# Signal
# Processing
# Systems

# WILEY SERIES IN TELECOMMUNICATIONS AND SIGNAL PROCESSING

John G. Proakis, Editor
Northeastern University

# Signal
# Processing
# Systems
## Theory and Design

N. Kalouptsidis

A Wiley-Interscience Publication
**JOHN WILEY & SONS, INC.**
New York · Chichester · Weinheim · Brisbane · Singapore · Toronto

Copyright © 1997 by John Wiley & Sons, Inc.

*Library of Congress Cataloging in Publication Data:*

Kalouptsidis, N.
    Signal processing systems : theory and design / N. Kalouptsidis.
        p.   cm. – (Wiley series in telecommunications and signal
    processing)
    "A Wiley-Interscience publication."
    Includes bibliographical references and index.
    ISBN 0-471-11220-8 (cloth : alk. paper)
    1. Signal processing. I. Title. II. Series.
TK5102.9.K34 1996
621.382'2–dc20                                        96-12447

Printed in the United States of America

10 9 8 7 6 5 4 3 2 1

*To Voula and Myrto*

# CONTENTS

# PREFACE

Signal processing systems are employed in a variety of applications and at an increasing rate. Signals such as speech, music, data, and video are compressed, encrypted, transmitted, stored, retrieved, filtered, enhanced, classified, or used for control. The scope of signal processing operations expands as performance requirements get more demanding. As an example, novel coding and compression systems are designed to cope with the storage, retrieval, and transmission of images in a multimedia environment. Furthermore, significant innovations in signal processing are introduced to serve emerging user needs, such as 3-D visualization, virtual reality, mobile robots, huge database retrieval, and signal content interpretation. The specific signal processing tasks selected for presentation in this book feature universal characteristics and generic value. They form primary building blocks upon which more sophisticated signal processing operations are developed. The methods that accomplish these tasks provide the fundamental knowledge base. Many other schemes can be accessed from the knowledge coded in this book.

The material covered in the book has been taught for several years at the University of Athens in a mixture of undergraduate courses, graduate courses, and seminars. It is assumed that the reader has a background on real analysis, linear algebra, and probability.

The book was written with two objectives in mind. The first was to motivate students with background in electrical engineering, computer science, physics, or mathematics to appreciate the usefulness and the versatility of signal-processing techniques. The second objective was to provide graduate students a solid background on the theory and design of signal processing systems. To achieve these goals within the given space limitations, a compromise betwen topic coverage and depth of analysis is necessary. An effort has been made to present theoretical concepts in a simple and natural manner. Mathematical tools are introduced as needed in tie with the application they intend to serve. Many mathematical facts are transferred to appendixes. Numerous examples are given. A good number of examples and exercises are simulations implemented in MATLAB. These simulations can form the core of a series of laboratory experiments and computer projects to be run in parallel with the course.

An undergraduate course can be based on Chapters 2–11. The entire book can be covered in a graduate one-year course.

I wish to thank G. Kapogiannopoulos for the preparation of the figures in the book. The comments and suggestions of N. Koutsoulis, M. Pouliakis, J. Tsinias, V. Tsoulkas, S. Theodoridis, K. Koutroumbas, M. Papadakis, and G. Gatt are gratefully acknowledged. I would particularly like to express my appreciation to E. Kofidis, P. Koukoulas, D. Linardatos, and A. Rondogiannis for valuable suggestions, in-depth reviews, and their excellent job in pointing out omissions, errors, and improvements.

Finally, I wish to thank my wife Voula for her endless support and encouragement, and my daughter Myrto for her love and understanding; without their help, this book would not have been written.

# 1

# INTRODUCTION

Signal processing systems convert signals produced by artificial or natural sources into a form suitable for the achievement of a specific purpose. Examples of signals include speech, audio, images, video, medical signals, geophysical data, economic time series, or sensor data. Typical purposes are data acquisition and display; efficient, reliable, and secure transmission and storage; enhancement of signal quality; prediction; control; classification; and others. Signal processing systems are useful in a wide range of application areas. Examples with significant impact are manufacturing, medical signal/image processing, and multimedia communications.

In manufacturing, signal processing systems are engaged in a multitude of tasks. They are involved in robotics with speech and vision capabilities; they are used in assembly line inspection where products are classified by a vision system for quality control. They are finally employed in communications on the workfloor and in nondestructive testing. Medical applications include digital hearing aids, medical diagnosis, computerized tomography, smart prostheses, or electrocardiogram monitoring. Multimedia communications are interactive communications between human and computer involving several media (text, still images, video, speech) and multiple senses (vision, hearing, tactil/force); they employ signal processing systems for efficient storage and transmission.

Signal processing systems are designed to perform a variety of functions. Among them are:

- Compression for speeding transmission and minimizing storage requirements.
- Modulation for efficient communication over digital and analog links.
- Error control coding for protecting the signal against the degradations occurring during transmission and storage.
- Equalization for combating distortion arising on real channels.
- Filtering for signal enhancement and noise removal.
- Encryption for privacy and authenticity.

1

- Control for modifying a plant so that it acquires a desired characteristic.
- Classification and clustering for content extraction and organization of pattern into homogeneous categories.
- Prediction for estimating future behaviour from past records.
- Identification for modeling of a plant or a signal source that operates in an uncertain environment.

*Signal compression and coding* is concerned with the economical representation of a signal. It strives to reduce the information bits needed to describe each signal value. Speech coding and image coding are used when speech or image data are involved. Compression algorithms are split into lossy and lossless schemes. In lossless compression the original signal is perfectly recovered from its compressed form. Compression rates of $2:1$ are achieved by lossless compression. Lossy compression causes distortion at the expense of better rates. The international JPEG (Joint Photographic Experts Group) standard achieves compression of images by $16:1$.

*Modulation* enables a better utilization of a communication channel. Modulation schemes catter for the simultaneous transmission of several signals through the same channel. Moreover they convert digital representations to analog waveforms suitable for transmission and storage. They constitute an essential component of every FAX and modem system.

*Error control coding* strengthens the signal with additional information to combat the noise incurring in transmission and storage. Communication channels such as telephone lines, coaxial cables, radio, satellite, and microwave links and storage media like CDROM degrade the signal by noise. Using the additional information bits provided by the error control scheme the receiver detects the errors and corrects them.

*Clustering and pattern recognition systems* extract content from the signal. By classifying the data into categories, they reveal their information structure. A speech recognition system converts spoken speech acquired by microphone into text suitable for printing or for giving commands to computers. In medical diagnosis, signals or images are classified according to diseases.

*Filtering* aims to suppress noise degradations in a signal or to enhance its characteristics. Sensor data, for instance, are always subject to measurement noise caused by the measuring device. Noise effects are removed by a filter. Slow variations in an image or a stock market bond index are enhanced by a filter that suppresses rapid variations.

*Control* takes place when action is applied to modify the behavior of a plant. Control is applied by an autopilot to keep the aircraft at a preassigned route. Control is used to maintain the temperature at a desired level or to keep a satellite in orbit around the earth. Control methods are finally employed in the design of robotic manipulators performing complex tasks.

*Encryption* schemes protect signals during transmission and storage. Originally intended for military applications, they have a widespread use in commercial applications involving sensitive data where access is limited to authorized parties.

*Identification* algorithms determine the structure of a system from input-output data. The unknown system could be a communication channel, a plant to be controlled in an uncertain environment, or a model of the earth to be used for geophysical exploration. Adaptive signal processing algorithms try to estimate time varying systems in real time.

*Prediction* methods are concerned with the estimation of the future behavior of a signal given data about its past. Prediction algorithms are involved in weather forecasting, stock market forecasting, and target tracking.

Signal processing is the theory and practice of algorithms and hardware that efficiently perform the above functions. The theory of signal processing relies on a rich blend of mathematical ideas and methods developed over the last 300 years. Signal processing practice relies on critical technologies such as very large scale integration (VLSI) and high-performance computers. In fact signal processing technology has been one of the primary enabling factors of the information and communication developments taking place in the twentieth century.

The phases involved in the design of a signal processing system can be grouped as follows:

1. Signal representation and modeling.
2. System representation and modeling.
3. Signal acquisition and synthesis.
4. Design formulation and optimization.
5. Efficient software/hardware implementation of the signal processing system.

Two critical factors that form an integral part of the design are:

6. Enabling technologies.
7. Application area.

The above phases provide a unified design framework. A more detailed account is presented next.

***Signal representation and modeling.*** A signal processed by a signal processing system must be represented in a proper mathematical format. There are several useful representations. The final choice depends on the specific class of signals and the signal processing function that is performed. Basic representations include time domain, frequency domain, representations in various orthogonal bases, higher-order statistical representations, and others. Very often the signal is thought of as the output of a source excited by another signal. Representation then may rely on the model of the source emitting the signal, which becomes a system representation issue.

***System representation and modeling.*** System representation provides the class of models within which the final structure is selected. Popular representations include time domain models, transformed models obtained by the application of transforms like the Fourier, the Laplace, and the $z$ transform, nonlinear models including Volterra representations and neural networks, time-varying and adaptive systems with parameters that vary with time to cope with the changes of the environment and the nonstationary characteristics of the input signal.

The choice of representation depends on the application area, the implementation technology, and the signal processing task to be accomplished. It is dictated by a trade-off of factors including availability of theory, implementation cost, complexity, and speed. Discrete models are currently the most popular because they feature precision, flexibility, programmability, and modularity much more so than analog representations. Furthermore mapping onto algorithms and architectures is much easier. Analog implementations are the preferred solution in applications with very low power, very low cost, very high bandwidths, or high parallelization requirements.

***Signal acquisition and synthesis.*** Current practice in signal processing relies on digital circuitry. Thus the signals to be processed must be converted to digital form. Data acquisition employs sampling and quantization. Sampling refers to the selection of certain signal values only. Quantization produces discrete approximations with a finite number of symbols. Eventually the output of the signal processing system will be restored to a continuous waveform such as speech, music, or video signal. Signal synthesis involves digital to analog conversion.

***Design formulation.*** The design formulation phase interprets the specific signal processing function (compression, filtering, classification, control, etc.) in a precise mathematical format. Then a system that executes this job efficiently is determined. This is often done by optimization. A performance criterion is chosen that assesses the effectiveness of each system in the model set. The system with optimum performance is specified using an optimization method and an algorithm. Least squares is a popular performance indicator.

***Software-hardware implementation.*** The structure determined by the design formulation is mapped onto an algorithm and subsequently onto an architecture. Signal processing algorithms involve some primary computational modules such as an FFT algorithm, numerical linear algebra algorithms such as singular value decomposition (SVD) or linear system solvers, and Kalman filtering.

In real time applications several signal processing functions must be carried out simultaneously with data acquisition. An adaptive signal processing system, for instance, performs a certain task within consecutive data blocks. This often places stringent complexity requirements, particularly with images or multidimensional data. Algorithms and architectures with high degree of parallelism are clearly desirable. The choice of architecture becomes a critical design issue. Single/multiple input multiple data (SIMD/MIMD) processors and VLSI special array processors are alternative platforms. The mapping of algorithms to architectures aims to find affordable or even optimal solutions.

**Enabling technologies.** Signal processing systems have successfully solved important practical problems in several application domains, partly due to the impressive evolution of microelectronics. The enabling technologies that have contributed to this success include VLSI, high-performance computers, ASICS (application specific integrated circuits) and computer-aided design tools.

**Application areas.** Signal processing systems are used in industrial, defense and scientific applications. Familiar application areas include medical image processing, geographical information systems, consumer multimedia, remote sensing, planetary research and education, control, computer-aided manufacturing, or robotics. As the design tools advance, the theory and practice of signal processing systems tend toward a cohesive and integrated interaction of phases 1–7.

**Book layout.** This book is almost exclusively concerned with phases 1–4. The book architecture is composed of tool development parts and signal processing design parts. Chapters primarily devoted to tool development are 2, 3, 5, 6, 7, 8, 9, 10, and 11. The main signal processing design themes are treated in Chapters 4, 12, 13, and 14. More precisely, signal representation and system modeling are presented in Chapters 2 and 3. Frequency domain representations are presented in Chapters 5 (periodic signals), 6 (deterministic signals), and 9 (stochastic signals). Series representations featuring a strong algebraic flavor are presented in Chapter 7. Descriptions in the complex domain utilizing the rich property of analyticity are considered in Chapter 8. Realization aspects with significant impact in estimation and control tasks are undertaken in Chapter 10. Stability is a crucial characteristic of a signal processing system and an important methodological tool. It is introduced in Chapter 4 and thoroughly studied in Chapter 11. The main signal processing design tasks are introduced in Chapter 4. Some are completed here; others are reconsidered at a later stage as they bear upon the tools and methods of subsequent chapters. In this respect Chapter 4 serves as a prelude to the main design themes of the book. Hopefully the challenge of the real and exciting problems discussed here will keep the interest of the reader alert for the rest of the book. Chapter 12 continues the discussion on filtering prediction and identification. Chapter 13 is devoted to control system design. Chapter 14 concludes with compression and coding.

There are two types of appendixes, local and global. Local appendixes appear at the end of the chapters. They usually convey proofs of theorems relevant to that chapter. Global appendixes appear at the end of the book and are useful to more than one chapters. Appendix I summarizes some facts from analysis. Appendixes II and III are written in a tutorial fashion and provide complementary material. Appendix II provides information on algebraic structures, mainly finite fields. Appendix III reviews some basic numerical algorithms.

The rest of the book is organized as follows: Chapter 2 is concerned with signal representation and modeling. A first acquaintance with signal models is provided. Deterministic, stochastic, multichannel, multidimensional, analog, discrete, and digital signals are discussed. Signal representations build upon approximations and decompositions into "simpler" constituents. Approximation requires limiting operations. Thus signal metrics and norms as well as convergence issues are discussed. A useful framework for the decomposition into "simpler" constituents is provided by orthogonal expansions. Hence signal representations in terms of orthogonal bases are discussed.

System representations are studied in Chapter 3. Input-output transformations, finite derivative, finite difference models, and state space representations are introduced. General system attributes such as causality, time invariance, memory, continuity, and fading memory are described. The fundamental ways of linking systems to form more powerful architectures are defined. Quantization and discrete simulation are introduced. The convolutional representation of linear systems and its nonlinear generalization to Volterra models are developed. Neural networks and radial basis function networks are also presented as nonlinear models and their universal approximation capabilities are highlighted.

Chapter 4 provides an introduction to the main design problems discussed in the book. Filtering prediction and identification are discussed in the context of least squares estimation. The LMS and the RLS algorithms are developed in connection with adaptive signal processing. The design of scalar and vector quantizers is studied and the $k$ means algorithm is derived. Pulse code modulation (PCM) and differential pulse code modulation (DPCM) are introduced as popular compression schemes. The basic elements of linear block codes for error control in transmission and storage are presented. The primary encryption concepts are highlighted and the RSA public key cryptosystem is developed. Finally, the perceptron algorithm and the backpropagation method are developed for pattern recognition applications.

Chapter 5 is concerned with the representation of periodic signals in terms of harmonically related exponential signals. Discrete, continuous time, digital and multidimensional signals are examined. The discrete Fourier transform (DFT) is introduced and its importance in signal processing algorithms is stressed. Fast Fourier transforms like the Cooley-Tukey FFT and the Good-Thomas FFT algorithms are developed for the implementation of the DFT.

Frequency domain representation of signals and systems is the subject of Chapter 6. The Fourier transform and its properties are developed for analog, discrete, and multidimensional signals. Convolutional and Volterra systems in the frequency domain are described. Frequency selective filtering, sampling, and modulation are important signal processing tasks discussed here.

Chapter 7 is concerned with the series representation of a discrete signal. The series formalism enables the use of algebraic tools in the study of signals. Discrete and digital signals generated by linear recurrent sources are analyzed, and their fine structure is revealed. Application to the analysis of linear feedback shift register sequences is discussed.

Signal and system analysis in the complex domain with the use of the $z$ and Laplace transforms is carried out in Chapter 8. These transforms convert signals into analytic functions of a complex variable whose rich structure enables the extraction of important information. An introduction to filter design is provided.

Chapter 9 develops spectral representations for stochastic wide sense stationary processes. The notion of power spectral density is introduced as the Fourier transform of the autocorrelation sequence. The integral of the power spectral density forms the spectral distribution of the signal. The spectral process of the signal itself is constructed, and the Fourier analysis and synthesis equations are derived. Higher-order spectral analysis via cumulants and polyspectra is also considered.

Chapter 10 is concerned with realization theory. It describes methods for the transfer between system representations. It searches for canonical forms with special characteristics and explores the uniqueness of system representations and their bearing on minimum memory requirements. The properties of controllability and observability are instrumental for this purpose. Controllability is a typical control property. It allows manouvering in every direction on the state space by proper inputs. Observability is a typical estimation property. It is essential in estimating the state evolution from output measurements. The strong duality of these concepts as well as the deeper duality between estimation and control is pointed out.

Stability of dynamical systems is the main topic of Chapter 11. The effects input and initial state perturbations have on the output are analyzed using Liapunov functions. Both continuous and discrete systems are considered. Finally, a simple example is used to navigate the reader toward the world of bifurcation and chaos, where very complex qualitative phenomena are observed.

Chapter 12 continues the discussion on prediction filtering and identification that was originated in Chapter 4. Infinite memory linear predictors are derived using the mean squared formulation. The basic elements of the Kolmogorov-Wiener theory are highlighted, and important insight into the structure of wide sense stationary processes is gained through the Wold decomposition. An alternative approach to filtering is offered by the Kalman filter. Kalman filters rely on state space representations, and they can cope with nonstationary signals and time varying characteristics. Identification is discussed in the context of convolutional and finite difference models.

Control system design methods are presented in Chapter 13. The transform design approach is performed in the $s$ or $z$ domain and relies on the transfer function description of systems. Pole placement design uses state space ideas. It determines the feedback gain that places the poles of the given plant at desired locations under full state information and computes the state estimator from input output measurements. Then it combines the two procedures into the controller structure. Regulator design also combines a control phase for the determination of the feedback gain under full state information and an estimation phase for the determination of the state. Estimation is performed by the Kalman filter. Control is accomplished by minimizing a quadratic function.

Chapter 14 continues the discussion on compression and error control coding originated in Chapter 4. Popular compression techniques such as transform coding and the discrete cosine transform are presented. Furthermore, the family of BCH codes is highlighted, and decoding algorithms are derived, including the Berlekamp-Massey algorithm. Finally, the Viterbi algorithm is described in connection with convolutional codes.

**Bibliographical notes.** An informative overview of signal processing perspectives is provided by the reports of the US National Science Foundation advisory committees. These are retrieved from the Internet address pubs@NSF.gov.

# 2

# SIGNAL MODELING: FIRST CONCEPTS

Signal representation and modeling are important factors in the design of signal processing systems. Successful design approaches depend upon judicious signal representations. It is for this reason that signal representation is a recurring theme throughout the book. This chapter provides a first acquaintance with signal modeling. The major signal categories described include deterministic, stochastic, multichannel, multidimensional, analog, discrete, and digital signals. The important deterministic classes of periodic and finite support signals are also presented. Finite support signals extend over a finite time horizon. They provide a good modeling environment because real signals are measured and processed in finite time intervals. Periodic signals model patterns of recurrent nature. If they have finite energy, they are represented by harmonically related sinusoids. The chapter initiates a study of two important signals, the matrix exponential signal and the Gaussian signal.

Two basic ideas bind together in a signal representation: approximation and decomposition into "simpler" components. Approximation involves a limiting operation, with consideration of signal metrics and norms as well as convergence issues. A useful framework for the decomposition into "simpler" constituents is provided by orthogonal expansions. This way signal representations are discussed in terms of orthogonal bases.

## 2.1 BASIC SIGNAL CATEGORIES

A *signal* expresses the variation of a variable with respect to another. Thus it is mathematically defined as a function $t \rightarrow x(t)$. The independent variable $t$ is commonly referred to as *time*, although it may not be physically so, and ranges in the *signal domain I*. $x(t)$ denotes the value of the signal at time $t$. All signal values are confined in the *signal range U*. The signal will be denoted by $x$. Abusing the notation, we often use the symbol $x(t)$ to denote both the value of the signal at $t$

and the function representing the signal. The true meaning is inferred from the text.

The above definition is extremely general. It includes:

- Electrical signals such as a current passing through a circuit element. The electric current describes the rate of change of charge with respect to time.
- Mechanical signals such as displacements and velocities of motions in mechanics. The position of a satellite, a space vehicle, or a robotic manipulator are such examples.
- Speech signals, music signals, or more generally signals in the audible range.
- Still images and video images.
- Medical signals such as the electrocardiogram (ECG) and the electroencephalogram (EEG).
- Economic time series such as the stock value or the currency exchange.
- Data sequences transmitted in communication networks.
- Sequences of zeros and ones processed by the central processing unit of a digital computer.

Depending on the nature of the signal domain $I$ and the signal range $U$, various categories result. *One-dimensional (1-D) signals* are characterized by a signal domain whose structure is specified by an 1-D parameter. If $I$ is a subinterval of the real numbers $R$, so that the independent parameter varies in a continuous range, the signal is called *continuous time* or *analog signal*. The most common cases are $I = R$ and $I = R^+$, the set of nonnegative real numbers. Simple examples of analog signals are the *unit step*

$$u_s(t) = \begin{cases} 1, & t \geq 0 \\ 0, & t < 0 \end{cases}$$

the *ramp signal*

$$u_r(t) = \begin{cases} t, & t \geq 0 \\ 0, & t < 0 \end{cases}$$

and the sinusoidal signal

$$x(t) = \sin \omega t, \qquad t \in R$$

In reality we are confronted with considerably more complicated situations. Figure 2.1 illustrates a segment of a speech signal. Human speech is generated as air from the lungs passes through the vocal tract. The vocal tract is a nonuniform acoustic tube expanding from the glottis to the lips. Moving the lips, jaw, tongue, and velum, the shape of the vocal tract changes, and as it is excited by airflow, it

**Figure 2.1.** Example of a speech signal.

**Figure 2.2.** The unit sample signal $\delta(n)$.

produces a speech signal. In this case the function $x(t)$ represents the resulting variation of acoustic pressure with time.

A *discrete (time) signal* results when the signal domain $I$ is a discrete set. We identify $I$ with a subinterval of the set of integers $Z$. The most common cases are $I = Z$ and $I = Z^+$, the set of nonnegative integers. Thus a discrete time signal is represented by a sequence. It is common in the signal processing literature to denote the time index by $n$ so that the time evolution is shown as $n \rightarrow x(n)$. The discrete unit step, the ramp signal, and the discrete sinusoid are obtained from their continuous time counterparts replacing $t$ with $n$, namely evaluating the signals only on integer values of time. An important example is the *unit sample signal*

$$\delta(n) = \begin{cases} 1, & n = 0 \\ 0, & \text{otherwise} \end{cases} \tag{2.1}$$

It is illustrated in Fig. 2.2.

Consider a positive number $T_s$ and an analog signal $x(t)$. If samples of $x(t)$ are taken $T_s$ units of time apart, the discrete signal $x_{T_s}(n) = x(nT_s)$ results. Discrete signals arising from analog signals this way are called *sampled signals*. $T_s$ is called the *sampling period* and $1/T_s$ the *sampling rate* or *sampling frequency*. Figures 2.3 and 2.4 illustrate an analog speech signal and a sampled version. Sampling is required whenever analog signals are to be processed by digital means. A detailed account is given in subsequent chapters.

The signal domain of a two-dimensional (2-D) signal is characterized by a two-dimensional parameter. Thus $I = I_1 \times I_2$, and we write $(t_1, t_2) \rightarrow x(t_1, t_2)$. If the signal is analog both $I_1$ and $I_2$ are subintervals of the real line $R$. In the case of

**Figure 2.3.** Example of a speech signal.

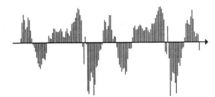

**Figure 2.4.** Discrete time signal obtained by sampling the speech signal of Fig. 2.3.

discrete 2-D signals, $I_1$ and $I_2$ are subintervals of $Z$. The 2-D unit step signal is

$$u_s(t_1, t_2) = u_s(t_1)u_s(t_2) = \begin{cases} 1, & \text{if } t_1 \geq 0 \text{ and } t_2 \geq 0 \\ 0, & \text{otherwise} \end{cases}$$

In a similar fashion the discrete 2-D unit step is obtained if the analog signal is evaluated at integers only. The 2-D unit sample is defined as

$$\delta(n_1, n_2) = \delta(n_1)\delta(n_2) = \begin{cases} 1, & \text{if } n_1 = 0 \text{ and } n_2 = 0 \\ 0, & \text{otherwise} \end{cases}$$

The above signals are depicted in Fig. 2.5.

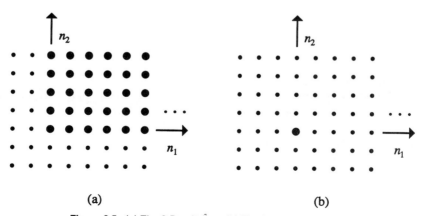

(a)  (b)

**Figure 2.5.** (a) The 2-D unit step. (b) The 2-D unit sample.

*Still images* are important examples of 2-D signals. Here the independent variables are spatial variables. The signal value may be the luminance of the object in a scene, such as a picture taken by an ordinary camera, the absorption of the body tissue in X-ray imaging, or the temperature profile of a region as in infrared imaging. Each value $x(n_1, n_2)$ of a discrete 2-D signal is referred to as *pixel* or *pel* (picture element). If the continuous time 2-D signal $x(t_1, t_2)$ is sampled with sampling periods $T_{s1}$ and $T_{s2}$, the discrete 2-D signal $x_{T_{s1}, T_{s2}}(n_1, n_2) = x(n_1 T_{s1}, n_2 T_{s2})$ results. A discrete image is obtained by sampling an analog signal. A television frame is represented by a discrete image of $512 \times 512$ pixels. An image on 35-mm film requires a spatial resolution of $1024 \times 1024$ pixels in the discrete image. Figure 2.6 represents an image and

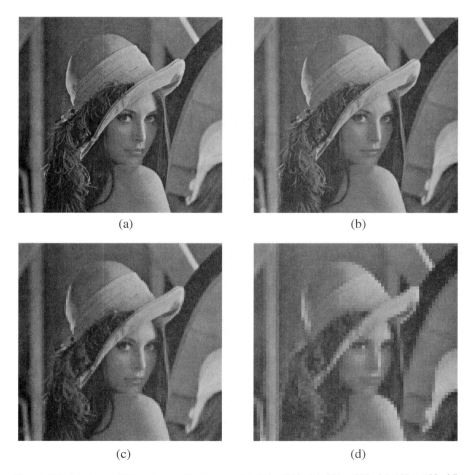

(a)  (b)

(c)  (d)

**Figure 2.6.** Lenna at different sampling rates. (a) $512 \times 512$. (b) $256 \times 256$. (c) $128 \times 128$. (d) $64 \times 64$.

sampled versions obtained at various sampling periods. Details tend to disappear as the spatial resolution decreases.

Our classification so far has been based on the signal domain. Further subdivisions are obtained if the signal range $U$ is considered. Let $k \geq 1$ be an integer. A *k-channel* signal is a map whose range $U$ is a subset of $R^k$; typically, $U = R^k$. We represent $x(t)$ in column format as

$$x(t) = \begin{bmatrix} x_1(t) \\ x_2(t) \\ \vdots \\ x_k(t) \end{bmatrix} = [\, x_1(t) \quad \cdots \quad x_k(t) \,]^T$$

where $x_1(t), x_2(t), \ldots, x_k(t)$ are the single-channel components of $x(t)$. Clearly the above definition applies with no changes to the discrete case. It is often appropriate to arrange the channels in matrix format. The multichannel signal then has the form

$$x(t) = \begin{bmatrix} x_{11}(t) & \cdots & x_{1k}(t) \\ \vdots & & \vdots \\ x_{m1}(t) & \cdots & x_{mk}(t) \end{bmatrix}$$

Multichannel signals appear when single-channel signals are grouped together. All currents passing through the elements of a circuit can be simultaneously viewed as a vector signal that provides the overall current information of the circuit. The motion of a particle in space is characterized by a six-dimensional signal consisting of three displacement coordinates and three velocity coordinates. Electrocardiograms concurrently taken from several points of the human body form a multichannel signal.

In several occasions we deal with complex valued signals, perhaps the most prominent of all being the exponential signal (Section 2.5). A *complex k-channel signal*, continuous or discrete, is one whose samples are $k$-tuples of complex numbers.

A *digital signal* or *message* is a discrete time signal whose signal range $U$ is a finite set. $U$ is endowed with some algebraic structure, usually that of a finite field. An introduction to the theory of finite fields that is adequate for the topics discussed in this book is provided in Appendix II.

The discrete unit step signal $u_s(n)$ and the unit sample signal $\delta(n)$ can be viewed as digital signals over the binary alphabet $\{0, 1\}$. Consider the finite field $U = GF(p) = \{0, 1, 2, \ldots, p - 1\}$, where $p$ is a prime number. The field operations are addition and multiplication modulo $p$. The signal

$$x(n) = n(\mathrm{mod}\ p) \tag{2.2}$$

is a digital signal on the alphabet $GF(p)$.

We already mentioned that digital processing of analog signals involves sampling. Digital devices, however, operate with digital signals. Since they have a finite amount of memory, they can handle only numbers represented by finite precision. Thus an irrational number like $\sqrt{2}$ must be represented by an approximation consisting of by as many bits as the memory capacity of the device allows. The process of reducing a discrete time signal to a digital signal by approximating each real sample with a finite number of bits, that is, with an element of a finite set $U$, is called *quantization*; the resulting signal is a quantized digital signal and the device yielding it is called a *quantizer*. Digital images are often quantized to 256 levels (8 bits). 0 corresponds to the darkest level and 255 to the brightest. Lenna is displayed in Fig. 2.7 (a) with spatial resolution of $512 \times 512$ pixels at 256 levels. Figures 2.7 (b), (c) and (d) display Lenna with 16

(a)                                     (b)

(c)                                     (d)

**Figure 2.7.** Lenna at different quantization levels.

levels (4 bits), 4 levels (2 bits) and 2 levels (1 bit) respectively. False contours appear as quantized levels decrease. Sampling followed by quantization constitutes the analog-to-digital conversion.

**Remark.** In the signal classification discussed so far, no special reference was made to the interesting class of continuous time signals with values in a discrete set. These signals are modeled by piecewise constant functions. The unit step is such example. As far as the content of this book is concerned, there is no reason to distinguish the above class from the family of continuous time signals. Digital signals with values in a finite alphabet, on the other hand, require separate treatment when their values are combined and manipulated by the rules of finite arithmetic systems, since the latter are in sharp contrast with the ordinary algebraic operations of real arithmetic.

The signals defined in previous sections are bound to model *deterministic* behaviors. Within a specific application environment all deterministic signals are defined on the same domain. In the continuous time case we set $I = R$ or $I = R^+$. In the discrete case we put $I = Z$ or $I = Z^+$. Regarding the signal range, we will restrict ourselves to three possibilities: $U \subset R^k$, (real signals), $U \subset C^k$ (complex signals), and $U \subset F^k$, $F$ finite field (digital signals). All the above options concern 1-D signals. Extension to 2-D signals is easily inferred. In summary, the large signal spaces, within which all subsequent deterministic exploration will take place, are listed in Table 2.1.

For simplicity, they are denoted by $L$, $\ell$, and $\ell(F)$, respectively. Extension to other classes is discussed when a need arises. Within each of the classes in the table, signals can be added, scaled, and multiplied in a natural way. Indeed, let us take $L$. For any two signals $x$, $y$, and any scalar $c$ (here scalars are the complex numbers), we define the sum $x + y$, the scalar muliplication $cx$ and the product $xy$, pointwise as

$$(x + y)(t) = x(t) + y(t) = (\, x_1(t) + y_1(t) \quad x_2(t) + y_2(t) \quad \cdots \quad x_k(t) + y_k(t)\,)^T$$
$$cx(t) = (\, cx_1(t) \quad cx_2(t) \quad \cdots \quad cx_k(t)\,)^T$$
$$(xy)(t) = x(t)y(t) = (\, x_1(t)y_1(t) \quad x_2(t)y_2(t) \quad \cdots \quad x_k(t)y_k(t)\,)^T$$

The pointwise operations appearing in the above expressions are the familiar ordinary operations in the ground field of complex numbers. The above

**TABLE 2.1    Model sets of deterministic signals**

| | |
|---|---|
| $L(R, C^k)$ | Analog $k$-channel complex-valued signals defined on $R$. |
| $\ell(Z, C^k)$ | Discrete $k$-channel complex-valued signals defined on $Z$. |
| $\ell(Z, F^k)$ | Digital $k$-channel sequences over the finite field $F$. |

definitions carry over to all classes of Table 2.1 with the obvious modifications reflecting the difference in the ground field.

The model sets $L$, $\ell$, and $\ell(F)$ equipped with the operations of addition and scalar multiplication become linear spaces of infinite dimension.

***Example 2.1  MATLAB implementation***
Implementation of pointwise operations in MATLAB is direct. Signals are restricted over a finite interval. Single-channel discrete signals are represented as vectors whose entries are the signal values. Analog signals are vectors formed by sampled values. The pointwise operations are then carried out by the pointwise vector operations $x + y$, $c * x$, and $x . * y$:

```
t = 0:.1:10;              % defines the time window;
x1 = 2*ones(1, length(t));  % this is the dc component of length t.
x2 = 4*sin(t);            % computes the sinusoid
x = x1 + x2;              % constructs the sum of a constant (dc)
                         % signal and a sinusoid.
plot(t,x1,t,x2,t,x)       % plots the three signals.          ∎
```

## 2.2  STOCHASTIC SIGNALS

In many cases the value of a signal cannot be completely determined. For instance, measurement of the same signal under the same environmental conditions yields different waveforms. The same signal transmitted twice by the same source through a communication channel results in different signals at the receiver. *Stochastic* signals provide a successful modeling approach for such cases. The value $x(n)$ of a stochastic or random discrete signal at time instant $n$ is a random variable $x(n) : \Omega \to R$, $\zeta \to x(n)(\zeta)$. Therefore it depends on the outcome $\zeta$ of a probabilistic experiment specified by the space $\Omega$, the set of events (subsets of $\Omega$ forming a $\sigma$ algebra), and the probability law $P$. Thus a random signal is modeled by a stochastic process. The behavior of the signal at time $n$ is governed by the distribution function of the random variable $x(n)$,

$$F_n(x) = P\{\zeta \in \Omega : x(n)(\zeta) \le x\} = P[x(n) \le x], \qquad x \in R$$

Here $x$ is the independent variable of the distribution function and should not be confused with the signal $x$. If the distribution function is differentiable with respect to $x$, as we usually assume in this book, the signal behavior at time $n$ is determined by the probability density function

$$f_n(x) = \frac{dF_n(x)}{dx}$$

$f_n(x)$ expresses the probability that the signal value at time $n$ lies in the infinitesimal interval

$$f_n(x)dx = P[x \leq x(n) \leq x + dx]$$

### 2.2.1  Finite family of distributions and densities

The family of density functions $f_n(x)$ as $n$ varies, does not suffice to characterize the statistical evolution of the signal. Suppose, for instance, that we are interested in a prediction problem where some information on the signal is available at time $n_1$, and we want to infer its behavior at some future time $n_2$. It is conceivable that in such application the conditional density $f_{n_2|n_1}(x_2|x_1)$ will be involved. This in turn relates to the second-order density $f_{n_1,n_2}(x_1, x_2)$. Thus we are led to consider the family of second-order distribution functions

$$F_{n_1,n_2}(x_1, x_2) = P[x(n_1) \leq x_1, x(n_2) \leq x_2]$$

More generally, if we try to infer conclusions on the behavior of the signal at time $n_k$ on the basis of signal information at times $n_1, n_2, \ldots, n_{k-1}$, the family of $k$th order distribution functions

$$F_{n_1,n_2,\ldots,n_k}(x_1, x_2, \ldots, x_k) = P[x(n_1) \leq x_1, x(n_2) \leq x_2, \ldots, x(n_k) \leq x_k]$$

will naturally show up. It can be proved that the family of distributions

$$\{F_{n_1,n_2,\ldots,n_k}(x_1, x_2, \ldots, x_k) : k \in Z^+, n_1, n_2, \ldots, n_k \in Z\}$$

characterizes the probabilistic behavior of the signal $x(n)$. The same description is valid for continuous time signals. The relevant family of distributions is

$$\{F_{t_1,t_2,\ldots,t_k}(x_1, x_2, \ldots, x_k) : k \in Z^+, t_1, t_2, \ldots, t_k \in R\}$$

Under very broad conditions it uniquely specifies the signal.

In the digital case each signal value is a random variable ranging in a discrete set $U$. In this case the family of densities consists of

$$f_{n_1,n_2,\ldots,n_k}(a_1, a_2, \ldots, a_k) = P[x(n_1) = a_1, x(n_2) = a_2, \ldots, x(n_k) = a_k]$$

where $k \in Z^+, n_1, n_2, \ldots, n_k \in Z, a_i \in U$.

Determination of the above families is a formidable task. Thus we either focus on specific classes of stochastic signals having the property that the family of distributions is spanned by considerably less information, or we seek approaches and formulations that involve only a few parameters of the above family. Such parameters are the mean and the autocorrelation function to be discussed shortly. An important example of the first type is the purely random signal. The discrete version is described next.

### *Example 2.2.    Purely random signal*

A *purely random process* describes a signal for which knowledge of its behavior at a time instant provides absolutely no information about its behavior at any other instant. Thus for any finite number of pairwise distinct time instants $n_1, n_2, \ldots, n_k$, the corresponding signal values are independent random variables:

$$f_{n_1, n_2, \ldots, n_k}(x_1, x_2, \ldots, x_k) = f_{n_1}(x_1) f_{n_2}(x_2) \cdots f_{n_k}(x_k)$$

Therefore the family of densities of a purely random signal is completely determined from the first-order family of densities $f_n(x)$, $n \in Z$. If these densities coincide for all $n$ in that $f_{n_1}(x) = f_{n_2}(x)$, for all $n_1, n_2$ and $x$, the signal is described by an *independently and identically distributed (IID)* random process.

Consider the uniform density on the interval $[a, b]$:

$$f(x) = \begin{cases} \dfrac{1}{b - a}, & x \in [a, b] \\ 0, & \text{otherwise} \end{cases} \qquad (2.3)$$

Figure 2.8 shows two realizations of a purely random process with uniform density on the interval $[0, 1]$. It was generated by these commands:

```
x = rand(1,300);
plot(x)
```
∎

### 2.2.2    Mean, correlation, and covariance functions

The mean of the discrete random signal $x(n)$ is given by the deterministic discrete signal

$$m_x(n) = E[x(n)] = \int_{-\infty}^{\infty} x f_n(x) dx$$

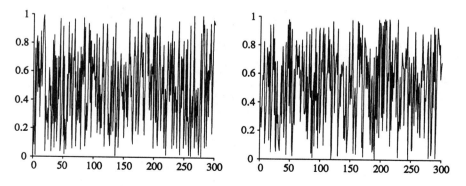

**Figure 2.8.** Realizations of a purely random signal.

If $x(n)$ is discrete with values in the finite subset $U$ of $R$, the mean is

$$m_x(n) = \sum_{a \in U} a P[x(n) = a]$$

In the sequel we write $m(n)$ for $m_x(n)$.

### *Example 2.3   Computation of the mean*

Consider a purely random Bernoulli process. Each signal value is either 0 or 1 with probability density

$$P[x(n) = 1] = p, \quad P[x(n) = 0] = 1 - p$$

Thus

$$m(n) = 0P[x(n) = 0] + 1P[x(n) = 1] = p$$

Next suppose that $y(n) = G[x(n)]$ where $x(n)$ is uniformly IID on the interval $[a, b]$. Then the mean is

$$m_y(n) = \int_a^b G(x) \frac{1}{b-a} dx = \frac{1}{b-a} \int_a^b G(x) dx$$

In particular, if $G(x) = x^2$, the mean is $\frac{1}{3}(b^2 + ab + a^2)$. The result can be verified by the MATLAB code

```
x = rand(1,300);
y = x.^2;
m = mean(y)
```
■

The mean of a process is associated with the first-order densities of the signal. Other parameters associated with $f_n(x)$, are the moments

$$m_k(n) = E[x^k(n)] = \int_{-\infty}^{\infty} x^k f_n(x) dx, \qquad k \geq 1$$

We easily check that the moments of the IID Bernoulli process are $m_k(n) = p$ for all $n \in Z$, $k \geq 1$. The central moments are

$$\mu_k(n) = E[x(n) - m(n)]^k = \int_{-\infty}^{\infty} (x - m(n))^k f_n(x) dx$$

Newton's identity gives

$$\mu_k(n) = \sum_{i=0}^{k} \binom{k}{i} (-1)^{k-i} E[x^i(n)] m^{k-i}(n) = \sum_{i=0}^{k} \binom{k}{i} (-1)^{k-i} m_i(n) m^{k-i}(n)$$

In particular, $\mu_1(n) = 0$, $\mu_2(n) = m_2(n) - m^2(n)$, $\mu_3(n) = m_3(n) - 3m(n)m_2(n)$ $+2m^3(n)$. The central moment $\mu_2(n)$ defines the *variance* of the signal at time $n$.

### Example 2.4  Moments of symmetric densities

A density function $f(x)$ is called *symmetric* if it is even, $f(x) = f(-x)$. In this case the graph of $f(x)$ is symmetric with respect to the $y$ axis. The uniform distribution on the interval $[-a, a]$ is symmetric. The odd moments of a random variable $x$ with a symmetric density are zero. Indeed,

$$E[x^{2k+1}] = \int_{-\infty}^{\infty} x^{2k+1} f(x) dx = \int_{-\infty}^{0} x^{2k+1} f(x) dx + \int_{0}^{\infty} x^{2k+1} f(x) dx$$

$$= -\int_{0}^{\infty} y^{2k+1} f(-y) dy + \int_{0}^{\infty} y^{2k+1} f(y) dy = 0$$

Let us compute the first four moments of a purely random process uniformly distributed on $[-a, a]$. Due to symmetry $m(n) = m_3(n) = 0$, and

$$m_2(n) = \int_{-a}^{a} x^2 f_n(x) dx = \frac{1}{2a} \int_{-a}^{a} x^2 dx = \frac{a^2}{3}$$

$$m_4(n) = \frac{1}{2a} \int_{-a}^{a} x^4 dx = \frac{a^4}{5}$$ ∎

The *(auto)correlation* and *(auto)covariance* functions relate to the second-order statistics of $x(n)$. They indicate the degree of correlation between two signal values. The autocorrelation function of a real analog signal is defined by the two-dimensional signal

$$R(t_1, t_2) = E[x(t_1)x(t_2)] = \int \int x_1 x_2 f_{t_1, t_2}(x_1, x_2) dx_1 dx_2$$

and the autocovariance function by

$$C(t_1, t_2) = E[(x(t_1) - m(t_1))(x(t_2) - m(t_2))]$$

The following relation is readily established

$$C(t_1, t_2) = R(t_1, t_2) - m(t_1)m(t_2) \tag{2.4}$$

We often work with a finite segment $s = [x(t_1) \quad x(t_2) \quad \cdots \quad x(t_k)]^T$ of the signal $x(t)$. The correlation matrix of $s$ is the matrix $E[ss^T]$ whose $ij$ entry is $R(t_i, t_j)$.

Definitions are entirely analogous in the discrete case. Extensions to complex-valued signals are direct. For instance, the autocorrelation function of a complex signal is defined by $R(t_1, t_2) = E[x(t_1)x^*(t_2)]$.

If $x(t)$ is a $k$-channel real signal the correlation function is specified by the $k \times k$ matrix valued 2-D signal

$$R(t_1, t_2) = E[x(t_1)x^T(t_2)]$$

Likewise the covariance function is

$$C(t_1, t_2) = E[(x(t_1) - m(t_1))(x(t_2) - m(t_2))^T]$$

**Example 2.5**    *Covariance of a purely random signal*
Let us consider a purely random discrete signal. Then

$$C(n_1, n_2) = \mu_2(n_1)\delta(n_1 - n_2) = \begin{cases} \mu_2(n_1), & n_1 = n_2 \\ 0, & \text{otherwise} \end{cases} \quad (2.5)$$

If $n_1 \neq n_2$, statistical independence of signal values gives $E[x(n_1)x(n_2)] = E[x(n_1)]E[x(n_2)]$. Therefore $C(n_1, n_2) = 0$. If $n_1 = n_2$, $C(n_1, n_1) = E[x^2(n_1)] - m^2(n_1) = \mu_2(n_1)$. The following code generates a purely random uniformly distributed process, computes the covariance $C(n, 0)$ and autocorrelation $R(n, 0)$ and plots the results in Fig. 2.9.

Theoretically the plots should look like a scaled unit sample (see Fig. 2.2). Deviations are due to the approximation of moments by sampled estimates. The MATLAB computation is based on the FFT algorithm. These topics are discussed in subsequent chapters.

```
x = rand(1,300);        % creates 300 samples of uniform purely random process
r = xcorr(x,'unbiased'); % computes the autocorrelation sequence via
                        % an unbiased estimate
c = xcov(x,'unbiased'); % computes the autocovariance sequence
subplot(121), plot(-N + 1: N-1, r), title('correlation')
subplot(122), plot(-N + 1: N-1, c), title('covariance')
                        % breaks the graph window into 1-by-2 grid and plots the correlation
                        % in the first box and the covariance in the second        ∎
```

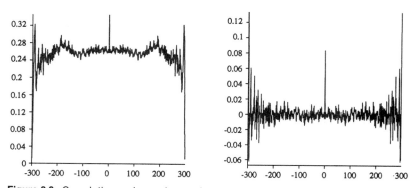

**Figure 2.9.** Correlation and covariance of a uniformly distributed independent process.

Note that if the covariance of a process $x(n)$ is given by Eq. (2.5), the signal values at two distinct times are uncorrelated; that is, the corresponding auto-covariance function is zero. A random signal with zero mean and autocorrelation given by Eq. (2.5) is called *white noise signal* for reasons that will be explained in Chapter 9.

Suppose that $x(n)$ is a $k_1$-channel signal and $y(n)$ is a $k_2$-channel signal. The *crosscorrelation* and *crosscovariance* of $x(n)$ and $y(n)$ are defined by

$$R_{xy}(n_1, n_2) = E[x(n_1)y^T(n_2)]$$

$$C_{xy}(n_1, n_2) = E\left[(x(n_1) - m_x(n_1))(y(n_2) - m_y(n_2))^T\right]$$

It holds that

$$R_{xy}(n_1, n_2) = R_{yx}^T(n_2, n_1)$$

In particular, if $x = y$, the crosscorrelation coincides with the autocorrelation and $R(n_1, n_2) = R^T(n_2, n_1)$. Another useful property of the autocorrelation is nonnegative definiteness. For any integers $k, n_1, \ldots, n_k$ and reals $a_1, \ldots, a_k$, it holds that

$$\sum_{i=1}^{k} \sum_{j=1}^{k} a_i a_j R(n_i, n_j) \geq 0$$

To see this, set $v = \sum_i a_i x(n_i)$. Then

$$\sum_{i=1}^{k} \sum_{j=1}^{k} a_i a_j R(n_i, n_j) = \sum \sum a_i a_j E[x(n_i)x^T(n_j)]$$

$$= E\left[\sum_i a_i x(n_i)\left(\sum_j a_j x(n_j)\right)^T\right] = E[vv^T] \geq 0$$

### 2.2.3 Higher-order moments of random signals

Occasionally one has to go beyond second-order densities to extract useful signal information. Important parameters conveying higher-order information are the higher order moments. Given an integer $k$, the $k$th-order moment of a real single channel signal $x(n)$ is specified by the $k$-dimensional signal

$$m_k(n_1, n_2, \ldots, n_k) = E[x(n_1)x(n_2) \cdots x(n_k)]$$

$$= \int \int \cdots \int x_1 x_2 \cdots x_k f_{n_1, n_2, \ldots, n_k}(x_1, x_2, \ldots, x_k) dx_1 dx_2 \cdots dx_k$$

Likewise, the $k$th-order central moment of $x(n)$ is

$$\mu_k(n_1, n_2, \ldots, n_k) = E[(x(n_1) - m(n_1))(x(n_2) - m(n_2)) \cdots (x(n_k) - m(n_k))]$$

where $m(n)$ denotes the mean of $x(n)$.

Very often a single realization of a stochastic signal is observed. Based on this information signal statistics (mean, covariance, higher-order moments) need to be estimated. *Ergodic processes* warrant that time averages of a single realization provide the statistical parameters of the process, asymptotically. The discussion of ergodic processes is beyond the scope of this book. Some comments are given in Chapter 4.

### 2.2.4  Random fields

The previous discussion extends to 2-D and multidimensional signals in a straightforward manner. A 2-D discrete random signal, also called *random field*, is determined by a sequence $x(n_1, n_2)$ where each value is a random variable. The mean, autocorrelation, and covariance functions are

$$m(n_1, n_2) = E[x(n_1, n_2)] = \int_{-\infty}^{\infty} x f_{n_1, n_2}(x) dx$$

$$R(n_1, n_2, \tilde{n}_1, \tilde{n}_2) = E[x(n_1, n_2) x(\tilde{n}_1, \tilde{n}_2)]$$

$$C(n_1, n_2, \tilde{n}_1, \tilde{n}_2) = R(n_1, n_2, \tilde{n}_1, \tilde{n}_2) - m(n_1, n_2) m(\tilde{n}_1, \tilde{n}_2)$$

Properties are discussed in the exercises. A purely random field is such that two different pixels are statistically independent. In this case

$$C(n_1, n_2, \tilde{n}_1, \tilde{n}_2) = \mu_2(n_1, n_2) \delta(n_1 - \tilde{n}_1) \delta(n_2 - \tilde{n}_2)$$

## 2.3  PERIODIC AND FINITE SUPPORT DETERMINISTIC SIGNALS

The signal classes given so far are very broad. Signal processing involves operations like differentiation, integration, or infinite summation. Not every signal can be processed this way. Depending on specific signal operations, properties like continuity, smoothness, integrability, finite energy, or periodicity have to be imposed. Two important deterministic signal classes are described next. More will be introduced as we go along.

### 2.3.1  Periodic signals

Periodicity expresses the repetitive occurrence of a pattern. It appears in many applications such as the oscillatory behavior of mechanical and electrical systems and the electrocardiogram. A continuous time signal $x(t)$ defined on $R$ is called *periodic* if there is a real number $T \neq 0$, called period of $x(t)$, such

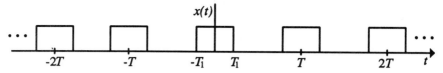

**Figure 2.10.** Periodic square wave.

that for all real $t$, $x(t + T) = x(t)$. It can be shown that if $x(t)$ is periodic with period $T$, all integer multiples of $T$ are periods as well, namely, $x(t + kT) = x(t)$ for all real $t$ and integers $k$. The least positive period of $x(t)$ is called *fundamental period* of $x(t)$. The sinusoidal signal $\sin(\omega t)$ is periodic with fundamental period $2\pi/\omega$. The periodic square wave is sketched in Fig. 2.10. It has fundamental period $T$ and is defined as

$$x(t) = \begin{cases} 1, & |t| \le T_1 \\ 0, & T_1 < |t| < \dfrac{T}{2}. \end{cases} \tag{2.6}$$

Everywhere else it repeats itself due to periodicity.

In the discrete case periodicity is defined in a similar fashion. The signal $x(n)$ is *periodic* if there is any integer $N \ne 0$ such that

$$x(n + N) = x(n), \qquad -\infty < n < +\infty \tag{2.7}$$

The smallest positive integer $N$ with the above property is called *fundamental period*.

### Example 2.6  Periodicity and sampling

Periodicity is not preserved under sampling. Consider the analog periodic sinusoid $\sin \omega t$. Let $T_s$ be the sampling period. The resulting discrete signal will be periodic if there is an integer $N$ such that $\sin \omega(n + N)T_s = \sin \omega n T_s$, or $\omega N T_s = 2k\pi$, for some integer $k$. Since $N$ is an integer, the above is satisfied if and only if $2\pi/\omega T_s = T/T_s$ is rational. Thus, if $\omega = 1$ (rad/sec) and $T_s = 0.1$ sec, the sampled sinusoid is not periodic. The following code will help the reader visualize the above:

```
t = 0:.1:10;
x1 = sin(10*t);        % x1 is not periodic
x2 = sin(2*pi*t);      % x2 is periodic
subplot(121), plot(t,x1), subplot(122), plot(t,x2)
```
■

The class of analog periodic signals is a linear subspace of $L$. In a similar fashion, analog periodic signals of given period constitute a linear subspace of $L$. Analogous statements hold for discrete time and digital signals (see Exercise 2.3).

The digital ramp signal of Eq. (2.2) is periodic because Eq. (2.7) in this case becomes $n + N = n(\text{mod } p)$ or $N = 0(\text{mod } p)$ or $N = kp$. Therefore the signal is periodic with fundamental period $p$.

### 2.3.2   Finite support signals

A signal $x(t)$ is a *finite support* signal if there are $T_1 \leq T_2$ such that $x(t) = 0$ for all $t \leq T_1$ and $t \geq T_2$; in other words, the signal is zero everywhere except possibly on a finite interval. The unit sample is an example of a finite support signal.

The set of analog finite support signals is a linear subspace of the basic signal model $L$, as is the set of finite support signals that vanish outside the same interval of length $T$. Similar statements hold for the discrete and the digital case. One interesting distinction is that in discrete time the space of signals that vanish outside an interval of length $N$ is finite-dimensional and can be identified with $C^N$. Indeed, let this interval be $0 \leq n \leq N - 1$. Then the signal $x$ can be identified with the vector of the $N$ values of $x$:

$$x = (\ x(0) \quad x(1) \quad \cdots \quad x(N - 1)\ )^T$$

that is

$$x_n = x(n - 1), \qquad 1 \leq n \leq N \tag{2.8}$$

Addition and scalar multiplication of such signals correspond to the usual addition and multiplication in $C^N$. Another consequence is that the latter space can be naturally identified with the space of discrete periodic signals of period $N$. Indeed, suppose, without loss of generality, that the interval consists of the first $N$ integers. Consider the map $\varphi$, which assigns the finite support signal $x$ to the periodic signal $x_p$ as

$$x_p(n) = \varphi(x)(n) = \sum_{k=-\infty}^{\infty} x(n - kN) \tag{2.9}$$

For each $n \in Z$ the above summation is finite because $x$ has finite support. The action of $\varphi$ is illustrated in Fig. 2.11.

We observe that $x_p$ is periodic of period $N$. Indeed,

$$x_p(n + N) = \sum_{k=-\infty}^{\infty} x(n + N - kN) = \sum_{k=-\infty}^{\infty} x(n - (k - 1)N) = x_p(n)$$

It is easy to check that $\varphi$ is linear. Moreover $\varphi$ is one-to-one. Let $\varphi(x) = \varphi(y)$. Then $x(n) = y(n)$ for all $n = 0, 1, \ldots, N - 1$, and since both signals are time limited on this interval, we have $x = y$. Finally, $\varphi$ is onto. Indeed, if $x_p$ is periodic of period $N$, then the signal

$$x(n) = \begin{cases} x_p(n), & 0 \leq n \leq N - 1 \\ 0, & \text{otherwise} \end{cases} \tag{2.10}$$

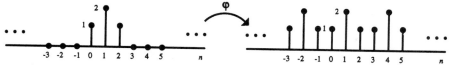

**Figure 2.11.** Identification of periodic signals of period N and signals of finite support of length N.

has finite support of length $N$ and $\varphi(x) = x_p$. Therefore $\varphi$ is an isomorphism. The signal $x_p(n)$, as defined in (2.9), is called *periodic extension* of $x(n)$.

### 2.3.3   2-D periodic and finite support signals

A discrete 2-D signal $x(n_1, x_2)$ is called *periodic* if there exist integers $N_1$ and $N_2$ such that for any $n_1$, $n_2$,

$$x(n_1, n_2) = x(n_1 + N_1, n_2) = x(n_1, n_2 + N_2)$$

The smallest such positive integers $N_1$ and $N_2$ are called *fundamental horizontal* and *vertical periods*, respectively. We observe that once $N_1 N_2$ samples over a rectangle of $Z^2$ are given, the remaining samples are determined by the periodicity condition. The most common case is the rectangle $I_1 \times I_2$:

$$I_1 = \{0 \le n_1 \le N_1 - 1\}, \quad I_2 = \{0 \le n_2 \le N_2 - 1\} \tag{2.11}$$

A periodic signal is illustrated in Fig. 2.12.

If there exist integer intervals $I_1, I_2$ of length $N_1$ and $N_2$, respectively, such that $x(n_1, n_2) = 0$ for all $n_1 \notin I_1$ or $n_2 \notin I_2$, $x(n_1, n_2)$ is a *finite support* signal. The most common case is the rectangle (2.11). A finite support signal as above can be identified with a matrix $X$ of $N_1$ rows and $N_2$ columns. For instance, if the intervals are given by (2.11), the entries of $X$ are $X_{mn} = x(m - 1, n - 1)$,

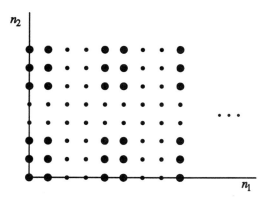

**Figure 2.12.** 2-D periodic signal with $N_1 = 4$ and $N_2 = 5$.

$1 \leq m \leq N_1$, $1 \leq n \leq N_2$. Pointwise addition and scalar multiplication correspond to matrix addition and matrix scalar multiplication. It follows that the set of signals having finite extent $N_1$, $N_2$ can be identified with the linear space of $N_1 \times N_2$ matrices. This space can also be shown to be isomorphic to the space of periodic sequences of periods $N_1$, $N_2$, as in the 1-D case.

A (still) discrete image is represented by a finite support signal and can be viewed as a matrix.

## 2.4  STATIONARY SIGNALS

A random signal $x(n)$, $n \in Z$, is called *stationary* if for every integer $N$, the shifted signal $x(n + N)$ has the same family of distributions. This means that every distribution remains invariant under a time shift. Assuming densities exist, for every integer $k$, $n_1, n_2, \ldots, n_k$, it holds that

$$f_{n_1+n,n_2+n,\ldots,n_k+n}(x_1, x_2, \ldots, x_k) = f_{n_1,n_2,\ldots,n_k}(x_1, x_2, \ldots, x_k)$$

In particular, the first-order densities satisfy $f_n(x) = f_{0+n}(x) = f_0(x)$. This states that the signal values follow the same distribution, namely they are identically distributed. For purely random processes this is necessary as well as sufficient for stationarity. Since first-order densities are identical, the mean $m(n)$ and all moments $m_k(n)$ are constant, independent of $n$.

Let us consider the implications of stationarity on second-order densities. We have

$$f_{n_1,n_2}(x_1, x_2) = f_{n_1-n_2+n_2,0+n_2}(x_1, x_2) = f_{n_1-n_2,0}(x_1, x_2)$$

Thus the family of second-order densities depends on the time difference $n_1 - n_2$ only. Hence the autocorrelation and autocovariance functions depend only on the difference $|n_1 - n_2|$. The one-dimensional signal

$$r(n) = R(n, 0) = E[x(n)x^T(0)]$$

completely determines the correlation function. Indeed,

$$R(n_1, n_2) = R(n_1 - n_2 + n_2, 0 + n_2) = R(n_1 - n_2, 0) = r(n_1 - n_2)$$

The sequence

$$r(n) = E[x(n + k)x^T(k)]$$

defines the autocorrelation sequence of the stationary process $x(n)$.

Shift invariance of the entire family of densities is hard to establish. Usually we are content with invariance of few parameters. The signal $x(n)$ is called *wide sense stationary (WSS)* if the mean is constant and the autocorrelation function

$R(n_1, n_2)$ depends only on the difference $|n_1 - n_2|$. A useful generalization is the following: $x(n)$ is stationary of order $k$ if for any $m \leq k$, and any integers $n$, $k_1$, $k_2, \ldots, k_m$ it holds that

$$E[x(n + k_1)x(n + k_2)x(n + k_3) \cdots x(n + k_m)x(n)] = c(k_1, k_2, \ldots, k_m)$$

Two processes $x(n)$, $y(n)$ are called *jointly wide sense stationary* if the augmented signal $(x^T(n) \ y^T(n))^T$ is wide sense stationary.

**Example 2.7** **Wide sense stationarity of sinusoids with uniform phase**
Let

$$x(t) = a\cos(\omega t + \phi)$$

$a$ and $\omega$ are constants while $\phi$ is a random variable with the uniform density $f_\phi(\phi)$ on $[-\pi, \pi]$. We show that $x(t)$ is wide sense stationary. We have

$$E[x(t)] = \int_{-\infty}^{\infty} a\cos(\omega t + \phi)f_\phi(\phi)d\phi = a \int_{-\pi}^{\pi} \cos(\omega t + \phi)\frac{1}{2\pi}d\phi = 0$$

Furthermore

$$E[x(t + \tau)x(\tau)] = E[a^2 \cos(\omega(t + \tau) + \phi)\cos(\omega\tau + \phi)]$$

$$= \frac{a^2}{4\pi} \int_{-\pi}^{\pi} [\cos(2\omega\tau + \omega t + 2\phi) + \cos\omega t]d\phi$$

As in the computation of the mean, the first term in the integrand gives integral zero. Hence

$$r(t) = \frac{1}{2}a^2 \cos \omega t \tag{2.12}$$

■

Stationarity of random fields is similarly defined. We consider wide sense stationarity only. The random field $x(n_1, n_2)$ is called WSS if $E[x(n_1, n_2)]$ is constant and

$$R(n_1, n_2, \tilde{n}_1, \tilde{n}_2) = R(n_1 - \tilde{n}_1, n_2 - \tilde{n}_2, 0, 0)$$

The autocorrelation sequence is specified by the 2-D sequence

$$r(n_1, n_2) = R(n_1 + k_1, n_2 + k_2, k_1, k_2)$$

A 2-D random signal is called *white noise* if it has zero mean and covariance

$$r(n_1, n_2, \tilde{n}_1, \tilde{n}_2) = m_2(n_1, n_2)\delta(n_1 - \tilde{n}_1, n_2 - \tilde{n}_2)$$

Thus any two distinct pixels are uncorrelated.

Cyclostationary processes are obtained when statistics are invariant under a common period. We confine ourselves to wide sense cyclostationarity. A signal $x(n)$ is called *wide sense cyclostationary (WSCS)* if there is a positive integer $N$ such that

$$m(n + N) = m(n), \quad R(n_1 + N, n_2 + N) = R(n_1, n_2) \qquad \text{for any } n_1, n_2$$

## 2.5  THE EXPONENTIAL SIGNAL

If a single signal had to be chosen for its importance, this would undoubtedly be the exponential signal. There are several reasons that the exponential signal plays such a prominent role. It provides the basis for explaining the behavior of periodic signals and the vehicle leading to the spectral characterization of signals. Moreover it generically characterizes the behavior of realizable linear time invariant systems as we shall see in later chapters.

### Example 2.8  Single-channel exponential signal
The complex exponential function

$$e^s = \sum_{n=0}^{\infty} \frac{1}{n!} s^n \tag{2.13}$$

is defined for all complex numbers $s$. This is readily established by the ratio test, which states that if there is $n_0$ such that for any $n \geq n_0$, $x(n) \neq 0$ and

$$\left| \frac{x(n+1)}{x(n)} \right| \leq L < 1 \tag{2.14}$$

then $\sum_{n=0}^{\infty} |x(n)| < \infty$. In the case of the exponential function, we obtain

$$\frac{\left| \frac{1}{(n+1)!} s^{n+1} \right|}{\left| \frac{1}{n!} s^n \right|} = \frac{1}{n+1} |s| \rightarrow 0$$

and the series (2.13) converges absolutely. It follows that the exponential function is well defined and continuous. Moreover it is analytic; namely its derivative with respect to $s$ exists everywhere and is obtained by termwise differentiation. Thus

$$\frac{d}{ds} e^s = \sum_{n=0}^{\infty} \frac{1}{n!} \frac{d}{ds} s^n = \sum_{n=1}^{\infty} \frac{1}{n!} n s^{n-1} = \sum_{n=1}^{\infty} \frac{1}{(n-1)!} s^{n-1} = e^s \tag{2.15}$$

Using the series definition (2.13), we can show (Exercise 2.4) the important formula

$$e^{s_1+s_2} = e^{s_1} \cdot e^{s_2} \tag{2.16}$$

Setting $s = j\omega$ in (2.13), we obtain the celebrated Euler's identity:

$$e^{j\omega} = \cos\omega + j\sin\omega \tag{2.17}$$

Let $s = \sigma + j\omega$. The *continuous time exponential signal* is the complex-valued signal

$$x(t) = e^{st} = e^{\sigma t}e^{j\omega t}, \qquad -\infty < t < \infty$$

Its real and imaginary components are readily obtained from Euler's identity

$$x(t) = e^{\sigma t}\cos\omega t + je^{\sigma t}\sin\omega t$$

**Invertibility.** For any $t \in R$ it holds $e^{st} \neq 0$ and $(e^{st})^{-1} = e^{-st}$. Indeed, from property (2.16) we have $e^{st} \cdot e^{-st} = e^{st-st} = e^0 = 1$.

**Differentiability.** From Eq. (2.15) we have $de^{st}/dt = se^{st}$

**Asymptotic behavior.** The graph of $e^{\sigma t}\cos\omega t$ is illustrated in Fig. 2.13. Note

$$\lim_{t\to\infty} e^{st} = 0, \quad \text{if and only if} \quad \Re s < 0 \tag{2.18}$$

If $\Re s > 0$, $e^{st}$ becomes unbounded as time increases. If $\Re s = 0$, the signal is bounded and $|e^{st}| = |e^{j\omega t}| = 1$.

**Periodicity.** The purely imaginary exponential signal

$$e^{j\omega t} = \cos\omega t + j\sin\omega t \tag{2.19}$$

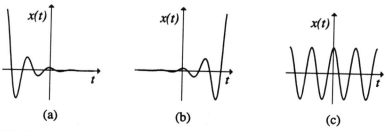

**Figure 2.13.** The real part of the complex exponential signal: (a) for $\Re s < 0$, (b) for $\Re s > 0$, and (c) for $\Re s = 0$.

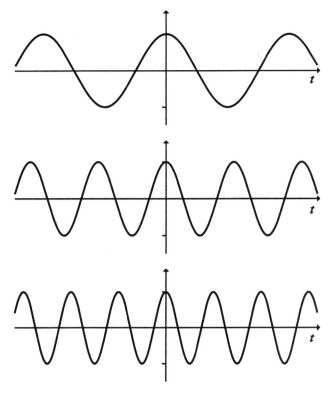

**Figure 2.14.** Graph of the cosine for three different frequencies.

is periodic. In fact this is the only case the complex exponential signal is periodic. The fundamental period is $T = 2\pi/\omega$; $\omega$ is the fundamental frequency. The signal $\cos \omega t$ is depicted in Fig. 2.14 for three frequency values, $\omega_1 < \omega_2 < \omega_3$. As frequency increases, the period decreases $T_1 > T_2 > T_3$. Thus as frequency becomes higher, the rate of oscillation increases and so does the variability of the signal. If the unit of time is sec, frequency units are rad/sec. We often use the scaling $\omega = 2\pi f$, where the unit for frequency $f$ is cycles per second or Hertz (Hz).

**The discrete exponential signal.** Let $a$ be a nonzero complex number written in polar form as $a = re^{j\Omega}$. The discrete complex exponential signal is defined as follows

$$x(n) = a^n = r^n e^{jn\Omega} = r^n \cos n\Omega + jr^n \sin n\Omega, \qquad -\infty < n < \infty$$

It is well known that $a^n \to 0$ if and only if $|a| < 1$. Thus the discrete exponential signal decays to zero if its parameter $a$ lies in the unit disc. The purely imaginary exponential signal

$$e^{j\Omega n} = \cos n\Omega + j \sin n\Omega$$

is obtained by sampling the continuous time exponential signal $e^{j\Omega t}$ with sampling period $T_s = 1$. It is periodic if $2\pi/\Omega$ is rational. The discrete periodic exponential signals of period $N$ have the form

$$x(n) = e^{j(2\pi/N)kn}, \qquad k = 0, \pm 1, \pm 2, \ldots \qquad (2.20)$$

**_Digital exponential signal._** Let $F$ be a finite field with $N$ entries and $a$ a nonzero element of $F$. The digital exponential signal is defined by

$$x(n) = a^n, \qquad -\infty < n < \infty$$

The reader should consult Appendix II at this point. The digital exponential takes values in the multiplicative cyclic subgroup generated by $a$. The finite structure of $F$ enforces periodicity. The fundamental period is the order of $a$. It is shown in Appendix II that the fundamental period is always a divisor of $N - 1$. As an example, consider the exponential signal in the field $GF(7)$. All possible fundamental periods are the divisors of 6, namely 2, 3, and 6.    ∎

### 2.5.1   Matrix exponential signal

In close analogy to the single-channel case we next discuss the multichannel exponential signal. Here we are contented with a first introduction of the concept and its properties. A complete description is supplied in Chapter 8.

**_Continuous time matrix exponential signal._** Let $A$ be a square matrix of order $k$ with complex entries. Motivated by (2.13), we define

$$e^A = \sum_{n=0}^{\infty} \frac{1}{n!} A^n = I + A + \frac{1}{2!}A^2 + \frac{1}{3!}A^3 + \cdots \qquad (2.21)$$

Thus $e^A$ is defined as the limit of the matrix sequence $S_N = \sum_{n=0}^{N} A^n/n!$. (For limits of matrix sequences, see Appendix I.) It can be shown that a matrix series $\sum_{n=0}^{\infty} C_n$ converges if the real series $\sum_{n=0}^{\infty} |C_n|$ converges (see Exercise 2.5). An induced matrix norm is used. Let $C_n = A^n/n!$. Then

$$\frac{|C_{n+1}|}{|C_n|} = \frac{\dfrac{1}{(n+1)!}|A^{n+1}|}{\dfrac{1}{n!}|A^n|} = \frac{1}{(n+1)}\frac{|A^{n+1}|}{|A^n|}$$

Since the norm is induced we have $|A^{n+1}|/|A^n| \leq |A|$. The ratio test implies that $e^A$ is defined for any matrix $A$. In particular, it is a real matrix if $A$ is real. The signal

$$x(t) = e^{At} = \sum_{n=0}^{\infty} \frac{t^n}{n!} A^n = I + tA + \frac{t^2}{2!} A^2 + \cdots, \qquad -\infty < t < \infty \qquad (2.22)$$

is the *continuous time matrix exponential signal*. Simple expressions are derived below in two special but important cases. The general case requires more effort and is developed in Chapter 8.

### Example 2.9   The nilpotent case

Contrary to the scalar case, the matrix series defining the exponential function may terminate after a finite number of terms. This occurs if there is an integer $N \geq 1$ such that $A^N = 0$. A matrix with this property is called *nilpotent*. In this case $A^n = 0$ for all $n \geq N$, and the exponential series reduces to the finite sum

$$e^{At} = I + tA + \frac{t^2}{2!} A^2 + \cdots + \frac{t^{N-1}}{(N-1)!} A^{N-1} = \sum_{n=0}^{N-1} \frac{t^n}{n!} A^n \qquad (2.23)$$

Suppose that

$$A = \begin{bmatrix} 0 & 1 \\ 0 & 0 \end{bmatrix}$$

We easily verify that $A^2 = 0$. Therefore

$$e^{At} = I + tA = \begin{bmatrix} 1 & 0 \\ 0 & 1 \end{bmatrix} + \begin{bmatrix} 0 & t \\ 0 & 0 \end{bmatrix} = \begin{bmatrix} 1 & t \\ 0 & 1 \end{bmatrix} \qquad \blacksquare$$

### Example 2.10   The diagonalizable case

Suppose that $A$ is diagonal with diagonal entries $\lambda_1, \ldots, \lambda_k$:

$$A = \text{diag}(\lambda_1, \lambda_2, \ldots, \lambda_k)$$

Then

$$A^n = \text{diag}(\lambda_1^n, \lambda_2^n, \ldots, \lambda_k^n)$$

and

$$e^{At} = \sum_{n=0}^{\infty} \frac{t^n}{n!} A^n = \text{diag}(e^{\lambda_1 t}, e^{\lambda_2 t}, \ldots, e^{\lambda_k t}) \qquad (2.24)$$

Now suppose that $A$ is diagonalizable; that is, it is similar to a diagonal matrix $D$. Let $P$ be an invertible matrix such that

$$A = PDP^{-1} \tag{2.25}$$

We observe that $A^2 = PDP^{-1}PDP^{-1} = PD^2P^{-1}$ and that generally

$$A^n = PD^nP^{-1} \tag{2.26}$$

Therefore

$$e^{At} = \sum_{n=0}^{\infty} \frac{t^n}{n!} PD^nP^{-1} = P\left(\sum_{n=0}^{\infty} \frac{t^n}{n!} D^n\right) P^{-1} = Pe^{Dt}P^{-1} \tag{2.27}$$

$e^{Dt}$ is computed by (2.24). The diagonal entries of $D$ are the eigenvalues of $A$, since similar matrices have the same eigenvalues. Moreover $A$ is diagonalizable, if the eigenvectors of $A$ form a basis. To see this, we write Eq. (2.25) as $AP = PD$. If $P$ is represented by its columns $P = (p_1 \quad p_2 \quad \cdots \quad p_k)$, we obtain

$$(Ap_1 \quad Ap_2 \quad \cdots \quad Ap_k) = P(\lambda_1 e_1 \quad \lambda_2 e_2 \quad \cdots \quad \lambda_k e_k)$$

where $e_i$ consists of 1 at the $i$th entry and zero otherwise. Hence $Ap_i = \lambda_i Pe_i = \lambda_i p_i$, $i = 1, 2, \ldots, k$, and the columns of $P$ are eigenvectors of $A$. $P$ is invertible if its columns form a basis in $C^k$. It is clear that not every matrix is diagonalizable. The $2 \times 2$ matrix of Example 2.9 is not diagonalizable, since the eigenvectors lie on the line $x_2 = 0$ and thus fail to span the entire space. More generally, every nonzero nilpotent matrix cannot be diagonalized (Exercise 2.7).

An important class of diagonalizable matrices is the set of matrices with pairwise distinct eigenvalues (see Exercise 2.30). A second example is the set of real symmetric matrices.

How typical is the diagonalizability property? It turns out that diagonalizable matrices are affluent. In fact they comprise a generic set in the class of square matrices. As a consequence nilpotent matrices, being nondiagonalizable, are exceptional. Let

$$A = \begin{bmatrix} 0 & 1 \\ 1 & 0 \end{bmatrix}$$

The characteristic polynomial of $A$ is $\phi(s) = \det(sI - A) = s^2 - 1$. The roots of $\phi(s)$ are the eigenvalues of $A$. These are $\lambda_1 = 1$, $\lambda_2 = -1$, and they are distinct. The eigenvectors of $\lambda_1 = 1$ are the nonzero solutions of the system $(I - A)x = 0$. Hence they constitute the nonzero points of the line $\{(x_1, x_2) : x_1 = x_2\}$. We pick one, say, $p_1 = (1 \quad 1)^T$. The eigenvectors of $\lambda_2 = -1$ are the nonzero solutions of $(-I - A)x = 0$ and coincide with the nonzero points of the line

$\{(x_1, x_2): x_1 + x_2 = 0\}$. One of them is $p_2 = (1 \quad -1)^T$. Then

$$P = (p_1 \quad p_2) = \begin{bmatrix} 1 & 1 \\ 1 & -1 \end{bmatrix}, \qquad P^{-1} = \begin{bmatrix} \dfrac{1}{2} & \dfrac{1}{2} \\ \dfrac{1}{2} & -\dfrac{1}{2} \end{bmatrix}$$

Equation (2.27) gives

$$e^{At} = \begin{bmatrix} \dfrac{1}{2}(e^t + e^{-t}) & \dfrac{1}{2}(e^t - e^{-t}) \\ \dfrac{1}{2}(e^t - e^{-t}) & \dfrac{1}{2}(e^t + e^{-t}) \end{bmatrix}$$

∎

The matrix counterpart of Equation (2.16) fails, that is, $e^{A+B} \neq e^A \cdot e^B$, unless $A$ and $B$ commute, that is, $AB = BA$ (see Exercise 2.8). As a consequence we have the following property:

**Invertibility.**  For any real $t$ the matrix $e^{At}$ is invertible and its inverse is given by

$$(e^{At})^{-1} = e^{-At}$$

Indeed, the matrices $At$ and $-At$ commute, and by (2.21), $e^0 = I$.

**Differentiation.**  The derivative of the exponential signal with respect to time is computed by termwise differentiation. Thus

$$\frac{d}{dt}e^{At} = \frac{d}{dt}\sum_{n=0}^{\infty}\frac{t^n}{n!}A^n = \sum_{n=0}^{\infty}\left(\frac{d}{dt}t^n\right)\frac{1}{n!}A^n = A\sum_{n=1}^{\infty}\frac{t^{n-1}}{(n-1)!}A^{n-1}$$

Hence

$$\frac{d}{dt}e^{At} = Ae^{At} \tag{2.28}$$

Equation (2.28) suggests that the matrix exponential signal is tightly connected with linear constant differential equations. These equations are extremely useful models for the description of evolutionary processes. They are discussed in detail in Chapter 8. The general form of a linear constant (or time invariant) differential equation is

$$\frac{d}{dt}X(t) = AX(t) \tag{2.29}$$

where $A$ is a given $k \times k$ matrix. We seek to determine a function $X(t)$ that satisfies (2.29). $X(t)$ is a matrix valued signal of dimensions $k \times m$, the most common case being $m = 1$. Inspection of Eq. (2.28) reveals that the matrix exponential is a solution of (2.29). Proceeding a step further, let $X_0$ be an arbitrary $k \times m$ matrix. Then the signal

$$X(t) = e^{At}X_0 \tag{2.30}$$

is a solution of (2.29). Indeed,

$$\frac{d}{dt}X(t) = \frac{d}{dt}(e^{At}X_0) = \left(\frac{d}{dt}e^{At}\right)X_0 = (Ae^{At})X_0 = A(e^{At}X_0) = AX(t)$$

Each solution of the form (2.30) takes the value $X_0$ at time $t = 0$:

$$X(0) = e^{A0}X_0 = IX_0 = X_0$$

We have seen that all functions (2.30) are solutions of (2.29). The converse is also true. Any solution of (2.29) is expressed as in (2.30). This is further discussed in Chapter 3. We thus conclude that the solutions of (2.29) are characterized by the matrix exponential signal.

### Example 2.11 Uniform motion

Consider the uniform motion of a particle moving on a line. If $y$ denotes the displacement of the particle from a given point, the acceleration $\ddot{y}$ will be zero. Let $x_1 = y$, $x_2 = \dot{y}$. Differentiation gives $\dot{x}_1 = \dot{y} = x_2$, $\dot{x}_2 = \ddot{y} = 0$. The vector $x = [x_1, x_2]^T$ satisfies (2.29), where

$$A = \begin{pmatrix} 0 & 1 \\ 0 & 0 \end{pmatrix}$$

The exponential matrix of $A$ was determined in Example 2.9. Thus

$$x(t) = \begin{pmatrix} x_1(t) \\ x_2(t) \end{pmatrix} = \begin{pmatrix} 1 & t \\ 0 & 1 \end{pmatrix} \begin{pmatrix} x_{10} \\ x_{20} \end{pmatrix} = \begin{pmatrix} x_{10} + x_{20}t \\ x_{20} \end{pmatrix}$$

We arrived at the familiar fact that velocity is constant and displacement varies linearly with time. ∎

### Example 2.12 Electrical circuit

Let us consider the constant resistance network of Fig. 2.15. It consists of an independent voltage source, a linear capacitor with capacitance $C$, an inductor with inductance $L$, and two resistors $R_1$ and $R_2$. We denote by $x_1(t)$ the voltage across the capacitor and by $x_2(t)$ the current passing through the inductor. Application of Kirchhoff's voltage law gives

$$u = x_1 + R_1 C \dot{x}_1$$
$$u = L\dot{x}_2 + R_2 x_2$$

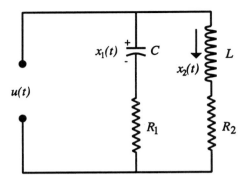

**Figure 2.15.** Electric circuit of Example 2.12.

Writing the above equations in matrix form, we obtain the differential equation

$$\dot{x} = Ax + bu$$

where

$$A = \begin{pmatrix} \dfrac{-1}{R_1 C} & 0 \\ 0 & \dfrac{-R_2}{L} \end{pmatrix}, \quad b = \begin{pmatrix} \dfrac{1}{R_1 C} \\ \dfrac{1}{L} \end{pmatrix}$$

Suppose that at time $t = 0$ the voltage across the capacitor is $x_{10}$ and the current through the inductor is $x_{20}$. The matrix $A$ is diagonal, so the exponential matrix $e^{At}$ is in accordance with Eq. (2.24):

$$e^{At} = \begin{pmatrix} e^{-t/R_1 C} & 0 \\ 0 & e^{-R_2 t/L} \end{pmatrix}$$

If the voltage source is switched off the circuit dynamics is given by (see Eq. 2.30):

$$x(t) = \begin{pmatrix} x_1(t) \\ x_2(t) \end{pmatrix} = e^{At} x_0 = \begin{pmatrix} e^{-t/R_1 C} & 0 \\ 0 & e^{-R_2 t/L} \end{pmatrix} \begin{pmatrix} x_{10} \\ x_{20} \end{pmatrix}$$

or

$$x_1(t) = e^{-t/R_1 C} x_{10}, \quad x_2(t) = e^{-R_2 t/L} x_{20}$$

As a consequence of the diagonal form of $A$ the above parameters are decoupled.   ∎

The issues of asymptotic behavior and periodicity are more delicate. Exercises 2.9 and 2.10 deal with the special case of diagonalizable matrices and thus provide a hint for the general case.

**Discrete matrix exponential signal.** Let $A$ be a $k \times k$ matrix with complex entries. The signal

$$x(n) = A^n, \quad 0 \leq n < \infty \tag{2.31}$$

constitutes the *discrete matrix exponential signal*. Note that time has been restricted to nonnegative values only. The discrete exponential signal can be defined for negative times provided $A$ is invertible so that $A^{-1}$, and hence $A^{-n}$, $n \geq 1$ exist.

**Recursivity.** We easily see that $x(n)$ is recursively obtained as

$$x(n+1) = Ax(n), \quad x(0) = I, \quad n \geq 0 \tag{2.32}$$

This property is analogous to the differentiation property of its continuous time counterpart. The asymptotic behavior is discussed in Chapter 11.

**Example 2.13  *MATLAB computation of the exponential signal***
The discrete time matrix exponential can be computed from Eq. (2.32) by the following code

```
function x = dexpon(A, x0, N)
x = [ ];
for i = 1:N      % specifies the observation interval
x = [x,x0];      % sets initial condition at x0
x0 = A*x0;       % recursively computes the previous equation
end
x = x';          % takes transpose
```

Faster implementations are obtained if the ltitr function of MATLAB is employed.

To determine the continuous time exponential, we proceed as follows: The solution of the differential equation (2.29), which starts at time $t_0$ (and not as previously at time 0) from $x_0$, is easily seen to be

$$x(t) = e^{A(t-t_0)}x_0 \tag{2.33}$$

Suppose that time is sampled with sampling period $T_s$. Let $x(t)$ be as above and $x(n) = x(nT_s)$. We will express $x(n)$ in terms of $x(n-1)$. For this purpose we apply Eq. (2.33) with $t_0 = (n-1)T_s$, $x_0 = x(n-1) = x((n-1)T_s)$.

Then $x(n) = x(nT_s) = e^{A(nT_s-(n-1)T_s)}x((n-1)T_s)$, or

$$x(n) = \Phi x(n-1), \quad \Phi = e^{AT_s} \tag{2.34}$$

The sequence $x(n)$ is computed with the aid of the discrete exponential function as follows:

```
function x = expon(A, x0, N, Ts)
Phi = expm(A*Ts)        % determines the exponential matrix of ATs
x = dexpon(Phi, x0, N)  % finds the discrete exponential          ∎
```

***Digital exponential signal.*** The digital exponential signal is defined by (2.31), however, the entries of $A$ belong to a finite field. The recursivity property (2.32) is clearly valid. In the single-channel case we saw that the digital exponential $x(n) = a^n$ is periodic if $a \neq 0$. This is no longer true in the matrix case. For instance, if $A$ is a nilpotent matrix, the exponential settles to zero after some initial transient period and hence is not periodic. We will later see that almost all exponential signals are periodic. A first result in this direction is the following:

**Theorem 2.1** Suppose that $A$ has entries in a finite field $F$ and is invertible. Then $A^n$, $n \in Z$ is periodic.

PROOF   The set of invertible matrices forms a multiplicative group with respect to matrix multiplication. It is a finite group and contains $A$ by assumption. Thus the order of $A$, $N$, is finite (see Appendix II) and $A^N = I$, $I$ being the identity matrix. Consequently $A^{n+N} = A^n A^N = A^n$.          ∎

### 2.5.2   2-D and m-D exponential signal

Let $s_1 = \sigma_1 + jw_1$, $s_2 = \sigma_2 + jw_2$. The continuous time 2-D exponential signal with parameters $s_1$ and $s_2$ is defined by

$$x(t_1, t_2) = e^{s_1 t_1 + s_2 t_2} = e^{\sigma_1 t_1 + \sigma_2 t_2 + j(w_1 t_1 + w_2 t_2)}$$

The purely imaginary exponential signal

$$x(t_1, t_2) = e^{j(w_1 t_1 + w_2 t_2)}$$

is periodic with periods $2\pi/w_1$ and $2\pi/w_2$. The discrete 2-D exponential signal is $x(n_1, n_2) = a_1^{n_1} a_2^{n_2}$. The purely imaginary signal has the form

$$x(n_1, n_2) = e^{j\Omega_1 n_1} e^{j\Omega_2 n_2} = e^{j(\Omega_1 n_1 + \Omega_2 n_2)}$$

Periodicity prevails if $\Omega_1 = 2\pi k_1/N_1$ and $\Omega_2 = 2\pi k_2/N_2$. Multidimensional extensions are easily constructed.

## 2.6  THE GAUSSIAN SIGNAL

Just as the exponential signal plays a prominent role in the analysis of deterministic signals, Gaussian signals are very important stochastic signals. A discrete random signal $x(n)$ with mean $m(n)$ and covariance $C(n_1, n_2)$ is called *Gaussian* or *normal* if for any $k$, $n_1$, $n_2$, ... , $n_k$, the random vector $x = (x(n_1) \quad x(n_2) \quad \cdots \quad x(n_k))^T$ has the normal density

$$f(x_1, x_2, \ldots, x_k) = \frac{1}{(2\pi)^{k/2}(\det C)^{1/2}} e^{-(x-m)^T C^{-1}(x-m)/2} \tag{2.35}$$

where

$$m = E[x] = (m(n_1) \quad m(n_2) \quad \cdots \quad m(n_k))^T, \quad C_{ij} = C(n_i, n_j)$$

A Gaussian signal is specified by the mean and covariance sequences. Therefore, if a Gaussian signal is wide sense stationary, it is also stationary. The above definition applies with the obvious notational changes to continuous time signals. Zero mean Gaussian white noise of variance 1 is generated by the command randn. A random field is called Gaussian if each vector formed by a finite number of pixels is Gaussian. A white noise Gaussian random field is shown in Fig. 2.16. It is generated by the following code:

```
[n1,n2] = meshdom(1:10,1:10);   % creates a 2-D grid
x = randn(10);                  % computes 2-D Gaussian white noise of variance 1.
subplot(121), mesh(x);          % yields a 2-D plot of x
subplot(122), mesh(xcorr2(x))   % plots the 2-D correlation
```

## 2.7  SIGNAL METRICS AND NORMS

As we will see in subsequent chapters, the performance of a signal processor, a controller or actuator, a filter or an encoder, is evaluated by comparing signals generated by any of the above systems and prespecified desired response signals. Such a comparison is effected by means of metrics (or distances). The metric helps us to define the proximity of two signals as well as the limit of a sequence of signals. The most common metrics are induced by norms on linear spaces. Of primary importance are finally the norms possessing the geometric features of Euclidean distance. These norms interpret the concept of signal energy and are produced by an inner product.

### 2.7.1  Definitions and examples

A set of signals $\mathcal{X}$ equipped with a distance function $d$ comprises a signal metric space. The set of real-valued functions bounded by a positive constant is a metric

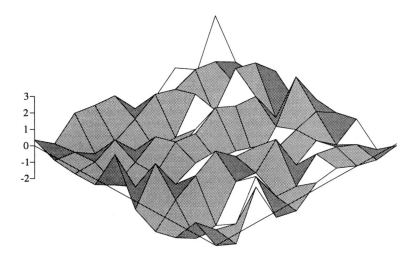

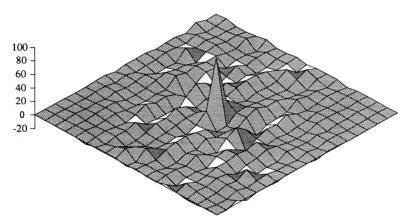

**Figure 2.16.** Plots of 2-D white Gaussian noise and its correlation.

space with distance function

$$d(x, y) = \sup_{t \in R} |x(t) - y(t)|$$

Most often $\mathcal{X}$ is a linear subspace of one of our basic signal models described in Section 2.1, and $d$ derives from a norm via the rule

$$d(x, y) = |x - y| \tag{2.36}$$

This brings us to the notion of signal normed spaces. Some important examples are described next. They constitute natural extensions of the $\ell_p$ norms on $R^k$, or $C^k$, described in Appendix I.

***Example 2.14    $\ell_p$ signal norms***
Let $p$ be a real number, $1 \leq p \leq \infty$, and

$$| x |_p = \left( \sum_{n=-\infty}^{+\infty} | x(n) |^p \right)^{1/p} \tag{2.37}$$

The set

$$\ell_p = \left\{ x \in \ell : | x |_p < \infty \right\} \tag{2.38}$$

is a normed space with norm given by (2.37). The proof is a direct consequence of the following inequalities stated without proof.

***Hölder's inequality.*** Let $1 \leq p, q \leq \infty$ such that $1/p + 1/q = 1$. Such numbers are called *conjugates*. For every signal $x$ and $y$, we have

$$\sum_{n=-\infty}^{\infty} | x(n) || y(n) | \leq \left( \sum_{n=-\infty}^{\infty} | x(n) |^p \right)^{1/p} \left( \sum_{n=-\infty}^{\infty} | y(n) |^q \right)^{1/q}$$

Holder's inequality is known as Schwarz inequality for $p = q = 2$.

***Minkowski's inequality.*** Given $1 \leq p \leq \infty$ and signals $x$ and $y$, it holds that

$$\left( \sum_{n=-\infty}^{\infty} | x(n) + y(n) |^p \right)^{1/p} \leq \left( \sum_{n=-\infty}^{\infty} | x(n) |^p \right)^{1/p} + \left( \sum_{n=-\infty}^{\infty} | y(n) |^p \right)^{1/p}$$

The $l_1$ norm is obtained for $p = 1$:

$$| x |_1 = \sum_{n=-\infty}^{\infty} |x(n)| \tag{2.39}$$

Signals with finite $l_1$ norm are called *absolutely summable*.
The $\ell_2$ norm is obtained for $p = 2$:

$$|x|_2 = \sqrt{\sum_{n=-\infty}^{\infty} |x(n)|^2} = \sqrt{\sum_{n=-\infty}^{\infty} x(n)x^*(n)} \tag{2.40}$$

Signals with finite $\ell_2$ norm are called *finite energy signals*. The signal energy is $|x|_2^2$.

By convention, the $\ell_p$ norm for $p = \infty$ yields the sup norm:

$$|x|_\infty = \sup_{-\infty < n < +\infty} |x(n)| \tag{2.41}$$

This is the supremum (least upper bound) amplitude of $x$. It is not always finite. The unit sample, the unit step, and the sinusoid have finite supremum amplitude because they are bounded. This is clearly not the case with the ramp signal.

It can be shown that for any $1 \leq p \leq q \leq \infty$, it holds that $\ell_p \subset \ell_q$. In particular,

$$\ell_1 \subset \ell_2 \subset \ell_\infty \tag{2.42}$$

(see Exercise 2.32). Thus an absolutely summable signal has finite energy and is bounded. Inclusions in (2.42) are proper. Finite support signals belong to all $\ell_p$ spaces. The unit step is not absolutely summable and does not have finite energy but is bounded and hence belongs to $\ell_\infty$. We will see in Section 4.1 that the sinc signal has finite energy but is not absolutely summable. ∎

### Example 2.15   $L_p$ signal norms

The continuous time counterparts of the $\ell_p$ signal spaces are the $L_p$ spaces. These are harder to analyze, and they require an excursion to Lebesgue integration. A detailed exposition would take us too far afield. To gain insight into the structure of these spaces, we first consider Riemann integrals up to the point where the need arises for an extension to the wider notion of Lebesgue integral. The continuous time analog of the discrete time norm (2.37) is

$$| x |_p = \left( \int_{-\infty}^{\infty} | x(t) |^p \, dt \right)^{1/p} \tag{2.43}$$

The above integral is interpreted as an improper Riemann integral. It then follows that for each $p : 1 \leq p \leq \infty$, the set

$$R_p = \left\{ x \in L : | x |_p < \infty \right\} \tag{2.44}$$

becomes a normed space with norm (2.43). For $p = 1$ we obtain the class of absolutely integrable signals, for $p = 2$ the finite energy signals, and for $p = \infty$ the class of bounded signals. In the latter case the norm is

$$| x |_\infty = \sup_{t \in R} | x(t) |$$

The spaces $R_p$ are different for different values of $p$ and are properly contained in $L$. They do, however, all contain a very important subspace: the set of finite support continuous signals. Indeed, the integral (2.43) of a finite support continuous signal becomes an ordinary finite Riemann integral, which exists because the integrand is a continuous function.  ∎

### 2.7.2 Convergence

Convergence in the large signal model set $\ell$ can be defined pointwise. A sequence of signals $x_N$ converges to a signal $x$ if for each $n$ the complex sequence $x_N(n)$ converges to $x(n)$. It is often necessary to work with more structured limiting procedures, as those induced by norms. In such cases we confine ourselves to normed subspaces of $\ell$, typically the $\ell_p$ spaces. Convergence in the $p$ norm means that $|x_N - x|_p \to 0$, $N \to \infty$. Convergence in the $p$ norm implies pointwise convergence but not conversely. The next theorem demonstrates the universal approximation capabilities of finite support signals.

**Theorem 2.2**   Given $\epsilon > 0$ and a signal $x$ with finite $p$ norm ($1 \le p < \infty$), there is a finite support signal $\tilde{x}$ such that $|x - \tilde{x}|_p < \epsilon$. Stated differently, there is a sequence of finite support signals $x_N$ converging to $x$ in the $p$th norm.

**PROOF**   Intuitively the idea is clear. A finite window is employed to look through the signal. As the window size gets bigger, better approximations result. Define the sequence of finite support signals

$$x_N(n) = \begin{cases} x(n), & |n| \le N \\ 0, & |n| > N \end{cases}$$

The sequence $x_N(n)$ is obtained by looking at the signal $x$ through a window of length $2N + 1$. We show that $x_N \to x$. Note that $x_N$ belongs to $\ell_p$. Moreover

$$|x_N - x|_p^p = \sum_{|n| \le N} |x_N(n) - x(n)|^p + \sum_{|n| > N} |x_N(n) - x(n)|^p$$

$$= \sum_{|n| > N} |x(n)|^p = |x|_p^p - \sum_{n=-N}^{N} |x(n)|^p$$

Hence $\lim_{N \to \infty} |x_N - x|_p^p = |x|_p^p - |x|_p^p = 0$.  ∎

The case $p = \infty$ has been excluded. The above result is not valid in $\ell_\infty$. To see this, take $x(n) = u_s(n)$, and note that $|x_N - x|_\infty = 1$.

The theorem can be phrased in more technical language as follows: Let $A$ be a subset of $\mathcal{X}$. The closure of $A$, $\bar{A}$, consists of all signals in $\mathcal{X}$ that can be obtained as limits of sequences from $A$:

$$\bar{A} = \left\{ x \in \mathcal{X} : \text{ there is } x_N \in A : x_N \to x \right\}$$

$\bar{A}$ embodies all the approximating power of $A$. If $\bar{A} = \mathcal{X}$, we say that $A$ is *dense* in $\mathcal{X}$. A dense set in $\mathcal{X}$ has a universal approximation capability in $\mathcal{X}$. Theorem 2.2 states that finite support signals are dense in $\ell_p$ for all $1 \leq p < \infty$. It can be shown that the closure of finite support signals with respect to the sup norm coincides with the class of signals that decay to zero.

Translation of Theorem 2.2 to continuous time signals presents serious difficulties. Consider the set of finite support signals once again, where this time we impose the natural assumption of continuity. We denote the corresponding set by $C_c$. It holds that $C_c \subset R_p$ for every $p$. Unfortunately, the limits of sequences of signals in $C_c$ with respect to the $p$ norm are not always Riemann integrable functions; that is, they do not belong to $R_p$. Rather, they are captured in the spaces of Lebesgue integrable functions, $L_p$. The space $L_p$ is given by (2.44), but the integral in (2.43) is replaced by the Lebesgue integral.

On several occasions we need to show that a sequence converges without explicit use of the limit. Fundamental or Cauchy sequences provide an important means for accomplishing this task. A sequence $x_N$ in a metric space is called *Cauchy or fundamental* if, for any $\epsilon > 0$, there is $N_0$ such that for all $N, M > N_0$ it holds $d(x_N, x_M) < \epsilon$. In $R^n$ a sequence is fundamental if and only if it is convergent. In a general normed space a convergent sequence is fundamental, but not vice versa. A counterexample is described in Exercise 2.21. This remark leads us to the following definition: A metric space $\mathcal{X}$ with the property that every Cauchy sequence in $\mathcal{X}$ converges in $\mathcal{X}$ is called *complete*. A complete normed space is called *Banach space*. Important examples of Banach spaces are provided by the following theorem, proved in the Appendix at the end of the chapter.

**Theorem 2.3**   Every normed space $\ell_p$, $1 \leq p \leq \infty$, is a Banach space. We mentioned previously that translation of the $\ell_p$ structure to the continuous time case does not lead to the $R_p$ spaces. The reason is that $R_p$ is not complete. In each of these spaces, fundamental sequences can be found that do not converge. It turns out that if $R_p$ is properly enlarged, it becomes complete, and hence a Banach space. The space resulting from the completion process is precisely the space $L_p$. Lebesgue integrals share all properties of Riemann integrals. They reduce to Riemann integrals when the latter exist. This is the case if the signal is piecewise continuous. A further feature of each $L_p$ space is that two functions that are identical almost everywhere—at all points of $R$ with the exception of a negligible set (a set of measure zero)—are identical as elements of $L_p$.

***Multidimensional signal metrics.***   The previous discussion extends to multidimensional signals in a natural way. The $\ell_p$ norm of a 2-D signal is

$$|x(n_1, n_2)|_p = \left( \sum_{n_1=-\infty}^{\infty} \sum_{n_2=-\infty}^{\infty} |x(n_1, n_2)|^p \right)^{1/p}$$

The linear space of 2-D signals with finite $\ell_p$ norm is a Banach space.

**Digital signals.** Metrics and norms of digital signals are of marginal interest in this book. One interesting metric used in the context of symbolic dynamics is defined as follows. Consider the class of digital signals defined on $Z^+$ with values in the integer field $GF(p)$. We define

$$d_p(x, y) = \sum_{n=0}^{\infty} \frac{1}{p^n} |x(n) - y(n)| \qquad (2.45)$$

where $|.|$ is the absolute value. The presence of $1/p^n$ warrants that the above distance is well defined for all digital signals. Indeed, since $|x(n) - y(n)| \le p - 1$, the series is dominated by $(p-1)\sum 1/p^n = p$. It is easy to see that it defines a valid metric. The following proposition states that if two signals are $1/p^k$ distance apart, their first $k$ values coincide.

**Proposition 2.1** Let $x, y$ be two signals defined on $Z^+$ with values in $GF(p)$ and $k$ be a positive integer. Then $d_p(x, y) \le 1/p^k$ if and only if $x(n) = y(n)$ for all $0 \le n \le k$.

**PROOF** Let $d_p(x, y) \le 1/p^k$. Suppose that $x(n) \ne y(n)$ for some $n < k$. Then $d_p(x, y) \ge |x(n) - y(n)|/p^n > 1/p^k$. Conversely, if $x(n) = y(n)$ for all $n \le k$, then

$$d_p(x, y) = \sum_{n=0}^{k} \frac{|x(n) - y(n)|}{p^n} + \sum_{n=k+1}^{\infty} \frac{|x(n) - y(n)|}{p^n} \le \sum_{n=k+1}^{\infty} \frac{p-1}{p^n} = \frac{1}{p^k}$$

∎

## 2.8 ORTHOGONAL EXPANSIONS

Among the norms introduced in the previous section, the continuous time $L_2$ norm and its discrete counterpart $l_2$ norm inherit additional important structure. Thus they allow the extension of most familiar concepts from Euclidean geometry to abstract signal spaces. In fact the generalization procedure becomes a routine once the notion of inner product is recognized.

A complex inner product associates with each pair of vectors $x, y$ from a complex linear space a complex number denoted $< x, y >$ so that the following properties are satisfied:

1. $< x, x > \ge 0$ and $< x, x > = 0 \Longleftrightarrow x = 0$.
2. $< x, y > = < y, x >^*$.
3. $< c_1 x + c_2 y, z > = c_1 < x, z > + c_2 < y, z >$.

A real inner product on a real vector space assigns a real number to each pair $x, y$, and property 2 simplifies to $< x, y > = < y, x >$. A real vector space $E$ equipped with a real inner product is called *Euclidean space* or *inner product space*. In the complex case, $E$ is called (complex) Euclidean space or *unitary space*. The above definition generalizes the standard Euclidean inner product on the plane

$$< x, y >= x_1 y_1 + x_2 y_2 = |x| \, |y| \cos \theta \qquad (2.46)$$

where $|x| = \sqrt{x_1^2 + x_2^2}$ is the length of $x$, and $\theta$ is the angle formed by $x$ and $y$. With the aid of the inner product various geometric notions are readily defined. Suppose that $E$ is a (complex) Euclidean space with inner product $<, >$. The *length* of a vector $x \in E$ is defined as

$$|x| = \sqrt{< x, x >} \qquad (2.47)$$

The latter expression defines a norm on $E$ (Exercise 2.23), and $E$ becomes a normed space. This space is not always a Banach space. If $E$ equipped with the norm induced by the inner product is a Banach space, it is called *Hilbert space*. The space $C^k$ together with the usual inner product $< x, y >= \sum_{i=1}^{k} x_i y_i^*$, is a complex Hilbert space. The distance induced by (2.47) is

$$d(x, y) = |x - y| = \sqrt{< x - y, x - y >} \qquad (2.48)$$

The inner product is a continuous mapping, where convergence is specified by the induced norm (2.47). Thus, if $x_N \to x$ (in the sense that $|x_N - x| \to 0$) and $y_N \to y$, then $< x_N, y_N > \to < x, y >$. The inner product and the norm induced by it are related by the celebrated *Cauchy-Schwarz inequality*:

$$| < x, y > | \leq |x||y| \qquad (2.49)$$

The above inequality becomes equality if and only if the two vectors $x$ and $y$ are linearly dependent, namely either one of them is zero or there is a scalar $c$ such that $x = cy$. In other words, $x$ and $y$ are colinear. Motivated by the plane equation (2.46), we define the angle formed by two vectors $x$ and $y$ in $E$ by

$$\cos \theta = \frac{< x, y >}{|x||y|} \qquad (2.50)$$

The Cauchy-Schwarz inequality certifies the standard trigonometric bound $|\cos \theta| \leq 1$.

### Example 2.16   *Discrete signal inner products*
Among the $\ell_p$ spaces introduced in the previous section, only the $\ell_2$ space inherits a natural inner product whose induced norm (2.47) coincides with the $\ell_2$ norm of Section 2.7. Let $x$ and $y$ be in $\ell_2$. Define the inner product by the rule

$$< x, y >= \sum_{n=-\infty}^{\infty} x(n) y^*(n) \qquad (2.51)$$

Hölder's inequality shows that (2.51) is a well-defined (finite) complex number. The inner product axioms are easily inferred. Thus $\ell_2$ is a Euclidean space and as such it provides a natural environment for studying geometric operations on signals.

In certain cases we find it convenient to work with slightly different inner products on $\ell_2$. Let $w(n)$ be a given sequence satisfying $0 < w(n) \le W$, for all $n \in Z$. We define the weighted inner product

$$< x, y >_w = \sum_{n=-\infty}^{\infty} w(n)x(n)y^*(n) \tag{2.52}$$

This is well defined because

$$\sum_{n=-\infty}^{\infty} |w(n)x(n)y^*(n)| \le W \sum_{n=-\infty}^{\infty} |x(n)y^*(n)| \le W|x|_2|y|_2$$

The meaning of (2.52) is that at each time $n$ the weighting factor $w(n)$ attaches different significance to the term $x(n)y(n)^*$. Among the most common weighting sequences $w(n)$ are the signals that decay to zero as $|n| \to \infty$. Such signals are usually referred to as *time windows*. ∎

### Example 2.17  Continuous time signal inner products
Among the $R_p$ spaces of Riemann integrable signals, only $R_2$ admits an inner product whose induced norm coincides with the $R_2$ norm. For any $x$ and $y$ in $R_2$, define the inner product

$$< x, y > = \int_{-\infty}^{\infty} x(t)y^*(t)dt \tag{2.53}$$

Hölder's inequality ensures that this is a finite number. The axioms of the inner product are easily verified. If Lebesgue integration is used in (2.53), the resulting space $L_2$ becomes a Hilbert space.

The set of periodic signals of period $T$ that are square integrable over a period forms an inner product space with inner product

$$< x, y > = \int_{T} x(t)y^*(t)dt$$ ∎

## 2.9  ORTHOGONAL FAMILIES

The angle between two vectors $x$ and $y$ was defined by Eq. (2.50). The most important case results when the two vectors form a right angle. This occurs when $< x, y > = 0$. We say that $x$ and $y$ are *orthogonal* if $< x, y > = 0$. A set of vectors $A$ is called *orthogonal*, if it consists of mutually orthogonal nonzero vectors. If in addition each vector $x$ in $A$ has unit length (i.e., $< x, x >= 1$), then $A$ is called *orthonormal*.

For any two orthogonal vectors $x$ and $y$, the *Pythagorean identity* holds

$$|x + y|^2 = |x|^2 + |y|^2 \tag{2.54}$$

Indeed,

$$|x + y|^2 = < x + y, x + y > = < x, x > + < x, y > + < y, x > + < y, y >$$

From orthogonality $< x, y >= < y, x >= 0$. Hence (2.54) is proved.

It is intuitively evident that nonzero orthogonal vectors are linearly independent. This is verified next.

**Theorem 2.4**   An orthogonal set $A$ is linearly independent.

**PROOF**   Let $x_1, x_2, \ldots, x_n$ vectors in $A$ and $\sum_{i=1}^{n} a_i x_i = 0$. Then

$$0 = < 0, x_j >= < \sum_{i=1}^{n} a_i x_i, x_j >= \sum_{i=1}^{n} a_i < x_i, x_j >= a_j < x_j, x_j >$$

The latter equality follows from the orthogonality of $A$. Since all vectors of $A$ are nonzero, $< x_j, x_j >\neq 0$, and thus $a_j = 0$. Therefore $A$ is linearly independent. ∎

### 2.9.1   Gram-Schmidt orthogonalization

Let $G = \{g_0, g_1, \ldots, g_k\}$ be a linearly independent set in an inner product space $H$. There exists an orthonormal set that spans the same space with $G$. Indeed, it suffices to construct an orthogonal set $\{f_0, f_1, \ldots, f_k\}$ having the same linear span with $G$. This set can then be converted to an orthonormal set by the normalization

$$\tilde{f}_i = \frac{1}{|f_i|} f_i = \frac{1}{\sqrt{< f_i, f_i >}} f_i \tag{2.55}$$

We set $f_0 = g_0$. The vector $f_1$ must lie on the plane spanned by $g_0, g_1$ and must be orthogonal to $f_0$. Hence we choose $f_1$ as the projection of $g_1$ on the line through the origin which is orthogonal to $f_0$. Thus $f_1 = g_1 - c_0 f_0$, and $< f_1, f_0 >= 0$. Then

$$0 = < g_1, f_0 > -c_0 < f_0, f_0 > \quad \text{or} \quad c_0 = \frac{< g_1, f_0 >}{< f_0, f_0 >}$$

Next we use induction. Suppose that the set $\{f_0, f_1, \ldots, f_m\}$ is orthogonal and spans the same space $M_m$ with $\{g_0, g_1, \ldots, g_m\}$. Let $f_{m+1}$ be the projection of $g_{m+1}$ on the line which is orthogonal to $M_m$. Thus

$$f_{m+1} = g_{m+1} - \sum_{i=0}^{m} c_i f_i, \quad < f_{m+1}, f_i >= 0, \quad i = 0, 1, \ldots, m$$

Taking into account that $< f_i, f_j >= 0$, $i \neq j$, we have

$$0 = < f_{m+1}, f_j >=< g_{m+1}, f_j > - \sum_{i=0}^{m} c_i < f_i, f_j >$$

$$= < g_{m+1}, f_j > -c_j < f_j, f_j >$$

or

$$c_j = \frac{< g_{m+1}, f_j >}{< f_j, f_j >}, \quad j = 0, 1, \ldots, m \qquad (2.56)$$

The Gram-Schmidt orthogonalization process is summarized below:

---

**GRAM-SCHMIDT ALGORITHM**

$$f_0 = g_0$$
$$f_{m+1} = g_{m+1} - \sum_{i=0}^{m} \frac{< g_{m+1}, f_i >}{< f_i, f_i >} f_i$$

---

Some important examples of orthogonal sets follow.

***Example 2.18   Orthonormal family of unit sample shifts***
Let $k$ be an integer. The discrete signal

$$\delta_k(n) = \begin{cases} 1, & n = k \\ 0, & n \neq k \end{cases} \qquad (2.57)$$

is obtained by shifting the unit sample $\delta(n)$ $k$ steps: $\delta_k(n) = \delta(n - k)$. Each $\delta_k(n)$ has finite energy (equal to 1) and thus belongs to $\ell_2$. The signals $\delta_k(n)$, $k \in Z$, form an orthonormal family in $\ell_2$ because

$$\sum_{n=-\infty}^{\infty} \delta_k(n) \delta_m(n) = \sum_{n=-\infty}^{\infty} \delta(n - k) \delta(n - m) = \begin{cases} 1, & k = m \\ 0, & k \neq m \end{cases} \qquad \blacksquare$$

***Example 2.19   Harmonically related continuous time exponential signals***
Let $\omega$ be a real number and $k$ be an integer. The exponential signal $e^{jk\omega t}$ has

fundamental period $2\pi/k\omega$. Let $T = 2\pi/\omega$. Then $T = k(2\pi/k\omega)$. Hence $T$ is a common period for all signals $e^{jk\omega t}$, $k \in Z$. The members of the above family are called *harmonically related* exponential signals because their fundamental frequencies are integer multiples of a single frequence $\omega$, and in music this results to harmonic tones. Harmonic exponential signals taken over any finite interval of duration $T = 2\pi/\omega$ form an orthogonal set. Indeed,

$$< e^{jk\omega t}, e^{jm\omega t} > = \int_T e^{j(k-m)\omega t} dt = \begin{cases} 0 & k \neq m \\ T & k = m \end{cases} \qquad (2.58)$$

∎

### Example 2.20   Harmonically related discrete exponentials

We saw in Section 2.5 that the periodic discrete time exponential signals of period $N$ have the form $f_k(n) = e^{jk(2\pi/N)n}$. The set of these signals is finite and can be described as

$$A = \{f_0(n), f_1(n), \ldots, f_{N-1}(n)\} \qquad (2.59)$$

Indeed, $f_N(n) = e^{jN2\pi n/N} = e^{j2\pi n} = 1 = f_0(n)$. In general, let $k \in Z$, and $q$, $r$ the quotient and remainder of $k$ upon division by $N$, $k = qN + r$, $0 \leq r \leq N - 1$. Then $f_k(n) = e^{j2\pi(qN+r)n/N} = e^{j2\pi rn/N} = f_r(n)$. $A$ constitutes an orthogonal set in the space of discrete periodic signals of period $N$, equipped with inner product

$$< x, y > = \sum_{n=0}^{N-1} x(n)y^*(n) \qquad (2.60)$$

To see this, we compute

$$< f_k, f_m > = \sum_{n=0}^{N-1} e^{jk2\pi n/N} e^{-jm2\pi n/N} = \sum_{n=0}^{N-1} e^{j(k-m)2\pi n/N}$$

The right-hand side is a partial sum of a geometric series. It is zero if $k \neq m$. If $k = m$, we have $< f_k, f_k > = N$.

∎

One notable difference between the discrete time periodic exponential family and the analog periodic exponential counterpart is that the former is a finite set, while the latter is an infinitely countable set. As a consequence discrete time harmonically related exponential signals lead to finite sum representations, whereas continuous time periodic exponentials give rise to infinite series representations. This remark becomes precise later on.

### 2.9.2   Finite orthogonal sets

Let $A$ be a subset of a linear space $E$. Let $[A]$ denote the *linear span* of $A$. An element of $[A]$ is a finite linear combination of vectors in $A$. The linear span of $A$ is the smallest linear subspace containing $A$. If $[A]$ is finite-dimensional and $f_0$, $f_1, \ldots, f_{N-1}$ is a basis in $[A]$, any vector $x \in [A]$ is expressed uniquely as

$$x = \sum_{k=0}^{N-1} a_k f_k \tag{2.61}$$

Now let us assume that $A$ is a finite orthogonal set in the inner product space $E$. Then $A = \{f_0, f_1, \ldots, f_{N-1}\}$, and $<f_i, f_j> = 0$, $i \neq j$. By Theorem 2.4, $A$ is linearly independent. Therefore any $x \in [A]$ is represented as in (2.61). Since $f_j$ are pairwise orthogonal, each term $a_k f_k$ corresponds to the projection of $x$ on the direction of $f_k$. Therefore the coefficient $a_k$ can be determined by the formula

$$a_k = \frac{<x, f_k>}{<f_k, f_k>}, \qquad k = 0, 1, 2, \ldots, N-1 \tag{2.62}$$

If the set $A$ is orthonormal, the latter becomes $a_k = <x, f_k>$. Indeed,

$$<x, f_m> = <\sum_k a_k f_k, f_m> = \sum_k a_k <f_k, f_m> = a_m <f_m, f_m> = a_m$$

Equations (2.62) and (2.61) are referred to as the analysis and synthesis equations, respectively:

> Synthesis equation:  $x = \sum_k a_k f_k$
>
> Analysis equation:  $a_k = <x, f_k> / <f_k, f_k>$

The synthesis equation is valid for any element of $[A]$, $x$. Now, if the space $E$ has finite dimension $N$, then necessarily $[A] = E$, and the synthesis equation holds for all elements of $E$. A direct consequence is the following important property: For any vector $x \in E$, it holds that

$$|x|^2 = \sum_k |a_k|^2 |f_k|^2 \tag{2.63}$$

Indeed,

$$|x|^2 = <x, x> = \sum_k \sum_m a_k a_m^* <f_k, f_m> = \sum_k |a_k|^2 |f_k|^2$$

The above analysis enables us to represent discrete periodic signals in terms of harmonically related exponentials. This topic is further discussed in Chapter 5.

### 2.9.3    Countable orthogonal sets

In contrast to the discrete periodic case, general orthogonal representations need an infinitely countable number of orthogonal directions. It thus appears that the synthesis equation will involve an infinite series. The limiting operation is now inevitable. The derivations become more elaborate. It is remarkable that much of the finite-dimensional theory remains valid in this more general context. A key role in this respect is played by the projection theorem.

Let us consider a countable orthogonal set

$$A = \left\{ f_k, k \in Z \right\} \tag{2.64}$$

in a Hilbert space $H$. The linear span of $A$ must be further expanded with limits of sequences of $[A]$. This leads to the closure of $A$, $\overline{[A]}$, introduced in Section 2.7.2. Each element of $\overline{[A]}$ is the limit of a sequence with entries in $[A]$. We call $\overline{[A]}$ the *linear closure* of $A$. It is easy to see that the linear closure of $A$ forms a linear subspace of $H$. The linear span $[A]$ of $A$ does not have finite dimension because, if it did, Theorem 2.4 would be violated. Since $[A] \subset \overline{[A]}$, the linear closure does not have finite dimension, either. The elements of the linear closure of $A$ are represented by series of orthogonal vectors of $A$. Let us introduce the concept of a series in $H$. If $x_n$ is a sequence in $H, n \in Z$, we write $x = \sum_{n=-\infty}^{\infty} x_n$, and say $x_n$ converges to $x$ in $H$, if the sequence of partial sums $x_{M,N} = \sum_{n=-M}^{N} x_n$ converges to $x$; that is, $|x_{M,N} - x| \to 0$, as $M, N \to \infty$. The next theorem extends the synthesis and analysis equations (2.62), (2.61) to countable orthogonal sets.

**Theorem 2.5**    Consider a countable orthogonal set $A$ of the form (2.64), in a Hilbert space $H$. Any element $x$ in $\overline{[A]}$ admits a unique representation

$$x = \sum_{k=-\infty}^{\infty} a_k f_k \tag{2.65}$$

where

$$a_k = \frac{< x, f_k >}{< f_k, f_k >}, \quad \sum_{k=-\infty}^{\infty} |a_k|^2 |f_k|^2 < \infty \tag{2.66}$$

Furthermore

$$|x|^2 = < x, x > = \sum_{k=-\infty}^{\infty} |a_k|^2 |f_k|^2 \tag{2.67}$$

The proof is supplied in the Appendix at the end of the chapter.    ∎

### 2.9.4  Universal approximation of orthogonal expansions

A critical question that is next addressed concerns the size of the linear closure of an orthogonal set $A$ in a Hilbert space $H$. The bigger $\overline{[A]}$ is, the more powerful the representation capability of $A$ in $H$. With these in mind we give the following definition:

An orthogonal set $A$ is called *complete* in $H$ if its linear closure covers $H$, that is, $\overline{[A]} = H$. We recollect from the finite-dimensional case that if $H$ has dimension $N$ and $A$ contains $N$ elements, then $[A] = H$ and $A$ is complete.

We next derive an important criterion for completeness. It will follow as a direct corollary of an extremely useful result known as *projection theorem*. The projection theorem is illustrated in Fig. 2.17. We are given a subspace $M$ and a point $x$. We want to determine the point $m_o \in M$ that lies the shortest distance from $x$. In other words, we want to solve the minimization problem

$$\min_{m \in M} |x - m| \tag{2.68}$$

The solution to this problem is the projection of $x$ onto $M$. It is characterized by the orthogonality principle

$$< x - m_o, m >= 0 \qquad \text{for all } m \in M \tag{2.69}$$

A precise formulation is given below.

**Theorem 2.6 Projection theorem**  Let $H$ be a Hilbert space and $M$ a closed subspace of $H$, that is $\overline{M} = M$. For any $x \in H$ there exists a unique point $m_o \in M$ such that

$$|x - m_o| \leq |x - m| \qquad \text{for all } m \in M$$

where $m_o$ is characterized by the orthogonality principle (2.69).

The proof is given in the Appendix at the end of the chapter.  ∎

With the aid of the projection theorem we deduce the following criterion of completeness of orthogonal sets.

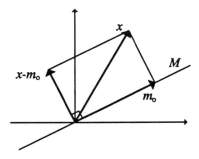

**Figure 2.17.**  Illustration of the projection theorem.

**Corollary 1**    An orthogonal set $A$ is complete in a Hilbert space $H$, if and only if, the only vector orthogonal to $A$ is the zero vector.

The proof is supplied in the Appendix at the end of the chapter.

***Example 2.21    Expansions of discrete energy signals in sample shifts***

Let us consider the orthonormal set of unit sample shifts. We can show by Corollary 1 that it is complete in $\ell_2$. Indeed, let $x$ be an energy signal that is orthogonal to all shifts of the unit sample. Then

$$0 = <x, \delta_k> = \sum_{n=-\infty}^{\infty} x(n)\delta_k(n) = \sum_{n=-\infty}^{\infty} x(n)\delta(n-k) = x(k)$$

Thus $x(k) = 0$ for all $k$. Consequently every energy signal $x(n)$ can be represented as

$$x(n) = \sum_{k=-\infty}^{\infty} x(k)\delta(n-k) \tag{2.70}$$

∎

Let $M^{\perp}$ denote the subspace of $H$ consisting of all vectors orthogonal to $M$. Thus $y \in M^{\perp}$ if $<y, x> = 0$ for all $x \in M$. It can be shown that $M^{\perp}$ is a closed subspace of $H$. Moreover $M \cap M^{\perp} = \{0\}$ and any $x \in H$ is uniquely written as $x = m_0 + \tilde{m}$, where $m_0$ belongs to $M$ and $\tilde{m}$ belongs to $M^{\perp}$. $m_0$ is the projection of $x$ onto $M$, and $\tilde{m}$ is the projection of $x$ onto $M^{\perp}$. These properties are summarized in the expression

$$H = M \oplus M^{\perp}$$

$M^{\perp}$ is called the *orthogonal complement* of $M$ in $H$.

### 2.9.5    The Hilbert space of a second-order stochastic signal

Consider the space of real random variables of zero mean and finite variance on a given probability space. This becomes a Hilbert space with inner product

$$<X, Y> = E[XY]$$

In fact it can be viewed as an $L_2$ space in a more abstract setup where expectation is interpreted as integration with respect to the probability measure. Two random variables are orthogonal if they are uncorrelated.

Let us consider a signal $x(n)$ whose values are second-order random variables, that is, they have finite variance. We say that $x(n)$ is a *second-order* stochastic process. Let $H_x(\infty)$ denote the linear closure of $A = \{x(n) : n \in Z\}$ in the above Hilbert space. $H_x(\infty)$ consists of linear combinations of signal values or limits of such linear combinations. Convergence is defined with respect to the distance induced by the inner product, namely

$$y_N \to y \quad \text{if and only if} \quad E[(y_N - y)^2] \to 0$$

and is referred to as *quadratic mean convergence*. The space $H_x(\infty)$ is a Hilbert space and encompasses all information generated by linear combinations of signal values or limits of such combinations. It is readily visualized by the theory of the previous sections if the signal is wide sense stationary white noise. In this case, by definition, the set $A$ is orthogonal. Every element $y$ in $H_x(\infty)$ is written

$$y = \sum_{k=-\infty}^{\infty} h(k)x(k), \quad \sum_{k=-\infty}^{\infty} |h(k)|^2 < \infty \tag{2.71}$$

The variance of $y$ is given by (see Eq. (2.67))

$$E[y^2] = <y, y> = \sum_{k=-\infty}^{\infty} |h(k)|^2 \gamma_2, \quad \gamma_2 = E[x^2(k)]$$

Finally Eq. (2.66) shows that

$$h(k) = \frac{E[yx(k)]}{\gamma_2}$$

The space $H_x(\infty)$ and its detailed structure is thoroughly studied in later chapters.

## BIBLIOGRAPHICAL NOTES

There are many good books devoted to signals and systems. Examples include Athans, Dertouzos, Spann, and Mason (1974), Lathi (1974), C. Liu and J. Liu (1975), Oppenheim, Willsky, and Young (1983), Oppenheim and Schafer (1989), Manolakis and Proakis (1989), Papoulis (1985), and Dickinson (1991). For an exposition of stochastic processes, the reader may consult Papoulis (1991), Wong and Hajek (1985), Priestley (1981), and Gardner (1986). Multidimensional signals are discussed in Jain (1989), Dudgeon and Mersereau (1984), and Lim (1990). For basic linear algebra concepts, the reader may consult Chen (1984), Hirch and Smale (1974), or standard linear algebra textbooks such as Strang (1980). A good introductory source for metrics, norms, and inner products is Kolmogorov and Fomin (1970). A readable account of the projection theorem and its manifold uses is contained in Luenberger (1969).

## PROBLEMS

**2.1.** Let $x(n)$ be the signal of Fig. 2.18. Plot each of the following signals:

(a) $x(n-3)$  (b) $x(2-n)$      (c) $x(2n)$
(d) $x(2n+1)$  (e) $x(n-1)\delta(n-2)$  (f) $x(n^2)$
Verify each case by MATLAB.

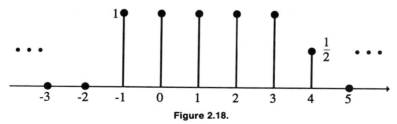

**Figure 2.18.**

**2.2.** Repeat Problem 2.1 for the signal $x(t)$ of Fig. 2.19 and the signals
(a) $x(t-2)$  (b) $x(3-t)$  (c) $x(3t)$
(d) $x(2t+1)$  (e) $x(t^2)$

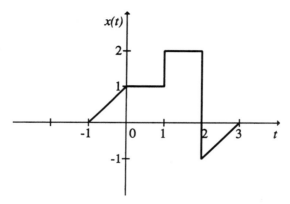

**Figure 2.19.**

**2.3.** Show that the set of analog signals of period $T$ form a linear subspace of $L$. Likewise show that the discrete periodic signals of period $N$ form a finite-dimensional linear subspace of $\ell$ of dimension $N$.

**2.4.** Prove the identity $e^{s_1+s_2} = e^{s_1}e^{s_2}$.

**2.5.** Show that the matrix series $\sum_{n=0}^{\infty} C_n$ converges absolutely if and only if the real series $\sum_{n=0}^{\infty} |C_n|$ converges.

**2.6.** Compute the analog matrix exponential signal $e^{At}$ for each of the two choices

$$A = \begin{pmatrix} 0 & 1 & 2 \\ 0 & 0 & 3 \\ 0 & 0 & 0 \end{pmatrix}, \quad A = \begin{pmatrix} 4 & 1 & 2 \\ 0 & 1 & 3 \\ 0 & 0 & 2 \end{pmatrix}$$

Verify your results with the MATLAB functions of Example 2.13.

**2.7.** Prove that every nonzero nilpotent matrix is not diagonalizable.

**2.8.** Prove that $e^{A+B} = e^A e^B$ if $A$ and $B$ commute, that is, $AB = BA$.

**2.9.** Let $A$ be a diagonalizable matrix. Prove that $\lim_{t \to \infty} e^{At} = 0$ if and only if the eigenvalues of $A$ have negative real part that is, $\Re\lambda(A) < 0$. Develop a MATLAB demonstrator as follows: Choose a diagonal matrix $D$ with negative real part diagonal elements and an invertible matrix $P$. Set $A = PDP^{-1}$. Compute $e^{At}$ with the function of Example 2.13, and plot the results.

**2.10.** Show that if $A$ is diagonalizable with purely imaginary eigenvalues and such that $\Im\lambda = k\omega$, $k \in Z$, the exponential signal $e^{At}$ is periodic with period $2\pi/\omega$. Derive a similar condition for the discrete exponential. Develop a MATLAB demonstrator as in Exercise 2.9.

**2.11.** Let $A$ be a diagonalizable matrix. Show that $\lim_{n \to \infty} A^n = 0$ if and only if all eigenvalues of $A$ are inside the unit circle, that is, $|\lambda(A)| < 1$. Develop a MATLAB demonstrator as in Exercise 2.9.

**2.12.** Create a MATLAB signal base incorporating the signals encountered in Chapter 2. The base consists of m files each one consisting of the function defining the signal. Use existing macros in MATLAB, or develop new ones to compute useful signal functions. For instance, the $\ell_p$ norm of a finite support signal $x$ is computed by the macro norm($x, p$).

**2.13.** Let $x(n_1, n_2)$ be the signal of Fig. 2.20. Plot each of the following signals and verify your results with MATLAB.
(a) $x(n_1 - 2, n_2 - 1)$   (b) $x(2 - n_1, 1 - n_2)$
(c) $x(2n_1, 2n_2)$   (d) $x(2n_1 + 1, n_2 + 2)$

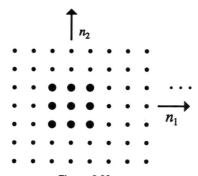

**Figure 2.20.**

**2.14.** When is the sum of two analog periodic signals periodic, and what is the fundamental period? Examine the same problem for the product of two signals.

**2.15.** Examine Problem 2.14 for the discrete case.

**2.16.** Check if each of the following signals is periodic. In the affirmative case determine the fundamental period.

$$x(t) = 3\cos(3t + \pi/5) \qquad x(t) = e^{j(\pi t - 2)}$$

$$x(n) = \sin(4\pi n/7 + 2) \qquad x(t) = e^{2\pi t}u_s(t)$$

$$x(n) = \cos(n/4)\cos(\pi n/4) \quad x(n) = \cos(\pi n^2/8)$$

**2.17.** Define the m-D discrete exponential signal. When is it periodic?

**2.18.** Compute the $p$ norms for $p = 1, p = 2, p = \infty$ for the signals that follow. Verify your results with MATLAB.

(a) $x(n) = u_r(n)$          (b) $x(n) = \dbinom{n}{k} u_s(n), \; k \in Z^+$

(c) $x(n) = \lambda^n u_s(n)$          (d) $x(n) = a^{|n|}$

**2.19.** Use MATLAB to generate an independent normally distributed signal $x(n)$ with variance $\lambda$ and mean $m$. Plot $x$. Use the commands mean and var to verify the values of $\lambda$ and $m$.

**2.20.** Compute the $p$ norms for $p = 1, p = 2, p = \infty$ for the analog signals that follow. Verify your results with MATLAB.

(a) $x(t) = u_r(t)$          (b) $x(t) = \dfrac{t^j}{j!} u_s(t), \; j \in Z^+$

(c) $x(t) = e^{\lambda t} u_s(t)$          (d) $x(t) = e^{a|t|}$

**2.21.** Let $C^2(I)$ be the space of continuous functions on the interval $I = [-1, 1]$ with norm

$$|x|_2 = \left( \int_{-1}^{1} |x(t)|^2 dt \right)^{1/2}$$

Prove that this space is not Banach by showing that the sequence

$$x_N(t) = \begin{cases} -1, & -1 \le t \le -\frac{1}{N} \\ Nt, & -\frac{1}{N} \le t \le \frac{1}{N} \\ 1, & \frac{1}{N} \le t \le 1 \end{cases}$$

is a Cauchy sequence that does not converge to a continuous function.

**2.22.** Compute the $p$ norms of the 2-D unit step, the 2-D unit sample, and the exponential signal.

**2.23.** This problem summarizes several technical details from Hilbert spaces that were used in this chapter. Let $H$ be a Hilbert space and $A$ a subset of $H$. Prove the following statements.

- Parallelogram law. For any $x, y \in H$,

$$|x + y|^2 + |x - y|^2 = 2|x|^2 + 2|y|^2.$$

- The inner product is a continuous function.
- The norm induced by the inner product is a valid norm.
- If the subspace $[A]$ has finite dimension, then it is closed, that is, $\overline{[A]} = [A]$.
- The orthogonal complement is always a closed subspace of $H$.

**2.24.** Show that the covariance of a 2-D real signal has the following properties:

$$R(n_1, n_2, \tilde{n}_1, \tilde{n}_2) = R(\tilde{n}_1, \tilde{n}_2, n_1, n_2)$$

$$\sum_{n_1} \sum_{n_2} \sum_{m_1} \sum_{m_2} a(n_1, n_2) R(n_1, n_2, m_1, m_2) a(m_1, m_2) \geq 0$$

for any finite support sequence $a(n_1, n_2)$.

**2.25.** Plot the following covariances of a 2-D WSS signal

$$r(n_1, n_2) = \sigma^2 \rho_1^{|n_1|} \rho_2^{|n_2|}, \qquad |\rho_1|, |\rho_2| < 1$$

$$r(n_1, n_2) = \sigma^2 \exp(-\sqrt{a_1 n_1^2 + a_2 n_2^2})$$

The first is separable, the second is not. If $a_1 = a_2 = a$, the second becomes $r(n_1, n_2) = \sigma^2 \rho^d$, $\rho = \exp(-|a|)$, and is called isotropic or circularly symmetric.

**2.26.** Prove that a second-order discrete stochastic signal is wide sense cyclostationary with period $L$ if and only if the so called $L$-blocked version $\mathbf{x}(n)$

$$\mathbf{x}(n) = (\, x(nL) \quad x(nL - 1) \quad \cdots \quad x(nL - L + 1)\,)^T$$

is wide sense stationary.

**2.27.** Suppose a signal has constant mean $m$. Prove that

$$\mu_2(n_1, n_2) = m_2(n_1, n_2) - m^2$$

Extend the result to central moments of order 3.

**2.28.** Let

$$x(n) = Ax(n - 1), \text{ where } A \text{ is a } k \times k \text{ matrix}$$

Suppose that the initial vector is a random vector $x_0$ with mean $m_0$ and covariance $C_0 = \text{cov}[x_0, x_0]$. Develop difference equations for the mean and the covariance of $x(n)$.

**2.29.** *Rademacher orthogonal functions.* Rademacher functions are piecewise constant periodic functions taking the values 1 and $-1$ in one period. They are generated as follows:

$$r_0(t) = 1, \qquad r_m(t) = r_{m-1}(2t) = r_1(2^{m-1}t), \qquad m \geq 1, \qquad 0 \leq t < 1$$

Derive plots of the first few Rademacher functions. Prove that they form an orthogonal set over the set of periodic functions of period 1.

**2.30.** Show by induction that if a matrix $A$ has pairwise distinct eigenvalues, it is diagonalizable. Show by an example that the converse is not true.

**2.31.** Show that a real symmetric matrix $A$ has real eigenvalues. It can be shown that $A$ is diagonalizable. Validate the statements by MATLAB simulations.

**2.32.** Prove (2.42).

**2.33.** Let $y(t) = x(t) \cos(\omega_0 t)$, where $x(t)$ is a stationary process with mean $m_x$ and autocorrelation $r_x(t)$. Prove that $m_y = m_x \cos(\omega_0 t)$ and that

$$r_y(t + \tau, t) = E[y(t + \tau)y(t)] = r_x(\tau)\left[\frac{1}{2}\cos(\omega_0 \tau) + \frac{1}{2}\cos(2\omega_0 t + \omega_0 \tau)\right]$$

Deduce that $y(t)$ is cyclostationary.

## APPENDIX

**PROOF OF THEOREM 2.3** Let $1 \leq p < \infty$. Let $x_N$ be a Cauchy sequence in $\ell_p$. Let $\epsilon > 0$ and $N_0 \in Z^+$ such that for all $N, M \geq N_0$, it holds that

$$|x_N - x_M|_p^p = \sum_{n=-\infty}^{\infty} |x_N(n) - x_M(n)|^p \leq \epsilon \tag{2.72}$$

For any $n \in Z$ it holds that $|x_N(n) - x_M(n)|^p \leq |x_N - x_M|_p^p \leq \epsilon$. Therefore for each $n$, the complex sequence $x_N(n)$ is a Cauchy sequence and hence convergent. Let $\lim_{N \to \infty} x_N(n) = x(n)$. We will show that $x(n)$ so defined belongs to $\ell_p$ and that

$$|x_N - x|_p \to 0 \tag{2.73}$$

Equation (2.72) implies that $\sum_{n=-K}^{K} |x_N(n) - x_M(n)|^p \leq \epsilon$. We fix $N$ and take the limit as $M \rightarrow \infty$. Then $\sum_{n=-K}^{K} |x_N(n) - x(n)|^p \leq \epsilon$. Now we take the limit as $K \rightarrow \infty$ to obtain

$$|x_N - x|_p^p = \sum_{n=-\infty}^{\infty} |x_N(n) - x(n)|^p \leq \epsilon$$

This proves (2.73). Finally, we have

$$|x|_p = |x - x_N + x_N|_p \leq |x_N - x|_p + |x_N|_p < \infty$$

which shows that $x \in \ell_p$. The case $p = \infty$ can be worked out along the same lines and is left to the reader. ∎

The following lemmas are needed for the proof of theorem 2.5.

**Lemma 1. Bessel's inequality**    Let $A$ be an orthonormal set of the form (2.64) in the Hilbert space $H$, and let $x \in H$. Then

$$\sum_{k=-\infty}^{\infty} |<x,f_k>|^2 \leq |x|^2 \qquad (2.74)$$

**PROOF**    Consider the partial sum $s_{M,N} = \sum_{k=-M}^{N} <x,f_k> f_k$. Then

$$0 \leq |x - s_{M,N}|^2 = |x|^2 - <x, s_{M,N}> - <s_{M,N}, x> + |s_{M,N}|^2 \qquad (2.75)$$

On the other hand,

$$<x, s_{M,N}> = <x, \sum_k <x,f_k> f_k> = \sum_k <x,f_k>^* <x,f_k>$$
$$= \sum_k |<x,f_k>|^2$$

Moreover

$$|s_{M,N}|^2 = <\sum_k <x,f_k> f_k, \sum_n <x,f_n> f_n>$$
$$= \sum_k \sum_n <x,f_k> <x,f_n>^* <f_k,f_n> = \sum_k |<x,f_k>|^2$$

Substituting the above into (2.75), we obtain

$$\sum_{k=-M}^{N} |<x,f_k>|^2 \leq |x|^2$$

and taking the limit as $N, M \rightarrow \infty$, we prove the inequality. ∎

**Lemma 2**    Let $A$ be an orthogonal set of the form (2.64). The series $\sum_{n=-\infty}^{\infty} a_n f_n$ converges if and only if the real series $\sum_{n=-\infty}^{\infty} |a_n|^2 |f_n|^2$ converges.

PROOF    Consider the partial sum $s_N = \sum_{n=0}^{N} a_n f_n$. Let $M > N \geq 0$. Then

$$|s_N - s_M|^2 = \left| \sum_{n=N+1}^{M} a_n f_n \right|^2 = \sum_{n=N+1}^{M} |a_n|^2 |f_n|^2$$

Therefore $s_N$ is a Cauchy sequence if and only if $a_N$ is a Cauchy sequence. Since both $H$ and $C$ are Hilbert spaces, $s_N$ is convergent if and only if $a_n$ is convergent. In a similar way we treat the partial sum $\sum_{n=-N}^{0} a_n f_n$.    ∎

PROOF OF THEOREM 2.5    Let $x \in \overline{[A]}$ and $\tilde{A} = \{\tilde{f}_n, n \in Z\}$ with $\tilde{f}_n = f_n / |f_n|$. $\tilde{A}$ is an orthonormal set spanning $\overline{[A]}$. Let

$$\hat{x} = \sum_{k=-\infty}^{\infty} < x, \tilde{f}_k > \tilde{f}_k = \sum_{k=-\infty}^{\infty} \frac{< x, f_k >}{< f_k, f_k >} f_k \qquad (2.76)$$

Bessel's inequality ensures that the series $\sum_k |< x, \tilde{f}_k >|^2$ converges. Lemma 2 implies that $\hat{x}$ is well defined in $H$. In fact $\hat{x} \in \overline{[A]}$ by definition of the series. We show next that

$$x - \hat{x} \perp \overline{[A]} \qquad (2.77)$$

Let $s_{M,N} = \sum_{k=-M}^{N} < x, \tilde{f}_k > \tilde{f}_k$. Then

$$< x - s_{M,N}, \tilde{f}_j > = < x, \tilde{f}_j > - \sum_{k=-M}^{N} < x, \tilde{f}_k > < \tilde{f}_k, \tilde{f}_j > = 0$$

Using continuity of the inner product, we obtain

$$< x - \hat{x}, \tilde{f}_j > = < x - \lim s_{M,N}, \tilde{f}_j > = \lim < x - s_{M,N}, \tilde{f}_j > = 0$$

We conclude that $x - \hat{x}$ is orthogonal to $\tilde{A}$. It follows that it is orthogonal to $[A]$. Now let $z$ be an element of $\overline{[A]}$ and $z_N \to z$. Then

$$< x - \hat{x}, z > = < x - \hat{x}, \lim z_N > = \lim < x - \hat{x}, z_N > = 0$$

We established (2.77). Note that $x, \hat{x} \in \overline{[A]}$. Since the linear closure of $A$ is a linear subspace, $x - \hat{x} \in \overline{[A]}$. Since (2.77) holds, $x - \hat{x}$ is orthogonal to itself and hence must be the zero vector.

Uniqueness of (2.65) is a direct consequence of the analysis equation (2.66). Equation (2.67) is established in the same way as (2.63).    ∎

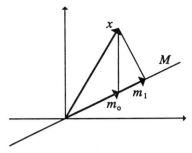

**Figure 2.21.** The triangle shows that if $m_o$ is not the projection of $x$ on $M$, another point $m_1$ can be found with shorter distance $|x - m_1| \leq |x - m_o|$.

PROOF OF THEOREM 2.6    We show first that a point $m_o$ in $M$ that minimizes (2.68) satisfies the orthogonality principle. The proof is a direct translation of the geometric argument depicted in Fig. 2.21. Indeed, suppose to the contrary, that there is $m \in M$ such that $< x - m_o, m > \neq 0$. Without loss of generality, we may take $|m| = 1$. Let $y = x - m_o$ and $m_1 = m_o + rm$, where $r = < x - m_o, m > = < y, m >$. Then $r \neq 0$ and

$$|x - m_1|^2 = |y - rm|^2 = |y|^2 - r < m, y > -r^* < y, m > +|r|^2 = |y|^2 - |r|^2$$

Thus $|x - m_1| < |y| = |x - m_o|$, contradicting minimality of $m_o$.

Next we show the converse. Suppose that $m_o$ satisfies the orthogonality principle (2.69). Take an element $m \in M$. Then $m_o - m$ belongs to $M$ because $M$ is a subspace. Orthogonality implies that $x - m_o \perp m_o - m$. Finally the Pythagorean identity (2.54) gives

$$|x - m|^2 = |(x - m_o) + (m_o - m)|^2 = |x - m_o|^2 + |m_o - m|^2 \qquad (2.78)$$

This implies that $|x - m| \geq |x - m_o|$ and establishes the claim. Incidentally, the above argument and (2.78) prove uniqueness of $m_o$.

Observe that so far we used only the notion of the inner product. The Banach structure inherent in a Hilbert space is invoked next to establish existence. If $x \in M$, we set $m_o = x$. Suppose that $x \notin M$. Let

$$v = \inf_{m \in M} |x - m| \qquad (2.79)$$

Let $m_k \in M$ with

$$|x - m_k| \to v \qquad (2.80)$$

We want to prove that $m_k$ is a Cauchy sequence. Application of the parallelogram law (Exercise 2.23) gives

$$|(m_j - x) + (x - m_i)|^2 + |(m_j - x) - (x - m_i)|^2 = 2|m_j - x|^2 + 2|x - m_i|^2$$

or

$$|m_j - m_i|^2 = 2|m_j - x|^2 + 2|x - m_i|^2 - 4|x - \frac{m_i + m_j}{2}|^2$$

For each $i,j$ the vector $(m_i + m_j)/2$ is in $M$ because $M$ is a linear subspace. By definition of $v$, $|x - (m_i + m_j)/2| \geq v$. Thus

$$|m_j - m_i|^2 \leq 2|m_j - x|^2 + 2|x - m_i|^2 - 4v^2$$

The latter inequality in conjunction with (2.80) prove that $m_k$ is a Cauchy sequence. Since $H$ is a Hilbert space, the sequence $m_k$ converges to a point $m_o$ in $H$. Now $m_k$ belongs to $M$. Hence $m_o$ belongs to $\overline{M}$. By assumption, $M$ is closed. Therefore $m_o \in M$. By continuity of the inner product, we obtain using (2.80), $|x - m_o| = v$, and the proof is complete. ∎

**PROOF OF COROLLARY 1**    Suppose that $A$ is complete. Let $x \in H$ such that $x \perp A$. Continuity and linearity of the inner product imply that $x \perp \overline{[A]} = H$. Thus $x$ must be the zero vector. Conversely, suppose that the only vector orthogonal to $A$ is the zero vector. Let $x \in H$ and $m_o$ the projection of $x$ onto $\overline{[A]}$. This is well defined because $\overline{[A]}$ is a closed subspace of $H$ and the projection theorem applies. By our assumption $x - m_o$ must be the zero vector, and thus $x \in \overline{[A]}$. The proof is complete. ∎

# 3

# BASIC SYSTEM REPRESENTATIONS

In this chapter some basic system representations are introduced. These include input output transformations, finite derivative and finite difference models, and state space representations. General system attributes such as causality, time invariance, memory, continuity, and fading memory are described. The fundamental ways of linking systems to form more powerful architectures are defined. Linear models are then isolated, and the convolutional representation is extracted. Particular emphasis on discrete simulation of each system representation is placed. The generalization of the convolutional format to the nonlinear case leads to Volterra models. Neural networks and radial basis function networks are also presented as nonlinear models and their universal approximation capabilities are highlighted.

## 3.1 SYSTEMS

Abstractly, a *system* can be thought of as a transformation between signals. A schematic representation is shown in Fig. 3.1. The term *filter* is used as a synonym. All signals that can be fed into $S$ are called *inputs*, or more emphatically, *admissible inputs*. The set of admissible inputs forms the domain of the operator $S$, and will be denoted by $\mathcal{U}$. The resulting signals are called *outputs*, and they belong to the range of $S$, $\mathcal{Y}$. Thus, if $S$ is excited by the input signal $u$, it produces the output signal $y = S(u)$. Depending on the nature of allowable inputs and possible outputs, we distinguish various system categories. If the input and the output are single-channel signals, $S$ is a single-input, single-output (SISO) system; otherwise, it is a *multichannel* or *multivariable* system. The *m-dimensional* systems are transformations between m-dimensional signals. *Continuous time systems* transform analog inputs to analog outputs, while *discrete systems* convert discrete inputs to discrete outputs. Likewise *digital systems* operate on digital signals.

Signal processing systems are built by combining basic units such as adders, multipliers, and memory devices. These are described next.

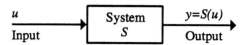

**Figure 3.1.** Pictorial description of system $S$.

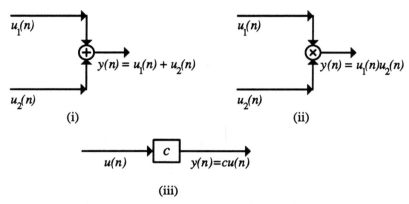

**Figure 3.2.** Graphical representation of (i) adder, (ii) multiplier, and (iii) scalor.

## *Example 3.1   Basic building blocks*

The basic pointwise operations between discrete signals can be interpreted as discrete systems. They are designated by the symbols of Fig. 3.2 and are realized by switching circuits. The adder and the multiplier are multichannel systems with 2 inputs and 1 output. The scalor multiplies an arbitrary input signal by a given scalar $c$. The *delay* system delays the input signal by one unit of time. It is depicted in Fig. 3.3(i) and is designated by the symbol $q$. The input output relationship of the delay system is

$$y(n) = q(u)(n) = u(n-1)$$

To determine the output at time $n$, we must have stored the input value of the previous time instant. This is why a delay system is a typical temporary memory system. It is physically realized by memory devices such as flip-flops.

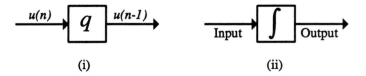

**Figure 3.3.** Graphical representation of (i) delay and (ii) integrator.

Pointwise operations between analog signals are represented by analog systems in a similar way. The integrator is another example. It is depicted in Fig. 3.3(ii) and is specified by

$$y(t) = \int_{-\infty}^{t} u(\tau)\, d\tau$$

The domain of the integrator is confined to integrable signals. The above analog systems are physically realized by operational amplifiers.

Pointwise digital operations are represented by digital systems. The same symbols of Fig. 3.2 are used for uniformity. The modulo-2 multiplier is physically realized by a digital device called AND gate. Electronic gates are simple circuits that either permit a flow of current or block it. The inverter $u \to u' = u + 1$ is another important gate.

A *combinatorial* circuit or *switching* function, with $k$ input terminals and $m$ output terminals, is a mapping $f : F^k \to F^m$, where $F$ is the binary field $\{0, 1\}$. We will establish that the AND gate and the inverter form a *complete* set of gates, in the sense that any combinatorial function can be realized by properly interconnecting gates of the above types. The function $f$ is written as

$$f = (f_1 \quad f_2 \quad \cdots \quad f_m)$$

Each switching function $f_i$, has $k$ input ports and one output port. It is specified by a list of its values. There are $2^k$ such values. Therefore $f_i$ can be identified with an element of $F^{2^k}$. If $k = 2$ and $m = 1$, there are 16 switching circuits, including the OR gate and the XOR gate. The OR gate is denoted by $\vee$ and is defined by the table

| $\vee$ | 0 | 1 |
|--------|---|---|
| 0 | 0 | 1 |
| 1 | 1 | 1 |

Very often the plus sign ($+$) is used to denote the OR operation, although it does not possess the usual properties of addition. The OR operation shares similar properties to the set theoretic union, while the AND operation corresponds to the intersection of sets. With this remark in mind the following relation is readily established.

$$x_1 \vee x_2 = \left[(x_1 \vee x_2)'\right]' = (x_1' \cdot x_2')'$$

The above expresses the OR gate in terms of inverters and the AND gate and can be used to develop an OR switching circuit. The *exclusive OR* (XOR) gate is the modulo $-2$ adder. It can be built from the expression

$$x_1 + x_2 = (x_1' \cdot x_2) \vee (x_1 \cdot x_2') \tag{3.1}$$

For each $1 \leq i \leq 2^k$, let $e_i$ be the vector that is one at position $i$ and zero elsewhere. Let $i = (i_1, i_2, \ldots, i_k)$ be the binary representation of $i$. $e_i$ can be identified with a combinatorial circuit with $k$ inputs and one output whose list of values is $e_i$. Let $e_i(u_1, \ldots, u_k)$ denote this circuit. It holds that

$$e_i(u_1, \ldots, u_k) = (u_1 + i_1 + 1)(u_2 + i_2 + 1) \cdots (u_k + i_k + 1) \tag{3.2}$$

Indeed, if $u_1 = i_1, \ldots, u_k = i_k$, we have

$$e_i(i_1, \ldots, i_k) = (i_1 + i_1 + 1)(i_2 + i_2 + 1) \cdots (i_k + i_k + 1) = 1$$

Let $j \neq i$. There exists $r$ such that $j_r \neq i_r$. Then $j_r + i_r = 1$ and $j_r + i_r + 1 = 0$. This is a factor of $e_i(j_1, \ldots, j_k)$; hence $e_i(j) = 0$ and the claim is proved. Since the functions $e_i(u_1, \ldots, u_k)$ form a basis, every function $f : F^k \to F$ is represented as

$$f(u_1, u_2, \ldots, u_k) = \sum_{i=1}^{2^k} f(i)e_i(u_1, u_2, \ldots, u_k) \tag{3.3}$$

Consequently every combinatorial function can be implemented by XOR and AND gates. Thus XOR and AND gates form a complete set. It follows from Eq.(3.1) that the inverter and the AND gate are also complete. If $f$ is given through a list of values, only those $i$ for which $f(i) = 1$ enter in (3.3). As Eq. (3.2) indicates, each $e_i$ is implemented by $k - 1$ AND gates coupled with inverters.

A further consequence of (3.3) is that every combinatorial function $f : F^k \to F$ can be represented as a polynomial function of several variables

$$f(u_1, \ldots, u_k) = a_0 + \sum_{j=1}^{k} a_j u_j + \sum_{j_1=1}^{k} \sum_{j_2=1}^{k} a_{j_1 j_2} u_{j_1} u_{j_2} + \cdots + \sum_{j_1, \ldots, j_k} a_{j_1 \cdots j_k} u_{j_1} \cdots u_{j_k} \tag{3.4}$$

It suffices to observe that each $e_i$ is written in the above form after all multiplications in (3.2) are carried out.

Pointwise m-D operations are interpreted as multidimensional systems in the obvious way. The 2-D shift is a discrete system, designated by $q_1 q_2$ and defined by

$$y(n_1, n_2) = u(n_1 - 1, n_2 - 1) \qquad \blacksquare$$

Physical processes require more elaborate representations as the next example indicates.

### Example 3.2 Satellite control
Geostatical satellites move at a certain distance from the earth on a circular orbit. Due to unpredictable meteorological conditions, deviations from the

nominal orbit occur. These deviations have an unstable behavior and need to be controlled. The controller is a signal processing system that processes the signals received by sensors and issues a command for corrective action. These actions translate into controlling forces generated by jet engines that restore the motion back to the nominal path. To set up the design problem, we must describe the system with inputs the control forces and outputs the position of the satellite. We view the satellite as a particle of mass $m$ moving in space under the influence of the gravitational field of the earth having mass $M$ and under the influence of a controlling force $u(t)$. The position of the particle is described in spherical coordinates by the vector $y(t) = (r(t), \theta(t), \phi(t))$, as shown in Fig. 3.4. The motion is governed by the following equations:

$$\ddot{r}(t) = r(t)\dot{\theta}^2(t)\cos^2\phi(t) + r(t)\dot{\phi}^2(t) - \frac{k}{r^2(t)} + \frac{u_r(t)}{m}$$

$$\ddot{\theta}(t) = -\frac{2\dot{r}(t)\dot{\theta}(t)}{r(t)} + 2\dot{\theta}(t)\dot{\phi}(t)\tan\phi(t) + \frac{u_\theta(t)}{mr(t)\cos\phi(t)}$$

$$\ddot{\phi}(t) = -\dot{\theta}^2(t)\cos\phi(t)\sin\phi(t) - \frac{2\dot{r}(t)\dot{\phi}(t)}{r(t)} + \frac{u_\phi(t)}{mr(t)}$$

where $u_r(t)$, $u_\theta(t)$, and $u_\phi(t)$ designate the spherical coordinates of $u(t)$. We observe that the input-output relationship $u \rightarrow y$ is considerably more complicated when compared to the pointwise operations of the previous example. Derivatives of the output variables are interrelated with inputs and outputs through a nonlinear mixing that involves trigonometric functions, squares, and the like. Furthermore this relationship is not well defined, unless additional information concerning the initial position and velocity of the system is

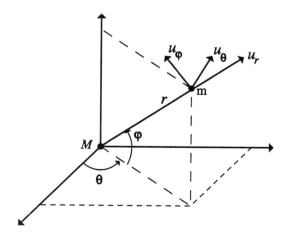

**Figure 3.4.** Illustration of satellite motion in spherical coordinates.

provided. It is a remarkable fact that despite the above complexity, a great deal of information can be extracted and utilized to design efficient control implementations.  ∎

## 3.2   CAUSALITY, MEMORY, AND TIME INVARIANCE

A 1-D system $S$ is *causal* or *nonanticipatory* if the effect never preceeds the cause. Thus for any input signal $u$ and any time instant $n$, the corresponding output at time $n$, $S(u)(n)$ depends only on previous input values $u(k)$, $k \leq n$. A precise statement goes as follows: For each integer $n$ and for any inputs $u(n)$, $v(n)$ such that $u(k) = v(k)$, for all $k \leq n$, it holds that $S(u)(n) = S(v)(n)$. Thus, at each time $n$, the output $S(u)(n)$ is uniquely determined by the past history of the input, $u(k)$, $k \leq n$. The definition is similar for continuous time systems. The basic building blocks of Example 3.1 are causal systems.

*Example 3.3   Dilation, decimators, and expandors*
Dilation of a signal causes expansion or contraction of the time scale. Let $a$ be a positive real. The dilation with parameter $a$ is represented by the system

$$y(t) = u(at) \tag{3.6}$$

Expansion results when $a < 1$; contraction when $a > 1$. Both cases are illustrated in Fig. 3.5. They constitute noncausal systems on $L(R)$. If the system domain is $L(R^+)$, contraction remains noncausal, while expansion becomes causal. Dilation makes sense for negative values of $a$. In particular, for $a = -1$, we obtain the reflection of a signal $y(t) = u(-t)$. Reflection is clearly a noncausal operation, since it reverses the input signal. Dilation in discrete time is referred to as

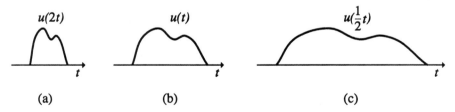

**Figure 3.5.** Illustration of dilation.

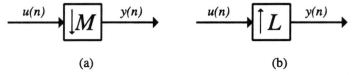

**Figure 3.6.** (a) Decimator, (b) Expandor.

*decimation.* Let $M$ be a positive integer. The *M-fold decimator* is defined by

$$y(n) = u(Mn)$$

and is depicted in Fig. 3.6(a). Only input values at multiples of $M$ are retained. As a consequence a certain amount of information is lost. Expansion in discrete time requires some caution to ensure time takes integer values. Given a positive integer $L$, the *L-fold expandor* or *L-fold interpolator* is illustrated in Fig. 3.6(b) and is defined by

$$y(n) = \begin{cases} u\left(\dfrac{n}{L}\right), & L \text{ divides } n \\ 0, & \text{otherwise} \end{cases}$$

All input values are retained but are filled with zeros every $L$ samples. In contrast to the decimator, the expandor preserves input information. ∎

Causality is a natural property when the system acts on input signals whose independent variable represents time. It imposes as an inevitable constraint, when real time operation of the system is involved. Causality does not bear the same significance in the 2-D case. Implementation issues however, lead us to consider causal 2-D systems. We will describe two pertinent concepts, causality (occasionally referred to as strong causality) and semicausality (causality in one direction). An illustration is provided in Fig. 3.7. Causality ensures that at each value $(n_1, n_2)$ the value of the output depends on previous values of the input $u(k_1, k_2), k_1 \leq n_1$, and $k_2 \leq n_2$ in both directions. In the semicausal case, $y(n_1, n_2)$ depends causally on one variable (in Fig. 3.7, the causal variable is $n_1$) and noncausally on the other. Thus it is computed from $u(k_1, k_2)$, with $k_1 \leq n_1$, and $k_2$ arbitrary.

A system is called *static* or *memoryless* if for any input signal and any time instant the corresponding output at that time depends only on the value of the input at the same time; otherwise, it is called *dynamic* or *system with memory.* The input-output relationship of a memoryless system has the form

$$y(n) = f(u(n), n) \tag{3.7}$$

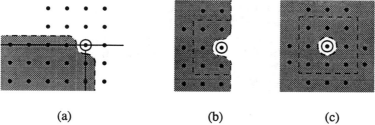

(a)          (b)          (c)

**Figure 3.7.** Illustration of (a) causality, (b) semicausality, and (c) noncausality in the 2-D case.

The explicit dependence of $f$ on time $n$ indicates that $f$ may be different for various values of time. Similar statements hold for continuous and digital systems. The pointwise operations are memoryless systems. In contrast, the integrator and the delay are systems with memory.

A static system is called *time invariant* if the associated function $f$ does not explicitly depend on time, that is, Eq. (3.7) takes the form

$$y(n) = f(u(n)) \tag{3.8}$$

Equation (3.8) means that if $u(n)$ is shifted by $k$ and is inserted into the system, the resulting output is simply the output of $u(n)$ shifted by the same amount $k$. We take this property as a definition of time invariance for the general case. We say that a continuous time system $S$ is *time invariant* if for any positive real $\ell$ and any input signal $u$ with corresponding output $y$, the $\ell$ input shift $u_\ell(t) = u(t - \ell)$ produces the same output $y$ but shifted by $\ell$, that is, $y_\ell(t) = y(t - \ell)$. In the general case and for emphatic purposes, $S$ is called *time varying*. Consider, for example, the $\tau$-shift system $S_\tau$: $y(t) = S_\tau(u(t)) = u(t - \tau)$. Then

$$y_\ell(t) = S_\tau[u(t - \ell)] = u(t - \ell - \tau) = u(t - \tau - \ell) = y(t - \ell)$$

which shows that $S_\tau$ is time invariant. In contrast, the system $y(t) = c(t)u(t)$, is time varying unless $c(t)$ is constant. The definition of time invariance in the discrete time case is virtually the same, except that $\ell$ is now an integer, which may be set equal to one without loss of generality. Decimators and interpolators are further important examples of time varying systems.

Strictly speaking, the characteristics of physical systems change with time because of aging and other deteriorating effects. In many cases of interest, however, variations are negligibly small in a given time interval, and hence the property of time invariance provides an acceptable description, which, as we will see later, greatly facilitates various aspects of analysis and design of systems.

A 2-D discrete system $S$ is *shift invariant* if for any integers $m_1$ and $m_2$ and any input $u(n_1, n_2)$, with output $y(n_1, n_2)$, it holds that

$$S(u(n_1 - m_1, n_2 - m_2)) = y(n_1 - m_1, n_2 - m_2)$$

**System functional.** The notion of time invariance warrants that once we know how to compute the output of a system at a specific time instant, then we can determine the output at any other time by a mere shift. Without loss of generality we take the specific instant to be the origin. Let $S$ be a time invariant system and $S_0$ the functional that focuses on the system output at time $t = 0$:

$$S_0 : \mathcal{U} \to R, \quad S_0(u) = S(u)(0)$$

$S_0$ is well defined. Conversely, $S$ is uniquely determined from $S_0$ due to time invariance. Indeed, to determine the output at time $t$, we simply shift the output at time zero by an amount equal to $t$. Let $q_t(u)(\tau) = u(\tau - t)$. Then

$$S(u)(t) = S(q_t(u))(0) = S_0(q_t(u)) \tag{3.9}$$

In the time invariant case the above one-to-one correspondence $S \leftrightarrow S_0$ enables us to work with the functional $S_0$ rather than the operator $S$. This facilitates the analysis in several cases.

If the time invariant system $S$ is also causal, the output at the origin depends only on the past history of the input, namely the input values corresponding to negative times. In this case we can further simplify our view of the system operator $S$ by restricting the functional $S_0$ on inputs defined on negative times only. In the discrete case the restriction $S_0$ has the following form: Let $\mathcal{U}(Z^-)$ denote the set of sequences defined over $Z^-$, the set of nonpositive integers. Then

$$S_0 : \mathcal{U}(Z^-) \to R, \qquad S_0(u) = S(u_e)(0)$$

where $u_e$ is any reasonable extension of $u$ to the set of all integers, for instance

$$u_e(n) = \begin{cases} u(n), & n \leq 0 \\ u(0), & n > 0 \end{cases}$$

The original system $S$ is reconstructed from the information provided by $S_0$ via the following generalization of rule (3.9):

$$S(u)(t) = S_0(P(q_t(u)))$$

$P$ truncates an element of $\mathcal{U}(Z)$ into an element of $\mathcal{U}(Z^-)$, namely $P(u)(n) = u(n)$, $n \leq 0$. The insertion of the truncation operator is needed to accommodate the difference of the domains of $S$ and $S_0$.

## 3.3 CONTINUITY AND FADING MEMORY

A continuous system is described by a continuous operator. Thus, if the input signals $u$ and $v$ are close, the corresponding outputs $S(u)$ and $S(v)$ are also close. To make the statement precise, a suitable notion of convergence must be specified in the input space $\mathcal{U}$ and in the output space $\mathcal{Y}$. Convergence is most commonly introduced with the aid of a metric. Consider, for instance, a discrete system $S$ whose input and output spaces are the set of bounded sequences on $Z$, $\ell_\infty$. We use the metric induced by the sup norm to define convergence of signals. Thus $S : \ell_\infty \to \ell_\infty$ is *continuous* if for any input $u \in \ell_\infty$ and for any $\epsilon > 0$ there exists $\delta > 0$ (depending on $u$ and $\epsilon$) such that, for any signal $v \in \ell_\infty$ with $d_\infty(u, v) = \sup_{n \in Z} |u(n) - v(n)| \leq \delta$, we have

$$d_\infty(S(u), S(v)) = \sup_{n \in Z} |S(u)(n) - S(v)(n)| \leq \epsilon.$$

$S$ is *uniformly continuous* if $\delta$ depends only on $\epsilon$ and not on $u$.

### Example 3.4 *Continuity of the shift and the decimator*

The unit shift $q : \ell_p \to \ell_p$ is uniformly continuous for any $1 \leq p \leq \infty$. This is a consequence of the readily established formula $|q(u) - q(v)|_p = |u - v|_p$. The decimator $D : \ell_\infty \to \ell_\infty$ is uniformly continuous. Indeed, notice first that

$$\{u(Mn) : n \in Z\} \subset \{u(n) : n \in Z\}$$

Let $\epsilon > 0$ and $\delta = \epsilon$. Then

$$|D(u) - D(v)|_\infty = \sup_{n \in Z} |u(Mn) - v(Mn)| \leq \sup_{n \in Z} |u(n) - v(n)| < \epsilon \qquad \blacksquare$$

The definition of continuity is readily carried over to the continuous time case. Suppose that the input and output spaces coincide with $C_\infty(R)$, the set of continuous bounded functions defined on $R$ and equipped with the sup norm. A system $S : C_\infty(R) \to C_\infty(R)$ is continuous if for any $u \in C_\infty(R)$ and any $\epsilon > 0$ there is $\delta > 0$ such that, for all $v \in C_\infty(R)$ satisfying $\sup_{t \in R} |u(t) - v(t)| \leq \delta$, it holds $\sup_{t \in R} |S(u)(t) - S(v)(t)| \leq \epsilon$. Exactly as in the discrete case, one shows that the delay system is continuous.

One useful type of convergence that does not arise from a metric is pointwise convergence on compact sets. Consider the space of continuous functions on $R$, $C(R)$. A sequence $u_N$ in $C(R)$ converges to $u$ in $C(R)$ if for any compact subset of $R$, $I$ and for any $\epsilon > 0$, there is $N_0$ such that $|u_N(t) - u(t)| \leq \epsilon$, for any $N \geq N_0$ and $t$ in $I$.

### Example 3.5 *Integrator*

Consider the integrator

$$y(t) = \int_0^t u(\tau)d\tau$$

The input space is the set of continuous functions defined on $R^+$, $C(R^+)$. The resulting outputs belong to $C(R^+)$. Furthermore the integrator is continuous with respect to pointwise convergence on compact sets. Indeed, suppose that $u_N(t) \to u(t)$, uniformly on the finite set $[0, T]$. Let $\epsilon > 0$ and $N_0$ such that $|u_N(t) - u(t)| \leq \epsilon/T$, for $N \geq N_0$ and $t$ in $[0, T]$. Then

$$|y_N(t) - y(t)| \leq \int_0^t |u_N(\tau) - u(\tau)|d\tau \leq \int_0^T \frac{\epsilon}{T}d\tau = \epsilon \qquad \blacksquare$$

### Examples 3.6 *Symbolic dynamics*

Consider the space of sequences defined on $Z^+$ with values in the finite field $GF(p)$, and the weighted metric

$$d_p(u, v) = \sum_n \frac{1}{p^n}|u(n) - v(n)|$$

We show that the $k$ advance shift $S(u)(n) = u(n + k)$ is continuous with respect to the metric $d_p$. Let $\epsilon > 0$. Pick $N_0$ such that $1/p^n < \epsilon$, for $n \geq N_0$. Let $\delta = 1/(p^{N_0+k})$. If $d_p(u, v) < \delta$, proposition 2.1 of Section 2.7.2 implies that $u(n) = v(n)$, for $0 \leq n \leq N_0 + k$. Therefore, if we ignore the first $k$ signal characters and shift the rest, we obtain $S(u)(n) = S(v)(n)$, for $0 \leq n \leq N_0$. Thus $d_p(S(u), S(v)) \leq 1/p^{N_0} < \epsilon$.    ■

It is easy to show that a discrete, time invariant, causal system $S : \ell_\infty \to \ell_\infty$ is continuous if and only if the associated functional $S_0$ is continuous; in other words, for any $u \in \ell_\infty$ and any $\epsilon > 0$ there is $\delta > 0$ such that for all $v \in \ell_\infty$

$$\sup_{n \leq 0} |u(n) - v(n)| < \delta \quad \text{implies} \quad |S(u)(0) - S(v)(0)| < \epsilon \tag{3.10}$$

Entirely analogous statements are valid for analog systems.

The computation of the output of a time invariant causal dynamic system $S$ requires the entire past input history. If the input values become of less value as we travel back in the remote past, we say that the system has fading memory. More precisely the system $S : \ell_\infty \to \ell_\infty$ has the fading memory property on a subset $K$ of $\ell_\infty$ if there is a decreasing sequence $w(n)$, defined on $Z^+$, such that $0 < w(n) \leq 1$, $\lim_{n \to \infty} w(n) = 0$ and for each $u \in K$ and $\epsilon > 0$ there is $\delta > 0$ so that for all $v \in K$

$$\sup_{n \leq 0} \{|u(n) - v(n)|w(-n)\} < \delta \quad \text{implies} \quad |S(u)(0) - S(v)(0)| < \epsilon \tag{3.11}$$

Comparison of (3.11) and (3.10) shows that the system $S$ has the fading memory property on the signal set $K$ if and only if the associated functional $S_0$ is continuous on the set of negative time truncations of $K$ (the set of functions of $K$ restricted on negative times) with respect to the weighted norm

$$|u|_w = \sup_{n \leq 0} |u(n)|w(-n) \tag{3.12}$$

*Example 3.7    Fading memory of the delay system*
We have seen that the delay system $q_k(u)(n) = u(n - k)$ is continuous on $\ell_\infty$. We show next it has the fading memory on $\ell_\infty$. Indeed, pick a positive decreasing sequence $w(n)$ such that $\lim_{n \to \infty} w(n) = 0$. Let $\epsilon > 0$ and $\delta = \epsilon w(k)$. For all $v \in \ell_\infty$ such that $|u(-n) - v(-n)|w(n) < \delta$, for all $n$, we have

$$|S(u)(0) - S(v)(0)| = |u(-k) - v(-k)| \leq \frac{\delta}{w(k)} = \epsilon \qquad ■$$

## 3.4  LINEARITY

Just as the exponential signal plays a prominent role in the study of signals, linear systems form a distinguished family of systems. In fact, we will see that strong relationships link linear systems and the exponential signal together. Linear systems are marked by the following attractive features: First, their specification needs considerably less information than arbitrary systems. Second, and more important, they carry a rich algebraic structure enabling the development of an extensive body of knowledge. A map between linear spaces is called linear if it preserves addition and scalar multiplication. Likewise a system $S$ is called *linear* if the input and output spaces are linear spaces and the transformation $S$ satisfies the *superposition property*

$$S(au + bv) = aS(u) + bS(v)$$

for any scalars $a$ and $b$ and any signals $u$ and $v$. The linear spaces involved in the above definitions are among the types discussed in Section 2.1. Clearly the delay is a linear system.

**Example 3.8    Static linear systems**
A static $l$-input, $m$-output system $y(t) = f(u(t), t)$ satisfies the superposition property if and only if, for each $t$, the assignment $u \rightarrow f(u, t)$ is linear. If we pick a base in $R^l$ and a base in $R^m$, the above expression is represented as

$$y(t) = F(t)\, u(t) \tag{3.13}$$

For each $t$, $F(t)$ is an $m \times l$ matrix. If the static system is also time invariant, Eq. (3.13) reduces to

$$y(t) = F\, u(t) \tag{3.14}$$

The adder and the scalor are linear systems while the multiplier is not. The latter equation is completely specified by the $ml$ entries of $F$. Thus it requires considerably less information than the general static structure. ■

## 3.5  SYSTEM CONNECTIONS

The architecture of a signal processing system consists of some basic building blocks, such as those described in Example 3.1, properly interconnected. Linear filters are built by interconnecting registers, adders, and scalors. Volterra filters use also multipliers. Neural networks are formed by the interconnection of some primary computational modules called *neurons*. The most important system interconnections are the cascade or series connection, the parallel connection, and the feedback connection. Knowledge of the properties of each of these system connecting operations greatly facilitates the analysis of complex architectures.

### 3.5.1  Series connection and invertibility

Two systems $S_1$ and $S_2$ are connected in series with $S_1$ preceding $S_2$ if the output

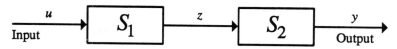

**Figure 3.8.** Cascade connection $S_2 S_1$.

of $S_1$ is exerted as input to $S_2$. The resulting system is called the *series* or *cascade connection* of $S_1$ and $S_2$. It is denoted by $S_2 S_1$ and is specified by the rule

$$y = (S_2 S_1)(u) = S_2(S_1(u))$$

A schematic representation is shown in Fig. 3.8. Apparently $S_2 S_1$ corresponds to the composition of the transformations $S_1$ and $S_2$. It is well defined provided that the range of $S_1$ is included in the domain of $S_2$. The series connection $S_r S_{r-1} \cdots S_1$ of any finite collection of systems $S_1, S_2, \ldots, S_r$ is defined in an obvious way. The series connection may be viewed as a control action of $S_1$ onto $S_2$ because the inputs of $S_2$ are manipulated or controlled by the signals produced by $S_1$.

The delay unit is capable of temporarily storing information. Longer storage capability can be achieved by cascading a number of delay systems. The shift register of length $k$ is described by

$$y(n) = qq \cdots q(u)(n) = q^k(u)(n) = u(n - k) \tag{3.15}$$

and is shown Fig. 3.9. The cascade connection is associative, that is, $(S_1 S_2)S_3 = S_1(S_2 S_3)$. Moreover it admits an identity element, denoted by $I$. $I$ lets the input pass undistorted, $I(u) = u$.

A system $S$ is called *left invertible* if there exists a system called *left inverse* or *left inverter* of $S$, designated by $S^{-l}$, that connected in series to $S$, as in Fig. 3.10, neutralizes the effect of $S$. The domain of the left inverse of $S$ includes the range of $S$ and $S^{-l} S = I$. Thus two different inputs yield different outputs if the system is left invertible (the transformation $S$ is one-to-one). If $y$ is the output of an invertible system $S$ we can reconstruct the input that produced $y$, by feeding $y$ into the inverse of $S$. $S$ is *right invertible* if there is a system $S^{-r}$ called *right inverse*

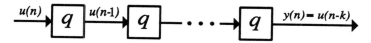

**Figure 3.9.** Shift register of length $k$.

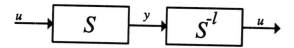

**Figure 3.10.** Definition of the left inverse.

of $S$ such that $S S^{-r} = I$. Finally, $S$ is *invertible* if a left inverse and a right inverse exist and coincide. The adder is not left invertible because there is no way to determine two signals from their sum. In contrast, it is right invertible. The scalor is invertible if the scaling factor $c$ is nonzero. Its inverse is again a scalor with factor $c^{-1}$. Several important applications involve inversion issues. For instance, communication system design aims to recover the transmitted signal from the received signal. In a way, it strives to develop a channel inverter that will counteract degradations incurred during transmission.

### 3.5.2  System banks and parallel connection

Let us consider systems $S_1, S_2, \ldots, S_k$ with the same domain. The multichannel system

$$y = (S_1 \times S_2 \cdots \times S_k)(u) = \begin{pmatrix} y_1 \\ y_2 \\ \vdots \\ y_k \end{pmatrix} = \begin{pmatrix} S_1(u) \\ S_2(u) \\ \vdots \\ S_k(u) \end{pmatrix}$$

is called system bank or *filter bank*. A two-channel filter bank is illustrated in Fig. 3.11(a). The same input signal applies to all systems. These systems operate concurrently (in parallel) and produce outputs $y_1, y_2, \ldots, y_k$, which constitute the multichannel output of the overall system. The parallel connection of two systems $S_1$ and $S_2$ denoted $S_1 + S_2$ is illustrated in Fig. 3.11(b). It is defined by

$$y = (S_1 + S_2)(u) = S_1(u) + S_2(u)$$

*Example 3.9  Multiplexers and demultiplexers*
A multiplexer serializes a multichannel signal. It serves as a buffer that matches the output of a multichannel system to a single-channel destination. It is

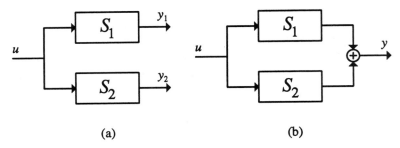

(a)                              (b)

**Figure 3.11.** (a) Filter bank and (b) parallel connection of two systems.

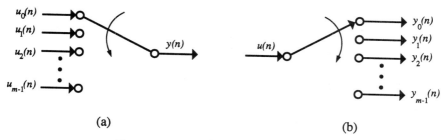

(a)                                    (b)

**Figure 3.12.** (a) Multiplexer; (b) Demultiplexer.

illustrated in Fig. 3.12(a). The input-output relationship is given by

$$y(n) = u_k\left(\frac{n-k}{m}\right) \qquad k = n(\mathrm{mod}\ m)$$

For instance, if $m = 2$, we have

$$(y(0) \quad y(1) \quad y(2) \quad \cdots) = (u_0(0) \quad u_1(0) \quad u_0(1) \quad u_1(1) \quad u_0(2) \quad u_1(2)\cdots)$$

The multiplexer can be realized by interconnecting registers, interpolators, and adders as in Fig. 3.13(a).

The demultiplexer distributes a serial stream into several channels. It serves as a buffer matching a single-channel source to a multichannel system. It is illustrated in Fig. 3.12(b). The input-output relationship is given by

$$y_k(n) = u(mn + k)$$

If $m = 2$, we have

$$\begin{pmatrix} y_0(0) & y_0(1) & y_0(2) & \cdots \\ y_1(0) & y_1(1) & y_1(2) & \cdots \end{pmatrix} = \begin{pmatrix} u(0) & u(2) & u(4) & \cdots \\ u(1) & u(3) & u(5) & \cdots \end{pmatrix}$$

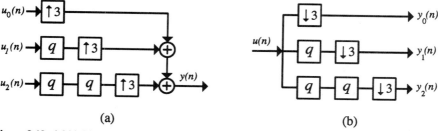

(a)                                    (b)

**Figure 3.13.** (a) Multiplexer built by registers, adders and interpolators; (b) Demultiplexer built by registers and decimators; in both cases $m = 3$.

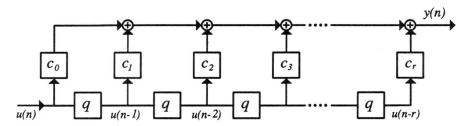

**Figure 3.14.** FIR filter architecture.

A demultiplexer architecture comprised of registers and decimators is depicted in Fig. 3.13(b). ∎

***Example 3.10    Time delay networks and FIR filters***
A finite impulse response (FIR) filter (the term is explained later on), or *linear time delay network*, is defined by

$$y(n) = c_0 u(n) + c_1 u(n-1) + \cdots + c_r u(n-r) = \sum_{k=0}^{r} c_k u(n-k) \qquad (3.16)$$

It can be implemented via the cascade and parallel connection of scalors, adders, and delays, as in Fig. 3.14. The cascade of $r$ delays expresses the time delay module and is implemented by a shift register. A polynomial time delay network (PTDN) forms the output signal by combining a finite number of input shifts in a polynomial fashion. It has the form

$$y(n) = p(u(n), u(n-1), \ldots, u(n-r)) \qquad (3.17)$$

where $p$ is a polynomial function of $r+1$ independent variables. The importance of these filters will be appreciated in Section 3.7.3, where their capability to approximate broad classes of nonlinear discrete systems is established.

A straightforward extension is obtained if the polynomial nonlinearity is replaced by a general nonlinear function $f$ of $r+1$ variables. The input-output expression is

$$y(n) = f(u(n), u(n-1), u(n-2), \ldots, u(n-r)) \qquad (3.18)$$

It is illustrated in Fig. 3.15. ∎

### 3.5.3  Feedback connection

Feedback indicates the presence of a loop in a system interconnection. Let us consider a rapidly moving vehicle such as an airplane taking off. The forces governing the motion are a control force $u(t)$ and the air resistance which is

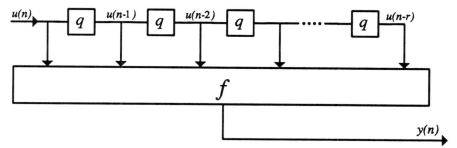

**Figure 3.15.** Time delay system architecture.

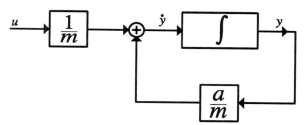

**Figure 3.16.** Diagram representation of Eq. (3.19).

assumed to be proportional to the vehicle's velocity, $y(t)$. The vehicle is viewed as a particle of mass $m$. Newton's law of motion states that

$$\frac{d}{dt}y(t) = \frac{a}{m}y(t) + \frac{1}{m}u(t) \tag{3.19}$$

This is an implicit equation in the sense that $y$ appears in both sides of the equation. William Thomson (Lord Kelvin) suggested the diagram of Fig. 3.16 as an implementation of Eq. (3.19). To verify that the diagram 3.16 represents relation (3.19), we denote the output of the integrator by $y(t)$. Then its input is $dy(t)/dt$, which, being the output of the adder, coincides with the right-hand side of Eq. (3.19). The basic feedback configuration is shown in Fig. 3.17.

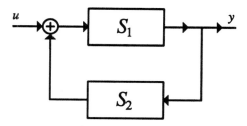

**Figure 3.17.** Feedback connection of systems $S_1$ and $S_2$.

Feedback is the natural mechanism in control design. Suppose that the system $S_1$ represents the system to be controlled; for instance, the satellite motion of Example 3.2. The controller is a system $S_2$ linked to the plant $S_1$ with the aim to steer the plant's output along a reference signal. If the controller is cascaded to the plant, it has no information on the system response and thus changes in the system environment cannot be taken into account and accommodated. The feedback connection does not suffer from this drawback. By its very nature, feedback acquires information from the system output, and upon the discrepancy with the desired signal issues a command for corrective action. Consider the system

$$y(n) = y(n-1) + bu(n), \qquad b = 1 \tag{3.20}$$

In the diagram representation of Fig. 3.17, $S_2$ is the delay and $S_1$ is the unit system. A natural extension leads to infinite impulse response (IIR) structures, (the term is explained later).

**Example 3.11   Recurrent time delay networks**
An *autoregressive (AR) filter* is specified by

$$y(n) = -\sum_{i=1}^{k} a_i y(n-i) + u(n)$$

The output appears in both sides of the above equation, and so a loop is expected in the diagram. Indeed, guided by Fig. 3.17, we obtain the structure of Fig. 3.18. If we cascade the FIR and AR architectures, we obtain

$$y(n) = -\sum_{i=1}^{k} a_i y(n-i) + \sum_{i=0}^{r} b_i u(n-i) \tag{3.21}$$

This is illustrated in Fig. 3.19. These filters are sometimes called *infinite impulse response (IIR) filters.*

In close analogy with the FIR counterparts, extension of the AR and IIR structures to a nonlinear set up is obtained by the form

$$y(n) = p(y(n-1), \ldots, y(n-k), u(n), u(n-1), \ldots, u(n-r)) \tag{3.22}$$

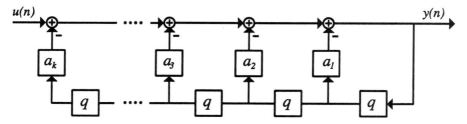

**Figure 3.18.** AR filter architecture.

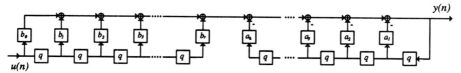

**Figure 3.19.** IIR filter architecture.

where $p$ is a polynomial function. A notable characteristic of the above form and of the feedback connection in general is that the output is not uniquely determined from the input. Determination of the output requires some further information, which is usually given via consecutive output values. We examine this further later on.  ■

## 3.6  CONVOLUTIONAL REPRESENTATIONS

Convolutional representations provide important linear system models. A motivation of the concept is given next. Roughly speaking the output of a discrete system at a certain instant is obtained by weighting the input values by constants and summing them up. Summation is replaced by integration in the continuous time case.

### 3.6.1  Motivation

We will first consider the implication of linearity on finite support inputs. The universal characteristics of finite support signals and continuity of the system operator enable the extension of the resulting expression to more general input classes.

***Finite support inputs and linearity.*** Let us consider a discrete system $S$ whose input space $\mathcal{U}$ is the set of finite support signals and the output space $\mathcal{Y}$ is the space of all sequences defined on $Z$, $\ell$. Let $u$ be an input signal and $N$ be an integer such that $u(n) = 0$, for all $|n| > N$. The unit sample shifts are finite support signals. We recall the expansion of $u(n)$ in terms of unit sample shifts

$$u(n) = \sum_{k=-N}^{N} \delta(n-k)u(k) = \sum_{k=-\infty}^{\infty} \delta(n-k)u(k) \qquad (3.23)$$

The superposition property gives

$$y(n) = \sum_{k=-\infty}^{\infty} h_k(n)u(k) \qquad (3.24)$$

where $h_k(n)$ designates the output signal resulting from the input $\delta(n - k)$:

$$h_k(n) = S(\delta(n - k)) \tag{3.25}$$

The family of response signals $\{h_k(n) : -\infty < k < \infty \}$ carries all information we need to know in order to specify any other system output resulting from a finite support input.

Equation (3.24) demonstrates that the behavior of a linear system is completely specified by the 2-D signal $h(n, k) = h_k(n)$, $k$, $n$ integers when the inputs are finite support signals.

***Causality and time invariance.*** If $S$ is causal, then for each $n$ no input values of the form $u(k)$, $k > n$ can be present in the sum (3.24). Therefore causality requires that $h_k(n) = 0$ for $n \in Z$ and $k > n$. Thus (3.24) reduces to

$$y(n) = \sum_{k=-\infty}^{n} h(n, k)u(k) \tag{3.26}$$

The two-dimensional signal $h(n, k)$ reduces to a lower triangular matrix of infinite size. The basic equation (3.24) becomes simpler when the system is time invariant. Let

$$h(n) = h_0(n) = S(\delta)(n) \tag{3.27}$$

Here $h(n)$ designates the output resulting from the unit sample input. Then time invariance gives $h_k(n) = h(n - k)$ and Eq. (3.24) becomes

$$y(n) = \sum_{k=-\infty}^{\infty} h(n - k)u(k) \tag{3.28}$$

The main input-output formulas derived above can be lifted to infinite support signals provided that the system inherits additional continuity properties. Two cases are considered to motivate the extension. First both input and output spaces coincide with an $\ell_p$ space where $1 \leq p < \infty$. The second case deals with $\ell_\infty$ signals.

**Inputs with finite p norm $1 \leq p < \infty$**

Suppose that the input space is $\ell_p$, the space of sequences of finite $p$ norm. We saw in Chapter 2 that finite support signals are dense in $\ell_p$ for $1 \leq p < \infty$. In fact, if we go through the proof of Theorem 2.2, we can observe that the signal $u_N(n) = \sum_{-N}^{N} \delta(n - k)u(k)$ coincides with the truncation of $u(n)$ on the interval $[-N, N]$. Therefore it converges to $u(n)$ in the $\ell_p$ norm as $N$ goes to infinity. Suppose that $S$ takes $\ell_p$ into $\ell_p$ and is continuous. Using the results of the previous paragraph, we conclude that $y_N(n) = S(u_N)(n) = \sum_{k=-N}^{N} h_k(n)u(k)$. Continuity implies (3.24) provided that the latter equation is viewed in the $\ell_p$ sense. Thus every linear continuous system acting on signals of finite $\ell_p$ norm is represented by Eq. (3.24). The case $p = \infty$ requires stronger continuity assumptions, as we show next.

## Fading memory systems with bounded inputs and outputs

The system $S$ is described by an operator $S : \ell_\infty \to \ell_\infty$. The argument used for the case $p \neq \infty$ does not apply because the set of finite support signals is not dense in $\ell_\infty$, as we saw in Section 2.7.2. This shortcoming is bypassed if a stronger continuity property on $S$ is imposed. Let us assume that $S : \ell_\infty \to \ell_\infty$ is a linear time invariant causal system with fading memory. Then $S$ is represented by

$$y(n) = \sum_{k=-\infty}^{n} h(n-k)u(k) = \sum_{k=0}^{\infty} h(k)u(n-k), \qquad (3.29)$$

while the corresponding series converges absolutely. Let $w(n)$ be the window implied by the fading memory.

We show that the set of finite support signals defined on $Z^-$ is dense in $\ell_\infty$ with respect to the windowed norm. Let $x \in \ell_\infty(Z^-)$ and $N$ a positive integer. Consider the truncated sequence $x_N(n) = x(n)$ for $-N \leq n \leq 0$, and 0 otherwise. Let $\epsilon > 0$. Since $w(n) \to 0$, there is $N_0$ such that $w(n) < \epsilon$, for $n \geq N_0$. Let $N \geq N_0$. Then

$$|x_N - x|_w = \sup_{n \leq -N} |x(n)|w(-n) \leq \sup_{n \leq -N_0} |x(n)|w(-n) \leq |x|_\infty \epsilon$$

Thus $|x_N - x|_w \to 0$. The rest of the proof is similar to the case $p \neq \infty$.

### 3.6.2  Discrete time convolutional representations

The convolutional representation of a discrete time invariant system is specified by

$$y(n) = \sum_{k=-\infty}^{\infty} h(k)u(n-k) = \sum_{k=-\infty}^{\infty} h(n-k)u(k) \qquad (3.30)$$

The behavior of such system is characterized by the signal $h(n)$, which for obvious reasons is called *unit sample response*. The less successful name *impulse response* is also used. The operation that combines two signals $u$ and $h$ to form the signal $y$ as in Eq. (3.30) is called *convolution* and is denoted

$$y = h * u$$

The input-output relationship of a causal convolutional system is given by Eq. (3.29), while Eq. (3.30) gives $h(n) = 0$, for $n < 0$. The right-hand side of (3.29) states that the output is formed by scaling input shifts. The weights are the same at each time due to time invariance. A signal that is zero for negative values of time is called *causal* because it lends itself to a causal linear time invariant system if it is employed as its unit sample response. Well posedness of the above representations require some further assumptions. For example, Eq. (3.30) reduces to a finite sum for each $n$ if the input history is limited to the interval $0 \leq k \leq n$. This is satisfied if the input is a causal signal, and the system is causal. The corresponding output is obtained by

$$y(n) = \sum_{k=0}^{n} h(n - k)u(k), \qquad 0 \leq n < \infty \qquad (3.31)$$

In more general situations the output at each time involves a series and extra conditions are needed. An important well posedness result is established next.

**Theorem 3.1 Well posedness** Suppose that $h(n)$ is absolutely summable, $h \in \ell_1$. The following statements hold:

1. For any $1 \leq p \leq \infty$ and any $u \in \ell_p$, the convolution $y = h * u$ is a well-defined signal in $\ell_p$ and the following bound is satisfied:

$$|h * u|_p \leq |h|_1 |u|_p \qquad (3.32)$$

2. The convolutional representation (3.29) defines a linear time invariant causal system with fading memory $S : \ell_\infty \to \ell_\infty$.

PROOF  Let $q$ be the conjugate of $p$: $1/p + 1/q = 1$. Then

$$|y(n)| \leq \sum_{k=-\infty}^{\infty} |h(n-k)||u(k)| = \sum_{k=-\infty}^{\infty} |h(n-k)|^{1/q}|h(n-k)|^{1/p}|u(k)|$$

Hölder inequality gives

$$|y(n)| \leq \left( \sum_{k=-\infty}^{\infty} |h(n-k)| \right)^{1/q} \left( \sum_{k=-\infty}^{\infty} |h(n-k)||u(k)|^p \right)^{1/p}$$

Thus

$$|y(n)|^p \leq \left( \sum_{k=-\infty}^{\infty} |h(n-k)| \right)^{p/q} \sum_{k=-\infty}^{\infty} |h(n-k)||u(k)|^p$$

$$\leq |h|_1^{p/q} \sum_{k=-\infty}^{\infty} |h(n-k)||u(k)|^p$$

Since $u \in \ell_p$, the sequence $n \to |u(n)|^p$ also belongs to $\ell_p$. Therefore

$$\sum_{n=-\infty}^{\infty} |y(n)|^p \le |h|_1^{p/q} \sum_{n=-\infty}^{\infty} \sum_{k=-\infty}^{\infty} |h(n-k)||u(k)|^p$$

We exchange sums in the right-hand side and use relation $(p/q) + 1 = p$ to obtain

$$\sum_{n=-\infty}^{\infty} |y(n)|^p \le |h|_1^{p/q} \sum_{k=-\infty}^{\infty} \left( \sum_{n=-\infty}^{\infty} |h(n-k)| \right) |u(k)|^p = |h|_1^{(p/q)+1} |u|^p = |h|_1^p |u|_p^p$$

and (3.32) is proved.

The bound (3.32) asserts that the system (3.29) is a well-defined operator on $\ell_\infty$. It is clearly linear time invariant and causal. The fading memory is discussed in Exercise 3.23. ∎

### Example 3.12  *Unit sample response of FIR systems*
The input output relationship of an FIR filter is

$$y(n) = \sum_{i=0}^{r} c_i u(n - i) \tag{3.33}$$

The above equation determines a linear time invariant causal system with fading memory on bounded inputs. The unit sample response is

$$h(n) = \sum_{i=0}^{r} c_i \delta(n - i) \tag{3.34}$$

If $n$ lies outside the range $0, 1, 2, \ldots, r$, it holds that $\delta(n - i) = 0$. Thus $h(n) = 0$. If $0 \le n \le r$, we have $h(n) = c_n$. Hence the unit sample response is a finite support signal, with values the filter coefficients. This property explains the name finite impulse response (FIR) system. ∎

### Example 3.13  *Exponential sample response and step input*
Let us consider a system of the form (3.28) with unit sample response

$$h(n) = a^n u_s(n), \qquad -\infty < n < \infty$$

and input $u(n) = u_s(n)$. Equation (3.31) gives

$$y(n) = \sum_{k=0}^{n} u_s(k) a^{n-k} u_s(n - k) = \sum_{k=0}^{n} a^{n-k} = \sum_{i=0}^{n} a^i, \qquad n \ge 0$$

Thus

$$y(n) = \begin{cases} \dfrac{a^{n+1} - 1}{a - 1}, & \text{if} \quad a \neq 1, \ n \geq 0 \\ n + 1, & \text{if} \quad a = 1, \ n \geq 0 \\ 0, & \text{if} \qquad n < 0 \end{cases}$$

The convolutional representation is implemented by the MATLAB command $y = \text{conv(h,u)}$.
We verify the previous calculations as follows:

```
n = 1:200;                              % time axis
step = ones(1,length(n));              % unit step input
h = (1/2).^n;                          % exponential impulse response
y = conv(h,step)                       % output
subplot(221), plot(step),
subplot(222), plot(h),
subplot(223), plot(y(1:length(n)))
```

The resulting output plot is illustrated in Fig. 3.20.  ■

### Examples 3.14 Decimators and interpolators
Two important examples of linear time varying systems are the decimator and the interpolator introduced in Section 3.2. The decimator is defined by $y(n) = u(Mn)$. Comparison with Eq. (3.24) shows that $h(n,k) = \delta(nM - k)$, since the only surviving input value is $u(Mn)$. Apparently the system is time varying.
In a similar fashion the $L$-fold interpolator

$$y(n) = \begin{cases} u\left(\dfrac{n}{L}\right), & \text{if } L \text{ divides } n \\ 0, & \text{otherwise} \end{cases}$$

is a linear time varying system with $h(n,k) = \delta(n - kL)$.  ■

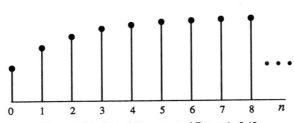

**Figure 3.20.** Plot of the output of Example 3.13.

### 3.6.3 Convolutional representation of digital systems

The analysis described above applies to digital systems. Equation (3.30) provides the convolutional representation of a digital system over finite support digital input signals with values in a field $F$. Causal inputs and causal impulse response lead to Eq. (3.31).

*Example 3.15* *Illustration of digital convolution*
Let $a, b$ be nonzero elements in $F$, and let

$$u(n) = b^n u_s(n), \qquad h(n) = a^n u_s(n), \qquad b \neq a$$

Then

$$y(n) = \sum_{k=0}^{n} b^{n-k} a^k = b^n \sum_{k=0}^{n} (ab^{-1})^k$$

Hence

$$y(n) = \begin{cases} \dfrac{a^{n+1} - b^{n+1}}{a - b}, & n \geq 0 \\ 0, & n < 0 \end{cases}$$
∎

### 3.6.4 Convolutional representation of analog systems

Continuous time linear systems admit similar representations with their discrete time counterpart, but they require considerably more effort. The main obstacle in pursuing the analysis applied for discrete systems is that no ordinary signal can produce a reasonably rich class of continuous time signals, by shifting, scaling and addition. Thus there is no continuous time analog of the unit sample signal to provide a useful decomposition of the form (3.23). Indeed, the natural counterpart of (3.23) is

$$u(t) = \int_{-\infty}^{\infty} \delta(t - \tau) u(\tau) d\tau \qquad (3.35)$$

Intuitively the role of $\delta(t - \tau)$ is to reject all values $u(\tau)$, $\tau \neq t$, and to allow only the value $u(t)$. Therefore $\delta(t)$ must be 1 for $t = 0$ and 0 otherwise. Substitution in the definition of the integral shows that the right-hand side of (3.35) is zero. Therefore (3.35) cannot be interpreted in the ordinary sense.

It turns out that if the definition of a continuous time signal is extended so as to include the so-called generalized functions, a signal with the above property can indeed be found. It is the unit impulse or Dirac function. The study of impulses is undertaken in chapter 6. In the meanwhile we content to define the convolutional representation of a continuous system by

$$y(t) = \int_{-\infty}^{\infty} h(t, \tau) u(\tau) d\tau$$

In the time invariant case the behavior of the system is characterized by the signal $h(t)$, called *impulse response*, and the operation of convolution

$$y(t) = \int_{-\infty}^{\infty} h(t-\tau)u(\tau)d\tau = \int_{-\infty}^{\infty} h(\tau)u(t-\tau)d\tau \qquad (3.36)$$

A causal system is specified by

$$y(t) = \int_{-\infty}^{t} h(t-\tau)u(\tau)d\tau = \int_{0}^{\infty} h(\tau)u(t-\tau)d\tau$$

If the input values are restricted to positive times the input-output equation takes the form of a finite integral

$$y(t) = \int_{0}^{t} h(t-\tau)u(\tau)d\tau = \int_{0}^{t} h(\tau)u(t-\tau)d\tau \qquad (3.37)$$

In analogy with the discrete case the following result holds:

**Theorem 3.2  Well posedness**   Suppose that $h(t)$ is absolutely integrable, that is, $h(t) \in L_1$. Then for each input in $L_p$, the convolution integral $y(t) = h(t) * u(t)$ is well defined and belongs to $L_p$, $1 \le p \le \infty$. Moreover the convolution bound (3.32) holds.   ∎

As a consequence the convolutional representation with absolutely integrable impulse response defines a causal linear time invariant system with fading memory on $L_\infty$, the space of bounded inputs.

Alternative frameworks for well posedness are possible. Consider for instance Eq. (3.37). The output is well defined at each time provided that some integrability assumptions on the input and the impulse response are met. It readily follows from Hölder's inequality that if $h(t)$ belongs to $L_q$ and $u(t)$ belongs to $L_p$, with $(p, q)$ a conjugate pair, the convolution integral $y(t)$ is a well-defined continuous function. An example of this type is the integrator

$$y(t) = \int_{0}^{t} u(\tau)d\tau$$

Comparison with Eq. (3.37) shows that the impulse response is the unit step signal, $h(t) = u_s(t)$. The unit step is bounded. The integrator output is well defined for all absolutely integrable inputs. The response resulting from the input $u(t) = e^{-at}u_s(t)$ is

$$y(t) = \int_{0}^{t} h(t-\tau)u(\tau)d\tau = \int_{0}^{t} e^{-a\tau}d\tau = \begin{cases} \dfrac{1}{a}(1 - e^{-at}), & t \ge 0 \\ 0, & t < 0 \end{cases}$$

### 3.6.5 Multichannel convolutional representations

Let $S$ be a linear system with $l$ input terminals and $m$ output terminals. For each $j = 1, 2, \ldots, l$, $\delta_j(n)$ denotes the $l$-channel signal that carries the unit sample $\delta(n)$ at the $j$ channel and is zero everywhere else,

$$\delta_j(n) = (0 \quad 0 \quad \cdots \quad \delta(n) \quad \cdots \quad 0 \quad 0)^T$$

Let $u(n) = (u_1(n) \quad u_2(n) \quad \ldots \quad u_l(n))^T$ be a finite support input that is zero outside $M \leq n \leq N$. Then

$$u(n) = \sum_{k=M}^{N} \sum_{j=1}^{l} \delta_j(n - k) u_j(k)$$

Linearity yields

$$y(n) = S(u)(n) = \sum_{k=M}^{N} \sum_{j=1}^{l} h_j(n, k) u_j(k) \tag{3.38}$$

where

$$h_j(n, k) = S(\delta_j(n - k)), \qquad j = 1, 2, \ldots, l \tag{3.39}$$

For each $n$, $k$, and $j$, $h_j(n, k)$ is a vector of length $m$. Equation (3.38) is written in matrix form as

$$y(n) = \sum_{k=M}^{N} h(n, k) u(k) = \sum_{k=-\infty}^{\infty} h(n, k) u(k) \tag{3.40}$$

where $h(n, k)$ denotes the 2-D matrix-valued signal

$$h(n, k) = (h_1(n, k) \quad h_2(n, k) \quad \ldots \quad h_l(n, k))$$

Time invariance and causality are dealt with as in the single-channel case. The corresponding expressions have exactly the same form except that the signals involved are matrix valued. Multichannel convolution can be obtained from ordinary convolutions as follows: We consider Eq. (3.38) for each output channel, and we interchange summation. Then

$$y_i(n) = \sum_{j=1}^{l} \sum_{k=M}^{N} h_{ij}(n, k) u_j(k), \qquad i = 1, 2, \ldots, m$$

In particular, the multichannel convolution takes the form

$$y_i(n) = \sum_{j=1}^{l} (h_{ij} * u_j)(n) \tag{3.41}$$

We conclude that the $m \times l$ multichannel convolution requires $ml$ ordinary convolutions and about $m(l - 1)$ additions. The $m \times l$ matrix-valued signal

$$h(n) = ( h_1(n) \quad \ldots \quad h_l(n) ) = ( S(\delta_1)(n) \quad \ldots \quad S(\delta_l)(n) )$$

consisting of the responses of the system to the $l$ vector unit sample signals completely specifies the input-output behavior of convolutional representation.

The continuous time multichannel convolutional representation has the same format as the single-channel counterpart except that the signals are now in matrix form.

### 3.6.6  Convolution as a Toeplitz product

Let $h(n)$ and $u(n)$ two discrete multichannel causal signals of duration $N$ and $M$, respectively, and $y(n) = h(n) * u(n)$ their convolution. If we stack up the consecutive values of $y(n)$, we obtain

$$y = \mathcal{H}u \tag{3.42}$$

where

$$y = \left( y^T(0) \quad y^T(1) \quad y^T(2) \quad \cdots \quad y^T(N + M - 2) \right)^T$$
$$u = \left( u^T(0) \quad u^T(1) \quad u^T(2) \quad \cdots \quad u^T(M - 1) \right)^T$$

and

$$\mathcal{H} = \begin{bmatrix} h(0) & 0 & 0 & \ldots & & 0 \\ h(1) & h(0) & 0 & \ldots & & 0 \\ h(2) & h(1) & h(0) & \ldots & & 0 \\ \vdots & \vdots & & & & \vdots \\ h(N-1) & h(N-2) & \ldots & h(0) & \ldots & \\ 0 & h(N-1) & & & 0 & \\ \vdots & 0 & h(N-1) & & \ddots & 0 \\ \vdots & & 0 & \ddots & & h(0) \\ & & & & & \vdots \\ & & & & h(N-1) & \\ 0 & \ldots & & \ldots & 0 & h(N-1) \end{bmatrix} \tag{3.43}$$

$\mathcal{H}$ is a $m(N + M - 1) \times lM$ block matrix whose entries are $m \times l$ matrices. It is solely determined from the samples of $h$. It is a *block Toeplitz* matrix; that is, all matrix entries located on a diagonal parallel to the main diagonal are equal.

Furthermore it is lower triangular. $\mathcal{H}$ is occasionally referred to as the *generator matrix*.

### Example 3.16   *Multichannel FIR filters*
Multichannel FIR systems are defined as in the scalar case by

$$y(n) = \sum_{i=0}^{r} c_i u(n - i) \tag{3.44}$$

The coefficients $c_i$ are $m \times l$ matrices. The unit sample response is

$$h(n) = \left[ \sum_{i=0}^{r} c_i \delta_1(n - i) \quad \sum_{i=0}^{r} c_i \delta_2(n - i) \quad \cdots \quad \sum_{i=0}^{r} c_i \delta_l(n - i) \right]$$

If $n$ lies outside the interval $0 \leq n \leq r$, $h(n) = 0$. On the other hand, for $0 \leq n \leq r$ we obtain

$$h(n) = (\, c_n e_1 \quad c_n e_2 \quad \cdots \quad c_n e_l \,) = c_n I_l = c_n$$

$I_l$ being the $l \times l$ identity matrix.  ■

### Example 3.17   *Convolutional encoders*
Multichannel FIR filters over finite fields characterize a very useful class of error control codes known as *convolutional codes*. An $(m, l, k)$ convolutional encoder is a multichannel FIR filter with $l$ inputs, $m$ outputs and filter order $k$, over a finite field, typically the binary field. Input and output buffers are added to match the input and output rates to the filter rate. The message or information sequence is converted by the demultiplexer from a serial stream to an $l$-channel string. The resulting $m$-channel output is serialized by a multiplexer to the coded sequence. Figure 3.21 illustrates a (3, 2, 1) convolutional encoder. Inspection of the block

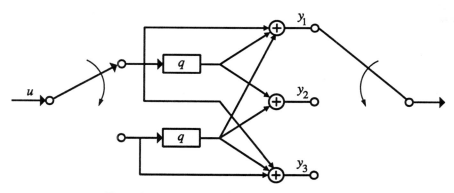

**Figure 3.21.** A (3 ,2, 1) binary convolutional encoder.

diagram leads to the following equations:

$$y_1(n) = u_1(n) + u_1(n-1) + u_2(n-1)$$
$$y_2(n) = u_1(n-1) + u_2(n-1)$$
$$y_3(n) = u_1(n) + u_2(n-1) + u_2(n)$$

The impulse response of the convolutional encoder is given by

$$h(0) = \begin{bmatrix} 1 & 0 \\ 0 & 0 \\ 1 & 1 \end{bmatrix}, \qquad h(1) = \begin{bmatrix} 1 & 1 \\ 1 & 1 \\ 0 & 1 \end{bmatrix}, \qquad h(n) = 0, n \neq 0,1$$

As a specific example, consider the message $u_1(n) = (101)$, $u_2(n) = (110)$, for $n = 0, 1, 2$ and zero otherwise. The filter register is initially empty, that is, $u(n) = 0$, for $n < 0$. The resulting coded sequence is

$$y_1(n) = (1\ 0\ 0\ 1), \quad y_2(n) = (0\ 0\ 1\ 1), \quad y_3(n) = (0\ 0\ 0\ 1)$$

for $0 \leq n \leq 3$, and zero elsewhere.    ∎

### 3.6.7  Two-dimensional linear systems

The 2-D convolutional representation is specified by the expression

$$y(n_1, n_2) = \sum_{k_1=-\infty}^{\infty} \sum_{k_2=-\infty}^{\infty} h(n_1 - k_1, n_2 - k_2) u(k_1, k_2) \qquad (3.45)$$

where

$$h(n_1, n_2) = S(\delta(n_1, n_2)) \qquad (3.46)$$

$h(n_1, n_2)$ is called *impulse response* or *2-D unit sample response*, and it characterizes the system. Eq. (3.45) defines the *2-D convolution* of $u$ and $h$:

$$y(n_1, n_2) = h(n_1, n_2) * *u(n_1, n_2) \qquad (3.47)$$

A causal shift invariant convolutional system has the form

$$y(n_1, n_2) = \sum_{k_1=-\infty}^{n_1} \sum_{k_2=-\infty}^{n_2} h(n_1 - k_1, n_2 - k_2) u(k_1, k_2)$$
$$= \sum_{k_1=0}^{\infty} \sum_{k_2=0}^{\infty} h(k_1, k_2) u(n_1 - k_1, n_2 - k_2)$$

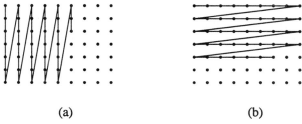

(a)                                             (b)

**Figure 3.22.** Image scanning by (a) columns and (b) by rows.

Semicausal systems in the direction of $n_1$ are given by

$$y(n_1, n_2) = \sum_{k_1=0}^{\infty} \sum_{k_2=-\infty}^{\infty} h(k_1, k_2) u(n_1 - k_1, n_2 - k_2)$$

Semicausal systems arise when the image is scanned by columns or rows; see Fig. 3.22.

The 2-D analog convolution has the form

$$y(t_1, t_2) = \int_{-\infty}^{\infty} \int_{-\infty}^{\infty} h(t_1 - \tau_1, t_2 - \tau_2) u(\tau_1, \tau_2) d\tau_1 d\tau_2 \tag{3.48}$$

***Example 3.18  Illustration of 2-D convolution***
Consider the 2-D system with impulse response

$$h(n_1, n_2) = a_1^{n_1} a_2^{n_2} u_s(n_1, n_2), \qquad a_1 \neq 1, \, a_2 \neq 1$$

We seek to determine the output to the 2-D unit step $u(n_1, n_2) = u_s(n_1, n_2)$. Both signals are separable, that is, $h(n_1, n_2) = h_1(n_1) h_2(n_2)$, $h_1(n_1) = a_1^{n_1} u_s(n_1)$, $h_2(n_2) = a_2^{n_2} u_s(n_2)$, and $u_s(n_1, n_2) = u_s(n_1) u_s(n_2)$. It is easy to see that

$$y(n_1, n_2) = (h_1(n_1) * u_s(n_1))(h_2(n_2) * u_s(n_2))$$

The above 1-D convolutions were determined in Example 3.13. Thus

$$y(n_1, n_2) = \left( \frac{a_1^{n_1+1} - 1}{a_1 - 1} \right) \left( \frac{a_2^{n_2+1} - 1}{a_2 - 1} \right) u_s(n_1, n_2) \qquad \blacksquare$$

## 3.7  VOLTERRA REPRESENTATIONS

Volterra representations constitute a natural extension of convolutional models to nonlinear systems. The output of a convolutional system is formed by weighting shifted versions of the input by the impulse response values and summing them up. This idea is translated in a nonlinear context if the shifted inputs are consecutively multiplied, weighted by coefficients, and then added up. The resulting expression in the causal case takes the form

$$y(n) = h_0 + \sum_{k=0}^{\infty} h_1(k)u(n-k) + \sum_{k_1=0}^{\infty} \sum_{k_2=0}^{\infty} h_2(k_1, k_2)u(n-k_1)u(n-k_2) + \cdots$$

$$+ \sum_{k_1=0}^{\infty} \sum_{k_2=0}^{\infty} \cdots \sum_{k_p=0}^{\infty} h_p(k_1, k_2, \ldots, k_p)u(n-k_1)u(n-k_2) \cdots u(n-k_p) \quad (3.49)$$

and is called *finite Volterra system* of degree $p$. The term "finite" emphasizes that the number of "nonlinearities" $p$ is finite. If $p$ goes to infinity, we obtain the general Volterra expansion

$$y(n) = h_0 + \sum_{m=1}^{\infty} \sum_{\mathbf{k}_m} h_m(k_1, k_2, \ldots, k_m)u(n-k_1)u(n-k_2) \cdots u(n-k_m) \quad (3.50)$$

The inner sum is a compact expression for

$$y_m(n) = \sum_{k_1=0}^{\infty} \cdots \sum_{k_m=0}^{\infty} h_m(k_1, k_2, \ldots, k_m)u(n-k_1)u(n-k_2) \cdots u(n-k_m)$$

and is called *homogeneous Volterra system* of degree $m$ because, if the input $u(n)$ produces the output $y_m(n)$ and $c$ is a constant, $cu(n)$ produces $c^m y_m(n)$. Each entry of the multiindex $\mathbf{k}_m = (k_1, k_2, \ldots, k_m)$ runs from 0 to $\infty$.

The $m$-dimensional sequence $h_m(k_1, k_2, \ldots, k_m)$ defines the *Volterra kernel* of degree $m$. The above representation is causal and time invariant. Time varying and noncausal Volterra models can be conceived as in the linear case.

***Well posedness.*** The output of the finite Volterra system (3.49) is always defined if the input is causal. Caution is required if the input is noncausal because each homogeneous term involves an infinite sum. If the general Volterra model (3.50) is considered, further complications are introduced. These issues can be approached in different ways, depending on the application at hand. Guided by the linear case we can make the following assumptions.

1. Each Volterra kernel $h_m(k_1, k_2, \ldots, k_m)$ is absolutely summable:

$$|h_m(k_1, k_2, \ldots, k_m)|_1 = \sum_{\mathbf{k}_m} |h_m(k_1, k_2, \ldots, k_m)| < \infty$$

2. The real sequence $|h_m(k_1, \ldots, k_m)|_1^{1/m}$, $m \geq 1$ is bounded; equivalently,

$$\limsup_{m \to \infty} |h_m|_1^{1/m} < \infty$$

The second condition controls the infinite number of nonlinearities in the Volterra expansion. It is automatically satisfied if the Volterra system is finite. To see the nature of the above conditions, consider a bounded input signal $u$ with finite sup norm $|u|_\infty$. The resulting output is bounded as follows:

$$|y(n)| \leq |h_0| + \sum_{m=1}^{\infty} \sum_{\mathbf{k}_m} |h_m(k_1, k_2, \ldots, k_m)||u(n - k_1)u(n - k_2) \cdots u(n - k_m)|$$

$$\leq \sum_{m=0}^{\infty} |u|_\infty^m \sum_{\mathbf{k}_m} |h_m(k_1, k_2, \ldots, k_m)| \leq \sum_{m=0}^{\infty} |h_m|_1 |u|_\infty^m$$

The right-hand side is a real power series with radius of convergence $\rho = (\limsup_{m \to \infty} |h_m|_1^{1/m})^{-1}$. The meaning of the above conditions is now apparent. Condition 1 asserts that each $|h_m|_1$ is finite. Condition 2 ensures that $\rho$ is positive. The conclusion is that inputs with amplitude bounded by $\rho$ produce well-defined and bounded outputs.

### 3.7.1  Multidimensional embedding

Every homogeneous Volterra system of degree $m$ can be lifted to a multi-dimensional linear system of dimension $m$ as follows: Each input signal $u(n)$ gives rise to the separable $m$-D signal

$$x(n_1, n_2, \ldots, n_m) = u(n_1)u(n_2) \cdots u(n_m)$$

This signal is convolved with the Volterra kernel $h_m(n_1, n_2, \ldots, n_m)$ via an m-D convolution

$$y(n_1, n_2, \ldots, n_m) = h_m(n_1, n_2, \ldots, n_m) * * \cdots * x(n_1, n_2, \ldots, n_m)$$

The output of the Volterra system $y_m(n)$ is then obtained by evaluation on the diagonal slice

$$y_m(n) = y(n, n, \ldots, n)$$

**Example 3.19**  *Output computation of a second degree Volterra system*
Consider the Volterra system of degree 2 with kernels

$$h_1(n) = a^n u_s(n), \quad h_2(n_1, n_2) = b_1^{n_1} b_2^{n_2} u_s(n_1)u_s(n_2)$$

The output produced by the unit step is obtained from the superposition of the outputs $y_1(n)$ and $y_2(n)$ of the homogeneous terms of degrees 1 and 2. $y_1(n)$ was calculated in Example 3.13. To determine $y_2(n)$, we use the embedding and the 2-D convolution of Example 3.18. Downloading the corresponding expression gives

$$y_2(n) = \frac{b_1^{n+1} - 1}{b_1 - 1} \frac{b_2^{n+1} - 1}{b_2 - 1} u_s(n)$$

Thus

$$y(n) = \frac{a^{n+1} - 1}{a - 1} + \frac{b_1^{n+1} - 1}{b_1 - 1} \frac{b_2^{n+1} - 1}{b_2 - 1}, \qquad n \geq 0$$

MATLAB simulation of a Volterra system of degree 2 is obtained as follows:

```
function y = volt2(h1,h2,u);
y1 = conv(h1,u);        % determines the output of the linear part;
u2 = u*u';              % forms the separable 2-D extension of the input, u is a column vector
y2 = conv2(h2,u2);      % computes the associated 2-D convolution
y = diag(y2);           % evaluates the diagonal slice
y = y1 + y;             % adds the responses of degrees 1 and 2.
```
∎

### 3.7.2 Symmetric, triangular, regular, finite support, and separable Volterra kernels

It is often convenient to work with Volterra expansions whose kernels possess additional properties. Five such cases are discussed next. Symmetric, triangular, and regular kernels can be imposed without harming generality, since any Volterra system is equivalently described by a Volterra system with symmetric triangular or regular kernels. FIR and separable kernels do not share the above property. Nevertheless, they provide reasonable approximations.

***Symmetric kernels.*** Let us first consider the second-degree homogeneous Volterra operator

$$y_2(n) = \sum_{k_1=0}^{\infty} \sum_{k_2=0}^{\infty} h_2(k_1, k_2) u(n - k_1) u(n - k_2) \tag{3.51}$$

The change of variables $k_1 \rightarrow k_2$, $k_2 \rightarrow k_1$ leads to

$$y_2(n) = \sum_{k_1=0}^{\infty} \sum_{k_2=0}^{\infty} h_2(k_2, k_1) u(n - k_1) u(n - k_2)$$

Therefore the kernel $h_2(k_1, k_2)$ and its transpose $h_2(k_2, k_1)$ represent the same input-output map. Let

$$h_{s,2}(k_1, k_2) = \tfrac{1}{2}(h_2(k_1, k_2) + h_2(k_2, k_1))$$

Clearly $h_{s,2}(k_1, k_2)$ is symmetric, that is, $h_{s,2}(k_1, k_2) = h_{s,2}(k_2, k_1)$. Furthermore $h_{s,2}(k_1, k_2)$ yields the same input-output map as $h_2(k_1, k_2)$. Indeed,

$$y_2(n) = \frac{1}{2} y_2(n) + \frac{1}{2} y_2(n)$$

$$= \frac{1}{2} \sum_{k_1=0}^{\infty} \sum_{k_2=0}^{\infty} h_2(k_1, k_2) u(n - k_1) u(n - k_2)$$

$$+ \frac{1}{2} \sum_{k_1=0}^{\infty} \sum_{k_2=0}^{\infty} h_2(k_2, k_1) u(n - k_1) u(n - k_2)$$

$$= \sum_{k_1=0}^{\infty} \sum_{k_2=0}^{\infty} h_{s,2}(k_1, k_2) u(n - k_1) u(n - k_2)$$

The above result is extended to the general case as follows: Let $\sigma$ denote a permutation (i.e., a rearrangement) of the indexes $k_1, k_2, \ldots, k_m$. Any such permutation leaves the input-output map of the $m$-degree homogeneous Volterra operator unaltered. The total number of permutations is $m!$. The symmetric kernel is then

$$h_{s,m}(k_1, \ldots, k_m) = \frac{1}{m!} \sum_{\sigma} h(k_{\sigma(1)}, k_{\sigma(2)}, \ldots, k_{\sigma(m)}) \qquad (3.52)$$

The sum runs over all permutations $\sigma$. As the preceding discussion indicates, the kernels of a given Volterra operator are not unique. Uniqueness is achieved if the kernels are assumed symmetric. The proof of this statement relies on the fact that the Volterra expansion is a Taylor series of the associated operator and is beyond the scope of this book. For further information the reader should refer to the bibliographical comments at the end of the chapter.

**Triangular kernels.** Let us consider a homogeneous Volterra operator of degree 2 with symmetric kernel $h_{s,2}(k_1, k_2)$. Due to symmetry the lower triangular values $h_{s,2}(k_1, k_2)$, $k_1 \geq k_2$ suffice to determine the entire matrix. Of course we could as well use the upper triangular part. This motivates the following definition: A Volterra kernel $h_{t,m}$ is called triangular if

$$h_{t,m}(k_1, k_2, \ldots, k_m) = 0 \quad \text{for} \quad k_1 < k_2 < \cdots < k_m$$

The corresponding homogeneous term becomes

$$\sum_{k_1=0}^{\infty} \sum_{k_2=0}^{k_1} \cdots \sum_{k_m=0}^{k_{m-1}} h_{t,m}(k_1, k_2, \ldots, k_m) u(n - k_1) u(n - k_2) \cdots u(n - k_m)$$

Given a symmetric Volterra system there is a unique triangular system that realizes the same input-output map, and conversely. The symmetric kernel associated with a given triangular kernel is constructed by Eq. (3.52). It is easy to establish the converse statement for second-degree systems. Indeed, all terms of the double sum not lying on the diagonal appear twice. Thus the triangular kernel is defined by

$$h_{t,2}(n_1, n_2) = \begin{cases} h_{s,2}(n_1, n_2), & n_1 = n_2 \\ 2h_{s,2}(n_1, n_2), & n_1 > n_2 \\ 0, & n_1 < n_2 \end{cases}$$

**Regular kernels.** Regular kernels remove the "discontinuities" of the triangular kernels at the diagonal. Using the change of variables $k_1 = i_1 + i_2 + \cdots + i_m$, $k_2 = i_2 + \cdots + i_m, \ldots, k_m = i_m$ where $0 \le i_j < \infty, j = 1, \ldots, m$, the corresponding homogeneous term becomes

$$\sum_{k_1=0}^{\infty} \sum_{k_2=0}^{k_1} \cdots \sum_{k_m=0}^{k_{m-1}} h_{t,m}(k_1, k_2, \ldots, k_m) u(n - k_1) \cdots u(n - k_m)$$

$$= \sum_{i_1=0}^{\infty} \sum_{i_2=0}^{\infty} \cdots \sum_{i_m=0}^{\infty} h_{t,m}(i_1 + \cdots + i_m, i_2 + \cdots + i_m, \ldots, i_m)$$

$$\cdot u(n - i_1 - \cdots - i_m) \cdots u(n - i_m)$$

$$= \sum_{i_1=0}^{\infty} \sum_{i_2=0}^{\infty} \cdots \sum_{i_m=0}^{\infty} h_{r,m}(i_1, i_2, \ldots, i_m) u(n - i_1 - \cdots - i_m) \cdots u(n - i_m)$$

where

$$h_{r,m}(i_1, i_2, \ldots, i_m) = h_{t,m}(i_1 + \cdots + i_m, i_2 + \cdots + i_m, \ldots, i_m)$$

$h_{r,m}$ extends smoothly over the region $i_1 \ge 0, \ldots, i_m \ge 0$ and is called *regular kernel*. Every Volterra system with triangular kernels is realized by a Volterra system with regular kernels and conversely.

The continuous time case is similar.

**Finite support Volterra systems.** In analogy with FIR models, a finite support Volterra system is a finite Volterra system whose kernels have finite support.

The resulting input-output expression has the form

$$y(n) = h_0 + \sum_{k=0}^{r_1} h_1(k)u(n-k) + \sum_{k_1=0}^{r_2} \sum_{k_2=0}^{r_2} h_2(k_1,k_2)u(n-k_1)u(n-k_2) + \cdots$$

$$+ \sum_{k_1=0}^{r_m} \cdots \sum_{k_m=0}^{r_m} h_m(k_1,k_2,\ldots,k_m)u(n-k_1)u(n-k_2)\cdots u(n-k_m) \quad (3.53)$$

The multiindex $r = (r_1 \cdots r_m)$ is the multiorder of the system and delineates the region over which the products of the shifted inputs affect the output. Finite support Volterra systems are identical to polynomial filters introduced in Example 3.10.

**Separable kernels.** In this class of Volterra systems each kernel is separable:

$$h_m(k_1,\ldots,k_m) = h_{m1}(k_1)h_{m2}(k_2)\cdots h_{mm}(k_m)$$

The corresponding homogeneous term becomes

$$y_m(n) = \sum_{k_m} h_{m1}(k_1)h_{m2}(k_2)\cdots h_{mm}(k_m)u(n-k_1)\cdots u(n-k_m)$$

$$= \left(\sum_{k_1} h_{m1}(k_1)u(n-k_1)\right)\left(\sum_{k_2} h_{m2}(k_2)u(n-k_2)\right)\cdots\left(\sum_{k_m} h_{mm}(k_m)u(n-k_m)\right)$$

or

$$y_m(n) = z_{m1}(n)z_{m2}(n)\cdots z_{mm}(n)$$

where

$$z_{mi}(n) = h_{mi}(n) * u(n), \qquad 1 \le i \le m$$

The corresponding architecture for a system of degree 3 is depicted in Fig. 3.23. The system is formed by a filter bank of linear filters followed by multipliers and adders. The impulse responses of the filters are the sequences $h_{mi}$.

### 3.7.3 Universal approximation capabilities of polynomial FIR filters

Polynomial FIR filters can simulate fairly general dynamical systems. More precisely every time invariant causal system with fading memory, $S : \ell_\infty \to \ell_\infty$, can be approximated by a polynomial FIR filter over uniformly bounded inputs.

**Theorem 3.3** Let $S : \ell_\infty \to \ell_\infty$ be a time invariant causal system and $M > 0$. Suppose that $S$ has fading memory on the set

$$\mathcal{U}_M = \{u \in \ell_\infty : |u|_\infty \le M\}$$

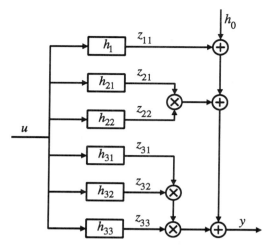

**Figure 3.23.** Architecture of a third-degree separable Volterra system.

Then for any $\epsilon > 0$ there exists a polynomial function $p : R^m \rightarrow R$ such that

$$|S(u)(n) - p(u(n-1), u(n-2), \ldots, u(n-m))| \leq \epsilon$$

for any $n \in Z$ and any $u$ satisfying $|u|_\infty \leq M$.

The proof is given in the Appendix at the end of the chapter. ∎

The family of polynomial FIR architectures is parametrized by the memory size $m$ and the weighting coefficients. The proof of Theorem 3.3 does not provide a constructive procedure for their determination. $m$ must be estimated from the fading memory property of $S$ and the window sequence $w(n)$. Once $m$ and the degree of the polynomial function $p$ are decided, the input-output expression of the polynomial system leads to a linear regression, and the coefficients can be learned by the methods of Chapter 4.

Besides its own significance the above theorem enables us to lift universal approximation capabilities of a static environment to a dynamic environment. Indeed, let $C$ be a dense subset of the set of continuous real-valued functions defined on a closed and bounded set $K$ of $R^m$. Then the class of architectures

$$y(n) = f(u(n), u(n-1), \ldots, u(n-m)), \quad m \in Z^+, \quad f \in C$$

approximates every time invariant causal system with fading memory on $K$. Indeed, by Theorem 3.3, any such system is approximated by a polynomial system. Then the polynomial function is approximated by a function $f$ in $C$. This idea is used later on to establish the universal approximation properties of time delay networks with a feed-forward multilayer neural network and radial basis nonlinearities.

## 3.8 FINITE DERIVATIVE REPRESENTATIONS

In this section the finite derivative model is introduced. Together with the finite difference model for discrete systems, and the state space representation discussed in the next sections, they constitute the finite recursive representations. This is so because computation of the output is recursive and the required memory is finite and constant. Finite recursive models offer two advantages. First, they facilitate the description of nonlinear processes. Modeling of real systems with recursive models is accomplished via direct translation of the physical laws governing the dynamics without intermediate complex manipulations. The second advantage is that they are naturally mapped to algorithms and architectures, making simulation of a real process direct and authentic.

### 3.8.1 Motivation

Classical mechanics and circuit theory provide natural examples for motivating finite derivative models. In these fields the dynamic behavior between inputs and outputs is governed by physical laws, which offer a more economical and compact representation of the dynamic evolution. In classical mechanics Newton's law in its simplest form asserts that the motion of a particle of mass $m$ moving under the action of a force $u(t)$ is governed by the equation

$$m\ddot{y}(t) = u(t) \tag{3.54}$$

where $y(t)$ denotes displacement of the particle from a given reference point. The second derivative, $\ddot{y}(t)$, is the acceleration of the particle. The crucial observation deduced from (3.54) is that the relevant motion parameters such as forces, displacements, and velocities are mathematically related via equations involving these variables as well as their derivatives. Exactly the same remark is valid in the case of circuits. Voltages and currents form relationships that typically involve a finite number of derivatives as well. Let us take, for instance, the following example.

### *Example 3.20   RLC circuit*
Consider the circuit depicted in Fig. 3.24. A resistor $R$, a capacitor $C$, and an

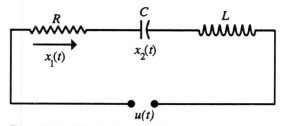

**Figure 3.24.** RLC circuit with independent voltage source.

inductor $L$ are connected in series to an independent voltage source. We view the circuit as a system with input the voltage supplied by the source $u(t)$ and output the current $y = x_1$. Indeed, let $x_2$ denote the voltage across the capacitor. Then $C\dot{x}_2 = x_1$. Kirchhoff voltage law gives $Rx_1 + x_2 + Lx_1 = u$. Differentiation of the latter equation and use of the capacitor equation gives $\ddot{y} = -R\dot{y}/L - y/CL + \dot{u}/L$. On the other hand, the system with input $u$ and output $y = x_2$ satisfies

$$\ddot{y} = -\frac{R}{L}\dot{y} - \frac{1}{CL}y + \frac{1}{LC}u \tag{3.55}$$

∎

### 3.8.2 Basic notions

The above remarks, and Eqs. (3.54) and (3.55) in particular, lead us to consider input-output relationships involving a finite number of input and output derivatives. Among these, we will focus attention on expressions solvable with respect to the higher-order output derivative

$$y^{(k)}(t) = F\left(y^{(k-1)}(t), \ldots, y^{(1)}(t), y(t), u(t), \ldots, u^{(r)}(t)\right) \tag{3.56}$$

Equation (3.56) constitutes the finite derivative representation of a system. To ensure that the system will supply an output when actuated by an input, assumptions on $F$ must be imposed. Moreover, to avoid the possibility of having infinitely many outputs from a single input, initial information about the output must be provided in conjunction with suitable additional assumptions on $F$. The precise technical framework for these questions is a subject of existence and uniqueness theory of differential equations. An important result in this direction and adequate for the purposes of the book is given next without proof.

***Well posedness.*** Suppose $F$ is continuously differentiable; that is, the partial derivatives of $F$ with respect to its arguments exist and are continuous. Let $u(t)$ be an input signal having derivatives of order $r$ and defined on an interval $I$. Then for any initial time $t_0$, in $I$, and any initial block vector $\mathbf{y}_0 = \begin{bmatrix} y_0^T & y_1^T & \cdots & y_{k-1}^T \end{bmatrix}^T$, there exists a unique signal $y(t)$ that satisfies (3.56) over a subinterval of $I$ and the *initial condition*

$$y(t_0) = y_0, \qquad y(t_0) = y_1, \qquad \ldots, \qquad y^{(k-1)}(t_0) = y_{k-1}$$

To show the explicit dependence on initial conditions we denote by $y(t, t_0, \mathbf{y}_0)$, the unique output trajectory produced by $t_0$ and $\mathbf{y}_0$.

***Induced transform.*** Well posedness conditions guarantee local existence of the output $y(t, t_0, \mathbf{y}_0)$ in an interval around $t_0$. Under further so-called completeness conditions the outputs can be extended over the reals. In this case the finite derivative model is mapped into a transform of the form $\mathcal{G} : L \times R \times R^k \to L$, $(u, t_0, \mathbf{y}_0) \to y(t, t_0, \mathbf{y}_0)$.

Transformations of the form $\mathcal{G}$ broaden the system scope introduced in Section 3.1, since they view the system output as the combined result of the input signal and its autonomous evolution specified by the initial data. We often find it convenient to separate the effects of the initial data from the input. This is accomplished in the following manner: Initial data $(t_0, \mathbf{y}_0)$ are kept fixed. The system $\mathcal{G}_{(t_0, \mathbf{y}_0)} : L \to L$, $\mathcal{G}_{(t_0, \mathbf{y}_0)}(u) = \mathcal{G}(u, t_0, \mathbf{y}_0)$ expresses the effects of inputs to outputs. It is called an *initialized system*. Initialized systems are thus represented by operators on signal spaces and thus are systems in the sense of Section 3.1. In a similar fashion we fix the input $u$, and we define the partial transform $\mathcal{G}_u : R \times R^k \to L$, $\mathcal{G}_u(t_0, \mathbf{y}_0) = \mathcal{G}(u, t_0, \mathbf{y}_0)$. This time the input is kept fixed, and the impact of the initial data on the output is modeled. Such systems are called *autonomous*, since their evolution is governed by its internal dynamics.

**Causality.** Equation (3.56) specifies a causal behavior for $t \geq t_0$, since it can be written in the equivalent integral form

$$y^{(k-1)}(t) = y_{k-1} + \int_{t_0}^{t} F\left(y^{(k-1)}(\tau), \ldots, y(\tau), u(\tau), \ldots, u^{(r)}(\tau)\right) d\tau, \qquad t \geq t_0$$

**Time Invariance.** Equation (3.56) determines a *time invariant representation* in the following sense: Let $y(t)$ denote the output produced by the input $u(t)$ and initial data $t_0$, $\mathbf{y}_0$. If $u(t)$ is arbitrarily shifted to $u(t + \tau)$, $\tau$ real, the resulting output is $y(t + \tau)$ provided that the system is initialized at $(t_0 - \tau, \mathbf{y}_0)$. The more general *time varying* case is defined by the model

$$y^{(k)}(t) = F\left(y^{(k-1)}(t), \ldots, y^{(1)}(t), y(t), u(t), \ldots, u^{(r)}(t), t\right) \qquad (3.57)$$

Well posedness of (3.57) is guaranteed if $F$ is continuously differentiable with respect to all arguments.

**Linearity.** The linear time varying form of Eq. (3.57) is

$$y^{(k)}(t) = -\sum_{j=0}^{k-1} a_j(t) y^{(j)}(t) + \sum_{i=0}^{r} b_i(t) u^{(i)}(t) \qquad (3.58)$$

which, in the time invariant case, reduces to

$$y^{(k)}(t) = -\sum_{j=0}^{k-1} a_j y^{(j)}(t) + \sum_{i=0}^{r} b_i u^{(i)}(t) \qquad (3.59)$$

The minus sign is inserted merely for notational purposes.

**Figure 3.25.** Mass-spring-dashpot mechanical system.

### Example 3.21    Mass-spring-dashpot system

We consider the mass-spring-dashpot mechanical system of Fig. 3.25. Forces affecting the motion are the external force $u(t)$, the friction force, and the restoring force of the spring. Let $y(t)$ be the displacement of the mass from a given reference point. The behavior of the spring is governed by Hooke's law, which states that the restoring force of the spring depends linearly on the displacement. The friction is proportional to velocity. Newton's law gives

$$\ddot{y} = -\frac{k}{m}y - \frac{\lambda}{m}\dot{y} + \frac{1}{m}u, \qquad k > 0, \lambda > 0 \tag{3.60}$$

If $\lambda = 0$, friction becomes negligible and Eq. (3.60) becomes $\ddot{y} = -ky/m + u/m$. This equation constitutes the *linear undamped oscillator* for reasons we will explain later. If friction is negligible and the spring is nonlinear, the motion is governed by the equation $m\ddot{y} + g(y) = u, yg(y) > 0$. In the presence of friction, it becomes $m\ddot{y} + f(y, \dot{y}) + g(y) = u$, with special case the nonlinear Lienard oscillator

$$m\ddot{y} + f(y)\dot{y} + g(y) = u \tag{3.61}$$

The mass-spring-dashpot system describes the motion of several real systems, such as the suspension system of a vehicle (see Fig. 3.26). Let $y_0$ be the distance of the vehicle from road surface when the vehicle is still, $y(t)$ the position from a given elevation level, and $u(t)$ the road elevation from the reference level. The motion is described by

$$M\ddot{y} + \lambda\dot{y} + ky = ku + \lambda\dot{u} \tag{3.62}$$

∎

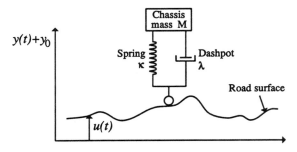

**Figure 3.26.** Suspension system.

## 3.9 FINITE DIFFERENCE MODELS

The discrete time counterpart of Eq. (3.56) relates a finite number of successive input and output shifts

$$y(n) = F(y(n-1), \ldots, y(n-k), u(n), u(n-1), \ldots, u(n-r)) \qquad (3.63)$$

The output is specified by $k$ preceding output values, $r$ preceding input samples and the current input sample. Equation (3.63) constitutes the finite difference model. The terminology originates from mathematical studies of such equations expressed in terms of differences. If $F$ is a polynomial function, we end up with the recurrent polynomial model discussed in Example 3.11. As in the continuous time case, the above model does not give rise to a single system in the sense of a well-defined transformation between inputs and outputs but rather to a family of systems. For a given input signal there are infinitely many outputs that satisfy Eq. (3.63), unless $k = 0$. The most intuitive and direct way to establish a legitimate input-output assignment is to specify $k$ consecutive values of the output, say, $y(n_0)$, $y(n_0 + 1)$, $\ldots$, $y(n_0 + k - 1)$. Then the entire future output profile $y(n)$, $n \geq n_0 + k$ is uniquely determined from the respective input via Eq. (3.63). Thus, unlike the continuous time case, well posedness is easy to establish.

**Well posedness.** Given a time instant $n_0$, a real block vector of size $k$ and entries of size $m \times 1$, $\mathbf{y}_0 = \begin{pmatrix} y_0^T & y_1^T & \cdots & y_{k-1}^T \end{pmatrix}^T$, and any input signal $u(n)$ defined on $[n_0 + k - r, \infty)$, there exists a unique output signal $y(n)$ that satisfies Eq. (3.63) for $n \geq n_0 + k$, and the so-called *initial condition* $y(n_0) = y_0$, $y(n_0 + 1) = y_1, \ldots$, $y(n_0 + k - 1) = y_{k-1}$. Sometimes we use the notation $y(n, n_0, \mathbf{y}_0)$ for the output trajectory produced by $n_0$ and $\mathbf{y}_0$.

Causality, time invariance, memory, and linearity are analogous to the continuous time case. The linear time varying model has the form

$$y(n) = -\sum_{j=1}^{k} a_j(n) y(n-j) + \sum_{i=0}^{r} b_i(n) u(n-i) \qquad (3.64)$$

In the time invariant case it becomes

$$y(n) = -\sum_{j=1}^{k} a_j y(n-j) + \sum_{i=0}^{r} b_i u(n-i) \qquad (3.65)$$

***Example 3.22    Euclid's algorithm***
Euclid's algorithm is introduced in Appendix II. The general recursion has the form

$$t^{(n-2)} = q^{(n)} t^{(n-1)} + t^{(n)} \qquad (3.66)$$

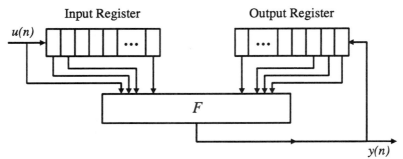

Input Register          Output Register

**Figure 3.27.** Block diagram realization of finite difference model.

If we solve with respect to $t^{(n)}$, the above equation takes the form

$$y(n) = -q\big(y(n-2), y(n-1)\big)y(n-1) + y(n-2) \qquad (3.67)$$

where $q$ is the quotient function and $y(n) = t^{(n)}$. Initial conditions are $y_{-1} = s$, $y_0 = t$. The algorithm is represented as an autonomous finite difference system.

∎

Equation (3.65) admits a finite parametrization. The input-output relationship is characterized by the parameters $k$, $r$, $a_j$, and $b_i$.

An attractive feature of finite difference models is that they are well suited for machine computation. A direct realization is illustrated by the block diagram of Fig. 3.27. The system memory is conveniently divided into the input register of length $r$ and the output register of length $k$. The contents of the registers are transferred to the possibly nonlinear module $F$, and the resulting output is fed back to the output register ready for the next clock cycle.

In the linear time invariant case the module $F$ of Fig. 3.27 consists of adders and multipliers. The block diagram specializes to the direct realization of Fig. 3.19.

Programming of Eq. (3.63) is straightforward for reasonable options of the function $F$. $F$ is usually available as a special function or called by a subroutine. Algorithms and architectures are available for a variety of such nonlinear functions.

***Example 3.23    MATLAB implementation of linear finite difference models***
The model (3.65) is implemented in MATLAB by the commands

```
y = filter(b,a,u)
y = filter(b,a,u,y0)
```

where $u$ is the input, $b$ is a vector with coefficients $b_0, \ldots, b_r$, and $a$ is a vector with coefficients $1, a_1, \ldots, a_k$. The second command treats nonzero initial conditions.

Here is an example.

```
a = [1,-3/4, 1/8];
b = [2,7];
n = 1:100;                        % time horizon
u = ones(1,length(n));           % unit step
y = filter(b,a,u);
plot(y(1:50))                     % plots the first 50 output values
```

## 3.10 STATE SPACE REPRESENTATIONS

### 3.10.1 Motivation

Despite their unquestionable success in the study of simple second-order dynamics of the oscillator type, finite derivative models have their own limitations as modeling tools. In the case of complex dynamics, they may be too restrictive or their development may require considerable effort. These problems are largely alleviated by the more general state space representation. To see how the state space methodology comes up as a natural and genuine modeling approach, we discuss an illustrative example.

*Examples 3.24   Communication satellite*
A simplified approximate description of the satellite (Example 3.2) is provided by the planar motion of a particle of mass $m$ moving in the gravitational field of the earth and under the action of controlling forces $u_1(t)$ and $u_2(t)$. Let $M$ denote the mass of the earth and $r(t)$ and $\theta(t)$ the polar coordinates of the displacement vector. The control problem of maintaining the radial displacement to a constant value pinpoints upon the system with input the force $u(t) = [u_1(t), u_2(t)]$ and output the radial displacement $y(t) = r(t)$. Can this system be modeled by a finite input-output derivative form? To answer this question, we apply Newton's law of motion to obtain $\ddot{r}(t) = r(t)\dot{\theta}^2(t) - k/r^2(t) + u_1(t)/m$, or in input-output notation

$$\ddot{y}(t) = y(t)\dot{\theta}^2(t) - \frac{k}{y^2(t)} + \frac{1}{m}u_1(t) \tag{3.68}$$

where $k = gM$. To obtain a finite input-output derivative model, we must eliminate the angular displacement $\theta(t)$. For this reason we consider the equation governing the angular motion:

$$\ddot{\theta}(t) = -\frac{2\dot{r}(t)\dot{\theta}(t)}{r(t)} + \frac{1}{mr(t)}u_2(t) \tag{3.69}$$

The acceleration $\ddot{\theta}(t)$ present in the latter equation suggests differentiation of Eq. (3.68). Substitution of Eq. (3.69) into Eq. (3.68) does not eliminate $\theta(t)$, and further differentiation is needed. Hopefully, after a finite number of differentiations, the angular displacement will be eliminated, and the system will take the desirable form (3.56). At this point, however, it is unclear if repeated differentiation a finite number of times will lead to elimination of $\theta(t)$. Even if this turns out to be true, the computations involved to determine the function $F$ of Eq. (3.56) could be prohibitive. Before more sophisticated apparatus is attempted, we must reconsider what precisely we have been after. We tried to obtain a mathematical model for the system with inputs the controlling forces and output the radial displacement, and in the process of doing so, we encountered some serious difficulties. A closer look reveals that these problems do not result from the poor modelability of the specific example but from our persistency to impose the finite derivative model. Indeed, although we failed to come up with a high-order differential equation, we certainly managed to write equations for our system that are really what we are looking for. The equations we alluded to are expressions (3.68) and (3.69), and the morale deduced from the above observations is that in order to bypass the aforementioned obstacles, the signal $\theta(t)$ and the associated Eq. (3.69) must be maintained. We can put our observation in broader context in the following manner:

*To model a mechanical system, first describe the overall motion and then use the resulting equations to identify a particular input-output relationship between specific motion parameters, as a constituent part of the motion.*

The first task comprises the main theme in Newtonian dynamics. It is accomplished by using the important fact that the motion of a set of $n$ particles moving in space is completely described by their displacements and velocities on a $6n$-dimensional space called *configuration space*. For instance, the satellite is characterized by the dispacement vector $[r(t), \theta(t)]^T$ and the velocity vector $[\dot{r}(t), \dot{\theta}(t)]^T$, by means of Eqs. (3.68) and (3.69). To describe it in the configuration space, we combine displacements and velocities into a single four-dimensional vector

$$x(t) = [x_1(t), x_2(t), x_3(t), x_4(t)]^T = [r(t), \dot{r}(t), \theta(t), \dot{\theta}(t)]^T$$

Equations (3.68) and (3.69) are written as a first-order vector differential equation on $R^4$,

$$\dot{x}_1(t) = x_2(t)$$

$$\dot{x}_2(t) = x_1(t)x_4^2(t) - \frac{k}{x_1^2(t)} + \frac{1}{m}u_1(t)$$

$$\dot{x}_3(t) = x_4(t)$$

$$\dot{x}_4(t) = \frac{-2x_2(t)x_4(t)}{x_1(t)} + \frac{1}{mx_1(t)}u_2(t) \tag{3.70}$$

The first and third equations merely state that the corresponding parameters are velocities. The overall motion is described in compact form as

$$\dot{x}(t) = f(x(t), u(t)) \tag{3.71}$$

where $\dot{x}(t) = [\dot{x}_1(t), \dot{x}_2(t), \dot{x}_3(t), \dot{x}_4(t)]^T$, and

$$f(x(t), u(t)) = \begin{bmatrix} x_2(t) \\ x_1(t)x_4^2(t) - \dfrac{k}{x_1^2(t)} + \dfrac{1}{m}u_1(t) \\ x_4(t) \\ \dfrac{-2x_2(t)x_4(t)}{x_1(t)} + \dfrac{1}{mx_1(t)}u_2(t) \end{bmatrix} \tag{3.72}$$

In the above expression $x(t)$ constitutes the *state* of the motion at time $t$. ∎

Once the equation of motion expressed in terms of the signal $x(t)$ is determined, we identify the means by which the specific motion parameter assigned as output is related to the state. In the satellite example we have $y(t) = r(t) = x_1(t)$. More generally the output being part of the motion is represented by an equation of the form

$$y(t) = g(x(t), u(t)) \tag{3.73}$$

Equations (3.71) and (3.73) constitute the *state space representation* of the system. The first equation describes the overall motion. The second equation specifies a particular function of the motion, in this case the radial displacement.

The situation we described above in the context of mechanical systems prevails with equal propensity in circuit theory. The state space representation proves a powerful modeling technique in this case too.

Guided by Eqs. (3.71) and (3.73), it is not difficult to guess the form of discrete time state space representations. A detailed description is given next.

### 3.10.2 Discrete state space representations

Discrete time state space representations are determined by the pair of equations

$$x(n+1) = f(x(n), u(n)) \tag{3.74}$$

$$y(n) = g(x(n), u(n)) \tag{3.75}$$

Three signal variables enter the above equations: the $l$-channel input signal $u(n)$, the $k$-channel state signal $x(n)$, and the $m$-channel output signal $y(n)$. The meaning of Eqs. (3.74) and (3.75) is that the relationship between $u$ and $y$ represents an evolving process whose overall dynamical evolution is governed by the first equation called *state equation* and substantiated by the state variable $x$. Like the finite difference model, the state equation (3.74) does not specify a well-defined transformation, $u \to x$, unless the state at a specific time instant is given. More precisely we readily verify the following claim:

Given a time instant $n_0$, a real $k$-dimensional vector $\mathbf{x}_0$ and any input signal $u$ defined on $n_0 \leq n < \infty$, there exists a unique state signal $x(n)$ defined on $n_0 \leq n < \infty$ that satisfies the state equation (3.74) and the initial condition $x(n_0) = \mathbf{x}_0$. Occasionally we want to stress the dependence on the initial data, and we write $\phi(n, n_0, \mathbf{x}_0, u)$ for $x(n)$. Thus

$$\phi(n_0, n_0, \mathbf{x}_0, u) = \mathbf{x}_0$$

$$\phi(n + 1, n_0, \mathbf{x}_0, u) = f(\phi(n, n_0, \mathbf{x}_0, u), u(n)), \qquad n \geq n_0$$

The resulting output is given by

$$y(n, n_0, \mathbf{x}_0, u) = g(\phi(n, n_0, \mathbf{x}_0, u), u(n)), \qquad n \geq n_0$$

The state space representation together with the initial state describe a causal system. Notice that the input values $u(n)$, $n < n_0$, prior to $n_0$, play no explicit role in the state and output computations, since their effect has been incorporated into the initial state, $\mathbf{x}_0$.

The dynamics of a system are incorporated into the state equation, since the output equation is static. Storage requirements are determined by the number of delay devices included in the state equation, which in turn are given by the dimension of the state space, $k$.

Equations (3.74) and (3.75) are time invariant because the defining laws $f$ and $g$ do not change with time. As a consequence the following shift invariant property holds: For any time $n_0$, initial state $\mathbf{x}_0$, and input $u(n)$, the state produced by the $j$-shift $v(n) = u(n + j)$ is $\phi(n + j, n_0 - j, \mathbf{x}_0, u)$, and the output is $y(n + j, n_0 - j, \mathbf{x}_0, u)$.

A time varying state space representation has the form

$$x(n + 1) = f(x(n), u(n), n) \tag{3.76}$$

$$y(n) = g(x(n), u(n), n) \tag{3.77}$$

The state space model (3.74)–(3.75) is linear if the functions $f$ and $g$ in Eqs. (3.74) and (3.75) are linear. In standard coordinates we have

$$x(n + 1) = Ax(n) + Bu(n) \tag{3.78}$$

$$y(n) = Cx(n) + Du(n) \tag{3.79}$$

where $A$ is $k \times k$, $B$ is $k \times l$, $C$ is $m \times k$, and $D$ is $m \times l$ matrix. In the time varying case the corresponding matrices vary with time.

Like finite difference models, linear time invariant state space representations are characterized by a finite number of parameters. These are $k, l, m, A, B, C,$ and $D$.

### 3.10.3   Variation of constants formula

Let us consider the linear time varying form of the state equation (3.78). Evaluation at consecutive time instants gives

$$\phi(n, n_0, \mathbf{x}_0, u) = \Phi(n, n_0)\mathbf{x}_0 + \sum_{k=n_0}^{n-1} \Phi(n, k+1)\, B(k)\, u(k), \quad n \geq n_0 + 1 \quad (3.80)$$

where

$$\Phi(n, n_0) = \begin{cases} A(n-1)A(n-2)\cdots A(n_0), & n \geq n_0 + 1 \\ I, & \text{for } n = n_0 \end{cases} \quad (3.81)$$

is the so-called *transition matrix*. The first term of the right-hand side of (3.80) satisfies

$$\phi(n, n_0, \mathbf{x}_0, 0) = \Phi(n, n_0)\, \mathbf{x}_0 \quad (3.82)$$

It represents the state evolution when the input is zero and provides the solution of the recursive equation

$$x(n+1) = A(n)x(n), \quad x(n_0) = \mathbf{x}_0 \quad (3.83)$$

In particular, the transition matrix is a solution of Eq. (3.83) with initial condition $\Phi(n_0, n_0) = I$. The second term of Eq. (3.80) satisfies

$$\phi(n, n_0, 0, u) = \sum_{k=n_0}^{n-1} \Phi(n, k+1)B(k)u(k) \quad (3.84)$$

This is the state response to the input $u(n)$ when the system is initially at rest, $\mathbf{x}_0 = 0$. Formula (3.80) is known as *variation of constants formula*, and it says that the overall state response is obtained from the superposition of the zero-input response (3.82) and the zero-state response (3.84).

Similar conclusions hold for the time varying form of Eq. (3.79). If we substitute formula (3.80) into (3.79), we obtain

$$y(n, n_0, \mathbf{x}_0, u) = C(n)\Phi(n, n_0)\mathbf{x}_0 + \sum_{k=n_0}^{n-1} C(n)\Phi(n, k+1)B(k)u(k) + D(n)u(n)$$

$$(3.85)$$

In the time invariant case, $A(n) = A$, and the transition matrix is given by the matrix exponential signal, $\Phi(n, n_0) = A^{n-n_0}$, $n \geq n_0$. Equations (3.80) and (3.85) reduce to

$$\phi(n, n_0, \mathbf{x}_0, u) = A^{n-n_0}\mathbf{x}_0 + \sum_{k=n_0}^{n-1} A^{n-k-1} Bu(k) \tag{3.86}$$

$$y(n, n_0, \mathbf{x}_0, u) = CA^{n-n_0}\mathbf{x}_0 + \sum_{k=n_0}^{n-1} CA^{n-k-1} Bu(k) + Du(n) \tag{3.87}$$

If the system is initially idle at time $n_0 = 0$, that is, $\mathbf{x}_0 = 0$, the output is given by

$$y(n) = \sum_{k=0}^{n-1} CA^{n-k-1} Bu(k) + Du(n) \tag{3.88}$$

### 3.10.4 Continuous time state space representations

Continuous time state space models are defined by the equations

$$\dot{x}(t) = f(x(t), u(t), t) \tag{3.89}$$

$$y(t) = g(x(t), u(t), t) \tag{3.90}$$

The space $R^k$ where the state $x(t)$ evolves defines the *state space*. The function $f$ is defined on $R^k \times R^l \times R$, and takes values in $R^k$. It defines the *state dynamics*. Finally the function $g : R^k \times R^l \times R \to R^m$ is the *output map*.

Like finite derivative models, well posedness of Eq. (3.89) is a much more delicate issue than the discrete time counterpart. Suitable assumptions must be postulated on $f$ to guarantee existence and uniqueness of solutions for given initial data. The following theorem is a standard result in the theory of ordinary differential equations. It is stated without proof.

**Theorem 3.4 Existence and uniqueness of solutions** Suppose that $f$ is continuously differentiable, namely its partial derivatives with respect to its arguments exist and are continuous. Then for any initial time $t_0$, initial state $\mathbf{x}_0$, and continuous input signal $u(t)$, there exists a unique state signal $x(t)$, defined on an interval $I$ containing $t_0$, that satisfies the initial condition $x(t_0) = \mathbf{x}_0$ and solves Eq. (3.89) on $I$. ∎

To show the dependence on initial state, we denote the corresponding state signal by $\phi(t, t_0, \mathbf{x}_0, u)$. Likewise we define the output signal by $y(t, t_0, \mathbf{x}_0, u) = g(\phi(t, t_0, \mathbf{x}_0, u), u(t), t)$. If $t_0$ and $\mathbf{x}_0$ are given, the input-output assignment $u \to y$ renders a well-defined system. Time invariant models result when $f$ and $g$ do not depend explicitly on time $t$. They have the form (3.71), (3.73).

Linear time invariant state space forms become

$$x(t) = Ax(t) + Bu(t) \tag{3.91}$$

$$y(t) = Cx(t) + Du(t) \tag{3.92}$$

If the parameters vary with time, linear time varying models are obtained.

### 3.10.5   Variation of constants formula for analog systems

Let us next try to develop an expression analogous to Eq. (3.80) for the solution of $\dot{x}(t) = A(t)x(t) + B(t)u(t)$. Although we cannot apply a recursive argument as in Section 3.10.3, we can use the insight and motivation of the discrete case to guess the following formula:

$$\phi(t, t_0, \mathbf{x}_0, u) = \Phi(t, t_0)\mathbf{x}_0 + \int_{t_0}^{t} \Phi(t, \tau)B(\tau)u(\tau)d\tau \tag{3.93}$$

The transition matrix $\Phi(t, t_0)$ satisfies the matrix differential equation

$$\frac{d}{dt}\Phi(t, t_0) = A(t)\Phi(t, t_0), \qquad \Phi(t_0, t_0) = I \tag{3.94}$$

Let $x(t)$ denote the right-hand side of (3.93). To establish the above formula, we verify that $x(t)$ satisfies the linear time varying state equation and the initial condition $x(t_0) = \mathbf{x}_0$. Hence, by the general uniqueness Theorem 3.4, it is the unique solution emanating at time $t_0$ from $\mathbf{x}_0$. The initial condition is trivially satisfied because $\Phi(t_0, t_0) = I$ and the integral evaluated at $t = t_0$ vanishes. Next we differentiate Eq. (3.93) and use the semigroup property of the transition matrix (Exercise 3.9), $\Phi(t, s)\Phi(s, \tau) = \Phi(t, \tau)$, as well as the rule for the derivative of products to obtain

$$\frac{d}{dt}x(t) = \frac{d}{dt}\Phi(t, t_0)\mathbf{x}_0 + \frac{d}{dt}\left(\Phi(t, t_0)\int_{t_0}^{t}\Phi(t_0, \tau)B(\tau)u(\tau)d\tau\right)$$

$$= A(t)\Phi(t, t_0)\mathbf{x}_0 + A(t)\Phi(t, t_0)\int_{t_0}^{t}\Phi(t_0, \tau)B(\tau)u(\tau)d\tau$$

$$\quad + \Phi(t, t_0)\Phi(t_0, t)B(t)u(t)$$

$$= A(t)\left[\Phi(t, t_0)\mathbf{x}_0 + \int_{t_0}^{t}\Phi(t, \tau)B(\tau)u(\tau)d\tau\right] + B(t)u(t)$$

$$= A(t)x(t) + B(t)u(t)$$

The claim is proved. The output equation $y(t) = C(t)x(t) + D(t)u(t)$ is written as

$$y(t) = C(t)\Phi(t, t_0)\mathbf{x}_0 + \int_{t_0}^{t} C(t)\Phi(t, \tau)B(\tau)u(\tau)d\tau + D(t)u(t) \qquad (3.95)$$

The first of the three terms in the right-hand side of Eq. (3.95) represents the response to zero input. The other two terms result when the system is initially at rest, that is, $x(t_0) = \mathbf{x}_0 = 0$.

If $A(t) = A$, the transition matrix is specified by $\dot{x} = Ax$, $x(t_0) = I$. Hence, as we saw in Section 2.5.1, it coincides with the exponential signal $\Phi(t, t_0) = e^{A(t-t_0)}$. Thus, for linear time–invariant state space representations, we have

$$\phi(t, t_0, \mathbf{x}_0, u) = e^{A(t-t_0)}\mathbf{x}_0 + \int_{t_0}^{t} e^{A(t-\tau)}Bu(\tau)d\tau \qquad (3.96)$$

$$y(t, t_0, \mathbf{x}_0, u) = Ce^{A(t-t_0)}\mathbf{x}_0 + \int_{t_0}^{t} Ce^{A(t-\tau)}Bu(\tau)d\tau + Du(t) \qquad (3.97)$$

### *Example 3.25*    *Constant resistance circuit*

Let us consider the constant resistance circuit of Fig. 3.28. It was initially discussed in Example 2.12. We choose as state variables the voltage across the capacitor and the current passing through the inductor. The output is as shown in Fig. 3.28.

Application of Kirchhoff's voltage law in the loops shown gives

$$u = x_1 + R_1 C\dot{x}_1, \quad u = L\dot{x}_2 + R_2 x_2, \quad y = \frac{-1}{R_1}x_1 + x_2 + \frac{1}{R_1}u$$

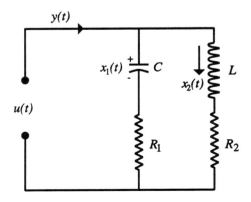

**Figure 3.28.** Constant resistance circuit.

In matrix form we obtain a linear time invariant state space representation with

$$A = \begin{pmatrix} \dfrac{-1}{R_1 C} & 0 \\ 0 & \dfrac{-R_2}{L} \end{pmatrix}, \quad B = \begin{pmatrix} \dfrac{1}{R_1 C} \\ \dfrac{1}{L} \end{pmatrix}, \quad C = \begin{pmatrix} \dfrac{-1}{R_1} & 1 \end{pmatrix}, \quad D = \dfrac{1}{R_1}$$

Suppose that the voltage across the capacitor at time $t_0 = 0$ is $x_{10}$ and that the initial current through the inductor is $x_{20}$. The exponential matrix $e^{At}$ was computed in Example 2.12. The variation of constants formula gives

$$x(t) = \begin{pmatrix} e^{-t/R_1 C} x_{10} \\ e^{-tR_2/L} x_{20} \end{pmatrix} + \int_0^t \begin{pmatrix} e^{-(t-\tau)/R_1 C} & 0 \\ 0 & e^{-(t-\tau)R_2/L} \end{pmatrix} \begin{pmatrix} \dfrac{1}{R_1 C} \\ \dfrac{1}{L} \end{pmatrix} u(\tau) d\tau$$

or

$$x_1(t) = e^{-t/R_1 C} x_{10} + \int_0^t e^{-(t-\tau)/R_1 C} \frac{1}{R_1 C} u(\tau) d\tau$$

$$x_2(t) = e^{-tR_2/L} x_{20} + \int_0^t e^{-(t-\tau)R_2/L} \frac{1}{L} u(\tau) d\tau$$

Suppose that the voltage source is a battery supplying a constant voltage $u(t) = V$. Then

$$x_1(t) = e^{-t/R_1 C} x_{10} + V - e^{-t/R_1 C} V$$

$$x_2(t) = e^{-tR_2/L} x_{20} + \frac{V}{R_2} - \frac{V}{R_2} e^{-tR_2/L}$$

If we let time go to infinity, we obtain the steady state values

$$\lim_{t \to \infty} x_1(t) = V, \quad \lim_{t \to \infty} x_2(t) = \frac{V}{R_2}, \quad \lim_{t \to \infty} y(t) = \frac{V}{R_2} \qquad \blacksquare$$

### 3.10.6   Linearization

Most practical systems are nonlinear. To make use of the rich linear theory, nonlinear systems are approximated by linear models. A common approximation relies on first-order Taylor expansion. In many cases this turns out to be adequate. We shall delve on this issue in Chapter 11 in connection with stability analysis. Consider the nonlinear system (3.71)–(3.73). Suppose that a given input $u^*(t)$ called *nominal* or *reference input* and an initial state $x_0^*$ produce the reference trajectory $x^*(t)$ and the reference output trajectory $y^*(t)$. Thus

$$\frac{d}{dt} x^*(t) = f(x^*(t), u^*(t)), \quad x^*(0) = x_0^*, \quad y^*(t) = g(x^*(t), u^*(t))$$

The first-order Taylor expansion of $f$ and $g$ are

$$f(x, u) - f(x^*(t), u^*(t)) = \frac{\partial f}{\partial x}\big|_{(x^*(t), u^*(t))}(x - x^*)$$

$$+ \frac{\partial f}{\partial u}\big|_{(x^*(t), u^*(t))}(u - u^*) + o(x, u)$$

$$g(x, u) - g(x^*(t), u^*(t)) = \frac{\partial g}{\partial x}\big|_{(x^*(t), u^*(t))}(x - x^*)$$

$$+ \frac{\partial g}{\partial u}\big|_{(x^*(t), u^*(t))}(u - u^*) + o(x, u)$$

If $x(t)$ and $y(t)$ are solutions of (3.71), (3.73) near $x^*(t)$, $y^*(t)$, respectively, the deviations $x(t) - x^*(t)$, $y(t) - y^*(t)$ are approximately described by the linear time varying system

$$\dot{x}(t) = A(t)x(t) + B(t)u(t) \tag{3.98}$$

$$y(t) = C(t)x(t) + D(t)u(t) \tag{3.99}$$

where

$$A(t) = \frac{\partial f}{\partial x}\big|_{(x^*(t), u^*(t))}, \quad B(t) = \frac{\partial f}{\partial u}\big|_{(x^*(t), u^*(t))} \tag{3.100}$$

$$C(t) = \frac{\partial g}{\partial x}\big|_{(x^*(t), u^*(t))}, \quad D(t) = \frac{\partial g}{\partial u}\big|_{(x^*(t), u^*(t))} \tag{3.101}$$

The above system is obtained from the Taylor expansion if we ignore terms of the form $o(x, u)$. We refer to the state space representation (3.98)–(3.99) as the *linearization* of (3.71)–(3.73) around the reference signals $u^*(t)$ and $x^*(t)$.

The linearized equations (3.98) and (3.99) are time varying in general. An important special case where they become time invariant results if the reference signals $x^*(t)$ and $u^*(t)$ are constant in time

$$x^*(t) = x^*, \quad u^*(t) = u^*$$

We then say that $x^*$ is an *equilibrium point* or a *stationary point* produced by $u^*$. It is easy to verify that such state and input values can be found from the equation

$$f(x^*, u^*) = 0 \tag{3.102}$$

In this case linearization becomes time invariant with parameters

$$A = \frac{\partial f}{\partial x}\big|_{(x^*, u^*)}, \quad B = \frac{\partial f}{\partial u}\big|_{(x^*, u^*)} \tag{3.103}$$

$$C = \frac{\partial g}{\partial x}\big|_{(x^*, u^*)}, \quad D = \frac{\partial g}{\partial u}\big|_{(x^*, u^*)} \tag{3.104}$$

Linearization of nonlinear discrete systems is defined in a similar way. Given the system

$$x(n + 1) = f(x(n), u(n)) \qquad (3.105)$$

$$y(n) = g(x(n), u(n)) \qquad (3.106)$$

a reference input $u^*(n)$ and initial state $x_0^*$, the linearization around the reference data is determined by the linear model

$$x(n + 1) = A(n)x(n) + B(n)u(n) \qquad (3.107)$$

$$y(n) = C(n)x(n) + D(n)u(n) \qquad (3.108)$$

The pertinent matrices are given by Eqs. (3.100) and (3.101) if we replace $t$ with $n$. Time invariant systems result when $x^*(n) = x^*$ and $u^*(n) = u^*$. Such values are determined from the system of algebraic equations

$$f(x^*, u^*) = x^* \qquad (3.109)$$

### Example 3.26   *Communication satellite*

Consider the motion of a communication satellite in space (Example 3.2). The position of the satellite is described in spherical coordinates $r(t)$, $\theta(t)$, and $\phi(t)$. The motion is controlled by thrusts $u_r(t)$, $u_\theta(t)$, and $u_\phi(t)$, as shown in Fig. 3.4. Let

$$x(t) = \begin{pmatrix} r(t) & \dot{r}(t) & \theta(t) & \dot{\theta}(t) & \phi(t) & \dot{\phi}(t) \end{pmatrix}^T$$

$$u(t) = \begin{pmatrix} u_r(t) & u_\theta(t) & u_\phi(t) \end{pmatrix}^T \qquad (3.110)$$

The state map is given by

$$f(x, u) = \begin{pmatrix} x_2 \\[1mm] x_1 x_4^2 \cos^2 x_5 + x_1 x_6^2 - \dfrac{k}{x_1^2} + \dfrac{u_r}{m} \\[1mm] x_4 \\[1mm] -\dfrac{2x_2 x_4}{x_1} + \dfrac{2x_4 x_6 \sin x_5}{\cos x_5} + \dfrac{u_\theta}{m x_1 \cos x_5} \\[1mm] x_6 \\[1mm] -x_4^2 \cos x_5 \sin x_5 - \dfrac{2x_2 x_6}{x_1} + \dfrac{u_\phi}{m x_1} \end{pmatrix} \qquad (3.111)$$

Consider the circular equatorial orbits

$$x^*(t) = (r_o \quad 0 \quad \omega t \quad \omega \quad 0 \quad 0)^T \tag{3.112}$$

To check if they qualify as solutions of the state equation associated with (3.111), we differentiate to obtain

$$\dot{x}^*(t) = (0 \quad 0 \quad \omega \quad 0 \quad 0 \quad 0)^T$$

Evaluation of (3.111) at (3.112) gives

$$f(x^*(t), u^*(t)) = \left(0 \quad r_o\omega^2 - \frac{k}{r_o^2} + \frac{u_r}{m} \quad \omega \quad \frac{u_\theta}{mr_o} \quad 0 \quad \frac{u_\phi}{mr_o}\right)^T$$

Equating the latter expressions we find

$$u_r = m\left(\frac{k}{r_o^2} - r_o\omega^2\right), \qquad u_\theta = 0, \qquad u_\phi = 0 \tag{3.113}$$

We conclude that circular equatorial orbits are feasible motions and can be generated by constant inputs (3.113). If the satellite is on such orbit, it stays in orbit. In the presence of disturbances, deviations will occur. We will later see that motion is unstable; hence, if no action is applied, the satellite will digress. To confirm this analysis, we will use linearization analysis. Linearization of (3.111) at reference signals (3.112) and (3.113) gives a time invariant state space model with matrices

$$A = \begin{pmatrix} 0 & 1 & 0 & 0 & 0 & 0 \\ \omega^2 + \dfrac{2k}{r_o^3} & 0 & 0 & 2\omega r_o & 0 & 0 \\ 0 & 0 & 0 & 1 & 0 & 0 \\ 0 & \dfrac{-2\omega}{r_o} & 0 & 0 & 0 & 0 \\ 0 & 0 & 0 & 0 & 0 & 1 \\ 0 & 0 & 0 & 0 & -\omega^2 & 0 \end{pmatrix}, \qquad B = \begin{pmatrix} 0 & 0 & 0 \\ \dfrac{1}{m} & 0 & 0 \\ 0 & 0 & 0 \\ 0 & \dfrac{1}{mr_o} & 0 \\ 0 & 0 & 0 \\ 0 & 0 & \dfrac{1}{mr_o} \end{pmatrix} \tag{3.114}$$

Note that the state reference trajectory is not a stationary point, yet the resulting linearization is time invariant. Consider the output

$$y = (y_1 \quad y_2 \quad y_3)^T = (r \quad \theta \quad \phi)^T = \begin{pmatrix} 1 & 0 & 0 & 0 & 0 & 0 \\ 0 & 0 & 1 & 0 & 0 & 0 \\ 0 & 0 & 0 & 0 & 1 & 0 \end{pmatrix} x \tag{3.115}$$

This is a linear equation and hence is not affected by linearization. We observe that the linearized motion in $(r, \dot{r}, \theta, \dot{\theta})$ coordinates does not affect and is not affected by the linearized motion in $(\phi, \dot{\phi})$ coordinates. This follows from the block diagonal form of the parameters $A$, $B$, and $C$. ∎

## 3.11 SAMPLERS AND QUANTIZERS

Digital processing of analog signals and digital manipulation of analog systems require faithful conversion of analog signals to digital signals and vice versa. Analog-to-digital (A-D) converters and digital-to-analog (D-A) converters are devices that effect such transformation. It is instructive to view the A-D converter as the series connection of a sampler and a quantizer, although physical devices are not built that way. A sampler transforms an analog signal into a discrete signal via the rule $u(t) \rightarrow u(t_n)$, $t_n$ is a real sequence. If $t_n$ forms an arithmetic progression containing the origin, $t_n = nT_s$, the sampler is called *uniform sampler* and is denoted by $\mathcal{S}_{T_s}$. A uniform sampler describes the sampling process mentioned in Chapter 2. We will adhere to uniform samplers throughout the book. A quantizer converts a discrete time signal into a digital signal with values in a given finite set. Both sampling and quantization amount to some sort of approximation. A sampler aims to represent an analog waveform with a sequence; a quantizer compresses the discrete information into a finite alphabet of symbols. Sampling theory seeks to characterize useful analog signal classes $\mathcal{U}$ so that, if $\mathcal{S}_{T_s}$ is restricted to $\mathcal{U}$, it has a left inverse that inacts the reconstruction of the analog waveform from its samples. In mathematics the problem of determining a function from its values is known as *interpolation*. Apparently sampling and interpolation are closely related. Invertibility of sampler is the subject of the sampling theorem discussed in Chapter 6, where in addition design procedures for the construction of the inverter are developed. A simple and widely used digital-to-analog converter is the zero-order hold (ZOH). It transforms a discrete time signal $u(n)$ into an analog signal $y(t)$ by holding the input values for an entire time interval

$$y(t) = (D|A)(u(n)) = u(n), \qquad nT_s \le t < (n+1)T_s \qquad (3.116)$$

The process is demonstrated in Fig. 3.29. More generally, the converter decides

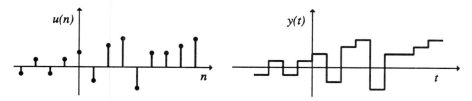

**Figure 3.29.** Zero-order hold.

on the value of the signal $u(t)$ on the interval $nT_s \leq t < (n+1)T_s$, using past and future signal values: $u(n), \ldots, u(n - M_1), u(n+1), \ldots, u(n + M_2)$. In the causal case $M_2 = 0$. Usually, polynomial interpolation is employed.

An $N$ point scalar quantizer is characterized by a mapping $Q : R \rightarrow C$. It converts each real number $x$ to an element of a finite set of reals

$$C = \{y_1, y_2, \ldots, y_N\}, \qquad y_i < y_{i+1}, \ i = 1, \ldots, N - 1$$

called *codebook*. The quantizer output values $y_i$ are referred to as *reproduction values*. $Q$ is written as the cascade $Q = \mathcal{D}\mathcal{E}$. The *encoder* $\mathcal{E} : R \rightarrow I$, $I = \{1, 2, \ldots, N\}$, assigns each real $x$ to the index $i$ associated with the reconstruction level $y_i$. The *decoder* $\mathcal{D} : I \rightarrow C$ converts each index $i$ to the corresponding output point $y_i$. In signal transmission the encoded index $\mathcal{E}(x)$ is transmitted, rather than the quantized value $Q(x) = y_i$. At the receiving end, the index $\mathcal{E}(x) = i$ is converted to the reconstruction level $y_i$ by the decoder, $\mathcal{D}(i) = y_i$, through a table-lookup procedure. Let $R = \log_2 N$. $R$ represents the number of bits required to represent each of the encoder outputs and thus defines the *transmission rate*, that is, the number of transmitted bits per sample. If the sampling period is $T_s$, the *bit rate* $R/T_s$ expresses the number of transmitted bits per second.

Mathematically a quantizer $Q$ is a simple function, namely its domain is partitioned into the $N$ sets, $R_i = \{x : Q(x) = y_i\} = Q^{-1}(y_i)$. $Q$ remains constant on each cell $R_i$. A quantizer is completely determined by the reconstruction levels $y_i$ and the corresponding partition cells $R_i$. The unbounded members of the partition are called *overload cells*, and the bounded sets are called *granular cells*.

The most popular quantizers are the *regular* quantizers. The cells of a regular quantizer are intervals containing the corresponding quantized values:

$$R_i = [x_{i-1}, x_i], \qquad x_{i-1} \leq y_i \leq x_i$$

The values $x_i$ are referred to as *transition or decision levels*. Thus $Q$ has the form

$$Q(x) = y_k \quad \text{if} \quad x_{k-1} \leq x < x_k, \qquad k = 1, 2, \ldots, N \tag{3.117}$$

Most often $x_0 = -\infty$ and $x_N = \infty$. If the input is bounded, the quantizer may be confined to a bounded interval of the real line. In this case all cells are granular cells, and the parameter $B = x_N - x_0$ defines the *dynamic range*. A regular quantizer is called *uniform* if the granular cells have constant length, $x_{i+1} = x_i + \Delta$.

The indicator function (or characteristic function) of a set $A$ is defined by

$$I_A(x) = \begin{cases} 1, & x \in A \\ 0, & \text{otherwise} \end{cases}$$

If $A$ is a half interval, $A = (-\infty, b)$, the indicator function $I_A$ is implemented by

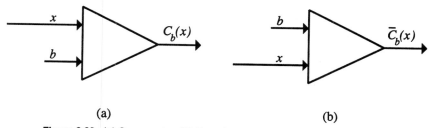

(a)            (b)

**Figure 3.30.** (a) Comparator. (b) Complementary binary threshold element.

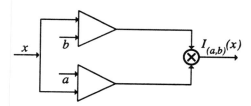

**Figure 3.31.** Implementation of the indicator function.

the comparator (also referred to as binary threshold element) of Fig. 3.30(a), where

$$C_b(x) = \begin{cases} 1 & x < b \\ 0 & x \geq b \end{cases}$$

The indicator of the half interval $(b, \infty)$ is implemented by the complementary binary threshold element $I_{(b,\infty)}(x) = 1 - C_b(x) = \bar{C}_b(x)$ as shown in Fig. 3.30(b). The indicator of a finite interval $(a, b)$ is realized by two comparators:

$$I_{(a,b)}(x) = C_b(x) - C_a(x) \tag{3.118}$$

or alternatively as $I_{(a,b)}(x) = C_b(x)\bar{C}_a(x)$. The multiplication can be implemented by an AND gate. Thus $I_{(a,b)}(x)$ can be built as in Fig. 3.31. Since a quantizer is a simple function, it is written as a linear combination of indicator functions

$$Q(x) = \sum_{i=1}^{N} y_i I_{R_i}(x) \tag{3.119}$$

In the case of regular quantizers, each indicator function of a granular cell is implemented by a pair of comparators as in Fig. 3.31. Thus we are led to the

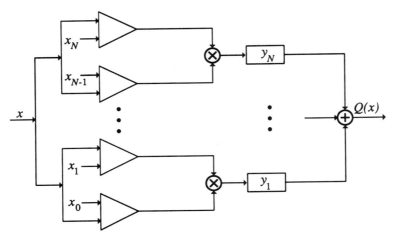

**Figure 3.32.** Architecture of a quantizer using comparators.

architecture of Fig. 3.32. If we substitute (3.118) into (3.119), we obtain

$$Q(x) = \sum_{i=1}^{N} y_i \big( C_{x_i}(x) - C_{x_{i-1}}(x) \big)$$

$$= \sum_{i=1}^{N-1} (y_i - y_{i+1}) C_{x_i}(x) + y_N C_{x_N}(x) - y_1 C_{x_0}(x)$$

which reduces the number of comparators by half. The efficient design of quantizers is treated in Chapter 4.

## 3.12 SIMULATION

A discrete time simulator of an analog system $S$ is a discrete system $S_{sim}$ which mitigates the behavior of $S$ in discrete time. $S_{sim}$ is a simulator of $S$ if the diagram

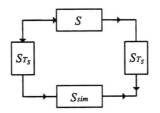

**Figure 3.33.** Discrete simulator.

of Fig. 3.33 approximately commutes

$$S_{T_s} S \approx S_{sim} S_{T_s} \tag{3.120}$$

If the above equality is restricted on a set of signals where the sampler has a right inverse $S_{T_s}^{-r}$, we have $S_{sim} = S_{T_s} S S_{T_s}^{-r}$.

**Example 3.27  Static systems**
Consider the static system $y(t) = f(u(t))$. The discrete system $y(n) = f(u(n))$ is a discrete simulator. Indeed,

$$S_{T_s} S(u)(t) = S(u)(nT_s) = f(u(nT_s)) = f(S_{T_s}(u)(t)) = S_{sim} S_{T_s}(u)(t)$$

and (3.120) holds. ∎

**Example 3.28  Integrator**
The sampled output of the integrator

$$y(t) = \int_{-\infty}^{t} u(\tau)d\tau$$

is

$$y(nT_s) = \int_{-\infty}^{nT_s} u(\tau)d\tau = y((n-1)T_s) + \int_{(n-1)T_s}^{nT_s} u(\tau)d\tau$$

Assuming sufficiently small spacing $T_s$, the integrand can be approximated by an expression that depends only on the end points $u((n-1)T_s)$ and $u(nT_s)$. Thus

$$y(nT_s) \approx y((n-1)T_s) + c(u((n-1)T_s), u(nT_s))$$

The discrete simulator has the form

$$y(n) = y(n-1) + c(u(n-1), u(n)) \tag{3.121}$$

The specification of $c(u(n-1), u(n))$ depends on the way $u(t)$ is approximated in the interval $(n-1)T_s \leq t \leq nT_s$. Common approximations are

$$y(n) = y(n-1) + T_s u(n-1) \tag{3.122}$$

$$y(n) = y(n-1) + T_s u(n) \tag{3.123}$$

$$y(n) = y(n-1) + \frac{T_s}{2}(u(n-1) + u(n)) \tag{3.124}$$

The first two schemes engage a zero-degree polynomial interpolation for the input in each interval $[(n-1)T_s, nT_s]$. They are known as the *rectangle rule*. The third algorithm relies on first-degree polynomial interpolation and is known as the *trapezoid rule*.    ∎

### 3.12.1   Discrete simulators for analog convolution

We develop a discrete simulator for the convolutional system

$$y(t) = \int_{-\infty}^{\infty} h(t-\tau)u(\tau)d\tau$$

as follows: We employ the zero-order hold as the sampler inverter and determine the simulator via the rule $S_{sim} = S_{T_s} \cdot S \cdot S_{T_s}^{-1}$. The output of the zero-order hold is $u(t) = u(n)$, if $nT_s \le t < (n+1)T_s$. Let $y(t)$ be the output of the analog convolutional system. The corresponding sampled version is

$$y(nT_s) = \int_{-\infty}^{\infty} h(\tau)u(nT_s - \tau)d\tau = \sum_{k=-\infty}^{\infty} \int_{kT_s}^{(k+1)T_s} h(\tau)u(nT_s - \tau)d\tau$$

Taking into account the piecewise constant form of $u(t)$, we have

$$y(n) = \sum_{k=-\infty}^{\infty} \left( \int_{kT_s}^{(k+1)T_s} h(\tau)d\tau \right) u(n-k)$$

Thus

$$y(n) = \sum_{k=-\infty}^{\infty} h_d(k)u(n-k), \quad h_d(n) = \int_{nT_s}^{(n+1)T_s} h(\tau)d\tau \tag{3.125}$$

The resulting simulator is a discrete convolutional system with unit sample response $h_d(n)$ determined from (3.125). If the integral cannot be computed in closed form, we usually assume that $h(t) \approx h(nT_s)$ in the interval $nT_s \le t \le (n+1)T_s$ so that

$$h_d(n) = T_s h(nT_s)$$

or

$$h_d(n) = T_s h((n+1)T_s)$$

Take the integrator as an example. The impulse response is $h(t) = u_s(t)$. Equation (3.125) gives $h_d(n) = T_s$, and

$$y(n) = T_s \sum_{k=-\infty}^{n} u(k) = y(n-1) + T_s u(n)$$

The simulator coincides with the rectangle integrator rule (3.123).

*Example 3.29   MATLAB simulation*
We compute the convolution of the causal signals

$$u(t) = u_s(t), \quad h(t) = e^{-2t}u_s(t)$$

using MATLAB simulation:

```
t = 0: .1:5;                              % sampling period Ts = 0.1
step = ones(1,length(t));                 % step input
h = exp(-2*t);                            % analog impulse response
hd = (1/2)*(1-exp(-0.2))*exp(-2*t);       % discrete impulse response
                                          % computed by the integral (3.125)
y = conv(hd,step);                        % simulated output
                                          % Alternatively we set
hd1 = 0.1*h;
y1 = conv(hd1,step);
plot(t,y(1:length(t)),t,y1(1:length(t)));
```

Let us next compute the output produced by a sum of two sinusoids at frequencies 50 and 120 Hz:

```
usine = sin(2*pi*50*t) + sin(2*pi*120*t);
ysine = conv(hd1,usine);
```

If we run the simulation of the same system with the exponential impulse response and look at the output plots, we will observe that the output is periodic tuned to the same input frequencies. We will examine this more closely in Chapter 5.

Consider finally the same system driven by Gaussian white noise. The simulation commands are

```
urand = randn(t);
yrand = conv(hd1,urand);                  %simulated output
my = mean(yrand);                         %mean
corry = xcorr(yrand,'unbiased');          %autocorrelation
N = length(yrand);
n = (-N + 1)*Ts:Ts:(N-1)*Ts;
plot(n,corry)
```

Analysis of the probabilistic properties of the output of a convolutional system is discussed in Chapter 9. ∎

### 3.12.2   Simulators of linear finite derivative models: Forward and backward difference approximations

Finite derivative models are simulated in a variety of ways. Some of them are based on the observation that finite derivative models are obtained as interconnections of a finite number of basic building blocks. Each block is simulated in some way and the same interconnection network is employed to link the corresponding building block simulators. Forward and backward difference simulators view the analog system as the interconnection of adders, scalors and differentiators. Integrating simulators, such as the bilinear simulator introduced in Chapter 11, view the analog system as the interconnection of adders, scalors and integrators. Adders and scalors, are static systems and are simulated in a straightforward manner (see Example 3.27). Forward difference simulators simulate the differentiator with the forward difference

$$\left(\frac{d}{dt}\right)_{\text{sim}}(u(n)) = \frac{u(n+1) - u(n)}{T_s} = \frac{1}{T_s}\Delta(u(n)) \tag{3.126}$$

Backward simulators simulate the differentiator with the backward difference

$$\left(\frac{d}{dt}\right)_{\text{sim}}(u(n)) = \frac{u(n) - u(n-1)}{T_s} = \frac{1}{T_s}\nabla(u(n)) \tag{3.127}$$

Higher-order derivatives are cascades of first-order derivatives. In the case of forward simulators, they are approximated by

$$\left(\frac{d^i}{dt^i}\right)_{\text{sim}}(u(n)) = \frac{1}{T_s^i}\Delta^i(u(n)) \tag{3.128}$$

Thus the forward difference simulator of the finite derivative model

$$y^{(k)}(t) = -\sum_{j=0}^{k-1} a_j y^{(j)}(t) + \sum_{i=0}^{r} b_i u^{(i)}(t)$$

has the form

$$\frac{1}{T_s^k}\Delta^k y(n) + \sum_{i=0}^{k-1}\frac{a_i}{T_s^i}\Delta^i y(n) = \sum_{j=0}^{r}\frac{b_j}{T_s^j}\Delta^j u(n), \qquad n \geq 0 \tag{3.129}$$

The latter expression is converted into the form (3.65); see Exercise (3.15). Likewise the backward difference simulator has the form

$$\frac{1}{T_s^k}\nabla^k y(n) + \sum_{i=0}^{k-1}\frac{a_i}{T_s^i}\nabla^i y(n) = \sum_{j=0}^{r}\frac{b_j}{T_s^j}\nabla^j u(n), \qquad n \geq k \tag{3.130}$$

**Example 3.30    First-order system**
Consider the first-order system

$$\dot{y}(t) + ay(t) = bu(t)$$

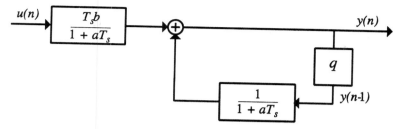

**Figure 3.34.** Backward difference simulator of Example 3.30.

The backward difference approximation gives,

$$\frac{1}{T_s}(y(n) - y(n-1)) + ay(n) = bu(n)$$

Assuming that $T_s \neq -1/a$ we obtain

$$y(n) - \frac{1}{1 + aT_s}y(n-1) = \frac{T_s b}{1 + aT_s}u(n) \qquad (3.131)$$

The above simulator is shown in Fig. 3.34. The step response is plotted in Fig. 3.36(a) for various values of the sampling period. Let us next consider the forward approximation

$$y(n+1) + (aT_s - 1)y(n) = bT_s u(n), \qquad n \geq 0 \qquad (3.132)$$

Figure 3.35 shows the corresponding architecture. Plots of the step response $y(n)$ for $a = 2$ and $T_s > 1$ are given in Fig. 3.36(b). An unbounded behavior is observed. Precise explanation for this phenomenon is given in Chapter 11. ∎

Forward and backward difference simulators are usually avoided because, even if (3.126) is a reasonable approximation, the high-order derivative approximation may be poor. Integrating simulators, on the other hand, have better performance. Integration smoothens the signal, and as a result the original approximation is not worsened. An important representative of this class is the

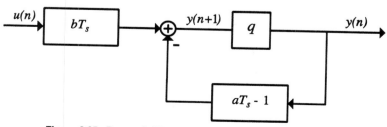

**Figure 3.35.** Forward difference simulator of Example 3.30.

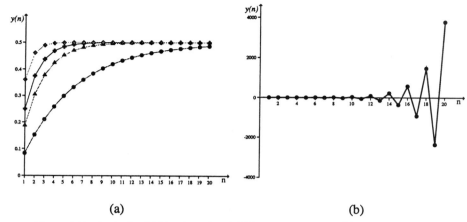

(a)    (b)

**Figure 3.36.** Analysis of (a) backward and (b) forward simulator of first-order system.

bilinear transformation method which uses the trapezoid rule for the simulation of the integrator. It is further discussed in Chapter 11.

### 3.12.3   Discrete simulators of linear state space models

Let us consider the continuous state space model

$$\dot{x}(t) = A(t)x(t) + B(t)u(t)$$

$$y(t) = C(t)x(t) + D(t)u(t)$$

We build the simulator $S_{sim} = S_{T_s} \cdot S \cdot S_{T_s}^{-1}$ using a zero-order hold for digital to analog conversion. Thus $u(t) = u(nT_s)$ for $nT_s \le t < (n+1)T_s$. The variation of constants formula gives

$$x((n+1)T_s) = \Phi((n+1)T_s, nT_s)x(nT_s)$$

$$+ \int_{nT_s}^{(n+1)T_s} \Phi((n+1)T_s, \tau)B(\tau)d\tau u(nT_s)$$

This leads to the following linear discrete state space representation:

$$x(n+1) = \Phi(n)x(n) + \Gamma(n)u(n)$$

$$y(n) = C(n)x(n) + D(n)u(n)$$

where

$$\Phi(n) = \Phi((n+1)T_s, nT_s), \quad \Gamma(n) = \int_{nT_s}^{(n+1)T_s} \Phi((n+1)T_s, \tau)B(\tau)d\tau \quad (3.133)$$

If the original system is time invariant, the resulting discrete simulator equation is time invariant too:

$$x(n+1) = \Phi x(n) + \Gamma u(n) \tag{3.134}$$

where

$$\Phi = e^{AT_s}, \quad \Gamma = \int_0^{T_s} e^{A\tau} d\tau B \tag{3.135}$$

### Example 3.31   Constant resistance network

Consider the circuit of Example 3.25. The state parameters of the discrete simulator are

$$\Phi = \begin{pmatrix} e^{-T_s/R_1 C} & 0 \\ 0 & e^{-T_s R_2/L} \end{pmatrix} \quad \Gamma = \begin{pmatrix} \dfrac{1}{R_1 C} \displaystyle\int_0^{T_s} e^{-t/R_1 C} dt \\ \dfrac{1}{L} \displaystyle\int_0^{T_s} e^{-t R_2/L} dt \end{pmatrix} = \begin{pmatrix} 1 - e^{-T_s/R_1 C} \\ \dfrac{1}{R_2}(1 - e^{-T_s R_2/L}) \end{pmatrix}$$

■

*Remarks on nonlinear system simulation.*   Nonlinear state space models are simulated in a variety of ways. Runge-Kutta methods are popular general purpose one-step algorithms of the form $x(n+1) = F(x(n), u(n))$ for the simulation of the state equation $\dot{x}(t) = f(x(t), u(t))$. The basic idea behind these methods is to preserve higher-order derivative information of $f$ through proper iterates of $F$. Runge-Kutta methods of order 3 and 5 are mostly preferred. Multistep methods such as Adams algorithm determine the state estimate $x(n+1)$ from several past estimates. The differential equation is replaced by an integral equation

$$x((n+k)T) = x((n-j)T) + \int_{(n-j)T}^{(n+k)T} F(x(\tau), u(\tau)) d\tau$$

and the integral is evaluated by polynomial interpolation.

### Example 3.32   SIMULINK simulation

Systems are conveniently described and analyzed by simulink, an extension of MATLAB. The operator specifying the system is represented by S (for System function). An S function is like any other MATLAB function. It can be expressed in MATLAB language as an M file, or a MEX file written in Fortran or C. Alternatively it can be represented by the block diagram editor of SIMULINK in terms of simpler block interconnections. Simpler blocks can either be copied from block libraries or built from the start. Once the S function representing the given system is defined in any of the above ways, analysis and design issues can be studied using MATLAB facilities and the toolboxes. Let us consider the backward simulator of Example 3.30. It has the form

$$y(n) = ay(n - 1) + bu(n)$$

and can be simulated in MATLAB in a variety of ways. The filter command is a standard option. Alternatively the architecture of Fig. 3.34 can be built using SIMULINK. From the standard subsystem blocks that make up the basic library, we open the discrete subsystem. The delay system and two copies of the gain are copied to the window called *untitled*. The delay is designated by $1/z$ for reasons explained in Chapter 8. We open the dialogue box of the gain, and we set the parameters $a$ and $b$. These variables are defined in the workspace. We set $a = 5/6$, $b = 1$. These correspond to $a = 2$, $T_s = 0.1$, $b = 12$ in the original analog model of Example 3.30. The output of the delay is fed to the gain $a$ and then added to the output of the gain $b$. The step function is chosen as input from the source library. The output port is an outport terminal taken from the connections list of blocks. We select **Parameters** from the **Simulation menu**. Several methods are available for the computation of the system output. They appear in the top of the SIMULINK control panel. We pick the linsim algorithm. We use the default values, and we set $[t, y]$ in the return variables field. These provide the time and output trajectories in the MATLAB workspace when the simulation is complete. Returning to the workspace, we produce output plots by

plot(t,y)

The same result is obtained if we save the file as **ltdn1** and use the simulation command

linsim('ltdn1',20)

The command linsim (linear simulation) specifies the simulation method used to compute the output. 20 designates the terminating simulation time. In the above architecture the model has a fixed input, the step function. The general system description appears in Fig. 3.37. Now the input is not specified. The block is called **ltdn** and is created from **ltdn1** by replacing the step function block with an inport block taken from the connections list of blocks.

Analysis of the system **ltdn** can be carried out from the command line. For this purpose the full syntax of the simulation algorithms is employed. All algorithms have identical calling syntax. The command

[t,x,y] = linsim('ltdn',tfinal,x0,options,u)

applies the simulation algorithm linsim to compute the time t, the state x, and the output y of the system **ltdn** between 0 and tfinal seconds. x0 specifies the initial state, which in this case is the initial content of the register. The argument options

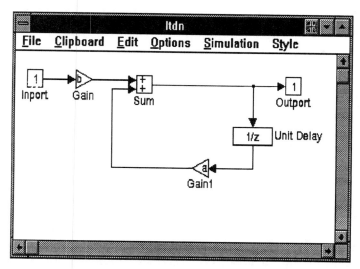

**Figure 3.37.** SIMULINK architecture of a first-order filter.

specifies simulation parameters such as relative error, minimum and maximum stepsize. If it is empty, the default parameters are employed. Finally u is the external input. It is either a string or a MATLAB expression. To derive the same simulation results with those of model **ltdn1**, we write

```
[t,x,y] = linsim('ltdn',20,16,[],'ones(length(t),1)');
plot(t,y)
```

**Simulation of satellite motion.**  Here we define the $S$ function by an M file. The reader is urged to derive a block diagram representation and to compare the simulation results. We open the new file **satel.m** where the description of the satellite equations is given. The input parameters $u1$, $u2$ and $u3$ are obtained from the original $u_r$, $u_\phi$, and $u_\theta$ upon scaling by the satellite mass $m$. The file has the form

```
function [sys,x0] = satel(t,x,u,flag)
% Satellite state equations as an m-file.
% State space equations:
if abs(flag) = = 1,
% if flag = 1, return state derivatives,
sys(1) = x(2);
sys(2) = x(1).*(x(4).^2).*(cos(x(5)).^2)
        + x(1).*(x(6).^2)-(39.88*10^13)./(x(1).^2) + u(1);
sys(3) = x(4);
sys(4) = (-2*x(2).*x(4))./(x(1)) + 2*x(4).*x(6).*tan(x(5))
        + u(2)./(x(1).*cos(x(5))));
```

```
sys(5) = x(6);
sys(6) = -(x(4).^2).*cos(x(5)).*sin(x(5))-(2*x(2).*x(6))./(x(1))
         + u(3)./x(1);
elseif flag = = 3
% If flag = 3, return system outputs, y
sys = [1 0 0 0 0 0;0 0 1 0 0 0;0 0 0 0 1 0]*x;
elseif flag = = 0,
% Return initial conditions
sys = [6,0,3,3,0,0]';
x0 = zeros(6,1);
else
%otherwise no need to return anything
sys = [];
end
```

The parameter $k$ is set to $k = GM = 39.88 * 10^{13}$. If flag $= 1$, the system variables sys(1), ..., sys(6) compute the state derivatives. If flag $= 3$, the system variables determine the outputs, and if flag $= 0$, sys yields parameter sizes and initial conditions.

We demonstrate by simulations that periodic solutions exist but they are unstable and call for control action. Periodic orbits along the space $\varphi = 0$, $\dot{\varphi} = 0$, have the form

$$x^*(t) = (r0, 0, w * t, w, 0, 0)$$

A periodic solution is a geostatic orbit if the satellite completes a full rotation around the earth in one day, that is, in 86,400 sec. This requires $w = 7.27 * 10^{-5}$. If external controls are set to zero, the desired distance of the satellite from the earth satisfies $w = \sqrt{k/r0^3}$, leading to the value $r0 = 42 * 10^6 m$. We set the initial condition at

$$x0 = (r0, 0, 0, w, 0, 0)$$

The Runge-Kutta method of order 5 is used to simulate the above model

```
[t,x,y] = rk45('satel',tfinal,x0);
```

where tfinal = 86,400 sec. Explicit plots are obtained by

```
subplot(221), plot(t,y(:,1)), title('radial distance')
subplot(222), plot(t,y(:,2)), title('theta')
subplot(223), plot(t,y(:,3)), title('phi')
```

Suppose next that the satellite is desired to rotate on a periodic orbit at the smaller distance $r1 = r0 = 36 * 10^6$ m. Full rotation in one day leads to the frequency value $w1 = 9.2454 * 10^{-5}$. Let

$$x0 = (r1, 0, 0, w1, 0, 0)$$

Simulations by the rk45 method show significant variations in the radial distance. Likewise theta is no longer a linear function of time. The enforcement of a periodic orbit in this case requires external control action. Guided by Eq. (3.113), we write

```
[t2,x2,y2] = rk45('satel',tfinal,x0,[],[t2,ones(length(t1),1)*k/(r0^2-r0*w^2),
                          zeros(length(t1),1),zeros(length(t1),1)]);
```

Trajectory plots show that a periodic orbit is accomplished. Next we perturb initial vectors and observe the outputs. Let

```
x0 = [r0 + 1000,0,0.2,w,0.3,0]';
```

There is an initial radial displacement by 1km and a deviation of 2 and 3 degrees in $\theta$ and $\phi$. Significant deviations are observed in the displacements plots. We conclude that instabilities are inherent in the motion. The design of stabilizing controllers is studied in Chapter 13.                                              ■

## 3.13  ARTIFICIAL NEURAL NETWORKS

Artificial neural networks are systems whose structure is inspired by the operation of the nervous system and the brain. The basic unit of a neural network is the neuron. A biological neuron is shown in Fig. 3.38. The nucleus of the neuron receives signals from other neurons through input paths called *dendrites* and processes them to form a new signal. If the resulting signal is sufficiently strong, it activates the firing of the neuron. An output signal is produced and transmitted through the output path, called *axon*, to other neurons. Transmission from the axon to dendrites of other neurons goes through a junction referred to as *synapse*, and the transferred signal is modified according to the synaptic strength of the junction. The brain consists of tens of billions of neurons. The main processing element of an artificial neural network imitates the operation of a biological neuron and is also referred to as a neuron. It is depicted

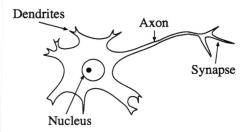

**Figure 3.38.** Illustration of a biological neuron.

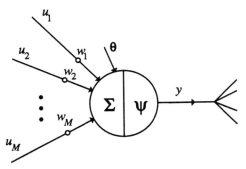

**Figure 3.39.** Artificial neuron.

in Fig. 3.39. It is a static shift invariant system with multiple inputs and one output described by

$$y(n) = \psi(x(n)) \tag{3.136}$$

$$x(n) = \sum_{j=1}^{M} w_j u_j(n) + \theta = w^T u(n) + \theta \tag{3.137}$$

The term $x(n)$ forms the summation of signals coming to the neuron, after they are weighted by the synaptic weights $w_j$. The function $\psi$ is called an *activation function*. It is typically a threshold function producing a nonzero value if the signal $x(n)$ is strong enough, or it can be a continuous function. The activation function is usually a sigmoid; that is, it features an S-shaped graph. A function $\psi : R \rightarrow R$ is called *sigmoid* if for some $A > 0$ and $B \leq 0$

$$\psi(u) = \begin{cases} A, & \text{if } u \rightarrow +\infty \\ B, & \text{if } u \rightarrow -\infty \end{cases}$$

If in addition $\psi$ is nondecreasing, it is a *squashing* function. Examples of sigmoid functions are the unit step, and the functions

$$\psi(u) = \frac{1}{1 + e^{-au}}, \quad \psi(u) = \frac{1 - e^{-au}}{1 + e^{-au}}, \qquad a > 0 \tag{3.138}$$

depicted in Fig. 3.40. The above examples are squashing functions. A single neuron having the unit step as activation function is known as *perceptron*. The perceptron can be viewed as a mapping that partitions the input space into two disjoint regions via a separating hyperplane. The hyperplane is given by

$$\{u \in R^M : w^T u + \theta = 0\}$$

and the disjoint regions are $\{u \in R^M : w^T u + \theta > 0\}$ and $\{u \in R^M :$

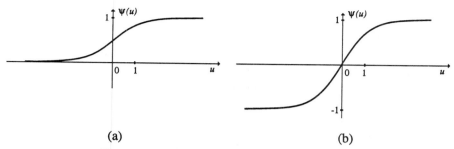

**Figure 3.40.** Common choices of the activation function.

$w^T u + \theta < 0$}. The first is given by all inputs of the perceptron whose resulting output is 1. The second consists of inputs yielding output 0. Due to the above property, perceptrons can be used for classification of patterns into two classes that are linearly separated. Pattern classification is discussed in the next chapter.

A neural network is formed by the series, parallel, and feedback interconnection of neurons. In addition it carries an input buffer where data are presented to the network and an output buffer that holds the response of the network to a given input. The neurons of a network are conveniently classified into input neurons, output neurons, and hidden or internal neurons. Input neurons receive the input signals, output neurons deliver the output signals, while the remaining neurons form the hidden neurons. The most important network architectures are delineated below.

### 3.13.1 Multilayer networks

Multilayer networks are represented by static nonlinear time invariant systems. Neurons are organized in a sequence of layers. Layers consisting of hidden neurons are called *hidden layers*. A network with one hidden layer, $m$ inputs, and $k$ outputs is shown in Fig. 3.41. Neurons on the same layer do not communicate with each other. Consider the $i$th neuron of the hidden layer. The input vector $u$ is modified by the weight vector $w_i$ and the threshold or *bias* vector $\theta_i$ associated with the neuron $i$. The resulting signal $w_i^T u + \theta_i$ is processed by the activation function $\psi$, which is the same for all neurons. In matrix form the outputs produced by the nodes of the hidden layer are written as

$$y(n) = \psi(Wu(n) + \theta) \tag{3.139}$$

where $y(n)$ contains the outputs of the $k$ neurons, $W$ is the $k \times m$ matrix of weight coefficients, $\theta$ is the $k \times 1$ vector of threshold values, and $\psi = (\psi, \psi, \dots, \psi)^T$ consists of $k$ copies of the activation function.

A multilayer neural network is formed by the cascade of single-layers. If there are $L = 3$ layers and the parameters of each single-layer subsystem are $W_i$, $\theta_i$,

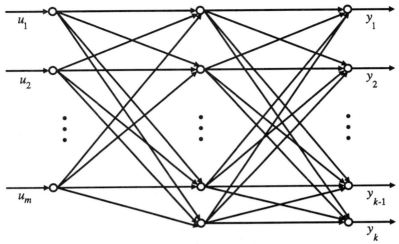

**Figure 3.41.** One hidden layer network.

$i = 1, 2, 3$, the input-output expression of the cascade is

$$y(n) = \psi[\theta_3 + W_3 * \psi(\theta_2 + W_2 * \psi(\theta_1 + W_1 * u))]$$

Often output layers involve no activation functions.

### 3.13.2 Universal approximation capabilities of multilayer neural networks

We demonstrate that any continuous function $f : R^m \rightarrow R^k$ can be uniformly approximated on compact sets by a multilayer neural network having a single hidden layer. All hidden nodes of the network have the same activation function, which can be of a very general nature.

**Theorem 3.5** Let $f$ denote a continuous real function defined on a compact set $I \subset R^m$ namely on a closed bounded subset of $R^m$ (see Appendix). Let $\psi$ be a continuous sigmoid and $\epsilon > 0$. Then there exists a positive integer $N$, a neural network $g(u)$ with one hidden layer and no activation function at the output layer

$$g(u) = \sum_{i=1}^{N} a_i \psi(w_i^T u + \theta_i) \tag{3.140}$$

such that $|f(u) - g(u)| < \epsilon$, for all $u \in I$. ∎

The proof is given in the Appendix at the end of the chapter.

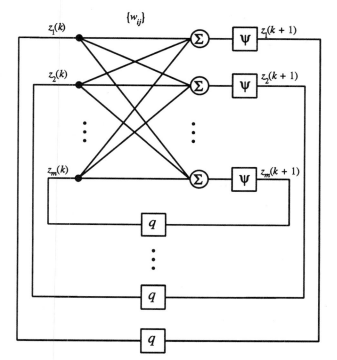

**Figure 3.42.** Recurrent network.

### 3.13.3 Recurrent networks

Recurrent networks engage memory facilities and thus are represented by dynamical systems. The simplest case is illustrated in Fig 3.42. A typical representative of the above architecture is the discrete *Hopfield, network*. Each input value, $u(n)$, is a column vector of length $m$. At time 0, $u(n)$ initializes $m$ processing units, each of which is described by the following operation:

$$z_i(k+1) = \psi\left( \sum_{j=1}^{m} w_{ij} z_j(k) \right)$$

where $\psi$ is an activation function. In vector form the above expression reads as

$$z(0) = u(n), \ z(k+1) = \psi(Wz(k)), \qquad k \geq 0 \qquad (3.141)$$

where $W$ is the matrix of the connection weights and $\psi$ is given by

$$\psi(x) = ( \psi(x_1) \quad \psi(x_2) \quad \cdots \quad \psi(x_m) )^T$$

The recurrent architecture of Fig. 3.42 is an autonomous dynamical system. Each input value is submitted as an initial condition. The network is triggered, and after a number of operations, it settles to a steady state vector which is then delivered to the output. If the weights are properly selected, the network has a finite set of steady state vectors $A$. Because of this feature the neural network acts as an associative memory unit. Each input value can be thought of as a noisy form of the stored patterns in $A$. The true pattern is retrieved from the network at the output. Character recognition is an application of this idea. The letters of the alphabet in a suitable coded form are the stored patterns. The aim of the network is to reveal the letters of a message from degraded versions.

## 3.14  RADIAL BASIS FUNCTION NETWORKS

Radial basis function networks look similar to single layer neural networks. They are illustrated in Fig. 3.43. The basic processing unit is the radial basis unit. Each such unit is characterized by a vector $w$. Its function is to produce a nonneglibible response, only if the input is close to $w$. Thus the output of the radial basis unit produced by the input $u$ has the form $r(d(u, w))$, where $r : R \to R$ is a function localized at the origin with fast decay to zero as its argument grows and $d$ is a measure of deviation of $u$ from $w$, typically a distance metric. The most common choice for $r$ is the Gaussian function

$$r(u) = e^{-\frac{1}{2}u^T C^{-1} u} \tag{3.142}$$

$C$ is a diagonal positive definite matrix. We employ the notation $< x, y >_Q = x^T Q y$. It is easy to see that $< x, y >_Q$ is a valid inner product when $Q > 0$. The induced squared norm is $|x|^2_Q = x^T Q x$. All inputs lying the same distance from $w$ produce the same output. Thus the radial basis unit is radially symmetric and this explains the name of the network. The input-output expression of the radial

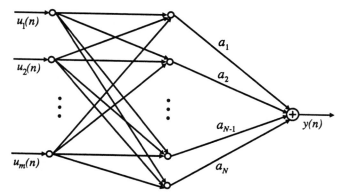

**Figure 3.43.** Radial basis function network.

basis function (RBF) network takes the form

$$y(n) = \sum_{i=1}^{N} a_i r(|u(n) - w_i|_Q^2) \tag{3.143}$$

It consists of $N$ radial basis units, each associated with a vector $w_i$. The outputs of these units are weighted by $a_i$ and summed. RBF networks approximate arbitrary continuous functions, just like neural networks with a single hidden layer. The difference is that the sigmoid activation function in a multilayer neural network produces nonzero responses in an unbounded interval, while the radial basis function is strongly localized.

### 3.14.1   Radial basis functions networks are universal approximators

RBF networks are universal approximators for quite general radial basis functions, like multilayer neural networks. Here we establish this important fact for Gaussian functions. Let $Q$ be a diagonal matrix with diagonal entries $\sigma_1, \sigma_2, \ldots, \sigma_m$, and let

$$|x|_Q^2 = x^T Q x = \sum_{i=1}^{m} \sigma_i x_i^2$$

**Theorem 3.6**   Consider the Gaussian function $g : R \to R$, $g(u) = e^{-u^2/2}$. Let $\mathrm{RBF}^m(g)$ denote the set of RBF networks of the form

$$\sum_{i=1}^{N} a_i g(|u - w_i|_Q), \qquad N \geq 1, \ a_i \in R, \ w_i \in R^m, \ \sigma_i > 0$$

Any real continuous function, $f$ defined on a compact subset $I$ of $R^m$ can be uniformly approximated by a network in $\mathrm{RBF}^m(g)$.
    The proof is given in the Appendix at the end of the chapter.   ∎

### 3.14.2   Time delay network and its universal character

The time delay network was introduced in Example 3.10. It is implemented by the equation

$$y(n) = f(u(n), u(n-1), \ldots, u(n-r)) \tag{3.144}$$

We have seen that if $f$ is a polynomial function, the corresponding structure is capable to approximate any time invariant causal system with fading memory. We have observed that the set of polynomials can be replaced by any dense set in the set of continuous functions on compact sets. Two such important sets were discussed above, multilayer neural networks and radial basis function networks. When the nonlinear function $f$ is replaced by a multilayer neural network the time delay neural network results. Likewise the time delay RBF network is obtained if $f$ is replaced by a radial basis function.

## BIBLIOGRAPHICAL NOTES

The basic continuous and discrete system representations are discussed in most books on signal and systems; see, for instance Athans, Dertouzos, Spann, and Mason (1974), Lathi (1974), Liu, C. and Liu, J. (1975), Oppenheim, Willsky, and Young (1983), Oppenheim and Schafer (1989), Manolakis and Proakis (1989), Papoulis (1985), and Dickinson (1991). Various aspects of digital systems are described in Kohavi (1978). The fading memory concept is introduced in Boyd (1985). The standard existence and uniqueness theorems of ordinary differential equations can be found in Brauer and Nohel (1969), Hirch and Smale (1974), and Arnold (1973). Multidimensional extensions of finite difference models have not been included due to space limitation. The interested reader may consult Lim (1990) and Dudgeon and Mersereau (1984). Neural networks and radial basis functions are discussed in Haykin (1994). Tutorial overviews are presented in Lippmann (1986) and Hush and Horne (1993). The universal approximation capability of polynomial filters is derived in Boyd (1985). A related approach is given in Sandberg (1991). The universal characteristics of neural networks for static systems is established in Cybenko (1989) and Hornik, Stinchcombe, and White (1989). The Stone-Weierstrass theorem is discussed in Dieudonne (1969).

## PROBLEMS

**3.1.** Check whether each of the following systems is dynamic, time invariant, causal, or linear:

(a) $y(n) = u(n-1)u(n-2)$,

(b) $y(n) = u(-n)$

(c) $y(n) = u(2n)$

(d) $y(t) = du(t)/dt$

(e) $y(n) + 3ny(n-1) + 2y(n-2) = 6u(n)$, $\;y(-2) = y(-1) = 0$

**3.2.** Prove that the time delay network (3.18) has the fading memory property on the set of inputs bounded by a given constant. The function $f$ is continuous.

**3.3.** Let $F$ be the binary field and $f : F^3 \to F$ be a combinatorial function specified by the table

| $(x_1, x_2, x_3)$ | 0 | 1 | 2 | 3 | 4 | 5 | 6 | 7 |
|---|---|---|---|---|---|---|---|---|
| $f(x_1, x_2, x_3)$ | 0 | 1 | 0 | 1 | 1 | 1 | 0 | 0 |

Write $f$ as a polynomial in three variables.

**3.4.** For any $k \in Z^+$, the output of a linear system produced by the input $t^k$ is $\cos kt$. Determine the output corresponding to the input $\pi + 6t^2 - 47t^5 + 8t^6$.

**3.5.** A convolutional system is specified by the relation

$$y(t) = \int_{-\infty}^{t} e^{-(t-\tau)} u(\tau - 2) d\tau$$

Find the impulse response. What is the response to the unit pulse: $u(t) = 1$ for $-1 \le t \le 2$ and zero otherwise?

**3.6.** Compute the convolution of a discrete rectangular pulse and a triangular pulse (see Fig. 3.44).

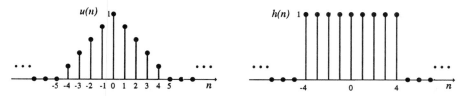

**Figure 3.44.**

**3.7.** Prove Theorem 3.2.

**3.8.** Compute and plot the outputs of the following systems using the **filter** command:
(a) $y(n) + ay(n - 1) = b_0 u(n) + b_1 u(n - 1)$, $a = 1/2, 2, 1, b_0, b_1$ arbitrary
(b) $y(n) + \frac{3}{4} y(n - 1) + \frac{1}{8} y(n - 2) = 7u(n - 1) + 3u(n - 2)$
for each of the inputs
(1) $u(n) = \delta(n)$         (3) $u(n) = \sin(2\pi/N)n, N = 300$
(2) $u(n) = u_s(n)$         (4) $u(n) = \text{rand}(1,N)$
Conduct experiments with other signals. Validate your results with SIMULINK.

**3.9.** Show that the transition matrix $\Phi(t, t_0)$, as defined in Eq. (3.94), has the following properties:
(a) $\Phi(t, t_0) = \Phi(t, s)\Phi(s, t_0)$         (b) $\det \Phi(t, t_0) = e^{\int_{t_0}^{t} tr[A(\tau)]d\tau}$

**3.10.** Use expression (3.24) to compute the output if

$$h_k(n) = \begin{cases} ka^n, & n = 0, 1, \ldots \\ 0, & \text{otherwise} \end{cases}$$

and the input is the causal rectangular pulse

$$u(n) = \begin{cases} 1, & \text{if } 0 \le n \le N \\ 0, & \text{otherwise} \end{cases}$$

**3.11.**   Show that the function

$$\psi(u) = \begin{cases} 0, & u < -\dfrac{\pi}{2} \\[2mm] \dfrac{1 + \cos(u + 3\pi/2)}{2}, & -\dfrac{\pi}{2} \leq u \leq \dfrac{\pi}{2} \\[2mm] 1, & u > \dfrac{\pi}{2} \end{cases}$$

is a squashing function and construct the graph.

**3.12.**   Show by examples that $S_2 S_1$ can be well defined, while $S_1 S_2$ does not exist. Show that even if both connections exist, $S_1 S_2 \neq S_2 S_1$.

**3.13.**   Let $S_1$ and $S_2$ be time invariant systems with state space representations

$$\dot{x}_i = f_i(x_i, u)$$

$$y_i = g_i(x_i, u), \qquad i = 1, 2$$

Show that the series connection $S_2 S_1$ is described in state space form by

$$\dot{x} = \begin{pmatrix} \dot{x}_1 \\ \dot{x}_2 \end{pmatrix} = \begin{pmatrix} f_1(x_1, u) \\ f_2(x_2, y_1) \end{pmatrix} = \begin{pmatrix} f_1(x_1, u) \\ f_2(x_2, g_1(x_1, u)) \end{pmatrix}$$

$$y = y_2 = g_2(x_2, y_1) = g_2(x_2, g_1(x_1, u))$$

Prove that the parallel connection is described in state space by

$$\dot{x} = \begin{pmatrix} \dot{x}_1 \\ \dot{x}_2 \end{pmatrix} = \begin{pmatrix} f_1(x_1, u) \\ f_2(x_2, u) \end{pmatrix}$$

$$y = y_1 + y_2 = g_1(x_1, u) + g_2(x_2, u)$$

To determine the state space form of the feedback connection, assume that $S_2$ has no direct input output link, that is $y_2 = g_2(x_2)$. Then show that the feedback state space representation is

$$y = g_1(x_1, u + y_2) = g_1(x_1, u + g_2(x_2))$$

$$\dot{x} = \begin{pmatrix} \dot{x}_1 \\ \dot{x}_2 \end{pmatrix} = \begin{pmatrix} f_1(x_1, u + g_2(x_2)) \\ f_2(x_2, g_1(x_1, u + g_2(x_2))) \end{pmatrix}$$

**3.14.**   Assume that the systems $S_1$, $S_2$ of Exercise 3.13 are linear and time invariant. Develop analogous state space representations for the cascade parallel and feedback connections.

**3.15.** Show that

$$\Delta^k = (q^{-1} - I)^k = \sum_{i=0}^{k} \binom{k}{i} (-1)^{k-i} (q^{-1})^i$$

$$\nabla^k = (I - q)^k = \sum_{i=0}^{k} \binom{k}{i} (-1)^i q^i$$

Show that (3.129) and (3.130) convert into finite difference expressions of the form (3.65).

**3.16.** Consider a state space model with parameters

$$A = \begin{pmatrix} -1 & 0 \\ 0 & -4 \end{pmatrix}, \quad B = \begin{pmatrix} 1 \\ 3 \end{pmatrix}$$
$$C = (2 \quad 4), \quad D = 1$$

Determine the parameters of the sampled system if the sampling period is $T_s = 0.1$.

**3.17.** The rotational motion of an asymmetric spinning space vehicle satisfies Euler's equation

$$\dot{x}_1 = \frac{I_2 - I_3}{I_1} x_2 x_3 + u_1$$

$$\dot{x}_2 = \frac{I_3 - I_1}{I_2} x_3 x_1 + u_2$$

$$\dot{x}_3 = \frac{I_1 - I_2}{I_3} x_1 x_2 + u_3$$

where $x_1$, $x_2$, and $x_3$ are the angular velocities about the three principal axes and $I_1 < I_2 < I_3$ are the corresponding moments of inertia. $u_1$, $u_2$, and $u_3$ are control torques generated by a reaction wheel mechanism. The goal is to maintain rotation of the space body at a constant angular velocity about one axis and cancel rotations about the other two axes. In this manner artificial gravity is accomplished about a single axis, while avoided are rotations about the other axes that are responsible for tumbling of people and equipment. The system is nonlinear due to the products $x_i x_j$ causing the gyroscopic coupling.

Compute the equilibria states when the input torques are zero. Determine the corresponding linearizations. Write an S function SIMU-LINK model for the nonlinear system. Determine the outputs by the MATLAB simulation algorithms, and compare results. Use the state space block to describe each linearization. Simulate each linearization by the linmod command and by the lsim command of the control toolbox. Use the command c2d to construct discrete simulators for each linearized system and repeat the simulation experiments.

**3.18.** Linearization is carried out in MATLAB by the comman linmod. Likewise the computation of equilibria is done by trim. Apply both commands to the block vdp representing the Van der Pol dynamical system

$$\dot{x}_1 = x_2$$
$$\dot{x}_2 = x_2(1 - x_1^2) - x_1$$

Experiment with other linear and nonlinear systems.

**3.19.** Use simulink to create a block for the finite support Volterra system

$$y(n) = 2u(n - 1) + u(n - 2) + 3u(n - 1)u(n - 3)$$

Carry out simulations using the inputs of the signal generator block, and plot the resulting outputs.

**3.20.** Simulate the RLC circuit of Example 3.25 in SIMULINK. Compare the plots of the state trajectories determined by the theory with those of simulation.

**3.21.** Consider the convolutional encoder of Fig. 3.45. Determine the coded sequence produced by the message $u = (1, 0, 1, 1, 1)$.

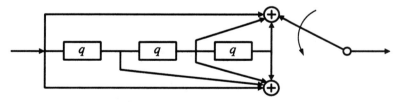

**Figure 3.45.**

**3.22.** Use the forward and backward simulators to analyze the second-order analog system

$$y^{(2)}(t) + 4y^{(1)}(t) + 3y(t) = u(t) + 5u^{(1)}(t)$$

in analogy with Example 3.30. First, simulate the analog system using the transfer function block of SIMULINK. Plot the outputs resulting from all inputs in the signal generator. Then determine the forward and backward simulators. Either use the filter command or create a block with the block editor, and compare the results with the analog computation.

**3.23.** Establish the fading memory property in Theorem 3.1. Hint. Take $w(n) = \sqrt{|h|_1\theta(n)}$, $\theta(n) = \sum_{k=n}^{\infty}|h(k)|$.

## APPENDIX

To establish the main approximation theorems of Chapter 3, we will make use of a celebrated extension of the Bolzano-Weierstrass theorem known as the Stone-Weierstrass theorem. The statement of this theorem follows without proof. Let $E$ be a metric space with metric $d$. $E$ is *compact* if every sequence in $E$ has a convergent subsequence in $E$. Thus, if $u_N$ belongs to $E$, there exists a subsequence $u_{N_k}$ and $u$ in $E$ such that $d(u_{N_k}, u) \to 0$. A set $A$ of real functionals $E \to R$ is an *algebra* if it is closed under addition, multiplication, and scalar multiplication. $A$ *separates points on $E$* if for every $u$, $v$ in $E$, $u \neq v$, there exists $G$ in $A$ such that $G(u) \neq G(v)$. $A$ *vanishes at no point of $E$* if for each $u$ in $E$ there exists $G$ in $A$ such that $G(u) \neq 0$.

**Theorem 3.7 Stone-Weierstrass theorem**  Suppose that $(E, d)$ is a compact metric space and $A$ an algebra of continuous functionals on $E$. Suppose that $A$ separates points on $E$ and vanishes at no point of $E$. Then, for every continuous functional $F : E \to R$ and every $\epsilon > 0$, there exists $G$ in $A$ such that $|F(u) - G(u)| < \epsilon$, for all $u \in E$. ∎

An immediate corollary is the following

**Corollary**  Suppose that $(E, d)$ is a compact metric space and $A$ a set of continuous functionals on $E$. Suppose that $A$ separates points on $E$. Then, for every continuous functional $F : E \to R$ and every $\epsilon > 0$, there exists a polynomial function $p : R^m \to R$ and $G_1, G_2, \ldots, G_m$ in $A$ such that

$$|F(u) - p(G_1(u), G_2(u), \ldots, G_m(u))| < \epsilon \qquad \text{for all } u \in E$$

PROOF   Let $\tilde{A}$ denote the collection of functionals

$$\tilde{A} = \{p(G_1(u), G_2(u), \ldots, G_m(u)), m \in Z^+, p \text{ polynomial}, G_i \in A\}$$

It is easy to check that $\tilde{A}$ is an algebra. It separates points on $E$ because it contains $A$. Finally, it vanishes at no point of $E$ because it contains the constant functionals. Thus the Stone-Weierstrass theorem applies and proves the claim. ∎

A direct application of the Stone-Weierstrass theorem is the following:

**Theorem 3.8**   Let $A$ denote the set of real trigonometric polynomials of period $T$:

$$f(t) = \sum_{k=-N}^{N} a_k e^{jk\omega t}, \ a_k = a_{-k}^*, \ \omega = \frac{2\pi}{T}$$

Then $A$ is dense in the set of continuous periodic functions of period $T$ with respect to pointwise convergence.

PROOF   $A$ is clearly a vector space. It is closed under multiplication because

$$\sum_{k=-N}^{N} a_k e^{jk\omega t} \sum_{l=-M}^{M} b_l e^{jl\omega t} = \sum_{r=-(N+M)}^{N+M} c_r e^{jr\omega t}$$

where $c_r = \sum_k a_k b_{r-k}$. It is easy to show that the symmetry constraint $c_r = c_{-r}^*$ is also satisfied. Let $t$, $s$ be in $[0, T]$ with $t \neq s$. Then $e^{j\omega t} \neq e^{j\omega s}$ and $A$ separates points. Finally, $A$ vanishes at no point because it contains the constant functions. Thus the assumptions of the Stone-Weierstrass theorem are satisfied, and the claim follows.   ∎

PROOF OF THEOREM 3.3   The Stone-Weierstrass theorem involves functionals rather than operators. For this reason we will work with the functional $S_0 : \ell_\infty(Z^-) \rightarrow R$, introduced in Section 3.2. As we recall, $S_0$ represents the system responses at time 0. Thanks to time invariance and causality, $S_0$ characterizes the system.

Let us consider the set of inputs

$$K = \{u \in \ell_\infty(Z^-) : |u|_\infty \leq M\}$$

$K$ consists of signals defined over the set of nonnegative integers that are bounded by the same constant $M$. Since $S$ has the fading memory on $K$, there is a decreasing sequence $w(n)$ defined on $Z^+$ such that $0 < w(n) \leq 1$, $\lim_{n\to\infty} w(n) = 0$ and $S$ is continuous on $K$ with respect to the weighted norm

$$|u|_w = \sup_{n \leq 0} |u(n)| w(-n)$$

We show next that $K$ is compact in $\ell_\infty(Z^-)$ with respect to the above norm. Indeed, let $u_N$ be a sequence in $K$. We form an array as follows: The first row contains the values of the signal $u_1$, the second row contains the values of the signal $u_2$, and so forth, the $N$th row contains the entries of the signal $u_N$. For each $N$, $u_N$ belongs to $K$. Hence each $u_N$ is bounded by $M$. In particular, the first column of the array $u_N(0)$ is a real sequence bounded by $M$. We pick a subsequence of $u_N(0)$ that converges, and we denote its limit $u^*(0)$. Clearly $|u^*(0)| \leq M$. Next we remove all rows of the array whose first entry is not a member of the above subsequence. For the sake of simplicity we denote the resulting rows by $u_N$, where now $u_N(0) \rightarrow u^*(0)$, and we take the second column of this array, $u_N(-1)$. This is a real sequence bounded by $M$; therefore it has a subsequence converging to $|u^*(-1)| \leq M$. We retain only the rows of the array whose second entry belongs to this subsequence. In this way we guarantee that the first two columns of the new array converge. Proceeding this way we construct an array having the property that the $n$th column converges to a limit denoted $u^*(-n)$. Since $u^*(-n)$ is bounded by $M$, it belongs to $K$. We continue to denote the rows of this array by $u_N$, although they form a subsequence of the original sequence. We prove that $u_N$ converges to $u^*$ with respect to the weighted norm. Let $\epsilon > 0$. Since $w(n) \rightarrow 0$, there exists $n_0$ such that $w(n_0) < \epsilon/2M$. Since $u_N$ belongs to $K$, we deduce that

$$\sup_{n \leq -n_0} |u_N(n) - u^*(n)| w(-n) \leq 2Mw(n_0) < \epsilon$$

Now recall $u_N(n) \rightarrow u^*(n)$, for all $n \leq 0$. Therefore for each $n$ in the finite set $-n_0 \leq n \leq 0$, there exists $N_0$ such that $|u_N(n) - u^*(n)| < \epsilon$, $N \geq N_0$. Hence

$$|u_N(n) - u^*(n)| w(-n) \leq |u_N(n) - u^*(n)| \leq \epsilon$$

Putting the above together, we establish that each sequence in $K$ has a convergent subsequence in $K$ and that $K$ is compact.

For each $k \in Z^+$, define the functional $G_k : \ell_\infty(Z^-) \rightarrow R$, $G_k(u) = u(-k)$. Each $G_k$ is continuous with respect to the weighted norm. Indeed, let $\epsilon > 0$. By definition, $w(k) > 0$. Take $\delta = \epsilon w(k)$. If $|u - v|_w < \delta$, it holds that

$$|G_k(u) - G_k(v)| = |u(-k) - v(-k)| = \frac{1}{w(k)} |u(-k) - v(-k)| w(k)$$

Thus

$$|G_k(u) - G_k(v)| \leq \frac{1}{w(k)} \sup_{n \leq 0} \{|u(n) - v(n)| w(-n)\} \leq \frac{1}{w(k)} w(k)\epsilon = \epsilon$$

It is easy to see that the set of the above functionals $G_k$ separates points. Indeed, if $u \neq v$, there exists $k$ such that $u(-k) \neq v(-k)$. The corresponding functional $G_k$ satisfies $G_k(u) \neq G_k(v)$.

We have completed all preparatory work for the application of the Stone-Weierstrass theorem. We take our compact metric space to be the set $K$ with the metric induced by the weighted norm. We also consider the set of functionals $G_k$ on $K$. We saw that these are continuous and separate points. All assumptions of the corollary of the Stone-Weierstrass theorem are satisfied. Therefore, for any $\epsilon > 0$, there exists an integer $m$, a polynomial function $p : R^m \rightarrow R$, and $G_{k1}, G_{k2}$,

..., $G_{km}$ such that

$$|S_0(u) - p(G_{k1}(u), G_{k2}(u), \ldots, G_{km}(u))| \leq \epsilon, \qquad u \in K$$

Using the one-to-one relationship of $S_0$ and the system $S$ discussed in Section 3.2, the above approximation translates to the original system setup as follows: Let $u$ be a signal defined on $Z$ and bounded by $M$. Let $n \in Z$. Then $Pq_{-n}u$ shifts the value $u(n)$ to the origin, truncates the signal to negative values while retains $u$ in $K$. Cleary it belongs to $K$. Hence

$$|S(u)(n) - p(u(n - k_1), \ldots, u(n - k_m))|$$

$$= |S_0(Pq_{-n}(u)) - p(G_{k1}(Pq_{-n}u), \ldots, G_{km}(Pq_{-n}u))| \leq \epsilon$$

The proof is complete.                                                      ∎

**PROOF OF THEOREM 3.5**    Without harm of generality we pick a sigmoidal function $\psi$ satisfying $A = 1$ and $B = 0$.

**Case 1:** $m = 1$
Suppose that $f$ is the indicator function of an interval, $f(u) = I_{[a,b]}(u)$. We pointed out in Section 3.11 that $f$ is expressed in terms of the unit step function as follows:

$$f(u) = u_s(u - a) - u_s(u - b)$$

We form the sequence $\psi_k(u) = \psi(ku)$, $k \in Z^+$. The sigmoid property implies that $\lim_{k \to \infty} \psi_k(u) = u_s(u)$, $u \neq 0$. Thus

$$f(u) = I_{[a,b]}(u) = \lim_{k \to \infty} [\psi(ku - ka) - \psi(ku - kb)]$$

This states that the indicator function of a finite interval can be approximated with a neural network of the form (3.140). If $f$ is a simple function, it can be written as a linear combination of indicator functions of intervals. Thus it can be approximated by a multilayer neural network as well. If, finally, $f$ is a continuous function on an interval, it can be uniformly approximated by simple functions and the claim is proved.

**Case 2:** $m > 1$
Theorem 3.8 is readily extended to multidimensional trigonometric polynomials. We approximate $f$ by a multidimensional trigonometric polynomial

$$\sum_{k \in I(N)} a_k e^{j\omega k^T u}, \qquad a_k = a^*_{-k}$$

We combine the pair $a_k e^{j\omega k^T u}$ and $a_{-k} e^{-j\omega k^T u}$ into the real unit $2\cos(\pi k^T u + \theta_k)$. Then the function is approximated by a sum of the form

$$\sum_k c_k \cos(\pi k^T u + \theta_k) \qquad (3.145)$$

The latter equation corresponds to a neural network with activation function given by the cosine. We have already seen that the theorem holds for the special case of real functions of a real variable. Hence the cosine can be approximated by a neural network with activation function $\psi$:

$$\cos(u) = \sum_i b_i \psi(c_i u + \phi)$$

Substitution into Eq. (3.145) proves the claim. ∎

PROOF OF THEOREM 3.6    We will apply the Stone-Weierstrass theorem. It is easy to see that $\mathrm{RBF}^m(g)$ forms a linear space. It is closed under multiplication because

$$\sum_{i=1}^{N} a_i e^{-\frac{1}{2}|u-w_i|^2_Q} \sum_{j=1}^{M} b_j e^{-\frac{1}{2}|u-v_j|^2_Q} = \sum_{i,j} a_i b_j e^{-\frac{1}{2}(|u-w_i|^2_Q + |u-v_j|^2_Q)}$$

Due to the parallelogram law (Exercise 2.23), the above can be expressed in the form

$$\sum_i \sum_j A_{ij} e^{-\frac{1}{2}|u-z_{ij}|^2_M}$$

where $A_{ij} = a_i b_j e^{-|w_i-v_j|^2_Q}$, $M = 2Q$, $z_{ij} = (w_i + v_j)/2$. The remaining requirements of the Stone-Weierstrass theorem are readily established and are left to the reader. ∎

# 4

---

# THEMES IN SYSTEM DESIGN

Signal processing systems are designed to perform a certain job efficiently. The task to be accomplished is often described by a certain input output characteristic and thus by a desired system $S_d$. The signal processing system is chosen from a prespecified class of systems $\Im_f$ so that it approximates $S_d$ and hence executes the desired job represented by $S_d$. The class $\Im_f$ delineates the structures that are actually buildable, the architectures that are feasible to implement. As with the desired system $S_d$, the class of admissible systems $\Im_f$ is specified after a thorough analysis and characterization of system families.

Analysis is concerned with the way specific input characteristics convey to the output. Determination of the output mean, covariance, higher-order moments, or densities in terms of input statistics is a typical analysis task. In a similar fashion analysis explores whether input properties such as finite energy, bounded amplitude, finite support, or stationarity carry to the output. Stability is an indispensable property of a signal processing system. A system is BIBO (bounded input, bounded output) stable if a bounded in amplitude input gives rise to a bounded output. An unstable amplifier is apt to burn if currents or voltages manifest overly large values. A stable computational algorithm guarantees that deviations of the input data, due to finite word length effects, do not lead to erroneous results at the output.

Efficiency refers to a particular performance criterion according to which systems of the class $\Im_f$ are assessed. The performance criterion, or cost function, is represented by an assignment $J(S)$ which reflects the deviation of $S$ from $S_d$. One approach is to specify an admissible performance range $B$ so that any system $S$ with $J(S) \in B$ is assessed as satisfactory and acceptable. A second approach is to determine an admissible system $S^*$ in $\Im_f$ that minimizes $J(S)$ and hence performs the given task better than all systems in $\Im_f$. In the latter case $J(S)$ is a number. We say that the triple $(S_d, \Im_f, J)$ specifies a *system design formulation*. Usually a design problem can be approached with more than one design formulations. A design formulation is successful when the minimization process provides implementable optimal solutions. There are cases where more than one successful design formulations are available, each offering some advantages and suffering some weaknessess. Adaption of a specific design depends on the relative importance of a multitude of factors pertinent to the particular application. Usually the final decision does not come out of a well-defined objective selection procedure but rather of engineering intuition, skill, and experience. In other cases the complexity of the system to be developed is such that no overall global design formulation is successful. Then the design problem is split into several design problems of smaller size and of manageable complexity.

Optimization is easier if the class of admissible systems admits a finite parametrization, namely there is a set $\Theta \subset R^d$, called *parameter space*, such that $\Im_f = \{S(\theta) : \theta \in \Theta \subset R^d\}$. Each member of $\Im_f$ is specified once the parameter is specified. If a finite parametrization is imposed, the design minimization reduces to a finite-dimensional minimization problem

$$\min_{\theta \in \Theta} J(S(\theta))$$

In the sequel the above general discussion is specialized to specific applications. A preliminary discussion of stability is given first.

### 4.1 STABILITY OF INPUT-OUTPUT MODELS

In broad terms, stability is concerned with the effects of perturbations of inputs, initial conditions (or initial states), and system parameters on the output. The output of a stable system remains bounded when inputs, initial conditions, and parameter variations are bounded. Small deviations from reference values do not cause large output deviations. A good example of nicely behaved system architectures are the time delay networks discussed in Section 3.5. Systems of the form

$$y(n) = f(u(n), u(n-1), \ldots, u(n-r), \theta)$$

where $\theta$ denotes a finite-dimensional vector and $f$ a piecewise continuous function, yield a stable behavior in the following sense: Given $M > 0$ and $N > 0$, there is $L > 0$ so that for any input bounded by $M$ and any $|\theta| \le N$, it holds that $|y(n)| < L$ for all $n$. Time delay neural networks with multilayer nonlinearities, time delay radial basis function networks, and polynomial filters are such examples. The story becomes much more interesting, once recurrent systems with the output appearing in the right-hand side of the above equation are considered. Stability of recurrent systems is undertaken in Chapter 11. The present section is essentially confined to convolutional systems. The convolutional model

$$y(n) = \sum_{k=-\infty}^{\infty} h(k)u(n-k) \tag{4.1}$$

is called *p-stable*, $1 \leq p \leq \infty$, if every input signal with finite $\ell_p$ norm produces an output signal with finite $\ell_p$ norm. If $p = \infty$, the system is called *BIBO stable*. In this case any input with bounded amplitude yields a bounded amplitude output. It is conceivable that the input and output spaces contain $\ell_p$. The system is called *unstable* if it is not stable. The definitions are the same for continuous time systems except that the $\ell_p$ spaces are replaced by the $L_p$ spaces.

The stability behavior of a convolutional model is characterized by the impulse response in the following manner:

**Theorem 4.1** The convolutional model (4.1) is BIBO stable if and only if $h(n)$ is absolutely summable, that is, $h(n) \in \ell_1$. In such case it is $p$ stable for any $1 \leq p \leq \infty$.

**PROOF** Let $|u(n)| \leq L$ and $h(n) \in \ell_1$. Then

$$\sum_{k=-\infty}^{\infty} |h(n-k)u(k)| \leq \sum_{k=-\infty}^{\infty} |h(n-k)|L \leq |h|_1 L$$

Hence the output of (4.1) is well defined and bounded. Conversely, suppose that (4.1) is BIBO stable and that the $\ell_1$ norm of $h(n)$ is infinite. Let $u(n) = \mathrm{sgn}h(-n)$. Clearly $u(n)$ is bounded. The resulting output $y(n)$ is also bounded. In particular $y(0)$ is finite. But

$$y(0) = \sum_k h(k)u(-k) = \sum_k h(k)\mathrm{sgn}h(k) = \sum_k |h(k)|$$

This is a contradiction. The proof is complete. ∎

If the system is causal, $h(n) = 0$, $n < 0$. Therefore a necessary and sufficient condition for stability is

$$\sum_{n=0}^{\infty} |h(n)| < \infty \qquad (4.2)$$

Some caution is needed when the system is described by the causal convolutional representation and the inputs are also causal. The particular input chosen to establish that $y(0)$ is unbounded does not apply. Condition (4.2) remains valid as a necessary and sufficient condition, but necessity is harder to establish (see Exercise 4.2).

### Example 4.1 Example of a stable and an unstable system
The impulse response of a system is $h(n) = 1/n^2$ if $n \geq 1$ and zero otherwise. $h(n)$ is absolutely summable. The partial sum $\sum_{n=2}^{N} 1/n^2$ is equal to the sum of the areas of $N - 1$ rectangles as illustrated in Fig. 4.1 (a). This area is bounded by the

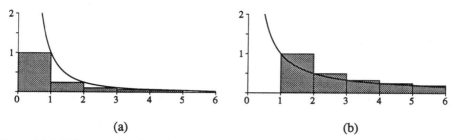

**Figure 4.1** (a) The sum of rectangular areas is bounded by the area surrounded by the graph $h(t) = 1/t^2$. (b) The sum of the rectangular areas exceeds the area under the graph $h(t) = 1/t$.

area of the curve $y = 1/t^2$ in the interval $[1, N]$. More precisely

$$\sum_{n=1}^{N} \frac{1}{n^2} \leq 1 + \int_1^N \frac{1}{t^2} dt = 2 - \frac{1}{N} \to 2$$

Therefore the system is stable. The unit-step response is sketched in Fig. 4.2 (a). The amplitude is bounded as expected.

Consider next the system with impulse response $h(n) = 1/n$, if $n \geq 1$ and zero otherwise. This signal is not absolutely summable. The $\ell_1$ norm is given by the harmonic series $\sum_{n=1}^{\infty} 1/n$, which diverges. Indeed, the partial sum of the series is given by the area of the rectangles of Fig. 4.1 (b). This area is bigger than the area surrounded by the graph of $y = 1/t$ on the interval $[1, N + 1]$. The latter is given by

$$\int_1^{N+1} \frac{1}{t} dt = \log(N + 1) \to \infty$$

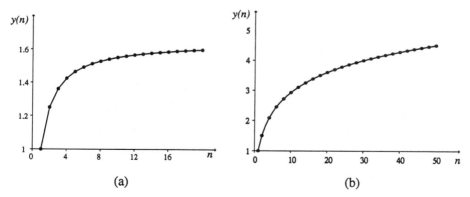

**Figure 4.2.** (a) Unit-step response of the system with impulse response $h(n) = 1/n^2$; (b) unit-step response of the system with impulse response $h(n) = 1/n$.

Hence the series diverges and the system is unstable. The step response is sketched in Fig. 4.2 (b). ∎

The concept of stability introduced above must be strengthened when time varying or nonlinear systems are considered. We say that the discrete linear causal (time varying) system $S$ is $\ell_p$ stable if there exist constants $k$ and $b$ such that

$$|S(u)|_p \leq k|u|_p + b \tag{4.3}$$

The $p$ gain of $S$ is defined as

$$\gamma_p = \inf\{k \in R : \text{there is } b \text{ such that (4.3) holds}\} \tag{4.4}$$

Intuitively the $p$ gain of $S$ describes the increase in the $p$ norm of the output signal. BIBO stability of linear time varying systems is described next.

**Theorem 4.2** Consider the time varying linear causal system

$$y(n) = \sum_{k=0}^{n} h(n, k)u(k) \tag{4.5}$$

The following statements are equivalent:

1. $\gamma_1 = \sup_{0 \leq n < \infty} \sum_{k=0}^{\infty} |h(n, k)| < \infty$.
2. The system (4.5) is $\ell_\infty$ stable in the sense of (4.3).
3. If the input to $S$ is bounded, that is, $u \in \ell_\infty$, the output $S(u)$ is bounded.

PROOF   Property 2 follows from property 1 as in the time invariant case. It is obvious that property 2 implies 3. The proof that property 1 follows from 3 requires the principle of uniform boundedness and is omitted. ∎

The definition of stability must be further modified if nonlinear input-output operators are considered. Both local and global approaches can be applied to shed new light into the structure of the far more complex nonlinear environment. We will not pursue this subject further except remark on Volterra systems. It is easy to see that if the well posedness assumptions of Section 3.7 are met, then all inputs bounded by the radius of convergence $\rho$ give rise to bounded outputs, and the system is BIBO stable. The converse statement is false.

Analog convolutional systems exhibit a similar behavior. The counterpart of Theorem 4.1 reads as follows:

**Theorem 4.3** The convolutional model

$$y(t) = \int_{-\infty}^{\infty} h(t - \tau)u(\tau)d\tau \tag{4.6}$$

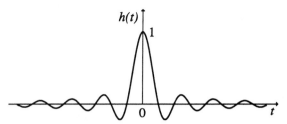

**Figure 4.3.** Impulse response of ideal lowpass filter.

is BIBO stable if and only if $h(t)$ is absolutely integrable, that is, $h(t) \in L_1$. In such case it is $p$ stable for any $1 \le p \le \infty$.

If the convolutional system is causal, a necessary and sufficient condition for stability is

$$\int_0^\infty |h(t)|\,dt < \infty$$

***Example 4.2    Instability of the ideal lowpass filter***
The signal

$$h(t) = \operatorname{sinc}(t) = \frac{\sin \pi t}{\pi t}$$

is plotted in Fig. 4.3. We will see in Chapter 6 that the system with impulse response $h(t)$ defines the ideal lowpass filter. $h(t)$ does not belong to $L_1$. Indeed, consider the integral

$$\Phi_n = \int_\pi^{n\pi} \frac{|\sin x|}{x}\,dx = \sum_{k=2}^n \int_{(k-1)\pi}^{k\pi} \frac{|\sin x|}{x}\,dx$$

In the interval $(k-1)\pi \le x \le k\pi$, we have $1/k\pi \le 1/x$. Thus

$$\int_{(k-1)\pi}^{k\pi} \frac{|\sin x|}{x}\,dx \ge \frac{1}{k\pi}\int_{(k-1)\pi}^{k\pi} |\sin x|\,dx \ge \left| \int_{(k-1)\pi}^{k\pi} \sin x\,dx \right| \frac{1}{k\pi}$$

$$= \frac{1}{k\pi}|\cos k\pi - \cos(k-1)\pi| = \frac{2}{k\pi}$$

Therefore

$$\Phi_n \ge \sum_{k=2}^n \frac{2}{k\pi} = \frac{2}{\pi}\sum_{k=2}^n \frac{1}{k}$$

We conclude that the $L_1$ norm of the sinc function is bounded below from the harmonic series that diverges. Hence the ideal lowpass filter is unstable. We observe that $h(t)$ is an energy signal, namely it belongs to $L_2$. Indeed, $|h(t)|^2 \leq 1/t^2$, and $1/t^2$ is absolutely integrable. The discrete case is similar. The signal

$$h(n) = \frac{\sin Wn}{\pi n}$$

is not absolutely summable but has finite energy. We will see in Chapter 6 that $h(n)$ defines the impulse response of the discrete ideal lowpass filter. ∎

The preceding stability results remain valid in the case of multidimensional systems, see Exercise (4.13).

## 4.2 LEAST MEAN SQUARES ESTIMATION

Let us consider two random vectors $X$ and $Y$ of size $n$ and $k$, respectively. We set $Z = \left( X^T \quad Y^T \right)^T$, and we assume that the joint density function $f_Z(z)$ exists. Suppose that we measure $Y$ and observe the value $y \in R^k$. Based on this information we want to estimate the corresponding value of $X$. A reasonable estimate is the vector $\hat{x} \in R^n$ that minimizes the mean squared error

$$J(x) = E\big[|X - x|^2 | Y = y\big] = E\big[(X - x)^T(X - x)| Y = y\big] \qquad (4.7)$$

over all $x \in R^n$. Take, for example, $X$ to be a block of signal values transmitted through a communication channel at a certain instant and $Y$ the signal block at the receiver. $Y$ is in general different from $X$ due to noise degradations caused by the channel. A receiver designed on the basis of least squares principle (4.7) will decode the specific received message $y$ into the vector $\hat{x}$ that minimizes (4.7). In a measurement context, $X$ represents the value of a signal and $Y$ the measurement of $X$ obtained by a measuring device. $Y$ differs from $X$ because of measurement noise inherent in the sensor. From the specific measurement $y$ we seek to estimate the true value of the signal. The guess $\hat{x}$ based on (4.7) ensures the minimum mean squared error. To determine the minimizing vector $\hat{x}$, we proceed as follows:

$$J(x) = E\big[X^T X | Y = y\big] - 2x^T E[X|Y = y] + x^T x$$

Setting the derivative with respect to $x$ equal to zero, we obtain

$$\frac{\partial J(x)}{\partial x} = 2x - 2E[X|Y = y] = 0$$

Hence

$$\hat{x} = E[X|Y = y] = \int_{-\infty}^{\infty} x f_{X|Y}(x|y)dx \qquad (4.8)$$

The Hessian $\partial^2 J(x)/\partial x^2 = 2I$ is positive definite. Thus the conditional mean of $X$ given $Y = y$ gives the unique minimum of (4.7).

### 4.2.1   Nonlinear least mean squares estimation

Let us define the conditional mean estimator $E[X|Y]$ as the function $R^k \to R^n$, $y \to E[X|Y = y]$. Among all nonlinear (measurable) functions $g : R^k \to R^n$, the conditional mean minimizes the mean squared error $E[|X - g(Y)|^2]$. Indeed, let $g$ be such function and $y \in R^k$. From the optimality of $\hat{x} = E[X|Y = y]$, we obtain

$$E[|X - g(Y)|^2|Y = y] = E[|X - g(y)|^2|Y = y] \geq E[|X - \hat{x}|^2|Y = y] \quad (4.9)$$

Given an arbitrary function $Q$ defined on $R^n \times R^k$, the following important identity holds:

$$E_Y[E[Q(X, Y)|Y]] = E[Q(X, Y)] \qquad (4.10)$$

where $E_Y$ denotes expectation taken with respect to the density of $Y$. Indeed, the left-hand side is

$$\int_{-\infty}^{\infty} \left[ \int_{-\infty}^{\infty} Q(x, y) f_{X|Y}(x|y)dx \right] f_Y(y)dy$$

$$= \int_{-\infty}^{\infty} \int_{-\infty}^{\infty} Q(x, y) f_{XY}(x, y)dxdy = E[Q(X, Y)]$$

If we take expectations on both sides of (4.9) and use (4.10), we obtain

$$E_Y[E[|X - g(Y)|^2|Y]] = E[|X - g(Y)|^2] \geq$$

$$E_Y[E[|X - E(X|Y)|^2|Y]] = E[|X - E(X|Y)|^2]$$

The conditional mean estimator is referred to as (nonlinear) least mean squares estimator. It has the following properties:

*Linearity.* For any matrix $A$ and any vector $b$ such that $AX + b$ makes sense, it holds that $E[AX + b|Y] = AE[X|Y] + b$. Likewise $E[X + Z|Y] = E[X|Y] + E[Z|Y]$. These follow from standard properties of the conditional mean.

**Orthogonality.** The error $\tilde{X} = X - E[X|Y]$ is uncorrelated with any function of $Y$. In fact, $E[g(Y)\tilde{X}^T] = 0$. Indeed, for any $y \in R^k$, we have

$$E[g(Y)\tilde{X}^T|Y = y] = g(y)E[\tilde{X}^T|Y = y] = g(y)(E[X|Y = y] - E[X|Y = y])^T = 0$$

The result follows if we take expectation and use (4.10).

### 4.2.2  The Gaussian case

A notable case where further significant information about the conditional mean $E[X|Y]$ can be extracted is the Gaussian case. Suppose that the augmented vector $Z = (X^T \quad Y^T)^T$ is Gaussian with mean $m_Z$ and covariance $C_{ZZ}$. The partitioned form of $Z$ implies the partitions

$$m_Z = (m_X^T \quad m_Y^T)^T, \quad C_{ZZ} = \begin{pmatrix} C_{XX} & C_{XY} \\ C_{YX} & C_{YY} \end{pmatrix} \tag{4.11}$$

The following theorem is proved in the Appendix at the end of the chapter.

**Theorem 4.4**  Suppose that $C_{ZZ}$ and $C_{YY}$ are invertible matrices. The conditional density $f_{X|Y}(x|y) = f_{XY}(x, y)/f_Y(y)$ is Gaussian with mean $m_X + C_{XY}C_{YY}^{-1}(y - m_Y)$ and variance $C_{XX} - C_{XY}C_{YY}^{-1}C_{YX}$.

Since $E[X|Y]$ is the mean of the above density, we can conclude that

$$E[X|Y] = m_X + C_{XY}C_{YY}^{-1}(Y - m_Y) \tag{4.12}$$

while the covariance of the error is

$$\text{cov}[\tilde{X}, \tilde{X}] = C_{XX} - C_{XY}C_{YY}^{-1}C_{YX} \tag{4.13}$$

Equation (4.12) expresses the conditional mean in terms of the second-order properties of $X$ and $Y$. A further important observation is that $E[X|Y]$ is a linear plus constant (affine) function of $Y$.

### 4.2.3  Linear least mean squares estimation

The conditional mean minimizes the mean squared error over all nonlinear functions. In general, it is a nonlinear function and often difficult to compute. The Gaussian case is a notable exception. A more tractable scenario is to confine minimization to functions of the form $g(Y) = AY + b$ and to try and determine the $A^*$ and $b^*$ that minimize

$$E[|X - AY - b|^2] \tag{4.14}$$

over all $A$ and $b$. This problem is readily solved by the projection theorem. The subspace $M$ is formed by the random vectors of the form $AY + b$. This space is finite-dimensional and thus is closed. Minimization of (4.14) amounts to finding among the points of $M$ the one lying the shortest distance from the given $X$. The optimum solution $A^*Y + b^*$ is the projection of $X$ onto $M$ and is determined by the orthogonality principle $< X - A^*Y - b^*, AY + b > = 0$ or

$$E[(X - A^*Y - b^*)^T(AY + b)] = 0 \qquad \text{for any } A, b \qquad (4.15)$$

Let $C_{XY} = E[XY^T] - E[X]E[Y^T]$ be the crosscovariance of $X$, $Y$ and $C_{YY} = E[YY^T] - E[Y]E[Y^T]$ the covariance of $Y$. Let also $E[X] = m_X$. Since Eq. (4.15) holds for every $A$, it will in particular hold for $A = 0$. In this case we have $(m_X - A^*m_Y - b^*)^T b = 0$ for any $b$. Hence

$$b^* = m_X - A^*m_Y \qquad (4.16)$$

If we replace (4.16) into (4.15) and set $b = 0$, the latter equation takes the form $E[Z^T AY] = 0$ for any $A$, where $Z = X_1 - A^*Y_1$, $X_1 = X - m_X$, $Y_1 = Y - m_Y$. Next we claim that this holds if and only if $E[ZY^T] = E[YZ^T] = 0$. Indeed,

$$E[Z^T AY] = E\left[\sum_i \sum_j a_{ij} Z_i Y_j\right] = \sum_i \sum_j a_{ij} E[Z_i Y_j]$$

The latter expression is zero for every $a_{ij}$ if and only if $E[Z_i Y_j] = 0$, and the claim is proved. The equation $E[ZY^T] = 0$ gives $E[X_1 Y^T] = A^*E[Y_1 Y^T]$. Note that $E[X_1 Y^T] = E[XY^T] - m_X m_Y^T = C_{XY}$. Likewise $E[Y_1 Y^T] = C_{YY}$. Therefore

$$A^* = C_{XY} C_{YY}^{-1} \qquad (4.17)$$

The linear least mean squares estimator will be denoted by $E^*[X|Y]$. It is given by

$$E^*[X|Y] = m_X + C_{XY} C_{YY}^{-1}(Y - m_Y) \qquad (4.18)$$

The above estimator depends only on the second-order properties of the random vectors $X$ and $Y$ and not on the entire family of probability density functions. Note also that if $X$ and $Y$ are uncorrelated, then $C_{XY} = 0$ and hence $E^*[X|Y] = E[X]$. The following properties are useful:

**Linearity.** $E^*[X + Z|Y] = E^*[X|Y] + E^*[Z|Y]$ and $E^*[AX|Y] = AE^*[X|Y]$. $A$ is a matrix of appropriate size.

**Error covariance.** The covariance of the error $\tilde{X} = X - E^*[X|Y]$, is given by the formula

$$\text{cov}[\tilde{X}, \tilde{X}] = C_{\tilde{X}\tilde{X}} = C_{XX} - C_{XY}C_{YY}^{-1}C_{YX} \tag{4.19}$$

Indeed, the orthogonality principle (4.15) together with

$$\text{cov}[AX + b, Y] = A\,\text{cov}[X, Y], \qquad \text{cov}[X, AY] = \text{cov}[X, Y]A^T$$

give

$$\text{cov}[\tilde{X}, \tilde{X}] = \text{cov}[\tilde{X}, X] = \text{cov}[X, X] - \text{cov}[A^*Y + b^*, X] = C_{XX} - C_{XY}C_{YY}^{-1}C_{YX}$$

The results of the previous section and in particular comparison of Eqs. (4.18) and (4.12) prove that the nonlinear least mean squares estimator $E[X|Y]$ and the linear least mean squares estimator $E^*[X|Y]$ are identical in the Gaussian case.

## 4.3   LEAST SQUARES FINITE MEMORY FILTERING AND PREDICTION

The ideas of the previous section provide a convenient framework for the analysis and study of filtering, prediction, and smoothing applications. Let $x(n)$ and $y(n)$ be two discrete random signals. We view $y(n)$ as the measurement signal that provides information about $x(n)$, the signal of actual interest. Measurements are contaminated by noise. The resulting degradations may be due to the measuring device, transmission errors, or uncontrolled artifacts. The basic goal of filtering is to process $y$ so that it acquires the desired characteristics carried by $x(n)$. A *filter* is a signal processing system that effects the above job. The filter acts on the measurement signal $y(n)$ and produces at the output the estimate of the signal $x(n)$, $\hat{x}(n)$. At each time $n$, the filter uses the samples $y(n - k)$ over the interval $k \in W$. System performance is assessed by a measure of the error $e = x - \hat{x}$. The available information involves some attributes of the signals $x$ and $y$, typically a finite number of samples. The size of $W$ may vary with $n$. Depending on the type of the window $W$, several filter models result. If $W$ consists of nonnegative integers, the filter is a causal system. Otherwise, it is referred to as *noncausal filter* or *smoother*. If $W$ is of infinite size, the so-called infinite horizon problem is obtained. Infinite horizon filtering is discussed in Chapter 12 in connection with Wiener filters. Within the finite horizon case we will consider windows of fixed length, independent of $n$, as well as windows of length $n$. The latter case appears when the measurement samples initiate at a certain time, which is commonly placed at the origin. The finite horizon case is easily casted in the framework of least squares estimation.

### 4.3.1   Growing memory window

Consider, first, a window of size $n$. Measurements start at time $n = 1$. They constitute a record of the stochastic process $y(n)$. At each time $n$, we form the vector of size $n$:

$$\mathbf{y}(n) = (\, y(n) \quad y(n-1) \quad \cdots \quad y(1)\,)^T$$

indicating the information available up to time $n$. The class of allowable filters consists of time varying systems of the form

$$\hat{x}(n) = g(\mathbf{y}(n), n) = g(y(n), \ldots, y(1), n) \tag{4.20}$$

Among the models of the form (4.20), the least mean squares estimator $\hat{x}(n) = E[x(n)|\mathbf{y}(n)]$ is optimal in the sense of minimizing the mean squared error $E[|x(n) - \hat{x}(n)|^2]$. The resulting filter architecture is nonlinear unless the given signals are jointly Gaussian. Furthermore it is time varying, even if the signals are jointly stationary because the dimension of $\mathbf{y}(n)$ grows with $n$.

Linear (more precisely, affine) structures are obtained at the expense of larger mean squared error if the class of allowable filters is restricted to

$$\hat{x}(n) = h^T(n)\mathbf{y}(n) + k(n) = \sum_{i=1}^{n} h_i(n)y(n-i+1) + k(n) \tag{4.21}$$

The problem admits a finite parametrization for fixed $n$. To determine the coefficients $h_i(n)$, $k(n)$ we invoke Eqs. (4.16) and (4.17), and we make the substitution $X \to x(n)$, $Y \to \mathbf{y}(n)$. Furthermore we set $h(n) = (\, h_1(n), \ldots, h_n(n)\,)^T$, and $c_{xy}(n) = (\, c_1(n) \quad \cdots \quad c_n(n)\,)$, where $c_i(n) = \text{cov} [x(n), y(n-i+1)]$. Let $C_y(n)$ be the matrix of size $n \times n$ with entries $\text{cov}[y(n-i), y(n-j)]$. The parameters of the filter at time $n$ are determined by the linear equations

$$h^T(n) = c_{xy}(n)C_y^{-1}(n) \quad k(n) = m_x(n) - \sum_{i=1}^{n} h_i(n)m_y(n-i+1) \tag{4.22}$$

If the signals $x(n)$ and $y(n)$ are jointly wide sense stationary, $c_i(n)$ and $[C_y(n)]_{ij}$ do not depend on $n$. Of course their size grows with $n$, and thus the filter coefficients depend on $n$, that is, the filter is a time varying system.

### 4.3.2 Fixed memory size

In many cases the complexity of the above time varying filter cannot be tolerated. To reduce computational requirements, we restrict the class of admissible systems to

$$\hat{x}(n) = g(y(n), y(n-1), \ldots, y(n-m+1), n) = g(\varphi_m(n), n) \tag{4.23}$$

where

$$\varphi_m(n) = (y(n)\ y(n-1)\ \ldots\ y(n-m+1))^T \tag{4.24}$$

The window size, $m$, is independent of the time index $n$. The nonlinear estimator becomes $\hat{x}(n) = E[x(n)|\varphi_m(n)]$. In general, it is time varying and nonlinear. If, however, $x(n)$ and $y(n)$ are jointly stationary, the filter becomes time invariant of the form

$$\hat{x}(n) = g(y(n), y(n-1), \ldots, y(n-m+1)) = g(\varphi_m(n))$$

In the affine case the situation is simpler. The class of allowable structures consists of

$$\hat{x}(n) = \sum_{i=1}^{m} h_i(n)y(n-i+1) + k(n) \tag{4.25}$$

As before the filter parameters are determined by the linear equations

$$h^T(n) = c_{xy}(n)C_y^{-1}(n), \quad k(n) = m_x(n) - \sum_{i=1}^{m} h_i(n)m_y(n-i+1) \tag{4.26}$$

where

$$C_y(n) = \mathrm{cov}[\varphi_m(n), \varphi_m(n)], \quad c_{xy}(n) = \mathrm{cov}[x(n), \varphi_m(n)] \tag{4.27}$$

The size of the above parameters is now constant. If $x(n)$ and $y(n)$ are jointly wide sense stationary, it is easy to see that $h^T(n)$ and $k(n)$ are constant with respect to $n$, and are determined by the linear system of equations

$$h^T = c_{xy}C_y^{-1}, \quad k = m_x - m_y \sum_{i=1}^{m} h_i \tag{4.28}$$

If $x(n)$ and $y(n)$ are zero mean, then $k = 0$, and $h$ is determined by the linear system of equations

$$Rh = d \tag{4.29}$$

where

$$R = E[\varphi_m(n)\varphi_m^T(n)], \quad d = E[\varphi_m(n)x(n)] \tag{4.30}$$

The matrix $R$ (or $C_y$) is symmetric Toeplitz and allows the development of fast solvers, such as the celebrated Levinson algorithm discussed below.

Implementation of the above method requires knowledge of $C_y$ and $c_{xy}$. In most cases the information that we have at our disposal is a record of samples $x(n)$, $y(n)$, $1 \leq n \leq N$. The statistical parameters are then estimated by the time averages

$$m_\varphi(N) = \frac{1}{N-m+1} \sum_{n=m}^{N} \varphi_m(n), \quad m_x(N) = \frac{1}{N} \sum_{n=1}^{N} x(n) \qquad (4.31)$$

$$C_y(N) = \frac{1}{N-m+1} \sum_{n=m}^{N} \left(\varphi_m(n) - m_\varphi(N)\right) \left(\varphi_m(n) - m_\varphi(N)\right)^T \quad (4.32)$$

$$c_{xy}(N) = \frac{1}{N-m+1} \sum_{n=m}^{N} \left(\varphi_m(n) - m_\varphi(N)\right)^T \left(x(n) - m_x(N)\right)$$

Time averages produce the corresponding moments if the underlying stochastic process is ergodic. A process $x(n)$ is called *mean square ergodic in the moments up to order m* if for all $1 \leq k \leq m$ and all integers $n_1, \ldots, n_k$,

$$\lim_{N \to \infty} E\left| \frac{1}{2N+1} \sum_{n=-N}^{N} x(n+n_1) \cdots x(n+n_k) - m(n_1, \ldots, n_k) \right|^2 = 0$$

The time averages converge to the moments in the quadratic mean. For the filtering problem, wide sense ergodicity in the second-order moments suffices. A process is called *ergodic in the wide sense* if it is ergodic in the moments of second order. Alternative concepts of ergodicity result if other convergence types are employed, for example, convergence with probability 1. Ergodicity tests are hard to establish and will not be discussed further.

**Total squared error.** The filter design problem is directly formulated in terms of the input output data samples via the total squared error design formulation. Without harming generality, we assume that the given signals have zero mean. In the opposite case, the means are subtracted. The class of allowable systems consists of FIR filters of given order $m$. Data $x(n)$, $y(n)$ are available in the interval $1 \leq n \leq N$. The performance criterion is the total squared error $\sum_{n=m}^{N} e^2(n)$. The optimum solution is easily obtained by the projection theorem. Alternatively, we can write the error as $e(n) = x(n) - \hat{x}(n) = x(n) - \varphi_m^T(n)\theta$, where $\theta$ is the vector of unknown coefficients and $\varphi_m(n)$ is given by Eq. (4.24). If we set the derivative with respect to $\theta$ equal to zero, we obtain $2\sum (\partial e(n)/\partial\theta)e(n) = 0$. The error depends linearly on $\theta$. Thus $\partial e(n)/\partial\theta = -\varphi_m(n)$ and $\sum \varphi_m(n)[x(n) - \varphi_m^T(n)\theta] = 0$. The optimum parameter $\theta(N)$ satisfies the linear system

$$R(N)\theta(N) = d(N) \qquad (4.33)$$

where

$$R(N) = \sum_{n=m}^{N} \varphi_m(n)\varphi_m^T(n), \qquad d(N) = \sum_{n=m}^{N} \varphi_m(n)x(n) \qquad (4.34)$$

**Exponentially forgetting window.** An extension of the total squared error is obtained if the error at each time is weighted. The cost function has the form

$$\sum_{n=m}^{N} w(n)e^2(n) \qquad (4.35)$$

The window $w(n)$ reflects the time varying significance of past and recent information; as such it is employed in the estimation of time varying dynamics. The exoponential forgetting window

$$w(n) = \lambda^{N-n}, \qquad 0 < \lambda \leq 1, \; n \leq N$$

stresses recent information and suppresses old data. Minimization of (4.35) using the exponentially forgetting window leads to the linear system

$$R(N)\theta(N) = d(N) \qquad (4.36)$$

where

$$R(N) = \sum_{n=m}^{N} \lambda^{N-n}\varphi_m(n)\varphi_m^T(n), \qquad d(N) = \sum_{n=m}^{N} \lambda^{N-n}\varphi_m(n)x(n) \qquad (4.37)$$

### 4.3.3   Prediction

Prediction is concerned with the estimation of a signal value from knowledge of past values. A predictor is a signal processing system that outputs estimates of future behavior. Prediction can be formulated as a filtering problem if we set $x(n) = y(n + d), d > 0$. The predictor output $\hat{y}(n)$ yields an estimate of the future signal value $y(n + d)$. In this sense predictors are always causal systems. The predictor memory can be finite or infinite. Infinite memory predictors are discussed in Chapter 12. Least squares finite predictors whose memory size is either fixed or grows with time are easily inferred from the results of the previous section. Let us consider the one-step-ahead prediction, $d = 1$. The nonlinear least mean squares predictor produces at the output the estimate $E[y(n)|y(n-1), y(n-2), \ldots, y(1)]$. The linear least squares predictor $E^*[y(n)|y(n-1), y(n-2), \ldots, y(1)]$ is simpler to compute, but it is time varying even if $y(n)$ is stationary. A simpler situation arises when FIR predictors

$$\hat{y}(n) = \sum_{i=1}^{m} a_i y(n-i)$$

are employed. Suppose that the signal $y(n)$ is zero mean wide sense stationary. Minimization of the mean squared error $E[e^2(n)]$, $e(n) = y(n) - \hat{y}(n)$, leads to the system $Ra = r$, where $R$ is given by (4.30) and

$$r = (r_1 \quad r_2 \quad \cdots \quad r_m)^T, \qquad r_i = E[y(n)y(n-i)]$$

The total least squared error case follows easily.

### Example 4.3   AR modeling

Suppose that a signal $x(n)$ is generated by the (autoregressive) AR model

$$x(n) + \sum_{i=1}^{m} a_i x(n-i) = w(n)$$

The process noise is zero mean with variance $\sigma$. The following MATLAB code simulates the model, generates a record of samples and uses the sample information to solve the normal equations. The resulting solution yields the AR coefficients.

```
w = (sig^0.5)*randn(1,L);          % produces L samples of the zero mean Gaussian
                                   % process noise of variance sig
x = filter(1,[1,-a],w);            % determines L samples of the AR signal with
                                   % AR coefficients given by the vector a

m = length(a);
R = zeros(m,m);
d = zeros(m,1);
for i = m:N-1                      % solves  the normal equations using N < L samples
  phi = x(i:-1:i-m + 1)';
  R = R + phi*phi';
  d = d + phi*x(i + 1);
end
theta = R\d;
xhat = filter([0,theta'],1,x);     % computes the predictor
                                   % output over the entire data set
epred = x-xhat;                    % computes the prediction error
plot([L-N + 1:L],epred(L-N + 1:L)) %plots the last N prediction error samples
```

Suppose next that the AR signal $x(n)$ is transmitted or stored and that the signal

$$y(n) = x(n) + v(n)$$

is received. A filter is next designed that removes the noise on the basis of sample information from both signals:

```
y = x + (vsig^0.5)*randn(1,L);          % zero mean Gaussian noise of variance
                                        % vsig is added
R = zeros(k,k);
d = zeros(k,1);
for i = k:N                             % solves the normal equations using N samples
   phi = y(i:-1:i-k + 1)';
   R = R + phi*phi';
   d = d + phi*x(i);
end
theta = R\d;
xfilt = filter([0,theta'],1,y);         % computes the filter
                                        % output over the entire data set
efilt = x-xfilt;                        % computes the estimation error
plot([L-N + 1:L],efilt(L-N + 1:L))      % plots the last N error samples
clg
plot([L-N + 1:L],epred(L-N + 1:L),[L-N + 1:L],efilt(L-N + 1:L))
                                        % plots the filtering and prediction errors   ∎
```

### 4.3.4 The Levinson algorithm

The Levinson algorithm and its variants constitute efficient methods for the solution of Toeplitz-like linear equations. It is of primary importance because Toeplitz equations occur frequently in signal processing applications. Let us consider the linear system of equations

$$Rc = d \tag{4.38}$$

$R$ is a symmetric Toeplitz matrix of order $p$ and $d$ a vector of length $p$:

$$R = \begin{pmatrix} r_0 & r_1 & \cdots & r_{p-1} \\ r_1 & r_0 & \cdots & r_{p-2} \\ \vdots & \vdots & \ddots & \vdots \\ r_{p-1} & r_{p-2} & \cdots & r_0 \end{pmatrix}, \qquad d = \begin{pmatrix} d_0 \\ d_1 \\ \vdots \\ d_{p-1} \end{pmatrix}$$

Let $R_m$ be the $m \times m$ matrix formed by deleting the last $p - m$ columns and rows of $R$. Let $\mathbf{d}_m$ denote the vector obtained by removing the last $p - m$ entries of $d$:

$$R_m = \begin{pmatrix} r_0 & r_1 & \cdots & r_{m-1} \\ r_1 & r_0 & \cdots & r_{m-2} \\ \vdots & \vdots & \ddots & \vdots \\ r_{m-1} & r_{m-2} & \cdots & r_0 \end{pmatrix}, \qquad \mathbf{d}_m = \begin{pmatrix} d_0 \\ d_1 \\ \vdots \\ d_{m-1} \end{pmatrix}$$

The Levinson algorithm recursively determines the solutions

$$R_m \mathbf{c}_m = \mathbf{d}_m$$

by repeated application of the general updating scheme of the Appendix at the end of the chapter. Apparently, after $p$ steps $\mathbf{c}_p = \mathbf{c}$. The following nesting structure is easily deduced:

$$R_{m+1} = \begin{pmatrix} R_m & \mathbf{r}_m \\ \mathbf{r}_m^T & r_0 \end{pmatrix}, \quad \mathbf{d}_{m+1} = \begin{pmatrix} \mathbf{d}_m \\ d_m \end{pmatrix} \tag{4.39}$$

where

$$\mathbf{r}_m^T = \begin{pmatrix} r_m & r_{m-1} & \cdots & r_1 \end{pmatrix} \tag{4.40}$$

If we apply identity (4.153) on the partitioning (4.39), we obtain

$$\mathbf{c}_{m+1} = \begin{pmatrix} \mathbf{c}_m \\ 0 \end{pmatrix} + \begin{pmatrix} \mathbf{b}_m \\ -1 \end{pmatrix} k_m^c \tag{4.41}$$

where

$$R_m \mathbf{b}_m = \mathbf{r}_m, \quad k_m^c = -\alpha_m^{-1} \beta_m^c \tag{4.42}$$

$$\alpha_m = r_0 - \mathbf{r}_m^T \mathbf{b}_m, \quad \beta_m^c = d_m - \mathbf{r}_m^T \mathbf{c}_m = d_m - \mathbf{d}_m^T \mathbf{b}_m \tag{4.43}$$

The computational complexity of standard linear solvers is $O(p^3)$. The above recursions will offer computational savings only if the vector $\mathbf{b}_m$ is efficiently determined. The vector $\mathbf{r}_m$ admits the following lower partitioning

$$\mathbf{r}_m^T + 1 = \begin{pmatrix} r_{m+1} & \mathbf{r}_m^T \end{pmatrix} \tag{4.44}$$

Likewise $R_{m+1}$ satisfies the lower partitioning

$$R_{m+1} = \begin{pmatrix} r_0 & \tilde{\mathbf{r}}_m^T \\ \tilde{\mathbf{r}}_m & R_m \end{pmatrix}, \quad \tilde{\mathbf{r}}_m = \begin{pmatrix} r_1 \\ \vdots \\ r_m \end{pmatrix} \tag{4.45}$$

Notice that

$$\tilde{\mathbf{r}}_m = J \mathbf{r}_m \tag{4.46}$$

where $J$ is the reversing order matrix

$$J = \begin{pmatrix} 0 & 0 & \cdots & 0 & 1 \\ 0 & 0 & \cdots & 1 & 0 \\ \vdots & & & & \\ 1 & 0 & \cdots & 0 & 0 \end{pmatrix}$$

If we apply identity (4.155) to the system $R_{m+1}\mathbf{b}_{m+1} = \mathbf{r}_{m+1}$ taking into account (4.45), we obtain

$$\mathbf{b}_{m+1} = \begin{pmatrix} 0 \\ \mathbf{b}_m \end{pmatrix} + \begin{pmatrix} -1 \\ \mathbf{a}_m \end{pmatrix} k_{m+1} \tag{4.47}$$

where

$$R_m \mathbf{a}_m = \tilde{\mathbf{r}}_m, \quad k_{m+1} = -\alpha_m^{-1}\beta_{m+1}, \quad \beta_{m+1} = r_{m+1} - \tilde{\mathbf{r}}_m^T \mathbf{b}_m \tag{4.48}$$

Computational reduction requires efficient procedures for the vector $\mathbf{a}_m$. Since each $R_m$ is symmetric and Toeplitz, it satisfies $JR_mJ = R_m$, $1 \leq m \leq p$. The above property together with Eq. (4.46) imply that

$$\mathbf{a}_m = J\mathbf{b}_m, \quad 1 \leq m \leq p-1$$

Thus $\mathbf{a}_m$ is directly obtained from $\mathbf{b}_m$ by the reversing order permutation $J$. The list of recursions obtained so far is complete. The computational cost is $O(m)$ for equations of the form (4.41) and $O(m)$ for the inner products involved in (4.43) and (4.48). The overall operations count is of order $O(p^2)$. Some further improvements result if we note that $\alpha_m$ can be recursively computed. Indeed, taking into account Eqs. (4.43), (4.44), (4.46), (4.47), and (4.48), we have

$$\alpha_{m+1} = r_0 - \mathbf{r}_{m+1}^T \mathbf{b}_{m+1} = r_0 - \begin{pmatrix} r_{m+1} & \mathbf{r}_m^T \end{pmatrix}\left( \begin{pmatrix} 0 \\ \mathbf{b}_m \end{pmatrix} + \begin{pmatrix} -1 \\ \mathbf{a}_m \end{pmatrix} k_{m+1} \right)$$

$$= r_0 - \mathbf{r}_m^T \mathbf{b}_m + (r_{m+1} - \mathbf{r}_m^T \mathbf{a}_m)k_{m+1} = \alpha_m + \beta_{m+1}k_{m+1}$$

Thus

$$\alpha_{m+1} = \alpha_m + \beta_{m+1}k_{m+1} = \alpha_m(1 - k_{m+1}^2) \tag{4.49}$$

The Levinson algorithm is summarized in Table 4.1.

*Remarks*

1. The Levinson algorithm is valid over arbitrary fields. In particular, it can be used for the solution of Toeplitz equations over finite fields.
2. The division operations require that each $\alpha_m$ be nonzero. This is related to the invertibility of the matrix $R$. Equation (4.163) of the Appendix implies that

$$\alpha_m = \frac{\det R_{m+1}}{\det R_m}$$

**Table 4.1  The Levinson algorithm**

$$\mathbf{d}_{m+1} = \begin{pmatrix} \mathbf{d}_m \\ d_m \end{pmatrix}$$

$$\mathbf{c}_{m+1} = \begin{pmatrix} \mathbf{c}_m \\ 0 \end{pmatrix} + \begin{pmatrix} \mathbf{b}_m \\ -1 \end{pmatrix} k_m^c$$

$$k_m^c = -\alpha_m^{-1} \beta_m^c$$

$$\beta_m^c = d_m - \mathbf{b}_m^T \mathbf{d}_m$$

$$\mathbf{r}_{m+1}^T = \begin{pmatrix} r_{m+1} & \mathbf{r}_m^T \end{pmatrix}$$

$$\mathbf{b}_{m+1} = \begin{pmatrix} 0 \\ \mathbf{b}_m \end{pmatrix} + \begin{pmatrix} -1 \\ \mathbf{a}_m \end{pmatrix} k_{m+1}$$

$$k_{m+1} = -\alpha_m^{-1} \beta_{m+1}$$

$$\beta_{m+1} = r_{m+1} - \mathbf{r}_m^T \mathbf{a}_m$$

$$\alpha_{m+1} = \alpha_m + \beta_{m+1} k_{m+1} = \alpha_m (1 - k_{m+1}^2)$$

It follows that $\alpha_m$ is nonzero if all main subdeterminants of $R$ are nonzero. This is the case if $R$ is a real positive definite matrix, as a consequence of the Sylvester criterion. Then $\alpha_m > 0$, or equivalently, by Eq. (4.49)

$$|k_m| < 1$$

Let us consider the Levinson algorithm in the context of fixed memory filtering. Then

$$R_m = E[\varphi_m(n)\varphi_m^T(n)], \qquad d_m = E[\varphi_m(n)x(n)]$$

Each intermediate vector $\mathbf{c}_m$ is precisely the optimum filter of order $m$. The linear system defining the vector $\mathbf{a}_m$ in Eq. (4.48) indicates that $\mathbf{a}_m$ is the one-step-ahead predictor minimizing the mean squared error $E[|e_m^f(n)|^2]$, where

$$e_m^f(n) = y(n) - \sum_{i=1}^m a_i y(n-i) = y(n) - \mathbf{a}_m^T \varphi_m(n-1)$$

Likewise $\mathbf{b}_m = \begin{pmatrix} b_1 & \cdots & b_m \end{pmatrix}^T$ is the optimum one-step-backward predictor that minimizes the mean squared error $E[|e_m^b(n)|^2]$, where

$$e_m^b(n) = y(n-m) - \sum_{i=1}^m b_i y(n-i+1) = y(n-m) - \mathbf{b}_m^T \varphi_m(n)$$

The parameter $\alpha_m$ coincides with the minimum mean squared prediction error. Indeed, the orthogonality principle gives

$$E[|e_m^f(n)|^2] = E[e_m^f(n)y(n)]$$

$$= E[y^2(n)] - \mathbf{a}_m^T E[\varphi_m(n-1)y(n)] = r_0 - \mathbf{a}_m^T \tilde{\mathbf{r}}_m = \alpha_m$$

Likewise $E[|e_m^b(n)|]^2 = a_m$. Using the Levinson recursion (4.41), it is easy to show that the minimum mean squared filtering error $J_m = E[e_m^2(n)]$ satisfies the recursion

$$J_{m+1} = J_m + \beta_m^c k_m^c = J_m - \alpha_m (k_m^c)^2$$

$J_m$ decreases with the filter order $m$.

### 4.3.5  Lattice realizations

The input-output expression of the optimum mean squared FIR filter of order $p$ is

$$\hat{x}(n) = \mathbf{c}_p^T \varphi_p(n) = x(n) - e_p(n) \tag{4.50}$$

where $e_p(n)$ is the filter error at time $n$. Order updates of the error are derived next. For this purpose we observe that the regressor vector satisfies the following shift invariant properties:

$$\varphi_{m+1}^T(n) = (\varphi_m^T(n) \quad y(n-m)), \quad \varphi_{m+1}^T(n) = (y(n) \quad \varphi_m^T(n-1)) \tag{4.51}$$

The first of the above partitions together with (4.41) give

$$e_{m+1}(n) = x(n) - (\varphi_m^T(n) \quad y(n-m)) \left( \begin{pmatrix} \mathbf{c}_m \\ 0 \end{pmatrix} + \begin{pmatrix} \mathbf{b}_m \\ -1 \end{pmatrix} k_m^c \right)$$

$$= (x(n) - \varphi_m^T(n)\mathbf{c}_m) + e_m^b(n)k_m^c$$

Hence

$$e_{m+1}(n) = e_m(n) + e_m^b(n)k_m^c \tag{4.52}$$

The backward error $e_m^b(n)$ is updated in a similar way. Using Eq. (4.47) and the lower partition (4.51), we obtain

$$e_{m+1}^b(n) = e_m^b(n-1) + e_m^f(n)k_{m+1} \tag{4.53}$$

Finally, the forward prediction error is determined by

$$e^f_{m+1}(n) = e^f_m(n) + e^b_m(n-1)k_{m+1} \tag{4.54}$$

Equations (4.50), (4.52), (4.53), and (4.54) provide the *lattice-ladder* realization of the FIR filter. The lattice section of the resulting architecture concerns the optimum predictor and is depicted in Fig. 4.4. The transversal parameters $c_i$ are replaced by the coefficients $k^c_m$ and $k_m$. The coefficients $k_m$ are called *reflection coefficients* due to their physical significance in scattering applications.

## 4.4  SYSTEM IDENTIFICATION

System identification is concerned with the determination of a system on the basis of given input output information. Most commonly this information is given in terms of a finite record of input output samples, $u(n)$, $y(n)$. Two approaches will be described, the *true model approach* and the *estimation or prediction error approach*. The true model approach assumes that the given input output signals are indeed generated by a system, called the *true system*. It then develops equations for the calculation of the true system using the input-output relationship and the available data. The estimation error approach starts from a given family of admissible models and seeks to determine one that best matches the given input output characteristics. In this case we end up with a system design

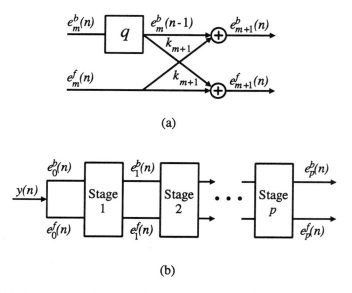

(a)

(b)

**Figure 4.4.** Lattice realization. Each stage in (b) is implemented by the architecture of (a).

formulation. The similarities and the differences in the two methods will become clear as we go along. We will be exclusively concerned with the identification of discrete time systems.

### 4.4.1  Noise free case

Let $u(n)$ and $y(n)$ denote the given input output signals. The true model approach assumes that there exists a system $S$ within the family of admissible models $\Im_f$ so that $y = S(u)$. It then tries to handle the latter equation so that the characteristics of $S$ are extracted. With prediction error identification we seek to determine an admissible system in $\Im_f$ that best interprets the input output information. Efficiency of an admissible system $S$ is assessed by comparing the output of $S$ produced by the given input $u$ to the given output $y$: $|S(u) - S_d(u)| = |S(u) - y|$. Reliance on the error explains the name of the method. We illustrate the above with the identification of static systems.

***Example 4.4    Identification of static systems***
***True model approach.***  Consider the single-input, single-output system

$$y(n) = f(u(n)) \tag{4.55}$$

The given input output samples are $u(n) = u_n$, $y(n) = y_n$, $n = 0, 1, 2, \ldots, N$. We choose the space of polynomial functions as the class of admissible systems. This space is sufficiently rich to provide good approximations for a fairly large class of functions $f$, while its elements lend themselves to efficient computation. Indeed, the Bolzano-Weierstrass theorem asserts that any continuous function is approximated arbitrarily well by a polynomial function of sufficiently high degree on a finite interval $[a, b]$. More precisely, if $f$ is continuous on $[a, b]$, for all $\epsilon > 0$, there exists $N$ and a polynomial function of degree $N$, $p(u)$ such that $|f(u) - p(u)| < \epsilon$ for all $u \in [a, b]$. The Stone-Weierstrass theorem described in Appendix 3A is an extension of the above result. Polynomial functions are efficiently evaluated by adders and scalors by the Hörner rule:

$$p(u) = (\cdots(((p_N u + p_{N-1})u + p_{N-2})u + p_{N-3})u + \cdots)u + p_0 \tag{4.56}$$

Each parenthesis has the form $p_i u + p_{i-1}$ and requires one multiplication and one addition. Since there are $N - 1$ parentheses, a total of $N$ multiplications and $N$ additions is needed. If $a_k$ denotes the content of the $k$ parenthesis, we have $a_0 = p_N$, $a_k = u a_{k-1} + p_{N-k}$, $k = 1, 2, \ldots, N$, $p(u) = a_N$.

Let us pursue the true model approach under the assumption that the true system is a polynomial function $p(u)$ of degree at most $N$. Equation (4.55) is rewritten as

$$p(u_i) = y_i, \qquad 0 \le i \le N \tag{4.57}$$

or in matrix form as $V\mathbf{p} = \mathbf{y}$, where $\mathbf{p}$ is the vector of the unknown polynomial coefficients, $\mathbf{y}$ is the vector of output samples, and $V$ is the so-called *Vandermonde matrix*

$$V = \begin{pmatrix} 1 & u_0 & u_0^2 & \cdots & u_0^N \\ 1 & u_1 & u_1^2 & \cdots & u_1^N \\ 1 & u_2 & u_2^2 & \cdots & u_2^N \\ \vdots & \vdots & \vdots & \ddots & \vdots \\ 1 & u_N & u_N^2 & \cdots & u_N^N \end{pmatrix} \tag{4.58}$$

It can be shown that the determinant of $V$ is

$$\det V = \prod_{\substack{j=0 \\ i>j}}^{N} (u_i - u_j)$$

It is nonzero if and only if $u_i \neq u_j$, $i \neq j$. Polynomial identification of (4.55) admits a finite parametrization. The unknown parameter vector consists of the polynomial coefficients and its size is given by the number of data points. This latter feature is often undesirable.

The polynomial $p(u)$ can alternatively be computed by Lagrange formula

$$p(u) = \sum_{i=0}^{N} y_i L_i(u) \tag{4.59}$$

$L_i(u)$ are named *Lagrange polynomials* and are determined by

$$L_i(u) = \frac{\prod_{j \neq i} (u - u_j)}{\prod_{j \neq i} (u_i - u_j)}, \qquad i = 0, 1, \ldots, N \tag{4.60}$$

Indeed, the denominators in (4.60) are nonzero because $u_i \neq u_j$ for $i \neq j$. Thus $L_i(u)$ are well defined. Each $L_i(u)$ has exactly $N$ roots. Therefore it has degree $N$. Hence $p(u)$ has degree at most $N$. Moreover

$$L_i(u_j) = \delta(i - j) = \begin{cases} 1, & \text{if } i = j \\ 0, & \text{if } i \neq j \end{cases}$$

We substitute the latter equation into (4.59), and we obtain $p(u_j) = \sum_{i=0}^{N} y_i L_i(u_j) = \sum_{i=0}^{N} y_i \delta(i - j) = y_j$. Hence the polynomial defined by (4.59) has degree at most $N$ and satisfies (4.57). If there was another polynomial $\tilde{p}(u)$ with the same properties, then $\tilde{p}(u_j) - p(u_j) = 0$. Therefore $\tilde{p}(u) - p(u)$ has degree at most $N$ and $N + 1$ roots. Necessarily, it is the zero polynomial.

***Prediction error identification by least squares.*** Suppose that the finite record of input-output samples $u_n = u(n)$, $y_n = y(n)$, $n = 0, \ldots, N$, is to be interpreted by a static time invariant single-input, single-output system of the form (4.55). The class of admissible systems consists of static systems admitting a finite parametrization

$$S_f = \{\hat{f}(u, \theta) : \theta \in R^d\} \tag{4.61}$$

Examples are polynomial functions, radial basis functions, and multilayer neural networks. We employ the total squared error as performance criterion

$$J(\theta) = \sum_{n=0}^{N} w(n) \left(y_n - \hat{f}(u_n, \theta)\right)^2 \tag{4.62}$$

with $w(n) > 0$. The value of $\theta$ that minimizes the above expression satisfies

$$\frac{d}{d\theta} J(\theta) = -2 \sum_{n=0}^{N} w(n) \frac{\partial \hat{f}}{\partial \theta}(u_n, \theta) \left(y_n - \hat{f}(u_n, \theta)\right) = 0$$

or

$$\sum_{n=0}^{N} w(n) \frac{\partial \hat{f}}{\partial \theta}(u_n, \theta) \hat{f}(u_n, \theta) = \sum_{n=0}^{N} w(n) \frac{\partial \hat{f}}{\partial \theta}(u_n, \theta) y_n \tag{4.63}$$

The above equations are nonlinear with respect to $\theta$. Algorithms for their solution are presented in Appendix III. Construction of a multilayer neural network, or a radial basis network that matches $y(n)$ at the output when presented with $u(n)$, fits the above framework. In both cases the input-output relationship has the form (4.55). The unknown parameters in the multilayer neural network configuration consist of the synaptic weights. Likewise $\theta$ consists of the centroids $w_i$ and the weights $a_i$ in a radial basis network architecture. In both cases $f$ depends nonlinearly on $\theta$.

The situation becomes simpler if the parametrization (4.61) is linear with respect to $\theta$. In this case there exist functions $\phi_i(u)$, $i = 0, 1, \ldots, d-1$ such that

$$\hat{f}(u, \theta) = \sum_{i=0}^{d-1} \phi_i(u) \theta_i$$

or in matrix form

$$\hat{f}(u, \theta) = \Phi^T(u)\theta, \quad \Phi^T(u) = (\phi_0(u) \quad \cdots \quad \phi_{d-1}(u)) \tag{4.64}$$

If $\theta(N)$ denotes the value of $\theta$ that minimizes (4.62), Eq (4.63) is rewritten as

$$Q(N)\theta(N) = s(N) \tag{4.65}$$

where

$$Q(N) = \sum_{n=0}^{N} w(n)\Phi(u_n)\Phi^T(u_n), \quad s(N) = \sum_{n=0}^{N} w(n)\Phi(u_n)y_n \qquad (4.66)$$

The matrix $Q(N)$ is invertible provided that the input data $u_n$ are sufficiently rich for the class of admissible systems, that is the $(N+1) \times d$ matrix $(d \ll N)$,

$$\Phi_N = \begin{pmatrix} \phi_0(u_0) & \phi_1(u_0) & \cdots & \phi_{d-1}(u_0) \\ \phi_0(u_1) & \phi_1(u_1) & \cdots & \phi_{d-1}(u_1) \\ \vdots & \vdots & \ddots & \vdots \\ \phi_0(u_N) & \phi_1(u_N) & \cdots & \phi_{d-1}(u_N) \end{pmatrix} \qquad (4.67)$$

has degree $d$:

$$\text{rank } \Phi_N = d \qquad (4.68)$$

Then $Q(N)$ is positive definite and hence invertible. Indeed, $Q(N) \geq 0$, since for any $\theta \in R^d$,

$$\theta^T Q(N)\theta = \sum_{n=0}^{N} w(n)(\theta^T \Phi(u_n))^2 \geq 0$$

If $\theta^T Q(N)\theta = 0$, for some $\theta$, then $\theta^T \Phi(u_n) = 0$ for all $0 \leq n \leq N$. Therefore $\theta$ belongs to the nullspace of $\Phi_N$, and Eq. (4.68) implies that $\theta = 0$.

***Polynomial identification.*** Polynomial identification results when $\phi_i(u) = u^i$. In contrast to Lagrange interpolation, parametrization (4.61) has constant dimension $d$ independent of $N$. Condition (4.68) holds. Indeed, let $\Phi_N\theta = 0$. Then

$$0 = \sum_{i=0}^{d-1} \phi_i(u_n)\theta_i = \sum_{i=0}^{d-1} u_n^i\theta_i, \qquad n = 0, \ldots, N$$

Consequently the polynomial with coefficients $\theta_i$ has $N+1$ roots, and since $N > d$, it must be the zero polynomial.

Notice that all entries $Q_{ij}(N)$ of (4.66) such that the sum of their indexes are the same, coincide. Such matrices are called *Hankel*. Their structure is illustrated in Fig. 4.5. All entries assigned to diagonals parallel to the antidiagonal are equal. Gauss elimination method solves Eq. (4.65) with $O(d^3)$ multiplications and additions. The Hankel structure enables the development of fast solvers such as the Levinson algorithm; see Exercise 4.20. ■

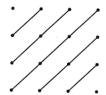

**Figure 4.5.** Structure of Hankel matrices.

## 4.4.2  Identification in the presence of disturbances

The noisy formulation aims to capture input-output relationships generated in an uncertain environment. The output $y$ is the result of the measurable input signal $u$ and a second signal $e$ that is not measurable and represents the uncertainty in the system operation. Thus the inputs of the unknown system are split into two groups. One input group is designated by $u$ and represents information that we have access to, inputs that we generate and measure. The second input group designated by $e$ represents everything that is unknown. Thus $e$ may incorporate all unmeasurable extraneous signals that affect the system, various modeling inaccuracies, errors in the data acquisition phase during which input-output data are collected. In a deterministic setup all signals involved are deterministic signals. In a stochastic formulation $e$ is modeled as a stochastic process. $u$ may be deterministic or stochastic.

Sometimes a more detailed description of the disturbances is required. One such case is illustrated in Fig. 4.6. The box $S$ represents the unknown system. The actual system output $z$ is not directly available. What is actually measured is $y$. The box $F$ models the data acquisition device. $w$ represents disturbances incurred during system operation and is called *process noise*. $v$ models the disturbances inherent in measuring devices. It is referred to as *measurement noise*. The system $F$ is often modeled by a static time invariant system $y(n) = F(z(n), v(n))$. The most popular case is the additive noise model

$$y(n) = z(n) + v(n)$$

***Correlation approach.*** The true model identification approach must incorporate a procedure for the removal of the disturbance. This is often accomplished by correlation techniques, and for this reason it is referred to as correlation approach. An illustration is given next.

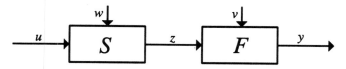

**Figure 4.6.** System operation and data acquisition in an uncertain environment.

***Example 4.5    Identification of stochactic static systems***
Consider a single-input, single-output static polynomial system of the form

$$y(n) = f(u(n)) + v(n) = \sum_{k=0}^{p} a_k u^k(n) + v(n)$$

The measurement noise $v(n)$ is a zero mean stationary process, independent of the input signal $u(n)$. We multiply both sides by $u^i(n)$ and take expectation, ignoring $n$. Independence of $u$ and $v$ implies that

$$E[yu^i] = \sum_{k=0}^{p} a_k E\left[u^{k+i}\right] + E[vu^i] = \sum_{k=0}^{p} a_k m_{k+i}$$

where $m_j = Eu^j$, the $j$th order moment of $u(n)$. Thus we obtain the Hankel system of equations

$$\begin{pmatrix} 1 & m_1 & \cdots & m_p \\ m_1 & m_2 & \cdots & m_{p+1} \\ \vdots & \vdots & \ddots & \vdots \\ m_p & m_{p+1} & \cdots & m_{2p} \end{pmatrix} \begin{pmatrix} a_0 \\ a_1 \\ \vdots \\ a_p \end{pmatrix} = \begin{pmatrix} m_{yu}(0) \\ m_{yu}(1) \\ \vdots \\ m_{yu}(p) \end{pmatrix}$$

where $m_{yu}(i) = E\left[y(n)u^i(n)\right]$.    ∎

**Prediction error identification.** The prediction error identification approach leads to a system design formulation in a manner entirely analogous to the filtering problem. Regarding the class of admissible systems, two useful options suggest themselves. The first option results if we mimick the error free approach. The class of admissible systems consists of models of the form $y = S(u)$. We then search for the one that minimizes a performance index of the error $e(n) = y(n) - S(u)(n)$. Much like filtering we may work with the least squares estimator $E[y(n)|u(n-1), u(n-2), \ldots, u(n-m)]$ or the linear least squares estimator $E^*[y(n)|u(n-1), u(n-2), \ldots, u(n-m)]$. The above class may fail to give satisfactory results in a stochastic setup because the output of the unknown system is not obtained only from $u$ but from the concomitant action of the noise term.

The second option considers a larger class of admissible models that filter both data sequences $u(n)$ and $y(n)$ to form an estimate $\hat{y}(n)$ of $y(n)$. We still come up with a filtering framework, albeit this time we deal with estimators having two inputs $u$ and $y$ and one output $\hat{y}$. $\hat{y}$ will be a noise free version of $y$. For instance, the best linear least squares estimator

$$\hat{y}(n) = E^*[y(n)|u(n-1), y(n-1), \ldots, u(1), y(1)] \tag{4.69}$$

is an affine function of the data. Hence it has the form

$$\hat{y}(n) = \sum_{i=1}^{n-1} h_i(n)y(n-i) + \sum_{i=1}^{n-1} g_i(n)u(n-i) + k(n) \qquad (4.70)$$

It is called *predictor* because it forms an estimate of $y(n)$ on the basis of past input-output information. The coefficients $h_i(n)$, $g_i(n)$, and $k(n)$ are determined by the formulas derived in Section 4.3.1. Explicit dependence on $n$ stresses the fact that the predictor is, in general, time varying. In applications with computationally intensive requirements, adoption of the above estimators is intractable because memory grows with time. Fixed memory filters offer a more reasonable choice. In the time-invariant case, these have the form

$$\hat{y}(n) = \sum_{i=1}^{k} h_i y(n-i) + \sum_{i=1}^{r} g_i u(n-i) = \varphi^T(n)\theta \qquad (4.71)$$

where the so-called regressor vector is given by

$$\varphi(n) = (y(n-1) \quad \cdots \quad y(n-k) \quad u(n-1) \quad \cdots \quad u(n-r))^T \qquad (4.72)$$

and

$$\theta = (h_1 \quad h_2 \quad \cdots \quad h_k \quad g_1 \quad \cdots \quad g_r)^T$$

The number of coefficients is now constant. Working as in the other least squares problems we arrive at a linear system of the form of Eqs. (4.29)–(4.30).

Estimation of the system parameters by prediction error identification can be viewed as a learning process within the class of admissible systems $\Im_f$. Think of $\Im_f$ as the a priori uncertainty about the system behavior. The selection of a specific system removes uncertainty and in this sense achieves learning. As the input is submitted to an admissible system, the output is formed and compared to the desired output. Learning is accomplished by picking the system that makes a given function of the error minimum.

Once learning is achieved and a specific admissible system is chosen to represent the input output data, the system capability to interpret data outside the data set used for learning becomes of interest. This is referred to as *generalization capability*, or *model validation*.

### Example 4.6 Linear and polynomial models
The family of admissible systems consists of FIR filters of given order $m$

$$\hat{y}(n) = \sum_{i=1}^{m} c_i u(n-i) \qquad (4.73)$$

Let $\theta^T = (\, c_1 \; \ldots \; c_m \,)$ be the vector of unknown coefficients. $\hat{y}(n)$ depends linearly on $\theta$

$$\hat{y}(n) = \varphi^T(n)\theta \tag{4.74}$$

where

$$\varphi^T(n) = (u(n-1) \; u(n-2) \; \ldots \; u(n-m)) \tag{4.75}$$

Let $e(n) = y(n) - \hat{y}(n)$. Minimization of the mean squared error $E[e^2(n)]$ leads to the linear system of equations

$$R\theta = d \tag{4.76}$$

where

$$R = E[\varphi(n)\varphi^T(n)], \quad d = E[\varphi(n)y(n)]$$

Analogous results are obtained if the total weighted squared error is employed. If the exponentially forgetting window is used, the optimum parameter $\theta(N)$ is determined by the linear system of equations

$$R(N)\theta(N) = d(N) \tag{4.77}$$

where

$$R(N) = \sum_{n=m+1}^{N} \lambda^{N-n} \varphi(n)\varphi^T(n)$$

$$d(N) = \sum_{n=m+1}^{N} \lambda^{N-n} \varphi(n)y(n)$$

Time delay polynomial models (see Example 3.10) also lead to equations of the form (4.74). The regressor vector carries all input shifts and their products. For a second-order polynomial model it has the form

$$\varphi^T(n) = (u(n-1) \; \ldots \; u(n-m) \; u^2(n-1) \; u(n-1)u(n-2) \; \ldots \; u^2(n-m)) \tag{4.78}$$

Suppose that $y(n)$, $u(n)$ are produced by an FIR filter of order $m$:

$$y(n) = \sum_{i=1}^{m} c_i^* u(n-i) = \varphi^T(n)\theta^* \tag{4.79}$$

If we replace Eq. (4.79) into the right-hand side of (4.76), we obtain

$$d = E[\varphi(n)y(n)] = E[\varphi(n)\varphi^T(n)]\theta^* = R\theta^*$$

Clearly $R$ is always nonnegative. If in addition $R > 0$, we can conclude that $\theta = \theta^*$. Condition $R > 0$ requires that the input be sufficiently rich to excite the unknown system modes. We will explore it further in Chapter 12. Let us consider a more general situation where the data sequences are determined by the linear regression

$$y(n) = \varphi^T(n)\theta^* + v(n) \tag{4.80}$$

Proceeding as in the noise free case, we substitute (4.80) into the right-hand side of (4.76) to obtain

$$R\theta = E[\varphi(n)y(n)] = E[\varphi(n)(\varphi^T(n)\theta^* + v(n))] = R\theta^* + E[\varphi(n)v(n)]$$

We assume that $R > 0$ and that the input $u(n)$ is uncorrelated with the zero mean process $v(n)$. Taking into account the form of the regressor vector (4.75), we conclude that $E[\varphi(n)v(n)] = 0$ and thus that $\theta = \theta^*$. Hence, if the estimates of $R$ and $d$ obtained from the finite record of samples are accurate, the estimated parameter $\theta$ will recover the true parameter $\theta^*$. A MATLAB simulation follows:

```
e = (sig^0.5)*randn(1,L);          % produces L samples of the zero mean Gaussian
                                   % process noise of variance sig
u = rand(1,L);                     % produces L samples of the measurable input
g = poly([1/2,1/4]);
v = filter(1,g,e);                 % noise in the true model
c = [1,3,4];                       % true FIR filter coefficients
y = filter([0,c],1,u) + v;         % output of the true model
n = length(c) + 1;
R = zeros(n,n);
d = zeros(n,1);
for i = n:N                        % solves the normal equations using N samples
   phi = u(i:-1:i-n + 1)';
   R = R + phi*phi';
   d = d + phi*y(i);
end
theta = R\d;
yhat = filter(theta',1,u);         % computes the predicted
                                   % output over the entire data set
ehat = y-yhat;                     % computes the prediction error
plot([L-N + 1:L],ehat(L-N + 1:L))  % plots the last N prediction error samples
```

## 4.5  ADAPTIVE SIGNAL PROCESSING

Adaptive signal processing systems execute a signal processing task in an uncertain and time varying environment. They share the following features:

- The system architecture is fairly simple. Adaptive systems usually operate in a real time environment with stringent computational complexity and storage requirements. In particular, they are finitely parametrizable.

- Adaptive systems are in general time varying systems. Parameter variations reflect the environment changes.
- An adaptive system is equipped with an adaptation mechanism that receives information from the environment and consolidates this knowledge into the system model. At the most primitive level, information is assessed by one or more data sequences. Adaptive algorithms estimate the parameters recursively. Updates are formed as new information data are received. In this sense adaptive schemes combine data acquisition and parameter estimation in a single mode. For this reason they are also called *on-line* methods in contrast to *off-line* or *batch* methods where data are first collected and stored and then processed.

An adaptive algorithm is a discrete system that processes the data sequence $z(n)$ and produces the sequence of estimates $\theta(n)$ at the output. Since each parameter estimate must be computed within consecutive measurement samples, parameter estimation must be carried out in real time. This places stringent requirements on the adaptive algorithm, and issues like computational complexity and storage become critical. To cope with these issues adaptive algorithms must be characterized by counts of operations and memory stages that remain constant and do not grow with time. This feature is offered by state space realizations. Hence an adaptive algorithm is described by the model

$$\phi(n + 1) = a(\phi(n), z(n))$$

$$\theta(n) = c(\phi(n), z(n)) \tag{4.81}$$

The algorithm is determined by the state map $a$, the output map $c$, and the initial seed $\phi_0$. The architecture of an adaptive filter is depicted in Fig. 4.7. The most common adaptive system is the FIR filter with time varying coefficients:

$$\hat{y}(n) = \sum_{i=1}^{p} a_i(n)u(n - i) = \varphi^T(n)\theta(n) \tag{4.82}$$

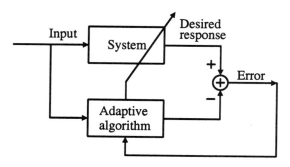

**Figure 4.7.** Block diagram of an adaptive filter.

Several adaptive schemes exist. Most of them rely on the basic principles outlined in Appendix III. The most distinguished representatives are the least mean square (LMS) algorithm and the recursive least squares (RLS) algorithm. The LMS algorithm is a steepest descent method, while the RLS algorithm has a Newton-Raphson philosophy (see Appendix III). These topics are described next.

### 4.5.1 The least mean square (LMS) algorithm

We assume that the output of the adaptive system depends linearly on the time varying parameters

$$\hat{y}(n) = \varphi^T(n)\theta(n) \tag{4.83}$$

The regressor vector is directly determined from the data. Examples are FIR predictors where $\varphi(n)$ is given by Eq. (4.75) or (4.72), and polynomial filtering where $\varphi(n)$ has the form of Eq. (4.78). The LMS algorithm adapts the parameter vector $\theta(n)$ by the steepest descent rule (see Appendix III):

$$\theta(n+1) = \theta(n) - \mu(n)\nabla J(\theta(n)) \tag{4.84}$$

In its primary form it uses the square of the difference between the desired value $y(n)$ and the predicted value $\hat{y}(n)$ as learning criterion:

$$J(\theta(n)) = \frac{1}{2}e^2(n) = \frac{1}{2}(y(n) - \varphi^T(n)\theta(n))^2$$

Then

$$\nabla J(\theta(n)) = -(y(n) - \varphi^T(n)\theta(n))\varphi(n)$$

and Eq. (4.84) takes the form

$$\theta(n+1) = \theta(n) + \mu(n)e(n)\varphi(n)$$

$$e(n) = y(n) - \varphi^T(n)\theta(n) \tag{4.85}$$

where $\mu(n)$ is the stepsize and $e(n)$ is the instantaneous error. The simplest variant of the LMS algorithm engages a small positive constant stepsize. It is instructive to observe the difference between the LMS algorithm and the steepest descent as applied to the solution of the normal equations (4.76). The steepest descent update for the cost

$$\frac{1}{2}E[e^2(n)] = \frac{1}{2}E[(y(n) - \varphi^T(n)\theta)^2]$$

reads

$$\theta(n+1) = \theta(n) + \mu E[e(n)\varphi(n)] \tag{4.86}$$

The term $E[e(n)\varphi(n)]$ is approximated by a time average. The LMS algorithm results if this time average is replaced by the instantaneous variable $e(n)\varphi(n)$. No rigorous justification of this relation is attempted. The convergence analysis of Eq. (4.86) is not difficult. In contrast, the LMS algorithm is much harder and will only be touched upon in Chapter 11.

A popular variant of the LMS algorithm is the normalized LMS. It employs a time varying gain given by

$$\mu(n) = \frac{\mu}{1 + \mu\varphi^T(n)\varphi(n)} \tag{4.87}$$

The above choice is justified in Chapter 11.

### 4.5.2   The recursive least squares (RLS) algorithm

The recursive least squares algorithm sequentially determines the linear regression estimate (4.77). The parameters of the normal equations satisfy the following update ($n$ is used instead of $N$):

$$R(n) = \lambda R(n-1) + \varphi(n)\varphi^T(n), \quad d(n) = \lambda d(n-1) + \varphi(n)y(n) \tag{4.88}$$

It is reasonable to expect that this simple updating pattern is reflected into $\theta(n)$. Indeed, the basic updating formulas (4.146) and (4.147) of the Appendix give

$$\theta(n) = \theta(n-1) + w(n)\alpha^{-1}(n)e(n) \tag{4.89}$$

$$e(n) = y(n) - \varphi^T(n)\theta(n-1) \tag{4.90}$$

$$w(n) = R^{-1}(n-1)\varphi(n) \tag{4.91}$$

$$\alpha(n) = \lambda + \varphi^T(n)R^{-1}(n-1)\varphi(n) = \lambda + \varphi^T(n)w(n) \tag{4.92}$$

where $e(n)$ expresses the error at time $n$ between the desired response or unknown system output $y(n)$ and the estimated output, using the previous estimate $\theta(n-1)$. The vector $w(n)$ is referred to as *Kalman gain*. Equation (4.89) is also written as

$$\theta(n) = \theta(n-1) + w(n)\tilde{e}(n), \quad \tilde{e}(n) = \frac{1}{\alpha(n)}e(n) \tag{4.93}$$

It follows from the alternative expression (4.145) that

$$\tilde{e}(n) = y(n) - \varphi^T(n)\theta(n) \tag{4.94}$$

where $\tilde{e}(n)$ is the error committed by the estimate at time $n$. Of course $\tilde{e}(n)$ cannot be computed by Eq. (4.94) because $\theta(n)$ is not available. For emphatic purposes $e(n)$ is referred to as the a priori error, and $\tilde{e}(n)$ as the a posteriori error.

It is reasonable to examine if the basic identity (4.146) applies once again for the computation of $w(n)$ via (4.91). Unfortunately, no direct updating expression for the regressor vector $\varphi(n)$ exists. $\varphi(n)$ does not relate to its shift by a simple additive law of the form (4.88). The easiest thing to do then is to update the inverse of $R(n)$, $P(n)$, in accordance with the matrix inversion lemma applied to (4.88). Then

$$P(n) = \frac{1}{\lambda}[P(n-1) - w(n)\alpha^{-1}(n)w^T(n)], \quad w(n) = P(n-1)\phi(n) \qquad (4.95)$$

The RLS algorithm consists of Eqs. (4.89), (4.90), (4.92), and (4.93). It has the state space form (4.81). The state consists of $P(n)$ and $\theta(n)$. Due to symmetry only the upper triangular part of $P(n)$ is needed. The memory requirements are proportional to $p(p-1)/2 + p$. The main computational load is caused by the update of $P(n)$, which requires $O(p^2)$ multiplications and additions. Savings by an order of magnitude in operations count and memory can be achieved by the fast RLS algorithms discussed in Exercise 4.18.

### *Example 4.7    Stock index forecasting*
The following MATLAB functions implement the LMS and the RLS algorithms. Figure 4.8 shows the values of the Athens stock exchange index over a

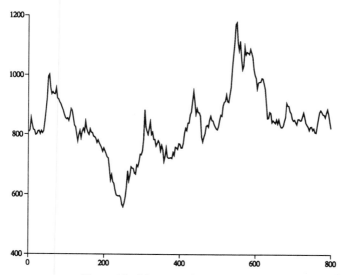

**Figure 4.8.** Athens stock exchange index.

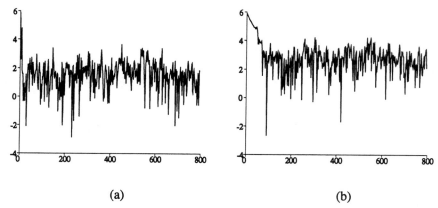

**Figure 4.9.** (a) RLS squared error; (b) LMS squared error.

period of three years. One-day-ahead prediction is performed by the LMS and the RLS algorithms. The total squared error for each experiment is plotted in Fig. 4.9. In both cases, the filter order is 6. The estimated outputs and the actual data series are depicted in Fig. 4.10. The RLS algorithm converges faster at the expense of increased computational complexity.

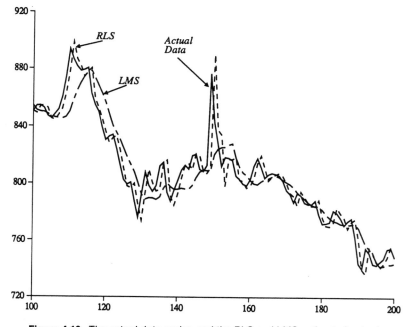

**Figure 4.10.** The actual data series and the RLS and LMS estimated outputs.

```
function [theta,Sqe] = lms(mi,M,N,u,y)
% implementation of the LMS algorithm
% theta     : parameter vector
% mi        : step size parameter
% M         : parameter vector length
% N         : length of the input data vector
% u         : input data vector
% y         : desired response

theta = zeros(1,M); % initilization
Sqe = [];           % squared error vector

for j = M:N
  phi = fliplr(u(j-M + 1:j));
  e = y(j)-theta*phi';
  Sqe = [Sqe e^2];
  theta = theta + mi*e*phi;
end;

function [theta,sqe] = rls(y,u,N,M,lamda,m)

% Implementation of the RLS algorithm
% theta     : parameter vector
% lambda    : forgetting factor
% M         : parameter vector length
% N         : length of data record
% u         : input data vector
% y         : desired response
% m         : initialization parameter

theta = zeros(1,M);
P = m*eye(M);
sqe = []; % squared error vector

for j = M:N
  phi = fliplr(u(j-M + 1:j));
  e = y(j)-theta*phi';
  sqe = [sqe e^2];
  w = (P*phi')/(lamda + phi*P*phi');
  theta = theta + w'*e;
  P = (1/lamda)*(P-w*phi*P);
end;
```

■

## 4.6  CODING AND COMPRESSION

Coding systems convert signals such as speech, images, and video into a compressed digital representation. Economy in bits is essential for a variety of signal processing applications such as storage, transmission, encryption, and pattern classification. Analog-to-digital conversion is a primary part of a coding system if the original signal is analog like speech or music. The coder tries to reduce the signal bit rate so that it meets the capacity requirements of the channel with acceptable quality of reproduction and with minimum implementation cost. For instance, a digitized speech at 64 kb/s is converted by a coder to a form suitable for a 9.6 kb/s channel. Coding techniques are divided into lossy and lossless schemes. In lossless compression, also called *noiseless coding* or *entropy coding*, the original signal is perfectly recovered from its compressed form. Compression rates of 2:1 are achieved. Lossless compression is used in storage and transmission of sensitive data where a single error can have catastrophic consequences. Computer programs and medical X rays are examples of such signals. Lossless coding algorithms such as Huffmann codes are variable length codes. They achieve compression by using fewer symbols for the most frequent letters and more symbols for the less frequent letters. In this way the average length of the codeword is smaller, and the storage capacity is reduced. Lossy compression achieves higher compression rates by allowing a loss in quality. Such losses are inevitable if the original signal is analog. Lossy coding systems enable a better exploitation of available resources. Cellular telephones, terestrial, and satellite broadcasting of high-definition television (HDTV) require lossy compression to reduce the large traffic of data. Multimedia communications involve high volumes of information that will exhaust the channel capacity without efficient compression.

Coding systems are grouped into waveform coders, transform coders, and model coders. Waveform coders operate directly on the discrete form of the signal or its sampled version if the original waveform is analog. Examples are pulse code modulation (PCM), differential pulse code modulation (DPCM), delta modulation, and vector quantization (VQ) in its primary form. VQ is a straightforward extension of scalar quantization. Transform coders use a transformed representation of the signal. Examples include the Karhunen-Loève and the discrete cosine transform (DCT). The inputs to a model coder are the parameters of a model that represents the signal source. An example is linear predictive coding with vector quantization (LPC-VQ) for speech coding. Speech is modeled as an autoregressive filter driven by either a white noise source (e.g., for unvoiced sounds like "sss" and "sh"), or a periodic pulse train (e.g. for voiced vowel sounds). The parameters of the model are determined by the Levinson algorithm and coded by a vector quantizer that relies on the so-called Itakura-Saito distortion rather than the average distortion. Waveform coding is introduced in this section. Transform coding and subband coding are discussed in Chapter 14.

### 4.6.1  Design of scalar quantizers

Quantizers were introduced in Section 3.11. Quantization design is discussed next. If the input to an A-D converter is a stochastic signal with probability densities $f_{t_1,\ldots,t_k}(x_1,\ldots,x_k)$, the corresponding probability densities of the uniformly sampled discrete signal are

$$f_{n_1,\ldots,n_k}(x_1,\ldots,x_k) = f_{n_1 T_s,\ldots,n_k T_s}(x_1,\ldots,x_k)$$

In particular, the mean and covariance sequences of the sampled signal are

$$m(n) = E[x(nT_s)] = m(nT_s), \quad C(n_1, n_2) = \text{cov}[x(n_1 T_s), x(n_2 T_s)]$$

It follows that if the analog waveform is stationary or wide sense stationary, the uniformly sampled signal inherits the same property. Suppose that the input to the quantizer $x(n)$ is stationary. All first-order densities of $x(n)$ collapse to a single density $f(x)$. We thus model the quantizer input as a random variable with density function $f(x)$. A reasonable performance criterion is the minimization of the mean squared quantization error

$$D = E[(x - Q(x))^2] \tag{4.96}$$

where $D$ is also called *average distortion*. Since $Q$ is characterized by the partition cells $R_i$ and the output levels $y_i$, $1 \le i \le N$, we can write $D = D(y_i, R_i)$ as

$$D = \int_{-\infty}^{\infty} (x - Q(x))^2 f(x) dx = \int_{\bigcup R_i} (x - Q(x))^2 f(x) dx$$

$$= \sum_{i=1}^{N} \int_{R_i} (x - y_i)^2 f(x) dx$$

In the case of regular quantizers, we have

$$D = \sum_{i=1}^{N} \int_{x_{i-1}}^{x_i} (x - y_i)^2 f(x) dx$$

Minimization of the average distortion $D$ is taken with respect to all partitions having a given number of cells $N$ and with respect to all reproduction values $y_i$. This is a nonlinear problem. In the case of regular quantization, the cells are uniquely specified by the decision levels $x_i$. In this case we may differentiate with respect to $x_i$ and $y_i$ and set the derivatives equal to zero. The resulting equations are nonlinear and require iterative algorithms for their solution such as the steepest descent and the Gauss-Newton algorithms described in Appendix III. In the sequel we treat the general nonregular case. We derive important necessary conditions for optimality using basic least squares estimation arguments. Simultaneous minimization of the average distortion with respect to the partition cells and the reconstruction levels is difficult. The situation becomes simpler if it is tackled in two steps. First, the reconstruction values are kept fixed but arbitrary, and optimization with respect to the partition cells is carried out. In the second step the partition is kept fixed, and the optimum codebook is found. The encoder-decoder decomposition of the quantizer provides an elegant interpretation of the two-step procedure: In the first step, the decoder structure is fixed and the optimal encoder is determined. In the second step, the encoder is given and the decoder is optimized. Stated differently, the original problem

$$\min_{R_i, y_i} D(y_i, R_i)$$

is set aside and the simpler problem

$$\min_{y_i} \min_{R_i} D(y_i, R_i)$$

is pursued. Let us fix the codebook $C = \{y_1, y_2, \ldots, y_N\}$. Let $Q$ be an arbitrary quantizer with codebook $C$ and $d(x, y) = (x - y)^2$. The following obvious inequality holds

$$d(x, Q(x)) \geq \min_{y_i \in C} d(x, y_i)$$

Hence

$$\int_{-\infty}^{\infty} d(x, Q(x)) f(x) dx \geq \int_{-\infty}^{\infty} \min_{y_i \in C} d(x, y_i) f(x) dx$$

The right-hand side of the above inequality provides a lower bound for the average distortion of any quantizer $Q$ with given codebook $C$. If a quantizer is found that attains this lower bound, it is optimum among all quantizers having the same codebook $C$. An optimum choice is

$$Q(x) = y_i, \quad \text{if } d(x, y_i) \leq d(x, y_j) \text{ for all } j \neq i$$

The above quantizer assigns $y_i$ to all input values that are closer to $y_i$ than any other reconstruction level. It constitutes the *nearest neighbor quantizer*.

The above argument remains valid for arbitrary positive distortion functions $d(x, y)$, not only the squared error. The partitions are readily determined if the squared error is used. Indeed, the partition cell of any $y_i$ different from the endpoints $y_1, y_N$ is the interval $(x_{i-1}, x_i)$, where $x_{i-1}$ is the midpoint of $y_{i-1}$ and $y_i$

$$x_{i-1} = \frac{y_{i-1} + y_i}{2} \tag{4.97}$$

To avoid pathological cases the reconstruction levels must reside in intervals of positive probability.

In the second step of the method, the partition is fixed, and the distortion is minimized with respect to the output levels $y_i$. We write the quantizer map as a linear combination of indicator functions (see Section 3.11)

$$Q(x) = \sum_{i=1}^{N} y_i I_{R_i}(x)$$

If the partition cells are given, the indicator functions $I_{R_i}$ are known. Minimization of

$$E[(x - Q(x))^2] = E[(x - \sum_{i=1}^{N} y_i I_{R_i}(x))^2]$$

with respect to $y_i$ is a linear least squares estimation problem and can be tackled with the method of Section 4.2.3. There the affine least squares estimate $\hat{X} = AY + b$ was derived. Here we are interested in the truly linear estimate $\hat{X} = AY$. Working as in the affine case, we obtain $A = R_{XY} R_{YY}^{-1}$, where

$$R_{XY} = E[XY^T], \quad R_{YY} = E[YY^T]$$

The difference between linear and affine estimation is that the former involves correlations rather than covariances. We apply the above to the quantization problem, using the correspondence $X \to x$, $Y \to (I_{R_1}, \ldots, I_{R_N})^T$ and $A \to (y_1, \ldots, y_N)$. $R_{YY}$ is diagonal. Indeed, since each $I_{R_i}$ is a discrete random variable, we have

$$E[I_{R_i}] = 1P[I_{R_i} = 1] + 0P[I_{R_i} = 0] = P(R_i) = P_i$$

Note that $I_{R_i} I_{R_j} = I_{R_i \cap R_j}$. If $i \neq j$, $R_i \cap R_j$ is the empty set, $I_{R_i \cap R_j} = 0$, and $E[I_{R_i \cap R_j}] = 0$. On the other hand, if $i = j$, $E[I_{R_i} I_{R_j}] = P_i$. In a similar fashion

$$(R_{XY})_i = E[x I_{R_i}] = \int_{R_i} x f(x) dx$$

Thus

$$y_i = \frac{E[x I_{R_i}]}{P_i} = E[x | R_i] = \frac{\int_{R_i} x f(x) dx}{\int_{R_i} f(x) dx}$$

The latter expression states that the quantizer output $y_i$ is the centroid, or the center of mass of $x$, constrained on the cell $R_i$. For regular quantizers it becomes

$$y_i = \frac{\int_{x_{i-1}}^{x_i} x f(x) dx}{\int_{x_{i-1}}^{x_i} f(x) dx} \tag{4.98}$$

Some interesting properties of the quantizer are discussed in Exercise 4.23.

### Example 4.8  Optimal quantizers for uniformly distributed inputs
Let us assume that the input to the quantizer $x$ has a uniform distribution on a symmetric interval of length $B$. The nearest neighbor rule implies the regularity

condition (4.97), while the centroid condition applied on the granular cells gives

$$y_i = \frac{x_i^2 - x_{i-1}^2}{2(x_i - x_{i-1})} = \frac{1}{2}(x_{i-1} + x_i) \tag{4.99}$$

Thus each reconstuction, level $y_i$ is located at the midpoint of the decision levels $x_{i-1}$ and $x_i$. Combining (4.97) and (4.99), we get $y_i - y_{i-1} = x_i - x_{i-1} = \Delta$. The resulting optimal quantizer is the uniform quantizer with quantization step

$$\Delta = \frac{B}{N} \tag{4.100}$$

The resulting average distortion is

$$D = \sum_{i=1}^{N} \frac{1}{B} \int_{x_{i-1}}^{x_i} (x - y_i)^2 dx = \frac{1}{B} \sum_{i=1}^{N} \frac{\Delta^3}{12} = \frac{N\Delta^3}{12B} = \frac{\Delta^2}{12}$$

If $\sigma_x^2$ is the variance of the quantizer input, the performance of a quantizer is often assessed by the signal-to-quantization error ratio (SNR):

$$SNR = 10 \log_{10} \frac{\sigma_x^2}{D}$$

expressed in decibels (dB). The dB scale is based on the correspondence

$$dB = 20 \log_{10} A$$

Indicative values are $0dB \leftrightarrow A = 1$, $20dB \leftrightarrow A = 10$, $40dB \leftrightarrow A = 100$, $-20dB \leftrightarrow A = 0.1$, while $1dB \leftrightarrow A \approx 1.12$. The use of logarithmic scale gives a better resolution when the range of $A$ is large or a small interval round zero.

In the case of uniformly distributed inputs the input variance is $B^2/12$. Thus the signal to noise ratio becomes

$$SNR = 10 \log_{10} \left(\frac{B}{\Delta}\right)^2 = 10 \log_{10} N^2 = 20 \log_{10} N$$

Let $R$ denote the quantizer rate $R = \log_2 N$. Then the signal-to-noise ratio is 6 dB per bit. In other words, the SNR increases 6 dB for each additional bit used for quantization. The above conclusion takes into account the granular noise only and neglects the overload noise. ∎

### 4.6.2 Compandors and nonuniform quantization

*Pulse code modulation (PCM)* is the simplest waveform coding method in which the signal value is quantized by a uniform quantizer. Despite the ease of

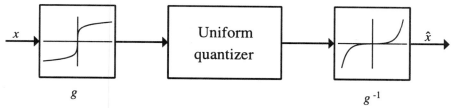

**Figure 4.11.** Nonuniform quantization via a compressor-expandor pair.

implementation, uniform quantizers offer inferior performance when compared to nonuniform quantization. We demonstrated in the previous example that uniform quantization results as an optimal procedure when the probability density of the input samples is uniform. Real signals like speech and images do not follow a uniform distribution. Speech signals present small amplitudes more frequently than large amplitudes. A nonuniform quantizer can offer improvements in performance if it provides more closely spaced levels at small signal amplitudes and more widely spaced levels at large signal amplitudes. It can be shown that a nonuniform quantizer can be realized by the cascade structure of Fig. 4.11. $g$ is called a *compressor* and its inverse $g^{-1}$ is called an *expandor*. The combined compressor-expandor pair is called *compandor*. $g$ features a small derivative for large input values and thus compresses large amplitudes. A typical example is the logarithm function $g(x) = \log x$ with derivative $1/x$. Two compandors that have been widely used in practice are the $\mu$ law and the $A$ law defined by

$$g_\mu(x) = \mathrm{sgn}(x)\frac{\log(1 + \mu|x|)}{\log(1 + \mu)}, \quad g_A(x) = \mathrm{sgn}(x)\frac{1 + \log A|x|}{1 + \log A}$$

with some modifications to avoid unbounded behavior at the origin. The parameters $\mu$ and $A$ reflect the amount of compression.

### 4.6.3 Predictive quantization

The primary goal of predictive quantization is to achieve compression by removing redundancy in the signal. Redundancy is related to predictability. The more predictable a signal is the more redundant information it carries. Existing popular compression techniques employ a linear predictor to extract the predictive signal attributes. The resulting prediction error sequence is freed from linear redundance and lends itself to more efficient digital representation. Predictive quantization relies on the following simple observation deduced from Fig. 4.12. The average distortion remains the same if a signal is subtracted before and after quantization. Indeed,

$$E[(x - \hat{x})^2] = E[(e + v - \hat{e} - v)^2] = E[(e - \hat{e})^2]$$

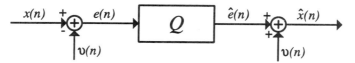

**Figure 4.12.** Difference quantization.

The signal $v(n)$ is typically obtained as the output of a fixed memory predictor operating on $x(n)$:

$$v(n) = E^*[x(n)|x(n-1), x(n-2), \ldots, x(n-m)] = \sum_{i=1}^{m} a_i x(n-i)$$

The power of the prediction error signal $e(n)$ is smaller than the signal power, and the signal to quantization noise ratio is improved. The predictor is determined off-line using very long records of the signal $x(n)$ and the techniques of Section 4.3.3. The resulting system is then built into the quantizer. Returning to Fig. 4.12, we observe that the sequence $v(n)$ must also be available at the decoder. This requires transmission of the past signal values $x(n-1), \ldots,$ $x(n-m)$ as side information. Our intention is to transmit differences rather than the signal itself to achieve better compression rates. So clearly we do not want to send signal samples. The difficulty is overcome if the predictive estimate $v(n)$ is formed by the previously estimated values $\hat{x}(n-k)$, rather than by the actual values $x(n-k)$. Thus $v(n)$ is given by

$$v(n) = \sum_{i=1}^{m} a_i \hat{x}(n-i)$$

The above approximation is reasonable because the primary goal of the predictive quantizer is to remove long-term redundancies rather than perform optimal sample-by-sample prediction. The resulting method is known as differential pulse code modulation (DPCM) and is illustrated in Fig. 4.13. The encoder produces the index $i$ as well as the reconstructed value $\hat{e}(n)$ needed for the determination of the predicted value $v(n)$. $P$ denotes the predictor used to compute $v(n)$. The index $i(n)$ is transmitted or stored. At the receiving end, the decoder $Q^{-1}$ reconstructs $\hat{e}(n)$ from the received index $i(n)$. Using the previously estimated values $\hat{x}(n-k)$ and the same predictor filter $P$, it determines the current estimate $\hat{x}(n)$.

Improved performance is offered by the DPCM variant of Fig. 4.14. The predictor is a two-channel FIR filter combining previously estimated values as well as previously estimated differences. It is expressed as

$$v(n) = \sum_{i=1}^{m} a_i \hat{x}(n-i) + \sum_{i=1}^{r} b_i \hat{e}(n-i)$$

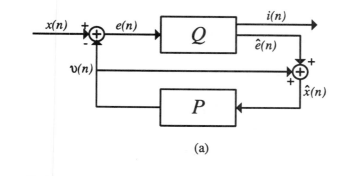

(a)

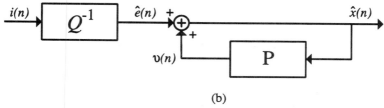

(b)

**Figure 4.13.** The DPCM method: (a) Encoder; (b) decoder.

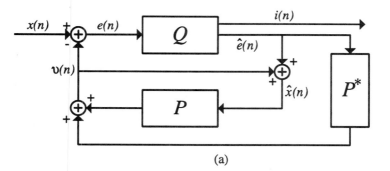

(a)

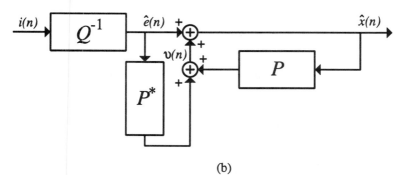

(b)

**Figure 4.14.** A DPCM variant: (a) Encoder; (b) decoder.

The AR part is represented by the box $P$, while the MA part is represented by the box $P^*$.

The off-line computation of the predictor coefficients in the above methods is best suited for stationary signals. Speech and images have nonstationary characteristics. Superior performance is accomplished if the predictor coefficients adapt to the time varying statistics of the signal. One such adaptive DPCM (ADPCM) system was adopted by CCITT as an international standard for speech coding at 32 kb/s. It employs the structure of Fig 4.14 with a 4-bit quantizer and a logarithmic compressor. The coefficients are adapted by the LMS algorithm.

### 4.6.4    Vector quantization

A scalar quantizer approximates each value of a single-channel signal by a number in a finite set. A vector quantizer of dimension $k$ operates on blocks of signal values $(x(n) \quad x(n-1) \quad \cdots \quad x(n-k+1))$ in its basic form. As a consequence it is described by a map $Q : R^k \rightarrow C$ where $C = \{y_1, y_2, \ldots, y_N\}$ and $y_i$ are vectors in $R^k$. The cells of the quantizer

$$R_i = \{x \in R^k : Q(x) = y_i\}$$

define a partition of $R^k$ into $N$ nonempty disjoint subsets. The forms and shapes of the cells in dimensions higher than one offer considerably more flexibility than the scalar case. This is one of the reasons that vector quantization offers superior performance as compared to scalar quantization. A statistical justification is provided by rate distortion theory, which is beyond the scope of this book.

The *resolution* or *rate* of a vector quantizer is $r = \log_2 N/k$, namely the number of bits per vector component. The cells of a *regular quantizer* are convex and contain the corresponding reproduction points, also called *code vectors* or *codewords*. A set $A$ is convex if for any points $x$, $y$ in $A$, the line segment $\lambda x + (1 - \lambda)y$, $0 \leq \lambda \leq 1$, is entirely contained in $A$. In the case of the real line, convex sets are intervals. Hence a regular quantizer reduces to a regular scalar quantizer. Implementation of regular scalar quantizers was discussed in Section 3.11. It relied on the representation

$$Q(x) = \sum_{i=1}^{N} y_i I_{R_i}(x) \tag{4.101}$$

and on the realization of each indicator function $I_{R_i}(x)$ by comparators. In higher dimensions the indicator function of a general convex set is not readily implemented, since convex sets can have intricate shapes. For this reason more restrictive structures are considered. A *polytopal* or *polyhedral quantizer* is a regular quantizer whose cells are polytopes. A polytope is a finite intersection of

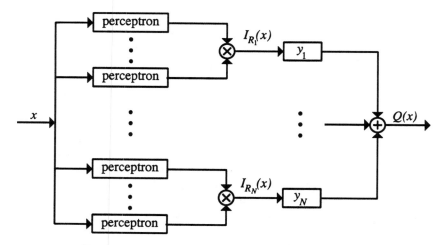

**Figure 4.15.** Realization of polytopal cells by neural networks.

half spaces of the form

$$\{x \in R^k : w_i^T x + \theta_i \geq 0\} \tag{4.102}$$

A polytopal quantizer is implemented by a multilayer neural network as follows: Let $R_j$ be one of the polytopal cells. Then $R_j = \cap_{i=1}^{L_j} H_i$, where $H_i$ is given by (4.102). A perceptron is associated to each halfspace $H_i$. It has weight vector $w_i$, threshold $\theta_i$, and the step function as activation function. The outputs of the $L_j$ perceptrons are broadcast to an AND gate. The resulting architecture implements the indicator function $I_{R_j}(x)$ and is depicted in Fig. 4.15.

A vector quantizer is decomposed into an encoder $\mathcal{E}$ and a decoder $\mathcal{D}$, $Q = \mathcal{D}\mathcal{E}$, exactly as in the scalar case. The design of vector quantizers using the least mean square error as distortion measure is entirely analogous to the scalar case. Suppose that the quantizer input is a random vector $x$ with density $f(x)$. Let $d(x, Q(x))$ be a distortion function, typically the Euclidean norm

$$d(x, Q(x)) = |x - Q(x)|_2^2 = (x - Q(x))^T (x - Q(x))$$

We seek to determine a quantizer $Q$ that minimizes the average distortion

$$D = E[d(x, Q(x))] = \int d(x, Q(x)) f(x) dx$$

It follows exactly as in the scalar case that if the code vectors $y_i$ are fixed, the optimal partition cells satisfy the *nearest neighbor rule*, namely

$$d(x, Q(x)) = \min_{y_i \in \mathcal{C}} d(x, y_i)$$

and hence

$$Q(x) = y_i \quad \text{if} \quad d(x, y_i) \leq d(x, y_j) \qquad \text{for all } j \qquad (4.103)$$

If the partition is fixed, Eq. (4.101) leads to a linear regression and the codevectors are determined by the centroid condition

$$y_i = E[x | I_{R_i}(x)] = \frac{E[x I_{R_i}(x)]}{P(R_i)}$$

The centroid condition requires knowledge of the input density, hardly available. Typically a collection $X$ of vectors $x_i$, $1 \leq i \leq M$, is given. Estimates of the input distribution function are calculated, and the centroids are estimated by numerical integration. Alternately, the design can be directly formulated in terms of the data set $X$ as follows: The quantizer seeks to minimize the average distortion

$$D = \sum_{i=1}^{M} d(x_i, Q(x_i)) = \sum_{i=1}^{N} \sum_{x_j \in R_i} d(x_j, Q(x_j)) = \sum_{i=1}^{N} \sum_{x_j \in R_i} d(x_j, y_i) \qquad (4.104)$$

over all partitions of the data set $X$ and over all codevectors $y_i$. Clearly

$$\min_{y_i, R_i} \sum_{i=1}^{M} d(x_i, Q(x_i)) \leq \min_{y_i} \min_{R_i} \sum_{i=1}^{M} d(x_i, Q(x_i))$$

We derive necessary conditions for the right-hand side of the above inequality. Fix the code $\mathcal{C} = \{y_1, \ldots, y_N\}$. Let

$$d(x, \mathcal{C}) = \min_{y_i \in \mathcal{C}} d(x, y_i)$$

Then $d(x_i, Q(x_i)) \geq d(x_i, \mathcal{C})$. Therefore

$$\sum_{i=1}^{M} d(x_i, Q(x_i)) \geq \sum_{i=1}^{M} d(x_i, \mathcal{C})$$

The lower bound is attained if $Q(x)$ satisfies the nearest neighbor condition (4.103). The cells are described by the rule

$$R_j = \{x \in X : d(x, y_j) \leq d(x, y_i), \text{ for al } i\} \qquad (4.105)$$

Next the partition cells are fixed and the $\ell_2$ norm is employed as distortion function. Minimization of

$$\sum_{i=1}^{N} \sum_{x_j \in R_i} (x_j - y_i)^T (x_j - y_i)$$

with respect to $y_i$ leads to

$$y_i = \frac{1}{k(R_i)} \sum_{x_j \in R_i} x_j \tag{4.106}$$

where $k(R_i)$ is the number of data vectors in $R_i$. In other words, $y_i$ is the mean of the cell $R_i$. The nearest neighbor condition and the centroid condition as described above are coupled. Hence they cannot be directly applied to determine the optimum quantizer. The obvious thing to do is to handle them in an iterative fashion. Given some initial values for the codevectors, the cells are determined by the nearest neighbor rule. Then new codevectors are computed by the centroid condition. The process is repeated until the average distortion drops below an acceptable level. This is the *k-means algorithm*, also referred to as *generalized Lloyd's algorithm*. It is summarized in Table 4.2.

It follows from the nearest neighbor rule and the centroid condition that the $k$ means algorithm is cost decreasing; that is, the average distortion satisfies $D_{m+1} \leq D_m$. $D_m$ stabilizes after a finite number of steps because the number of partitions of $X$ is finite and monotonicity of the cost prevents the algorithm from returning to an already used partition.

Initialization of the $k$ means algorithm can be done in several ways. The simplest is to take the first $N$ vectors from the data set as codevectors. If these are highly correlated, a better way is to pick vectors periodically. Clustering techniques can also be employed (see also the discussion of Section 4.9.4). One potential problem is that a codevector $y_i$ is associated with an empty cell $R_i$. It may happen that the nearest neighbor rule distributes all training vectors to the remaining codewords. To cope with the empty cell problem and yet maintain the given number of cells $N$, splitting is applied. The cell with the highest number of training vectors is assigned two codevectors and then split into two cells. The empty cell is deleted.

Another issue that needs to be addressed is the occurrence of ties in the nearest neighbor rule. It may happen that a training vector lies the same distance from

---

**Table 4.2    The $k$ means algorithm**

---

1    Initialization step, $m = 1$. Specify initial codevectors $y_1^1, \ldots, y_N^1$. Give initial distortion a large value.
2    For $m = 1 : M_{max}$. Given the code vectors $y_i^m$, create new cluster sets $R_i^m$ by the nearest neighbor rule (4.105).
3    Compute new code vectors $y_i^{m+1}$ using the centroid condition (4.106).
4    Compute the average distortion

$$D_{m+1} = \sum_{i=1}^{N} \sum_{x_j \in R_i} (x_j - y_i)^T (x_j - y_i)$$

If $(D_m - D_{m+1})/D_m$ is small, stop.

---

more than two codewords. One way to handle this problem is to assign such a vector arbitrarily to one of the equidistant codewords and after the algorithm terminates, reassign the training vector to another equidistant codeword and iterate the algorithm once again.

### 4.6.5 Image vector quantization

The application of vector quantization to images is straightforward. Each still image is viewed as a block matrix. The entries are rectangular blocks of fixed size, typically $4 \times 4$. Each such block is an input to the quantizer. If the intensity image is partitioned into $4 \times 4$ blocks the quantizer has dimension 16. The training sequence consists of the blocks of several images. Images represented by

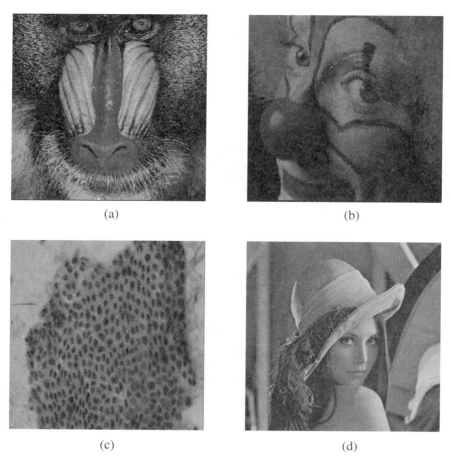

(a)

(b)

(c)

(d)

**Figure 4.16.** Images used for vector quantization.

(a)

(b)

**Figure 4.17.** (a) Original image and (b) image obtained by the vector quantizer of size 64 trained by the three images of Fig. 4.16.

8 bits per pixel can be compressed without perceptible loss of picture quality down to less than 1 bit per pixel by vector quantization schemes.

***Example 4.9    Illustration of image vector quantization***
The images of Fig.. 4.16 have spatial resolution $256 \times 256$ pixels and 256 (8 bits) quantization levels. They are partitioned into $4 \times 4$ blocks. The $k$ means algorithm with the mean square distortion measure is used to design a quantizer of dimension 16. The training set consists of the blocks of the 3 images (top left, top right, and bottom left). Lenna is used for performance assessment (bottom right). Vector quantizers of dimension 16 and size 64, 32, and 16 are constructed. The rate of a vector quantizer of size 64 is $\log_2 64/16 = 0.375$ bits per pixel. It compresses the original image from 8 bits to 0.375 bits per pixel. Likewise a quantizer of size 16 has a rate of $4/16 = 0.25$ bits per pixel. Vector quantized Lenna at 0.375 bits per pixel is shown in Fig. 4.17. ■

## 4.7  ERROR CONTROL CODING

### 4.7.1  The digital communication system

Error control coding reinforces a signal with additional information to combat the errors occurring during transmission and storage. The general idea is illustrated in Fig. 4.18. The source coder converts the original waveform (speech, music, ASCII data, image, video) into a sequence of bits. If the waveform is analog, sampling and quantization are performed by the source encoder as part of the analog to digital conversion. The aim of the source encoder is to produce an economical representation of the signal along the lines of the

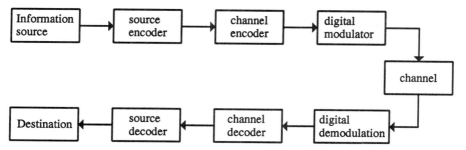

**Figure 4.18.** The digital communication system.

previous section. The channel encoder fragments the information sequence into blocks of length $k$ and converts each block $u$ into a codeword $v$ of length $n$. $n$ is always greater than $k$. The difference $n - k$ reflects the surplus information intended to ensure reliable transmission and storage. The coded sequence $v$ is transformed by the modulator (or writing unit) prior to transmission or storage into an analog waveform of given duration. If the coded sequence is binary, the modulator outputs one analog waveform of duration $T$ seconds for 0 and another analog waveform of duration $T$ for 1. Analog transmission is required by coaxial cables, optical fibers, and microwave links. It is also required by storage media like magnetic tapes and optical memory units.

Transmission media are classified as guided (twisted pair, coaxial cable, optical fiber) and unguided (propagation through air, vacuum, seawater). With guided media the electromagnetic waves are guided along an enclosed physical path, and transmitters/receivers are directly attached to the medium. In an unguided medium transmission and reception are achieved by means of an antenna. In the omnidirectional case the transmitting antenna broadcasts a signal in all directions. The signal is received by many antennas. In the directional configuration, connection is point to point. The transmitting antenna emits a focused electromagnetic beam, and the receiving antenna receives it. Point-to-point unguided transmission is effected in the microwave range. Examples include terrestrial links and satellite communication. In satellite communication, data are not transfered directly from transmitter to receiver but are relayed via satellite.

Transmission through a channel is subject to various impairments. Attenuation and noise are the most important. As an electromagnetic signal is transmitted along a medium, it gradually becomes weaker at greater distances: This is referred to as *attenuation*. In digital communication systems attenuation is coped with regenerative repeaters. Repeaters are placed at intermediate points. Each repeater receives the signal, recovers the digital stream, and transmits a new digital signal. In this way accumulation of distortion is avoided. Noise represents unwanted electromagnetic energy. It takes the form of thermal noise, intermodulation noise, crosstalk, and impulse noise. Thermal noise results from thermal agitation of electrons in a conductor. It is present in all electronic devices and transmission media and is a function of temperature. It is often modeled as a white noise process. Intermodulation noise is mainly caused by nonlinearities in the transmitter, receiver, or transmission system. Crosstalk is an unwanted coupling between signal paths. Impulse noise consists of spikes of short duration and relatively high amplitude. It is caused by external electromagnetic disturbances, such as lightning and faults and flaws in the communication system. It is a primary source of error in digital transmission.

As a result of channel degradation the signal arriving at the receiving end or retrieved from the storage medium is no longer the same. To recover the transmitted signal, the receiver performs several sophisticated signal processing functions. The demodulator (or reading unit) converts each analog waveform of duration $T$ into a character from a finite alphabet. The channel decoder converts the *received sequence* $r$ into the estimated sequence $\hat{u}$. Finally, a source decoder converts the sequence $\hat{u}$ into the original signal form and delivers it to the user.

### 4.7.2 Channel encoding

Channel encoders that have attracted most attention fall within two classes, block codes and convolutional codes. An $(n, k)$ *block encoder* over the field with $N$ elements, $F$, transforms each information block $u = (u_0, u_1, \ldots, u_{k-1})$, $u_i \in F$, into the codeword $v = (v_0, v_1, \ldots, v_{n-1})$ by a static time invariant digital system.* Hence the input output expression of the channel encoder is

$$v = G(u)$$

Block channel encoders are implemented by combinatorial circuits. A standard requirement is that the encoder is a 1–1 mapping, ensuring that each codeword results from a unique message. Since there are $N^k$ possible message blocks the number of codewords is also $N^k$. The set of $N^k$ codewords of length $n$ is called an $(n, k)$ block code and is denoted by $C$. Since $G$ is 1–1, it has a left inverse $\tilde{G}$. The inverse circuit $\tilde{G}$ is used at the receiving end to reconstruct the message from the codeword, $u = \tilde{G}(v)$. *Systematic coders* form a convenient class of block coders that feature an extremely simple inverse circuit $\tilde{G}$. A systematic coder has the form

$$v = G(u) = (\tilde{v} \quad u)$$

The message appears unaltered in the codeword, occupying the last $k$ bits. The job of the encoder is to choose the $n - k$ digits of $\tilde{v}$, called *parity check* digits, so that the transmitted information is immune to noise. The inverse of the systematic encoder simply picks the last $k$ digits and is described by the

---

* Following the standard practice in the coding literature, we use the symbol $n$ for code blocklength rather than time. Moreover we represent vectors by rows.

projection

$$\tilde{G}(v) = v \begin{pmatrix} 0 \\ I_k \end{pmatrix}$$

In contrast to block codes, convolutional encoders are encoders with memory. In a $(n, k, m)$ *fixed encoder* the codeword at time $i$, $v_i$, is a vector of length $n$ that depends on the present and $m$ past message blocks of length $k$. It has the form

$$v_i = G(u_i, u_{i-1}, \ldots, u_{i-m})$$

The parameter $m$ specifies the memory of the encoder. Since $G$ can always be represented by a polynomial function, the encoder has the structure of a polynomial filter over a finite field. It is implemented by a register and an interconnection of AND and XOR gates. *Convolutional codes* result when $G$ is linear. They are represented by FIR filters. Decoding of convolutional codes is discussed in Chapter 14.

### 4.7.3   Channel decoding

Next we concentrate on the channel decoder. We assume that both the codeword and the received sequence are binary. We say that *binary coding* is used and that the demodulator makes *hard decisions*. Since the channel encoder is 1–1, the decoder can equivalently determine the estimated code sequence $\hat{v}$ rather than $\hat{u}$. We view $v$ and $r$ as random vectors. Given a specific realization of $r$, we seek to determine an estimate of $v$. This reminds us of least squares estimation. Least squares is not a natural choice, however, because the resulting estimate $\hat{v} = E[v|r]$ does not yield a binary sequence. Maximum likelihood decoding is an alternative. Given $r$, we search for the codeword that most likely produced $r$. Thus we pick $\hat{v}$ as the codeword that maximizes the a posteriori probability

$$\max_{v \in \mathcal{C}} P(v|r)$$

Since $P(r)$ is known, Bayes's rule

$$P(v|r) = \frac{P(r|v)P(v)}{P(r)}$$

enables us to study the equivalent problem $\max_{v \in \mathcal{C}} P(r|v)P(v)$. If all codewords are equally likely, we can equivalently maximize

$$\max_{v \in \mathcal{C}} P(r|v)$$

A decoder that estimates $\hat{v}$ by maximizing $P(r|v)$ over all codewords is called *maximum likelihood decoder.* If the channel output in a given time interval depends only on the transmitted signal in that interval, and not on any previous transmission, the channel is said to be *memoryless*. In this case the cascade of the modulator, the channel, and the demodulator form a *discrete memoryless channel* that is completely specified by the transition probabilities $P(r_i|v_i)$. In the binary case the channel is specified by the probabilities $P(0|1)$ and $P(1|0)$. If $P(0|1) = P(1|0) = p$, the binary channel is called *symmetric*. In this case both error types are equiprobable. Binary symmetric channels occur when binary coding is used, the demodulator makes hard decisions, and the distribution of the noise is symmetric. For discrete memoryless channels maximum likelihood becomes $P(r|v) = \prod_i P(r_i|v_i)$. To convert the product into a sum, we use the logarithm. Clearly the optimum estimated codeword will not be affected. Thus we examine

$$\max_{v \in \mathcal{C}} \sum_{i=0}^{n-1} \log P(r_i|v_i)$$

Suppose that in addition the channel is symmetric. If $r_i \neq v_i$, $P(r_i|v_i) = p$, while if $r_i = v_i$, $P(r_i|v_i) = 1 - p$. The Hamming distance

$$d(r, v) = \#\{i : r_i \neq v_i\}$$

is given by the number of places where $r$ and $v$ differ. Then

$$\log P(r|v) = \sum_{i: r_i \neq v_i} \log P(r_i|v_i) + \sum_{i: r_i = v_i} \log P(r_i|v_i)$$

$$= d(r, v) \log p + [n - d(r, v)] \log(1 - p)$$

$$= d(r, v) \log \frac{p}{1 - p} + n \log(1 - p)$$

If the probability of transmission error satisfies $p < 1/2$, it holds $\log(p/(1 - p)) < 0$. Furthermore the term $n \log(1 - p)$ does not depend on $v$. We conclude that for binary symmetric channels, the maximum likelihood decoding rule reduces to the *minimum (Hamming) distance decoding rule*:

$$\text{Choose } \hat{v} : d(r, \hat{v}) = \min_{v \in \mathcal{C}} d(r, v)$$

Both block and convolutional codes strive to achieve efficient transmission and storage. Key issues that account for efficiency are reliability, namely low probability of decoding error, high speed, and low implementation cost. The speed of information transmission is indicated by the code rate $R$. In the case of block codes, the code rate is given by $R = k/n$, namely the number of informa-tion bits per transmitted symbol. If the modulator uses a waveform of duration

$T$ seconds for each bit, the information transmission rate is $R/T$ bits per second. . Good block error control codes require long blocklength. The reason is that errors occur in clustered form. Some message segments contain more than the average number of errors, and some others less. In his landmark paper Shannon (1948), C. Shannon showed that there is a channel parameter $C$ in bits per second, called *capacity*, so that if the transmission rate is less than $C$, an error control code exists that achieves arbitrarily small decoding error probability. The proof of the above result does not offer constructive methods for finding such good codes. The objective of error control coding is to develop implementable coding schemes that approximate the features anticipated by Shannon's work.

### 4.7.4   Error detection, correction, and minimum distance

Suppose that the codeword $v$ is transmitted and that $r$ is received. The following simple error detection rule is employed: An error is declared if $r$ is not a codeword. This strategy warrants error detection if $r$ is not a codeword. On the other hand, if $r$ is a codeword different from the transmitted word $v$, an erroneous detection decision results. We will see below that the probability of false detection is small. We will also see that the above rule is much simpler to implement than error correction via maximum likelihood or minimum distance decoding.

The minimum distance of a block code $C$ is a critical parameter of the encoder reflecting its error detection and correction capability. It is defined by

$$d_{\min} = \min\{d(v,z), v, z \in C, v \neq z\} \qquad (4.107)$$

Suppose that $v$ is transmitted or stored and that $r$ is received. Let

$$e = r - v$$

be the error occurred in transmission or storage. The Hamming weight of $e$ is the number of nonzero entries, $w(e) = d(r,v)$. If $w(e) \leq d_{\min} - 1$, $r$ differs from $v$ in no more than $d_{\min} - 1$ places. Hence $r$ cannot be a codeword different from $v$; if it were, $d_{\min} - 1 \geq d(r,v) \geq d_{\min}$, contradiction. Thus all errors with at most $d_{min} - 1$ nonzero entries are detected. For this reason we say the error detection capability of the code is $d_{\min} - 1$.

Suppose next that $w(e) = d(r,v) = t$ so that $t$ errors occurred during transmission of $v$. Let $z$ be an arbitrary codeword different from $v$. The Hamming distance satisfies all properties of a distance metric. The triangle inequality gives

$$d(z,v) - d(r,v) \leq d(z,r)$$

Since $z \neq v$, it holds that

$$d_{\min} - t \leq d(z,v) - d(r,v) \leq d(z,r)$$

If the minimum distance satisfies

$$t < d_{\min} - t \quad \text{or} \quad d_{\min} \geq 2t + 1,$$

then $t < d(z, r)$. This states that all codewords $z$ different from $v$ have distance from $r$ greater than $t$. Hence the minimum distance decoder will correctly decode $r$ into $v$, the actual transmitted word.

In summary, a block code with minimum distance $d_{\min}$ will detect all error patterns with fewer than $d_{\min} - 1$ errors and will correct all error patterns with $[(d_{\min} - 1)/2]$ or fewer errors. The parameter $[(d_{\min} - 1)/2]$ reflects the error correcting capability of the code. For given $n$ and $k$, block codes with minimum distance as large as possible are desired. BCH codes provide a family of codes within which minimum distance can be designed to satisfly a given lower bound. They are treated in Chapter 14.

### 4.7.5  Linear block codes

Let $F$ denote a finite field, and $F^k$ the set of $k$-tuples $(u_0, u_1, \ldots, u_{k-1})$, $u_i \in F$. $F^k$ becomes a vector space with the pointwise addition and scalar multiplication. A block $(n, k)$ encoder $G : F^k \to F^n$ is called *linear* if $G$ is a linear map. The resulting codebook $C$ is a linear subspace of $F^n$. Using standard coordinates the encoder is represented as

$$v = uG \tag{4.108}$$

where $G$ is a matrix of size $k \times n$. If $G$ is represented by its rows, Eq. (4.108) is rewritten as

$$v = u_0 g_0 + u_1 g_1 + \cdots + u_{k-1} g_{k-1}$$

The codebook $C$ is the row space of $G$. The rows are linearly independent vectors because $G$ is 1–1. The matrix $G$ is called a *generator matrix* for $C$.

Linear block coding is readily implemented by a matrix by vector multiplication circuit. The circuit involves about $kn$ scalors and adders. If $F$ is the binary field, each adder corresponds to modulo 2 addition and is implemented by an XOR gate. The scalor with parameter $g$ denotes a link if $g = 1$, and no link if $g = 0$.

The linear block encoder must be properly constrained if it is going to appear in systematic form. Since $v = (\hat{v} \quad u)$, the analogous partitioning $G = (P \quad L)$ leads to $(\hat{v} \quad u) = uG = (uP \quad uL)$. Therefore $\hat{v} = uP$, $u = uL$. We conclude that $L$ is the identity matrix of order $k$. The generator matrix of a linear code in systematic form is given by

$$G = (P \quad I_k)$$

The identity matrix leaves the message block unaltered in the codeword. $P$ has size $k \times (n - k)$ and is responsible for the generation of the parity check digits contained in $\hat{v}$. $P$ is the design parameter that specifies how the redundant information will be chosen to secure reliable transmission and storage.

### 4.7.6   Parity matrix, syndrome computation, detection, and correction

Suppose that the codeword $v$ is transmitted and that $r$ is received. Let $e = r - v$ be the resulting error. The detection rule checks if $r$ is a codeword and declares an error if the answer is not affirmative. This requires browsing over the words of the codebook $C$ and seeing if $r$ is one of them. An efficient implementation of the error detection scheme relies on the alternate characterization of the codebook as the null space of a parity matrix. Suppose that $G$ is in systematic form. Then

$$u = (\hat{v} \quad u)\begin{pmatrix} 0 \\ I_k \end{pmatrix} = v\begin{pmatrix} 0 \\ I_k \end{pmatrix}$$

and

$$v = uG = v\begin{pmatrix} 0 \\ I_k \end{pmatrix}(P \quad I_k) = v\begin{pmatrix} 0 & 0 \\ P & I_k \end{pmatrix}$$

Thus

$$v\left(I_n - \begin{pmatrix} 0 & 0 \\ P & I_k \end{pmatrix}\right) = v\begin{pmatrix} I_{n-k} & 0 \\ -P & 0 \end{pmatrix} = 0$$

We conclude that $v$ is a codeword if and only if it satisfies

$$vH^T = 0 \tag{4.109}$$

$$H = \begin{pmatrix} I_{n-k} & -P^T \end{pmatrix} \tag{4.110}$$

In the binary case $P = -P$. The matrix $H$ has size $(n - k) \times n$ and is called *parity check matrix* of $C$.

The characterization of the code by the parity check matrix $H$ makes error detection straightforward. We compute the vector of length $n - k$,

$$s = rH^T \tag{4.111}$$

where $s$ defines the *syndrome* of the received word $r$. If $s \neq 0$, $r$ is not a codeword, and an error has been committed. The error detection circuit amounts to a matrix by vector multiplication and involves $(n - k)n$ scalar multiplications and additions. Note that

$$s = rH^T = (v + e)H^T = vH^T + eH^T = 0 + eH^T = eH^T \tag{4.112}$$

The syndrome cannot be computed by the latter equation since the error $e$ is not known. It can of course be determined from (4.111). Equation (4.112) is a linear system of equations. In the binary case it has $2^k$ solutions. The set of solutions is the linear variety containing all error patterns

$$e = r - v, \qquad v \in C$$

where $r$ is the particular solution (4.111) and $C$ is the subspace of codewords specified by Eq. (4.109). The minimum distance decoder determines the estimated codeword by finding the error pattern with minimum weight in the above linear variety

$$d(\hat{v}, r) = w(\hat{e}) = \min\{d(r, v), v \in C\} = \min\{|e|, eH^T = s\}$$

The decoding algorithm involves the following steps

1. Compute the syndrome $s = rH^T$.
2. Construct the combinatorial circuit that transforms each input $s$ to the solution $\hat{e}$ of the system $s = eH^T$ that has minimum weight
3. Decode $r$ to $\hat{v} = r - \hat{e}$.

Steps 1 and 3 are fairly simple. In contrast, the implementation of step 2 is complicated for large blocklengths $n$. To obtain simpler decoding structures the class of codes must be further narrowed. We will return to this subject in Chapter 14 in connection with BCH codes.

### Example 4.10  Binary Hamming codes

Hamming codes are linear codes that are capable of correcting a single error. The easiest way to introduce these codes is by means of the parity matrix $H$. The size of $H$ is of the form $m \times (2^m - 1)$; that is, the blocklength is $2^m - 1$, and the parity checks are $m$. The columns of $H$ are all nonzero binary vectors of length $m$. For instance, the parity matrix of the linear (7,4) Hamming code is given by

$$H = \begin{pmatrix} 0\,0\,0\,1\,1\,1\,1 \\ 0\,1\,1\,0\,0\,1\,1 \\ 1\,0\,1\,0\,1\,0\,1 \end{pmatrix}$$

The columns of $H$ are the binary representations of all decimals between 1 and 7. If a single error occurs at position $j$, the error pattern has the form $e = [0, 0, \ldots, 1, \ldots, 0]$. The syndrome becomes $s = rH^T = eH^T = h_j$, where $h_j$ is the $j$th column of $H$. Hence the syndrome provides the binary representation of the location of the error. We conclude that all single errors are corrected.  ∎

Two types of errors are encountered in transmission and storage: burst and random errors. Random errors are distributed randomly in the received sequence and are primarily associated with memoryless channels. The deep space channel and many satellite channels are representative examples. Burst errors, on the other hand, occur in clusters that contain hundreds or thousands of bits and are associated with channels with memory. For example, in radio channels error bursts are caused by signal fading due to multipath transmission. In wire and cable links, bursts are caused by impulsive switching noise and crosstalk. Burst errors in the compact disc (CD) audio system are caused by dropouts, the result of surface degradation from fingerprints and scratches on the disc.

A burst of length $l$ is a vector whose nonzero components are confined to $l$ consecutive digits. It can be shown that a linear block code over $GF(N)$ that is able to correct all bursts of length $t$ or less must have at least $2t$ parity symbols. This result is known as *Rieger bound* (see Exercise 4.35).

Good codes correcting a small number of bursts have been found by computer search. Longer bursts are efficiently handled by interleaving, discussed in Chapter 14.

## 4.8   ENCRYPTION

Encryption systems (or cryptographic systems or cryptosystems) aim to achieve secure transmission and storage of information. They are of primary importance in communication networks where shared access is provided. The main goals of a modern cryptographic system are security or privacy and authenticity. Privacy ensures that unauthorized disclosure of information is prevented. Authenticity or integrity guarantees the authenticity of the message to the receiver and the prevention of unauthorized modification of information. Secure communication and storage are vulnerable to active and passive threats. Passive threats (eavesdropping) refer to interception of information. Active threats involve deliberate modification of messages. For example, the message *Credit Smith's account with $100* modifies to *Credit Jone's account with $10,000*. With the exception of optical fibers, communication links like telephone twisted pair, coaxial cable, microwave, or satellite are vulnerable to eavesdropping. Encryption systems are designed to prevent passive threats and to detect active threats.

### 4.8.1   Cryptosystems

We confine ourselves to digital structures. A cryptosystem consists of an *enciphering transformation* and a *deciphering transformation*. It is illustrated in Fig. 4.19. The enciphering transformation converts a message sequence called *plaintext* or *cleartext* into an unintelligent form, the *ciphertext*, with the aid of the *key*. The deciphering transformation recovers the original message from the ciphertext with the use of the key. The enciphering transformation has the form

$$E : \mathcal{U} \times \mathcal{K} \to \mathcal{Y}, \qquad (m, k) \to E(m, k) = E_k(m) = c$$

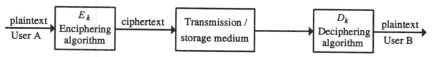

**Figure 4.19.** Encryption system.

The input space $\mathcal{U}$ consists of message sequences with values in a finite alphabet, for instance, 0 and 1. Each message $m$ is transformed to the ciphertext sequence $c$ with the aid of the key $k$ and the enciphering algorithm $E_k$. The key space $\mathcal{K}$ is finite. The ciphertext $c$ is transmitted or stored in the unprotected medium, while the key $k$ is securely delivered to the receiver. The receiver reconstructs the message from the ciphertext and the key by means of the deciphering transformation

$$D : \mathcal{Y} \times \mathcal{K} \to \mathcal{U}, \qquad (c, k) \to D(c, k) = D_k(c) = m$$

For each $k \in \mathcal{K}$ the enciphering transformation $E_k$ is invertible and the inverse is the deciphering transformation:

$$E_k^{-1} = D_k$$

Both the enciphering and the deciphering algorithms must lend themselves to efficient implementation. For operational reasons the use of the same system at the transmitter and receiver is often recommended. This mandates that $D_k = E_k$, that is, $E_k^{-1} = E_k$. The security of a crypto system is assessed by the computational effort needed to determine the key or message from known message-ciphertext information. The system transformations $E$ and $D$ are also assumed known. Experience shows that physical devices cannot be kept secret. In most cases they are installed in places of public access. Therefore the enciphering and deciphering algorithms can be revealed. The interception of the ciphertext is a common passive threat, as was pointed out above. Therefore the ciphertext is always assumed available. Finally, acquisition of segments of the transmitted message is almost impossible to avoid. These correspond to silent periods in a speech signal, recurring keywords in a computer program like begin, end, if then, and others. Thus the security of the cryptographic system is assessed by the computational complexity required to solve the equation

$$E(m, k) = c \qquad (4.113)$$

The key $k$ is the unknown, while $E$, $c$, and portions of $m$ are known. In summary, for each $k$ the algorithms $E_k$ and $D_k = E_k$ must be easy to implement, while Eq. (4.113) must be computationally hard to solve with respect to $k$ given parts of $m$, $c$ and the function $E$. Apparently the cryptanalyst easily acquires the message by simply applying the deciphering algorithm $D_k$ on the ciphertext, if he/she manages to determine the key $k$.

In the above sense a secure cryptosystem achieves authenticity as well between trustworthy communicating parties. The key $k$ employed by user A to transmit information to user B acts like a digital signature for A. When user B deciphers the ciphertext, he essentially validates that the message is indeed sent by A because he is the only one knowing the secret key. A third sender not knowing $k$ would have enciphered a message $m'$ with a different key $k'$. When user B deciphers with the key $k$, he produces the unintelligible message $D_k(E_{k'}(m'))$. (Intelligible sequences are very few in the set of all sequences.) Thus it is impossible for a third party to forge A's signature, except B who has access to the secret key. Protection from B is achieved by public key cryptosystems discussed below.

In analogy with the division of error control codes into block and convolutional codes, encryption systems are divided into *block ciphers* and *stream ciphers*. Block ciphers partition the message sequence into large blocks and transform each block into a ciphertext block with the aid of the key. Stream ciphers process the plaintext in small segments. Each ciphertext segment depends on previous key or message segments.

***Example 4.11   Transposition systems***
Transposition systems are block ciphers that reshuffle bits or characters in the message. Suppose that message blocks have length $d$. Each key is a permutation of $d$ items. The key space consists of $d!$ elements. Each key is represented by an orthogonal matrix $P$ consisting of 0 and 1. Equation (4.113) takes the form

$$c_i = Pm_i, \qquad i = 1, 2, \ldots$$

Suppose now that $d = 3$ and that the message block $m_i = (m_{i1} \quad m_{i2} \quad m_{i3})^T$ is reindexed to $(m_{i1} \quad m_{i3} \quad m_{i2})^T$. The corresponding key is

$$P = \begin{pmatrix} 1 & 0 & 0 \\ 0 & 0 & 1 \\ 0 & 1 & 0 \end{pmatrix}$$

A generic block of length $d$, namely a block with pairwise distinct entries, suffices to determine the key. More generally, a linear cipher $c_i = Pm_i$, featuring an arbitrary linear invertible transformation $P$ as key, is handled by collecting $d$ known input-output blocks into the matrix equation

$$PM = C, \quad M = (m_1 \quad m_2 \quad \cdots \quad m_d), \quad C = (c_1 \quad c_2 \quad \cdots \quad c_d)$$

If $M$ is invertible, the key is found from

$$P = CM^{-1}$$

The complexity of the above linear system is proportional to $O(d^3)$, the complexity of a linear system solver.

Transposition ciphers are too simple to be of any real use as stand-alone devices. They are nevertheless engaged as parts of more complex architectures. Transposition ciphers are readily analyzed under a ciphertext attack only, that is, without knowledge of plaintext. We simply observe that the statistical properties of the language of the plaintext are preserved in the ciphertext. For example the letter e in English is the most frequent letter occurring 13% of the time. The code is then analyzed by estimating one-, two-, and three-dimensional densities of the ciphertext characters and by matching the observed statistics to the known statistics of the plaintext language. ∎

A *block cipher* is described by a static system of the form

$$c_i = f(m_i, k) \qquad (4.114)$$

In this expression $m_i$ denotes the plaintext block at time $i$. It has size $d$, and its entries are characters of the alphabet $F_m$. $c_i$ denotes the ciphertext block at time $i$. It has size $d$ and its entries are characters of the alphabet $F_c$. The same key $k$ is used for the encipherment of blocks. $k$ has size $n$, and its digits belong to the alphabet $F_k$. $f$ is the enciphering transformation. For fixed $k$ the assignment $f_k : F_m^d \rightarrow F_c^d, m \rightarrow c = f_k(m) = f(m, k)$ must be invertible so that deciphering is feasible. Each such $f_k$ is called *substitution* and is implemented by a combinatorial circuit. In fact $f_k$ can also be represented by a matrix by vector product; however, the dimensions are now prohibitively large. The design of a block cipher aims to determine suitable laws $f$ so that the resulting linear equations are not accompanied by other linear constraints that will help reduce complexity forcing the cryptanalyst to no better choice than exchaustive search over the key space. The latter is controlled by the large number of keys.

A popular cryptosystem of the form (4.114) is DES (Data Encryption Standard), developed by IBM. DES acts on 64-bit message blocks with a 56-bit key. The function $f$ is a cascade of permutations and substitutions. Hardware implementations achieve rates of several million bits per second. The system has been used for authentication in Treasury electronic fund transfer applications and for PIN encryption for automated teller machines (ATM).

Stream ciphers process small segments of cleartext sequentially. A popular recipe is the Vernam cipher

$$c_i = m_i + k_i \qquad (4.115)$$

Each segment of the ciphertext $c_i$ results from the modulo 2 addition of the message segment $m_i$ and the key vector $k_i$. The key sequence $k_i$ is produced by a key generator algorithm and the initial finite key. Deciphering uses the same key generator and the exlusive OR to recover the message:

$$c_i + k_i = m_i + k_i + k_i = m_i$$

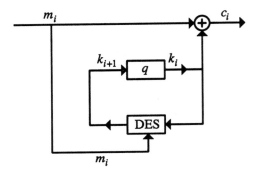

**Figure 4.20.** DES in output feedback mode.

The enciphering and deciphering operations are identical. Security of the above system is assessed by the ability of the key generator to produce purely random-like sequences, so as to mask the statistical redundancy of the message sequence. Under the known plaintext attack, portions of the transmitted message are known. Since $c_i$ is available, a number of consecutive samples of the key sequence $k_i$ are accessible, by (4.115). Hence the key generator must be sufficiently complex so that the computation of the secret key from partial knowledge of the key sequence is prevented.

More complex stream ciphers can be built by combining well-designed block ciphers and Vernam ciphers. In general, a stream cipher is described by a nonlinear state space model

$$c_i = g(m_i, k_i)$$

$$k_{i+1} = h(k_i, m_i)$$

The initial state is formed by the secret key. Deciphering is simple if the output map $g$ has the additive format of the Vernam cipher. The function $h$ can be implemented using the DES module as illustrated in Fig. 4.20.

### 4.8.2   Public key cryptosystems

The cryptosystem of Fig. 4.19 requires that the key be securely distributed. Thus secure channels are not avoided, albeit the key information is much smaller than the messages and thus easier to accommodate. Public key cryptosystems alleviate the key distribution problem by using two keys for each user one public and one private. The idea is illustrated in Fig. 4.21. User A has a public key corresponding to the enciphering transformation $E_A$ and a private key corresponding to the deciphering transformation $D_A$. Public keys of all users are publically displayed in a directory. Suppose that user A wants to communicate with user B. He/she finds the public key of B, $E_B$, in the public directory and enciphers the message

$$c = E_B(m) \tag{4.116}$$

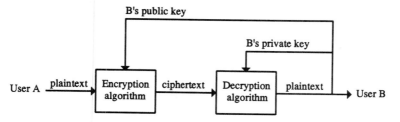

**Figure 4.21.** Public key cryptosystem.

User B deciphers the received ciphertext $c$ using its own private key $D_B$:

$$m = D_B(c) \tag{4.117}$$

The public enciphering transformation is the inverse of the private deciphering transformation $D_B$.

The above communication does not require secret exchange of keys. In fact each of the communicating parties does not know the secret key of the other. The enciphering and deciphering algorithms must be easy to compute. In addition each $E_A$ must be designed so that the message $m$ is computationally impossible to derive from (4.116). In a similar way even if a portion of the message $m$ is known, Eq. (4.117) is impossible to solve with respect to the secret key $D_B$.

The above procedure provides for secrecy but cannot be used for authenticity, since any user can send messages to B using B's public key. Authenticity is accomplished as follows: User A enciphers the message $m$ by its own private key and sends the resulting ciphertext to B:

$$c = D_A(m) \tag{4.118}$$

Upon receipt, user B deciphers the received ciphertext $c$ using A's public key

$$m = E_A(c) \tag{4.119}$$

The received message is authentic because it is only A who can encipher with the secret key $D_A$. The transaction achieves authenticity but is insecure because any user has access to A's public key and hence can recover the message from (4.119).

Secrecy and authenticity are simultaneously achieved by cascading the preceding functions. User A enciphers the message $m$ by its private key $D_A$ and then secretly transmits it to B using B's public key

$$c = E_B[D_A(m)] \tag{4.120}$$

Upon receipt, user B deciphers $c$ by its private key and validates the message by the public key of A:

$$m = E_A[D_B(c)] \tag{4.121}$$

A practical public key cryptosystem offering secrecy and authenticity is the RSA scheme named after Rivest, Shamir, and Adleman. This is described below.

***Example 4.12  RSA public key cryptosystem***
The encryption transformations $E$ and $D$ in the RSA scheme rely on exponentiation over finite arithmetic systems. The basic idea is that exponentiation can be efficiently computed, but the inverse operation is computationally much harder. First, two large prime numbers $p$ and $q$ are chosen and their product $n = pq$ is determined. The Euler's totient function (see Appendix II) $\phi(n)$ is easily computed from the prime factors of $n$:

$$\phi(n) = (p - 1)(q - 1) \tag{4.122}$$

An integer $d$ relatively prime to $\phi(n)$ is chosen in the interval

$$\max(p, q) + 1 \le d \le n - 1$$

where $d$ serves as the secret key. Euclid's algorithm is used to find an integer $e$ so that

$$ed = 1(\text{mod } \phi(n)) \quad \text{or} \quad ed - k\phi(n) = 1 \tag{4.123}$$

(see Appendix II). The pair $(e, n)$ forms the public key. Enciphering and deciphering are performed by exponentiation over integers modulo $n$. Messages are treated as positive integers less than $n$, namely as elements of $Z_n$. Then

$$c = m^e(\text{mod } n) \tag{4.124}$$

$$m = c^d(\text{mod } n) \tag{4.125}$$

The above equations are mutual inverses provided that

$$m = m^{ed}(\text{mod } n)$$

or

$$m^{ed-1} = 1(\text{mod } n) \tag{4.126}$$

Euler's theorem (see Appendix II) implies that $m^{\phi(n)} = 1(\text{mod } n)$ for any $m \in Z_n$. Hence Eq. (4.126) holds because of (4.123). To determine the secret key $d$ from (4.123), $\phi(n)$ must be known. The computation of $\phi(n)$ from (4.122) requires the prime factors $p$ and $q$. These are secret, although their product $n$ is known. Thus cracking the code boils down to factoring a number $n$ into its prime factors. Existing algorithms factor a 200 digits integer in billions of years at the rate of one step per microsecond.

Exponentiation is efficiently performed given $e$, $d$, and $n$. The storage requirements of 200-decimal integers (664 bits) are about 2 kilobits per user. In particular, the public keys $n$ and $e$ require about 1.3 kilobits per user for storage. The conventional cryptosystems of Fig. 4.19 like DES require much less storage. The DES requires only 56 bits for the key. Enciphering and deciphering a 664-bit number require about $10^3$ multiplications. Present hardware implementations of the RSA scheme are too slow for communication links capable of high bit rates.

When the RSA is used for both secrecy and authenticity, a complication arises due to the cascade of exponentiations over different moduli:

$$c = E_B[D_A(m)] = [m^{d_A} (\bmod\ n_A)]^{e_B} (\bmod\ n_B)$$

If $n_A > n_B$, the block $m^{d_A} (\bmod\ n_A)$ might not be in the range $[0, n_B - 1]$, and if it is reduced modulo $n_B$, recovery of the original signal is not feasible. This issue is settled at a slight increase of complexity as follows:

User A compares the public keys $n_A$ and $n_B$. If $n_A < n_B$, he/she proceeds in the usual way. The message is signed by $D_A(m)$ and enciphered with B's public key: $c = E_B[D_A(m)]$. If $n_A > n_B$, the message is first enciphered by B's public key and then signed by A: $c' = D_A[E_B(m)]$. Upon receipt, B will compare $n_A$ and $n_B$. If $n_A < n_B$, he/she will decipher and validate: $m = E_A[D_B(c)]$. If $n_A > n_B$, he/she will first validate and then decipher: $m = D_B[E_A(c')]$.  ∎

## 4.9 PATTERN RECOGNITION

Pattern recognition is concerned with the classification of patterns into two or more categories. Features or attributes of patterns are extracted and measured. Features provide a more economical representation and serve as main inputs to the classification system. A quantitative characterization of the categories is supplied. On the basis of the above information, a pattern classifier is designed that assigns patterns into categories.

In computer-aided medical diagnosis, medical signals or images form the original patterns. A classifier automatically decides whether the signal or image corresponds to a healthy or pathological case. Consider, for instance, the classification of cells into suspected deseases on the basis of histological examinations. Cell features are measured, including nuclei and cytoplasm attributes. Examples are area, perimeter, diameter, nuclei cytoplasms ratio, and histograms. Besides the above features that aim to describe the geometric shape of the cell, optical features such as color and brightness are also determined. The results are formatted into a vector that represents the cell and is used for classification. Character recognition systems provide further examples. The classes are the letters of an alphabet or alphanumeric characters. A pattern classifier recognizes characters that have been degraded by noise. A third important application is encountered in computer vision. A pattern classifier in

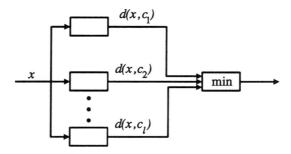

**Figure 4.22.** Nearest neighbor classifier.

an assembly line uses the features extracted from camera pictures to split products into defective and nondefective items.

### 4.9.1 Pattern classifiers

A pattern classifier, much like a quantizer, assigns vectors of $R^p$ called *patterns* to letters of a finite alphabet called *labels*. Each label associates to a category, cluster, or class. Categories are disjoint sets of patterns $c_1, c_2 \ldots, c_l$. An example of a pattern classifier is presented in Fig. 4.22. The underlying assumption is that patterns are grouped into categories in accordance with a similarity (or dissimilarity) measure $d(x, y)$. Here the measure is a distance metric. If $x$ and $y$ are in the same cluster, $d(x, y)$ is small, and the patterns are similar. The distance of a pattern $x$ from a closed set $c$ is defined by

$$d(x, c) = \inf\{d(x, y), y \in c\}$$

If $c$ is finite, the infimum becomes a minimum. The classifier of Fig. 4.22 is a bank of static systems. Each of them computes the distance of the input from one of the given categories. Then the minimum of these distances is computed, and the input is assigned to the class that is closer. This is the *nearest neighbor rule*. Often each category is represented by a single vector, the cluster representative.

An alternative approach is to specify a class of pattern classifier models offering a universal approximation capability (e.g., multilayer neural networks) or meeting some a priori assumptions concerning the categories (e.g., linear separation). Within the above class the pattern classifier that optimizes a certain performance measure on the basis of given patterns is searched. In this context the perceptron algorithm and the backpropagation algorithm are next introduced.

### 4.9.2 The perceptron algorithm

Let us consider two linearly separated categories. Linear separation enables us to narrow down the model set of pattern classifiers to the perceptron family

$u_s(w^T x + \theta)$; $u_s(x)$ is the unit step. Thus the family of allowable structures consists of a single neuron with the step function as activation function. Let $x(n)$, $1 \leq n \leq N$, be a collection of known patterns, and $A$, $B$ the two known categories. Let $d(n)$ be the desired response

$$d(n) = \begin{cases} 1, & \text{if } x(n) \in A \\ 0, & \text{if } x(n) \in B \end{cases}$$

describing the labels of the given patterns. Now $d(n)$ is known because $x(n)$, $A$ and $B$ are all known. In analogy with Example 4.4, we seek to determine the weights so that some function of the error $e(n) = d(n) - y(n)$ is minimized. The situation is complicated by the presence of the nondifferentiable nonlinear activation function.

Since $A$ and $B$ are linearly separated closed subsets of $R^p$, there exist $w^* \in R^p$, $\theta^* \in R$, such that $w^{*T} x + \theta^* > 0$ for all $x$ in $A$ and $w^{*T} x + \theta^* < 0$ for all $x \in B$. Consider the augmented vectors $\tilde{x}^T = (x^T, 1)$, the corresponding sets $\tilde{A}, \tilde{B}$, and $\tilde{w}^T = (w^{*T}, \theta^*)$. Linear separation states that there is a vector $\tilde{w}$ in $R^{p+1}$ such that $\tilde{w}^T \tilde{x} > 0$ for all $\tilde{x}$ in $\tilde{A}$ and $\tilde{w}^T \tilde{x} < 0$ for all $\tilde{x} \in \tilde{B}$. The development of the perceptron algorithm will rely on the latter condition. The tilde notation is abandoned for simplicity. In geometric terms we seek a vector $w$ such that the line orthogonal to $w$ contains $A$ in one of the two half planes and $B$ in the other half plane. It is clear that if there is one line separating $A$ and $B$, there is an infinite number of separating lines. To determine $w$, we set up an optimization problem. The performance of each vector $w$ is assessed by the sum of distances of incorrectly classified samples. It is graphically illustrated in Fig. 4.23. The expression of the perceptron criterion is

$$J(w) = -\sum_{x \in X_A(w)} w^T x + \sum_{x \in X_B(w)} w^T x \qquad (4.127)$$

The set $X_A(w)$ consists of the samples of $A$ misclassified by the line defined by $w$. Likewise $X_B(w)$ consists of samples in $B$ misclassified by $w$. Clearly $J(w) \geq 0$,

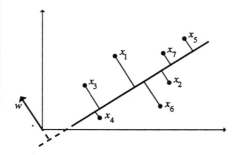

**Figure 4.23.** Illustration of the perceptron criterion.

while $J(w) = 0$ if and only if both $X_A(w)$ and $X_B(w)$ are empty. A better representation of the cost is obtained if $B$ is replaced by $-B$. Then the set of misclassified samples $X(w)$ is written

$$X(w) = X_A(w) \cup X_{-B}(w) = \{x \in X : w^T x \le 0\}$$

where $X = A \cup (-B)$. The perceptron criterion takes the form

$$J(w) = - \sum_{x \in X(w)} w^T x \qquad (4.128)$$

We will use the steepest descent approach and, in particular, the algorithm

$$w(n+1) = w(n) + \gamma s(n) \qquad (4.129)$$

$$s(n) = \sum_{x(n) \in X(w(n))} x(n) = -\nabla J(w(n)) \qquad (4.130)$$

The cost (4.128) is not a continuous function, and the second equality in Eq. (4.130) does not hold everywhere. Hence convergence results of the steepest descent algorithm cannot be invoked. A direct analysis of the asymptotic behavior of the algorithm is described below. Let $w^*$ be a separating vector, so that the subspace orthogonal to $w^*$ separates $A$ and $B$. The same property is shared by all vectors of the form $aw^*$, $a > 0$. We show that (4.129) converges to a vector of the form $aw^*$ in a finite number of steps. To this end we show that there is $N$ such that $X(w(N))$ is empty. Then the algorithm locks at $w(N)$, and $w(N)$ is the desired solution, as there are no misclassified samples. Suppose that the conclusion does not hold and that $X(w(n))$ is nonempty for any $n$. Then for any $n$, $s(n) \ne 0$. Moreover

$$|w(n+1) - aw^*|^2 = |w(n) - aw^* + \gamma s(n)|^2$$

$$= |w(n) - aw^*|^2 + \gamma^2 |s(n)|^2 + 2\gamma w^T(n)s(n) - 2a\gamma w^{*T}s(n)$$

Since $w^T(n)s(n) \le 0$, we have

$$|w(n+1) - aw^*|^2 \le |w(n) - aw^*|^2 + \gamma^2 |s(n)|^2 - 2a\gamma w^{*T}s(n)$$

Since $w^*$ is a separating vector and $s(n)$ is nonzero, $w^{*T}s(n) > 0$. Equation (4.130), shows that $s(n)$ takes values on the finite set $S$ of all possible nonzero sums of samples. Let

$$\beta = \max\{|s|^2, s \in S\}, \quad \rho = \min\{w^{*T}s : s \in S\}, \quad a = \frac{\beta\gamma}{\rho}$$

Then

$$|w(n+1) - aw^*|^2 \le |w(n) - aw^*|^2 - \beta\gamma^2$$

Repetitive application of this inequality gives

$$|w(n+1) - aw^*|^2 \le |w(1) - aw^*|^2 - n\beta\gamma^2$$

Hence after $n > N_0 = |w(1) - aw^*|^2/\beta\gamma^2$ steps the right-hand side becomes negative, a contradiction.

A simpler variant results if the sum of misclassified samples is replaced by a single misclassified sample. The algorithm becomes

$$w(n+1) = w(n) + x(n), \quad x(n) \in X(w(n)) \tag{4.131}$$

Convergence of the above scheme is proved similarly.

With reference to the original problem setup specified by the categories $A$, $B$, weights $w$, and bias $\theta$, the algorithm reads as follows: The values of the weights and biases are set to small random numbers. At time $n$ the input is presented, and the output with the current weight and bias values is calculated and compared with the desired response. If the output is correctly classified, the weights and bias remain unchanged. If the input belongs to class $A$, namely the desired response is 1 and the resulting output is 0, the update (4.131) takes the form

$$w(n+1) = w(n) + x(n), \quad \theta(n+1) = \theta(n) + 1$$

On the other hand, if $d(n) = 0$ and the output is 1, then $-x(n) \in X_{-B}(w(n))$ and (4.131) becomes

$$w(n+1) = w(n) - x(n), \quad \theta(n+1) = \theta(n) - 1$$

The above cases can be combined in the following expressions:

$$w(n+1) = w(n) + [d(n) - y(n)]x(n)$$

$$\theta(n+1) = \theta(n) + [d(n) - y(n)]$$

An entire pass of the training set is called an *epoch*. Usually several epochs are required before the algorithm converges.

We have shown that if two classes are linearly separable, there exists a perceptron architecture that achieves pattern classification. Parameters are trained by the perceptron algorithm.

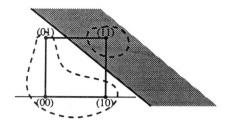

**Figure 4.24.** Classes of points for the AND gate.

*Example 4.13    AND gate*

The AND gate is described by the table

| $x$ | $y$ | $xy$ |
|---|---|---|
| 0 | 0 | 0 |
| 0 | 1 | 0 |
| 1 | 0 | 0 |
| 1 | 1 | 1 |

All possible input pairs are presented in Fig. 4.24. The pair $(1, 1)$ corresponds to 1 while all other pairs correspond to 0. Let $A$ denote the singleton $(1, 1)$, and let $B$ be the set of $(0, 0)$, $(0, 1)$, $(1, 0)$. It follows from Fig. 4.24, that these sets are linearly separable. Therefore the perceptron algorithm will determine a separating line. One such line is $x_2 + x_1 - 3/2 = 0$. A MATLAB implementation using the neural networks toolbox is presented next. The input vectors are registered as columns of the matrix $P$:

$$P = \begin{pmatrix} 0 & 0 & 1 & 1 \\ 0 & 1 & 0 & 1 \end{pmatrix}$$

The desired response is expressed in terms of the row $T$:

$$T = (0 \quad 0 \quad 0 \quad 1)$$

```
[R,Q] = size(P);                    % number of inputs R, number of data vectors Q
[S,Q] = size(T);                    % number of neurons S
W = rands(S,R);                     % initialize weights to small random numbers
b = rands(S,1);                     % initialize biases to small random numbers
df = 1;                             % frequency of displayed epochs
me = 20;                            % maximum number of epochs
tp = [df,me];
[W,b,epochs] = trainp(W,b,P,T,tp);  % training by the perceptron algorithm
y = hardlim(W*P,b);                 % assesses network performance
```

The trainp command implements the update equation (4.131) and returns the final weights and bias values $W$ and $b$. The variable epochs gives the number of epochs employed during training.

Let us next consider the XOR gate with values

| $x$ | $y$ | $x+y$ |
|---|---|---|
| 0 | 0 | 0 |
| 0 | 1 | 1 |
| 1 | 0 | 1 |
| 1 | 1 | 0 |

The pairs $(0,0)$, $(1,1)$ yield 0 and are assigned to set $B$, while $(1,0)$, $(0,1)$ produce 1 and are assigned to $A$. It is easy to check that these sets are not linearly separable. ∎

Nonlinearly separable categories like XOR can be dealt with by the back-propagation algorithm treated next.

### 4.9.3 Back-propagation algorithm

Let us consider a multilayer neural network (see Fig. 3.41) with $K_0$ inputs and $L$ hidden layers, each containing $K_i$ neurons. All neurons are similar and have the structure of Fig. 3.39. The nonlinearity is a smooth sigmoid function. At each time $n$ the input value $u(n)$ is presented, and the results produced by each neuron are transmitted from layer to layer. The output $y(n)$ is formed at the rightmost layer and has $K_L$ entries. Given the parameters $L$, $K_i$, $i = 0, 1, \ldots, L$, the neuron structure of Fig. 3.39, and the communication links specified by the architecture of Fig. 3.41, we seek to determine the weights of each neuron so that the output approximates a given desired form. The desired behavior is specified by the multichannel signal $d(n) = \big( d_1(n) \quad \ldots \quad d_{K_L}(n) \big)^T$. The network will be trained so that it learns to produce the desired output $d(n)$ in response to the excitation $u(n)$.

Input and desired response samples $u(n)$ and $d(n)$ are given over the interval $1 \le n \le N$. For each level $l$ and each node $i$ of layer $l$, we seek to determine the vector of weights of neuron $i$, $w_i^{(l)}$, that minimizes the total squared error

$$J(w) = \frac{1}{2} \sum_{n=1}^{N} \sum_{i=1}^{K_L} e_i^2(n) = \frac{1}{2} \sum_{n=1}^{N} |e(n)|^2$$

where

$$e(n) = \big( e_1(n) \quad \ldots \quad e_{K_L}(n) \big)^T = d(n) - y(n)$$

If we set derivatives equal to zero, we end up with a set of nonlinear equations. Thus we must resort to iterative algorithms. The input-output expression of the $j$

node at level $l$ is

$$y_j^{(l)}(n) = f(x_j^{(l)}(n)) \tag{4.132}$$

$$x_j^{(l)}(n) = w_j^{(l)T} u^{(l)}(n) \tag{4.133}$$

Vector $u^{(l)}(n)$ denotes the input at time $n$ for each node of level $l$. $x_j^{(l)}(n)$ designates the linear combination of inputs that is forwarded to $\psi$, and vector $w_j^{(l)}$ includes the weights from all nodes of level $l - 1$ to node $j$. The length of these vectors is $K_{l-1}$. It holds that $u^{(l)}(n) = y^{(l-1)}(n)$.

The back propagation algorithm is a steepest descent method. The descent direction is the cost gradient

$$\frac{1}{2} \sum_{n=1}^{N} \frac{\partial}{\partial w_j^{(l)}} |e(n)|^2 \tag{4.134}$$

The computation of the gradient is carried out as follows: Differentiation gives

$$\frac{1}{2} \frac{\partial}{\partial w_j^{(l)}} |e(n)|^2 = \frac{1}{2} \frac{\partial}{\partial x_j^{(l)}(n)} |e(n)|^2 \frac{\partial}{\partial w_j^{(l)}} x_j^{(l)}(n) \tag{4.135}$$

Linearity of Eq. (4.133) implies that

$$\frac{\partial}{\partial w_j^{(l)}} x_j^{(l)}(n) = u^{(l)}(n) \tag{4.136}$$

We set

$$\frac{1}{2} \frac{\partial}{\partial x_j^{(l)}(n)} |e(n)|^2 = \delta_j^{(l)}(n) \tag{4.137}$$

Then Eq. (4.135) becomes

$$\frac{1}{2} \frac{\partial}{\partial w_j^{(l)}} |e(n)|^2 = \delta_j^{(l)}(n) u^{(l)}(n) \tag{4.138}$$

The parameters $\delta_j^{(l)}(n)$ are recursively computed. The computation starts from the last layer $l = L$ and proceeds from right to left. This backward direction explains the name of the method. At the first step we have

$$\delta_j^{(L)}(n) = \frac{1}{2} \frac{\partial}{\partial x_j^{(L)}(n)} \sum_{i=1}^{K_L} e_i^2(n) = \frac{1}{2} \frac{\partial}{\partial x_j^{(L)}(n)} e_j^2(n) = e_j(n) \frac{\partial}{\partial x_j^{(L)}(n)} e_j(n)$$

or

$$\delta_j^{(L)}(n) = -\epsilon_j^{(L)}(n)\psi'(x_j^{(L)}(n))$$

$$\epsilon_j^{(L)}(n) = e_j(n) = d_j(n) - \psi(x_j^{(L)}(n))$$

Let $l < L$. We assume that $\delta_i^{(l+1)}(n)$ have been computed for $i = 1, 2, \ldots, K_{l+1}$. Differentiation gives

$$\delta_j^{(l)}(n) = \frac{1}{2}\sum_{i=1}^{K_{l+1}} \frac{\partial|e(n)|^2}{\partial x_i^{(l+1)}(n)} \frac{\partial x_i^{(l+1)}(n)}{\partial x_j^{(l)}(n)} = \sum_{i=1}^{K_{l+1}} \delta_i^{(l+1)}(n) \frac{\partial x_i^{(l+1)}(n)}{\partial x_j^{(l)}(n)}$$

Moreover

$$\frac{\partial x_i^{(l+1)}(n)}{\partial x_j^{(l)}(n)} = \frac{\partial\left(w_i^{(l+1)^T} y^{(l)}(n)\right)}{\partial x_j^{(l)}(n)} = w_i^{(l+1)^T} \frac{\partial y^{(l)}(n)}{\partial x_j^{(l)}(n)}$$

But

$$y^{(l)}(n) = \left(\psi(x_1^{(l)}(n)) \quad \psi(x_2^{(l)}(n)) \quad \cdots \quad \psi(x_{K_l}^{(l)}(n))\right)^T$$

Hence

$$\frac{\partial y^{(l)}(n)}{\partial x_j^{(l)}(n)} = \left(0 \quad \cdots \quad 0 \quad \psi'(x_j^{(l)}(n)) \quad 0 \quad \cdots \quad 0\right)^T$$

Thus

$$\frac{\partial x_i^{(l+1)}(n)}{\partial x_j^{(l)}(n)} = w_{ij}^{(l+1)}\psi'(x_j^{(l)}(n))$$

Combining the above relations, we have

$$\delta_j^{(l)}(n) = \epsilon_j^{(l)}(n)\psi'(x_j^{(l)}(n))$$

$$\epsilon_j^{(l)}(n) = \sum_{i=1}^{K_{l+1}} \delta_i^{(l+1)}(n)w_{ij}^{(l+1)}$$

A significant simplification results if the instantaneous error replaces the average error, much as in the perceptron algorithm and the LMS algorithm:

$$s(n) = \frac{\partial|e(n)|^2}{\partial w}$$

**Table 4.3** Back-propagation algorithm

$$w_j^{(l)}(n+1) = w_j^{(l)}(n) + \gamma \delta_j^{(l)}(n) u^{(l)}(n)$$

Forward computations at step $l$

$$u^{(l)}(n) = \left[ u_1^{(l)}(n) \dots u_{K_l}^{(l)}(n) \right]^T$$

$$u_j^{(l)}(n) = \psi\left( x_j^{(l-1)}(n) \right)$$

$$x_j^{(l-1)}(n) = w_j^{(l-1)T}(n) u^{(l-1)}(n)$$

Backward computations at step $l$

$$\delta_j^{(L)}(n) = \epsilon_j^{(L)}(n) \psi'(x_j^{(L)}(n))$$

$$\epsilon_j^{(L)}(n) = d_j(n) - \psi(x_j^{(L)}(n))$$

For $l = L - 1, \dots, 1$

$$\delta_j^{(l)}(n) = \epsilon_j^{(l)}(n) \psi'(x_j^{(l)}(n))$$

$$\epsilon_j^{(l)}(n) = \sum_{i=1}^{K_{l+1}} \delta_i^{(l+1)}(n) w_{ij}^{(l+1)}(n)$$

The back-propagation algorithm with the above modification is summarized in Table 4.3. A convenient choice for the activation function is $\psi(x) = 1/(1 + e^{-x})$. In this case the derivative becomes

$$\psi'(x) = \psi(x)(1 - \psi(x))$$

### *Example 4.14* *XOR*

We consider XOR synthesis with neural networks. The perceptron is now replaced with a neural network having 1 hidden layer and 5 neurons featuring the tansig activation function. The input vectors are registered as columns of the matrix $P$:

$$P = \begin{pmatrix} 0 & 0 & 1 & 1 \\ 0 & 1 & 0 & 1 \end{pmatrix}$$

The desired response is expressed in terms of the row $T$:

$$T = (0 \quad 1 \quad 1 \quad 0)$$

The program returns the network weights, biases, number of epochs and the total squared error through training.

```
[R,Q] = size(P);        % number of inputs R, number of data vectors Q
S1 = 5;                 % number of hidden neurons
```

```
[S2,Q] = size(T);              % number of output neurons S2
[W1,b1] = rands(S1,R);         % initializes weights and biases from input to
                               % the hidden layer by small random numbers
[W2,b2] = rands(S2,S1);        % initialize weights and biases from hidden
                               % layer to output by small random numbers
df = 40;                       % frequency of displayed epochs
me = 600;                      % maximum number of epochs
eg = 0.02;                     % acceptable error level
lr = 0.02;                     % learning rate
tp = [df,me,eg,lr];            % training by back-propagation
[W1,b1,W2,b2,ep,tr] = trainbp(W1,b1,'tansig',W2,b2,'purelin',p,t,tp);
y1 = tansig(W1*P,b1);
y2 = purelin(W2*y1,b2)         % assesses network performance          ∎
```

The speed of back propagation is improved by the incorporation of momentum. Momentum ignores small features in the error. Thus the network does not get trapped in shallow local minima. Learning with momentum relies on the update

$$w_j^{(l)}(n+1) - w_j^{(l)}(n) = m(w_j^{(l)}(n) - w_j^{(l)}(n-1)) + (1-m)\left(\gamma\delta_j^{(l)}(n)u^{(l)}(n)\right)$$

The weights change uses a fraction of the last weight change and a fraction of the change proposed by the back propagation rule. Typically the momentum constant is set to 0.95.

### Example 4.15   Character recognition
Multilayer neural networks and the back-propagation algorithm can be used in character recognition. Consider the following 14 letters

A  E  F  H  I  K  L  M  N  T  V  X  Y  Z

Each letter is described by a $7 \times 5$ matrix with entries 0 or 1. A and E are indicated in Fig. 4.25.

A is represented by the matrix

$$A = \begin{pmatrix} 0 & 0 & 1 & 0 & 0 \\ 0 & 1 & 0 & 1 & 0 \\ 1 & 0 & 0 & 0 & 1 \\ 1 & 1 & 1 & 1 & 1 \\ 1 & 0 & 0 & 0 & 1 \\ 1 & 0 & 0 & 0 & 1 \\ 1 & 0 & 0 & 0 & 1 \end{pmatrix}$$

A neural network with 35 input nodes, 14 output nodes (as many as the letters), and 20 hidden nodes is proposed. Each letter is presented at the input in the given

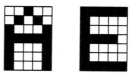

**Figure 4.25.** Representation of letters A and E.

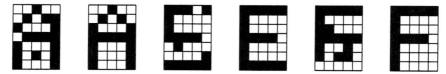

**Figure 4.26.** Distorted forms of letters A, E, and F, and the reconstructed letters produced by the network.

order: A E F $\cdots$ Z. The desired response consists of 14-dimensional vectors $(1 \quad 0 \quad 0 \quad \cdots \quad 0), (0 \quad 1 \quad 0 \quad \cdots \quad 0), \cdots, (0 \quad 0 \quad 0 \quad \cdots \quad 1)$, corresponding to the given letters. The network is trained by back-propagation and its ability to recognize distorted letters is illustrated in Fig. 4.26. Degradations modify certain entries of the letter matrix. ∎

### 4.9.4 Clustering

There are applications where the categories are not a priori known. In these cases the pattern classifier is called to group patterns without prior class knowledge. We refer to this process as *clustering*. Clustering algorithms reveal the homogeneity in the data by extracting the similarities and dissimilarities in the data set.

Suppose that the number of categories is known. Suppose also that it is reasonable to employ the average of a cluster as a representative. Then clustering can be accomplished by the $k$ means algorithm. The code vectors are the cluster representatives and the cells become the clusters. Let $x_i$, $1 \leq i \leq N$, denote the data patterns and $l$ the number of categories. $l$ vectors $w_j$ are initially assigned as cluster representatives. The distance of each remaining pattern $x$ from all representatives $w_j$, $d(x, w_j)$, is computed, and $x$ is assigned to the nearest representative. Once an epoch is completed, a partition of clusters is formed. New cluster representatives are determined by the centroid condition, and the average distortion is computed. The process is repeated until no significant change in the distortion is noticed.

In some applications (particularly when digital data are involved) the notion of a single cluster representative formed by averaging is not natural. An alternate approach is to represent each cluster $c_i$ by a set $r(c_i)$. The set $r(c_i)$ may contain more than one representatives, including the possibility $r(c_i) = c_i$. The $k$ means algorithm is readily modified to handle this more general case. It simply replaces the distance $d(x, w_i)$ with $d(x, r(c_i))$. The distance of a point $x$ to a set $A$ containing $k(A)$ elements can be defined in several ways. Popular choices are

$$d(x, A) = \min\{d(x, a), a \in A\}$$

$$d(x, A) = \sum_{a \in A} |x - a|$$

$$d(x, A) = \frac{1}{k(A)} \sum_{a \in A} |x - a|$$

The problem becomes harder if the number of categories is not known. A heuristic and computationally simple technique is to specify a threshold value and sweep the data, creating new categories each time the minimum distance of the current pattern from the preceding patterns exceeds the theshold. Hierarchical clustering algorithms are computationally more demanding alternatives.

## 4.10 CONTROL SYSTEMS DESIGN

Control aims to improve the performance of a system by proper input action. It is accomplished by means of feedback. The structure of the control system is illustrated in Fig. 4.27. $S$ designates the system that we want to control. It is commonly referred to as the *plant*. The plant $S$ is modified by feedback so that it acquires a certain desired behavior, that is, the output $y$ attains a desired characteristic. Control is exerted by the *controller* or *compensator* in the presence of the process noise $w$ and the measurement noise $v$. In *asymptotic tracking*, the controller attempts to maintain the output $y$ near the reference signal $r$, $y(t) \approx r(t)$. The configuration of Fig. 4.28 is known as *unity feedback*, and it

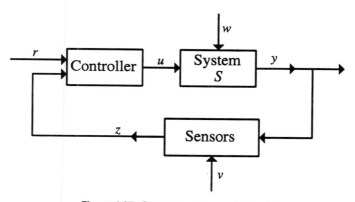

**Figure 4.27.** Structure of the control system.

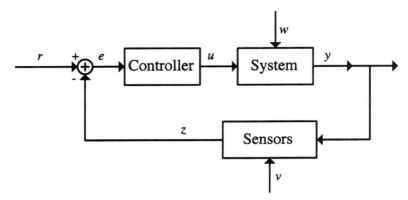

**Figure 4.28.** Unity feedback system.

will be proved very useful for asymptotic tracking. Here the controlled variable $y$ is measured and the measurement $z$ is compared to the reference signal $r$. The controller processes the error $e = r - z$ and the resulting output, $u$, steers the system so as to reduce the error $e$. The requirement $y = r$ cannot be satisfied due to physical limitations. Thus controller performance is evaluated by the asymptotic decay of the error $\lim_{t \to \infty} e(t) = 0$, or the minimization of a suitable error norm under reasonable consumption of input energy. A primary requirement in control design is stability of the closed loop system depicted in Fig. 4.28 so that signal amplitudes remain bounded when the inputs are bounded. The controller must achieve satisfactory tracking performance under the influence of parasitic signals and in the presence of plant's parameter variations. Control design is described in Chapter 13.

## BIBLIOGRAPHICAL NOTES

Good sources for stability of input-output operators are Desoer and Vidyasagar (1975) and Vidyasagar (1978). Least-squares estimation, with applications to filtering and prediction, are treated in many books, examples are Priestley (1981), Gardner (1986), Papoulis (1991), and Anderson and Moore (1979). The Levinson algorithm and its variants are covered in Marple (1987). Adaptive filters are treated in Honig and Messerchmitt (1984), Haykin (1991), Widrow and Stearns (1985), and Kalouptsidis and Theodoridis (1993). Recommended reading on prediction filtering and identification is supplied in Chapter 12. References on compression and coding are given in Chapter 14. The material on encryption is based on the book of Denning (1983). There are many excellent books on control. Consult the bibliographical notes at the end of Chapter 13. Pattern recognition topics are provided in Duda and Hart (1973) and Fukunaga (1990). Clustering algorithms are discussed in Jain and Dubes (1988). For the back-propagation algorithm, the reader may consult Haykin (1994) and the extensive list of references therein.

## PROBLEMS

**4.1.** Prove that the static system

$$y(n) = f(u(n))$$

is BIBO stable if $f$ is continuous. If in addition $f$ belongs to class $O(u)$, namely there exists $K$ such that $|f(u)| \leq K|u|$ for all $u$, the system is $p$ stable for all $p$.

**4.2.** Suppose that the discrete causal convolutional system is BIBO stable in the following sense: For any $L > 0$ there is $M > 0$ such that all inputs bounded by $L$ produce outputs bounded by $M$. Show that $h(n)$ is absolutely summable.

*Hint.* For each $M > 0$ the inputs $u_M(n) = \mathrm{sgn}\, h(M - n)$ are bounded by 1. Thus all corresponding outputs are bounded by some $L$. Pick sufficiently large $M$ such that $\sum_{n=0}^{M} |h(n)| > L$. Evaluate $y(M)$ to obtain a contradiction. It can be shown, using a theorem from functional analysis called uniform boundedness, that the above concept of stability is equivalent to BIBO stability.

**4.3.** Use the MATLAB command conv to simulate the systems of Example 4.1. Observe the stability behavior through the output plots of various inputs.

**4.4.** Prove that the time varying system

$$y(n) = \sum_{k=0}^{\infty} h(n, k) u(k)$$

is $\ell_1$ stable if

$$\sup_{0 \leq k < \infty} \sum_{n=0}^{\infty} |h(n, k)| < \infty$$

**4.5.** Show that the linear time varying system

$$y(n) = c_0(n)u(n) + c_1(n)u(n - 1) + \cdots + c_p(n)u(n - p), \qquad n \geq 0$$

is BIBO stable if the time varying coefficients are bounded. In particular, if

$$c_i(n) = c_i(n - 1) + w_i(n),$$

the system is BIBO stable provided that the disturbances $w_i(n)$ are absolutely summable. Provide a MATLAB validation.

**4.6.**  Show that the cascade and parallel connection of discrete BIBO stable systems is BIBO stable. Is this true for the feedback connection?

Exercises 4.7–4.12 deal with the analysis of output statistics of static systems.

**4.7.**  Consider the static system with $k$ inputs and $m$ outputs

$$y(n) = f(u(n)) \qquad (4.139)$$

and the family of first-order input distributions

$$P_{u,n}(u) = P\{\zeta : u(n)(\zeta) \le u\} = P[u(n) \le u]$$

Show that the family of first-order output distributions is given by

$$P_{y,n}(y) = P[u(n) \in D_y] = \int_{D_y} P_{u,n}(u)\,du$$

where

$$D_y = \{u \in R^k : f(u) \le y\}$$

Derive a similar expression for the higher-order output distributions. Deduce that if the input is stationary, the output is stationary.

**4.8.**  If (4.139) is single-input, single-output and $f$ is a squashing function (see Section 3.13), prove that

$$P_y(y) = \begin{cases} 0, & y < 0 \\ P_u(f^{-1}(y)), & 0 < y < 1 \\ 1, & y \ge 1 \end{cases}$$

Specialize the result to the activation functions of Section 3.13. Calculate the probability density $p_y(y)$ in terms of the input density by differentiating the distribution function. If $f(u) = \text{sgn}(u)$ and the input is zero mean Gaussian with variance $\sigma^2$, show that $p_y(1) = 1 - \Phi(0)$, $p_y(-1) = \Phi(0)$, where $\Phi(x)$ is the error integral

$$\Phi(x) = \int_{-\infty}^{x} e^{-z^2/2\sigma^2}\,dz$$

**4.9.**  It is known from probability theory that if $f$ is differentiable, and for each $y$ there is at most one $u$ that solves the equation $f(u) = y$, then

$$p_y(y) = \frac{p_u(f^{-1}(y))}{\left| \det\left( \dfrac{\partial f}{\partial u} \bigg|_{f^{-1}(y)} \right) \right|} \qquad (4.140)$$

Consider the 2-input, 2-output system that converts Cartesian coordinates to polar coordinates $r = \sqrt{x^2 + y^2}$, $\phi = \arctan y/x$. Prove that

$$p_{r\phi}(r, \phi) = rp_{xy}(r\cos\phi, r\sin\phi), \qquad r > 0$$

If $x$ and $y$ are uncorrelated zero mean normal random variables with variance $\sigma^2$, then

$$p_{r\phi}(r, \phi) = \frac{r}{2\pi\sigma^2} e^{-r^2/2\sigma^2}, \qquad r > 0, \quad |\phi| < \pi$$

Conclude that the random variables $r$ and $\phi$ are independent, $\phi$ is uniformly distributed in the interval $[-\pi, \pi]$, and $r$ follows the *Rayleigh distribution*

$$p_r(r) = \frac{r}{\sigma^2} e^{-r^2/2\sigma^2} u_s(r)$$

**4.10.** High-order output densities can be determined with the aid of (4.140). If $f(u_i) = y_i$, $1 \leq i \leq l$, have a unique solution, show that

$$p_{y, n_1, \dots, n_l}(y_1, \dots, y_l) = \frac{p_{u, n_1 \dots, n_l}(u_1, \dots, u_l)}{\prod_{i=1}^{l} |f'(u_i)|} \qquad (4.141)$$

In particular, show that if the input $u(n)$ is a purely random process, the output of (4.139) is also a purely random process.

**4.11.** Formula (4.140) enables the computation of the first-order density of the output, if $k = m$. If $m < k$, form the new system

$$\tilde{y} = \begin{pmatrix} y \\ \hat{y} \end{pmatrix} = \begin{pmatrix} f(u) \\ \hat{f}(u) \end{pmatrix}$$

by augmenting $f$ with $k - m$ additional variables, and set

$$\hat{y}_{m+1} = u_{m+1}, \dots, \hat{y}_k = u_k$$

unless physical motivation dictates a different choice. The density of the augmented system $\tilde{y}$ is determined by (4.140). Then the density of the first block of $\tilde{y}$, $y$, is found using the standard formula

$$p_y(y) = \int p_{\tilde{y}}(y, \hat{y}) d\hat{y}$$

Apply the above method to analyze the additive noise model $y = u + \xi$. $y$ represents the measured value of the signal $u$ and $\xi$ the measurement noise. $u$ and $\xi$ have joint density $p_{u,\xi}(u,\xi)$. Setting $\hat{y} = \xi$, show that

$$p_y(y) = \int p_{u,\xi}(y - \hat{y}, \hat{y})d\hat{y}$$

Deduce that the density of the sum of two independent random variables is given by the convolution of their densities. Show that if $u$ and $\xi$ are jointly Gaussian, $y$ is Gaussian as well.

**4.12.** Apply the method of Exercise 4.11 for the multiplicative noise model $y = u\xi$. Show that

$$p_y(y) = \int \frac{1}{|\hat{y}|} p_{u,\xi}\left(\frac{y}{\hat{y}}, \hat{y}\right)d\hat{y}$$

If $u$ and $\xi$ are independent and uniformly distributed in $[0,1]$, deduce

$$p_y(y) = \int_y^1 \frac{1}{\hat{y}}d\hat{y} = \begin{cases} -\ln y, & 0 < y < 1 \\ 0, & \text{otherwise} \end{cases}$$

**4.13.** Prove that the 2-D convolutional model

$$y(n_1, n_2) = \sum_{k_1=-\infty}^{\infty} \sum_{k_2=-\infty}^{\infty} h(n_1 - k_1, n_2 - k_2)u(k_1, k_2)$$

is BIBO stable if and only if $h(n_1, n_2)$ is absolutely summable. In particular show that the 2-D FIR system is $p$ stable for all $1 \le p \le \infty$. Specialize the above to a separable system.

**4.14.** In analogy with the program given in Example 4.6, write a MATLAB code to compute the optimum predictor (4.71) using the extended regressor of Eq. (4.72). Verify that the corresponding prediction error is white. Compare the performance of the above predictor and $E^*[y(n)|u(n-1), \ldots, u(n-m)]$ in terms of the minimum mean squared error.

**4.15.** Using the basic identities of Appendix 4A for the inversion of partitioned matrices, derive a Levinson-type algorithm for the solution of nonsymmetric Toeplitz equations.

**4.16.** A block matrix $R$ with entries square matrices $r_{ij}$ is called *block Toeplitz* if $r_{i+k,j+k} = r_{ij}$ for all $k$. Let $R_m$ denote the submatrix of $R$ obtained by retaining the first $m$ block rows and columns. Show that a block Toeplitz matrix satisfies $JR_mJ = R_m^{bT}$, where $J$ is the block order reversing matrix

$$J = \begin{pmatrix} 0 & 0 & \cdots & 0 & I \\ 0 & 0 & \cdots & I & 0 \\ \vdots & & & & \\ I & 0 & \cdots & 0 & 0 \end{pmatrix}$$

and $R_m^{bT}$ stands for the block transpose of $R_m$; that is the $ij$ block entry is given by the $ji$ entry of $R_m$. Note that the block transpose does not coincide with the ordinary transpose of $R_m$. Derive a block Levinson algorithm for the block system $Rc = d$, using the identities for the inversion of partitioned matrices. Show that multichannel filtering and prediction can be solved with the above algorithm. Program the resulting block Levinson variant in MATLAB. Use the command clock to compare performance with the conventional solution.

**4.17.** Let $\mathcal{H}_p(n)$ be the linear space spanned by the random variables $y(n)$, $y(n - 1)$, ..., $y(n - p)$. Prove in connection with Section 4.3 that the backward errors

$$e_0^b(n), \ e_1^b(n), \ e_2^b(n), \ldots, e_p^b(n)$$

form an orthogonal base in $\mathcal{H}_p(n)$ and that the parameters $k_m^c$ appearing in the Levinson algorithm express the projection of $x(n)$ in the above basis. Furthermore show that the reflection coefficients $k_m$ represent the signal value $y(n + 1)$ in the above orthogonal basis.

**4.18.** Let $w_{p+1}(N + 1)$ be the augmented Kalman gain

$$w_{p+1}(N + 1) = R_{p+1}^{-1}(N)\varphi_{p+1}(N + 1)$$

Use the inversion for partitioned matrices identities of the Appendix and the nesting properties (4.39), (4.45) to show the recursions

$$w_{p+1}(N + 1) = \begin{pmatrix} 0 \\ w_p(N) \end{pmatrix} + \begin{pmatrix} 1 \\ -a_p(N) \end{pmatrix} \hat{k}(N + 1)$$

$$w_{p+1}(N + 1) = \begin{pmatrix} w_p(N + 1) \\ 0 \end{pmatrix} + \begin{pmatrix} -b_p(N) \\ 1 \end{pmatrix} \tilde{k}(N + 1)$$

The Kalman gain is updated as follows: First $w_{p+1}(N + 1)$ is computed from the first recursion. Then $w_p(N + 1)$ is determined from the second recursion. The forward and backward predictors are updated by the RLS recursion. Derive a complete list of recursions, and show that complexity is reduced to $O(p)$ operations per adaptation step. Simulate the resulting fast RLS algorithm by MATLAB, and make comparisons with the RLS and LMS algorithms.

**4.19.** Let

$$x(n) + \frac{13}{12}x(n - 1) + \frac{9}{24}x(n - 2) + \frac{1}{24}x(n - 3) = w(n),$$

where $w(n)$ is white noise. Using the code of Example 4.3, simulate the source and generate data $x(n)$, $1 \le n \le N$. Show that the true parameters are recovered. Discuss the filtering problem $y(n) = x(n) + v(n)$ as in Example 4.3. Use the RLS and the LMS algorithms described in Example 4.7 to recover the AR coefficients from data $(x(n), y(n))$.

**4.20.** If $H$ is a Hankel matrix and $J$ is the order exhange permutation, show that $JH$ and $HJ$ are Toeplitz and hence that the Levinson algorithm is applicable for the solution of Hankel equations. Program the algorithm in MATLAB, and use it to solve the least squares polynomial identification of Example 4.4.

**4.21.** Consider the second-order polynomial system

$$y(n) = a_0 + a_1 u(n) + a_2 u^2(n) + v(n)$$

Compute the coefficients under the assumption that $u(n)$ is zero mean Gaussian, independent of $v(n)$. Verify your results by MATLAB simulations.

**4.22.** This exercise relates to Example 4.4. Given polynomials $p(u)$, $q(u)$, of degree at most $d - 1$, prove that the function

$$< p, q >= \sum_{n=0}^{N} w(n) p(u_n) q(u_n) = \mathbf{p}^T Q(N) \mathbf{q}$$

defines a valid inner product, where $\mathbf{p}, \mathbf{q}$ are the vectors of the polynomial coefficients and $Q(N)$ is defined by (4.66).

Prove that if $p_0(u)$, $p_1(u)$, ..., $p_k(u)$ is a set of orthogonal monic polynomials with respect to the above inner product and $\deg p_i(u) = i$, then they satisfy the two-term recurrent equation

$$p_{n+1}(u) = (u + a_n) p_n(u) + b_n p_{n-1}(u) , \qquad n \geq 0$$

with initial conditions

$$p_0(u) = 1, \quad p_{-1}(u) = 0, \quad b_0 = 1$$

or equivalently in vector form

$$\mathbf{p}_{n+1} = \begin{pmatrix} 0 \\ \mathbf{p}_n \end{pmatrix} + a_n \begin{pmatrix} \mathbf{p}_n \\ 0 \end{pmatrix} + b_n \begin{pmatrix} \mathbf{p}_{n-1} \\ 0 \\ 0 \end{pmatrix}$$

**4.23.** Show that a quantizer satisfying the centroid condition has the following properties:
(a) $E[Q(u)] = E[u]$.
(b) The quantization error is uncorrelated with the quantizer output, $E[Q(u)(u - Q(u))] = 0$. In contrast, the quantizer input and the quantization error are correlated. Indeed, show that
(c) $E[(u - Q(u))u] = \sigma_u^2 - \sigma_{Q(u)}^2$.

**4.24.** A simple repetitive $(n, 1)$ code simply repeats a single information bit $n$ times. Show that that these codes are low bit rate codes with minimum distance $n$.

**4.25.** A simple parity check $(n, n - 1)$ code adds one bit so that the total number of ones is even. Show that these codes are high-rate codes with minimum distance 2.

**4.26.** Consider the $(7, 4)$ linear code with generator matrix

$$G = \begin{pmatrix} 1 & 1 & 0 & 1 & 0 & 0 & 0 \\ 0 & 1 & 1 & 0 & 1 & 0 & 0 \\ 1 & 1 & 1 & 0 & 0 & 1 & 0 \\ 1 & 0 & 1 & 0 & 0 & 0 & 1 \end{pmatrix}$$

Find the codeword resulting from the message $u = 1011$. Construct the parity check matrix. Compute the syndrome $s = s_1 s_2 s_3$. Let $r = 1001001$ be the received sequence. Show that the minimum distance decoding determines the transmitted message. Determine the error detecting and the error correcting capability of the code.

**4.27.** Show that for any binary $(n, k)$ linear code with minimum distance greater or equal than $2t + 1$ the number of parity check digits satisfies the Hamming bound

$$n - k \geq \log_2 \left( 1 + \binom{n}{1} + \binom{n}{2} + \cdots + \binom{n}{t} \right)$$

**4.28.** Consider the $(8, 4)$ linear code generated by the matrix

$$G = \begin{pmatrix} 0 & 1 & 1 & 1 & 1 & 0 & 0 & 0 \\ 1 & 1 & 1 & 0 & 0 & 1 & 0 & 0 \\ 1 & 1 & 0 & 1 & 0 & 0 & 1 & 0 \\ 1 & 0 & 1 & 1 & 0 & 0 & 0 & 1 \end{pmatrix}$$

Develop an encoder circuit and a syndrome computation circuit. Show that $d_{min} = 4$. Decode the received sequence $r = 00000100$.

**4.29.** Implement the $k$ means algorithm in MATLAB. Generate a set of 12 IID Gaussian two-dimensional vectors with zero mean and unit variance. Classify the patterns into four classes.

**4.30.** Consider the autoregressive source

$$x(n) = 0.9x(n - 1) + w(n)$$

$w(n)$ is white noise of unit variance.

Form 60000 samples of the $m$-dimensional vectors $(x(n) \cdots x(n - m + 1))$, for $1 \leq m \leq 7$. Construct vector quantizers of dimensions $m$ at a rate of 1 bit per sample using the $k$ means algorithm. Draw a plot of the signal to quantization noise ratio versus the quantizer dimension.

**4.31.** Demonstrate by experiment that the basic VQ method has a blockiness and sawtooth effect. Hint. Construct a closeup of the eye and the shoulder of Lenna and its coded version.

**4.32.** Consider the RSA encryption scheme with public keys $n = 55$ and $e = 7$. Encipher the plaintext $M = 10$. Break the cipher by finding $p$, $q$, and $d$. Decipher the ciphertext $C = 35$.

**4.33.** Consider the enciphering transformation $f_k(m) = k_1 m + k_0 (\mathrm{mod}\ 26)$. Suppose that it is known that the plaintext letter E (4) is encrypted to F (5) and that H (7) is encrypted to W (22). Find the keys $k_1$ and $k_0$.

**4.34.** In connection with Example 4.14, prove that the XOR gate leads to nonlinearly separable categories. Check the MATLAB code of Example 4.13 for this case.

**4.35.** Prove the Rieger bound. Hint. Prove the following claims. (a) There is no codeword that is a burst of length $2t$. (b) Any two vectors that are zero except in their first $2t$ components yield different syndromes. (c) There are at least $N^{2t}$ vectors with different syndromes.

**4.36.** Let $A = \{(1, 1), (1, 3), (2, 2)\}$ and $B = \{(5, 4), (5, 5), (6, 4)\}$. Are the sets linearly separable? Run the perceptron algorithm. Repeat for the sets $A = \{(1, 1), (5, 5), (2, 2)\}$ and $B = \{(1, 3), (5, 4), (6, 4)\}$. Work out the above cases with the backpropagation algorithm.

## APPENDIX

In this appendix some results about inversion of matrices are reviewed.

*Matrix inversion for modified matrices.* Let us consider the linear system

$$AX_o = E \quad \text{or} \quad X_o = A^{-1}E \tag{4.142}$$

Suppose that the parameters $A$ and $E$ are modified and the new system

$$(A + BCD)X_n = E + BF \tag{4.143}$$

is obtained. We want to express $X_n$ in terms of the previous estimate $X_o$. The matrices involved have arbitrary dimensions provided that the above expression is well defined. We write (4.143) as $AX_n + BCDX_n = E + BF$. We multiply both sides by $A^{-1}$ (assuming that it exists) to obtain $X_n + A^{-1}BCDX_n = A^{-1}E + A^{-1}BF$, or

$$X_n = A^{-1}E + A^{-1}BK \tag{4.144}$$

$$K = F - CDX_n \tag{4.145}$$

We substitute (4.144) into (4.145) to obtain $K = F - CDA^{-1}E - CDA^{-1}BK$, or $(I + CDA^{-1}B)K = F - CDA^{-1}E$. Putting together the latter expression and (4.144), we arrive at

$$X_n = A^{-1}E + A^{-1}BK \tag{4.146}$$

$$K = (I + CDA^{-1}B)^{-1}[F - CDA^{-1}E] \tag{4.147}$$

If we set $E = I$, $F = 0$ in the above formulas, the resulting solution $X_n$ is the inverse of $A + BCD$. It is given by

$$(A + BCD)^{-1} = A^{-1} - A^{-1}B[I + CDA^{-1}B]^{-1}CDA^{-1} \tag{4.148}$$

If $C$ is invertible, we write $I = CC^{-1}$, and the above formula becomes

$$(A + BCD)^{-1} = A^{-1} - A^{-1}B[C^{-1} + DA^{-1}B]^{-1}DA^{-1} \tag{4.149}$$

**Matrix inversion for partitioned matrices.** As above we assume that $X_o$ given by (4.142) is at our disposal. This time we want to use this estimate in order to determine the solution of an augmented problem $RX_n = Q$ written in partitioned form

$$\begin{pmatrix} A & B \\ C & D \end{pmatrix} \begin{pmatrix} X \\ \tilde{X} \end{pmatrix} = \begin{pmatrix} E \\ F \end{pmatrix} \tag{4.150}$$

Equivalently we have the pair of equations

$$AX + B\tilde{X} = E, \quad CX + D\tilde{X} = F$$

We solve the first equation with respect to $X$, and substitute into the second

$$X = A^{-1}E - A^{-1}B\tilde{X} \tag{4.151}$$

$$CA^{-1}E + [D - CA^{-1}B]\tilde{X} = F$$

The latter equation becomes

$$\tilde{X} = [D - CA^{-1}B]^{-1}[F - CA^{-1}E] \tag{4.152}$$

We combine (4.151) and (4.152) as follows:

$$\begin{pmatrix} X \\ \tilde{X} \end{pmatrix} = \begin{pmatrix} A^{-1}E \\ 0 \end{pmatrix} + \begin{pmatrix} -A^{-1}B \\ I \end{pmatrix} \tilde{X} \tag{4.153}$$

$$\tilde{X} = [D - CA^{-1}B]^{-1}[F - CA^{-1}E] \tag{4.154}$$

The above expressions require that $A^{-1}$ as well as $(D - CA^{-1}B)^{-1}$ exist. Suppose next that the previous estimate is $X_o = D^{-1}F$. This means that previous computations used data that are stored in the southeast part of the matrix and the lower part of the right-hand side. In this case the roles of $X$ and $\tilde{X}$ are reversed. Thus the second block row is solved with respect to $\tilde{X}$, and the following equations result

$$\begin{pmatrix} X \\ \tilde{X} \end{pmatrix} = \begin{pmatrix} 0 \\ D^{-1}F \end{pmatrix} + \begin{pmatrix} I \\ -D^{-1}C \end{pmatrix} X \tag{4.155}$$

$$X = (A - BD^{-1}C)^{-1}[E - BD^{-1}F] \tag{4.156}$$

We compute the inverse of $R$ using the above formulas. Let

$$R^{-1} = \begin{pmatrix} X & Y \\ \tilde{X} & \tilde{Y} \end{pmatrix}$$

Then

$$R\begin{pmatrix} X \\ \tilde{X} \end{pmatrix} = \begin{pmatrix} I \\ 0 \end{pmatrix}, \quad R\begin{pmatrix} Y \\ \tilde{Y} \end{pmatrix} = \begin{pmatrix} 0 \\ I \end{pmatrix} \tag{4.157}$$

We solve the first using (4.153) and (4.154). Notice first that if $\Delta = D - CA^{-1}B$, Eq.(4.149) gives

$$\Delta^{-1} = (D - CA^{-1}B)^{-1} = D^{-1} + D^{-1}C[A - BD^{-1}C]^{-1}BD^{-1} \tag{4.158}$$

Thus (4.153) and (4.154) imply that

$$\tilde{X} = -\Delta^{-1}CA^{-1}$$

$$X = A^{-1} + A^{-1}B[D - CA^{-1}B]^{-1}CA^{-1} = A^{-1} + A^{-1}B\Delta^{-1}CA^{-1}$$

Likewise we have

$$Y = -A^{-1}B\Delta^{-1}$$

$$\tilde{Y} = \Delta^{-1}$$

We arrive at the important identity

$$\begin{pmatrix} A & B \\ C & D \end{pmatrix}^{-1} = \begin{pmatrix} A^{-1} + A^{-1}B\Delta^{-1}CA^{-1} & -A^{-1}B\Delta^{-1} \\ -\Delta^{-1}CA^{-1} & \Delta^{-1} \end{pmatrix} \tag{4.159}$$

In a similar fashion we obtain

$$\begin{pmatrix} A & B \\ C & D \end{pmatrix}^{-1} = \begin{pmatrix} \Gamma^{-1} & -\Gamma^{-1}BD^{-1} \\ -D^{-1}C\Gamma^{-1} & D^{-1} + D^{-1}C\Gamma^{-1}BD^{-1} \end{pmatrix} \tag{4.160}$$

$$\Gamma = A - BD^{-1}C \tag{4.161}$$

Identity (4.160) can be rewritten in the format

$$R^{-1} = \begin{pmatrix} I & 0 \\ -D^{-1}C & \Gamma \end{pmatrix} \begin{pmatrix} \Gamma^{-1} & 0 \\ 0 & \Gamma^{-1} \end{pmatrix} \begin{pmatrix} I & -BD^{-1} \\ 0 & D^{-1} \end{pmatrix} \tag{4.162}$$

The above equation expresses the inverse of $R$ as a product of a block lower triangular matrix times a block diagonal matrix times an upper triangular matrix (block LDU decomposition). It is established by direct evaluation of the right-hand side. The determinant of $R^{-1}$ is

$$\det R^{-1} = \det \Gamma (\det \Gamma^{-1})^2 \det D^{-1} = \det \Gamma^{-1} \det D^{-1} \tag{4.163}$$

PROOF OF THEOREM 4.4   Since $Z$ is Gaussian, $Y$ is also Gaussian, and

$$f_{X|Y}(x|y) = L \exp\left( -\frac{1}{2}(z - m_Z)^T C_{ZZ}^{-1}(z - m_Z) + \frac{1}{2}(y - m_Y)^T C_{YY}^{-1}(y - m_Y) \right)$$

where

$$L = \frac{(\det C_{YY})^{1/2}}{(2\pi)^{n/2}(\det C_{ZZ})^{1/2}} \tag{4.164}$$

Next we use the matrix inversion for partitioned matrices formula (4.160) to substitute $C_{ZZ}^{-1}$ in the above expression. We observe that $C_{ZZ}^{-1}$ is symmetric and is

given by

$$\begin{pmatrix} \Gamma^{-1} & -\Gamma^{-1}C_{XY}C_{YY}^{-1} \\ -C_{YY}^{-1}C_{YX}\Gamma^{-1} & C_{YY}^{-1} + C_{YY}^{-1}C_{YX}\Gamma^{-1}C_{XY}C_{YY}^{-1} \end{pmatrix}$$

where $\Gamma = C_{XX} - C_{XY}C_{YY}^{-1}C_{YX}$. Thus the argument of the exponential becomes $-\frac{1}{2}$ times

$$\begin{pmatrix} x - m_X \\ y - m_Y \end{pmatrix}^T \begin{pmatrix} \Gamma^{-1} & -\Gamma^{-1}C_{XY}C_{YY}^{-1} \\ -C_{YY}^{-1}C_{YX}\Gamma^{-1} & C_{YY}^{-1} + C_{YY}^{-1}C_{YX}\Gamma^{-1}C_{XY}C_{YY}^{-1} \end{pmatrix} \begin{pmatrix} x - m_X \\ y - m_Y \end{pmatrix}$$

$$-(y - m_Y)^T C_{YY}^{-1}(y - m_Y)$$

$$= (x - m_X)^T \Gamma^{-1}(x - m_X) - 2(y - m_Y)^T C_{YY}^{-1}C_{YX}\Gamma^{-1}(x - m_X)$$

$$+(y - m_Y)^T C_{YY}^{-1}C_{YX}\Gamma^{-1}C_{XY}C_{YY}^{-1}(y - m_Y)$$

It is easy to verify that the latter expression can be written as the quadratic form

$$\left(x - (m_X + C_{XY}C_{YY}^{-1}(y - m_Y))\right)^T \Gamma^{-1}\left(x - (m_X + C_{XY}C_{YY}^{-1}(y - m_Y))\right)$$

Using formula (4.163), we find that

$$\det C_{ZZ}^{-1} = \frac{\det \Gamma^{-1}}{\det C_{YY}}$$

The theorem is proved if the above equation is replaced into (4.164). ∎

# 5

# HARMONIC ANALYSIS OF SIGNALS AND SYSTEMS

This chapter is concerned with the representation of periodic signals in terms of harmonically related exponential signals. Discrete, continuous time, digital, and multidimensional signals are examined. The periodic properties of outputs of convolutional systems are analyzed. The transform that converts a discrete signal into its spectral coefficients is known as discrete Fourier transform (DFT). Several computational tasks ultimately lead to a DFT computation. An important example is convolution. Fast Fourier transforms (FFT) are efficient algorithms for the calculation of the DFT. The Cooley-Tukey FFT and the Good-Thomas FFT algorithms are discussed. The study of spectral representations is continued in subsequent chapters.

## 5.1 HARMONIC ANALYSIS OF DISCRETE PERIODIC SIGNALS

We saw in Chapter 2 that the space of discrete periodic signals of period $N$ has dimension $N$. The set of $N$ periodic harmonically related exponentials of period $N$, $e^{-j(2\pi/N)kn}$, $0 \leq k \leq N - 1$, is orthogonal in this space. Hence every periodic signal $x$ of period $N$ is written as

| | | |
|---|---|---|
| Synthesis equation: | $$x(n) = \sum_{k=0}^{N-1} a_k e^{j(2\pi/N)kn}$$ | (5.1) |
| Analysis equation: | $$a_k = \frac{1}{N} \sum_{n=0}^{N-1} x(n) e^{-j(2\pi/N)kn}$$ | (5.2) |

The analysis equation follows from Eq. (2.62) of Chapter 2. Summation is carried over any interval of length $N$. The synthesis equation provides the *Fourier expansion* of discrete periodic signals. The coefficients $a_k$ are the *Fourier coefficients* or *spectral lines* of $x(n)$. The sequence of spectral lines can be viewed as a discrete periodic signal of period $N$. The Parseval identity Eq.(2.63) takes the form

$$\frac{1}{N}\sum_{n=0}^{N-1}|x(n)|^2 = \sum_{k=0}^{N-1}|a_k|^2 \tag{5.3}$$

and expresses the preservation of power.

**Example 5.1    *A sum of periodic sinuosoids***
The signal

$$x(n) = 1 + 2\sin\frac{2\pi}{N}n + \cos\frac{6\pi}{N}n, \qquad N > 3$$

is periodic with period $N$. Using Euler's identity, we obtain

$$x(n) = 1 + \frac{1}{j}e^{j(2\pi/N)n} - \frac{1}{j}e^{-j(2\pi/N)n} + \frac{1}{2}e^{j(6\pi/N)n} + \frac{1}{2}e^{-j(6\pi/N)n}$$

Simple inspection yields

$$a_0 = 1, \quad a_1 = \frac{1}{j} = -j, \quad a_{-1} = a_{N-1} = -\frac{1}{j} = j, \quad a_3 = \frac{1}{2}, \quad a_{-3} = a_{N-3} = \frac{1}{2}$$

and $a_k = 0$, for the remaining values of $k$ in $0 \leq k \leq N-1$. The Fourier coefficients are complex numbers, even if $x(n)$ is real. The real and imaginary parts are sketched in Fig. 5.1 for $N = 8$, while amplitude and phase are sketched in Fig. 5.2. ∎

The next signal cannot be directly handled by inspection. The Fourier coefficients are calculated via the analysis equation.

**Example 5.2    *Discrete periodic square wave***
Let $x(n)$ be the discrete periodic square wave:

$$x(n) = \begin{cases} 1, & |n| \leq N_1 \\ 0, & -\frac{N}{2} \leq n < -N_1, N_1 < n < \frac{N}{2}. \end{cases} \tag{5.4}$$

where $N$ is even. The analysis equation gives

$$a_k = \frac{1}{N}\sum_{n=-N_1}^{N_1} e^{-j(2\pi/N)kn} = \frac{1}{N}\left\{2\,\Re\left(\sum_{n=0}^{N_1} e^{j(2\pi/N)kn}\right) - 1\right\}$$

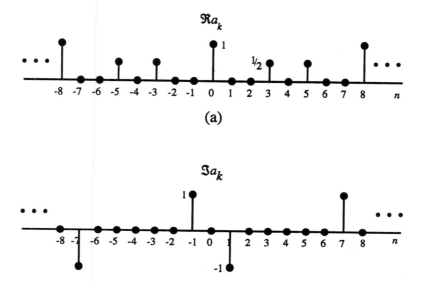

**Figure 5.1.** Fourier coefficients of signal in Example 5.1: (a) Real part; (b) imaginary part.

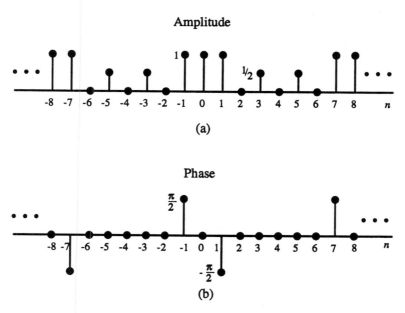

**Figure 5.2.** Amplitude and phase of the Fourier coefficients of the signal discussed in Example 5.1.

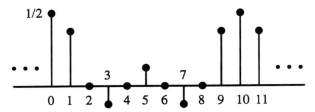

**Figure 5.3.** Fourier coefficients for the periodic square wave when $N_1 = 2$ and $N = 10$.

Straightforward computations yield

$$a_0 = \frac{2N_1 + 1}{N}, \quad a_k = \frac{1}{N} \frac{\sin\left(2\pi k(N_1 + \frac{1}{2})/N\right)}{\sin(2\pi k/2N)}, \qquad k = 1, \ldots, N - 1$$

The signal takes the values 0 and 1 equally often if $N - (2N_1 + 1) = 2N_1 + 1$ or $N = 4N_1 + 2$. The case $N_1 = 2$, $N = 10$ is illustrated in Fig. 5.3.   ■

## 5.2  DISCRETE FOURIER TRANSFORM

The discrete Fourier transform (DFT) translates a periodic signal from the time domain to the frequency domain. Using the isomorphism between periodic signals and signals of finite extent, we may view the DFT as the transformation of a finite support signal, $x(n)$, into the finite support signal formed by its Fourier coefficients. More precisely, let $N$ be an integer and

$$w_N = e^{-j(2\pi/N)}$$

The subscript $N$ is omitted when no confusion arises. The DFT of length $N$ converts a signal of extent $N$, $x$, into the signal of finite extent $N$, $X$, via the rule

$$X(k) = \sum_{n=0}^{N-1} x(n) w_N^{kn}, \qquad k = 0, 1, \ldots, N - 1 \tag{5.5}$$

Comparison of Eqs. (5.5) and (5.2) shows that $X(k) = Na_k$. If we view a signal of extent $N$ as an $N$-dimensional vector, the DFT is represented by the linear transformation

$$X = W_N x \tag{5.6}$$

where $W_N$ is the $N \times N$ matrix with entries $(W_N)_{kn} = w_N^{kn}$, $0 \le k, n \le N - 1$. The columns of $W_N$ consist of the harmonically related exponential signals, which as we have seen, are orthogonal. Thus $W_N$ is a unitary matrix, and its inverse is

given by

$$W_N^{-1} = \frac{1}{N} W_N^*$$

where $W_N^*$ is the conjugate transpose. The latter equation is a compact way of writing the inversion formula

$$x(n) = \frac{1}{N} \sum_{k=0}^{N-1} X(k) w_N^{-kn}, \qquad n = 0, 1, \ldots, N-1 \tag{5.7}$$

Computation of the DFT can be carried out with the aid of definition (5.5) or (5.6). It requires $N^2$ complex multiplications and $N(N-1)$ complex additions provided that $W_N$ is available. We will see that the special structure of $W_N$ enables the development of efficient algorithms collectively known as fast Fourier transform (FFT). The importance of this remark stems from the fact that a wide range of signal processing computations ultimately leads to the computation of DFT.

The inverse DFT is specified by Eq. (5.7). If we take conjugates, we obtain

$$x^*(n) = \frac{1}{N} \sum_{k=0}^{N-1} X^*(k) w_N^{kn}$$

Thus, to compute the inverse DFT, we form the conjugate of $X$, find its DFT, and apply conjugation and scaling by $1/N$.

### Example 5.3  *MATLAB computation of the DFT*
The discrete Fourier transform is implemented in MATLAB by the command fft. Efficiency is maximized if $N$ is a power of 2. Returning to Example 5.1, we derive the plots of Fig. 5.1 and Fig. 5.2 by the code

```
N = 8;
n = 0:N-1;
x = ones(1,N) + 2*sin(2*pi/N*n) + cos(6*pi/N*n);
X = fft(x)/N;
subplot(221), plot(real(X)), subplot(222), plot(imag(X))
subplot(223), plot(abs(X)),  subplot(224), plot(angle(X))
```
■

## 5.3  DISCRETE FOURIER TRANSFORM OF DIGITAL SIGNALS

In analogy with Eq. (5.5), the discrete Fourier transform of a digital signal of length $N$ with values in the finite field $F$ is defined by

$$X(k) = \sum_{n=0}^{N-1} x(n) w^{kn} \tag{5.8}$$

We can no longer invoke orthogonality arguments to invert the above equation, since tools like the projection theorem are not valid. We proceed algebraically as follows: Let $y(m) = \sum_{k=0}^{N-1} X(k)w^{-km}$. Taking into account Eq. (5.8), we obtain

$$y(m) = \sum_k \sum_n x(n)w^{kn}w^{-km} = \sum_n x(n) \sum_k w^{k(n-m)} \qquad (5.9)$$

The sum $\sum_{k=0}^{N-1} w^{kr}$ is the partial sum of a geometric series with ratio $w^r$. We readily establish the following formula:

$$\sum_{k=0}^{N-1} w^{kr} = \begin{cases} \dfrac{w^{rN} - 1}{w^r - 1}, & w^r \neq 1 \\ N(\mathrm{mod}\ p), & w^r = 1 \end{cases}$$

where $p$ is the characteristic of $F$. To arrive at the familiar inversion formula, we assume that the order of $w$ equals the signal length $N$. Thus $w^N = 1$. Returning to Eq. (5.9), we see that the inner sum equals zero if $n \neq m$. Therefore $y(m) = N(\mathrm{mod}\ p)x(m)$, and

$$x(n) = \frac{1}{N(\mathrm{mod}\ p)} \sum_{k=0}^{N-1} X(k)w^{-kn}, \quad w^N = 1 \qquad (5.10)$$

Just as the DFT of a real vector is a complex vector, the DFT of a digital signal with values in $F$ can be defined in a larger field. Indeed, let $\hat{F}$ be a field containing $F$ as a subfield ($\hat{F}$ is an extension field of $F$) and $w \in \hat{F}$ of order $N$. Then Eq. (5.8) maps signals of length $N$ with values in $F$, into signals of length $N$ with values in $\hat{F}$. It is also necessary that the length $N$ coincides with the order of $w$.

### Example 5.4    *Digital exponential*
Consider the exponential signal

$$x(n) = a^n$$

where $a$ is a primitive element in the field $F$. Let $w$ be another primitive element. Then the signal blocklength is equal to the order of $w$. Let $k_0$ be the unique integer such that $a^{-1} = w^{k_0}$. It exists because $w$ is primitive in $F$. The DFT of $x(n)$ is

$$X(k) = \sum_{n=0}^{N-1} (aw^k)^n = \frac{a^N w^{kN} - 1}{aw^k - 1} = 0 \qquad \text{for } k \neq k_0$$

and

$$X(k_0) = N(\mathrm{mod}\ p)$$

There is only one nonzero spectral line at frequency $k_0$.    ∎

## 5.4 FOURIER ANALYSIS OF ANALOG PERIODIC SIGNALS

The set of continuous time harmonically related exponentials of period $T$ is a countable orthogonal set in the Hilbert space $L_2(T)$ of periodic functions of period $T$ that are Lebesgue square integrable over an interval of length $T$. Therefore any signal in the linear closure of the set of harmonically related exponentials can be decomposed as

$$
\begin{aligned}
x(t) &= \sum_{k=-\infty}^{\infty} a_k e^{jk\omega t} \\[2mm]
a_k &= \frac{1}{T} \int_T x(t) e^{-jk\omega t} dt
\end{aligned}
\left.\vphantom{\begin{aligned}x\\a\end{aligned}}\right\}
\quad T = \frac{2\pi}{\omega}
$$

$$(5.11)$$
$$(5.12)$$

The complex coefficients $a_k$ are called *Fourier coefficients* or *spectral lines*. Each $a_k$ corresponds to the projection of the periodic signal $x(t)$ on the $k$th orthogonal direction $e^{jk\omega t}$ and indicates the spectral content of $x(t)$ at frequency $k\omega$. Integration is carried over an arbitrary interval of length $T$. Application of Parseval identity (2.67) to periodic signals gives

$$
\frac{1}{T} \int_T |x(t)|^2 dt = \sum_{k=-\infty}^{\infty} |a_k|^2
\qquad (5.13)
$$

The latter equation expresses the preservation of power in the time and frequency domain. Each term $|a_k|^2$ provides the power at harmonic $k\omega$ rad/s. The sum of powers of all harmonics yields the total signal power.

Fourier himself used cosines and sines rather than exponentials in his exposition of this series. Conversion from one format to the other is easily done as follows: Let $x(t)$ be a real signal. Then

$$
x(t) = x^*(t) = \sum_{k=-\infty}^{\infty} a_k^* e^{-jk\omega t} = \sum_{k=-\infty}^{\infty} a_{-k}^* e^{jk\omega t}
$$

Therefore $a_k = a_{-k}^*$. The Fourier expansion of the above signal is

$$
x(t) = a_0 + \sum_{k=1}^{\infty} a_k e^{jk\omega t} + \sum_{k=1}^{\infty} a_{-k} e^{-jk\omega t} = a_0 + \sum_{k=1}^{\infty} 2\Re[a_k e^{jk\omega t}]
$$

Let $a_k = A_k e^{j\theta_k}$. Then

$$
x(t) = a_0 + 2 \sum_{k=1}^{\infty} A_k \cos(k\omega t + \theta_k)
\qquad (5.14)
$$

On the other hand, if $a_k = B_k + jC_k$, it holds that

$$x(t) = a_0 + 2\sum_{k=1}^{\infty}[B_k \cos k\omega t - C_k \sin k\omega t] \tag{5.15}$$

From a practical point of view the latter formulae are more natural since only positive frequencies can be measured. The use of exponential signals is more convenient from a mathematical point of view. In the following example the signal is formed by linear combination of sinusoidal signals. It clearly belongs to the linear span of the harmonic exponential family, and therefore it admits representation (5.11).

**Example 5.5   Sum of sinusoids**
Let

$$x(t) = 1 + 2\cos\omega t + \sin 2\omega t$$

Then

$$x(t) = 1 + e^{j\omega t} + e^{-j\omega t} + \frac{1}{2j}e^{j2\omega t} - \frac{1}{2j}e^{-j2\omega t}$$

The Fourier coefficients are

$$a_0 = a_1 = a_{-1} = 1, \quad a_2 = \frac{1}{2j}, \quad a_{-2} = -\frac{1}{2j}$$

and $a_k = 0$ for $k \neq 0, \pm 1, \pm 2$. The signal $x(t)$ is plotted in Fig. 5.4, while the real and imaginary part of the spectral coefficients are depicted in Fig. 5.5.   ∎

The same answer is obtained if the Fourier coefficients are computed by the analysis equation (5.12). If a signal is a linear combination of a finite number of

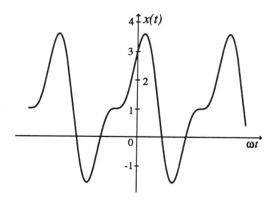

**Figure 5.4.** Plot of signal $x(t)$ of Example 5.5.

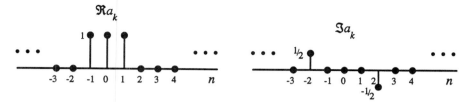

**Figure 5.5.** Real and imaginary part of Fourier coefficients of signal $x(t)$ of example 5.5.

sinusoidal signals having the same period, the Fourier coefficients are derived by simple inspection and are represented by a finite support discrete signal. The significance of the analysis and synthesis equations is mainly appreciated in cases where simple inspection proves to be of no help, yet the Fourier coefficients are readily determined by the analysis equation (5.12). An important example of this type is the periodic square wave, drawn in Fig. 2.10.

### Example 5.6  The periodic square wave

The difference between this example and the preceding one is that $x(t)$ no longer resides in the linear span of $A$, the set of harmonically related exponentials. Apparently there is nothing in the graph of the periodic square wave to remind us of a sinusoidal oscillatory behavior. It turns out, however, that the periodic square wave belongs to the linear closure of $A$ as most periodic signals do. Before we certify this claim, let us assume that the square wave can be synthesized by exponentials. The Fourier series coefficients are given by Eq. (5.12). Let $\omega = 2\pi/T$. To facilitate computations, we integrate over the period $-T/2 \le t \le T/2$:

$$a_0 = \frac{1}{T} \int_{-T/2}^{T/2} x(t)dt = \frac{1}{T} \int_{-T_1}^{T_1} dt = \frac{2T_1}{T}$$

while for $k \neq 0$, we find that

$$a_k = \frac{1}{T} \int_{-T_1}^{T_1} e^{-jk\omega t} dt = -\frac{1}{jk\omega T} e^{-jk\omega t} \Big|_{-T_1}^{T_1} = \frac{\sin k\omega T_1}{k\pi}$$

If $T = 4T_1$, the signal takes the value 1 for half the period and the value zero for the rest of the period. In this case the Fourier coefficients are shown in Fig. 5.6. Note $a_k = (\sin k\pi/2)/k\pi, \ k \neq 0$. Hence

$$a_k = \begin{cases} \dfrac{1}{2}, & k = 0 \\ 0, & k \text{ even} \\ \pm\dfrac{1}{k\pi}, & k \text{ odd} \end{cases} \tag{5.16}$$

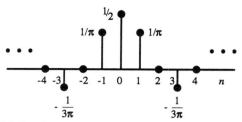

**Figure 5.6.** Fourier coefficients for the periodic square wave, $T = 4T_1$.

The total signal power is

$$\frac{1}{T}\int_T x^2(t)dt = \frac{1}{T}\int_{-T_1}^{T_1} x(t)dt = a_0 = \frac{2T_1}{T} = \frac{1}{2}$$

The main lobe contains all frequencies between the first left and right zero crossings. Therefore the total power of the main lobe is

$$|a_0|^2 + |a_1|^2 + |a_{-1}|^2 = \frac{1}{4} + \frac{2}{\pi^2}$$

Hence the first two harmonics together with the mean value occupy a significant part of the total signal power.

Now we inquire whether $x(t)$ is expandable in Fourier series. Sufficient evidence that this is plausible is deduced by a computer experiment. Figure 5.7 shows plots of the partial sum $\hat{x}(t) = \sum_{k=-N}^{N} a_k e^{jk\omega t}$ for several values of $N$. The coefficients $a_k$ are given by Eq. (5.16). The following code has been used. The arguments are the vector of the Fourier coefficients $A$, the fundamental

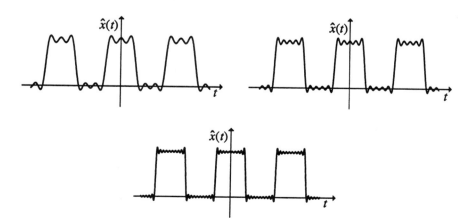

**Figure 5.7.** Plot of the partial sum of the Fourier series associated with the periodic square wave.

frequency $\omega$, the time vector $t$, and the number of terms $N$.

```
function x = blpa(A,omega,t,N)
  x = A(1)*ones(1,length(t));      % A = [a0 a1 a2 ... ]
  plot(t,x); pause
  for k = 1:N,
    z = exp(j*k*omega*t) + exp(-j*k*omega*t);
    x = x + A(k + 1)*z;
    plot(t,x); pause
  end;
```

The above function assumes that the Fourier coefficients are available. In the case of the periodic square wave, they are computed by

```
k = 1:N;
A = [1/2,sin(k*pi/2)./(k*pi)];
```

If the available information consists of signal samples, determination of the Fourier coefficients is accomplished as follows:

### Example 5.7  Computation of Fourier series with the DFT

Let $x(t)$ be a periodic signal of period $T$ with Fourier series (5.11). We sample $x(t)$ with sampling period $T_s = T/N$. The positive integer $N$ provides the number of samples in one period. We wish to compute the Fourier coefficients from the samples $x(mT_s)$. Evaluation of Eq. (5.11) at the sampling times gives

$$x(mT_s) = \sum_{k=-\infty}^{\infty} a_k e^{jk(2\pi/T)mT_s} = \sum_{k=-\infty}^{\infty} a_k e^{jk(2\pi/N)m} = \sum_{k=-\infty}^{\infty} a_k w_N^{-km} \qquad (5.17)$$

We divide $k$ with $N$. Let $r$ be the quotient and $n$ the remainder. Then $k = n + rN$, $0 \le n \le N - 1$. Substitution into Eq. (5.17) and use of $w_N^{rN} = (w_N^N)^r = 1$ gives

$$x(mT_s) = \sum_{n=0}^{N-1} \sum_{r=-\infty}^{\infty} a_{n+rN} w_N^{-m(n+rN)} = \sum_{n=0}^{N-1} w_N^{-mn} \sum_{r=-\infty}^{\infty} a_{n+rN} \qquad (5.18)$$

We set

$$\hat{a}_n = \sum_{r=-\infty}^{\infty} a_{n+rN} \qquad (5.19)$$

The signal $\hat{a}_n$ forms the periodic extension of $a_n$. Using Eq. (5.19), we rewrite Eq. (5.18) as

$$x(mT_s) = \sum_{n=0}^{N-1} \hat{a}_n w_N^{-mn} = N\ IDFT(\hat{a}_n), \qquad m = 0, 1 \ldots, N - 1. \qquad (5.20)$$

where IDFT stands for the inverse discrete Fourier transform. Application of the DFT gives

$$\hat{a}_n = \frac{1}{N}\sum_{m=0}^{N-1} x(mT_s)w_N^{mn} = \frac{1}{N}\, DFT(x(mT_s)) \tag{5.21}$$

Thus the coefficients $\hat{a}_n$ are determined from the DFT of the sampled signal. The true coefficients $a_n$ cannot, in general, be exactly reproduced from their periodic extension $\hat{a}_n$. Perfect reconstruction is achieved if $x(t)$ is *bandlimited*, namely the Fourier coefficients form a finite support signal: $a_k = 0$, $|k| > M$. A bandlimited periodic signal is a trigonometric polynomial of the form

$$x(t) = \sum_{k=-M}^{M} a_k e^{j(2\pi k/T)t} \tag{5.22}$$

The *bandwidth* of $x(t)$, $2M2\pi/T$, provides the band of frequencies outside of which the information content of the signal is negligible. We show in the next section that the class of bandlimited periodic signals is dense in the set of continuous functions over an interval of length $T$. The preceding discussion and the isomorphism of periodic and finite support signals described in Section 2.3.2 demonstrate that if $N \geq 2M + 1$ (i.e., the sampling rate exceeds the signal bandwidth), it holds that

$$a_n = \begin{cases} \hat{a}_n, & 0 \leq n \leq M \\ \hat{a}_{M+n}, & -M \leq n \leq 0 \end{cases} \tag{5.23}$$

In this case the Fourier coefficients are computed via the DFT by Eq. (5.21).

The above scheme is readily implemented in MATLAB. Consider, for instance, the signal

$$x(t) = 1 + \cos(2\pi 50t) + 6\cos(2\pi 200t)$$

The frequency content is limited to two frequencies at 50 Hz and 200 Hz. The signal bandwidth is 400 Hz. Let $\omega_0 = 2\pi 50$. The only nonzero spectral lines are $a_0, a_1, a_{-1}, a_4$, and $a_{-4}$. Thus $N \geq 2*4+1 = 9$.

```
N = 16;
T = 1/50;              % fundamental period
Ts = T/N;              % sampling period
t = [0:N-1]*Ts;        % time interval
x = ones(1,N) + cos(2*pi*50*t) + 6*cos(2*pi*200*t);
plot(t,x)
X = fft(x)/N;
w0 = 2*pi*50;          % fundamental frequency
w = [-N/2: N/2-1]*w0;  % frequency axis
X1 = fftshift(X);      % shifts X to the center
subplot(221), plot(w,real(X1)), subplot(222), plot(w,imag(X1))
subplot(223), plot(w,abs(X1)), subplot(224), plot(w,angle(X1))
```
■

The computer experiment of Example 5.6 is justified by the following theorem:

**Theorem 5.1** The family of harmonically related exponential signals of period $T$ is complete over $L_2(T)$.

Hence every periodic signal of period $T$ that is square (Lebesgue) integrable over an interval of length $T$ is represented by a Fourier series. The proof of the theorem is provided in the Appendix at the end of the chapter.

## 5.5 FOURIER ANALYSIS OF 2-D PERIODIC SIGNALS

### 5.5.1 Fourier expansions

Let $N_1$, $N_2$ be positive integers and

$$w_{N_1} = e^{-j2\pi/N_1}, \quad w_{N_2} = e^{-j2\pi/N_2} \tag{5.24}$$

For any $0 \le k_1 \le N_1 - 1$, $0 \le k_2 \le N_2 - 1$, the exponential signals $f_{k_1,k_2}(n_1, n_2) = w_{N_1}^{-k_1 n_1} w_{N_2}^{-k_2 n_2}$ form an orthogonal set in the space of 2-D periodic signals with periods $N_1$, $N_2$ and inner product

$$< x(n_1, n_2), y(n_1, n_2) > = \sum_{n_1=0}^{N_1-1} \sum_{n_2=0}^{N_2-1} x(n_1, n_2) y^*(n_1, n_2) \tag{5.25}$$

Indeed, it is easy to verify that

$$<f_{k_1,k_2}, f_{m_1,m_2}> = \begin{cases} N_1 N_2, & \text{if } k_1 = m_1 \text{ and } k_2 = m_2 \\ 0, & \text{otherwise} \end{cases} \tag{5.26}$$

Since the above space has finite dimension, every 2-D periodic signal with periods $N_1$, $N_2$ is represented by

$$x(n_1, n_2) = \sum_{k_1=0}^{N_1-1} \sum_{k_2=0}^{N_2-1} a(k_1, k_2) w_{N_1}^{-k_1 n_1} w_{N_2}^{-k_2 n_2} \tag{5.27}$$

$$a(k_1, k_2) = \frac{1}{N_1 N_2} \sum_{n_1=0}^{N_1-1} \sum_{n_2=0}^{N_2-1} x(n_1, n_2) w_{N_1}^{k_1 n_1} w_{N_2}^{k_2 n_2} \tag{5.28}$$

The Fourier coefficients $a(k_1, k_2)$ constitute the *spectral lines* of the signal. They are complex numbers, in general. The value $a(k_1, k_2)$ yields the contribution of the signal at the spatial frequency $(k_1 2\pi/N_1, k_2 2\pi/N_2)$.

## 5.5.2  Two-dimensional DFT

The two-dimensional discrete Fourier transform converts the matrix of values of a periodic signal in one period $\mathcal{X}$ into the matrix of Fourier coefficients $\mathbf{X}$ by $X(k_1, k_2) = N_1 N_2 a(k_1, k_2)$. From Eq. (5.28) we have

$$X(k_1, k_2) = \sum_{n_1=0}^{N_1-1} \sum_{n_2=0}^{N_2-1} x(n_1, n_2) w_{N_1}^{k_1 n_1} w_{N_2}^{k_2 n_2} = \sum_{n_1=0}^{N_1-1} w_{N_1}^{k_1 n_1} \sum_{n_2=0}^{N_2-1} x(n_1, n_2) w_{N_2}^{k_2 n_2} \quad (5.29)$$

Let $W_N$ be the matrix that represents the one dimensional DFT of length $N$; see Eq. (5.6). Then Eq. (5.29) is written in matrix form as

$$\mathbf{X} = W_{N_1} \mathcal{X} W_{N_2} \tag{5.30}$$

Hence

$$\mathcal{X} = W_{N_1}^{-1} \mathbf{X} W_{N_2}^{-1} = \frac{1}{N_1 N_2} W_{N_1}^* \mathbf{X} W_{N_2}^* \tag{5.31}$$

It is easy to verify that Eq. (5.31) leads to (5.27).

***Example 5.8    Spectral lines of a 2-D sinusoid***
Let

$$x(n_1, n_2) = \cos\left(\frac{4\pi n_1}{N} + \frac{6\pi n_2}{N}\right), \qquad N > 3$$

This signal is illustrated in Fig 5.8. Using Euler's identity we have

$$x(n_1, n_2) = \frac{1}{2}\left[e^{j(4\pi n_1/N + 6\pi n_2/N)} + e^{-j(4\pi n_1/N + 6\pi n_2/N)}\right] \tag{5.32}$$

Comparison of (5.27) with (5.32) gives

$$a(k_1, k_2) = \begin{cases} \dfrac{1}{2}, & \text{if } k_1 = 2, k_2 = 3 \\[2mm] \dfrac{1}{2}, & \text{if } k_1 = -2 = N - 2, k_2 = -3 = N - 3 \\[2mm] 0, & \text{otherwise} \end{cases}$$

The 2-D DFT is implemented in MATLAB by the command fft2. The above

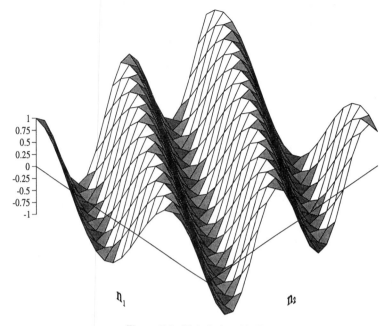

**Figure 5.8.**  Plot of signal in Example 5.8.

results are obtained as follows:

```
N = 16;
n1 = 0:N-1;
n2 = 0:N-1;
l1 = ones(size(n1));
l2 = ones(size(n2'));
x = cos(4*pi*n1'*l1/N + 6*pi*l2*n2/N);
mesh(x); pause
a = fft2(x)/N^2;
subplot(121);mesh(real(a)); subplot(122);mesh(imag(a));
```

■

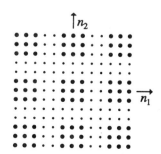

**Figure 5.9.**  Plot of signal in Example 5.9.

*Example 5.9 Spectral lines of the 2-D periodic square wave*
We consider the 2-D periodic square wave of Fig. 5.9:

$$x(n_1, n_2) = \begin{cases} 1, & \text{if } |n_1| \le M_1, |n_2| \le M_2 \\ 0, & \text{if } M_1 < |n_1| < \dfrac{N_1}{2} \text{ or } M_2 < |n_2| < \dfrac{N_2}{2} \end{cases} \tag{5.33}$$

We compute the Fourier coefficients from Eq. (5.28):

$$a(k_1, k_2) = \frac{1}{N_1 N_2} \sum_{n_1=-M_1}^{M_1} \sum_{n_2=-M_2}^{M_2} e^{-j(2\pi/N_1)k_1 n_1} e^{-j(2\pi/N_2)k_2 n_2}$$

$$= \frac{1}{N_1} \sum_{n_1=-M_1}^{M_1} e^{-j(2\pi/N_1)k_1 n_1} \frac{1}{N_2} \sum_{n_2=-M_2}^{M_2} e^{-j(2\pi/N_2)k_2 n_2}$$

Using Example 5.2, we have

$$a(0, 0) = \left(\frac{2M_1 + 1}{N_1}\right) \left(\frac{2M_2 + 1}{N_2}\right)$$

$$a(k_1, 0) = \frac{2M_2 + 1}{N_1 N_2} \frac{\sin(2\pi k_1(M_1 + \frac{1}{2})/N_1)}{\sin(\pi k_1/N_1)}$$

$$a(0, k_2) = \frac{2M_1 + 1}{N_1 N_2} \frac{\sin(2\pi k_2(M_2 + \frac{1}{2})/N_2)}{\sin(\pi k_2/N_2)}$$

$$a(k_1, k_2) = \frac{1}{N_1 N_2} \frac{\sin(2\pi k_1(M_1 + \frac{1}{2})/N_1)}{\sin(\pi k_1/N_1)} \frac{\sin(2\pi k_2(M_2 + \frac{1}{2})/N_2)}{\sin(\pi k_2/N_2)}$$

The spectral lines are illustrated in Fig. 5.10. ∎

*Example 5.10 Spectral lines of an image*
Consider now the image of Fig. 5.11. The spectral lines of the corresponding 2-D signal are shown in Fig. 5.12. ∎

### 5.5.3 Representation of 2-D DFT via the Kronecker product

The two-dimensional DFT is a linear transformation. Thus it is represented by a matrix $W$ of size $N_1 N_2 \times N_1 N_2$. This matrix is nicely expressed in terms of the Kronecker product. If $A$ and $B$ are matrices of sizes $M_1 \times M_2$ and $N_1 \times N_2$, respectively, their Kronecker product has size $M_1 N_1 \times M_2 N_2$ and is defined as

$$A \otimes B = \begin{pmatrix} a_{11}B & a_{12}B & \cdots & a_{1M_2}B \\ a_{21}B & a_{22}B & \cdots & a_{2M_2}B \\ \vdots & \vdots & \ddots & \vdots \\ a_{M_1 1}B & a_{M_1 2}B & \cdots & a_{M_1 M_2}B \end{pmatrix}$$

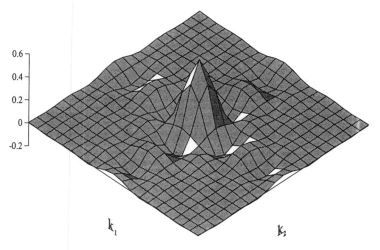

**Figure 5.10.** Spectral lines of the signal discussed in Example 5.9.

The Kronecker product is a block matrix with matrix entries of size $N_1 \times N_2$. The 2-D DFT relates to the Kronecker product in the following way: We assign the one-dimensional array $x$ to the matrix of the signal values $x(n_1, n_2)$, $\mathcal{X}$, using lexicographic order by rows. The first row of the matrix is registered first, followed by the second row, then the third row, and so on. Thus

$$x = (\, x(0,0) \quad x(0,1) \quad \cdots \quad x(0, N_2 - 1) \quad x(1,0) \quad \cdots \quad x(N_1 - 1, N_2 - 1)\,)^T$$

In a similar way the matrix $\mathbf{X}$ of the Fourier coefficients is converted into the

**Figure 5.11.** The image in Example 5.10.

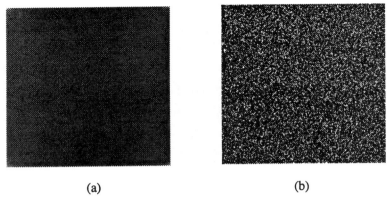

(a)                                        (b)

**Figure 5.12.** (a) The logarithm of the magnitude, and (b) the phase of the spectral lines of the image in Example 5.10.

vector $\hat{x}$. The assignment

$$\{0, 1, \ldots, N_1 N_2 - 1\} \ni i \to (n_1, n_2) \in \{0, 1, \ldots, N_1 - 1\} \times \{0, 1, \ldots, N_2 - 1\}$$
$$i = n_1 N_2 + n_2$$

converts the 1-D index $i$ to the 2-D index $(n_1, n_2)$ obtained from the remainder and the quotient of the division of $i$ by $N_2$. This indexing assigns the entries of the vector $x$ to the entries of the matrix $\mathcal{X}$. It holds that

$$\hat{x} = (W_{N_1} \otimes W_{N_2})(x) \tag{5.34}$$

Indeed, we write $x$ and $\hat{x}$ in block form as

$$x^T = \begin{pmatrix} x_0^T & x_1^T & \cdots & x_{N_1-1}^T \end{pmatrix}, \quad \hat{x}^T = \begin{pmatrix} X_0^T & X_1^T & \cdots & X_{N_1-1}^T \end{pmatrix}$$

where

$$x_{n_1}^T = \begin{pmatrix} x(n_1, 0) & x(n_1, 1) & \cdots & x(n_1, N_2 - 1) \end{pmatrix}$$

Let $x_{n_1}(n_2)$ denote the $n_2 + 1$ entry of the above vector, that is, $x_{n_1}(n_2) = x(n_1, n_2)$. Equating the $k_1$th block of (5.34) and taking into account the definition of the Kronecker product, we find that

$$X_{k_1} = \sum_{n_1=0}^{N_1-1} w_{N_1}^{n_1 k_1} W_{N_2} x_{n_1}$$

Zooming at the $k_2$ entry, we obtain

$$X_{k_1}(k_2) = X(k_1, k_2) = \sum_{n_1=0}^{N_1-1} w_{N_1}^{n_1 k_1} \sum_{n_2=0}^{N_2-1} w_{N_2}^{n_2 k_2} x(n_1, n_2)$$

The latter expression is valid, and the proof of Eq. (5.34) is complete. We can alternatively work with lexicographic ordering by columns. The details are left to the reader.

***Example 5.11    2-D DFT of 2 × 2 two-dimensional signals***
We compute the 2-D DFT of $2 \times 2$ two-dimensional signals. We have $N_1 = N_2 = 2$ and $w_2^0 = 1$, $w_2^1 = e^{-j\pi} = -1$. Hence

$$W_2 = \begin{pmatrix} 1 & 1 \\ 1 & -1 \end{pmatrix}$$

The 2-D signal

$$\mathcal{X} = \begin{pmatrix} x(0,0) & x(0,1) \\ x(1,0) & x(1,1) \end{pmatrix}$$

is transformed via (5.30) into the matrix $\mathbf{X} = W_2 \mathcal{X} W_2$, or

$$\begin{pmatrix} X(0,0) & X(0,1) \\ X(1,0) & X(1,1) \end{pmatrix} =$$

$$\begin{pmatrix} x(0,0) + x(0,1) + x(1,0) + x(1,1) & x(0,0) - x(0,1) + x(1,0) - x(1,1) \\ x(0,0) + x(0,1) - x(1,0) - x(1,1) & x(0,0) - x(0,1) - x(1,0) + x(1,1) \end{pmatrix}$$

$$(5.35)$$

The Kronecker product is

$$W_2 \otimes W_2 = \begin{pmatrix} 1 & 1 & 1 & 1 \\ 1 & -1 & 1 & -1 \\ 1 & 1 & -1 & -1 \\ 1 & -1 & -1 & 1 \end{pmatrix}$$

Lexicographic ordering by rows gives

$$x = \begin{pmatrix} x(0,0) & x(0,1) & x(1,0) & x(1,1) \end{pmatrix}^T$$

Therefore

$$(W_2 \otimes W_2)(x) = \begin{pmatrix} x(0,0) + x(0,1) + x(1,0) + x(1,1) \\ x(0,0) - x(0,1) + x(1,0) - x(1,1) \\ x(0,0) + x(0,1) - x(1,0) - x(1,1) \\ x(0,0) - x(0,1) - x(1,0) + x(1,1) \end{pmatrix}$$

The latter expression provides the lexicographic ordering by rows of the matrix (5.35).    ∎

### 5.5.4    Computational complexity of the 2-D DFT

The 2-D DFT requires

$$C_0 = N_1^2 N_2^2 \tag{5.36}$$

multiplications and additions, assuming that the coefficients $w_{N_1}^{k_1 n_1}$, $w_{N_2}^{k_2 n_2}$ have been computed and stored. Significant savings result from the separable structure (5.30) of the DFT. The product of two matrices $A$, $B$ of sizes $M \times N$ and $N \times L$, respectively, is written

$$Y = AB = A(b_1 \quad b_2 \quad \cdots \quad b_L) = (Ab_1 \quad Ab_2 \quad \cdots \quad Ab_L)$$

Each product $Ab_i$ requires $MN$ operations. Therefore the computation of $Y$ requires $MNL$ operations. The product of three matrices $A$, $B$, $C$ of sizes $M \times N$, $N \times L$, $L \times K$, $Y = ABC$ is written $Y = AY_1$, $Y_1 = BC$. The first product calls for $MNK$ operations, while the second product needs $NLK$ operations. The total is $NK(M + L)$ operations. Application of the above to (5.30) leads to

$$C_1 = N_1 N_2 (N_1 + N_2) \tag{5.37}$$

operations. Operations are reduced by an order of magnitude. If $N_1 = N_2 = N$, the general linear form of the DFT requires, according to (5.36), $N^4$ operations, while (5.37) needs $2N^3$ operations. To get a feeling of the computational savings, consider an image of $1024 \times 1024 \approx 10^3 \times 10^3$ pixels. Then

$$C_0 = (10^3)^4 = 10^{12}, \quad C_1 = 2(10^3)^3 = 2 \times 10^9$$

## 5.6    RESPONSE OF CONVOLUTIONAL SYSTEMS TO PERIODIC INPUTS

### 5.6.1    Fourier coefficients of responses produced by periodic inputs

Consider the convolutional system

$$y(n) = \sum_{k=-\infty}^{\infty} h(n - k)u(k) \tag{5.38}$$

with absolutely summable impulse response $h(n)$. We feed the system with the exponential input $u(n) = e^{j\omega n}$. The resulting output exists and is bounded. It is given by

$$y(n) = \sum_{k=-\infty}^{\infty} h(k)e^{j(n-k)\omega} = H(\omega)e^{jn\omega} \tag{5.39}$$

where

$$H(\omega) = \sum_{k=-\infty}^{\infty} h(k)e^{-j\omega k}$$

Since $h(n)$ is absolutely summable, $H(\omega)$ exists. We will see in Chapter 6 that $H(\omega)$ is the Fourier transform of $h(n)$. It constitutes the *frequency response* of the system. Equation (5.39) states that the output is obtained from the input exponential scaled by $H(\omega)$.

Let us next consider a periodic input with Fourier expansion

$$u(n) = \frac{1}{N} \sum_{k=0}^{N-1} U(k)w_N^{-kn}, \qquad w_N = e^{-j2\pi/N}$$

where $U(k)$ denotes the DFT of $u(n)$. For each $k$ the exponential $(w_N^{-k})^n$ produces the output $H(2\pi k/N)w_N^{-kn}$. By linearity the output to $u(n)$ is

$$y(n) = \frac{1}{N} \sum_{k=0}^{N-1} H(k)U(k)w_N^{-kn} \tag{5.40}$$

where, with a slight abuse of notation, we write

$$H(k) = H(\omega)|_{\omega=k2\pi/N} = \sum_{n=-\infty}^{\infty} h(n)e^{-j2\pi kn/N} \tag{5.41}$$

Thus the response of a BIBO stable convolutional system to a periodic input is periodic with the same period. It occupies the same bandwidth with harmonics $2\pi k/N, 0 \le k \le N - 1$. The DFT coefficients of the output $Y(k)$ are given by the pointwise product

$$Y(k) = H(k)U(k) \tag{5.42}$$

where $U(k)$ is the input DFT and $H(k)$ consists of the $N$ values of the frequency response sampled at frequencies $2\pi k/N$. If the system is a causal FIR system and the impulse response has finite extent $N$, equal to the period of $u(n)$, then $H(k)$ coincides with the discrete Fourier transform of $h(n)$. In this case the convolution becomes

$$y(n) = \sum_{k=0}^{n} h(n-k)u(k) = \sum_{k=0}^{N-1} h(k)u(n-k) \tag{5.43}$$

Furthermore the DFT coefficients of $y(n)$ are obtained from the pointwise product of the DFT's $U(k)$ and $H(k)$.

***Example 5.12    Output spectral lines produced by an exponential impulse response and a sinusoidal input***
Consider the BIBO stable convolutional system with impulse response $h(n) = a^n u_s(n)$, $|a| < 1$ and the periodic input

$$u(n) = \cos\frac{2\pi n}{N} = \frac{1}{2}e^{j(2\pi/N)n} + \frac{1}{2}e^{-j(2\pi/N)n}$$

Then

$$H(k) = \sum_{n=0}^{\infty} a^n e^{-jk(2\pi/N)n} = \sum_{n=0}^{\infty}(ae^{-jk2\pi/N})^n = \frac{1}{1 - ae^{-j2\pi k/N}}$$

Thus

$$y(n) = \frac{1}{N}\left(Y(1)e^{j2\pi n/N} + Y(-1)e^{-j2\pi n/N}\right)$$

where

$$Y(1) = H(1)U(1) = \frac{1}{2}\frac{N}{\left(1 - ae^{-j2\pi/N}\right)}$$

$$Y(-1) = H(-1)U(-1) = \frac{1}{2}\frac{N}{\left(1 - ae^{j2\pi/N}\right)} \qquad \blacksquare$$

The preceding analysis carries over to continuous time systems. If the BIBO stable convolutional system

$$y(t) = \int_{-\infty}^{\infty} h(t - \tau)u(\tau)d\tau \qquad (5.44)$$

is excited by the exponential input $u(t) = e^{j\omega t}$, it yields the output $y(t) = H(\omega)e^{j\omega t}$ where

$$H(\omega) = \int_{-\infty}^{\infty} h(t)e^{-j\omega t}dt$$

is the Fourier transform of $h(t)$, called *frequency response*. If the system is fed with a finite energy periodic input

$$u(t) = \sum_{k=-\infty}^{\infty} U(k)e^{jk\omega_0 t}, \qquad \omega_0 = \frac{2\pi}{T}$$

the output $y(t)$ is a finite energy signal over any interval of length $T$, with Fourier series

$$y(t) = \sum_{k=-\infty}^{\infty} Y(k)e^{jk\omega_0 t}$$

The spectral lines are determined by

$$Y(k) = H(k)U(k) \tag{5.45}$$

where $H(k) = H(k\omega_0)$, that is

$$H(k) = \int_{-\infty}^{\infty} h(t)e^{-jk\omega_0 t} dt \tag{5.46}$$

***Example 5.13    Output spectral lines produced by an exponential impulse response and a square wave input***
The system with impulse response

$$h(t) = e^{-at}u_s(t), \qquad a > 0$$

is fed with the periodic square wave (Example 5.6). The input spectral lines are given by Eq. (5.16). Hence

$$H(k) = \int_0^{\infty} e^{-at}e^{-kj\omega_0 t} dt = \frac{1}{a + k\omega_0 j}$$

The Fourier coefficients of the output are

$$Y(k) = \frac{1}{a + k\omega_0 j} U(k) = \begin{cases} \dfrac{1}{2a}, & \text{if } k = 0 \\ 0, & \text{if } k \text{ even} \\ \pm \dfrac{1}{k\pi(a + k\omega_0 j)}, & \text{if } k \text{ odd} \end{cases}$$

A plot of the output spectral lines is given in Fig. 5.13.    ■

### 5.6.2    Cyclic convolution

Let us consider two periodic signals of period $N$, $u(n)$ and $h(n)$. We want to determine the periodic signal $y(n)$ having the property that its DFT $Y(k)$ is given by the product $H(k)U(k)$. Motivated by the preceding section, we give the following definition: The *cyclic* or *circular convolution* of $u(n)$ and $h(n)$ is defined by

$$y(n) = h(n) \odot u(n) = \sum_{k=0}^{N-1} h(n-k)u(k) \tag{5.47}$$

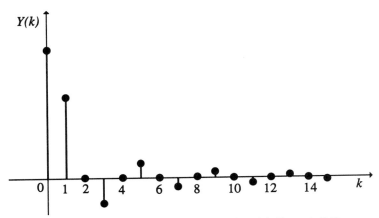

**Figure 5.13.** Spectral lines of the output (real part) in Example 5.13.

where $y(n)$ is periodic of period $N$. The isomorphism of periodic signals of period $N$ and signals of finite support $N$, discussed in Section 2.3.2, translates the cyclic convolution to finite support signals. In the latter case definition (5.47) applies with the understanding that for values of $n - k$ outside the interval $0 \leq n \leq N - 1$, the corresponding signal value $h(n - k)$ is not set to zero but is folded back to the interval by mod $N$ reduction. Thus the cyclic convolution of $N$ point signals has the form

$$y(n) = \sum_{k=0}^{N-1} h((n - k)_N) u(k) \qquad (5.48)$$

where $(n)_N$ denotes the remainder of the division of $n$ with $N$. To stress the difference between cyclic convolution and ordinary convolution, the latter is called *linear convolution* because of its property to characterize linear systems.

***Example 5.14    Linear and cyclic convolution of signals of length 2***
We consider the finite extent signals with samples $u(0)$, $u(1)$ and $h(0)$, $h(1)$, respectively. The linear convolution $y'(n) = h(n) * u(n)$ is

$$y'(0) = h(0)u(0), \quad y'(1) = h(0)u(1) + h(1)u(0), \quad y'(2) = h(1)u(1)$$

The cyclic convolution is

$$y(0) = h(0)u(0) + h(1)u(1), \quad y(1) = h(1)u(0) + h(0)u(1) \qquad \blacksquare$$

It is straightforward to verify that circular convolution is commutative for single-channel signals, assosiative and distributive over addition and scalar multiplication. The analysis of Section 5.6.1 shows that the cyclic convolution

$y = h \odot u$ is converted by the DFT to the pointwise product

$$Y(k) = H(k)U(k), \quad k = 0, 1, \ldots, N - 1 \tag{5.49}$$

The above equation is employed in the fast computation of cyclic convolution. For this purpose it uses a fast Fourier transform for the transfers to and from the frequency domain. More details are given in Chapter 8.

Like the DFT, the cyclic convolution is an important computational concept because it lends itself to efficient algorithms and because various other tasks can be reduced to cyclic convolutions. Next we will see how to determine ordinary convolutions from cyclic convolutions.

### 5.6.3   Connection of cyclic convolution and linear convolution

Suppose that $h'(n)$ and $u'(n)$ are zero outside $0 \le n \le N - 1$. Let $y'(n)$ be their linear convolution:

$$y'(n) = \sum_{k=0}^{N-1} h'(n - k)u'(k) \qquad 0 \le n \le 2N - 2$$

Let $h(n)$ and $u(n)$ denote the $N$ periodic extensions. Recall that $u'(n) = u(n)$, for $0 \le n \le N - 1$ and that $u'(n) = 0$, otherwise. Then the circular convolution of $h(n)$ and $u(n)$ is

$$y(n) = \sum_{k=0}^{n} h(n - k)u(k) + \sum_{k=n+1}^{N-1} h(n - k)u(k)$$

The index $n - k$ in the first sum is in the range $0 \le n \le N - 1$. Therefore $(n - k)_N = n - k$. The index in the second sum is negative; hence $n - k = -N + N + n - k$, and $(n - k)_N = N + n - k$. Thus

$$y(n) = y'(n) + \sum_{k=n+1}^{N-1} h'(N + n - k)u'(k) = y'(n) + y'(N + n), \qquad 0 \le n \le N - 1 \tag{5.50}$$

because $h'(N + n - k) = 0$, for $k \le n$. Formula (5.50) states that the cyclic convolution is computed from the sum of the linear convolution and its $N$ right shift.

The above relationship forms the basis for the efficient computation of linear convolution. Indeed, consider two signals: $u(n)$ and $h(n)$ of length $M$ and $L$, respectively. The linear convolution $y'(n)$ has length $M + L - 1$. Pick $N > M + L - 2$ and augment $u$ and $h$ with zeros until their size is $N$. Form the cyclic convolution of the resulting zero padded signals. Since $y'(N + n) = 0$, for all $n \ge 0$, Eq. (5.50) yields $y(n) = y'(n)$, $0 \le n \le N - 1$ and in particular

$y(n) = y'(n)$ for $0 \leq n \leq M + L$. Therefore any cyclic convolution algorithm can be employed for the computation of linear convolution. A popular method converts the signals into the frequency domain via the FFT, multiplies their transforms, and converts the resulting product to the time domain by the FFT.

## 5.7  HARMONIC PROCESS

Harmonic processes provide a natural setup for periodicity in the class of wide sense stationary stochastic processes. A continuous time harmonic process is defined by

$$x(t) = \sum_{i=1}^{k} A_i \cos(\omega_i t + \phi_i) \tag{5.51}$$

The parameters $k$ and $\omega_i$, $i = 1, 2, \ldots, k$, are constant. $A_i$ are independent random variables, while $\phi_i$ are independent uniformly distributed on $(-\pi, \pi)$ random variables. Furthermore, for any $i$ and $j$, $A_i$ and $\phi_j$ are independent. The above model is reminiscent of Eq. (5.14), which gives the Fourier expansion of a real periodic signal. In Example 2.7 we studied the special case $k = 1$ and showed that the signal is wide sense stationary with zero mean and autocorrelation function $r(t) = \frac{1}{2} E[A_1^2] \cos \omega_1 t$. Taking expectations, we deduce that $E[x(t)] = 0$. In a similar fashion the autocorrelation function is

$$r(t) = E[x(t+\tau)x(\tau)] = E\left[\sum_i A_i \cos(\omega_i(t+\tau) + \phi_i) \sum_j A_j \cos(\omega_j\tau + \phi_j)\right]$$

or

$$r(t) = \sum_i \sum_j E[A_iA_j]E\left[\cos(\omega_i(t+\tau) + \phi_i)\cos(\omega_j\tau + \phi_j)\right]$$

If $i \neq j$, $\phi_i$ and $\phi_j$ are independent random variables. Hence arbitrary (measurable) functions of these random variables are also independent. Thus

$$E\left[\cos(\omega_i(t+\tau) + \phi_i)\cos(\omega_j\tau + \phi_j)\right] = E[\cos(\omega_i(t+\tau) + \phi_i)]E\left[\cos(\omega_j\tau + \phi_j)\right]$$

The right-hand side is zero provided $i \neq j$. Therefore

$$r(t) = \sum_i E[A_i^2]E[\cos(\omega_i(t+\tau) + \phi_i)\cos(\omega_i\tau + \phi_i)]$$

The results obtained for the case $k = 1$ imply that

$$r(t) = \frac{1}{2}\sum_{i=1}^{k} E[A_i^2] \cos \omega_i t \tag{5.52}$$

where $r(t)$ is periodic if the frequencies, are integer multiples of a single frequency $\omega_i = n_i\omega$. In this case $r(t)$ is a bandlimited periodic signal and the Fourier series expansion of $r(t)$ is finite. The Fourier coefficients are

$$a_k = \begin{cases} \frac{1}{4}E[A_i^2], & k = \pm n_i \\ 0, & \text{otherwise} \end{cases}$$

The discrete case is similar. A discrete harmonic process has the form

$$x(n) = \sum_{i=1}^{k} A_i \cos(\omega_i n + \phi_i) \tag{5.53}$$

where, as before, $k$ and $\omega_i$ are constant, $A_i$, $\phi_i$ are independent, and $\phi_i$ are uniformly distributed random variables. The process $x(n)$ is zero mean wide sense stationary with autocorrelation

$$r(n) = \frac{1}{2}\sum_{i=1}^{k} E[A_i^2] \cos \omega_i n \tag{5.54}$$

It is periodic if $\omega_i = (2\pi/N)n_i$, $n_i$, $N$ integers. The Fourier coefficients are given as in the continuous time counterpart.

## 5.8 THE FAST FOURIER TRANSFORM

The fast Fourier transform (FFT) declares a family of efficient algorithms that compute the DFT transform. Their great potential in digital signal processing was demonstrated by Cooley and Tukey in 1965. The Cooley-Tukey FFT algorithm relies on a divide-and-conquer technique and has a long history. One of its variants, the decimation in frequency FFT, dates back to Gauss. Two versions of the Cooley-Tukey FFT algorithm are presented in the sequel.

### 5.8.1 Decimation in time FFT

Let $x(n)$ be a signal of duration $0 \leq n \leq N - 1$, $N$ even. We consider the signals $x_1(n)$ and $x_2(n)$ consisting of the even and odd values of $x(n)$, respectively:

$$x_1(n) = x(2n), \quad x_2(n) = x(2n + 1), \qquad n = 0, 1, \ldots, \frac{N}{2} - 1$$

Both have duration $N/2$. The DFT of $x(n)$ can be written as follows:

$$X(k) = \sum_{n \text{ even}} x(n)w_N^{nk} + \sum_{n \text{ odd}} x(n)w_N^{nk}$$

or

$$X(k) = \sum_{n=0}^{N/2-1} x(2n)w_N^{2nk} + \sum_{n=0}^{N/2-1} x(2n+1)w_N^{(2n+1)k} \qquad (5.55)$$

Note $w_N^2 = e^{-j4\pi/N} = w_{N/2}$. Therefore Eq. (5.55) becomes

$$X(k) = \sum_{n=0}^{N/2-1} x_1(n)w_{N/2}^{nk} + w_N^k \sum_{n=0}^{N/2-1} x_2(n)w_{N/2}^{nk}$$

Hence

$$X(k) = X_1(k) + w_N^k X_2(k), \qquad k = 0, 1, \ldots, \frac{N}{2} - 1 \qquad (5.56)$$

The latter equation states that the first $N/2$ values of the DFT can be evaluated from the DFT of the decimating signals $x_1(n)$ and $x_2(n)$ of length $N/2$. Periodicity of the DFT gives $X_1(k + N/2) = X_1(k)$, $X_2(k + N/2) = X_2(k)$. Moreover $w_N^{k+N/2} = w_N^k w_N^{N/2} = w_N^k e^{-j\pi} = -w_N^k$. The above remarks lead to

$$X\left(k + \frac{N}{2}\right) = X_1(k) - w_N^k X_2(k), \qquad 0 \le k \le \frac{N}{2} - 1 \qquad (5.57)$$

The transforms $X_1(k)$ and $X_2(k)$ are combined into the so-called *butterfly* to supply $X(k)$. The structure of the butterfly is shown in Fig. 5.14. It transforms two numbers $A$ and $B$ into $X$ and $Y$ according to the rule

$$X = A + w_N^k B, \quad Y = A - w_N^k B$$

It utilizes one multiplier and one adder-subtractor. The process described so far is illustrated in Fig. 5.15 for $N = 8$. According to Eqs. (5.56) and (5.57), the evaluation of $X(k)$ and $X(k + N/2)$ in terms of $X_1(k)$ and $X_2(k)$ involves $N/2$

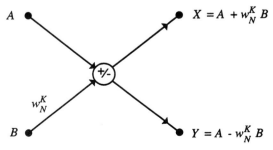

**Figure 5.14.** Diagram of the butterfly operation.

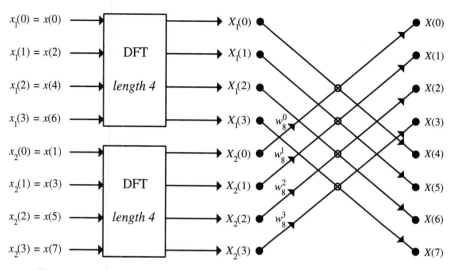

**Figure 5.15.** Computation of 8-point DFT via two 4-point DFTs and four butterflies.

butterflies, whereas each requires one complex multiplication. Some butterflies, like the one corresponding to $w_8^0 = 1$, require no multiplication. Thus the overall complexity does not exceed $N/2$ multiplications and additions. If the DFT's $X_1(k)$ and $X_2(k)$ are directly evaluated, we have a total of at most $2(N/2)^2 + N/2 = N^2/2 + N/2$ operations. Hence a 50% saving is achieved by Eqs. (5.56) and (5.57) over the direct computation of the DFT.

The computational reduction becomes much more significant if $N$ is a power of 2, $N = 2^L$, and the above decimation process is repeated. In this case the computation of each $X_1(k)$ and $X_2(k)$ is carried out by the butterfly structure, utilizing four $N/4$–point DFT's, two for each one. Continuing this way, after $L - 1$ steps we end up with $N/2$ DFT's each of length 2. It is easy to see that a 2-point DFT is merely an adder. Indeed, $w_2^0 = 1$, $w_2^1 = -1$, and $X(0) = x(0) + x(1)$, $X(1) = x(0) - x(1)$. The 8-point FFT is illustrated in Fig. 5.16.

**Bit reversal.** As Fig. 5.16 illustrates, the signal $x(n)$ is submitted to the FFT in shuffled order

$$( x(0) \quad x(4) \quad x(2) \quad x(6) \quad x(1) \quad x(5) \quad x(3) \quad x(7) )$$

Thus an indexing procedure is needed that will permute the data to the form appropriate for the FFT algorithm. The indexing identity is revealed if indices are represented in binary form. Table 5.1 indicates the natural order and the shuffled order for length 8.

We observe that shuffled numbers are bit reversed. In general, if $N = 2^L$, the

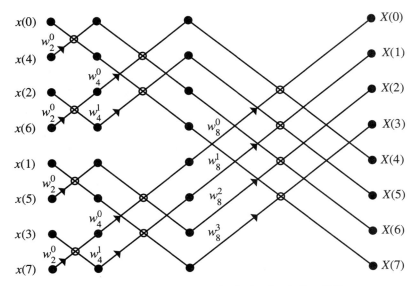

**Figure 5.16.** Structure of 8-point decimation in time FFT.

integer $n$ with binary representation $n = (b_L b_{L-1} \cdots b_1)$ is converted into the bit-reversed order $r(n) = (b_1 b_2 \cdots b_L)$, and the corresponding value $x(n)$ is stored in the array position $r(n)$.

***Computational complexity and memory requirements.*** Direct computation of the DFT as a matrix by vector multiplication requires $N^2$ complex multiplications and additions. Memory requirements for the storage of the entries of $W_N$ are of the same order of magnitude. Computational complexity of the FFT is much lower. Savings are impressive when $N = 2^L$. In this case there are $L = \log_2 N$ levels, and at each level $N/2$ butterflies are engaged. Each butterfly executes at most one multiplication. Hence the number of complex multipli-

**TABLE 5.1   Bit reversal**

| Regular Order | Binary Form | Shuffled Order | Shuffled Binary Form |
|---|---|---|---|
| 0 | 000 | 0 | 000 |
| 1 | 001 | 4 | 100 |
| 2 | 010 | 2 | 010 |
| 3 | 011 | 6 | 110 |
| 4 | 100 | 1 | 001 |
| 5 | 101 | 5 | 101 |
| 6 | 110 | 3 | 011 |
| 7 | 111 | 7 | 111 |

cations is at most $N \log_2 N/2$. The number of additions is analogous. Reduction from $N^2$ to $N \log_2 N$ is close to an order of magnitude. Indeed, direct computation of the DFT of length $N = 1000 \approx 2^{10}$ requires $10^6$ operations. The FFT calls for $10^4/2$ complex operations only.

An important feature of the FFT that drastically limits memory requirements is in-place computation. This means that the outputs at each level can be stored at the same locations the inputs were stored. Indeed, the input at each stage is not needed at subsequent stages. Consequently storage amounts to $2N$ memory locations for the butterflies and $2N$ locations for the bit reversal.

### 5.8.2  Decimation in frequency FFT

The decimation in time FFT algorithm is based on the splitting of the signal into its even and odd values. The decimation in frequency FFT partitions the signal into its first $N/2$ and its last $N/2$ values. It turns out that this partitioning leads to bit reversal of the DFT vector, and thus the resulting algorithm is called decimation in frequency FFT.

Let

$$x_1(n) = x(n), \qquad x_2(n) = x\left(n + \frac{N}{2}\right), \qquad n = 0, 1, \ldots, \frac{N}{2} - 1$$

The DFT becomes

$$X(k) = \sum_{n=0}^{N/2-1} x(n) w_N^{nk} + \sum_{n=N/2}^{N-1} x(n) w_N^{nk} = \sum_{n=0}^{N/2-1} [x_1(n) + e^{-j\pi k} x_2(n)] w_N^{nk}$$

Now

$$e^{-j\pi k} = \begin{cases} 1, & \text{if } k \text{ even} \\ -1, & \text{if } k \text{ odd} \end{cases}$$

This leads us to consider the even and odd values of $X(k)$ separately. Thus

$$X(2k) = \sum_{n=0}^{N/2-1} [x_1(n) + x_2(n)] w_{N/2}^{nk}$$

$$X(2k+1) = \sum_{n=0}^{N/2-1} [x_1(n) - x_2(n)] w_N^{n} w_{N/2}^{nk}$$

If we set

$$f(n) = x_1(n) + x_2(n), \quad g(n) = [x_1(n) - x_2(n)] w_N^{n}, \qquad n = 0, 1, \ldots, \frac{N}{2} - 1$$

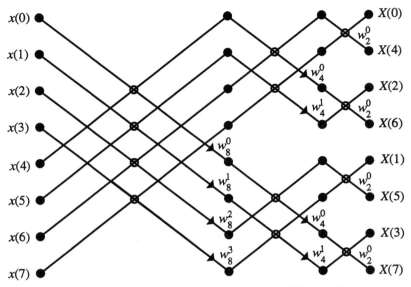

**Figure 5.17.** Structure of decimation in frequency FFT for length 8.

we obtain

$$X(2k) = F(k) \quad X(2k+1) = G(k), \qquad k = 0, 1, \ldots, \frac{N}{2} - 1$$

The computation of the $N$-point DFT reduces to two $N/2$-point DFT's. If $N = 2^L$, the above procedure is repeated. The algorithm terminates after $L$ steps. The procedure is illustrated in Fig. 5.17 for $N = 8$. The butterfly structure is depicted in Fig. 5.18. Input data are presented in normal order, but the transformed vector is read out in bit-reversed order. The same bit reversal permutation applies. The number of operations is about $N \log_2 N/2$ complex multiplications and additions.

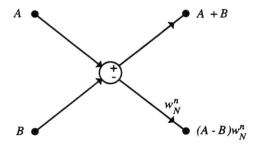

**Figure 5.18.** Butterfly for the decimation in frequency FFT.

### 5.8.3   Cooley-Tukey FFT algorithms for signals of arbitrary length

Suppose that $N = N_1 N_2$. We will effectively view the Cooley-Tukey FFT as mapping a 1-D array into a 2-D array. We replace the time index $n$ by two indexes $n_1$, $n_2$ obtained by dividing $n$ with $N_1$:

$$n = n_2 N_1 + n_1, \qquad 0 \le n_1 \le N_1 - 1, \ 0 \le n_2 \le N_2 - 1$$

In a similar fashion we replace the frequency index $k$ by the indexes $k_1$, $k_2$ obtained by the division of $k$ by $N_2$:

$$k = k_1 N_2 + k_2, \qquad 0 \le k_1 \le N_1 - 1, \ 0 \le k_2 \le N_2 - 1$$

The above indexing procedures correspond to the lexicographic ordering by columns and rows. In the new index space the DFT has the following form:

$$X(k_1, k_2) = \sum_{n_1=0}^{N_1-1} \sum_{n_2=0}^{N_2-1} w_N^{(k_1 N_2 + k_2)(n_2 N_1 + n_1)} x(n_1, n_2)$$

Let $w_N^{N_2} = w_1$, $w_N^{N_1} = w_2$. Then $w_1^{N_1} = w_N^N = 1$. Likewise $w_2^{N_2} = 1$. Furthermore $w_N^{(k_1 N_2 + k_2)(n_2 N_1 + n_1)} = w_1^{n_1 k_1} w_2^{n_2 k_2} w_N^{n_1 k_2}$. Thus

$$X(k_1, k_2) = \sum_{n_1=0}^{N_1-1} w_1^{n_1 k_1} \left( w_N^{n_1 k_2} \sum_{n_2=0}^{N_2-1} w_2^{n_2 k_2} x(n_1, n_2) \right) \qquad (5.58)$$

The innermost sum computes the $N_2$ point Fourier transform of each row of the array $x(n_1, n_2)$. The resulting matrix is multiplied entry by entry by $w_N^{n_1 k_2}$, and the $N_1$-point Fourier transform follows for each column. Complexity is reduced even if the smaller size DFT's are conventionally computed. The inner sum involves $N_2$ point DFT's for each row, that is, $N_1 N_2^2$ operations. Entry by entry multiplication requires $N_1 N_2 = N$ multiplications, and the outer Fourier transform needs $N_2 N_1^2$ operations. Therefore the total number is $N_1 N_2^2 + N_2 N_1^2 + N = N(N_1 + N_2 + 1)$, which compares favorably to $N^2$.

The above procedure is repeated if the factors $N_1$, $N_2$ are further factored.

*Extension to finite fields.* The FFT schemes presented so far extend to finite fields without difficulty. The basic recursions remain valid, the only difference being that the ordinary operations are replaced by the operations of the field under consideration.

### 5.8.4    Computation of the 2-D DFT from the 1-D FFT

Computation of the 2-D DFT with the aid of (5.30) relies on the direct computation of the one-dimensional DFT. Computational complexity is significantly reduced if the 1-D FFT is employed. We represent the 2-D signal by its columns $\mathcal{X} = \begin{pmatrix} x_0 & x_1 & \cdots & x_{N_2-1} \end{pmatrix}$. Equation (5.30) takes the form

$$\mathbf{X} = W_{N_1} \begin{pmatrix} x_0 & x_1 & \cdots & x_{N_2-1} \end{pmatrix} W_{N_2} = V W_{N_2}$$

where

$$V = \begin{pmatrix} W_{N_1} x_0 & W_{N_1} x_1 & \cdots & W_{N_1} x_{N_2-1} \end{pmatrix}$$

The columns of $V$ are computed using the FFT on the columns of $\mathcal{X}$. If we represent $V$ by its rows, we find that

$$V = \begin{pmatrix} v_0^T \\ v_1^T \\ \vdots \\ v_{N_1-1}^T \end{pmatrix}, \qquad \mathbf{X} = \begin{pmatrix} v_0^T W_{N_2} \\ v_1^T W_{N_2} \\ \vdots \\ v_{N_1-1}^T W_{N_2} \end{pmatrix}$$

$\mathbf{X}$ is computed by applying the FFT on the rows of $V$. If $N_1$ and $N_2$ are powers of 2, the computation of $V$ requires $N_2 N_1 \log_2 N_1 / 2$ operations, while the computation of $\mathbf{X}$ requires $N_1 N_2 \log_2 N_2 / 2$ operations. The total count is

$$C_3 = \frac{N_1 N_2}{2} (\log_2 N_1 + \log_2 N_2) = \frac{N_1 N_2}{2} (\log_2 N_1 N_2) \tag{5.59}$$

For an image of $1024 \times 1024$ pixels, we have $C_3 \cong \frac{1}{2} 10^6 (\log_2(2^{20})) = 10^7$. Additional savings are achieved by the 2-D FFT.

### 5.8.5    Two-dimensional FFT

Suppose that $N_1$, $N_2$ are even integers and that

$$x_{00}(n_1, n_2) = x(2n_1, 2n_2), \quad x_{01}(n_1, n_2) = x(2n_1, 2n_2 + 1)$$
$$x_{10}(n_1, n_2) = x(2n_1 + 1, 2n_2), \quad x_{11}(n_1, n_2) = x(2n_1 + 1, 2n_2 + 1)$$

Then

$$X(k_1, k_2) = \sum_{n_1=0}^{N_1/2-1} \sum_{n_2=0}^{N_2/2-1} x(2n_1, 2n_2) w_{N_1}^{2n_1 k_1} w_{N_2}^{2n_2 k_2}$$

$$+ \sum_{n_1=0}^{N_1/2-1} \sum_{n_2=0}^{N_2/2-1} x(2n_1, 2n_2 + 1) w_{N_1}^{2n_1 k_1} w_{N_2}^{(2n_2+1)k_2}$$

$$+ \sum_{n_1=0}^{N_1/2-1} \sum_{n_2=0}^{N_2/2-1} x(2n_1 + 1, 2n_2) w_{N_1}^{(2n_1+1)k_1} w_{N_2}^{2n_2 k_2}$$

$$+ \sum_{n_1=0}^{N_1/2-1} \sum_{n_2=0}^{N_2/2-1} x(2n_1 + 1, 2n_2 + 1) w_{N_1}^{(2n_1+1)k_1} w_{N_2}^{(2n_2+1)k_2}$$

It holds that $w_N^2 = w_{N/2}$. Hence

$$X(k_1, k_2) = X_{00}(k_1, k_2) + X_{01}(k_1, k_2) w_{N_2}^{k_2} + X_{10}(k_1, k_2) w_{N_1}^{k_1} + X_{11}(k_1, k_2) w_{N_1}^{k_1} w_{N_2}^{k_2}$$

Each $X_{ij}(k_1, k_2)$, $i, j = 0, 1$ has size $N_1/2 \times N_2/2$ and is periodic with horizontal period $N_1/2$ and vertical period $N_2/2$. Exactly as in the 1-D case, the relation $w_{N_1}^{N_1/2} = w_{N_2}^{N_2/2} = -1$ yields

$$X(k_1 + \frac{N_1}{2}, k_2) = X_{00}(k_1, k_2) + X_{01}(k_1, k_2) w_{N_2}^{k_2}$$

$$- X_{10}(k_1, k_2) w_{N_1}^{k_1} - X_{11}(k_1, k_2) w_{N_1}^{k_1} w_{N_2}^{k_2}$$

$$X(k_1, k_2 + \frac{N_2}{2}) = X_{00}(k_1, k_2) - X_{01}(k_1, k_2) w_{N_2}^{k_2}$$

$$+ X_{10}(k_1, k_2) w_{N_1}^{k_1} - X_{11}(k_1, k_2) w_{N_1}^{k_1} w_{N_2}^{k_2}$$

$$X(k_1 + \frac{N_1}{2}, k_2 + \frac{N_2}{2}) = X_{00}(k_1, k_2) - X_{01}(k_1, k_2) w_{N_2}^{k_2}$$

$$- X_{10}(k_1, k_2) w_{N_1}^{k_1} + X_{11}(k_1, k_2) w_{N_1}^{k_1} w_{N_2}^{k_2}$$

Thus the DFT is computed by means of four 2-D DFT's of size $N_1/2 \times N_2/2$. The basic processing unit is the two-dimensional butterfly. It has 4 inputs and 4 outputs and is illustrated in Fig. 5.19. The computation of the 2-D DFT needs $N_1 N_2/4$ butterflies. Each executes three multiplications. The total count of multiplications is $3N_1 N_2/4$. If $N_1 = N_2 = N$ and $N$ is a power of 2, the process is repeated $\log_2 N$ times, and the multiplications count is at most $C_4 = \frac{3}{4} N^2 \log_2 N$. The two-dimensional FFT achieves 25% computational reduction in comparison to the algorithm that uses the 1-D FFT; see

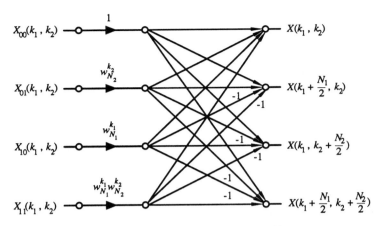

**Figure 5.19.** 2-D butterfly, for $N_1 = N_2 = N$.

Eq. (5.59). Indexing requires bit reversal as in the one-dimensional FFT. The index $(n_1, n_2)$ converts to the index $(r(n_1), r(n_2))$ where the permutation $r$ indicates the bit reversal.

### 5.8.6  The Good-Thomas FFT algorithm

The Good-Thomas FFT algorithm converts the one-dimensional DFT, into a multidimensional DFT using the Chinese remainder theorem for integers. The Cooley-Tukey FFT algorithm achieves superior performance when the block-length $N$ is a power of a prime. In contrast, the Good-Thomas algorithm requires that $N$ is a product of relatively prime integers. We describe the method for the two-dimensional case. Let

$$N = N_1 N_2, \qquad gcd(N_1, N_2) = 1$$

Coprimeness of $N_1$ and $N_2$ guarantees the existence of $r_1$ and $r_2$ such that

$$r_1 N_1 + r_2 N_2 = 1 \tag{5.60}$$

The Chinese remainder theorem (see Appendix II) asserts that each integer $n$ in $0 \le n \le N - 1$ is written

$$n = n_1 r_2 N_2 + n_2 r_1 N_1 \,(\mathrm{mod}\, N) \tag{5.61}$$

where $n_1 = n(\mathrm{mod}\, N_1)$ and $n_2 = n(\mathrm{mod}\, N_2)$. We use representation (5.61) to write the DFT of an $N$ point signal $x$ as

$$X(k) = \sum_{n_1=0}^{N_1-1} \sum_{n_2=0}^{N_2-1} x(n_1, n_2) w^{k(n_1 r_2 N_2 + n_2 r_1 N_1)} \tag{5.62}$$

The frequency index $k$ is converted into a two-dimensional index in such a way that the exponent in (5.62) is amenable to further manipulations. For this purpose we assume the following representation:

$$k = k_2 N_1 + k_1 N_2 (\text{mod } N) \tag{5.63}$$

Then $w^{kn} = w^{k_2 n_2 r_1 N_1^2} w^{k_1 n_1 r_2 N_2^2}$, where we make use of the properties

$$w^{k_2 n_1 r_2 N_1 N_2} = w^{k_2 n_1 r_2 N} = 1 = w^{k_1 n_2 r_1 N_1 N_2}$$

Let

$$w_1 = w^{r_2 N_2^2}, \qquad w_2 = w^{r_1 N_1^2} \tag{5.64}$$

$w_1$ is an $N_1$ root of unity. Indeed, $w_1 = w^{N(r_2 N_2)} = 1$. Likewise $w_2$ is an $N_2$ root of unity. Now Eq. (5.62) is rewritten as

$$X(k_1, k_2) = \sum_{n_1=0}^{N_1-1} \sum_{n_2=0}^{N_2-1} x(n_1, n_2) w_1^{n_1 k_1} w_2^{n_2 k_2} \tag{5.65}$$

This is a true 2-D DFT provided that Eq. (5.63) is well defined and invertible. Thus we need to establish that the pair $(k_1, k_2)$ is uniquely specified in $0 \le k_1 \le N_1 - 1$ and $0 \le k_2 \le N_2 - 1$. For this purpose we multiply both sides of (5.63) with $r_2$ and use (5.60):

$$r_2 k = k_2 r_2 N_1 + k_1 r_2 N_2 = k_2 r_2 N_1 + k_1 (1 - r_1 N_1) = k_1 + m N_1$$

Hence $k_1 = r_2 k (\text{mod } N_1)$. Likewise, if we multiply (5.63) with $r_1$, we will find that $k_2 = r_1 k (\text{mod } N_2)$. It follows that the indexing map $k \rightarrow (k_1, k_2)$ is well defined and invertible.

The derivation of the algorithm is complete. Let us take, for example, $N = 15$. Then $N_1 = 3$, $N_2 = 5$, $r_1 = 2$, $r_2 = -1$. The one-dimensional array

$$(0 \quad 1 \quad 2 \quad 3 \quad 4 \quad 5 \quad 6 \quad 7 \quad 8 \quad 9 \quad 10 \quad 11 \quad 12 \quad 13 \quad 14)$$

is mapped onto the two-dimensional array

$$\begin{pmatrix} 0 & 6 & 12 & 3 & 9 \\ 10 & 1 & 7 & 13 & 4 \\ 5 & 11 & 2 & 8 & 14 \end{pmatrix}$$

The index map scans the $3 \times 5$ array by moving on the diagonals. Likewise $k$ is recovered from $(k_1, k_2)$ by $k = 5k_1 + 3k_2$.

In summary, the Good-Thomas algorithm determines the discrete Fourier transform by converting it into a multidimensional DFT, by means of (5.65). The resulting multidimensional DFT is computed by means of the methods of Sections 5.8.4 and 5.8.5. The above procedure achieves significant computational savings for reasons similar to those explained in Section 5.8.3. Besides Eq. (5.65), the algorithm requires two indexing schemes. The first relates to the time domain. It converts $n$ into the pair $(n_1, n_2)$, of the remainders of $n$ by $N_1$ and $N_2$. The second indexing map recovers the frequency index $k$ from the pair $(k_1, k_2)$ via Eq. (5.63).

## BIBLIOGRAPHICAL NOTES

Representations of discrete and continuous time periodic signals in terms of harmonically related exponentials are discussed in C. Liu and J. Liu (1975), Oppenheim, Willsky, and Young (1983), and Manolakis and Proakis (1989). 2-D signals are treated in Dudgeon and Mersereau (1984) and Lim (1990). The history of the FFT algorithm is provided in the paper of Heideman, Johnson, and Burrus (1984). FFT algorithms are analyzed in Blahut (1985), Burrus and Parks (1985), and Oppenheim and Schafer (1989). Other orthogonal transforms are described in Ahmed and Rao (1975). For a mathematical treatment of Fourier series and harmonic analysis, the reader is referred to Dym and McKean (1972) and Katznelson (1976).

## PROBLEMS

**5.1.** Determine the Fourier coefficients for the following signals. Provide MATLAB validation.
   (a) $e^{j5t}$      (b) $\cos 3t + \sin 8t$      (c) signal of Fig. 5.20
   (d) $x(t)$ periodic with period 2 and $x(t) = e^{-3t}$, $-1 < t < 1$.
   (e) signal of Fig. 5.21.

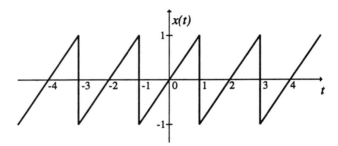

**Figure 5.20.**

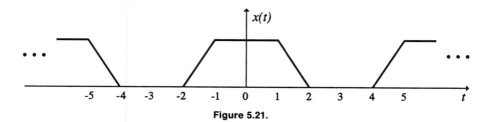

**Figure 5.21.**

**5.2.** Determine the Fourier coefficients of the following discrete signals. Provide MATLAB validation.
(a) $x(n) = \cos \omega n$, $\omega = 2\pi m/N$, $m$ and $N$ are relatively prime, $(m < N)$
(b) $x(n) = \sin \omega n$, with the same assumptions as in (a)

**5.3.** Show that the Kronecker product has the properties:
(a) $(A + B) \otimes C = A \otimes C + B \otimes C$
(b) $(A \otimes B) \otimes C = A \otimes (B \otimes C)$
(c) $(A \otimes B)^T = A^T \otimes B^T$
(d) $(A \otimes B)^{-1} = A^{-1} \otimes B^{-1}$
(e) $(A \otimes B)(C \otimes D) = (AC) \otimes (BD)$
(f) $\det(A \otimes B) = (\det A)^n (\det B)^m$, where $A$ is $m \times m$ and $B$ is $n \times n$
(g) $\text{Tr}(A \otimes B) = \text{Tr}(A)\,\text{Tr}(B)$
(h) Let $\text{vec}(A)$ denote the vector obtained from $A$ by lexicographic ordering by rows. Show that $\text{vec}(ABC) = (C^T \otimes A)\text{vec}(B)$

Verify the above properties by the MATLAB command kron applied on matrices of your choice.

**5.4.** (a) A multichannel signal $x(n)$ is called periodic if there exists an integer $N$ such that $x(n + N) = x(n)$ for all $n$. Let

$$x(n) = (\, x_1(n) \quad x_2(n) \quad \cdots \quad x_m(n)\,)$$

Show that $x(n)$ is periodic with period $N$ if and only if all signal components are periodic with period $N$. Use this as a starting point to prove that

$$x(n) = \sum_{k=0}^{N-1} A_k e^{jkn2\pi/N}, \quad A_k = \frac{1}{N}\sum_{k=0}^{N-1} x(n)e^{-jkn2\pi/N}$$

$A_k$ are vectors of the same length with $x(n)$.
(b) Derive a similar conclusion for multichannel analog signals.
(c) Let $x(n)$ be a single-channel signal with period $N$. Compute the Fourier coefficients of the multichannel signal

$$(\, x(n) \quad x(n-1) \quad \cdots \quad x(n-r)\,), \qquad 0 < r < N$$

**5.5.** If $x(n_1, n_2)$ is a 2-D periodic signal with periods $N_1$, $N_2$ and Fourier coefficients $X(k_1, k_2)$, find the Fourier coefficients of the signals
(a) $x(n_1 - m_1, n_2 - m_2)$, $m_1$, $m_2$ given integers
(b) $x(n_2, n_1)$ assuming that $N_1 = N_2$
(c) $x^*(n_1, n_2)$

**5.6.** Let $x(n_1, n_2)$ as in Problem 5.5. What is the period of the 1-D sequence $x(n, n)$? If $N_1$, $N_2$ are relatively prime, find the Fourier coefficients of $x(n, n)$.

**5.7.** Compute the 2-D DFT of the signal $a^{n_1 - n_2}$, $0 \le n_1, n_2 < N = N_1 = N_2$.

**5.8.** Prove the following properties of the discrete Fourier transform:
1. Linearity: $ax(n) + by(n) \leftrightarrow aX(k) + bY(k)$
2. Cyclic shift: $x(R_N(n-1)) \leftrightarrow w^k X(k)$, where $R_N(k)$ is the remainder of $k$ when divided by $N$ and $w = e^{-j2\pi/N}$
3. Modulation: $w^n x(n) \leftrightarrow X(R_N(k+1))$
4. Real signals: $X(k) = X^*(N-k)$ if $x(n)$ is real.

**5.9.** Consider a finite energy signal over an interval of length $T$, $x(t)$. Using the projection theorem, prove that among the bandlimited periodic signals of bandwidth $2N + 1$:

$$\sum_{k=-N}^{N} a_k e^{jk\omega t}, \qquad \omega = \frac{2\pi}{T}$$

the signal whose coefficients $a_k$ are the Fourier coefficients of $x(t)$, minimizes the mean squared error

$$\int_T \left| x(t) - \sum_{k=-N}^{N} a_k e^{jk\omega t} \right|^2 dt$$

Show that the minimum error is decreasing with $N$.

**5.10.** Determine the spectral lines of

$$x(n_1, n_2) = \cos\left(\frac{2\pi n_1}{N}\right) + \cos\left(\frac{4\pi n_1}{N} + \frac{2\pi n_2}{N}\right) + \sin\left(\frac{6\pi n_2}{N}\right)$$

**5.11.** Explain how the decimation in time and frequency FFT schemes can be obtained from Eq. (5.58) and the general length case.

**5.12.** Show that the 2-D butterfly can be computed with 3 complex multiplications and 8 complex additions. Use the remark that the butterfly is a $2 \times 2$ DFT.

**5.13.** Verify the calculations of Examples 5.12 and 5.13 by MATLAB.

**5.14.** Verify Eq. (5.34) using MATLAB and the kron command.

**5.15.** Let $x(t) = \sum_{k=-\infty}^{\infty} a_k e^{jk\omega_0 t}$, $a_0 = 0$, $a_k = a^*_{-k}$. If the real and imaginary parts of $a_k$ are zero mean independent Gaussian random variables, show that $x(t)$ is a harmonic process of the form (5.51), where the coefficients $A_k$ follow the Rayleigh distribution (see Exercise 4.9).

**5.16.** Consider a nonprimitive element $b$ of a finite field $F$. Compute the DFT of $x(n) = b^n$.

**5.17.** Consider the signal $x(n) = (n+1)(n+2)/2$ in a finite field. Show that $x(n)$ takes values in the subfield of integers. Calculate the DFT over $GF(2^3)$.

**5.18.** Show that the FFT algorithms are valid in finite fields.

**5.19.** Show that the circular convolution can be written as a product $y = Cu$, where $y$ and $u$ are the $N$-point vectors of signal samples, and $C$ is the circular matrix

$$C = \begin{pmatrix} h(0) & h(N-1) & \cdots & h(1) \\ h(1) & h(0) & \cdots & h(2) \\ \vdots & \vdots & \ddots & \vdots \\ h(N-1) & h(N-2) & \cdots & h(0) \end{pmatrix}$$

The rows of $C$ are obtained by the cyclic shifts of the first row.
Show that every circular matrix $C$ is diagonalized by the DFT matrix $W_N$ and that the diagonal elements (eigenvalues of $C$) are formed by the DFT of the first row of $C$. For the proof use Eq. (5.49).

**5.20.** Show that the Cooley-Tukey decimation in time FFT interconnection network can be written as a cascade of exhange permutations and the bit reversal. For $1 \leq k \leq L$ and $N = 2^L$, the $k$-exchange permutation is

$$p_k(n) = p_k(b_L b_{L-1} \cdots b_1) = (b_L \cdots \hat{b}_k \cdots b_1)$$

where $\hat{b}_k = 1 + b_k$. Prove that the cascade can be realized by a single-exchange permutation and the shift permutation

$$\zeta^k(n) = (b_{L-k} b_{L-k-1} \cdots b_{L-k+1})$$

by showing that

$$p_k = \zeta^{-(L-k+1)} p_1 \zeta^{L-k+1}$$

Use the above analysis to deduce the architectures of Fig. 5.22 and Fig. 5.23.

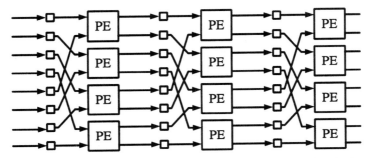

**Figure 5.22.** Perfect shuffle exchange multistage network for the decimation in time FFT of length 8.

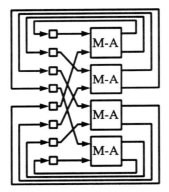

**Figure 5.23.** Perfect shuffle exchange single stage network for the decimation in time FFT of length 8.

**5.21.**    Draw a diagram showing the flow of computations for the 2-D FFT.

**5.22.**    Construct an example that shows the indexing map of the Good-Thomas algorithm for the frequency indexes.

## APPENDIX

**PROOF OF THEOREM** 5.1    We will prove that if a finite energy periodic function $x(t)$ is orthogonal to the set of harmonically related exponentials $A$, that is,

$$\int_0^T x(t)e^{-jk\omega t}dt = 0, \qquad k \in Z \tag{5.66}$$

it is zero almost everywhere, and hence it is the zero function in the space $L_2(T)$.

We will rely on Theorem 3.8 of the Appendix of Chapter 3. To take advantage of continuity, we consider the function

$$y(t) = \int_0^t x(\tau)d\tau - C$$

The constant $C$ will be chosen to secure that $y(t)$ is orthogonal to $A$ too. Indeed, note that $y(0) = -C$. Setting $k = 0$ in (5.66), we have $y(T) = -C$. Since $x(t)$ is absolutely integrable over $T$, $y(t)$ is absolutely continuous and $\dot{y}(t) = x(t)$ almost everywhere. Thus

$$\int_0^T \dot{y}(t)e^{-jk\omega t}dt = 0, \qquad k \in Z$$

Integration by parts gives for $k \neq 0$,

$$y(t)e^{-jk\omega t}|_0^T - (-jk\omega) \int_0^T y(t)e^{-jk\omega t}dt = 0$$

Thus

$$\int_0^T y(t)e^{-jk\omega t}dt = 0, \qquad k \in Z, k \neq 0$$

We wish to extend the above so as to cover the case $k = 0$. To ensure that $\int_0^T y(t)dt = 0$, we pick

$$C = \frac{1}{T}\int_0^T \int_0^t x(\tau)d\tau dt$$

Then

$$\int_0^T y(t)e^{-jk\omega t}dt = 0, \qquad k \in Z$$

We established that the continuous function $y(t)$ on the interval $[0, T]$ is orthogonal to the harmonic exponential family. Next we approximate $y(t)$ with a trigonometric polynomial using Theorem 3.8 of the Appendix of Chapter 3. For any given $\epsilon > 0$, let $z(t) = \sum_{k=-N}^{N} a_k e^{jk\omega t}$, $a_k = a_{-k}^*$ so that $|y(t) - z(t)| < \epsilon$ for all $t \in [0, T]$. $z(t)$ belongs to the linear span of $A$. Therefore $y(t)$ is orthogonal to $z(t)$. This fact together with the Cauchy-Schwarz inequality give

$$|y|_2^2 = <y, y> = <y - z, y> = \int_0^T (y(t) - z(t))y^*(t)dt \le \epsilon \int_0^T |y(t)|dt$$

$$\le \epsilon\sqrt{T}\left(\int_0^T |y(t)|^2 dt\right)^{1/2} \le \epsilon\sqrt{T}|y|_2$$

If $|y|_2 \neq 0$, then $|y|_2 \le \epsilon\sqrt{T}$ for all $\epsilon$. Thus $y = 0$. Differentiating, we obtain $\dot{y}(t) = x(t) = 0$ almost everywhere, and the proof is complete. ∎

# 6

---

# FREQUENCY ANALYSIS OF SIGNALS AND SYSTEMS

Representation of signals in the frequency domain is finding numerous applications in signal processing. Frequency domain modeling makes certain analysis and design properties easier to visualize and handle. Important examples are frequency selective filtering, sampling, and modulation. The signal is expressed in the frequency domain by its Fourier transform. The Fourier transform provides a representation of general, not necessarily periodic, signals in terms of exponential signals and in this way uncovers their spectral content. In contrast to the periodic case where the spectrum is concentrated on harmonically related frequencies, the spectral information is now spread over a continuum range of frequencies.

## 6.1  FOURIER ANALYSIS OF ANALOG SIGNALS

### 6.1.1  Motivation and definitions

To gain some insight into the nature of the Fourier transform, we will outline the argument Fourier used to extend his series expansions of periodic signals to aperiodic waveforms. A rigorous description is taken up in subsequent sections. Let us consider a finite support continuous time signal $x(t)$ and its periodic extension $x_T(t)$:

$$x_T(t) = \sum_{r=-\infty}^{\infty} x(t - rT), \qquad T > 2S \tag{6.1}$$

Plots are shown in Fig. 6.1. We observe that as $T$ increases towards infinity, $x_T(t)$ approaches $x(t)$. The Fourier series expansion of $x_T(t)$ is

$$x_T(t) = \sum_{k=-\infty}^{\infty} a_k e^{jk\omega_0 t}, \qquad \omega_0 = \frac{2\pi}{T} \tag{6.2}$$

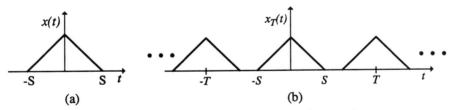

**Figure 6.1.** A finite support signal and its periodic extension.

The spectral components are given by

$$a_k = \frac{1}{T} \int_T x_T(t) e^{-jk\omega_0 t} dt = \frac{1}{T} \int_{-S}^{S} x(t) e^{-jk\omega_0 t} dt = \frac{1}{T} \int_{-\infty}^{\infty} x(t) e^{-jk\omega_0 t} dt$$

We define the function of the real variable $\omega \to X(\omega)$:

$$X(\omega) = \int_{-\infty}^{\infty} x(t) e^{-j\omega t} dt \tag{6.3}$$

Then $a_k = \frac{1}{T} X(k\omega_0)$, and (6.2) becomes

$$x_T(t) = \frac{1}{T} \sum_{k=-\infty}^{\infty} X(k\omega_0) e^{jk\omega_0 t} \tag{6.4}$$

If we take limits on both sides of (6.4), and we recall that $x(t) = \lim_{T \to \infty} x_T(t)$, we obtain

$$x(t) = \frac{1}{2\pi} \lim_{\omega_0 \to 0} \sum_{k=-\infty}^{\infty} X(k\omega_0) e^{jk\omega_0 t} \omega_0$$

We recognize the presence of an integral in the right-hand side of the above equation. Thus

$$x(t) = \frac{1}{2\pi} \int_{-\infty}^{\infty} X(\omega) e^{j\omega t} d\omega \tag{6.5}$$

The latter expression represents the aperiodic signal $x(t)$ over a continuum range of periodic exponential signals. The spectral content over an infinitesimal interval $[\omega, \omega + d\omega]$ is $X(\omega)$. Equation (6.3) defines the Fourier transform $X(\omega)$ of the signal $x(t)$ and is called the *analysis equation*. It describes the information content of the signal in the frequency domain. Equation (6.5) defines the inverse Fourier transform and is called the *synthesis equation*. It demonstrates how the signal is recovered in the time domain. The above equations are valid over a very broad class of signals. Precise definitions and a rigorous description are given later.

For each frequency $\omega$, $X(\omega)$ is a complex number. Frequently we refer to it as the *signal spectrum*. It is represented in polar coordinates by the *magnitude spectrum* and the *phase spectrum*, $X(\omega) = |X(\omega)|e^{j \arg X(\omega)}$. In Cartesian coordinates, it is expressed in terms of its real and imaginary parts $X(\omega) = \Re X(\omega) + j\Im X(\omega)$. The frequency change $\omega \to f = \omega/2\pi$ expresses frequency in Hertz (cycles per second). In this case the Fourier equations take the form

$$X(f) = \int_{-\infty}^{\infty} x(t)e^{-j2\pi ft}\,dt, \qquad x(t) = \int_{-\infty}^{\infty} X(f)e^{j2\pi ft}\,df \tag{6.6}$$

$X(\omega)$ is complex valued even if $x(t)$ is real. Expressions for the real and imaginary parts are obtained if we insert Euler's identity into the definition of the Fourier transform:

$$\Re X(\omega) = \int_{-\infty}^{\infty} (\Re x(t)\cos\omega t + \Im x(t)\sin\omega t)\,dt$$

$$\Im X(\omega) = \int_{-\infty}^{\infty} (\Im x(t)\cos\omega t - \Re x(t)\sin\omega t)\,dt$$

In particular, if $x(t)$ is real, the above become

$$\Re X(\omega) = \int_{-\infty}^{\infty} x(t)\cos\omega t\,dt, \qquad \Im X(\omega) = -\int_{-\infty}^{\infty} x(t)\sin\omega t\,dt \tag{6.7}$$

The Fourier transform of a matrix valued signal $x(t)$ with entries $x_{ij}(t)$ is

$$X(\omega) = \int_{-\infty}^{\infty} x(t)e^{-j\omega t}\,dt \qquad \text{or} \qquad [X(\omega)]_{ij} = X_{ij}(\omega)$$

### 6.1.2 Existence of Fourier transform

The Fourier transform does not exist for all signals. Even when it does exist, the synthesis equation (6.5) may not be valid. Precise conditions that ensure the validity of Fourier representation are given next. Equations (6.3) and (6.5) can be interpreted in a variety of ways, depending on the limit involved and the type of integration. Equality in the main equations may extend to the entire real line or to a typical subset. For instance, equality in (6.5) may be confined to points $t$ such that $x(t)$ is continuous. Experience gained in the long study of the above problems shows that the best method of attack is to consider the following main questions:

1. When is the Fourier transform $X(\omega)$ well defined?

2. When is the inverse Fourier transform

$$\hat{x}(t) = \frac{1}{2\pi} \int_{-\infty}^{\infty} X(\omega) e^{j\omega t} d\omega \tag{6.8}$$

well defined?

3  When does it hold $x(t) = \hat{x}(t)$?

First, we address the above questions in the environment of $L_1$, the space of absolutely integrable functions. Then we focus attention on the space of rapidly decreasing functions where all three questions are nicely answered. Subsequently we consider the space of square integrable functions $L_2$. Finally, we briefly discuss the Fourier transform of impulsive signals in the context of tempered distributions. The following theorem gives sufficient conditions for the existence of Fourier transform and provides an answer to question 1.

**Theorem 6.1**   Suppose that $x(t)$ is absolutely integrable in the Lebesgue sense

$$|x|_1 = \int_{-\infty}^{\infty} |x(t)| dt < \infty$$

The Fourier transform $X(\omega)$ exists for any $\omega \in R$. Moreover it is uniformly continuous bounded function satisfying $|X(\omega)| \leq |x|_1$, $\omega \in R$.

**PROOF**   Note that

$$|X(\omega)| \leq \int_{-\infty}^{\infty} |x(t)||e^{-j\omega t}| dt = |x|_1$$

Thus $X(\omega)$ is a bounded function. Let $\omega$, $v$ in $R$. Then

$$X(\omega + v) - X(v) = \int_{-\infty}^{\infty} x(t) \left[ e^{-j(\omega+v)t} - e^{-jvt} \right] dt$$

Now we want to take the limit as $\omega \to 0$ and insert it into the integral. This is permissible by the dominated convergence theorem provided that the integrand is bounded by an absolutely integrable function and also converges almost everywhere. In our case these assumptions are met. Indeed,

$$|x(t)||e^{-j(\omega+v)t} - e^{-jvt}| \leq |x(t)||e^{-j\omega t} - 1| \leq 2|x(t)|$$

and the integrand is bounded by the integrable function $2|x(t)|$. Thus

$$|X(\omega + v) - X(v)| \leq \int_{-\infty}^{\infty} |x(t)||e^{-j\omega t} - 1| dt$$

Inserting the limit into the integral, and noting that $e^{-j\omega t} \to 1$, as $\omega \to 0$, we establish continuity. Analogous results hold for multichannel signals (see Exercise 6.4). ∎

### 6.1.3 Computation of Fourier transform

The Fourier transform can be computed by means of the DFT. The procedure is analogous to that used for the Fourier series coefficients (see Example 5.7). The algorithm we will describe uses a finite record of signal samples $x(nT_s)$ and produces a finite record of Fourier transform samples $X(kW_s)$. $T_s$ and $W_s$ denote the sampling periods in the time and frequency domain, respectively. In effect, a time as well as a frequency window are imposed on the signal and its transform. The time window is $T_1 \leq t \leq T_2$, and the frequency window is $0 \leq \omega \leq W$. Let $N$ be the number of sampled points in the time and frequency domain. Then $T_2 - T_1 = NT_s$, $W = NW_s$. We apply the rectangle approximation on the Fourier integral (6.3) to obtain

$$X(\omega) \approx \int_{T_1}^{T_2} x(t)e^{-j\omega t}dt = \sum_{n=0}^{N-1} \int_{T_1+nT_s}^{T_1+(n+1)T_s} x(t)e^{-j\omega t}dt$$

$$\approx T_s e^{-j\omega T_1} \sum_{n=0}^{N-1} x(T_1 + nT_s)e^{-j\omega nT_s}$$

Evaluation at equidistant frequencies gives

$$X(k) = X(kW_s) \approx T_s e^{-jkW_sT_1} \sum_{n=0}^{N-1} x(T_1 + nT_s)e^{-jknT_sW_s}$$

The latter expression reminds us of the DFT. Indeed, let $x(n) = x(T_1 + nT_s)$, $0 \leq n \leq N - 1$, and $T_sW_s = 2\pi/N$. Then

$$X(k) = X(kW_s) \approx T_s e^{-jkW_sT_1} DFT(x(n))$$

In a similar fashion we handle the inverse Fourier transform. The above algorithm must be used with caution especially in connection with the phase spectrum. The inevitable use of time and frequency windows causes errors. Problems are acute near the highest frequency $W$. They are alleviated if plots are visualized in frequency intervals $[0, W_1]$ with $W_1 \ll W$. Computation of the analog Fourier transform is further discussed in Section 8.8.5; see also Exercise 6.17.

*Example 6.1   MATLAB computation of Fourier transform*
A MATLAB implementation is described below. The variable $x$ denotes the signal in the time domain. $T_1$, $T_2$ are the endpoints of the time window over

which the signal values are defined. $N$ denotes the number of frequency points. Preferably $N$ is a power of 2 to maximize efficiency of the Cooley-Tukey FFT algorithm. $T_s$ and $W_s$ are the sampling periods in the time and frequency domain, respectively. The time window is $t = [T_1 : T_s : T_2 - T_s]$. Moreover $T_s = (T_2 - T_1)/N$, $W_s = 2\pi/(N * T_s)$. The sampled frequencies are $w = [-N/2 + 1 : N/2] * W_s$. The choice of $N$ and a variant of the algorithm are described later in connection with the Poisson summation formula.

```
function [w,X] = strft(x)
x1 = ['[y,Ts,N,T1] = ',x,';'];   % handles signals entered by a string; see Example 6.2
eval(x1);
t = T1:Ts:T1 + (N-1)*Ts;
Ws = 2*pi/(N*Ts);
w = [-(N/2) + 1:(N/2)]*Ws;
X = Ts*fft(y);
X = exp(-j*w*T1).*X;
subplot(221), plot(t,y);
subplot(222), plot(w,abs(fftshift(X)));
subplot(223), plot(w,angle(fftshift(X)));
```                                                                     ∎

The use of the above function is elaborated next.

### Example 6.2    Causal exponential
We easily check that the causal exponential signal

$$x(t) = e^{at}u_s(t), \qquad a \in R$$

is absolutely integrable provided that $a < 0$. The Fourier transform is

$$X(\omega) = \int_0^\infty e^{at - j\omega t} dt = \frac{1}{j\omega - a} = \frac{-a - j\omega}{a^2 + \omega^2}$$

The magnitude and phase are

$$|X(\omega)| = \frac{1}{\sqrt{a^2 + \omega^2}}, \qquad \arg X(\omega) = \tan^{-1}\left(\frac{\omega}{a}\right)$$

Plots of magnitude and phase are shown in Fig 6.2. The magnitude $|X(\omega)|$ attains its maximum value $-1/a$ at $\omega = 0$ and attenuates at higher frequencies as $|\omega| \to \infty$. We confirm the conclusions of Theorem 6.1 that $X(\omega)$ is uniformly continuous and bounded. In fact $X(\omega) \to 0$ as $|\omega| \to \infty$. The last property is known as *Riemann-Lebesgue lemma* and is established later. The causal exponential signal is generated by the following MATLAB function.

```
function [x,Ts,N,T1] = cexp(T2,N,a)
T1 = 0;
Ts = T2/N;
t = 0:Ts:(N-1)*Ts;
x = exp(a*t);
```

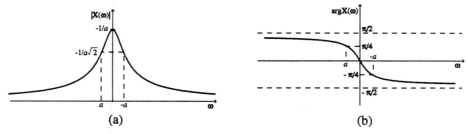

**Figure 6.2.** Magnitude and phase spectrum of the signal in Example 6.2.

The command strft('cexp(T2,N,a)') plots the signal in the time domain and the magnitude and phase spectrum in the frequency domain. ∎

### Example 6.3    Even exponential
The signal

$$x(t) = e^{a|t|}, \qquad a < 0$$

is sketched in Figure 6.3. It is easy to show that $x(t) \in L_1$. The Fourier transform is

$$X(\omega) = \int_{-\infty}^{0} e^{-at}e^{-j\omega t}\,dt + \int_{0}^{\infty} e^{at}e^{-j\omega t}\,dt = \frac{-2a}{a^2 + \omega^2}$$

The spectrum is real. It is shown in Fig. 6.3. The signal is constructed by the following code:

```
function [x,Ts,N,T1] = evexp(T1,T2,N,a)
Ts = (T2-T1)/N;
t = T1:Ts:(N-1)*Ts + T1;
x = exp(-a*abs(t));
```

The command strft('evexp(T1,T2,N,a)') plots the signal in the time as well as in the frequency domain. ∎

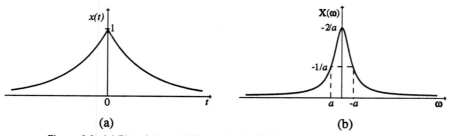

**Figure 6.3.** (a) Plot of signal of Example 6.3; (b) the spectrum of the signal.

***Example 6.4    Rectangular pulse***
Consider the rectangular pulse of duration $P$:

$$x(t) = \begin{cases} 1, & \text{if } |t| < P \\ 0, & \text{otherwise} \end{cases}$$

Clearly $x(t)$ is absolutely integrable. The Fourier transform is

$$X(\omega) = \int_{-P}^{P} e^{-j\omega t} dt = \frac{2\sin \omega P}{\omega} \tag{6.9}$$

The rectangular pulse and its spectrum are shown in Fig. 6.4. $X(\omega)$ is real and $X(0) = 2P$. Indeed, de l'Hôpital's rule gives

$$X(0) = \lim_{\omega \to 0} \frac{2\sin \omega P}{\omega} = \lim_{\omega \to 0} 2P\cos \omega P = 2P$$

The spectrum decreases to zero as we pass to higher frequencies. The *zero crossings* of $X(\omega)$ are the frequencies where $X(\omega)$ becomes zero. They are given by $\omega = \lambda\pi/P, \lambda \in Z, \lambda \neq 0$. ∎

### 6.1.4    Properties of Fourier transform

Throughout this subsection we assume that $x(t)$ is an absolutely integrable single-channel complex-valued signal. According to the previous theorem, the Fourier transform $X(\omega)$ exists. We will use the notation

$$x(t) \xrightarrow{\mathcal{F}} X(\omega) = \mathcal{F}(x(t))$$

***Complex conjugate.***  Conjugation in the time domain converts to reflection and

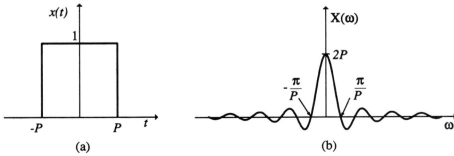

**Figure 6.4.** Rectangular pulse: (a) time domain; (b) frequency domain.

conjugation in the frequency domain:

$$x^*(t) \xrightarrow{\mathcal{F}} X^*(-\omega), \quad x^*(-t) \xrightarrow{\mathcal{F}} X^*(\omega)$$

Indeed, let $z(t) = x^*(t)$. Then

$$Z(\omega) = \int_{-\infty}^{\infty} x^*(t) e^{-j\omega t} \, dt = \left( \int_{-\infty}^{\infty} x(t) e^{j\omega t} \, dt \right)^* = X^*(-\omega)$$

The second relation is proved in a similar way.

**Linearity.** For any constants $c_1$ and $c_2$ and any signals $x_1(t)$ and $x_2(t)$ in $L_1$, we have

$$c_1 x_1(t) + c_2 x_2(t) \xrightarrow{\mathcal{F}} c_1 X_1(\omega) + c_2 X_2(\omega)$$

The proof is a direct consequence of the linearity of the integral.

**Real, imaginary, even, and odd parts.** Given a signal $x(t)$ in $L_1$ with transform $X(\omega)$, we want to determine the signals whose Fourier transforms are the real and imaginary parts of $X(\omega)$. To this end we define the conjugate even and conjugate odd part of $x(t)$ as follows:

$$x_e(t) = \frac{1}{2}(x(t) + x^*(-t)), \qquad x_o(t) = \frac{1}{2}(x(t) - x^*(-t)) \qquad (6.10)$$

Clearly $x_e(t)$ and $x_o(t)$ are absolutely integrable. The conjugate even part of $x(t)$ is invariant under reflection followed by conjugation, while the odd part of $x(t)$ is a conjugate odd function of $t$. Thus $x_e(t) = x_e^*(-t)$, $x_o(t) = -x_o^*(-t)$. Moreover $x(t) = x_e(t) + x_o(t)$. Then

$$x_e(t) \xrightarrow{\mathcal{F}} \Re X(\omega), \qquad x_o(t) \xrightarrow{\mathcal{F}} j\Im X(\omega)$$

The proof is a direct consequence of the two preceding properties.

**Time and frequency shifting.** For any real $t_0$ and $\omega_0$ it holds that

$$x(t - t_0) \xrightarrow{\mathcal{F}} e^{-j\omega t_0} X(\omega), \qquad e^{j\omega_0 t} x(t) \xrightarrow{\mathcal{F}} X(\omega - \omega_0) \qquad (6.11)$$

Let $z(t) = x(t - t_0)$. Clearly $z(t) \in L_1$ and $|z|_1 = |x|_1$. Then

$$Z(\omega) = \int_{-\infty}^{\infty} x(t - t_0) e^{-j\omega t} \, dt = \int_{-\infty}^{\infty} x(\tau) e^{-j\omega(\tau + t_0)} \, d\tau = e^{-j\omega t_0} X(\omega)$$

The frequency shifting property is similarly proved. We observe that shifting in the time domain is converted into multiplication of the spectrum by the spectral factor $e^{-j\omega t_0}$. Note that the spectrum of a shifted in time signal has the same magnitude as the original signal, whereas its phase undergoes a linear change:

$$|Z(\omega)| = |e^{-j\omega t_0} X(\omega)| = |X(\omega)|, \quad \arg Z(\omega) = \arg X(\omega) - \omega t_0$$

As an example, let $z(t) = x(t) \cos \omega_c t$. The Fourier transform is readily computed using the frequency shifting property and linearity. Indeed ,

$$Z(\omega) = \mathcal{F}\left(\frac{1}{2} e^{j\omega_c t} x(t)\right) + \mathcal{F}\left(\frac{1}{2} e^{-j\omega_c t} x(t)\right) = \frac{1}{2}(X(\omega - \omega_c) + X(\omega + \omega_c))$$

$$(6.12)$$

The above property forms the basis of amplitude modulation (AM), which we discuss further in Section 6.6.

**Scaling.** For any real $a$ it holds that

$$x(at) \xrightarrow{\mathcal{F}} \frac{1}{|a|} X\left(\frac{\omega}{a}\right) \tag{6.13}$$

Indeed, let $a > 0$ and $z(t) = x(at)$. Then

$$Z(\omega) = \int_{-\infty}^{\infty} x(at) e^{-j\omega t} \, dt = \frac{1}{a} \int_{-\infty}^{\infty} x(\tau) e^{-j\frac{\omega}{a}\tau} \, d\tau = \frac{1}{a} X\left(\frac{\omega}{a}\right)$$

A similar argument applies if $a < 0$. The time scaling property is illustrated in Fig. 6.5. If $|a| > 1$, the signal is compressed in the time domain; hence it changes more rapidly and therefore is expanded in the frequency domain. If $|a| < 1$, the signal expands in the time domain; hence it changes more slowly and the spectrum is compressed.

**Symmetries for real signals.** The Fourier transform of a real signal is not always real as Example 6.2 indicates. It does, however, possess the following symmetry:

$$x(t) = x^*(t) \xrightarrow{\mathcal{F}} X(\omega) = X^*(-\omega)$$

The spectrum is a complex conjugate even function of the frequency. As a consequence we have

$$\Re X(\omega) = \Re X(-\omega), \quad \Im X(\omega) = -\Im X(-\omega)$$

$$|X(\omega)| = |X(-\omega)|, \quad \arg X(\omega) = -\arg X(-\omega)$$

The real part and the magnitude of $X(\omega)$ are even functions, while the imaginary part and the phase are odd functions of the frequency.

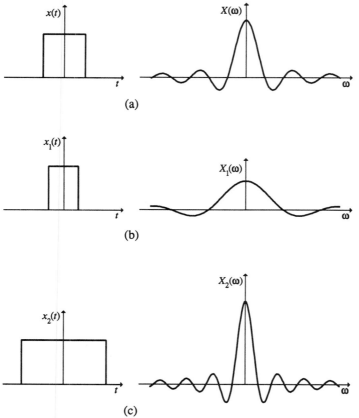

**Figure 6.5.** Illustration of scaling: (a) signal $x(t)$ with spectrum $X(\omega)$; (b) signal $x_1(t) = x(at), |a| > 1$; (c) $x_2(t) = x(at), |a| < 1$.

**Convolution.** Let $x(t)$ and $h(t)$ be two signals in $L_1$ with Fourier transforms $X(\omega)$ and $H(\omega)$, respectively. Let $y(t) = h(t) * x(t)$ denote their convolution. Then the Fourier transform $Y(\omega)$ exists for all $\omega \in R$ and is given by

$$h(t) * x(t) \xrightarrow{\mathcal{F}} H(\omega)X(\omega) \qquad (6.14)$$

It follows from the convolution bound of Theorem 3.2 that $y(t)$ is in $L_1$. Hence $Y(\omega)$ is well defined. Note that

$$Y(\omega) = \int_{-\infty}^{\infty} y(t)e^{-j\omega t}\,dt = \int_{-\infty}^{\infty} \left( \int_{-\infty}^{\infty} h(t-\tau)x(\tau)d\tau \right) e^{-j\omega t}\,dt$$

$$= \int_{-\infty}^{\infty} \left( \int_{-\infty}^{\infty} h(t-\tau)e^{-j\omega t}\,dt \right) x(\tau)d\tau$$

A rigorous justification of the last equality is provided by Fubini's theorem. The change of variables $t - \tau = v$ in the inner integral gives

$$
\begin{aligned}
Y(\omega) &= \int_{-\infty}^{\infty} \left( \int_{-\infty}^{\infty} h(v) e^{-j\omega(v+\tau)} \, dv \right) x(\tau) d\tau \\
&= \int_{-\infty}^{\infty} \left( \int_{-\infty}^{\infty} h(v) e^{-j\omega v} \, dv \right) e^{-j\omega\tau} x(\tau) d\tau \\
&= \int_{-\infty}^{\infty} H(\omega) e^{-j\omega\tau} x(\tau) \, d\tau = H(\omega) X(\omega)
\end{aligned}
$$

The convolution theorem enables us to determine the spectrum of the output signal of a convolutional system in terms of the input spectrum and the impulse response spectrum. The Fourier transform of the impulse response defines the *frequency response* of the system. It was first encountered in Chapter 5. The convolution property ensures that the convolution representation is converted into pointwise multiplication. This important result has numerous applications, as we will later see.

### Example 6.5   *Deterministic autocorrelation function*
Given an absolutely integrable signal $x(t)$ with spectrum $X(\omega)$, we wish to compute the signal $r(t)$ with Fourier transform the energy spectrum $|X(\omega)|^2$. Since $|X(\omega)|^2 = X(\omega) X^*(\omega)$, the complex conjugate property together with the convolution theorem asserts that

$$
r(t) = x(t) * x^*(-t) = \int_{-\infty}^{\infty} x(t - \tau) x^*(-\tau) d\tau = \int_{-\infty}^{\infty} x(t + \tau) x^*(\tau) d\tau
$$

The signal $r(t)$ is the *deterministic autocorrelation function* of $x(t)$. Being the convolution $x(t) * x^*(-t)$, of two integrable signals, it is also absolutely integrable. The same argument shows that the *deterministic crosscorrelation* of two absolutely integrable signals $x(t)$ and $y(t)$,

$$
r_{xy}(t) = x(t) * y^*(-t) = \int_{-\infty}^{\infty} x(t + \tau) y^*(\tau) d\tau \tag{6.15}
$$

has Fourier transform

$$
R_{xy}(\omega) = X(\omega) Y^*(\omega) \tag{6.16}
$$

∎

**Differentiation.** If $x(t) \in L_1$ and $tx(t) \in L_1$, then $X(\omega)$ is differentiable and

$$
-jtx(t) \xrightarrow{\mathcal{F}} \frac{d}{d\omega} X(\omega) \tag{6.17}
$$

An analogous assertion describes differentiation in the time domain. Suppose that the derivative $\dot{x}(t)$ of $x(t)$ exists and belongs to $L_1$. Then

$$\dot{x}(t) \xrightarrow{\mathcal{F}} j\omega X(\omega) \tag{6.18}$$

Repeated application of (6.18) and (6.17) gives

$$x^{(k)}(t) \xrightarrow{\mathcal{F}} (j\omega)^k X(\omega), \quad (-j)^k t^k x(t) \xrightarrow{\mathcal{F}} \frac{d^k}{d\omega^k} X(\omega) \tag{6.19}$$

**PROOF**   Note that

$$\frac{X(\omega+v)-X(\omega)}{v} = \int_{-\infty}^{\infty} \frac{e^{-jvt}-1}{v} e^{-j\omega t} x(t) dt$$

Now we want to take the limit as $v \to 0$ and insert it into the integral. This is permissible by the dominated convergence theorem provided that the integrand is bounded by an absolutely integrable function and also converges almost everywhere. In our case these assumptions are met. Indeed,

$$\left| \frac{e^{-jvt}-1}{v} e^{-j\omega t} x(t) \right| = \frac{|e^{-jvt}-1|}{|v|} |x(t)| \le |t||x(t)|, \quad v \neq 0$$

Thus the integrand is bounded with respect to $v$ by $|t||x(t)|$, and the latter is integrable by assumption. Furthermore

$$\lim_{v \to 0} \frac{e^{-jvt}-1}{v} e^{-j\omega t} x(t) = \frac{d}{dv} e^{-jvt}|_{v=0} e^{-j\omega t} x(t) = -jt e^{-j\omega t} x(t)$$

Hence the dominated convergence theorem applies and gives

$$\frac{dX(\omega)}{d\omega} = \int_{-\infty}^{\infty} -jt e^{-j\omega t} x(t) dt = \mathcal{F}(-jtx(t))$$

The shifting property implies that the Fourier transform of $\frac{1}{h}[x(t+h)-x(t)]$ is $X(\omega)(e^{j\omega h}-1)/h$. The dominated convergence theorem gives

$$j\omega X(\omega) = \lim_{h \to 0} \int_{-\infty}^{\infty} \frac{1}{h}[x(t+h)-x(t)]e^{-j\omega t} dt = \int_{-\infty}^{\infty} \dot{x}(t)e^{-j\omega t} dt \quad \blacksquare$$

**Theorem 6.2  Riemann-Lebesgue lemma**   Let $x(t)$ be an absolutely integrable signal. Then $X(\omega) \to 0$ as $|\omega| \to \infty$.

**PROOF**   Suppose first that $\dot{x}(t)$ exists and belongs to $L_1$. The differentiation property together with the boundedness of the Fourier transform of absolutely integrable functions imply that

$$|X(\omega)| = \frac{1}{|\omega|}|\mathcal{F}(\dot{x})| \le \frac{1}{|\omega|}|\dot{x}|_1 \to 0, \qquad |\omega| \to \infty$$

In general, given $\epsilon > 0$, we pick a function $y$ such that $y$, $\dot{y}$ are in $L_1$ and $|x - y|_1 < \epsilon$. Such $y$ can always be found. Indeed, $x$ is first approximated by a signal of compact support $z$ as in Theorem 2.2. Then $z$ is approximated by a trigonometric polynomial $y$ over the support of $z$ by Theorem 5.1. $y$ is the desired function. Then

$$|X(\omega)| \le |X(\omega) - Y(\omega)| + |Y(\omega)| \le |x - y|_1 + |Y(\omega)| \le \epsilon + \frac{1}{|\omega|}|\dot{y}|_1$$

The right-hand side decays to zero as $|\omega| \to \infty$. ∎

In Theorem 6.1 we proved that $X(\omega)$ is a bounded function if $x(t)$ belongs to $L_1$. The Riemann-Lebesgue lemma strengthens this further by asserting that $X(\omega) \to 0$, as $|\omega| \to \infty$. It would be highly desirable if we could push it a little further and establish that $X(\omega) \in L_1$. If this turned out to be true, we would have automatically answered the second question posed in Section 6.1.2, namely well posedness of the inverse Fourier transform. Indeed, if $X(\omega)$ is absolutely integrable, Theorem 6.1 shows that $\hat{x}(t)$ is a well-defined bounded continuous function. Unfortunately, the condition $X(\omega) \in L_1$ does not follow from the assumption that $x(t) \in L_1$. Consider, for instance, the causal exponential signal discussed in Example 6.2. The Fourier transform in this case behaves like $1/\omega$, which, in analogy with the harmonic series, diverges. Likewise the rectangular pulse treated in Example 6.4 belongs to $L_1$, but its Fourier transform does not (see Example 4.2). On the other hand, the signal of Example 6.3 as well as its Fourier transform are absolutely integrable.

A very useful space of signals that are absolutely integrable in both time and frequency domain is the space of rapidly decreasing functions $S$ introduced next.

***Rapidly decreasing signals.*** Let $S$ denote the space of infinitely differentiable functions $x(t)$ such that for any nonnegative integers $m$, $k$ it holds $t^m x^{(k)}(t) \to 0$ as $|t| \to \infty$. It is easy to show that $S$ is a vector space. Moreover $S$ is contained in $L_p$ for any $1 \le p \le \infty$. Indeed, if $x(t) \in S$, $x(t)$ is continuous and satisfies $x(t) \to 0$, $|t| \to \infty$. It can be further shown that $S$ is dense in every $L_p$ space, $1 \le p \le \infty$. Let $x(t) \in S$. Since $S$ is contained in $L_1$, the Fourier transform $X(\omega)$ exists. It is easy to see that the $k$th derivative $x^{(k)}(t)$ belongs to $S$ and that $t^m x(t) \in S$ for any nonnegative integers $m$, $k$. Therefore the differentiation property in the frequency domain implies that $X(\omega)$ is infinitely differentiable. The Riemann-Lebesgue lemma enables us to conclude that $X(\omega) \in S$. A typical example of a rapidly decreasing signal is the Gaussian signal discussed below.

***Example 6.6    The Gaussian signal***

The Gaussian signal defines the probability density function of the Gaussian distribution. It has the form

$$x(t) = \frac{1}{\sigma\sqrt{2\pi}} e^{-t^2/2\sigma^2}$$

Since $x(t)$ is a probability density, it is absolutely integrable. In fact

$$\frac{1}{\sigma\sqrt{2\pi}} \int_{-\infty}^{\infty} e^{-t^2/2\sigma^2} \, dt = 1 \tag{6.20}$$

and $x(t)$ is infinitely differentiable because it is the exponential of a polynomial. De l' Hôpital's rule implies that it is a rapidly decreasing function. Thus $x(t) \in \mathcal{S}$ and $X(\omega) \in \mathcal{S}$. The differentiation property in the time as well as in the frequency domain enables us to conclude that

$$\frac{dX(\omega)}{d\omega} = -j\mathcal{F}(tx(t)) = j\mathcal{F}\left(\sigma^2 \frac{dx}{dt}\right) = -\sigma^2 \omega X(\omega) \tag{6.21}$$

This is a first-order linear time varying differential equation. The solution is easily computed by integration: $X(\omega) = X(0)e^{-\sigma^2\omega^2/2}$. Due to (6.20) the spectrum at zero frequency is equal to 1, $X(0) = 1$. Thus

$$X(\omega) = e^{-\sigma^2\omega^2/2} \tag{6.22}$$

We observe that the spectrum of a Gaussian signal is also a Gaussian signal with variance the inverse of the variance of the time domain representation.

The following *m*-file implements the Gaussian signal:

```
function [x,Ts,N,T1] = gaussian(T1,T2,N,sigma)
Ts = (T2-T1)/N;
t = T1:Ts:(N-1)*Ts + T1;
x = exp(-t.^2/2*(sigma)^2)/(sigma*sqrt(2*pi));
```                                                ■

## 6.1.5    Inverse Fourier transform

We have seen that $\mathcal{S}$ is a subset of $L_1$ and that if $x(t)$ is in $\mathcal{S}$, $X(\omega)$ is also in $\mathcal{S}$. Hence $X(\omega)$ is absolutely integrable. Thus, for all signals $x(t) \in S$, both $X(\omega)$ and $\hat{x}(t)$ are well defined. At a more general level, if $x(t) \in L_1$ and in addition $X(\omega) \in L_1$, $\hat{x}(t)$ exists for all real $t$ and is a bounded continuous function. Given these assumptions we proceed to the third question, and we inquire whether the inverse Fourier transform is capable of recovering the signal in the time domain:

$$x(t) = \hat{x}(t) \tag{6.23}$$

It is conceivable that (6.23) cannot hold everywhere because $\hat{x}(t)$ is always bounded and continuous, while $x(t)$ does not have to be so. On the other hand, if $x(t)$ is a rapidly decreasing function, it is bounded continuous, and we have reasons to believe that (6.23) is valid everywhere. Let us substitute Eq. (6.3) into (6.8). Assuming that we can interchange integration order, we obtain

$$\hat{x}(t) = \frac{1}{2\pi} \int \int x(\tau)e^{-j\omega\tau} d\tau e^{j\omega t} d\omega = \frac{1}{2\pi} \int x(\tau) \left( \int e^{j\omega(t-\tau)} d\omega \right) d\tau$$

Unfortunately, the integral $\int e^{j\omega(t-\tau)} d\omega$ does not exist as an ordinary integral but it can be interpreted as a distribution; we will later see that it is given by the Dirac function. A different argument is employed to establish the theorem that follows. The basic idea is to approximate the underlying Dirac function by a Gaussian signal.

**Theorem 6.3**   For any rapidly decreasing signal $x(t)$, the Fourier transform $X(\omega)$ is also rapidly decreasing. Moreover the inversion formula

$$x(t) = \frac{1}{2\pi} \int_{-\infty}^{\infty} X(\omega)e^{j\omega t} d\omega$$

holds for every real $t$.

**PROOF**   Let $x(t)$ and $y(t)$ be two functions in $L_1$ with Fourier transforms $X(\omega)$ and $Y(\omega)$, respectively. The following equality holds:

$$\int x(s) Y(s)ds = \int X(s)y(s)ds \tag{6.24}$$

Indeed, both integrals are well defined. Since $X(s)$, $Y(s)$ are bounded, they belong to $L_\infty$, and Hölder's inequality applies. We substitute $Y(s)$ in the left side by (6.3), and use Fubini's theorem to obtain

$$\int x(s) Y(s)ds = \int x(s) \int y(v)e^{-jsv} dvds$$

$$= \int y(v) \int x(s)e^{-jsv} dsdv$$

$$= \int y(v)X(v)dv$$

Suppose that $x(t)$ is a rapidly decreasing signal. Fix $t$, and consider the shift $x(s+t)$. Replacing $y(s)$ with the Gaussian signal in Eq. (6.24) gives

$$\int x(s+t)e^{-s^2\sigma^2/2}ds = \frac{1}{\sigma\sqrt{2\pi}} \int X(s)e^{jts-(s^2/2\sigma^2)}ds$$

If we set $v = \sigma s$ in the left side integral, we obtain

$$\int x\left(\frac{v}{\sigma} + t\right)e^{-v^2/2}dv = \frac{1}{\sqrt{2\pi}}\int X(s)e^{jts-(s^2/2\sigma^2)}ds$$

Let $\sigma \to \infty$. Then $x(v/\sigma + t) \to x(t)$ and $e^{jts-(s^2/2\sigma^2)} \to e^{jts}$. The dominated convergence theorem enables us to insert the limit inside the integrals:

$$x(t)\int_{-\infty}^{\infty} e^{-v^2/2}dv = \frac{1}{\sqrt{2\pi}}\int_{-\infty}^{\infty} X(s)e^{jts}ds = \sqrt{2\pi}\hat{x}(t)$$

Since the Gaussian signal is a probability density, the integral in the left side is equal to $\sqrt{2\pi}$. The inversion formula is established. ∎

The above theorem asserts that the Fourier transform is one-to-one and onto the space of rapidly decreasing functions. Using a similar analysis we can show more generally that if both $x(t)$ and $X(\omega)$ are absolutely integrable, the inversion formula is valid at every continuity point of $x(t)$.

### 6.1.6 Duality, modulation, and Parseval identity

*Duality.* The duality property expresses the striking similarity of the analysis and synthesis equations. Let $x(t)$ be a rapidly decreasing signal with Fourier transform $\mathcal{F}(x) = X$. Then, if we set $y(t) = X(t)$, the Fourier transform of $y(t)$ is $Y(\omega) = 2\pi x(-\omega)$. Indeed,

$$Y(\omega) = \int_{-\infty}^{\infty} X(v)e^{-j\omega v}dv = 2\pi x(-\omega)$$

Let $R$ denote the reflection operator, $R(x)(\omega) = x(-\omega)$. Application of the Fourier transform twice gives a scaled reflection of the original waveform

$$\mathcal{F}^2(x)(\omega) = 2\pi R(x)(\omega) = 2\pi x(-\omega) \tag{6.25}$$

In a similar way we see that the inverse Fourier transform $\mathcal{F}^{-1}(X)$ is obtained from the Fourier transform of the reflection of $X$:

$$\mathcal{F}^{-1}(X) = \frac{1}{2\pi}\mathcal{F}R(X) \tag{6.26}$$

The above conclusions are valid for continuous absolutely integrable signals whose Fourier transform is absolutely integrable as well. Take, for example, $y(t) = 2/(t^2 + 1)$. We know from Example 6.3 that $y(t) = X(t)$, where $X(\omega)$ is the Fourier transform of the signal $x(t) = e^{-|t|}$. Clearly $x(t)$ is continuous, it belongs to $L_1$, and $X(\omega)$ belongs to $L_1$. Thus, by duality, we conclude that $Y(\omega) = 2\pi e^{-|\omega|}$.

***Modulation.*** Let us consider two rapidly decreasing signals $x(t)$ and $h(t)$ with Fourier transforms $X(\omega)$ and $H(\omega)$, respectively. Since $x(t)$ and $h(t)$ are absolutely integrable, their convolution is also absolutely integrable with Fourier transform given by the product of Fourier transforms according to the convolution property. Since $X(\omega)$ and $H(\omega)$ belong to $S$, their product belongs to $S$. The duality property implies that the convolution is a rapidly decreasing function. We also expect from duality that convolution in the frequency domain converts into multiplication in the time domain. More precisely it holds that

$$x(t)h(t) \xrightarrow{\ \mathcal{F}\ } \frac{1}{2\pi} X(\omega) * H(\omega) \tag{6.27}$$

Indeed,

$$\mathcal{F}(X * H) = \mathcal{F}(X)\mathcal{F}(H) = \mathcal{F}^2(x)\mathcal{F}^2(h) = (2\pi)^2 R(xh) = 2\pi\mathcal{F}^2(xh)$$

If we take the inverse Fourier transform, the result follows. Equation (6.27) is referred to as *modulation*.

***Parseval identity.*** The Parseval identity expresses the preservation of energy in the frequency domain. For any rapidly decreasing function $x(t)$, it holds that

$$\int_{-\infty}^{\infty} |x(t)|^2 dt = \frac{1}{2\pi} \int_{-\infty}^{\infty} |X(\omega)|^2 d\omega \tag{6.28}$$

More generally, for any $x(t)$, $y(t)$ in $S$, we have

$$\int_{-\infty}^{\infty} x(t)y^*(t)dt = \frac{1}{2\pi} \int_{-\infty}^{\infty} X(\omega)Y^*(\omega)d\omega \tag{6.29}$$

Indeed, from Example 6.5 we have

$$r(0) = \int_{-\infty}^{\infty} |x(t)|^2 dt = \frac{1}{2\pi} \int_{-\infty}^{\infty} |X(\omega)|^2 e^{j\omega 0} d\omega = \frac{1}{2\pi} \int_{-\infty}^{\infty} |X(\omega)|^2 d\omega$$

Equation (6.29) follows in a similar way.

### 6.1.7 The $L_2$ theory and Plancherel theorem

Using the space of rapidly decreasing functions $S$ as a base, we can extend the concept of Fourier transform to energy signals. We have pointed out that $S$ is dense in $L_p$. In particular, it is dense in $L_1$ as well as in $L_2$. Let $x(t) \in L_2$. Pick a sequence $x_n(t)$ in $S$ such that $x_n(t)$ tends to $x(t)$ in the $L_2$ sense. Then $x_n(t)$ is a Cauchy sequence. Since the Fourier transform is linear and preserves lengths—thanks to Parseval identity it is an isometry, apart from the scaling factor $1/2\pi$—

the sequence of Fourier transforms $X_n = \mathcal{F}(x_n)$ is also a Cauchy sequence. Indeed,

$$|x_n - x_m|_2 = \frac{1}{\sqrt{2\pi}}|\mathcal{F}(x_n - x_m)|_2$$

$$= \frac{1}{\sqrt{2\pi}}|\mathcal{F}(x_n) - \mathcal{F}(x_m)|_2 = \frac{1}{\sqrt{2\pi}}|X_n - X_m|_2$$

Since $L_2$ is a Hilbert space every Cauchy sequence converges. Therefore there exists an element $X$ in $L_2$ such that $X = \lim_{n\to\infty} X_n$. The assignment $x \to X$ defines the Fourier transform on $L_2$.

**Technical remarks.** $X$ does not depend on the particular sequence $x_n$ chosen. Indeed, let $\tilde{x}_n \to x$ and $\tilde{X}_n \to \tilde{X}$. Then

$$|\tilde{X} - X|_2 = |\lim \tilde{X}_n - \lim X_n|_2 = \lim |\tilde{X}_n - X_n|_2$$

$$= \sqrt{2\pi} \lim |\tilde{x}_n - x_n|_2 = \sqrt{2\pi}|x - x|_2 = 0$$

It is straightforward to show that apart from the scaling factor $1/2\pi$, the Fourier transform is a linear isometry onto $L_2$, namely it is a linear one-to-one and onto transformation that preserves the inner product of $L_2$. Note finally that if $x$ belongs to $L_1 \cap L_2$, the Fourier transform defined above coincides with the original Fourier transform of $x$ viewed as an element of $L_1$. Indeed, recall that $\mathcal{S}$ is dense in $L_1 \cap L_2$. Let $x_n$ in $\mathcal{S}$ with $x_n \to x$ in the $L_2$ sense. It can be shown that there is a subsequence $x_{n_k}$ of $x_n$ that converges to $x$ almost everywhere. Since $x_{n_k}$ and $x$ are in $L_1$ and the Fourier transform is continuous, we have $X_{n_k}(\omega) \to X(\omega)$. Then $X_{n_k}(\omega) \to X(\omega)$ in the $L_2$ sense also, and the claim is established. An intuitively appealing construction of the $L_2$ Fourier transform and its inverse is via the truncations

$$X_T(\omega) = \int_{-T}^{T} x(t)e^{-j\omega t}dt, \qquad \hat{x}_W(t) = \frac{1}{2\pi}\int_{-W}^{W} X(\omega)e^{j\omega t}d\omega$$

It can be shown that

$$\lim_{T\to\infty} |X_T - X|_2 = 0, \qquad \lim_{W\to\infty} |\hat{x}_W - x|_2 = 0$$

**Example 6.7   Rectangular frequency window**
Let us consider a signal whose Fourier transform is a window of width $W$:

$$X(\omega) = \begin{cases} 1, & \text{if } |\omega| < W \\ 0, & \text{otherwise} \end{cases} \tag{6.30}$$

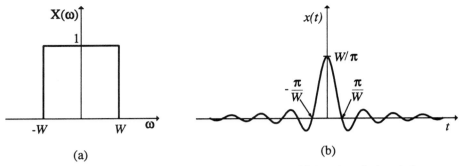

**Figure 6.6.** (a) Window function in the frequency domain; (b) time domain description.

It is easy to see that $X(\omega)$ is square integrable. Hence it is the Fourier transform of a signal $x(t)$ that can be computed by the synthesis equation (6.5):

$$x(t) = \frac{1}{2\pi} \int_{-W}^{W} e^{j\omega t} d\omega = \frac{\sin Wt}{\pi t} \tag{6.31}$$

The two signals are sketched in Fig. 6.6. Note that $x(t)$ is an energy signal but is not absolutely integrable (see Example 4.2).

Examples 6.7 and 6.4 are dual. They are easily derived from the sinc function sinc $x = \sin \pi x / \pi x$. ■

***Example 6.8    Characteristic function of a random variable***
Let $u$ be a random variable with probability density function $f(u)$. The Fourier transform of $f(-u)$ is called *characteristic function*. It is denoted $\phi_u(\omega)$, or when no confusion arises, $\phi(\omega)$. Thus

$$\phi(\omega) = \int_{-\infty}^{\infty} f(-u)e^{-j\omega u} du = \int_{-\infty}^{\infty} f(u)e^{j\omega u} du \tag{6.32}$$

From the definition of the expected value, we have

$$\phi(\omega) = E[e^{j\omega u}] \tag{6.33}$$

Since $\int_{-\infty}^{\infty} f(u)du = 1$ and $f(u) \geq 0$, the density $f(u)$ belongs to $L_1$, and the Fourier transform always exists. The inversion formula, however, is not applicable, unless further assumptions on the density are imposed, for instance, that $f(u) \in L_2$. A more general definition views the characteristic function as the Stieltjes integral of the probability distribution function of $u$. In this case the probability distribution is uniquely determined from the characteristic function.

If $u$ obeys the uniform distribution on the interval $[-a, a]$, the characteristic function is given by $\phi(\omega) = 2a$ sinc$(\omega a/\pi)$ according to Example 6.4. If $u$ is

Gaussian with mean $m$ and variance $\sigma^2$, Example 6.6 shows that the characteristic function is

$$\phi(\omega) = e^{jm\omega - \omega^2 \sigma^2 / 2}$$

The existence of higher-order moments of $u$ are related to the degree of differentiability of its characteristic function. If all moments of the random variable $u$ exist and are finite, expansion of the exponential function in (6.33) in power series gives

$$\phi(\omega) = E\left[\sum_{n=0}^{\infty} \frac{(j\omega)^n}{n!} u^n\right] = \sum_{n=0}^{\infty} \frac{(j\omega)^n}{n!} E[u^n]$$

Therefore

$$E[u^n] = \left(\frac{1}{j}\right)^n \frac{d^n}{d\omega^n} \phi(\omega)|_{\omega=0} \qquad \blacksquare$$

### 6.1.8  Impulsive signals in time and frequency domain

We pointed out in Section 3.6.4 that the main obstacle in carrying out the discrete time analysis to analog representations is that Eq. (3.35) is not satisfied by ordinary inputs. This delicate issue can be settled only if the class of signals is suitably extended. Comparison of (3.35) with (3.36) shows that

$$u(t) = \delta(t) * u(t) \tag{6.34}$$

Consequently $\delta(t)$ is the impulse response of the identity system. Although $\delta(t)$ does not exist as an ordinary signal, the identity system is perfectly defined. This observation encourages us to identify $\delta(t)$ by the operator $u(t) \rightarrow u(t)$. Since this is a time invariant system, it suffices to specify the output of the identity system only at time $t = 0$, in accordance with the discussion of Section 3.2. Hence $\delta(t)$ is determined by the functional

$$\delta : \mathcal{U} \rightarrow R, \quad \delta(u) = u(0) \tag{6.35}$$

It is easy to see that $\delta$ is a linear functional. Moreover continuity of $\delta$ is anticipated if $\mathcal{U}$ is equipped with a proper notion of convergence. Before we give precise definitions, we note that $\delta$ is not the only generalized function of interest. Other interesting examples are linear combinations, shifts, and derivatives of $\delta$ functions. The space of linear and continuous functionals suffices to cover all such cases. Precise statements are given next.

The domain of generalized functions is called *test space*. It is denoted by $\mathcal{D}$ instead of the $\mathcal{U}$ used above. The choice of the test space is dictated by the application. A popular candidate is the space of infinitely differentiable functions of finite support. Thus a signal $u(t)$ belongs to $\mathcal{D}$ if derivatives of all orders $u^{(k)}(t)$ exist and are continuous on $R$ and if there is $T > 0$ such that $u(t) = 0$ for any $|t| \geq T$. The above test space equipped with ordinary addition and scalar multiplication is a linear space. Convergence in $\mathcal{D}$ is defined as follows: A sequence $u_N$ converges to $u$ if there is an interval $I$ such that $u_N^{(k)}(t) \to u^{(k)}(t)$ uniformly in $I$ for all integers $k$ and furthermore if all functions $u_N$ are zero outside $I$. Every linear and continuous functional $\lambda : \mathcal{D} \to R$ is called *generalized function* or *distribution*. Addition and scalar multiplication of generalized functions are defined by

$$(\lambda_1 + \lambda_2)(u) = \lambda_1(u) + \lambda_2(u), \quad (a\lambda)(u) = a\lambda(u)$$

### Example 6.9    The Dirac or Delta function
The generalized function $\delta : \mathcal{D} \to R$, $\delta(u) = u(0)$ is called the *Dirac function*. It is easy to check that it is linear and continuous on the test space $\mathcal{D}$. More generally, the functional $\delta_t(u) = u(t)$ defines the shift of $\delta$ by $t$, for $t \in R$.    ∎

**Ordinary signals as distributions.** Ordinary signals must be embedded in the space of distributions if the latter is to be a useful concept. We assume that $x(t)$ is absolutely integrable on each finite interval. We say it is locally integrable, and we write $x(t) \in L_{1e}$. We assign the distribution $\lambda_x$ to $x(t)$ as

$$\lambda_x(u) = \int_{-\infty}^{\infty} u(t)x(t)dt \tag{6.36}$$

The above integral is well defined because the test function $u(t)$ is zero outside a finite interval $I$ and the functions $x(t)$, $u(t)$ are integrable on $I$, $x(t)$ by assumption and $u(t)$ by continuity. It can be shown that $\lambda_x$ is linear and continuous on $\mathcal{D}$. Hence it forms a generalized function. With the aid of the assignment $x \to \lambda_x$, ordinary locally integrable signals are represented as distributions. This says that the signal $x(t)$ is identified with the functional of the system having impulse response $x(t)$. It can be shown that $x(t)$ is uniquely recovered from $\lambda_x$, namely the correspondence $x \to \lambda_x$ is one-to-one. Generalized functions obtained this way are called *regular distributions* or simply *signals*. The remaining distributions, such as the Dirac function, are called *singular distributions*. Linear combinations of shifts of the Dirac function are termed *impulsive signals* because of their bursty character. Motivated by (6.36), we often use the nonlegitimate expression

$$\lambda(u) = \int_{-\infty}^{\infty} \lambda(t)u(t)dt \tag{6.37}$$

to represent singular distributions in integral form and to identify them with $\lambda(t)$. All basic operations and transforms of signals can be carried over to distributions. The most important are considered in the Appendix at the end of the chapter.

Let us next discuss the representation of impulsive signals in the frequency domain. The starting point for the rigorous definition of the Fourier transform of generalized functions is the Parseval identity rewritten as

$$\int_{-\infty}^{\infty} u(t)x^*(t)dt = \frac{1}{2\pi} \int_{-\infty}^{\infty} U(\omega)X^*(\omega)d\omega$$

Let us view $u$ as test function. We identify $x(t)$ with the regular distribution $\lambda_x$ and the Fourier transform $X(\omega)$ with the distribution $\Lambda_X$. Then the above definition takes the form $\Lambda_{X^*}(U) = 2\pi\lambda_{x^*}(u)$. It is reasonable to define the Fourier transform of the distribution $\lambda$ as the distribution $\Lambda$ that satisfies the relation

$$\Lambda(U) = 2\pi\lambda(u) \tag{6.38}$$

The above definition will be successful if every element of the test space $\mathcal{D}$ is uniquely expressed as the Fourier transform of an element of the same space. In other words, the assignment $\mathcal{F} : \mathcal{D} \to \mathcal{D}$ is one-to-one and onto. The test space of infinitely differentiable finite support functions does not have this property (why?). This difficulty is bypassed if the set $\mathcal{S}$ of rapidly decreasing functions is used as test space. A sequence $u_N \in \mathcal{S}$ converges to the function $u \in \mathcal{S}$ if the sequence $u_N^{(k)}(t)$ converges to $u^{(k)}(t)$, absolutely and uniformly, on every finite interval. A continuous linear mapping $\lambda : \mathcal{S} \to C$, where $C$ denotes the complex numbers, is called a *tempered distribution*. In the sequel we consider complex rather than real generalized functions because signals in the frequency domain are complex in general. Since the Fourier transform is one-to-one and onto on $\mathcal{S}$, the Fourier transform of the tempered distribution $\lambda : \mathcal{S} \to C$ is uniquely defined as the tempered distribution $\Lambda : \mathcal{S} \to C$ that satisfies (6.38).

**Remark.** Every tempered distribution can be identified by a distribution in the ordinary sense of the definition but not conversely. Indeed, the space of infinitely differentiable functions of compact support $\mathcal{D}$ is clearly contained in $\mathcal{S}$. It can be shown that $\mathcal{D}$ is dense in $\mathcal{S}$. Every tempered distribution $\lambda : \mathcal{S} \to C$ defines the ordinary distribution $\tilde{\lambda} : \mathcal{D} \to C$ by the rule $\tilde{\lambda}(u) = \lambda(u)$.

It can be shown that tempered distributions are distributions satisfying a certain growth condition at infinity. In particular, every function $x$ in $L_p$, $1 \leq p \leq \infty$, determines a regular tempered distribution. Indeed, let $\lambda_x(u)$ be defined by Eq. (6.36). Hölder's inequality gives

$$|\lambda_x(u)| \leq \int_{-\infty}^{\infty} |u(t)x(t)|dt \leq |u|_q|x|_p, \quad \frac{1}{q} + \frac{1}{p} = 1$$

where $|u|_q$ is finite because $u$ is rapidly decreasing and thus belongs to all $L_p$ spaces. This shows that $\lambda_x$ is a linear and continuous functional on $\mathcal{S}$.

Suppose that $x(t)$ is absolutely integrable. Then the Fourier transform $X(\omega)$

exists. According to the above discussion, $x(t)$ can be viewed as a tempered distribution $\lambda_x$. Let $\Lambda_X$ be its Fourier transform. According to Theorem 6.1, $X(\omega)$ is bounded; that is, it belongs to $L_\infty$. Hence $X(\omega)$ is also assigned to a tempered distribution. Equation (6.38) shows that this distribution coincides with $\Lambda_X$ and the two notions are consistent. The same argument works for the Fourier transform of energy signals.

***Example 6.10***   ***The Fourier transform of the delta function***
The Dirac function $\lambda = \delta$ can be viewed as the tempered distribution $\delta : S \to C$, $u \to \delta(u) = u(0)$. It can be shown that $\delta$ is linear and continuous on $S$. Let $u$ be a rapidly decreasing function. Then

$$\Lambda(u) = 2\pi\delta(\mathcal{F}^{-1}u) = 2\pi\big(\mathcal{F}^{-1}u\big)(0) = \int_{-\infty}^{\infty} u(t)dt$$

Thus $\Lambda = 1$ in the sense of generalized functions. The spectrum of the impulse covers the entire range of frequencies. ∎

***Example 6.11***   ***The Fourier transform of periodic signals***
Let us consider the constant function $x(t) = 1$. From duality and the previous example, an impulse is expected in the frequency domain. Since $x(t)$ is bounded, it belongs to $L_\infty$ and thus gives rise to the tempered distribution

$$\lambda_x(u) = \int_{-\infty}^{\infty} u(t)dt$$

The Fourier transform is according to (6.38):

$$\Lambda_x(u) = 2\pi\lambda_x(\mathcal{F}^{-1}u) = 2\pi \int \mathcal{F}^{-1}u\,dt = 2\pi u(0) = 2\pi\delta(u)$$

Hence

$$1 \xrightarrow{\mathcal{F}} 2\pi\delta(\omega) \tag{6.39}$$

The shift property (see Appendix 6A) is easily extended to tempered distributions. Application of the shift property in the frequency domain gives

$$e^{j\omega_0 t} \xrightarrow{\mathcal{F}} 2\pi\delta(\omega - \omega_0) \tag{6.40}$$

From the above we readily find that

$$\cos\omega_0 t \xrightarrow{\mathcal{F}} \pi[\delta(\omega - \omega_0) + \delta(\omega + \omega_0)], \quad \sin\omega_0 t \xrightarrow{\mathcal{F}} \frac{\pi}{j}[\delta(\omega - \omega_0) - \delta(\omega + \omega_0)] \tag{6.41}$$

More generally, the Fourier transform of a square integrable periodic signal has the form

$$\sum_{k=-\infty}^{\infty} c_k e^{jk\omega_0 t} \xrightarrow{\mathcal{F}} \sum_{k=-\infty}^{\infty} 2\pi c_k \delta(\omega - k\omega_0) \tag{6.42}$$

The spectrum of a periodic signal with period $T$ consists of impulses $\delta(\omega - k\omega_0)$, equally spaced at distance $\omega_0 = 2\pi/T$ apart, with height $2\pi$ times the corresponding Fourier series coefficients.

As an example take the periodic square wave of Fig. 2.10. The Fourier series coefficients were computed in Example 5.6. Hence

$$X(\omega) = \sum_{k=-\infty}^{\infty} 2\frac{\sin k\omega_0 T_1}{k} \delta(\omega - k\omega_0)$$

∎

**Example 6.12    The Fourier transform of the step signal**
The step signal $u_s(t)$ can be written in terms of its even and odd parts as $u_s(t) = \frac{1}{2} + \frac{1}{2}\text{sgn}(t)$, where the signum function is defined by

$$\text{sgn}(t) = \begin{cases} 1, & \text{if } t \geq 0 \\ -1, & \text{if } t < 0 \end{cases}$$

Since the signum is bounded, it can be viewed as a tempered distribution. The following relation is given without proof:

$$\text{sgn}(t) \xrightarrow{\mathcal{F}} \frac{2}{j\omega} \tag{6.43}$$

Using (6.39) and (6.43), we obtain

$$U_s(\omega) = \pi\delta(\omega) + \frac{1}{j\omega} \tag{6.44}$$

∎

**Example 6.13    Real causal signals and the Hilbert transform**
If $x(t)$ is real and causal, the real part of its spectrum $\Re X(\omega)$ determines the imaginary part $\Im X(\omega)$ through the Hilbert transform. A nonrigorous explanation goes as follows: Consider the even and odd parts of $x(t)$. Since $x(t)$ is real and causal, Eq. (6.10) implies that $x(t) = 2x_e(t)u_s(t)$. The modulation property together with Eq. (6.44) give

$$X(\omega) = \frac{1}{\pi}\Re X(\omega) * \left(\pi\delta(\omega) + \frac{1}{j\omega}\right) = \Re X(\omega) + \frac{1}{\pi}\Re X(\omega) * \frac{1}{j\omega}$$

Thus

$$X(\omega) = \Re X(\omega) + j\Im X(\omega) = \Re X(\omega) + \frac{1}{j\pi}\int_{-\infty}^{\infty} \frac{\Re X(v)}{\omega - v}\,dv$$

or

$$\Im X(\omega) = -\frac{1}{\pi} \int_{-\infty}^{\infty} \frac{\Re X(v)}{\omega - v} dv \qquad (6.45)$$

The latter expression states that if $x(t)$ is a real causal signal, the imaginary part of its spectrum is determined from its real part. In exactly the same way, we obtain the equation

$$\Re X(\omega) = \frac{1}{\pi} \int_{-\infty}^{\infty} \frac{\Im X(v)}{\omega - v} dv \qquad \blacksquare$$

### 6.1.9   The Poisson summation formula

Consider the rapidly decreasing signal $x(t)$ with Fourier transform $X(\omega)$ and reals $T, W > 0$. Let

$$x_T(t) = \sum_{n=-\infty}^{\infty} x(t - nT), \quad X_W(\omega) = \sum_{k=-\infty}^{\infty} X(\omega - kW) \qquad (6.46)$$

It can be shown that both functions are well defined and periodic with periods $T$ and $W$, respectively. Their Fourier series expansions are

$$x_T(t) = \frac{1}{T} \sum_{k=-\infty}^{\infty} X(kW_T)e^{jkW_T t}, \qquad W_T = \frac{2\pi}{T} \qquad (6.47)$$

$$X_W(\omega) = \frac{2\pi}{W} \sum_{n=-\infty}^{\infty} x(nT_W)e^{-jnT_W \omega}, \qquad T_W = \frac{2\pi}{W} \qquad (6.48)$$

Equation (6.47) states that the Fourier series coefficients of the periodic extension of $x(t)$, $x_T(t)$, are obtained from the Fourier transform of $x(t)$ sampled every $W_T = 2\pi/T$ frequency points and scaled by $1/T$. In a similar fashion the Fourier series coefficients of the periodic extension of $X(\omega)$, $X_W(\omega)$, are given by the signal samples $x(nT_W)$ scaled by $T_W$.

If $x(t)$ has finite support, Eq. (6.47) forms the essence of the argument we used in the motivation of Fourier transform discussed in Section 6.1.1. In the general case we proceed as follows: Let $a_k$ be the Fourier series coefficients of $x_T(t)$. Then

$$a_k = \frac{1}{T} \int_0^T e^{-jkW_T t} x_T(t) dt = \frac{1}{T} \sum_{n=-\infty}^{\infty} \int_0^T e^{-jkW_T t} x(t + nT) dt$$

$$= \frac{1}{T} \sum_{n=-\infty}^{\infty} \int_{nT}^{(n+1)T} e^{-jkW_T t} x(t) dt = \frac{1}{T} X(kW_T)$$

A similar argument proves Eq. (6.48).

*Computation of Fourier transform.* The Poisson summation formula sheds further light into the computation of the Fourier transform treated in Section 6.1.3. In Example 5.7 we discussed how to compute the Fourier series coefficients $a_k$ of a periodic function from its samples by means of the DFT. We constructed the discrete periodic extension $\hat{a}_k = \sum_{r=-\infty}^{\infty} a_{k+rN}$, and showed that $\hat{a}_k = (1/N)\,\mathrm{DFT}(x(nT_s))$. In the case under consideration, $x_T(t)$ is the periodic extension given by (6.47), $x_T(nT_s)$ are the given samples, and $a_k = \frac{1}{T}X(kW_s)$, $W_s = 2\pi/T$ are the Fourier series coefficients. The periodic extension of $a_k$ is

$$\hat{a}_k = \sum_{r=-\infty}^{\infty} a_{k+rN}$$

$$= \frac{1}{T}\sum_{r=-\infty}^{\infty} X((k+rN)W_s) = \frac{1}{T}\sum_{r=-\infty}^{\infty} X(kW_s + rNW_s)$$

Equation (6.46) implies that $\hat{a}_k$ is obtained by sampling the periodic extension of $X(\omega)$ of period $NW_s$ at frequencies $\omega = kW_s$: $\hat{a}_k = \frac{1}{T}X_{NW_s}(kW_s)$. Hence

$$X_{NW_s}(kW_s) = \sum X(kW_s - rNW_s) = \frac{T}{N}\mathrm{DFT}\big(x_{NT_s}(nT_s)\big) \qquad (6.49)$$

Determination of the Fourier transform values $X(kW_s)$ from the samples of its periodic extension $X_{NW_s}(kW_s)$ is possible provided that $X(\omega)$ is bandlimited, that is $X(\omega) = 0$ for $|\omega| > W$, and moreover $W \le NW_s - W$ so that $X(\omega)$ does not overlap with the shifts $X(\omega + rNW_s)$. Thus

$$N \ge \frac{2W}{W_s} \quad \text{or} \quad N \ge \frac{WT}{\pi} \qquad (6.50)$$

In summary the Fourier transform samples $X(kW_s)$ of a bandlimited signal of bandwidth $W$ are determined by (6.49), which takes the form

$$X(kW_s) = \frac{T}{N}\mathrm{DFT}(x_{NT_s}(nT_s))$$

provided that (6.50) holds. The samples $x_{NT_s}(nT_s))$ are determined by sampling the periodic extension $x_T(t) = \sum_{m=-\infty}^{\infty} x(t - mT)$, $T = 2\pi/W_s$. Of course the infinite sum is approximated by a finite sum, and so an error is unavoidable. If $x(t)$ is time limited to $T_1 \le t \le T_2$, we pick $T \ge T_2 - T_1$ to secure $x_T(t) = x(t)$. It can be shown that no signal exists that is both time limited (finite support) and bandlimited. Therefore approximation errors always occur.

## 6.2 FOURIER ANALYSIS OF DISCRETE SIGNALS

### 6.2.1 Definition and examples

Let $x(n)$ be a complex-valued sequence defined on $Z$. The (discrete time) Fourier transform of $x(n)$ is a complex function of a real variable defined by

$$X(\omega) = \sum_{n=-\infty}^{\infty} x(n)e^{-j\omega n}, \quad \omega \in R \tag{6.51}$$

The signal $x(n)$ is recovered from $X(\omega)$ via the *inverse Fourier transform*

$$x(n) = \frac{1}{2\pi} \int_{2\pi} X(\omega)e^{j\omega n} d\omega \tag{6.52}$$

We will rigorously analyze the above equations mimicking the continuous time case. Our job, however, is much simpler now. We will show that absolute summability of the signal suffices to warrant affirmative answers to all three questions regarding well posedness of the Fourier transform. Indeed, let us assume that $x(n) \in \ell_1$, that is.

$$\sum_{n=-\infty}^{\infty} |x(n)| < \infty$$

Then the series (6.51) converges absolutely and uniformly for all $\omega$. Indeed,

$$|X(\omega)| \leq \sum_{n=-\infty}^{\infty} |x(n)||e^{-j\omega n}| = \sum_{n=-\infty}^{\infty} |x(n)| = |x|_1$$

A known result from analysis confirms that the series (6.51) converges absolutely and uniformly to a continuous function $X(\omega)$. Thus the Fourier transform $X(\omega)$ is a bounded continuous function. The function $X(\omega)e^{j\omega n}$ is continuous and hence Riemann integrable on every finite interval. Therefore the signal $\hat{x}(n) = \frac{1}{2\pi} \int_{2\pi} X(\omega)e^{j\omega n} d\omega$ is well defined. Convergence of (6.51) implies that

$$X(\omega)e^{j\omega k} = \sum_{n=-\infty}^{\infty} x(n)e^{-j\omega n}e^{j\omega k}$$

converges absolutely and uniformly. Therefore

$$\hat{x}(n) = \frac{1}{2\pi} \int_{2\pi} X(\omega)e^{j\omega n} d\omega = \frac{1}{2\pi} \sum_{k=-\infty}^{\infty} x(k) \int_{0}^{2\pi} e^{-j\omega(k-n)} d\omega = x(n)$$

$X(\omega)$ is periodic with period $2\pi$. Indeed,

$$X(\omega + 2\pi) = \sum_{n=-\infty}^{\infty} x(n)e^{-j(\omega+2\pi)n} = \sum_{n=-\infty}^{\infty} x(n)e^{-j\omega n}e^{-j2\pi n} = X(\omega)$$

Thus the range of frequencies is confined to an interval of length $2\pi$. We recognize in Eq. (6.51) the Fourier expansion of the periodic function $X(-\omega)$. The signal in the time domain $x(n)$ coincides with the Fourier coefficients of $X(-\omega)$. Equation (6.52) defining the inverse Fourier transform expresses the Fourier series coefficients of the periodic function $X(-\omega)$—compare (6.52) with Eqs. (5.11) and (5.12).

### Example 6.14  Computation of the discrete time Fourier transform

Since $X(\omega)$ is periodic with period $2\pi$, it suffices to compute the points $X(kW_s)$ in one period. If $N$ is the number of these points, we have $W_s = 2\pi/N$. The computation of $X(kW_s)$ from $x(n)$ is dual to the computation of the Fourier coefficients of a periodic function discussed in Example 5.7. Proceeding the same way we find that

$$X(kW_s) = \sum_{n=0}^{N-1} \hat{x}_N(n)w^{kn} = \mathrm{DFT}(\hat{x}_N(n)) \tag{6.53}$$

where $\hat{x}_N(n)$ is the periodic extension of $x(n)$ of period $N$

$$\hat{x}_N(n) = \sum_{m=-\infty}^{\infty} x(n + mN)$$

If $x(n)$ has finite extent $N_0$, that is, $x(n) = 0$, $|n| > N_0$, and the sampling rate in the frequency domain exceeds the time extent of $x(n)$, $2\pi/W_s = N \geq 2N_0$, the above formula becomes

$$X(kW_s) = \mathrm{DFT}(x(n)) \tag{6.54}$$

If the signal has infinite support, approximation errors are inevitable. A MATLAB implementation is described below. $N_1$, $N_2$ are the endpoints of the time window over which the values of the signal $x$ are defined. The number of frequency points $N$ is chosen as the smallest power of 2 satisfying $N > N_2 - N_1$.

```
function [w,X] = dtft(x)
x1 = ['[y,N1,N2] = ',x,';'];
eval(x1);
X = fft(y);
N = N2-N1 + 1;
Ws = 2*pi/N;                      % sampling period in the  frequency domain
w = [-(N/2) + 1:(N/2)]*Ws;       % frequency window
subplot(221), plot(N1:N2,y);
subplot(222), plot(w,abs(fftshift(X)));
subplot(223), plot(w,angle(fftshift(X)));
```

■

### Example 6.15  Causal exponential

The causal exponential signal

$$x(n) = a^n u_s(n), \qquad |a| < 1, \; a \in C$$

is absolutely summable. The Fourier transform is

$$X(\omega) = \sum_{n=-\infty}^{\infty} a^n u_s(n) e^{-j\omega n} = \sum_{n=0}^{\infty} (ae^{-j\omega})^n = \frac{1}{1 - ae^{-j\omega}}$$

where $X(\omega)$ is a continuous complex function of $\omega$. The magnitude and phase spectra are depicted in Fig. 6.7, for $a > 0$. They are given by

$$|X(\omega)|^2 = \frac{1}{1 + |a|^2 - 2\Re(ae^{-j\omega})},$$

$$\arg X(\omega) = \tan^{-1}\left(\frac{-\Re a \sin \omega + \Im a \cos \omega}{1 - \Re a \cos \omega - \Im a \sin \omega}\right)$$

The following *m* file implements the discrete causal exponential signal.

```
function [x,N1,N2] = dtcexp(N2,a)
N1 = 0;
n = 0:N2-1;
x = a.^[0:N2];
```

The command dtft('dtcexp(N2,a)') plots the signal in the time and frequency domain.  ∎

### Example 6.16  Discrete even exponential

Let

$$x(n) = a^{|n|}, \qquad |a| < 1, \; a \in R$$

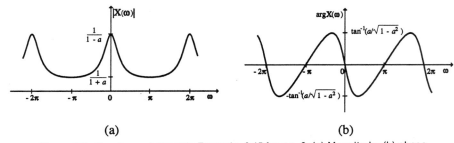

(a)                                        (b)

**Figure 6.7.** Spectrum of signal in Example 6.15 for $a > 0$: (a) Magnitude; (b) phase.

Clearly $x(n)$ belongs to $\ell_1$. Then

$$X(\omega) = \sum_{n=-\infty}^{\infty} a^{|n|} e^{-j\omega n} = \sum_{n=-\infty}^{-1} a^{-n} e^{-j\omega n} + \sum_{n=0}^{\infty} a^n e^{-j\omega n}$$

The change of variables $k = -n$ in the first sum gives

$$X(\omega) = \sum_{k=1}^{\infty} (a e^{j\omega})^k + \sum_{n=0}^{\infty} (a e^{-j\omega})^n = \sum_{k=0}^{\infty} (a e^{j\omega})^k - 1 + \sum_{n=0}^{\infty} (a e^{-j\omega})^n$$

Both series are geometric series. Thus

$$X(\omega) = \frac{1}{1 - a e^{j\omega}} - 1 + \frac{1}{1 - a e^{-j\omega}} = \frac{1 - a^2}{1 + a^2 - 2a \cos \omega}$$

In contrast to the previous example, $X(\omega)$ is a real function of $\omega$ if $a$ is real. Indeed, $x(n)$ is even, and as we will shortly see, even signals have real transforms, exactly as in the analog case. $X(\omega)$ is plotted in Fig. 6.8 via the code:

```
function [x,N1,N2] = dtevexp(N1,N2,a)
x = a.^[abs(N1:N2)];
dtft('dtevexp(N1,N2,a)'
```
■

***Example 6.17   Unit sample shifts***
Consider the unit sample

$$x(n) = \delta(n)$$

Since $x(n) = 0$, for any $n \neq 0$, $X(\omega) = x(0)e^{j0\omega} = 1$. The spectrum of the unit sample extends to all frequencies. More generally, the Fourier transform of the shifted unit sample $x(n) = \delta(n - k)$ is $X(\omega) = e^{-jk\omega}$. The magnitude spectrum is $\mid X(\omega) \mid = 1$, and the spectrum phase is linear in one period, $\arg X(\omega) = -k\omega$.
■

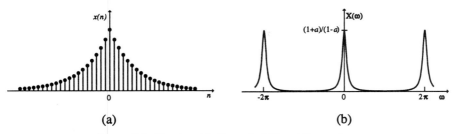

**Figure 6.8.** The signal in Example 6.16 and its spectrum.

*Example 6.18   Rectangular pulse*
Let

$$x(n) = \begin{cases} 1, & \text{if } |n| \le N_1 \\ 0, & \text{if } |n| > N_1 \end{cases}$$

The plot of $x(n)$ is provided in Fig. 6.9. The Fourier transform is

$$X(\omega) = \sum_{n=-N_1}^{N_1} e^{-j\omega n} = \frac{\sin \omega(N_1 + \frac{1}{2})}{\sin(\omega/2)}$$

Application of de l'Hôpital's rule gives

$$X(0) = \left(N_1 + \frac{1}{2}\right) \frac{\cos \omega(N_1 + \frac{1}{2})}{\frac{1}{2}\cos(\omega/2)}\bigg|_{\omega=0} = 2N_1 + 1$$

$X(\omega)$ becomes zero at frequencies $\omega = 2\lambda\pi/(2N_1 + 1)$, $\lambda$ is an integer in the range $0 < \lambda < 2N_1 + 1$. The larger the pulse bandwidth $2N_1 + 1$, the larger is the number of frequencies at which $X(\omega)$ is zero and the smaller is the bandwidth of the main lobe (see Fig. 6.9). The spectrum tends to the Dirac function as $N_1 \to \infty$. ∎

**Extension to finite energy signals.** There are examples of important signals, such as the impulse response of an ideal lowpass filter,

$$h(n) = \frac{\sin Wn}{\pi n}$$

which are not absolutely summable and thus the previous results do not apply. Although the signal $h(n)$ does not belong to $\ell_1$, it belongs to $\ell_2$. Recall from Section 2.7 that $\ell_1 \subset \ell_2$. It is therefore reasonable to inquire if Eqs. (6.51) and (6.52) are valid in $\ell_2$. The $\ell_2$ theory was essentially developed in Section 2.9. We proved there that if $x(n) \in \ell_2$, the series (6.51) converges in $\ell_2$ and (6.52) holds.

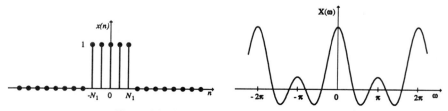

**Figure 6.9.** Rectangular pulse and its spectrum.

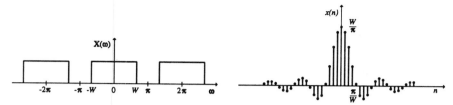

**Figure 6.10.** Fourier transform in Example 6.19 and its time reconstruction.

*Example 6.19   Periodic square wave in the frequency domain*
Let $x(n)$ be a signal whose Fourier transform is given by the periodic square wave
of Fig. 6.10. Thus

$$X(\omega) = \begin{cases} 1, & \text{if } |\omega| \le W \\ 0, & \text{if } W < |\omega| \le \pi \end{cases} \tag{6.55}$$

$X(\omega)$ has finite energy on an interval of length $2\pi$, but is not a continuous
function. Thus $x(n)$ is not absolutely summable, but it has finite energy. It is
reconstructed with the aid of Eq. (6.52):

$$x(n) = \frac{1}{2\pi} \int_{-W}^{W} e^{j\omega n} d\omega = \frac{\sin Wn}{\pi n} \tag{6.56}$$

A behavior similar to that of Example 6.7 is noticed.   ∎

Low frequencies for analog signals are described by intervals of small width
centered around the origin, while high frequencies correspond to values far away
from zero to the left or to the right, namely to frequencies $|\omega| > M$, where $M$ is
large. Periodicity of the Fourier transform imposes a different setup in the
discrete case. In the interval $-\pi < \omega < \pi$, low frequencies correspond to intervals around $\omega = 0$, while high frequencies are placed near the points $-\pi$ and $\pi$.

### 6.2.2   Properties of the discrete time Fourier transform

The discrete time Fourier transform has similar properties to the Fourier
transform of analog signals. The derivation of these properties is entirely
analogous and is left to the reader. Few exceptions are discussed in greater
length in the sequel. The most important properties are listed in Table 6.1. $x(n)$,
$h(n)$ are complex signals with Fourier transforms $X(\omega)$ and $H(\omega)$. One difference
concerns the differentiation property. A discrete signal cannot be differentiated
with respect to time, since it takes discrete values. The scaling property is also
different. The scaled signal $x(Mn)$ is defined only if $M$ is an integer. The scaling
property is analyzed by the properties of decimation and interpolation.

**TABLE 6.1  Properties of the discrete time Fourier transform**

| Property | Time Domain | Frequency Domain |
|---|---|---|
| Conjugation in time | $x^*(n)$ | $X^*(-\omega)$ |
| Conjugation in frequency | $x^*(-n)$ | $X^*(\omega)$ |
| Reflection | $x(-n)$ | $X(-\omega)$ |
| Linearity | $ax(n) + bh(n)$ | $aX(\omega) + bH(\omega)$ |
| Real part | $x_e(n) = \frac{1}{2}[x(n) + x^*(-n)]$ | $\Re X(\omega)$ |
| Imaginary part | $x_o(n) = \frac{1}{2}[x(n) - x^*(-n)]$ | $j\Im X(\omega)$ |
| Time shift | $x(n - n_0)$ | $e^{-j\omega n_0} X(\omega)$ |
| Frequency shift | $e^{j\omega_0 n} x(n)$ | $X(\omega - \omega_0)$ |
| Real signal | $x(n) = x^*(n)$ | $\Re X(\omega) = \Re X(-\omega)$ |
| | | $\Im X(\omega) = -\Im X(-\omega)$ |
| | | $\mid X(\omega) \mid = \mid X(-\omega) \mid$ |
| | | $\arg X(\omega) = -\arg X(-\omega)$ |
| Convolution | $h(n) * x(n)$ | $H(\omega)X(\omega)$ |
| Modulation | $x(n)h(n)$ | $\dfrac{1}{2\pi}\displaystyle\int_{2\pi} X(\omega - \nu)H(\nu)d\nu$ |
| Differentiation | $(-j)^k n^k x(n)$ | $\dfrac{d^k}{d\omega^k} X(\omega)$ |
| Parseval theorem | $\displaystyle\sum_{n=-\infty}^{\infty} \mid x(n) \mid^2$ | $\dfrac{1}{2\pi}\displaystyle\int_{2\pi} \mid X(\omega) \mid^2 d\omega$ |
| | $\displaystyle\sum_{n=-\infty}^{\infty} x(n)h^*(n)$ | $\dfrac{1}{2\pi}\displaystyle\int_{2\pi} X(\omega)H^*(\omega)d\omega$ |
| Scaling: | | |
| Decimation | $x_M(n) = x(Mn), M > 1$ | $\dfrac{1}{M}\displaystyle\sum_{k=0}^{M-1} X\left(\dfrac{\omega}{M} - \dfrac{2\pi k}{M}\right)$ |
| Interpolation | $x_L(n) = x(n/L)$ | $X(L\omega)$ |
| | if $n = 0 \pmod L$, $L > 1$ | |
| | otherwise $x_L(n) = 0$ | |

***Decimation and polyphase decomposition.*** Consider the discrete signal $x(n)$ and a positive integer $M$. The decimated signal

$$x_M(n) = x(Mn), \qquad -\infty < n < \infty \tag{6.57}$$

is obtained by sampling $x(n)$ every $M$ samples. One important way decimation appears in practice is in multirate filtering. In modern applications a variety of systems requires transmission and storage of signals at different sampling rates. Suppose that $x(n)$ is obtained from the analog signal $s(t)$ with sampling rate $F = 1/T_s$, $x(n) = s(nT_s)$. If the sampling rate is decreased to $\tilde{F} = 1/\tilde{T}_s$, the resulting discrete signal is $s(n\tilde{T}_s) = s(nMT_s) = x(nM) = x_M(n)$, where the integer $M$ is given by $M = \tilde{T}_s/T_s = F/\tilde{F} > 1$. The sampling rate reduction corresponds to decimation. To determine the effect of decimation in the frequency domain, we consider the polyphase decomposition of $x(n)$. Let $M = 2$. Define $x_0(n) = x(2n)$, $x_1(n) = x(2n + 1)$. Then

$$X(\omega) = \sum_{n=-\infty}^{\infty} x(n)e^{-j\omega n} = \sum_{n \text{ even}} x(n)e^{-j\omega n} + \sum_{n \text{ odd}} x(n)e^{-j\omega n}$$

$$= \sum_{n=-\infty}^{\infty} x_0(n)e^{-j\omega 2n} + \sum_{n=-\infty}^{\infty} x_1(n)e^{-j\omega(2n+1)}$$

$$= X_0(2\omega) + e^{-j\omega}X_1(2\omega)$$

In general, the $M$ *polyphase components* of $x(n)$ are defined by

$$x_k(n) = x(Mn + k), \qquad 0 \le k \le M - 1$$

$X(\omega)$ is expressed in terms of the spectra $X_k(\omega)$ via the *polyphase decomposition*

$$X(\omega) = \sum_{k=0}^{M-1} e^{-jk\omega}X_k(M\omega) \tag{6.58}$$

The decimated signal $x_M(n)$ coincides with the zero polyphase component $x_0(n)$. We evaluate (6.58) at the frequencies: $\omega - 2\pi l/M$, $0 \le l \le M - 1$, and we sum up to obtain

$$\sum_{l=0}^{M-1} X\left(\omega - \frac{2\pi l}{M}\right) = \sum_{l=0}^{M-1} \sum_{k=0}^{M-1} e^{-jk(\omega-2\pi l/M)}X_k(M\omega - 2\pi l)$$

Invoking periodicity of the discrete time Fourier transform, we have

$$\sum_{l=0}^{M-1} X\left(\omega - \frac{2\pi l}{M}\right) = \sum_{k=0}^{M-1} e^{-jk\omega}X_k(M\omega) \sum_{l=0}^{M-1} e^{j2\pi kl/M}$$

Application of the identity

$$\frac{1}{M} \sum_{l=0}^{M-1} e^{j2\pi kl/M} = \begin{cases} 1 & \text{if } k = 0 \ (\text{mod } M) \\ 0 & \text{otherwise} \end{cases} \tag{6.59}$$

gives

$$X_0(M\omega) = \frac{1}{M} \sum_{l=0}^{M-1} X\left(\omega - \frac{2\pi l}{M}\right)$$

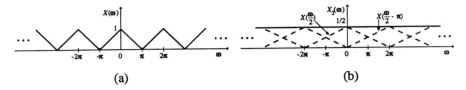

**Figure 6.11.** (a) Initial spectrum $X(\omega)$ and (b) spectrum of the decimated signal, $X_M(\omega)$, $M = 2$.

or

$$X_M(\omega) = X_0(\omega) = \frac{1}{M}\sum_{k=0}^{M-1} X\left(\frac{\omega}{M} - \frac{2\pi}{M}k\right) \tag{6.60}$$

The spectrum of the decimated signal is obtained by stretching the original spectrum $X(\omega)$ by $M$, shifting it by $2\pi, 4\pi, \ldots, 2\pi(M-1)$, and adding all up. It is illustrated in Fig. 6.11.

***Interpolation.*** While decimation corresponds to sampling rate reduction, interpolation achieves increase of the sampling rate. Suppose that the analog signal $s(t)$ is sampled at two different rates $F$ and $\tilde{F}$, $F < \tilde{F}$ and $\tilde{F}/F = L$ is a positive integer. Let $x(n)$ and $\tilde{x}(n)$ be the resulting discrete signals. Then $\tilde{x}(n) = s(n/\tilde{F}) = s(nF/F\tilde{F}) = x(n/L)$. Sampling rate increase by factor $L$ requires the values $x(n/L)$, which makes sense only if $n$ is a multiple of $L$. The remaining values must be interpolated. As in Example 3.3 we define the signal

$$x_L(n) = \begin{cases} x\left(\dfrac{n}{L}\right), & \text{if } n = 0 \ (\text{mod} L) \\ 0, & \text{otherwise} \end{cases} \tag{6.61}$$

The spectrum of $x_L(n)$ is

$$X_L(\omega) = \sum_{n=-\infty}^{\infty} x_L(n)e^{-j\omega n} = \sum_{k=-\infty}^{\infty} x\left(\frac{kL}{L}\right)e^{-j\omega kL} = X(\omega L)$$

$X_L(\omega)$ is obtained from the original spectrum $X(\omega)$ via frequency scaling by $L$. An example is depicted in Fig. 6.12 for $L = 3$.

### 6.2.3  Signals with impulsive spectra

Purely imaginary exponential signals of the form $e^{j\omega_0 n}$ do not have Fourier transform in the ordinary sense. They can, however, be visualized in the frequency domain if interpreted as distributions. Due to periodicity these

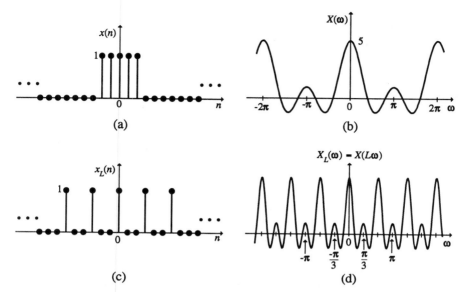

**Figure 6.12.** (a) Signal $x(n)$, (b) spectrum $X(\omega)$, (c) interpolation $x_L(n)$, and (d) spectrum $X_L(\omega)$.

distributions are periodic of period $2\pi$. No formal account will be presented. Instead we sketch the underlying ideas via a few examples. Consider the discrete signal $x(n)$. We view its Fourier transform as a tempered distribution in the following sense: Given a rapidly decreasing function $u \in S$ with Fourier transform $\hat{u} \in S$,

$$X(\omega)(\hat{u}) = \sum_{n=-\infty}^{\infty} x(n)e^{-j\omega n}(\hat{u}) = \sum_{n=-\infty}^{\infty} x(n) \int_{-\infty}^{\infty} e^{-j\omega n}\hat{u}(\omega)d\omega$$

$$= 2\pi \sum_{n=-\infty}^{\infty} x(n)u(-n)$$

where $u(n)$ are the samples of the analog signal $u(t)$. All signals $x(n)$ for which the latter expression is finite give rise to well-defined (in a distributional sense) spectra.

**Example 6.20    The Fourier transform of periodic signals**
Let $x(n) = 1$. Then

$$X(\omega)(\hat{u}) = 2\pi \sum_{n=-\infty}^{\infty} u(n)$$

Recall the Poisson summation formula

$$\sum_{n=-\infty}^{\infty} u(t - nT) = \frac{1}{T} \sum_{n=-\infty}^{\infty} \hat{u}\left(n\frac{2\pi}{T}\right) e^{jn2\pi t/T}$$

Setting $t = 0$, $T = 1$, we obtain $\sum_n u(n) = \sum_n \hat{u}(2\pi n)$. Thus $X(\omega)(\hat{u}) = 2\pi \sum_n \hat{u}(2\pi n)$, and

$$X(\omega) = \sum_{k=-\infty}^{\infty} \delta(\omega - 2\pi k)$$

Shifting in the frequency domain implies that

$$e^{j\omega_0 n} \xrightarrow{\mathcal{F}} 2\pi \sum_{k=-\infty}^{\infty} \delta(\omega - \omega_0 - 2\pi k) \tag{6.62}$$

It is now easy to compute the Fourier transform of a periodic signal. Recall that the exponential signal $e^{j\omega_0 n}$ is periodic if $\omega_0 = 2\pi m/N$. Using the linearity of Fourier transform, we have

$$\sum_{m=0}^{N-1} a_m e^{j2\pi mn/N} \xrightarrow{\mathcal{F}} 2\pi \sum_{m=0}^{N-1} a_m \sum_{k=-\infty}^{\infty} \delta\left(\omega - \frac{2\pi m}{N} - 2\pi k\right)$$

or

$$\sum_{m=0}^{N-1} a_m e^{j2\pi mn/N} \xrightarrow{\mathcal{F}} \sum_{l=-\infty}^{\infty} 2\pi a_l \delta\left(\omega - \frac{2\pi l}{N}\right) \tag{6.63}$$

The latter equality results from the periodicity of $a_m$ and the identification of each integer $l$ with the pair $(k, m)$ of the quotient and remainder of its division by $N$. Consider the $N$ periodic extension of the unit sample

$$x(n) = \sum_{k=-\infty}^{\infty} \delta(n - kN)$$

The Fourier coefficients are $a_k = 1/N$. Hence

$$X(\omega) = \frac{2\pi}{N} \sum_{k=-\infty}^{\infty} \delta\left(\omega - \frac{2\pi k}{N}\right) \qquad\blacksquare$$

## 6.3 MULTIDIMENSIONAL FOURIER TRANSFORM

The Fourier theory of the previous sections extends to multidimensional signals. The development is entirely analogous to the 1-D case. We will trace the main points leaving most details to the reader. Let $x(t_1, t_2, \ldots, t_k)$ denote a $k$-dimensional absolutely integrable signal

$$\int_{R^k} |x(t_1, t_2, \ldots, t_k)| \, dt_1 \, dt_2 \cdots dt_k < \infty$$

The Fourier transform is defined by

$$X(\omega_1, \omega_2, \ldots, \omega_k) = \int_{R^k} x(t_1, t_2, \ldots, t_k) e^{-j \sum_{i=1}^{k} t_i \omega_i} \, dt_1 \, dt_2 \cdots dt_k$$

It is a bounded continuous function. Under conditions similar to the 1-D case, the signal is recovered in the time domain by

$$x(t_1, t_2, \ldots, t_k) = \left( \frac{1}{2\pi} \right)^k \int_{R^k} X(\omega_1, \omega_2, \ldots, \omega_k) e^{j \sum_{i=1}^{k} t_i \omega_i} \, d\omega_1 \, d\omega_2 \cdots d\omega_k$$

The discrete case is analogous. The Fourier transform of $x(n_1, \ldots, n_k)$ is

$$X(\omega_1, \ldots, \omega_k) = \sum_{n_1 = -\infty}^{\infty} \cdots \sum_{n_k = -\infty}^{\infty} x(n_1, \ldots, n_k) e^{-j \sum_{i=1}^{k} n_i \omega_i} \tag{6.64}$$

Signal reconstruction is achieved by

$$x(n_1, \ldots, n_k) = \frac{1}{(2\pi)^k} \int_{-\pi}^{\pi} \cdots \int_{-\pi}^{\pi} X(\omega_1, \ldots, \omega_k) e^{j \sum_{i=1}^{k} n_i \omega_i} \, d\omega_1 \cdots d\omega_k \tag{6.65}$$

The multidimensional discrete Fourier transform is a periodic function with period $2\pi$ in each dimension:

$$X(\omega_1, \ldots, \omega_i, \ldots, \omega_k) = X(\omega_1, \ldots, \omega_i + 2\pi, \ldots, \omega_k), \qquad 1 \leq i \leq k$$

The Fourier transform of a separable signal is separable and is given by

$$x(n_1, \ldots, n_k) = x_1(n_1) \cdots x_k(n_k) \xrightarrow{\mathcal{F}} X(\omega_1, \ldots, \omega_k) = X_1(\omega_1) \cdots X_k(\omega_k)$$

**Example 6.21**   *2-D unit sample shifts*
Let

$$x(n_1, n_2) = \delta(n_1, n_2) = \delta(n_1) \delta(n_2)$$

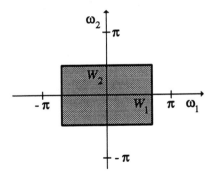

**Figure 6.13.** Frequency response of 2-D rectangular ideal lowpass filter.

Then $X(\omega_1, \omega_2) = x(0, 0)e^{-j(\omega_1 0 + \omega_2 0)} = 1$. More generally,

$$\delta(n_1 - k_1, n_2 - k_2) \xrightarrow{\mathcal{F}} e^{-j(k_1\omega_1 + k_2\omega_2)}$$ ∎

### *Example 6.22    2-D frequency window*

Suppose that the Fourier transform of $x(n_1, n_2)$ is the periodic extension of

$$X(\omega_1, \omega_2) = \begin{cases} 1, & |\omega_1| \leq W_1 < \pi, |\omega_2| \leq W_2 < \pi \\ 0, & \text{elsewhere in } [-\pi, \pi] \times [-\pi, \pi] \end{cases}$$

As we will see in Section 6.5 the above function defines the frequency response of the ideal lowpass filter and is illustrated in Fig. 6.13. Due to separability, calculations are reduced to those of Example 6.19:

$$x(n_1, n_2) = \frac{1}{2\pi} \int_{-W_1}^{W_1} e^{j\omega_1 n_1} d\omega_1 \frac{1}{2\pi} \int_{-W_2}^{W_2} e^{j\omega_2 n_2} d\omega_2 = \frac{\sin W_1 n_1}{\pi n_1} \frac{\sin W_2 n_2}{\pi n_2}$$

The signal $x(n_1, n_2)$ is sketched in Fig. 6.14. ∎

### *Example 6.23    Characteristic function of a random vector*

Let $u = (u_1, \ldots, u_k)^T$ be a random vector and $f(u) = f(u_1, \ldots, u_k)$ be the probability density function of $u$. The $k$-dimensional Fourier transform of $f(-u)$ defines the *characteristic function* of $u$, $\phi(\omega) = \phi(\omega_1, \ldots, \omega_k)$. Thus

$$\phi(\omega) = \int_{-\infty}^{\infty} f(-u)e^{-j\omega^T u} du = \int_{-\infty}^{\infty} \cdots \int_{-\infty}^{\infty} f(u_1, \ldots, u_k)e^{j\sum \omega_i u_i} du_1 \cdots du_k \tag{6.66}$$

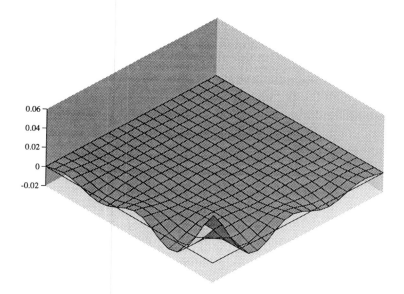

**Figure 6.14.** The 2-D frequency window in the time domain.

From the definition of the expected value, we have

$$\phi(\omega) = E[e^{j\omega^T u}] \tag{6.67}$$

As in the scalar case the characteristic function characterizes the probability distribution of a random vector. If the exponential function is expanded in power series, Eq. (6.67) becomes

$$\phi(\omega) = \sum_{n=0}^{\infty} \frac{j^n}{n!} E[(\omega^T u)^n]$$

For $n = 1$ we obtain

$$\frac{\partial \phi(\omega)}{\partial \omega}\Big|_{\omega=0} = jE[u] = jm_u \tag{6.68}$$

The quadratic term is

$$\frac{j^2}{2!} E[(\omega^T u)^2] = -\frac{1}{2} E[\omega^T u u^T \omega]$$

Therefore

$$\frac{\partial^2 \phi(\omega)}{\partial \omega^2}\Big|_{\omega=0} = -E[u u^T] \tag{6.69}$$

∎

***Example 6.24   Gaussian vectors***
Suppose that $u$ is a Gaussian vector $N(m, C)$ with probability density function

$$f(u) = \frac{1}{(2\pi)^{k/2}(\det C)^{1/2}} e^{-(u-m)^T C^{-1}(u-m)/2} \qquad (6.70)$$

We will use the characteristic function to prove that the mean value is $m$ and the covariance matrix is $C$. Suppose first that $m = 0$. Then

$$\phi(\omega) = \frac{1}{(2\pi)^{k/2}(\det C)^{1/2}} \int_{-\infty}^{\infty} e^{j\omega^T u - u^T C^{-1}u/2} du$$

We complete the squares in the exponent

$$j\omega^T u - \frac{1}{2}u^T C^{-1}u = -\frac{1}{2}[u^T C^{-1}u - 2j\omega^T CC^{-1}u - \omega^T C\omega] - \frac{1}{2}\omega^T C\omega$$

This is further written as

$$-\frac{1}{2}(u - jC\omega)^T C^{-1}(u - jC\omega) - \frac{1}{2}\omega^T C\omega$$

Thus

$$\phi(\omega) = e^{-\frac{1}{2}\omega^T C\omega} \int_{-\infty}^{\infty} \frac{1}{(2\pi)^{k/2}(\det C)^{1/2}} e^{-(u-jC\omega)^T C^{-1}(u-jC\omega)/2} du$$

The integrand in the above expression is the Gaussian density $N(jC\omega, C)$. Thus the integral equals 1, and the characteristic function becomes

$$\phi(\omega) = e^{-\omega^T C\omega/2} \qquad (6.71)$$

If the mean of $u$ is nonzero, the above takes the form

$$\phi(\omega) = e^{jm^T \omega - \omega^T C\omega/2} \qquad (6.72)$$

Equation (6.68) gives

$$m_u = \frac{1}{j}\frac{\partial \phi(\omega)}{\partial \omega}\Big|_{\omega=0} = \frac{1}{j}e^{jm^T \omega - \omega^T C\omega/2}[jm - C\omega]|_{\omega=0} = m$$

Likewise Eq. (6.69) gives

$$C = E[uu^T] - mm^T = \text{cov}[u, u]$$

Next we show that gaussianity is preserved under linear-affine transformations. Let $A$ be an $m \times k$ matrix. The vector $z = Au$ is Gaussian with density

$N(Am, ACA^T)$. Indeed, the characteristic function of $z$ is

$$\phi_z(\omega) = E[e^{j\omega^T z}] = E[e^{j\omega^T Au}] = \phi_u(A^T\omega) = e^{jm^T A^T\omega - \omega^T ACA^T\omega/2}$$

Two important consequences are the following: If $u$, $y$ are jointly Gaussian vectors—that is, the augmented vector $w^T = [u^T y^T]$ is Gaussian with mean $m_w$ and covariance $C_w$—the sum $z = u + y$ is $N(m_u + m_y, C_u + C_y + C_{uy} + C_{yu})$.

Indeed, we write $z = Aw$ with $A = (I \quad I)$, $I$ being the identity matrix. The statement follows from the partitions

$$m_w = \begin{pmatrix} m_u \\ m_y \end{pmatrix}, \qquad C_w = \begin{pmatrix} C_u & C_{uy} \\ C_{yu} & C_y \end{pmatrix} \tag{6.73}$$

The second consequence is that if $w^T = [u^T y^T]$ is Gaussian then $y$ is $N(m_y, C_y)$, where $m_y$ and $C_y$ are obtained from partition (6.73). The proof follows if we set $A = (0 \quad I)$. If finally $y = u + a$, and $a$ is a constant vector, $y$ is $N(m + a, C)$. Indeed,

$$\phi_y(\omega) = E[e^{j\omega^T(u+a)}] = e^{j\omega^T a}E[e^{j\omega^T u}] = e^{j\omega^T a}\phi_u(\omega)$$

Thus

$$\phi_y(\omega) = e^{j\omega^T(m+a) - j\omega^T C\omega/2} \qquad \blacksquare$$

### *Example 6.25   Perceptron*

Consider the perceptron system introduced in Section 3.13. The activation function is the hard limiter. If the input $u$ is a Gaussian vector with density $N(m, C)$, the random variable $w^T u + \theta$ is Gaussian with density $N(w^T m + \theta, w^T Cw)$. Hence

$$P[y = 1] = \int_0^\infty \frac{1}{(2\pi w^T Cw)^{1/2}} e^{-(z - w^T m - \theta)^2/(2w^T Cw)} dz \qquad \blacksquare$$

## 6.4  SYSTEM ANALYSIS IN THE FREQUENCY DOMAIN

In this section input-output representations are converted into the frequency domain. Analysis and design of signal processing systems is greatly benefited from such a spectral viewpoint, as subsequent sections demonstrate.

### 6.4.1   Convolutional systems in the frequency domain

As we recall from Chapter 3, the convolutional representation of a system is

given by

$$y(n) = h(n) * u(n) = \sum_{k=-\infty}^{\infty} h(n-k)u(k) = \sum_{k=-\infty}^{\infty} h(k)u(n-k) \quad (6.74)$$

Suppose that the impulse response $h(n)$ is absolutely summable, $h(n) \in \ell_1$. The Fourier transform $H(\omega)$ exists and forms the frequency response of the system. As we saw in Section 5.6, it characterizes the outputs to exponential inputs. The system is stable, and the convolution bound (Theorem 3.2) implies that each absolutely summable or finite energy input signal gives rise to an absolutely summable or finite energy output. Under those circumstances the Fourier transforms of the relevant signals exist, and the convolution property of the Fourier transform converts Eq. (6.74) into

$$Y(\omega) = H(\omega)U(\omega) \quad (6.75)$$

Analogous conclusion holds for the continuous time convolutional system

$$y(t) = h(t) * u(t) = \int_{-\infty}^{\infty} h(t-\tau)u(\tau)d\tau$$

If $h(t)$ is absolutely integrable and $u(t)$ belongs to $L_1$ or $L_2$, the output inherits the same property and the convolution property of the Fourier transform leads to (6.75), interpreted in the analog domain.

In a similar fashion the multidimensional convolution

$$y(n_1, \ldots, n_k) = h(n_1, \ldots, n_k) * \cdots * u(n_1, \ldots, n_k) \quad (6.76)$$

converts to

$$Y(\omega_1, \ldots, \omega_k) = H(\omega_1, \ldots, \omega_k)U(\omega_1, \ldots, \omega_k) \quad (6.77)$$

In polar coordinates (6.75) becomes

$$|Y(\omega)| = |H(\omega)||U(\omega)|, \quad \arg Y(\omega) = \arg H(\omega) + \arg U(\omega) \quad (6.78)$$

The additive form of the second equation enables the graphical description of the output phase spectrum via the superposition of the input phase spectrum and the frequency response phase spectrum. A similar property holds for the logarithm of the magnitude

$$\log|Y(\omega)| = \log|H(\omega)| + \log|U(\omega)| \quad (6.79)$$

Plots of phase and magnitude in dB are called *Bode plots*. The use of logarithmic scale gives a better resolution when the frequency band of interest is large, or when it confines to a small interval round zero.

***Example 6.26   Output spectrum of a convolutional system***
The unit sample response of a convolutional system is

$$h(n) = a^n u_s(n), \qquad |a| < 1$$

We want to compute the output spectrum produced by the input $u(n) = b^{|n|}$, $|b| < 1$. According to Examples 6.15 and 6.16, we have

$$H(\omega) = \frac{1}{1 - ae^{-j\omega}}, \quad U(\omega) = \frac{1 - b^2}{1 + b^2 - 2b\cos\omega}$$

Hence the output spectrum is

$$Y(\omega) = \frac{1 - b^2}{(1 - ae^{-j\omega})(1 + b^2 - 2b\cos\omega)}$$

MATLAB simulations and plots are easily derived.   ∎

### 6.4.2   Linear time varying systems in the frequency domain

Consider the linear time varying system

$$y(n) = \sum_{k=-\infty}^{\infty} h(n,k)u(k) \tag{6.80}$$

We will assume that for all integers $n$, the $\ell_1$ norms of $h(n,k)$ are finite and uniformly bounded:

$$\sup_{n \in Z} \sum_{k=-\infty}^{\infty} |h(n,k)| < \infty \tag{6.81}$$

It then follows, as in Theorem 4.2 (see also Exercise 4.4), that the system is $\ell_1$ stable. Therefore for each absolutely summable input $u(n)$, the output is also absolutely summable, and its Fourier transform is

$$Y(\omega_1) = \sum_{n=-\infty}^{\infty} y(n)e^{-j\omega_1 n} = \sum_n \left( \sum_k h(n,k)u(k) \right) e^{-j\omega_1 n}$$

We replace $u(k)$ by its inverse Fourier transform to obtain

$$Y(\omega_1) = \frac{1}{2\pi} \sum_n \left( \sum_k h(n,k) \int_0^{2\pi} U(\omega_2)e^{j\omega_2 k} d\omega_2 \right) e^{-j\omega_1 n}$$

or

$$Y(\omega_1) = \frac{1}{2\pi} \int_0^{2\pi} K(\omega_1, \omega_2) U(\omega_2) d\omega_2$$

$$K(\omega_1, \omega_2) = \sum_n \sum_k h(n, k) e^{-j\omega_1 n} e^{j\omega_2 k}$$

Notice that $K(\omega_1, \omega_2)$ is related to the 2-D Fourier transform of $h(n, k)$ by

$$K(\omega_1, \omega_2) = H(\omega_1, -\omega_2)$$

Condition (6.81) must be strengthened to warrant that $H(\omega_1, \omega_2)$ exists in the ordinary sense. For instance, it may be replaced by the assumption that $h(n, k)$ is absolutely summable as a 2-D signal. Nevertheless, (6.81) suffices to secure that $H(\omega_1, \omega_2)$ exists as a distribution. This is the case with time invariant systems.

***Example 6.27    Time invariant case***
In the time invariant case the kernel $h(n, k)$ is a function of the time difference, $h(n, k) = h(n - k, 0)$. Condition (6.81) holds if the impulse response $h(n) = h(n, 0)$ is absolutely summable. Using the substitution of variables $m = n - k$, we have

$$H(\omega_1, \omega_2) = \sum_n \sum_k h(n - k) e^{-j\omega_1 n} e^{-j\omega_2 k}$$

$$= \sum_n \sum_m h(m) e^{-j\omega_1 n} e^{-j\omega_2(n-m)}$$

$$= \sum_n e^{-j(\omega_1 + \omega_2)n} \sum_m h(m) e^{j\omega_2 m}$$

It follows from the Poisson summation formula and the discussion of Section 6.2.3 that

$$\sum_n e^{-j(\omega_1 + \omega_2)n} = 2\pi \sum_n \delta(\omega_1 + \omega_2 + 2\pi n) \tag{6.82}$$

Then

$$H(\omega_1, \omega_2) = 2\pi \sum_n \delta(\omega_1 + \omega_2 + 2\pi n) H(-\omega_2)$$

Therefore

$$K(\omega_1, \omega_2) = 2\pi \sum_n \delta(\omega_1 - \omega_2 + 2\pi n) H(\omega_2)$$

and

$$Y(\omega_1) = \frac{1}{2\pi} \int_0^{2\pi} 2\pi \sum_n \delta(\omega_1 - \omega_2 + 2\pi n) H(\omega_2) U(\omega_2) d\omega_2$$

Since integration is confined to the interval $[0, 2\pi]$, the train of impulses contributes only the term $\delta(\omega_1 - \omega_2)$. Thus we arrive at the familiar formula

$$Y(\omega_1) = H(\omega_1) U(\omega_1)$$  ∎

### 6.4.3   Periodically time varying linear systems

Motivated by the two expressions of convolution (6.74), we express (6.80) as

$$y(n) = \sum_{k=-\infty}^{\infty} r(n - k, k) u(k) \tag{6.83}$$

where $r(n, k) = h(n + k, k)$, and

$$y(n) = \sum_{k=-\infty}^{\infty} p(n, k) u(n - k) \tag{6.84}$$

where $p(n, k) = h(n, n - k)$. In this manner we explicitly introduce a shift either in $h$ or in $u$. The 2-D signals $r(n, k)$ and $p(n, k)$ are occasionally referred to as *Green functions*. Equation (6.80) defines a *linear periodically time varying (LPTV)* system if there exists $N$ such that $h(n + N, k + N) = h(n, k)$. This is weaker than 2-D periodicity of $h(n, k)$. In terms of the Green functions, it takes the form $p(n + N, k) = p(n, k)$ for all $k$ and $r(n, k + N) = r(n, k)$ for all $n$. To determine the behavior of a LPTV system in the frequency domain, we view $p(n, k)$ as a function of $n$ with $k$ fixed, and we denote by $\hat{p}(m, k)$ the corresponding Fourier coefficients. Substituting the Fourier series expansion of $p(n, k)$ into Eq. (6.84), we obtain

$$y(n) = \sum_{k=-\infty}^{\infty} \sum_{m=0}^{N-1} \hat{p}(m, k) e^{j2\pi mn/N} u(n - k)$$

Thus the output spectrum has the form

$$Y(\omega) = \sum_{m=0}^{N-1} \sum_{k=-\infty}^{\infty} \hat{p}(m, k) \mathcal{F}\left\{ e^{j2\pi mn/N} u(n - k) \right\}$$

Using the shift property in time as well as in frequency domain, we have

$$Y(\omega) = \sum_{m=0}^{N-1} \sum_{k=-\infty}^{\infty} \hat{p}(m, k) e^{-jk(\omega - 2\pi m/N)} U\left( \omega - \frac{2\pi m}{N} \right)$$

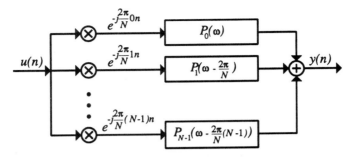

**Figure 6.15.** Architecture for LPTV systems.

or

$$Y(\omega) = \sum_{m=0}^{N-1} P_m\left(\omega - \frac{2\pi m}{N}\right) U\left(\omega - \frac{2\pi m}{N}\right)$$

where for each $m$, $P_m(\omega)$ denotes the Fourier transform of $\hat{p}(m,k)$ viewed as a sequence with respect to $k$. A diagram representation is shown in Fig. 6.15. Analogous results are obtained if representation (6.83) is used. They are summarized in Exercise 6.31.

### 6.4.4  Frequency analysis of Volterra systems

Volterra representations are translated in the frequency domain with the aid of multidimensional embedding. Due to the well posedness assumptions stated in Section 3.7, the Fourier transforms of the kernels $H_m(\omega_1, \ldots, \omega_m)$ are bounded continuous functions. If the input is absolutely summable with Fourier transform $U(\omega)$, the m-D separable extension $x_m(n_1, n_2, \ldots, n_m) = u(n_1)u(n_2)\cdots u(n_m)$ is separable and absolutely summable with Fourier transform $X_m(\omega_1, \ldots, \omega_m) = U(\omega_1)\cdots U(\omega_m)$. Using the notation of Section 3.7.1 and the convolution property we obtain

$$Y(\omega_1, \ldots, \omega_m) = H_m(\omega_1, \ldots, \omega_m)U(\omega_1)\cdots U(\omega_m) \qquad (6.85)$$

To complete the calculation, we need to express the Fourier transform of the 1-D signal $y_m(n) = y(n, n, \ldots, n)$ in terms of the m-D transform $Y(\omega_1, \ldots, \omega_m)$. Let us for simplicity take $m = 2$. The inverse 2-D Fourier transform gives

$$y_2(n) = y(n, n) = \frac{1}{4\pi^2}\int_0^{2\pi}\int_0^{2\pi} e^{jn(\omega_1+\omega_2)} Y(\omega_1, \omega_2)d\omega_1 d\omega_2$$

The change of variables $\omega = \omega_1 + \omega_2$ leads to

$$y_2(n) = \frac{1}{4\pi^2}\int_0^{4\pi} e^{jn\omega}\int_0^{2\pi} Y(\omega - \omega_2, \omega_2)d\omega_2 d\omega$$

Comparison of the right-hand side with the inverse Fourier transform of $y_2(n)$ shows that

$$Y_2(\omega) = \frac{1}{2\pi} \int_0^{2\pi} Y(\omega - \omega_2, \omega_2) d\omega_2$$

The above analysis can easily be generalized. The Fourier transform of $y_m(n) = y(n, n, \ldots, n)$ is

$$Y_m(\omega) = \left(\frac{1}{2\pi}\right)^{m-1} \int_0^{2\pi} \cdots \int_0^{2\pi} Y\left(\omega - \sum_{i=2}^m \omega_i, \omega_2, \ldots, \omega_m\right) d\omega_2 \cdots d\omega_m$$

If Eq. (6.85) is inserted into the latter equation, we obtain the frequency domain representation of the homogeneous $m$-degree Volterra system. Superposition of the homogeneous terms gives the general case. In particular, a second-degree Volterra system is expressed in the frequency domain as

$$Y(\omega) = H_1(\omega)U(\omega) + \frac{1}{2\pi} \int_0^{2\pi} H_2(\omega - \omega_2, \omega_2) U(\omega - \omega_2) U(\omega_2) d\omega_2 \quad (6.86)$$

**Response to exponential inputs.** If the input to the Volterra system is the exponential $u(n) = e^{j\omega n}$, the resulting output is

$$y(n) = \sum_{m=1}^{\infty} \sum_{\mathbf{k}_m} h_m(k_1, k_2, \ldots, k_m) e^{j\omega(n-k_1)} \cdots e^{j\omega(n-k_m)}$$

$$= \sum_{m=1}^{\infty} e^{j\omega m n} \sum_{\mathbf{k}_m} h_m(k_1, k_2, \ldots, k_m) e^{j\omega(-\sum_{i=1}^m k_i)}$$

Thus

$$y(n) = \sum_{m=1}^{\infty} e^{jmn\omega} H_m(\omega, \omega, \ldots, \omega) \quad (6.87)$$

$H_m(\omega, \omega, \ldots, \omega)$ is the m-D Fourier transform of $h_m(k_1, \ldots, k_m)$ evaluated on the diagonal slice. For second-order Volterra systems, Eq. (6.87) becomes

$$y(n) = e^{j\omega n} H_1(\omega) + e^{j\omega 2n} H_2(\omega, \omega)$$

In particular, the periodic input $u(n) = e^{j2\pi n/N}$ produces the output

$$y(n) = e^{j2\pi n/N} H_1\left(\frac{2\pi}{N}\right) + e^{j4\pi n/N} H_2\left(\frac{2\pi}{N}, \frac{2\pi}{N}\right)$$

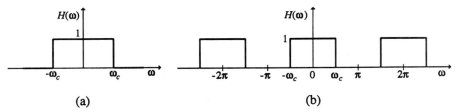

**Figure 6.16.** Frequency response of ideal lowpass filter: (a) analog case; (b) discrete case.

In the linear case a single tone at the input generates a single tone at the output. In contrast, the second-order Volterra system produces two tones at the output.

## 6.5   FREQUENCY SELECTIVE FILTERS

Frequency selective filters preserve or amplify the spectral content of the input within a frequency band and reject the spectral content at another. The bass amplifier in audio systems accenuates the low frequencies and attenuates the high frequencies. The human ear detects only the audio frequency "portion" of a signal, that is, the part of the spectrum that is located below 20 kHz. Frequency selective filters are also used to separate signals with nonoverlaping spectra. For instance, a high-frequency disturbance degrading a low-frequency signal (surface noise in audio recording) can be removed by a filter that will eliminate the high-frequency signal portion and will pass the low-frequency content. Frequency selective filtering is neatly described in the framework of convolutional systems. The input-output relationship of a continuous time convolutional system is $y(t) = h(t) * u(t)$. The convolutional property of the Fourier transform gives $Y(\omega) = H(\omega)U(\omega)$. $H(\omega)$ defines the frequency response of the system, $U(\omega)$ is the input spectrum, and $Y(\omega)$ is the output spectrum. Preservation of the input spectrum in the interval $[-\omega_c, \omega_c]$ and rejection of the remaining part are accomplished by the frequency response

$$H(\omega) = \begin{cases} 1, & |\omega| \leq \omega_c \\ 0, & |\omega| > \omega_c \end{cases} \qquad (6.88)$$

plotted in Fig. 6.16 (a). The frequency band $[-\omega_c, \omega_c]$ is called *passband*, while the subset $|\omega| > \omega_c$ is called *stopband*. $\omega_c$ is referred to as *cutoff frequency*. The filter with frequency response given as in (6.88) is called *ideal lowpass filter* because low frequencies are allowed and high frequencies are rejected. Lowpass filters differentiate slow from fast variations of the input signal. The impulse response was found in Example 6.7 to be $h(t) = (\omega_c/\pi)\text{sinc}(\omega_c t/\pi)$. Since $h(t)$ extends to negative instants of time, the filter is noncausal. In Example 4.2 we saw that $h(t)$ is not absolutely integrable. Therefore the ideal lowpass filter is unstable. A primary goal of filter design is to develop realizable stable structures that

(a)
(b)

**Figure 6.17.** Image filtering by lowpass filter: (a) original image; (b) filtered image.

adequately approximate the above ideal behavior. Elements of this theory are discussed in Chapter 10.

Discrete filters (or digital filters) are similarly defined. The discrete ideal lowpass filter is depicted in Fig. 6.16 (b). The frequency response in one period is

$$H(\omega) = \begin{cases} 1, & |\omega| \leq \omega_c \\ 0, & \pi > |\omega| > \omega_c \end{cases} \tag{6.89}$$

The filter impulse response was calculated in Example 6.19. The filter is unstable and noncausal. Digital filter design aims to develop efficient architectures that approximate the ideal characteristic. When a lowpass filter is applied on a time series such as the price index in the stock market, it isolates long-term from short-term fluctuations.

Two-dimensional lowpass filters are similarly defined. The frequency response of a 2-D discrete lowpass filter has the form

$$H(\omega_1, \omega_2) = \begin{cases} 1, & (\omega_1, \omega_2) \in R \\ 0, & \text{otherwise} \end{cases} \tag{6.90}$$

where $R$ is a neighborhood of the origin contained in $[-\pi, \pi] \times [-\pi, \pi]$, such as the rectangle of Example 6.22. Filtering of the image shown in Fig. 6.17 (a) leads to the image of Fig. 6.17 (b). Fast variations have been suppressed.

Other types of frequency selective filters, besides lowpass filters, are the *highpass filters* and the *bandpass filters*. These are depicted in Figs. 6.18 (a) and (b). The highpass filter rejects slow variations (low frequencies) and preserves high frequencies. The bandpass filter allows the frequencies within a certain band and rejects the others. The width of the passband is called

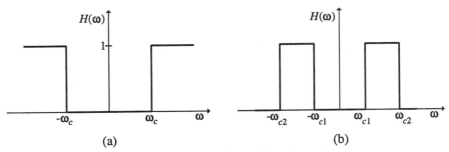

**Figure 6.18.** (a) Highpass filter; (b) bandpass filter.

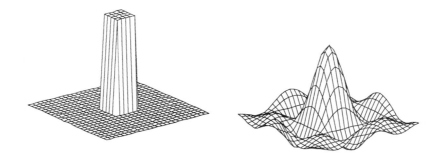

**Figure 6.19.** 2-D lowpass filter in the frequency and space domain.

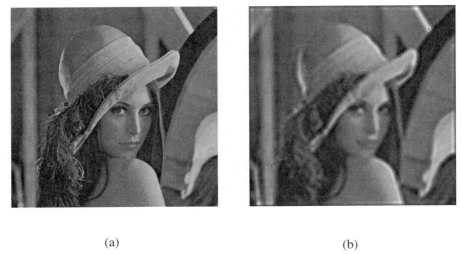

(a)                                                    (b)

**Figure 6.20.** (a) Image of $512 \times 512$ pixels; (b) lowpass filtered version using the filter of Example 6.28.

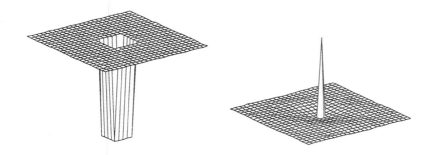

**Figure 6.21.** Highpass approximation in the frequency and space domain.

*bandwidth.* More generally, frequency selective filters in 1-D and multidimensional cases, in analog or discrete form, are specified by a frequency response whose magnitude has piecewise constant characteristics.

### Example 6.28   Lowpass and highpass filtering of images

Consider the 2–D ideal lowpass filter of Example 6.22. We choose $W_1 = W_2 = W = 3/25\pi$, and we approximate the Fourier transform $H(\omega_1, \omega_2)$ by an FIR filter whose impulse response $\hat{h}(n_1, n_2)$ coincides with the ideal lowpass frequency response for the first $25 \times 25$ points. The frequency response magnitude and the impulse response are shown in Fig. 6.19. The lowpass characteristic is visible. The blurring effect of lowpass filtering on a digital image is shown in Fig. 6.20.

A highpass filter is sketched in the frequency and spatial domain in Fig. 6.21. Highpass image filtering is illustrated in Fig. 6.22. Details and local contrast tend

(a)                                    (b)

**Figure 6.22.** (a) Image of $512 \times 512$ pixels; (b) highpass filtered version using the filter of Example 6.28.

to be accentuated. The filtered image appears to be sharper. The above example is implemented by the following MATLAB code:

```
load lenna
HI = zeros(25,25);
HI(12:14,12:14) = ones(3,3);       % ideal lowpass filter
HIs = fsamp2(HI);                  % approximation
mesh(HI), pause;                   % displays frequency response
mesh(HIs), pause;                  % displays impulse response

freqz2(HIs), pause;
Y = filter2(HIs,X);                % filters the image
colormap(gray(256));
subplot(121), imshow(X), axis off;
subplot(122), imshow(Y), axis off;

Hh = ones(15,15);
Hh(7:9,7:9) = zeros(3,3);          % highpass filter
Hhs = fsamp2(Hh);
mesh(Hh), pause;                   % displays frequency response
mesh(Hhs), pause;                  % displays impulse response
freqz2(Hhs), pause;
Z = filter2(Hhs,X);                % filters the image
colormap(gray(256));
subplot(121), imshow(X), axis off;
subplot(122), imshow(Z), axis off;
```
∎

### Example 6.29   Audio signals

Audio signals like speech and music are perceived directly by the human ear. Their bandwidth extends from 20 Hz to 20 kHz. In many applications a narrower bandwidth is used. Evidently as bandwidth increases, quality of sound is enhanced. Telephone voice is of moderate quality. It extends from 300 to 3400 Hz. The reduction in bandwidth leads to a reduction in the cost of the facility. To guard the transmitted signal against interference from other signals occupying the same transmission medium, the bandwidth is expanded from 0 Hz to 4 kHz. Thus telephone voice is obtained by passing the speech signal through a lowpass filter with a cutoff frequency of 4 kHz. Voice of teleconference quality requires approximately 7 kHz of bandwidth. High fidelity sound raises the bandwidth requirements to 15 kHz. Compact disks (CD) use 20 kHz for each of two channels for stereo.                                                              ∎

### Example 6.30   Video

A video signal corresponds to a sequence of images. Each image is converted to a one-dimensional signal by scanning. An electron beam sweeps across a photosensitive plate upon which a visual image is optically focused and produces an analog electric signal proportional to the brightness level of the image. Transmission of a black-and-white TV signal requires a bandwidth of about 4 MHz. Thus the TV transmitter includes a filter that passes frequency components in the range 0 to 4.2 MHz. Cable TV and TV broadcast utilize a bandwidth of 6 MHz to accommodate color and audio information and to guard against interference from other TV signals. ∎

## 6.6 MODULATION AND MULTIPLEXING

A modulation system modifies a given signal $c(t)$, called *carrier* by the input signal $u(t)$. Amplitude modulation (AM) modifies the amplitude of the carrier in accordance with $y(t) = c(t)u(t)$. In frequency modulation (FM) the input signal varies the frequency of the carrier (see Exercise 6.25). Modulation is employed in signal transmission for two main reasons. First, the signal spectrum must be shifted to a higher-frequency range in order to ensure effective transmission. Direct transmission of speech or music through an unguided medium requires antennas with many kilometers in diameter. The second reason is that modulation permits frequency division multiplexing, that is, shared access of the transmission medium. A communication channel can be viewed as a bandpass filter, that is, a convolutional system whose frequency response is bandlimited to a specific range of frequencies. When an audio signal such as speech or music is transmitted through the atmosphere, it is rapidly attenuated. Transmission over longer distances through the above channel is accomplished if the spectrum of the transmitted signal is in the range 0.3 to 3 MHz. Conversion of the lowpass audio signal into such a bandpass waveform is effected by modulation.

Modulation schemes are required to reduce the size of transmitting antennas. The size of a high-quality antenna is $1/10$ of the electromagnetic wavelength. Transmission of an acoustic signal of bandwidth 10 kHz would require an antenna of size 3 km, since the wavelength of the signal is $3 \times 10^8/10^4 = 3 \times 10^4$m. Shifting the signal spectrum to megacycles ensures reasonable antenna size.

### 6.6.1 Transmission media

Let us briefly outline the characteristics of the major transmission media. Twisted pair is the most common medium linking residential telephone sets and the local telephone exchange or "end office." Both analog and digital transmission are handled. Analog transmission requires amplifiers every 5 or 6 km to combat attenuation. Digital transmission requires repeaters every 2 or 3 km. The channel bandwidth is about 250 kHz. Coaxial cable offers superior performance at a higher cost. It supports both analog and digital transmission. The bandwidth extends to about 500 MHz, and the data rate is about 500 Mbps. Coaxial cable faces increasing competition from optical fiber and microwave links. Its present primary use is in cable TV.

Optical fiber systems operate in the range of about $10^{14}$ to $10^{15}$ Hz with a bandwidth of 2 GHz = $2 \cdot 10^9$ Hz and a data rate of 2 Gbps. Besides their significantly greater capacity, they feature greater repeater spacing (10–100 km) and greater immunity to external electromagnetic fields. Thus they are less susceptible to interference, impulse noise, or crosstalk. Finally, they do not radiate energy, causing little interference and offering a high degree of security from eavesdropping. They enjoy an increasingly widespread use.

Radio encompasses frequencies in the range of 3 kHz to 300 GHz. Frequencies ranging from about 2 to 40 GHz are referred to as microwave frequencies. Microwave enables point-to-point transmission. Terrestrial microwave transmission is achieved by dish-shaped antennas located at sufficient height above ground level. Terrestrial microwave systems are engaged in long haul communications for voice and TV. Attenuation is less severe than coaxial cable allowing a repeaters or amplifiers spacing of 10 to 100 km. On the other hand, microwave frequencies are subject to interference, and hence the allocation of frequency bands is strictly regulated.

Communication satellites are relay microwave stations linking two or more ground based stations. The satellite provides point-to-point link between distant stations. It also serves broadcast communication from one earth station to several ground stations. The satellite receives signals on one frequency band (uplink), amplifies or repeats the signal and transmits it on another frequency band (downlink). Satellites are used for TV distribution, long distance telephone transmission and other applications.

A mathematical model that captures additive signal impairments, attenuation, and the bandwidth characteristics of the transmission media outlined above is given by the model

$$r(t) = c(t) * u(t) + e(t) \tag{6.91}$$

where $u(t)$ is the channel input, $e(t)$ is the additive noise, $c(t)$ is the channel impulse response. In the frequency domain, $C(\omega)$ is approximately a bandpass filter whose passband represents the transmission bandwidth limitations. Finally, $r(t)$ is the channel output. Equation (6.91) is illustrated in Fig. 6.23. A special but important case is when $c(t)$ takes the form $c(t) = a\delta(t)$; $\delta(t)$ is the Dirac function. Then Eq. (6.91) takes the form

$$r(t) = au(t) + e(t)$$

The term $a$ represents the attenuation incurred during transmission.

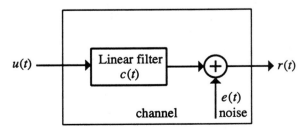

**Figure 6.23.** Channel representation as a linear filter.

Nonlinearities and time varying effects are also present in certain transmission media. Mobile cellular radio channels exhibit a time variant behavior, due to multipath propagation.

### 6.6.2  Modulation principle

We assume for simplicity that transmission is error free and that the channel filter $C(\omega)$ is bandpass, equal to 1 in the passband. Let $c(t) = e^{j(\omega_c t + \theta_c)}$ be the carrier signal in an AM system. The shift property of the Fourier transform gives $Y(\omega) = e^{j\theta_c} U(\omega - \omega_c)$. The transmitter modulates the carrier signal $c(t)$ with $u(t)$ and transmits the signal $y(t)$. The *carrier frequency* $\omega_c$ indicates the center of the channel bandwidth through which efficient transmission is achieved.

The demodulator at the receiving end recovers the transmitted signal $u(t)$ by $u(t) = e^{-j(\omega_c t + \theta_c)} y(t)$. The modulator is implemented with two multipliers and two carrier signals, one for the real and one for the imaginary part. A simpler structure is obtained if only one of the two parts is used. This leads to sinusoidal amplitude modulation defined by the carrier, $c(t) = \cos(\omega_c t + \theta_c)$. Suppose that $\theta_c = 0$. The carrier spectrum is

$$C(\omega) = \pi[\delta(\omega - \omega_c) + \delta(\omega + \omega_c)]$$

Using (6.12), the spectrum of the modulated signal is

$$Y(\omega) = \frac{1}{2}[U(\omega - \omega_c) + U(\omega + \omega_c)]$$

The process of modulation is illustrated in Fig. 6.24. $2\omega_M$ is the signal bandwidth.

Demodulation is always possible if the complex exponential carrier is employed. In contrast, demodulation with sinusoidal carrier requires $\omega_c > \omega_M$. Indeed, if $\omega_c \leq \omega_M$, the two shifted spectra $U(\omega - \omega_c)$ and $U(\omega + \omega_c)$ overlap

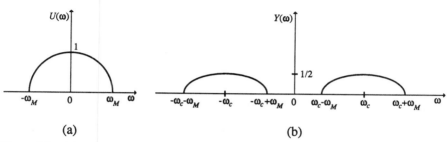

**Figure 6.24.** Sinusoidal amplitude modulation : (a) input spectrum; (b) spectrum of modulated signal.

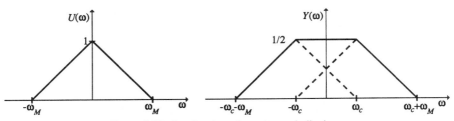

**Figure 6.25.** Overlapping of spectra and aliasing.

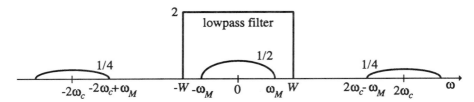

**Figure 6.26.** Demodulation with sinusoidal carrier.

and as a result the input spectrum is significantly altered, making the reconstruction of the original signal impossible (Fig. 6.25). If $\omega_c > \omega_M$, demodulation is accomplished as follows: The modulated signal $y(t)$ is modulated again with the same carrier and then is filtered by a lowpass filter of amplitude 2 and bandwidth $[-W, W]$. The cutoff frequency satisfies $\omega_M < W < 2\omega_c - \omega_M$. The process is illustrated in Fig. 6.26. The demodulator is shown in Fig. 6.27.

### 6.6.3 Frequency division multiplexing

Frequency division multiplexing (FDM) relies on the observation that if two signals have nonoverlaping spectra, they can be simultaneously transmitted

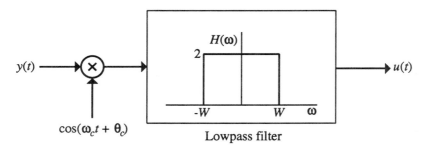

**Figure 6.27.** Demodulator.

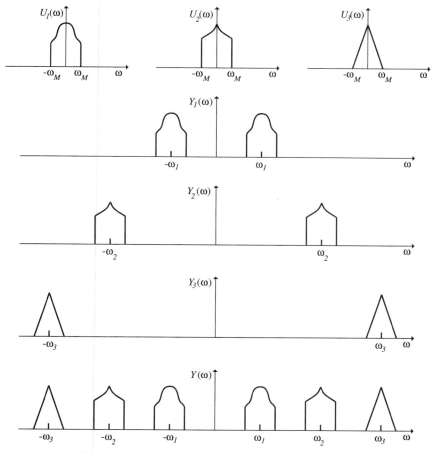

**Figure 6.28.** Description of frequency division multiplexing.

through a communication channel and split at the receiver by bandpass filters. If the signals have overlapping spectra, the previous idea works provided that modulation is applied prior to transmission. In this way many signals can exploit the channel simultaneously. The number of signals that can be served depends on the bandwidth of the signals and the bandwidth of the channel.

Frequency division multiplexing with sinusoidal amplitude modulation is illustrated in Fig. 6.28. $N$ signals $u_1(t)$, $u_2(t)$, ... , $u_N(t)$ are modulated with carrier frequencies $\omega_1, \omega_2, \ldots, \omega_N$, and the sum is transmitted. Each of the signals $u_1(t), u_2(t), \ldots, u_N(t)$, is reconstructed at the receiver by a bandpass filter and a demodulator. The bandpass filter isolates the spectrum of each $u_i(t)$, from the spectrum of the sum $y(t)$ and the demodulator shifts the signal back to its original band (see Fig. 6.29).

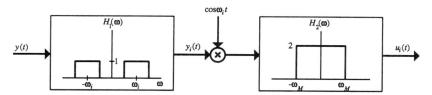

**Figure 6.29.** Reconstruction of multiplexed signals with a bandpass filter and demodulator.

FDM has been successfully used in telephony, AM broadcasting, and TV distribution, including broadcast TV and cable TV. With the domination of digital transmission of speech signals over telephone channels, the use of frequency division multiplexing has declined in favor of time division multiplexing described below. Commercial AM broadcasting extends from 535 to 1605 kHz and uses sinusoidal amplitude modulation. Each AM station is assigned a frequency $\omega_i$. Simultaneous transmission of radio stations is carried out by FDM. The receiver tunes to the station frequency $\omega_i$ and applies the procedure of Fig. 6.29.

Besides AM modulation, frequency and phase modulation are alternative modulation methods jointly termed *angle modulation* methods. In frequency modulation the frequency of the carrier is modified by the signal. Likewise a phase modulator modifies the phase of the carrier by the signal. Angle modulation systems are much harder to analyze that amplitude modulation (see Exercise 6.25). Their main property is that they expand the signal bandwidth several times while offering a high degree of noise immunity. For this reason FM systems are used in high fidelity broadcasting and point-to-point transmission where the transmitter power is limited. Commercial FM radio broadcasting utilizes the frequency band 88 to 108 MHz.

Dozens of TV signals are simultaneously transmitted through coaxial cable (cable TV) or through the atmosphere (broadcast) by frequency division multiplexing. Each TV signal is allocated a bandwidth of 6 MHz. A special form of amplitude modulation known as *vestigial sideband AM* (see Exercise 6.24) is employed for the modulation of the black-and-white video signal. Color information is modulated by a second carrier system. It is shifted to higher frequencies so that there is no overlap. Finally, the spectrum of the audio portion of the TV signal is shifted further to the right by frequency modulation. The composite signal occupies a bandwidth less than 6 MHz.

### 6.6.4   Time division multiplexing (TDM)

Besides FDM, shared use of a channel is accomplished by *time division multiplexing*. As the term implies time is divided into slots and each slot is allocated to a different signal. TDM relies on *pulse amplitude modulation* (PAM) or *switched sampling*. A switch is closed at regular intervals and is kept closed for a fixed duration. Pulse amplitude modulation uses the periodic rectangular pulse $p_T(t)$ as a carrier signal. The PAM modulated signal is $y(t) = P_T(t)u(t)$ where $p_T(t)$ has period $T$ and duration $\Delta$. In accordance with Example 5.6, the Fourier series of $p_T(t)$ is

$$p_T(t) = \sum_{k=-\infty}^{\infty} a_k e^{jk\omega_p t}, \qquad \omega_p = \frac{2\pi}{T}$$

The Fourier coefficients are

$$a_0 = \frac{\Delta}{T}, \quad a_k = \frac{\sin(k\omega_p \Delta/2)}{\pi k}$$

The PAM signal $y(t)$ is expressed as

$$y(t) = \sum_{k=-\infty}^{\infty} a_k e^{jk\omega_p t} u(t)$$

and is illustrated in Fig. 6.30. Suppose that the input $u(t)$ is bandlimited with bandwidth $2\omega_M$. The frequency shifting property leads to the following expression for the spectrum of the PAM signal:

$$Y(\omega) = \sum_{k=-\infty}^{\infty} a_k U(\omega - k\omega_p)$$

$Y(\omega)$ is obtained from the superposition of shifted and scaled input spectra.

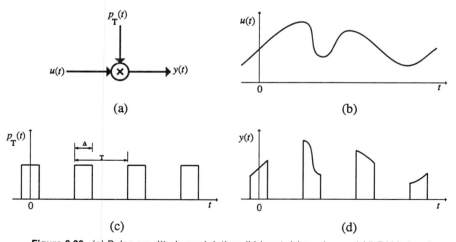

**Figure 6.30.** (a) Pulse amplitude modulation, (b) input, (c) carrier, and (d) PAM signal.

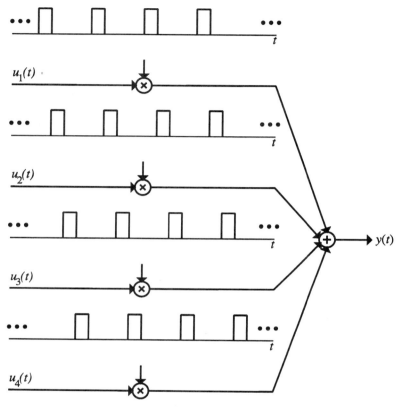

**Figure 6.31.** Time division multiplexing.

Extraction of $U(\omega)$ from $Y(\omega)$ is possible if

$$\omega_M < \omega_p - \omega_M \quad \text{or} \quad 2\omega_M < \omega_p = \frac{2\pi}{T} \tag{6.92}$$

The above inequality does not allow spectra overlapping. When Eq. (6.92) is satisfied, reconstruction of $u(t)$ from $y(t)$ is achieved by a lowpass filter of cutoff frequency $\omega_c$ in the interval $\omega_M < \omega_c < \omega_p - \omega_M$ and scaling by $1/a_0$. Condition (6.92) imposes a constraint on the period of the pulse and the bandwidth of the modulating signal and is independent of the pulse duration. Pulse amplitude modulation is directly related to the sampling theorem proved in the next section and provides a practical approach to sampling.

Time division multiplexing utilizes pulse amplitude modulation, as illustrated in Fig. 6.31. Signals $u_1(t), \ldots, u_N(t)$ are modulated by shifted rectangular pulses of period $T$ and duration $\Delta$. Each pulse carrier is off during a period of duration $T - \Delta$, and the resulting PAM signal is zero. During this period other signals are

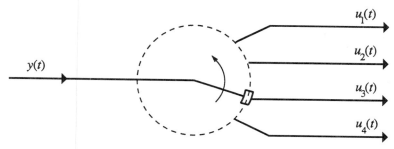

**Figure 6.32.** Reconstruction in TDM.

transmitted. The number of simultaneously served signals are $T/\Delta$. Separation of signals at the receiver is done by a switch as shown in Fig. 6.32.

## 6.7 SAMPLING

Let $x(t)$ be an analog signal with finite energy having Fourier transform $X(\omega)$. If $x(t)$ is sampled with sampling period $T_s$, the sampled signal

$$x_s(n) = x(nT_s)$$

results. Clearly $x_s(n)$ belongs to $\ell_2$. To distinguish analog and discrete signals in the frequency domain, we designate discrete frequencies by $\Omega$, and we reserve $\omega$ for analog frequencies. We explore the relation between the discrete time Fourier transform $X_s(\Omega)$ of the sampled signal $x_s(n)$ and the Fourier transform of the original analog signal $x(t)$. The sampled signal $x_s(n)$ is reconstructed from its Fourier transform by

$$x_s(n) = x(nT_s) = \frac{1}{2\pi} \int_{-\pi}^{\pi} X_s(\Omega) e^{j\Omega n} d\Omega \tag{6.93}$$

Likewise the analog signal is obtained from its Fourier transform by

$$x(t) = \frac{1}{2\pi} \int_{-\infty}^{\infty} X(\omega) e^{j\omega t} d\omega \tag{6.94}$$

In particular,

$$x(nT_s) = \frac{1}{2\pi} \int_{-\infty}^{\infty} X(\omega) e^{j\omega n T_s} d\omega \tag{6.95}$$

Comparison of (6.95) and (6.93) suggests the change of variables $\omega T_s = \Omega$. Thus

Eq. (6.95) becomes

$$x(nT_s) = \frac{1}{2\pi T_s} \int_{-\infty}^{\infty} X\left(\frac{\Omega}{T_s}\right) e^{j\Omega n} d\Omega = \frac{1}{2\pi T_s} \sum_{k=-\infty}^{\infty} \int_{(2k-1)\pi}^{(2k+1)\pi} X\left(\frac{\Omega}{T_s}\right) e^{j\Omega n} d\Omega$$

The change of variable $\Omega = \Omega - 2k\pi$ leads to

$$x(nT_s) = \frac{1}{2\pi T_s} \int_{-\pi}^{\pi} \sum_{k=-\infty}^{\infty} X\left(\frac{\Omega - 2k\pi}{T_s}\right) e^{j\Omega n} d\Omega \tag{6.96}$$

Comparison of Eqs. (6.93) and (6.96) gives

$$X_s(\Omega) = \frac{1}{T_s} \sum_{k=-\infty}^{\infty} X\left(\frac{\Omega + 2k\pi}{T_s}\right) \tag{6.97}$$

Hence the discrete time Fourier transform of the sampled signal is obtained from the Fourier transform of the analog signal via the following procedures:

1. Scaling, $X(\Omega) \rightarrow Y(\Omega) = X(\Omega/T_s)$.
2. Shift, $Y(\Omega) \rightarrow Y(\Omega + 2k\pi)$.
3. Superposition of the shifted spectra and scaling by $1/T_s$.

The spectrum of a bandlimited analog signal is illustrated Fig. 6.33 (a). It is zero outside the interval $[-\frac{\omega_0}{2}, \frac{\omega_0}{2}]$. Scaling is depicted in Figure 6.33 (b). Superposition of the shifts by $2\pi$ and $-2\pi$, is shown in Fig. 6.33 (c). In panel (c) shifts do not overlap because the left endpoint of the bandwidth of $Y(\Omega - 2\pi)$, exceeds the right endpoint of the bandwidth of $Y(\Omega)$:

$$\frac{\omega_0}{2} T_s \leq 2\pi - \frac{\omega_0}{2} T_s \quad \text{or} \quad \omega_0 \leq \frac{2\pi}{T_s}$$

If the above condition is satisfied, the analog spectrum is embedded in the spectrum of the discrete signal. If $\omega_0 > 2\pi/T_s$, the shifted spectra overlap, as illustrated in Fig. 6.33(d). In the latter case the analog signal is aliased and its reconstruction from the sampled signal is not feasible. Thus, if $x(t)$ is bandlimited (i.e., $X(\omega) = 0$ for $|\omega| > \omega_0/2$) and the sampling rate exceeds the bandwidth $\omega_0$ of $x(t)$, the spectrum $X(\omega)$ is preserved distortionless in the spectrum of its samples, $X_s(\Omega)$. Under these conditions the analog signal can be recovered from its samples. Indeed, Eq. (6.94) becomes due to the bandlimited nature of the signal

$$x(t) = \frac{1}{2\pi} \int_{-\pi/T_s}^{\pi/T_s} X(\omega) e^{j\omega t} d\omega \tag{6.98}$$

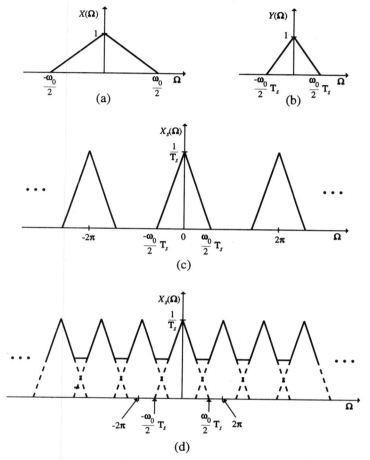

**Figure 6.33.** Sampled signal in the frequency domain: (a) Analog signal; (b) scaling; (c) shift, superposition, and scaling; (d) overlapping of spectra.

while Eq. (6.97) gives

$$X_s(\omega T_s) = \frac{1}{T_s} X(\omega), \qquad \frac{-\pi}{T_s} < \omega \le \frac{\pi}{T_s} \tag{6.99}$$

If Eq. (6.99) is substituted in (6.98), we find that

$$x(t) = \frac{T_s}{2\pi} \int_{-\pi/T_s}^{\pi/T_s} X_s(\omega T_s) e^{j\omega t} d\omega = \frac{T_s}{2\pi} \int_{-\pi/T_s}^{\pi/T_s} \sum_{n=-\infty}^{\infty} x_s(n) e^{-j\omega T_s n} e^{j\omega t} d\omega$$

or

$$x(t) = \sum_{n=-\infty}^{\infty} x(nT_s) \frac{T_s}{2\pi} \int_{-\pi/T_s}^{\pi/T_s} e^{j\omega(t-nT_s)} d\omega = \sum_{n=-\infty}^{\infty} x(nT_s) \frac{\sin[(\pi/T_s)(t-nT_s)]}{(\pi/T_s)(t-nT_s)}$$

or

$$x(t) = \sum_{n=-\infty}^{\infty} x(nT_s) \operatorname{sinc}\left(\frac{t-nT_s}{T_s}\right) \tag{6.100}$$

We have thus proved the following important result:

**Theorem 6.4 Sampling theorem** If $x(t)$ is a bandlimited signal (i.e., $X(\omega) = 0$ for $|\omega| > \omega_0/2$) and the sampling rate satisfies the *Nyquist criterion*

$$\omega_0 \leq \frac{2\pi}{T_s}$$

the signal is reconstructed from its samples according to formula (6.100).

The frequency $2\pi/T_s$ is called *Nyquist frequency*. The sampling theorem certifies that a bandlimited signal is faithfully represented by its samples if it is sampled at a sufficiently high rate, higher than the signal bandwidth.

The sampling theorem admits a straightforward extension to the multidimensional case. The 2-D analog of Eq. (6.100) is

$$x(t_1, t_2) = \sum_{n_1=-\infty}^{\infty} \sum_{n_2=-\infty}^{\infty} x(n_1 T_{1s}, n_2 T_{2s}) \operatorname{sinc}\left(\frac{t_1 - n_1 T_{1s}}{T_{1s}}\right) \operatorname{sinc}\left(\frac{t_2 - n_2 T_{2s}}{T_{2s}}\right)$$

The above representation is valid provided that $X(\omega_1, \omega_2) = 0$ for $|\omega_1| \geq W_1/2$, $|\omega_2| \geq W_2/2$ and $W_1 \leq 2\pi/T_{1s}$, $W_2 \leq 2\pi/T_{2s}$.

Equation (6.100) represents an interpolation formula. Using the shift and scaling properties of the Fourier transform and Example 6.7 we obtain the frequency domain form

$$X(\omega) = \sum_{n=-\infty}^{\infty} x(nT_s) e^{-jT_s\omega n} T_s H_{\mathrm{lp}}(\omega)$$

or

$$X(\omega) = T_s X_s(T_s\omega) H_{\mathrm{lp}}(\omega) \tag{6.101}$$

where $H_{\mathrm{lp}}(\omega)$ is the frequency response of the ideal lowpass filter with cutoff frequency $\pi/T_s$. Equation (6.101) states that the analog bandlimited signal is reconstructed from its samples by lowpass filtering.

Practical sampling is achieved by the switched sampling scheme described is Section 6.6.4 or by zero-order sampling. Zero-order sampling relies on the interpolation formula (6.100). Let $h_0(t)$ denote the causal rectangular pulse of duration $T_s : h_0(t) = 1$ for $0 \leq t \leq T_s$ and zero elsewhere. The analog signal produced by a zero-order hold is expressed as

$$\hat{x}(t) = \sum_{n=-\infty}^{\infty} x(nT_s)h_0(t - nT_s) \tag{6.102}$$

We seek to determine a reconstruction filter with frequency response $H(\omega)$ that recovers the analog waveform $x(t)$ from $\hat{x}(t)$. Then

$$H(\omega)\hat{X}(\omega) = X(\omega) \tag{6.103}$$

Equation (6.102) is written in the frequency domain as

$$\hat{X}(\omega) = \sum x(nT_s)e^{-nT_s\omega}H_0(\omega) = X_s(T_s\omega)H_0(\omega)$$

Example 6.4 and the time shift property give

$$H_0(\omega) = e^{j\omega T_s/2}\left[\frac{2\sin(\omega T_2/2)}{\omega}\right]$$

Combining (6.103) and (6.101), we obtain

$$H(\omega) = \frac{X(\omega)}{\hat{X}(\omega)} = \frac{T_s H_{\mathrm{lp}}(\omega)}{H_0(\omega)}$$

The Riemann-Lebesgue lemma asserts that most signals are approximately bandlimited. Audio signals extend from 20 Hz to 20 kHz. The corresponding Nyquist rate is 40 kHz. Digital speech of high quality is produced at rates higher than 8 kHz, or 8000 speech samples per second. CD digital audio systems use sampling rates higher than 44.1 kHz and PCM with 16 bits per sample.

To achieve exact bandlimited behavior digital processing systems utilize a lowpass filter for preprocessing as in Fig. 6.34. In voice telephone transmission the cutoff frequency of the lowpass filter is set at 4 kHz. If the sampling rate of the A/D is sufficiently high aliasing effects are canceled. The 2-D case is entirely analogous. Antialiasing in an imaging system is realized by a lens and the scanning aperture that converts an optical image into an electric signal.

### Example 6.31  *Digital telephone transmission*
Transmission of speech signals over telephone lines is predominantly digital. The speech encoder consists of an antialiasing lowpass filter and a A/D converter. The filter limits the frequency content of the speech signal below 3400 Hz. The

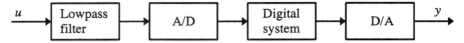

**Figure 6.34.** Digital processing system.

sampling rate is 8 kHz or higher, in accordance with the Nyquist criterion. An 8 bit quantizer is then employed. PCM and DPCM are the most widely used methods. In North America PCM with nonuniform quantization and the $\mu$ law with $\mu = 255$ have been adopted. In international communications the $A$ law compandor has been adopted with $A = 87.6$. PCM achieves a rate of 64 kb/s. Adaptive PCM has been adopted by CCITT as an international standard for speech coding at 32 kb/s. It employs a 4 bit quantizer and a logarithmic compressor. The coefficients are adapted by the LMS algorithm.

Each subscriber is connected to a switching center (exchange) by the local loop or subscriber loop. Switching centers that directly support subscribers are known as *end offices*. Direct connection between every pair of end offices is impractical, and intermediate exchanges are used. Branches that link exchanges are referred to as *trunks*. Trunks carry digitized speech signals of many subscribers using time division multiplexing. ∎

## BIBLIOGRAPHICAL NOTES

Frequency analysis of signals and systems is covered in most books devoted to signals and systems that are cited in the list of references. A detailed account is given in Papoulis (1962) and Bracewell (1986). A rigorous treatment of Fourier transform can be found in Dym and McKean (1972) and Rudin (1970). Frequency analysis of multidimensional signals is provided in Dudgeon and Mersereau (1984) and Lim (1990). The presentation of distributions follows the introductory exposition of Kolmogorov and Fomin (1970). The frequency domain representation of Volterra systems is discussed in Schetzen (1980). Some delicate issues are rigorously treated by Boyd (1985). Time varying systems in the frequency domain are described in Zadeh and Desoer (1963).

## PROBLEMS

**6.1.** Following the analysis of Section 6.1.3, develop an algorithm and a MATLAB implementation of the inverse Fourier transform.

**6.2.** Compute the Fourier transform of the following signals:
(a) $(e^{-at} \cos \omega_0 t) u_s(t)$, $a > 0$
(b) $e^{-a|t|} \sin bt$, $a > 0$
(c) $(\sin \pi t / \pi t)(\sin 2\pi(t - 1) / \pi(t - 1))$
(d) $x(t)$, as in Fig. 6.35.

(e) $e^{-2t}(u_s(t+3) - u_s(t-2))$
(g) $x(t)$, as in Fig. 6.36.

Verify your answers with **MATLAB** simulation, and sketch the signals in the time and frequency domain.

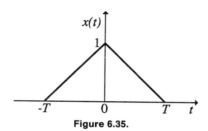

**Figure 6.35.**

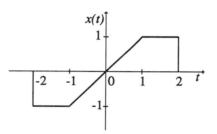

**Figure 6.36.**

**6.3.** The following are Fourier transforms of analog signals. Compute each signal in the time domain.
(a) $2\sin(4(\omega - 2\pi))/(\omega - 2\pi)$
(b) $\cos(4\omega + \pi/3)$
(c) $X(\omega)$ of Fig. 6.37.

Verify your answers with **MATLAB** simulation and sketch the signals in the time and frequency domain.

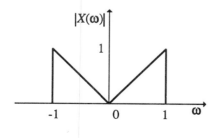
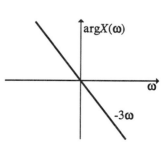

**Figure 6.37.**

**6.4.** Prove Theorem 6.1 for multichannel as well as multidimensional signals.

**6.5.** The Fourier transform $X(\omega)$ of a real even signal satisfies $\ln |X(\omega)| = -|\omega|$. Find $x(t)$. Repeat the calculation under the assumption that $x(t)$ is an odd signal.

**6.6.** Compute the convolution of the signals that follow, using the convolution property of the Fourier transform:
(a) $x(t) = te^{-3t}u_s(t)$, $h(t) = e^{-3t}u_s(t)$
(b) $x(t) = e^{-t}u_s(t)$, $h(t) = e^{t}u_s(-t)$

Verify your answers with MATLAB simulation, and sketch the signals in the time and frequency domain.

**6.7.** Show that the Hilbert transform can be viewed as a convolutional system, and find the frequency response. Organize a MATLAB experiment indicating that the real part only (or the imaginary part only) suffices to retrieve the Fourier transform of a real causal signal. Does the result extend to 2-D signals?

**6.8.** Compute the Hilbert transform of $\cos 4t$ and $\sin 2t$.

**6.9.** Let $y(t) = x(t)p(t)$. The spectrum of $x(t)$ is the triangular pulse of Fig. 6.38 with $T = 1$. Compute the spectrum of $y(t)$ for each $p(t)$:
(a) $p(t) = \cos(t/2)$
(b) $p(t) = \cos 2t$

Verify your answers with MATLAB simulation, and sketch the signals in the time and frequency domain.

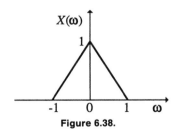

**Figure 6.38.**

**6.10.** Let

$$x(t) = \frac{\sin(3Wt/2)}{\pi t}, \quad p(t) = \cos 2Wt + 4\cos 8Wt$$

Let $h(t)$ be a periodic signal with fundamental frequency $W$. Show that the signal $y(t) = h(t) * (p(t)x(t))$ is periodic, and compute its spectral lines.

**6.11.** Determine the Fourier transform for each of the discrete signals
(a) $(\frac{1}{4})^n u_s(n+3)$
(b) $|a|^n \sin \omega_0 n, |a| < 1$
(c) $3^n u_s(-n)$
(d) $\cos(18\pi n/7) + \sin(2n)$
(e) $n(\frac{1}{2})^{|n|}$
(f) $x(n)$, as shown in Fig. 6.39.

Verify your answers with MATLAB simulation, and sketch the signals in the time and frequency domain.

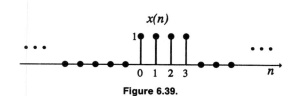

$x(n)$

**Figure 6.39.**

**6.12.** The following are Fourier transforms of discrete time signals. Determine the signals in the time domain.
(a) $X(\omega) = \sum_{k=-\infty}^{\infty} (-1)^k \delta(\omega - \pi k/2)$
(b) $X(\omega) = \sin^2 \omega$
(c) $X(\omega) = 1 - 3e^{-j4\omega} + 5e^{-j6\omega}$
(d) $X(\omega)$, as shown in Fig. 6.40.

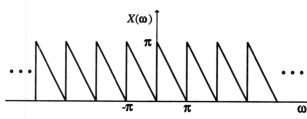

$X(\omega)$

$\pi$

$-\pi$      $\pi$        $\omega$

**Figure 6.40.**

**6.13.** Prove the properties of Table 6.1.

**6.14.** Calculate the convolution of the signals that follow with the aid of the convolution property of Fourier transform:
(a) $u(n) = (-1)^n$, $h(n) = (\frac{1}{4})^n u_s(n)$
(b) $u(n) = (n+1)(\frac{1}{2})^n u_s(n)$, $h(n) = (\frac{1}{2})^n u_s(n)$.

Verify your answers with MATLAB simulation, and sketch the signals in the time and frequency domain.

**6.15.** The deterministic crosscorrelation of real signals $x(n)$, $y(n)$ is

$$\phi_{xy}(n) = \sum_{k=-\infty}^{\infty} x(n+k)y(k)$$

Show that $\Phi_{xx}(\omega) \geq 0$ for each $\omega$. Express $\Phi_{xy}(\omega)$ in terms of $X(\omega)$ and $Y(\omega)$.

**6.16.** Determine the 2-D Fourier transform of the following signals:
(a) $x(n_1, n_2) = \delta(n_1 - 2)\delta(n_2 + 3)$
(b) $x(n_1, n_2) = (\frac{1}{2})^{n_2} \cos(2\pi n_1/3)u_s(n_2)$
(c) $x(n_1, n_2) = \cos(\pi n_2/3 + 2\pi n_1/5)$.

**6.17.** Consider the causal exponential signal $x(t) = e^{at}u_s(t)$. Compute the phase and magnitude of the finite time window approximation

$$x_{T_2}(\omega) = \int_0^{T_2} x(t)e^{-j\omega t}\,dt$$

Analyze the errors as $T_2$ goes to infinity. Determine the magnitude and phase estimates obtained by the algorithm of Section 6.1.3. Derive Taylor series approximations with respect to $T_s$. Analyze their behavior near $W$.

**6.18.** Let $y(n) = x(n)p(n)$. The spectrum of $x(n)$ is the periodic triangular pulse of Fig. 6.41. Compute the spectrum of $y(n)$ for the following choices of $p(n)$:
(a) $p(n) = \cos(\pi n)$
(b) $p(n) = \cos(\pi n/2)$.

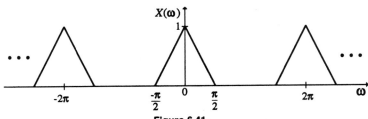

**Figure 6.41.**

**6.19.** Using the Fourier transform, compute the output of the analog systems with input $u(t) = \cos 2\pi t + \sin 6\pi t$ and impulse responses:
(a) $h(t) = \sin 4\pi t/\pi t$
(b) $h(t) = (\sin 4\pi t)(\sin 8\pi t)/\pi t^2$
(c) $h(t) = (\sin 4\pi t)(\cos 8\pi t)/\pi t$.

**6.20.** Consider the analog system with impulse response $h(t) = (\sin 2\pi t)/\pi t$. Determine the output for each of the inputs:
(a) $u(t)$ of Fig. 6.42.
(b) $u(t) = 2/(1 + t^2)$.

Check your answers with MATLAB.

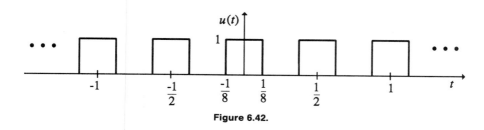

**Figure 6.42.**

**6.21.** Plot the logarithm of the magnitude and the phase for each of the following signals described in the frequency domain
(a) $1 + \frac{1}{2}e^{-j\omega}$
(b) $1 - 2e^{-j\omega}$
(c) $1/[(1 - \frac{1}{4}e^{-j\omega})(1 + \frac{1}{2}e^{-j\omega})]$

Check your answers with MATLAB.

**6.22.** In sinusoidal amplitude modulation the input to the modulator with bandwidth $2\omega_M$ is converted to a signal of bandwidth $4\omega_M$. Single sideband modulation requires a bandwidth of only $2\omega_M$ and thus offers a better channel utilization. The upper and lower sidebands are illustrated in Fig. 6.43. The spectrum of the modulating signal is shown in Fig. 6.43(a). Sinusoidal AM with carrier frequency $\omega_c$ leads to the modulated signal of Fig. 6.43(b). The two upper sidebands are depicted in Fig. 6.43(c), and the two lower sidebands are indicated in Fig. 6.43(d). Show that the lower sidebands are eliminated by the highpass filter

$$H(\omega) = u_s(-\omega - \omega_c) + u_s(\omega - \omega_c) = \begin{cases} 1, & |\omega| > \omega_c \\ 0, & \text{otherwise} \end{cases}$$

as in Fig. 6.44. Prove that the time domain representation of the signal $y_2(t)$ with spectrum plotted in Fig. 6.43(c) is given by

$$y_2(t) = u(t) \cos \omega_c t - \hat{u}(t) \sin \omega_c t$$

where $\hat{u}(t)$ is the Hilbert transform of $u(t)$. Prove that demodulation of the SSB AM signal can be carried out as in the case of sinusoidal AM.

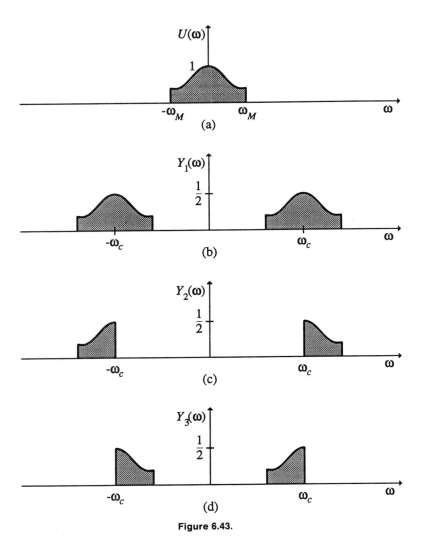

**Figure 6.43.**

**6.23.** Describe sinusoidal modulation and pulse amplitude modulation in discrete time and derive similar conclusions with the analog case.

**6.24.** *Vestigial sideband AM.* The characteristics of the highpass filter in single sideband AM modulation described in Exercise 6.22 (see also Fig. 6.44) are very difficult to implement in practice. At the expense of a slight increase of bandwidth a portion of the unwanted sideband called a *vestige* is allowed. The generation of the vestigial sideband (VSM) AM signal is shown in Fig. 6.45. The demodulator is the same as in sinusoidal AM. It is depicted in Fig. 6.46. Prove that the message signal is undistorted in the interval $|\omega| \le W$, if the VSB filter $H(\omega)$ satisfies

$$H(\omega - \omega_c) + H(\omega + \omega_c) = \text{constant}, \qquad |\omega| \le W$$

Deduce that $H(\omega)$ must have the characteristic of Fig. 6.47.

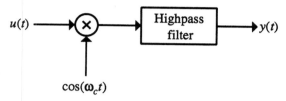

**Figure 6.44.** Single sideband (SSB) AM signal.

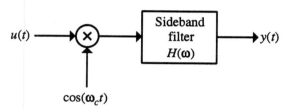

**Figure 6.45.** Vestigial sideband AM signal.

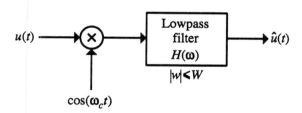

**Figure 6.46.** Vestigial sideband AM demodulator.

**6.25.** *Frequency modulation FM.* In frequency modulation the input signal modifies the frequency of the carrier. The FM modulator is described by the following state space model:

$$\dot{\theta}(t) = \omega_c + k_f u(t)$$
$$y(t) = \cos \theta(t)$$

The modulated signal $y(t)$ results from the carrier signal $\theta(t)$ whose derivative is linearly modified by the input $u(t)$. The nonlinear form of the output equation makes the analysis of the FM modulator more complicated. Prove that the output produced by the input $u(t) = E \cos \omega_m t$ is

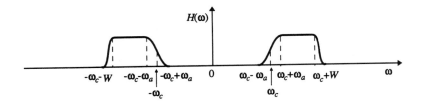

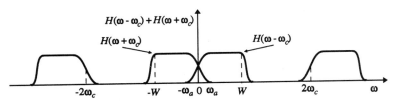

**Figure 6.47.** Characteristic of the VSB filter.

given by

$$y(t) = \cos \omega_c t \cos\left(\frac{k_f E}{\omega_m}\sin \omega_m t\right) - \sin \omega_0 t \sin\left(\frac{k_f E}{\omega_m}\sin \omega_m t\right)$$

Prove that $y(t)$ is periodic with spectral lines at integer multiples of $\omega_m$. The factor $m = k_f E/\omega_m$ is called *modulation index*. Narrowband FM results when $m$ is small. In this case prove that $y(t)$ is approximately given by $y_a(t) = \cos \omega_c t - m(\sin \omega_m t)\sin \omega_c t$. Plot the spectrum of $y_a(t)$. It can be shown that the lines with essential spectral content cover a bandwidth of $10\omega_m$ Hz. In FM broadcasting the carrier frequencies are separated by 200 kHz.

**6.26.** Let $h_l(n)$ be the impulse response of a lowpass discrete time filter. Show that the signal $h_h(n) = (-1)^n h_l(n)$ is the impulse response of a highpass filter.

**6.27.** The frequency response of a 2-D ideal lowpass filter with circular passband is

$$H(\omega_1, \omega_2) = \begin{cases} 1, & \omega_1^2 + \omega_2^2 \leq R^2 < \pi^2 \\ 0, & \text{elsewhere in the interval } [-\pi,\pi]\times[-\pi,\pi] \end{cases}$$

Using

$$\omega^2 = \omega_1^2 + \omega_2^2, \quad \phi = \tan^{-1}\frac{\omega_2}{\omega_1}, \quad \theta = \tan^{-1}\frac{n_2}{n_1}$$

show that

$$h(n_1, n_2) = \frac{1}{2\pi} \int_0^R \omega J_0\left(\omega\sqrt{n_1^2 + n_2^2}\right) d\omega = \frac{R}{2\pi} \frac{J_1\left(R\sqrt{n_1^2 + n_2^2}\right)}{\sqrt{n_1^2 + n_2^2}}$$

where $J_0(x)$, $J_1(x)$ are the Bessel functions of the first kind of order 0 and 1, respectively. Derive plots using MATLAB.

**6.28.** Conduct a MATLAB experiment to demonstrate that the phase spectrum of an image contains more intelligibility than the magnitude spectrum. *Hint*: Load an image; compute the phase and magnitude spectrum and their inverse Fourier transform. Compare the results with the original image.

**6.29.** Derive a set of linear equations for the computation of the image intensity pixels given that the phase spectrum is known. What is the complexity for $256 \times 256$ level images?

**6.30.** Derive formula (6.60) for the decimator output spectrum using the results of Section 6.4.2. Repeat for the interpolator.

**6.31.** Use representation (6.83) and periodicity of the Green function $r(n, k)$ to derive the formula

$$Y(\omega) = \sum_{m=0}^{N-1} \left(\sum_i \hat{r}(i, m)e^{-j\omega i}\right) U\left(\omega - \frac{2\pi m}{N}\right) = \sum_m r_m(\omega) U\left(\omega - \frac{2\pi m}{N}\right)$$

where $\hat{r}(i, m)$ denote the Fourier coefficients of $r(i, k)$. Develop a diagram representation for a periodically time varying linear system.

**6.32** Write a MATLAB program that computes the response of a Volterra system of degree 2 with finite support kernels produced by an exponential input. Use the program to validate Eq. (6.87).

**6.33.** Following the analysis of the Appendix at the end of the chapter, show that scaling of a distribution $\lambda$ by a scalar $a$ is defined by

$$C_a\lambda(u) = \frac{1}{|a|}\lambda(C_{a^{-1}}(u))$$

where $C_a(x(t)) = x(at)$. In particular, show that $C_a(\delta) = \delta/|a|$.

**6.34.** Following the analysis of the Appendix at the end of the chapter, show that the product of an infinitely differentiable function $f(t)$ and a distribution $\lambda$ is defined by

$$f\lambda_x(u) = \int_{-\infty}^{\infty} u(t)[f(t)x(t)]dt = (f\lambda)(u) = \lambda(fu)$$

In particular, show that $f\delta = f(0)\delta$.

## APPENDIX

Some important operations of generalized functions are defined in this appendix.

**Shift.** To define the shift of a distribution, we first consider regular distributions. Let $x(t)$ be a locally integrable signal and $\lambda_x$ the corresponding generalized function. Let $\tau \in R$ and $z(t) = q^\tau x(t) = x(t - \tau)$. Clearly $z(t)$ is locally integrable, and the corresponding generalized function is

$$\lambda_z(u) = \int_{-\infty}^{\infty} u(t)z(t)dt = \int_{-\infty}^{\infty} u(t)x(t - \tau)dt = \int_{-\infty}^{\infty} u(t + \tau)x(t)dt$$

or $\lambda_z(u) = \lambda_x(q^{-\tau}u)$. Motivated by the above computation, we define the $\tau$ shift of the distribution $\lambda$ by

$$q^\tau \lambda(u) = \lambda(q^{-\tau}u) \qquad (6.104)$$

The shift of the Dirac function is computed as follows:

$$q^\tau \delta(u) = \delta(q^{-\tau}u) = u(0 + \tau) = u(\tau) \qquad (6.105)$$

**Differentiation.** The generalized function of the derivative of a signal $x(t)$ is

$$\lambda_{\dot{x}}(u) = \int_{-\infty}^{\infty} u(t)\dot{x}(t)dt = u(t)x(t)\Big|_{-\infty}^{\infty} - \int_{-\infty}^{\infty} \dot{u}(t)x(t)dt$$

The first term is zero because the test signal $u(t)$ is zero outside a finite interval. The second term is well defined because $u(t)$ is differentiable. Thus $\lambda_{\dot{x}}(u) = -\lambda_x(\dot{u})$. We observe that the right-hand side is meaningful, even if the original signal $x(t)$ is not differentiable. Therefore, when an ordinary signal is viewed as a distribution, it can be differentiated. On the basis of the above discussion, we define the derivative $\dot{\lambda}$ of a distribution $\lambda$ as $\dot{\lambda}(u) = -\lambda(\dot{u})$. More generally, the derivative of order $k$ is

$$\lambda^{(k)}(u) = (-1)^k \lambda(u^{(k)}) \qquad (6.106)$$

It can be shown that the latter equation defines a distribution.

The derivative of the Dirac function is $\dot{\delta}(u) = -\delta(\dot{u}) = -\dot{u}(0)$. In general,

$$\delta^{(k)}(u) = (-1)^k \delta(u^{(k)}) = (-1)^k u^{(k)}(0)$$

The unit step $u_s(t)$ is not differentiable because of the jump at the origin.

Clearly it is locally integrable. The corresponding distribution is

$$\lambda_{u_s}(u) = \int_{-\infty}^{\infty} u(t)u_s(t)dt = \int_0^{\infty} u(t)dt$$

The derivative of the unit step is in accordance with (6.106),

$$\dot{\lambda}_{u_s}(u) = -\int_0^{\infty} \dot{u}(t)dt = u(0) = \delta(u)$$

where we made use of the assumption that $u(t)$ has finite support. We have proved the useful identity

$$\frac{d}{dt}u_s = \delta \tag{6.107}$$

**Convolution.** The convolution of a generalized function and an ordinary signal often arises. Let $R$ denote the reflection operator $R(u)(t) = u(-t)$, and $q^t$ the $t$ shift $q^t(u)(\tau) = u(\tau - t)$. The convolution is written

$$y(t) = h(t) * u(t) = \int_{-\infty}^{\infty} h(\tau)(q^t Ru)(\tau)d\tau = \lambda_h(q^t Ru)$$

The last equality requires that $h$ is locally integrable and $u$ is a test signal. Motivated by this observation we proceed to the following definition. Let $\lambda$ be a distribution and $u(t)$ a test signal. The convolution $y = \lambda * u$ is the ordinary signal

$$y(t) = (\lambda * u)(t) = \lambda(q^t Ru) \tag{6.108}$$

Notice that $y(t)$ is always an ordinary signal rather than a generalized function. A consequence of the above definition is $\delta * u = u$. Indeed,

$$(\delta * u)(t) = \delta(q^t Ru) = (q^t R)(u)(0) = u(t)$$

Equation (6.108) can be extended to more general signals, but the proof is beyond the scope of the book.

**Unified framework of discrete and analog signals via distributions.** Distributions provide a unified framework for the description of analog and discrete signals. We have seen how to represent an analog locally integrable signal by a generalized function. In a similar fashion a discrete signal $x(n)$ is represented by the distribution

$$\Gamma_x = \sum_{n=-\infty}^{\infty} x(n)\delta(t - n), \quad \Gamma_x(u) = \sum_{n=-\infty}^{\infty} x(n)u(n)$$

for each test function $u$. Since $u$ has finite support, the sum is finite and $\Gamma_x$ is well defined. A natural consequence of the above representation is that the distribution associated with the unit sample signal coincides with the Dirac function. Indeed,

$$\Gamma_{\delta(n)}(u) = \sum_{n=-\infty}^{\infty} \delta(n)u(n) = u(0) = \delta(u)$$

# 7

---

# SERIES REPRESENTATION OF
# DISCRETE CAUSAL SIGNALS

The series representation of a discrete causal signal views the underlying sequence as a series. The series formalism enables the use of algebraic tools in the study of signals. It is particularly useful in the characterization of signals represented by rational series, namely series written as ratios of polynomials. We will demonstrate that these signals are obtained by the superposition of combinatorial exponentials. Along the same spirit, discrete and digital signals generated by linear recurrent sources are analyzed, and their fine structure is revealed. Application of these ideas to the analysis of linear feedback shift register sequences is discussed. Convolution is represented by series multiplication. Cyclic convolution corresponds to multiplication modulo a certain polynomial. Based on these representations fast convolution and fast Fourier transform algorithms are discussed. When the indeterminate is replaced by a complex variable and the series is evaluated on the set of complex numbers, the $z$ transform results. That topic is discussed in the next chapter.

## 7.1 BASIC PROPERTIES

Let $x(n)$ denote a signal defined on the set of nonnegative integers $Z^+$ and taking values in a field $F$. The *series* representation in an indeterminate variable $q$ of the signal $x(n)$ is

$$x(q) = \sum_{n=0}^{\infty} x(n)q^n \tag{7.1}$$

We view the above as a formal expression in $q$, namely as an alternative way of writing the sequence $x(n)$, that enables us to introduce algebraic notions like series multiplication and rationality.

*Example 7.1   Finite support signals are polynomials in q*
A signal is represented by a polynomial if and only if it is a finite support signal. Thus, if $x(n) = 0$ for $n \geq M + 1$, then $x(q) = \sum_{n=0}^{M} x(n)q^n$. The unit sample shift $x(n) = \delta(n - k)$ is such an example with $x(q) = q^k$.                                            ∎

### 7.1.1   The convolution property and the representation of convolutional systems

The series representation paints the convolution operation with an algebraic color. Let $h(n)$ and $u(n)$ be two signals with values in the field $F$. The series representation of the causal convolution

$$y(n) = (h * u)(n) = \sum_{i=0}^{n} h(n - i)u(i) \tag{7.2}$$

is the product

$$y(q) = h(q)u(q) \tag{7.3}$$

Indeed

$$y(q) = \sum_{n=0}^{\infty} \sum_{i=0}^{n} h(n - i)u(i)q^n = \sum_{i=0}^{\infty} \left( \sum_{n=i}^{\infty} h(n - i)q^n \right) u(i)$$

The change of variables $n - i \leftarrow j$ in the inner sum leads to

$$y(q) = \sum_{i=0}^{\infty} \left( \sum_{j=0}^{\infty} h(j)q^{i+j} \right) u(i) = \sum_{i=0}^{\infty} \left( \sum_{j=0}^{\infty} h(j)q^j \right) u(i)q^i = h(q)u(q)$$

As we recall from Chapter 3, the input-output representation of a causal convolutional system driven by a causal input is given by (7.2). Let $h(q)$ and $u(q)$ be the series of the impulse response and the input. The convolution property (7.3) states that the output series is the product of the input series and the impulse response series.

To emphasize that a convolutional representation of a system is a transformation $u \rightarrow y$ characterized by $h(q)$, we use the notation

$$y(n) = h(q)u(n), \quad h(q) = \sum_{n=0}^{\infty} h(n)q^n$$

This notation is heavily used in Chapter 12.

## 7.1.2 Rational series

We say that the series $x(q)$ associated with the signal $x(n)$ is *rational*, if there are real polynomials $a(q)$ and $b(q)$ such that

$$\left(\sum_{n=0}^{N} a(n)q^n\right)\left(\sum_{n=0}^{\infty} x(n)q^n\right) = \sum_{n=0}^{M} b(n)q^n$$

and we write $x(q) = b(q)/a(q)$. The signal series $x(q)$ is called *proper rational* (*strictly proper rational*) if the degree of $b(q)$ does not exceed (is strictly less than) the degree of $a(q)$.

### Example 7.2   Exponential signal
The series representation of the exponential signal $x(n) = a^n$ is given by the strictly proper rational expression

$$x(q) = \sum_{n=0}^{\infty} a^n q^n = \frac{1}{1 - aq}$$

Indeed,

$$(1 - aq)\sum_{n=0}^{\infty} a^n q^n = \sum_{n=0}^{\infty} a^n q^n - \sum_{n=0}^{\infty} a^{n+1} q^{n+1} = 1 \qquad \blacksquare$$

### Example 7.3   Series representation of periodic signals
Suppose that $x(n)$ is periodic on the interval $0 \le n < \infty$ with period $N$. The series representation of $x(n)$ is strictly proper rational and is given by

$$x(q) = \frac{1}{1 - q^N} \sum_{n=0}^{N-1} x(n)q^n \qquad (7.4)$$

Conversely, if the latter expression is valid, the signal is periodic with period $N$. Indeed, division of $n$ by $N$ gives unique $k$, $r$, $0 \le r \le N - 1$, such that $n = kN + r$. Thus

$$x(q) = \sum_{n=0}^{\infty} x(n)q^n = \sum_{r=0}^{N-1}\sum_{k=0}^{\infty} x(kN + r)q^{kN+r} = \sum_{r=0}^{N-1} x(r)q^r \sum_{k=0}^{\infty} (q^N)^k$$

Exactly as in the treatment of the exponential signal, we can show that the geometric series $\sum_{k=0}^{\infty}(q^N)^k$ is equal to $1/(1 - q^N)$. The converse follows easily. More generally, if the series of a signal $x(n)$ satisfies

$$x(q) = \frac{1}{1 - q^N} x_0(q) \qquad (7.5)$$

for some polynomial $x_0(q)$ of degree less than $N$, then $x(n)$ is periodic of period $N$ and the coefficients of $x_0(q)$ give the truncation of $x(n)$ within one period.

Consider the causal periodic square wave of period $N$. In a single period the signal equals one for $N_1$ consecutive times and zero for the remaining $N - N_1$ instants. According to formula (7.4) the series is given by

$$\frac{1}{1 - q^N} \sum_{n=0}^{N_1 - 1} q^n = \frac{1}{1 - q^N} \frac{q^{N_1} - 1}{q - 1} \qquad \blacksquare$$

### 7.1.3 Series derivative

The derivative of the series $x(q) = \sum_{n=0}^{\infty} x(n)q^n$ is defined as

$$\dot{x}(q) = \frac{d}{dq}x(q) = \sum_{n=1}^{\infty} nx(n)q^{n-1} = \sum_{n=0}^{\infty} (n+1)x(n+1)q^n \qquad (7.6)$$

The signal associated with $\dot{x}(q)$ is $(n+1)x(n+1)$. More generally, the $m$th derivative of $x(q)$ is

$$\frac{d^m}{dq^m}x(q) \leftrightarrow (n+1)(n+2)\cdots(n+m)x(n+m) \qquad (7.7)$$

***Example 7.4    Combinatorial exponential signal***
We have seen that the exponential $a^n$ is represented by $1/(1 - aq)$. Formal differentiation of $1/(1 - aq)$ gives $a/(1 - aq)^2$. Thus

$$\sum_{n=0}^{\infty} (n+1)a^{n+1}q^n = \frac{a}{(1 - aq)^2} \quad \text{or} \quad \sum_{n=0}^{\infty} (n+1)a^n q^n = \frac{1}{(1 - aq)^2}$$

Consequently the signal $(n+1)a^n$ is represented by the series $1/(1 - aq)^2$. Differentiating twice, we obtain

$$\sum_{n=0}^{\infty} \frac{1}{2}(n+1)(n+2)a^n q^n = \frac{1}{(1 - aq)^3}$$

Repeated differentiation leads to the expressions

$$x(n) = \binom{n+m}{m}a^n, \quad x(q) = \frac{1}{(1 - aq)^{m+1}} \qquad (7.8)$$

$\blacksquare$

### 7.1.4   Multichannel series and the matrix exponential signal

If $x(n)$ is an $m \times k$ matrix valued signal with entries $x_{ij}(n)$, its series expansion in a single indeterminate $q$ is given by Eq. (7.1). Thus $x(q)$ can be viewed as an $m \times k$ matrix whose $ij$ entry is the series $x_{ij}(q)$ of the signal $x_{ij}(n)$.

**Example 7.5   A matrix valued signal**
Let

$$x(n) = \begin{bmatrix} u_s(n) & \left(\dfrac{1}{2}\right)^n \\ 0 & (3)^n \end{bmatrix}$$

The nonzero entries are exponential signals. Hence

$$x(q) = \begin{bmatrix} \dfrac{1}{1-q} & \dfrac{1}{1-q/2} \\ 0 & \dfrac{1}{1-3q} \end{bmatrix}$$   ∎

The series of a multichannel signal is not always derived from its entries so readily. An important example is the matrix exponential signal $x(n) = A^n$. Comparison with the scalar case leads us to examine the matrix rational function $(I - qA)^{-1}$ as a possible candidate. Thus

$$\sum_{n=0}^{\infty} A^n q^n (I - qA) = \sum_{n=0}^{\infty} A^n q^n - \sum_{n=0}^{\infty} A^{n+1} q^{n+1} = I$$

Hence

$$x(n) = A^n, \quad x(q) = (I - qA)^{-1}$$

## 7.2   PARTIAL FRACTION EXPANSION

### 7.2.1   Single-channel signals

Let $x(n)$ be a signal with values in a field $F$ and $x(q)$ its series representation. Suppose that $x(q)$ is rational:

$$x(q) = \frac{b(q)}{a(q)} \tag{7.9}$$

We assume that the constant term of $a(q)$ is one. Following the Appendix at the end of the chapter, we factor the denominator as

$$a(q) = (1 - \lambda_1 q)^{p_1} (1 - \lambda_2 q)^{p_2} \cdots (1 - \lambda_m q)^{p_m} \qquad (7.10)$$

where $1/\lambda_1$, $1/\lambda_2$, $\ldots$, $1/\lambda_m$ are the distinct roots of $a(q)$ in an extension field, each occurring $p_1$, $p_2$, $\ldots$, $p_m$ times, respectively. Each factor $(1 - \lambda_i q)^{p_i}$ is reminiscent of the denominator in the series of the combinatorial exponential signal. This section aims to show that $x(n)$ is obtained by the superposition of such exponentials.

**Theorem 7.1**   Let $x(q)$ be given by Eq. (7.9) with the denominator factor as in Eq. (7.10). Then $x(q)$ is uniquely decomposed as

$$x(q) = r_{00} + q^1 r_{01} + \cdots + q^{p_0} r_{0p_0}$$

$$+ \frac{1}{1 - \lambda_1 q} r_{11} + \frac{1}{(1 - \lambda_1 q)^2} r_{12} + \cdots + \frac{1}{(1 - \lambda_1 q)^{p_1}} r_{1p_1} + \cdots$$

$$\vdots$$

$$+ \frac{1}{1 - \lambda_m q} r_{m1} + \frac{1}{(1 - \lambda_m q)^2} r_{m2} + \cdots + \frac{1}{(1 - \lambda_m q)^{p_m}} r_{mp_m}$$

or compactly

$$x(q) = \sum_{j=0}^{p_0} q^j r_{0j} + \sum_{i=1}^{m} \sum_{j=1}^{p_i} \frac{1}{(1 - \lambda_i q)^j} r_{ij} \qquad (7.11)$$

The coefficients $r_{0j}$, $0 \leq j \leq p_0$ correspond to the quotient polynomial obtained from the division of $b(q)$ by $a(q)$. In particular, if $x(q)$ is strictly proper, these coefficients are zero. The remaining coefficients belong to the extension field spanned by the roots of $a(q)$. They are computed by the formula

$$r_{ij} = \frac{(-1)^{p_i - j}}{(p_i - j)! \lambda_i^{p_i - j}} \frac{d^{p_i - j}}{dq^{p_i - j}} \left[ (1 - \lambda_i q)^{p_i} x(q) \right] \Bigg|_{q = 1/\lambda_i} \qquad (7.12)$$

The proof is supplied in the Appendix at the end of the chapter.   ∎

A series of the form $1/(1 - \lambda_i q)^j$ is called *partial fraction*. Equation (7.11) is referred to as partial fraction expansion of $x(q)$. The first index $i$ enumerates the nonzero roots of the denominator, and the second index the multiplicities of each pole. In particular, if $\lambda_i^{-1}$ is a simple root (i.e., $p_i = 1$), for each $i$ and $x(q)$ is strictly proper, then we set $r_i = r_{i1}$ and obtain

$$x(q) = \sum_{i=1}^{m} \frac{1}{1 - \lambda_i q} r_i, \quad r_i = (1 - \lambda_i q) x(q)|_{q = 1/\lambda_i} \qquad (7.13)$$

Each term of the form $q^i$ in Eq. (7.11) corresponds to the $i$ shift of the unit sample signal, in accordance with Example 7.1. Moreover the fraction $1/(1 - \lambda_i q)^j$ describes the combinatorial exponential signal, according to Example 7.4. Thus we have established the representation

$$x(n) = \sum_{j=0}^{p_0} \delta(n-j)r_{0j} + \sum_{i=1}^{m} \sum_{j=1}^{p_i} \binom{n+j-1}{j-1} \lambda_i^n r_{ij} \qquad (7.14)$$

If $x(q)$ is strictly proper and the roots $\lambda_i^{-1}$ are simple, we obtain the simpler form

$$x(n) = \sum_{i=1}^{m} \lambda_i^n r_i \qquad (7.15)$$

The coefficients $r_{ij}$ can alternatively be computed by repeated divisions. Indeed, consider the rational function

$$\frac{b(q)}{(1 - q\lambda)^m}, \qquad \deg b = k < m$$

We perform the successive divisions

$$b(q) = a_0(q)(1 - q\lambda) + r_m$$

$$a_0(q) = a_1(q)(1 - q\lambda) + r_{m-1}$$

$$\vdots$$

$$a_{m-2}(q) = a_{m-1}(q)(1 - q\lambda) + r_1$$

Observe that $\deg a_0(q) = k - 1$, $\deg a_1(q) = k - 2$, ..., $\deg a_{m-2}(q) = k - (m-1)$, $\deg a_{m-1}(q) = k - m < 0$. Thus $a_{m-1}(q) = 0$. If we divide the first equation by $(1 - q\lambda)^m$, the second by $(1 - q\lambda)^{m-1}$, and so forth, the $m$th equation by $(1 - q\lambda)$, we establish the claim.

**Example 7.6   A rational signal**
Let

$$x(q) = \frac{3q - 2}{\frac{1}{8}q^2 - \frac{3}{4}q + 1}$$

The roots of the denominator are $\lambda_1^{-1} = 2$ and $\lambda_2^{-1} = 4$. Thus $x(q)$ is factored as

$$x(q) = \frac{3q - 2}{(1 - \frac{1}{2}q)(1 - \frac{1}{4}q)} \qquad (7.16)$$

Partial fraction expansion gives

$$x(q) = \frac{1}{1 - \frac{1}{2}q} r_1 + \frac{1}{1 - \frac{1}{4}q} r_2 \tag{7.17}$$

where

$$r_1 = \left(1 - \frac{1}{2}q\right) x(q)\Big|_{q=2} = \frac{3q - 2}{1 - \frac{1}{4}q}\Big|_{q=2} = 8$$

$$r_2 = \left(1 - \frac{1}{4}q\right) x(q)\Big|_{q=4} = \frac{3q - 2}{1 - \frac{1}{2}q}\Big|_{q=4} = -10$$

Hence

$$x(n) = 8\left(\frac{1}{2}\right)^n - 10\left(\frac{1}{4}\right)^n, \quad n \geq 0 \qquad \blacksquare$$

**Example 7.7    A rational signal with multiple pole**
Let

$$x(q) = \frac{q^3}{(1 - q)^2}$$

If we perform division, we find that

$$x(q) = q + 2 + \frac{3q - 2}{(1 - q)^2}$$

Partial fraction expansion gives

$$x(q) = q + 2 + \frac{r_{11}}{1 - q} + \frac{r_{12}}{(1 - q)^2}$$

where

$$r_{12} = x(q)(1 - q)^2\Big|_{q=1} = (q + 2)(1 - q)^2 + (3q - 2)\Big|_{q=1} = 1$$

$$r_{11} = (-1)\frac{d}{dq}\left(x(q)(1 - q)^2\right)\Big|_{q=1} = -3$$

Thus

$$x(n) = \delta(n - 1) + 2\delta(n) - 3u_s(n) + (n + 1)u_s(n) \qquad \blacksquare$$

**Root cancellations.** If a root of the denominator, $1/\lambda_i$, is also a root of the numerator, some of the coefficients $r_{ij}$ vanish. For instance, if $1/\lambda_i$ is a simple root of $a(q)$ and $b(q)$, then $r_i = 0$, by Eq. (7.13).

### 7.2.2 Multivariable partial fraction expansion

Let us consider a $K \times L$ matrix-valued rational series

$$x(q) = \begin{bmatrix} x_{11}(q) & x_{12}(q) & \cdots & x_{1L}(q) \\ x_{21}(q) & x_{22}(q) & \cdots & x_{2L}(q) \\ \vdots & \vdots & & \vdots \\ x_{K1}(q) & x_{K2}(q) & \cdots & x_{KL}(q) \end{bmatrix}$$

Let $a_{kl}(q)$ denote the denominator polynomial of $x_{kl}(q)$, and $a(q)$ their least common multiple. Then $x(q)$ is factored as follows:

$$x(q) = \frac{1}{a(q)} \begin{bmatrix} b_{11}(q) & b_{12}(q) & \cdots & b_{1L}(q) \\ b_{21}(q) & b_{22}(q) & \cdots & b_{2L}(q) \\ \vdots & \vdots & & \vdots \\ b_{K1}(q) & b_{K2}(q) & \cdots & b_{KL}(q) \end{bmatrix}$$

Let $1/\lambda_1, 1/\lambda_2, \ldots, 1/\lambda_m$ be the nonzero roots of $a(q)$ with multiplicities $p_1, p_2, \ldots, p_m$. We assume that the constant term of $a(q)$ is nonzero, so $a(q)$ can be factored as in (7.10). We apply partial fraction expansion to each rational function $b_{kl}(q)/a(q)$ to obtain

$$\frac{b_{kl}(q)}{a(q)} = \sum_{j=0}^{p_0(kl)} r_{0j}^{(kl)} q^j + \sum_{i=1}^{m} \sum_{j=1}^{p_i} \frac{1}{(1 - \lambda_i q)^j} r_{ij}^{(kl)} \tag{7.18}$$

Let $p_0 = \max(p_0(kl))$. For each $i, j$ let $R_{ij}$ be the matrix whose $kl$-entry is $r_{ij}^{(kl)}$. Then Eq. (7.18) is written in matrix form

$$x(q) = \sum_{j=0}^{p_0} R_{0j} q^j + \sum_{i=1}^{m} \sum_{j=1}^{p_i} \frac{1}{(1 - \lambda_i q)^j} R_{ij} \tag{7.19}$$

The matrices $R_{ij}$ are determined by the matrix analog of Eq. (7.12):

$$R_{ij} = \frac{(-1)^{p_i-j}}{(p_i - j)! \lambda_i^{p_i-j}} \frac{d^{p_i-j}}{dq^{p_i-j}} \left[ (1 - \lambda_i q)^{p_i} x(q) \right] \Big|_{q=1/\lambda_i}$$

Conversion in the time domain gives

$$x(n) = \sum_{j=0}^{p_0} \delta(n-j) R_{0j} + \sum_{i=1}^{m} \sum_{j=1}^{p_i} \binom{n+j-1}{j-1} \lambda_i^n R_{ij} \qquad (7.20)$$

**Real signals.** If $x(n)$ is real, the coefficients $R_{0j}$ are real matrices because they are obtained from the division of $b(q)$ by $a(q)$. On the other hand, the parameters $\lambda_i$ are, in general, complex, and hence the corresponding coefficients $R_{ij}$ are complex. If $1/\lambda_k$ is complex, its complex conjugate $1/\lambda_k^*$ is also a root of the same multiplicity. Suppose that $\lambda_k^* = \lambda_s$. It follows from (7.20) that $R_{kj}^* = R_{sj}$, $j = 1, 2, \ldots, p_k = p_s$. As a result all imaginary terms in (7.20) vanish and produce a real signal.

**Digital signals.** Suppose that $x(n)$ is a digital rational signal of the form (7.9) with values in the field $GF(p)$. Consider the irreducible factorization of $a(q)$, as described in the Appendix at the end of the chapter, $a(q) = f_1^{k_1}(q) \cdots f_m^{k_m}(q)$. Each irreducible factor $f_i(q)$ has coefficients in $GF(p)$ and degree $d_i$. Let us drop the subscript $i$ for simplicity. If $1/\lambda$ is a root of $f$, $f$ is the minimal polynomial of $1/\lambda$ (see Appendix II). The remaining roots of $f$ are $1/\lambda^p$, $1/\lambda^{p^2}$, $\ldots$, $1/\lambda^{p^{d-1}}$. The expansion of $x(n)$ involves terms of the form

$$r_1 \lambda^n + r_2 \lambda^{pn} + r_3 \lambda^{p^2 n} + \cdots + r_d \lambda^{(p^{d-1})n}$$

For each $n$, $x(n)$ is a field integer; hence $x^p(n) = x(n)$. Using Lemma 5 of Appendix II, we obtain

$$\left[ \sum_{i=1}^{d} r_i \lambda^{np^{i-1}} \right]^p = \sum_{i=1}^{d} r_i^p \lambda^{np^i}$$

Therefore

$$r_2 = r_1^p, \quad r_3 = r_2^p = r_1^{p^2}, \quad \ldots, \quad r_d = r_1^{p^{d-1}}$$

All coefficients associated with the conjugate roots of $\lambda$ are determined from $r_1$.

**Example 7.8     A digital rational signal**
Consider a binary-valued signal with rational transform

$$x(q) = \frac{1}{q^4 + q + 1}$$

The denominator is irreducible. The roots are $\lambda$, $\lambda^2$, $\lambda^4$, and $\lambda^8$ in the field $GF(2^4)$. Note that $\lambda^{-1} = \lambda^{14}$, $\lambda^{-2} = \lambda^{13}$, $\lambda^{-4} = \lambda^{11}$, and $\lambda^{-8} = \lambda^7$. Hence we write

$$q^4 + q + 1 = (1 + \lambda^7 q)(1 + \lambda^{11} q)(1 + \lambda^{13} q)(1 + \lambda^{14} q)$$

Partial fraction expansion gives

$$x(q) = \frac{r_1}{1 + \lambda^{14} q} + \frac{r_2}{1 + \lambda^{13} q} + \frac{r_3}{1 + \lambda^{11} q} + \frac{r_4}{1 + \lambda^7 q}$$

where

$$r_1 = x(q)(1 + \lambda^{14} q)|_{q=\lambda^1} = [(1 + \lambda^8)(1 + \lambda^{12})(1 + \lambda^{14})]^{-1} = \lambda^{14}$$

$$r_2 = x(q)(1 + \lambda^{13} q)|_{q=\lambda^2} = [(1 + \lambda^9)(1 + \lambda^{13})(1 + \lambda^{16})]^{-1} = \lambda^{13} = r_1^2$$

$$r_3 = x(q)(1 + \lambda^{11} q)|_{q=\lambda^4} = [(1 + \lambda^{11})(1 + \lambda^2)(1 + \lambda^3)]^{-1} = \lambda^{11} = r_2^2$$

$$r_4 = x(q)(1 + \lambda^7 q)|_{q=\lambda^8} = [(1 + \lambda^{19})(1 + \lambda^{21})(1 + \lambda^{22})]^{-1} = \lambda^7 = r_3^2$$

We conclude that

$$x(n) = \lambda^{7n+7} + \lambda^{11n+11} + \lambda^{13n+13} + \lambda^{14n+14}$$

Although each term in the right-hand side is an element of $GF(2^4)$, the sum is 0 or 1. To see this, check the equation $x^2(n) = x(n)$. ∎

### 7.2.3 Analysis of the matrix exponential signal

**Complex exponential.** We saw that the series of the matrix exponential signal $A^n$, $n \geq 0$, is $(I - qA)^{-1}$; $A$ is a square $k \times k$ matrix. Let us recall the basic inversion formula

$$(I - qA)^{-1} = \frac{1}{\det(I - qA)} \, \text{adj}(I - qA) \qquad (7.21)$$

The denominator is factored as

$$a(q) = \det(I - qA) = (1 - \lambda_1 q)^{p_1}(1 - \lambda_2 q)^{p_2} \cdots (1 - \lambda_m q)^{p_m}$$

where $\lambda_1, \lambda_2, \ldots, \lambda_m$, are the nonzero eigenvalues of $A$. The $ij$ entry of the $k \times k$ matrix $\text{adj}(I - qA)$ is $(-1)^{i+j}$ times the determinant resulting after the $i$ column and the $j$ row of $(I - qA)$ are removed; hence it is a polynomial of degree at most $k - 1$. We conclude that $(I - qA)^{-1}$ is a rational function. Partial fraction expansion gives

$$A^n = \sum_{j=0}^{p_0} \delta(n-j) R_{0j} + \sum_{i=1}^{m} \sum_{j=1}^{p_i} \binom{n+j-1}{j-1} \lambda_i^n R_{ij} \qquad (7.22)$$

$$R_{ij} = \frac{(-1)^{p_i-j}}{(p_i-j)!\lambda_i^{p_i-j}} \frac{d^{p_i-j}}{dq^{p_i-j}} \left[ (1-\lambda_i q)^{p_i} (I-qA)^{-1} \right] \Bigg|_{q=1/\lambda_i} \qquad (7.23)$$

The coefficients $R_{0j}$ are zero if 0 is not included in the eigenvalues of $A$. In the opposite case the polynomial $a(q)$ has degree $k - p_0$, and $x(q)$ is not proper. If $A$ has nonzero simple eigenvalues, Eq. (7.22) becomes

$$A^n = \sum_{i=1}^{m} \lambda_i^n R_i, \quad R_i = (1-\lambda_i q)(I-qA)^{-1} \Bigg|_{q=1/\lambda_i} \qquad (7.24)$$

The representation of the matrix exponential in terms of combinatorial exponentials leads to a complete characterization of the signals produced by the linear source $x(n+1) = Ax(n)$. Suppose that $A$ is a $k \times k$ matrix and that $x_0$ is an arbitrary initial $k \times l$ matrix. The solution emanating at time $n = 0$ from $x_0$ is given by $x(n) = A^n x_0$. Thus

$$x(n) = \sum_{j=0}^{p_0} \delta(n-j) R_{0j} x_0 + \sum_{i=1}^{m} \sum_{j=1}^{p_i} \binom{n+j-1}{j-1} \lambda_i^n R_{ij} x_0 \qquad (7.25)$$

The coefficients satisfy Eq. (7.23). Formula (7.25) is of fundamental importance. It confirms that linear combinations of unit sample shifts and combinatorial exponentials signals are the only type of signals that can be generated by a linear signal source. Signals like $1/n$ or $\log n$ cannot be observed at the output of a linear generator.

A further remark we deduce from (7.25) is that the eigenvalues of $A$ play a neuralgic role in the qualitative behavior of the solutions of the linear difference equation. Representation (7.25) complements the preliminary analysis carried out in Section 2.5.1.

### Example 7.9  A real matrix exponential signal
Let

$$A = \begin{bmatrix} \dfrac{1}{2} & 1 \\ 0 & 3 \end{bmatrix}$$

Since $A$ is upper triangular, the eigenvalues are the diagonal entries. Both are nonzero and simple. Thus $A^n = (1/2)^n R_1 + (3)^n R_2$. Now

$$I - qA = \begin{bmatrix} 1 - \frac{1}{2}q & -q \\ 0 & 1 - 3q \end{bmatrix}$$

$$(I - qA)^{-1} = \frac{1}{(1 - \frac{1}{2}q)(1 - 3q)} \begin{bmatrix} 1 - 3q & q \\ 0 & 1 - \frac{1}{2}q \end{bmatrix}$$

Thus

$$R_1 = \left(1 - \frac{1}{2}q\right)(I - qA)^{-1}\bigg|_{q=2} = \begin{bmatrix} 1 & \dfrac{q}{1 - 3q} \\ 0 & \dfrac{1 - q/2}{1 - 3q} \end{bmatrix}_{q=2} = \begin{bmatrix} 1 & -\dfrac{2}{5} \\ 0 & 0 \end{bmatrix}$$

$$R_2 = (1 - 3q)(I - qA)^{-1}\big|_{q=1/3} = \begin{bmatrix} 0 & 2/5 \\ 0 & 1 \end{bmatrix}$$

Therefore

$$A^n = \begin{bmatrix} \left(\dfrac{1}{2}\right)^n & -\dfrac{2}{5}\left(\dfrac{1}{2}\right)^n + \dfrac{2}{5}\,3^n \\ 0 & 3^n \end{bmatrix}, \quad n \geq 0 \qquad \blacksquare$$

**Digital exponential.** If the entries of $A$ are in $GF(p)$, formula (7.22) remains valid, except that the polynomial $\det(I - qA)$ is now factored over an extension of $GF(p)$. We illustrate the procedure with the following example:

**Example 7.10  A digital matrix exponential signal over GF(2)**
Let

$$A = \begin{pmatrix} 0 & 1 \\ 1 & 1 \end{pmatrix}$$

Then

$$I - qA = \begin{pmatrix} 1 & q \\ q & 1 + q \end{pmatrix}, \quad (I - qA)^{-1} = \frac{1}{q^2 + q + 1}\begin{pmatrix} 1 + q & q \\ q & 1 \end{pmatrix}$$

The characteristic polynomial $q^2 + q + 1$ is irreducible over $GF(2)$. We form the prime factorization in the extension field $GF(2^2)$. We pick $\lambda$ such that $\lambda^2 = \lambda + 1$. The eigenvalues of $A$ are $\lambda$ and $\lambda^2$. Note that $\lambda^3 = 1$. Hence $q^2 + q + 1 = (1 + \lambda q)(1 + \lambda^2 q)$. Thus

$$A^n = (\lambda^n)R_1 + (\lambda^{2n})R_2$$

where

$$R_1 = \frac{1}{1+\lambda^2 q}\begin{bmatrix} q+1 & q \\ q & 1 \end{bmatrix}\Bigg|_{q=\lambda^2} = \begin{bmatrix} \lambda^2 & 1 \\ 1 & \lambda \end{bmatrix}$$

$$R_2 = \frac{1}{1+\lambda q}\begin{bmatrix} q+1 & q \\ q & 1 \end{bmatrix}\Bigg|_{q=\lambda} = \begin{bmatrix} \lambda & 1 \\ 1 & \lambda^2 \end{bmatrix}$$

Thus

$$A^n = \begin{bmatrix} \lambda^{n+2} + \lambda^{2n+1} & \lambda^n + \lambda^{2n} \\ \lambda^n + \lambda^{2n} & \lambda^{n+1} + \lambda^{2n+2} \end{bmatrix}, \quad n \geq 0$$

It is easy to check that each entry is either 0 or 1.                                    ■

### 7.2.4  Convolution of rational signals

The convolution of rational signals is rational and hence admits partial fraction expansion. This is a consequence of the convolution property, which asserts that the series of convolution is the product of series.

***Example 7.11    Output series of a convolutional system***
The unit sample response of a convolutional system is

$$h(n) = a^n u_s(n)$$

We want to compute the output produced by the unit step $u(n) = u_s(n)$. We have $u(q) = 1/(1-q)$, $h(q) = 1/(1-aq)$. The output is the rational signal

$$y(q) = \frac{1}{(1-q)(1-aq)}$$

Partial fraction expansion gives

$$y(n) = \frac{a}{a-1}a^n u_s(n) + \frac{1}{1-a}u_s(n), \qquad a \neq 1 \tag{7.26}$$

We check the above calculations by the following **MATLAB** exercise. Set $a = \frac{1}{2}$.

```
n = 0:300;
u = ones(1,length(n));      % unit step
y = 2*u-(1/2).^n;           % output determined by Eq. (7.26)
h = (1/2).^n;               % impulse response
y1 = conv(h,u);             % output of the convolutional system
plot(n,y,n,y1(1:n + 1))
```
■

### 7.2.5   Interconnections of convolutional systems and invertibility

The product of series is associative and distributive over addition. In the single-channel case it is commutative. Moreover the unit sample signal acts as an identity. These properties are easy to establish and are left as exercises. A restatement of the associative property is that the impulse response of the cascade connection of two convolutional systems with impulse responses $h_1(n)$ and $h_2(n)$ is given by the convolution $h_2(n) * h_1(n)$. Similarly the impulse response of the parallel connection is $h_2(n) + h_1(n)$.

A single-input single-output system with unit sample response $h(n)$ has a causal convolutional inverse if and only if there is a causal signal $\tilde{h}(n)$, so that $h(q)\tilde{h}(q) = 1$, or $h(n) * \tilde{h}(n) = \delta(n)$. Specifically,

$$h(0)\tilde{h}(0) = 1$$
$$h(0)\tilde{h}(1) + h(1)\tilde{h}(0) = 0$$
$$\vdots$$
$$h(0)\tilde{h}(n) + h(1)\tilde{h}(n-1) + \cdots + h(n)\tilde{h}(0) = 0$$

The above equations are solvable with respect to $\tilde{h}(n)$ if and only if $h(0) \neq 0$, namely 0 is not a zero of $h(q)$. Then $\tilde{h}(0) = 1/h(0)$ and recursively

$$\tilde{h}(n) = -\frac{1}{h(0)}\left[\sum_{k=1}^{n} h(k)\tilde{h}(n-k)\right]$$

The series of the impulse response of the inverse system is $\tilde{h}(q) = h^{-1}(q)$. If $h(q)$ is rational, $h(q) = b(q)/a(q)$, then $h^{-1}(q) = a(q)/b(q)$.

**Example 7.12   *Invertible and noninvertible systems***
Let $h(n) = \delta(n-1)$. Then $h(q) = q$. The system has no causal inverse because $h(0) = 0$. In contrast, the rational system

$$h(q) = \frac{q-2}{q^2 + 2q + 1}$$

satisfies $h(0) = -2$. The inverse exists and is given by

$$h^{-1}(q) = \frac{1 + 2q + q^2}{q - 2} \qquad \blacksquare$$

## 7.3   SERIES REPRESENTATION OF LINEAR FINITE DIFFERENCE MODELS

This section is concerned with the series representation of linear finite difference models. We start with the left and right shift properties.

**Left shift property.** Let $l$ be a positive integer. Suppose that a matrix-valued signal $x(n)$ is specified over the interval $-l \leq n \leq \infty$. Consider the $l$-shift $x^{(l)}(n) = x(n-l)$, $n \geq 0$. Then

$$x^{(l)}(q) = q^l x(q) + \sum_{i=1}^{l} x(-i) q^{l-i}$$

If $l = 1$, the above formula becomes $x^{(1)}(q) = qx(q) + x(-1)$.

The proof is a direct consequence of the definitions. Indeed,

$$x^{(l)}(q) = \sum_{n=0}^{\infty} x(n-l) q^n = \sum_{i=-l}^{\infty} x(i) q^{i+l} = q^l \left[ \sum_{i=-l}^{-1} x(i) q^i + \sum_{i=0}^{\infty} x(i) q^i \right]$$

The *right shift property* is analogous. The series of the right shift $x^{(-l)}(n) = x(n+l)$ satisfies

$$q^l x^{(-l)}(q) = x(q) - \sum_{i=0}^{l-1} x(i) q^i$$

In particular, $qx^{(-1)}(q) = x(q) - x(0)$.

***Example 7.13    A linear finite difference source***
A discrete single-channel signal $x(n)$ is produced by the source

$$x(n) = -\frac{5}{6} x(n-1) + \frac{1}{6} x(n-2)$$

and initial data $x(-2) = 0$, $x(-1) = 1$. The shift property gives

$$x(q) = -\frac{5}{6} [qx(q) + x(-1)] + \frac{1}{6} [q^2 x(q) + x(-2) + x(-1)q]$$

We group together all terms involving $x(q)$ and solve with respect to $x(q)$ to obtain

$$x(q) = \frac{1}{1 + \frac{5}{6}q - \frac{1}{6}q^2} \left[ -\frac{5}{6} x(-1) + \frac{1}{6} x(-2) + \frac{1}{6} x(-1)q \right]$$

$$= \frac{1}{1 + \frac{5}{6}q - \frac{1}{6}q^2} \left[ -\frac{5}{6} + \frac{1}{6} q \right]$$

The roots of the denominator are -1 and 6. Hence

$$x(n) = -\frac{6}{7}(-1)^n + \frac{1}{42} \left( \frac{1}{6} \right)^n, \quad n \geq 0 \qquad \blacksquare$$

The procedure outlined in the previous example is fairly general as is indicated next.

### 7.3.1   Series representation of linear time invariant difference models

Consider the difference model

$$y(n) = -\sum_{l=1}^{k} A_l y(n - l) + \sum_{l=0}^{r} B_l u(n - l) \qquad (7.27)$$

Let $\mathbf{y}_0$ be the vector specifying the $k$ initial output samples

$$\mathbf{y}_0 = \left( y(-k)^T \quad y(-k + 1)^T \quad \cdots \quad y(-1)^T \right)^T \qquad (7.28)$$

Let $\mathbf{u}_0$ denote the vector of initial input values at times $-r, -r + 1, \ldots, -1$. We consider the $q$ series of both sides of (7.27). The shift property gives

$$y(q) = -\sum_{l=1}^{k} A_l \left[ q^l y(q) + \sum_{i=1}^{l} y(-i) q^{l-i} \right]$$

$$+ B_0 u(q) + \sum_{l=1}^{r} B_l \left[ q^l u(q) + \sum_{i=1}^{l} u(-i) q^{l-i} \right]$$

or

$$\left( I + \sum_{l=1}^{k} A_l q^l \right) y(q) + \sum_{i=1}^{k} \left( \sum_{l=i}^{k} A_l q^{l-i} \right) y(-i)$$

$$= \left( \sum_{l=0}^{r} B_l q^l \right) u(q) + \sum_{i=1}^{r} \left( \sum_{l=i}^{r} B_l q^{l-i} \right) u(-i) \qquad (7.29)$$

Let us define the polynomial matrices

$$A(q) = I + \sum_{l=1}^{k} A_l q^l$$

$$B(q) = \sum_{l=0}^{r} B_l q^l$$

$$A_i(q) = \sum_{l=i}^{k} A_l q^{l-i}, \qquad 1 \le i \le k$$

$$B_i(q) = \sum_{l=i}^{r} B_l q^{l-i}, \qquad 1 \le i \le r$$

Then Eq. (7.29) becomes

$$y(q) = -\sum_{i=1}^{k} A^{-1}(q)A_i(q)y(-i) + \sum_{i=1}^{r} A^{-1}(q)B_i(q)u(-i) + A^{-1}(q)B(q)u(q)$$

The latter equation is compactly written as

$$y(q) = -h^y(q)\mathbf{y}_0 + h^u(q)\mathbf{u}_0 + h(q)u(q) \tag{7.30}$$

where

$$h^y(q) = A^{-1}(q)[A_k(q)\, A_{k-1}(q)\, \cdots\, A_1(q)]$$

$$h^u(q) = A^{-1}(q)[B_r(q)\, B_{r-1}(q)\, \cdots\, B_1(q)]$$

$$h(q) = A^{-1}(q)B(q) \tag{7.31}$$

An alternate useful formula is obtained if the right-hand side member of Eq. (7.29) is rearranged so that it has a series format. The following expression is readily established:

$$y(q) = -A^{-1}(q)A_y(q) + A^{-1}(q)B_u(q) + A^{-1}(q)B(q)u(q) \tag{7.32}$$

$$A_y(q) = \sum_{i=0}^{k-1}\left(\sum_{l=i+1}^{k} A_l y(i-l)\right)q^i, \quad B_u(q) = \sum_{i=0}^{r-1}\left(\sum_{l=i+1}^{r} B_l u(i-l)\right)q^i$$

Formula (7.30) characterizes the output of the system in terms of initial data and the input. Some remarkable conclusions can be drawn from this equation. First of all, the output is decomposed into a sum of three terms: The first term on the right side of (7.30) depends only on the initial output data. It provides the output when the input is zero. The second term involves only initial input values and vanishes when the input is causal. The third term captures the main effect of the input and represents the output when initial conditions are zero. The function $h(q)$ is the *transfer function* of the finite difference model and characterizes the system only if initial conditions are zero. It is useful to interpret Eq. (7.30) with the aid of Fig. 3.19. The first term results at the output when the input register is set to zero, no input data enter the processor, and only the output register is loaded with initial content. The second term is obtained at the output when the output register is set to zero, the input is inactive, and the input register is initially loaded. The third term results when both registers are initially zero. For obvious reasons the polynomial $A(q)$ is called *characteristic polynomial* of the recursive equation.

**Example 7.14   First-order system**

Let us consider the single-input single-output system

$$y(n) + ay(n-1) = b_0 u(n) + b_1 u(n-1)$$

We have

$$A_1(q) = A_1 = a \qquad\qquad B_1(q) = B_1 = b_1$$

$$A(q) = 1 + A_1 q = 1 + aq \qquad B(q) = B_0 + B_1 q = b_0 + b_1 q$$

Thus

$$h^y(q) = \frac{a}{1 + aq} \qquad h^u(q) = \frac{b_1}{1 + aq} \qquad h(q) = \frac{b_0 + b_1 q}{1 + aq}$$

and the output is

$$y(q) = -\frac{a}{1+aq} y(-1) + \frac{b_1}{1+aq} u(-1) + \frac{b_0 + b_1 q}{1 + aq} u(q) \qquad\blacksquare$$

**Characterization of unit sample response.** If the initial vectors $y_0$ and $u_0$ are zero, the input-output description of the system has a convolutional form, and in addition the transfer function $h(q)$ is rational. Therefore the unit sample response $h(n)$ is determined via partial fraction expansion.

$$h(n) = \sum_{j=0}^{p_0} \delta(n-j) R_{0j} + \sum_{i=1}^{m} \sum_{j=1}^{p_i} \binom{n+j-1}{j-1} \lambda_i^n R_{ij} \qquad (7.33)$$

Thus a signal qualifies as unit sample response of a linear finite difference model only if it is a linear combination of combinatorial exponential signals.

Alternatively, the impulse response can be computed as the output of (7.27) resulting from $\delta(n)$. Then

$$h(n) + \sum_{j=1}^{k} A_j h(n-j) = \sum_{i=0}^{r} B_i \delta(n-i), \qquad (7.34)$$

Let $M = \min(k, r)$. Calculation at instants $n = 0, 1, \ldots, r$ gives

$$h(0) = B_0$$

$$h(1) = -A_1 h(0) + B_1$$

$$h(2) = -A_1 h(1) - A_2 h(0) + B_2 \qquad\qquad (7.35)$$

$$\vdots$$

$$h(r) = -A_1 h(r-1) - A_2 h(r-2) - \cdots - A_M h(r-M) + B_r$$

These expressions specify the first $r + 1$ values of $h(n)$. For $n \geq r + 1$ the unit sample input does not affect the computation of $h(n)$ in Eq. (7.34), and thus

$$h(n) = -\sum_{j=1}^{k} A_j h(n - j), \qquad n \geq r + 1 \tag{7.36}$$

We can easily infer from (7.36) and (7.35) that the impulse response is not a finite support signal unless all $A_i$ are zero. This explains why linear time invariant finite difference models are referred to as infinite impulse response (IIR) filters.

**Example 7.15    Computation of impulse response**
Let

$$y(n) + \frac{3}{4} y(n - 1) + \frac{1}{8} y(n - 2) = u(n) + u(n - 1)$$

We have

$$A(q) = 1 + \frac{3}{4} q + \frac{1}{8} q^2, \quad B(q) = 1 + q$$

Thus

$$h(q) = A^{-1}(q) B(q) = \frac{1 + q}{1 + \frac{3}{4} q + \frac{1}{8} q^2} \tag{7.37}$$

Therefore the impulse response is

$$h(n) = -2\left(-\frac{1}{2}\right)^n + 3\left(-\frac{1}{4}\right)^n, \qquad n = 0, 1, \ldots \qquad \blacksquare$$

**Example 7.16    Multichannel impulse response**
Let

$$y(n) + Ay(n - 1) = u(n), \qquad A = \begin{bmatrix} 1 & 3 \\ 0 & 2 \end{bmatrix}$$

At the individual channel level we have

$$y_1(n) + y_1(n - 1) + 3y_2(n - 1) = u_1(n)$$
$$y_2(n) + 2y_2(n - 1) = u_2(n)$$

It holds that $k = 1$, $r = 0$, $A_1 = A$, and $B_0 = I_2$. Thus

$$A(q) = I_2 + A_1 q = \begin{bmatrix} 1 + q & 3q \\ 0 & 1 + 2q \end{bmatrix}, \qquad B(q) = B_0 = I_2$$

Furthermore $h(q) = A^{-1}(q)B(q) = A^{-1}(q)$, $\det A(q) = (1+q)(1+2q)$, and

$$A^{-1}(q) = \frac{1}{\det A(q)} \begin{bmatrix} 1+2q & -3q \\ 0 & 1+q \end{bmatrix} = \begin{bmatrix} \dfrac{1}{1+q} & \dfrac{-3q}{(1+q)(1+2q)} \\ 0 & \dfrac{1}{1+2q} \end{bmatrix}$$

Hence the impulse response is

$$h(n) = \begin{bmatrix} (-1)^n & -3(-1)^n + 3(-2)^n \\ 0 & (-2)^n \end{bmatrix}, \qquad n \geq 0$$

If the system is initially at rest and the input is $u(n) = [\delta(n-1) \quad \delta(n)]^T$, the resulting output is $y(q) = h(q)u(q)$. Therefore $u(q) = [q \quad 1]^T$ and

$$y(q) = \begin{bmatrix} \dfrac{q}{1+q} - \dfrac{3q}{(1+q)(1+2q)} \\ \dfrac{1}{1+2q} \end{bmatrix}$$

We conclude that

$$y(n) = \begin{bmatrix} \delta(n) - 4(-1)^n + 3(-2)^n \\ (-2)^n \end{bmatrix}, \qquad n \geq 0 \qquad \blacksquare$$

**Responses due to nonzero initial conditions.** The series $h^y(q)$ and $h^u(q)$ in Eq. (7.31) are rational and have the same denominator with the transfer function $h(q)$. Therefore they are expressed by the same combinatorial exponentials as $h(q)$, differing only in the coefficients $R_{ij}$.

**Example 7.17 Nonzero initial conditions**
Consider the system of Example 7.15. We have $A_2(q) = A_2 = \frac{1}{8}$, $A_1(q) = A_1 + A_2 q = \frac{3}{4} + \frac{1}{8}q$. Thus

$$h^y(q) = A^{-1}(q)[A_2(q) \quad A_1(q)] = \frac{1}{1 + \frac{3}{4}q + \frac{1}{8}q^2} \begin{bmatrix} \frac{1}{8} & \frac{3}{4} + \frac{1}{8}q \end{bmatrix}$$

The denominator is factored as $A(q) = (1 + \frac{1}{2}q)(1 + \frac{1}{4}q)$. Therefore $h^y(q)$ becomes in the time domain

$$h^y(n) = \left(-\frac{1}{2}\right)^n r_1 + \left(-\frac{1}{4}\right)^n r_2$$

where

$$r_1 = \left(1 + \frac{1}{4}q\right)^{-1} \left(\frac{1}{8} \quad \frac{3}{4} + \frac{1}{8}q\right)\Big|_{q=-2} = \left(\frac{1}{4} \quad 1\right)$$

$$r_2 = \left(1 + \frac{1}{2}q\right)^{-1} \left(\frac{1}{8} \quad \frac{3}{4} + \frac{1}{8}q\right)\Big|_{q=-4} = \left(-\frac{1}{8} \quad -\frac{1}{4}\right)$$

Let $y(-2) = 0$ and $y(-1) = 1$. Then $\mathbf{y}_0^T = (0 \quad 1)$. If the input is zero, the output is

$$y(n) = -\left(-\frac{1}{2}\right)^n + \left(\frac{1}{4}\right)\left(-\frac{1}{4}\right)^n, \quad n \geq 0$$

In a similar fashion we find $B_1(q) = 1$, $h^u(q) = A^{-1}(q)$, and

$$h^u(n) = 2(-1/2)^n - (-1/4)^n, \quad n \geq 0 \qquad \blacksquare$$

### 7.3.2   Extension of solutions

Suppose that the input to a linear time invariant single-input single-output finite difference model (7.27) is defined for all integers. Given an initial output vector $\mathbf{y}_0$, we inquire whether the resulting output is defined for all integers as well. If $A_k \neq 0$, the input-output expression can be solved with respect to $y(n-k)$, enabling the backward in time computation. The series representation of the positive and negative time trajectories can be analyzed separately and then patched together to synthesize the entire signal. We illustrate the idea for a simple first-order system, leaving the general case to the reader.

*Example 7.18   First order system*
Let

$$y(n) = ay(n-1) + bu(n), \qquad a \neq 0, \; y(-1) = y_0 \tag{7.38}$$

Let $y^+(n) = y(n)$, $n \geq 0$, denote the positive time trajectory, and $y^-(n) = y(-n-2)$, $n \geq 0$, denote the negative time trajectory. In a similar fashion let $u^+(n) = u(n)$ and $u^-(n) = u(-n-1)$, $n \geq 0$. Proceeding as in Example 7.14, we find that

$$y^+(q) = \frac{a}{1-aq}y_0 + \frac{b}{1-aq}u^+(q), \quad y^+(n) = a^{n+1}y_0 + ba^n * u^+(n)$$

If we solve Eq. (7.38) with respect to $y(n - 1)$, we obtain

$$y(-n - 1) = \frac{1}{a}y(-n) - \frac{b}{a}u(-n), \quad n \geq 0$$

$$y(-n - 2) = \frac{1}{a}y(-n - 1) - \frac{b}{a}u(-n - 1), \quad n \geq 0$$

or

$$y^-(n) = \frac{1}{a}y^-(n - 1) - \frac{b}{a}u^-(n), \quad n \geq 0$$

We arrived at an equation of the form (7.38) with $a$ replaced by $a^{-1}$ and $b$ replaced by $-ba^{-1}$. Hence

$$y^-(q) = \frac{a^{-1}}{1 - a^{-1}q}y_0 - \frac{ba^{-1}}{1 - a^{-1}q}u^-(q)$$

$$y^-(n) = a^{-n-1}y_0 - ba^{-n-1} * u^-(n)$$

Patching the negative and the positive pieces together, we obtain

$$y(n) = a^{n+1}y_0 + (ba^n u_s(n) * u(n)u_s(n))$$
$$- (ba^{n+1}u_s(-n - 2) * u(n + 1)u_s(-n - 2)) \qquad (7.39)$$

∎

## 7.4 LINEAR FEEDBACK SHIFT REGISTERS

Let us consider the autoregressive part of Eq. (7.27) over a finite field:

$$y(n) = \sum_{i=1}^{k} h_i y(n - i) \qquad (7.40)$$

A linear feedback shift register (LFSR) is a device that implements the above equation. It is illustrated in Fig. 7.1. The contents of the register are initially loaded with the initial vector $(y(-k) \quad y(-k + 1) \quad \cdots \quad y(-1))$. At each cycle the content of each register stage is shifted to the right, while the first stage is fed by a linear combination of the previous values of the register stages.

### 7.4.1 Periodic structure of digital rational signals

Partial fraction expansion enables us to extract important information about the periodicity properties of digital rational signals. This is illustrated in the following theorem:

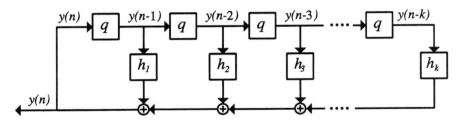

**Figure 7.1.** Linear feedback shift register.

**Theorem 7.2**    Let $x(n)$ be a strictly proper rational signal with values in the finite field $GF(p^m)$. All poles are nonzero and have multiplicities $m_1, \ldots, m_l$. The orders of the poles are $N_1, \ldots, N_l$. Then the period of $x(n)$ is given by

$$\mathrm{lcm}(N_1 p^r, N_2 p^r, \ldots, N_l p^r)$$

where the integer $r$ is such that $p^{r-1} \le \max m_i \le p^r$.

The proof is given in the Appendix at the end of the chapter. ∎

If $x(q)$ is rational but not strictly proper, the unit sample shifts are present in the partial fraction representation of $x(n)$ and at least one of the coefficients $r_{0j}$ is nonzero. The term $\sum_{j=0}^{p_0} \delta(n - j) r_{0j}$ describes a finite support signal that appears as a transient characteristic in the signal. After $p_0 + 2$ time instants, it dies out. Consequently the signal is not periodic, but eventually periodic, namely $x(n + N) = x(n)$, for $n \ge p_0 + 2$. The sequence repeats itself indefinitely once an initial time interval is neglected. The output sequences of linear feedback shift registers are strictly proper rational. Hence their periodicity structure is fully explained by the previous theorem.

***Example 7.19    Linear feedback shift register of length 4***
Let us consider the linear feedback shift register (LFSR) of Fig. 7.2. It implements the equation

$$y(n) = y(n - 1) + y(n - 2) + y(n - 4)$$

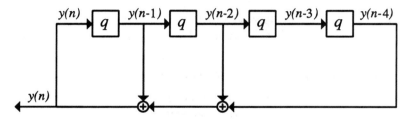

**Figure 7.2.** Linear feedback shift register in Example 7.19.

The characteristic polynomial is $a(q) = 1 + q + q^2 + q^4$. It is factored as $a(q) = (1 + q)(1 + q^2 + q^3)$. The roots of both factors are embedded in the extension field $GF(2^3)$ generated by the polynomial $q^3 + q + 1$ and the primitive element $a^3 = a + 1$. The roots of $a(q)$ are $1$, $a^3$, $a^5$, and $a^6$. The inverses of the roots are $\lambda_1 = 1$, $\lambda_2 = a$, $\lambda_3 = a^2$, and $\lambda_4 = a^4$. Hence

$$a(q) = (1 + q)(1 + aq)(1 + a^2 q)(1 + a^4 q)$$

The output of the linear feedback shift register has the form

$$y(n) = r_1 + r_2 a^n + r_3 a^{2n} + r_4 a^{4n}$$

Let us determine the possible periods of the output signals of the LFSR. The constant sequence $y(n) = r_1$ corresponding to the root $\lambda_1 = 1$ has period 1. As we recall from Example II.2 of Appendix II all nonzero elements of $GF(2^3)$ different from 1 are primitive. Thus their order is 7. Hence each sequence of the LFSR has period 1 if $r_2 = r_3 = r_4 = 0$, and 7 otherwise. As an example suppose that the initial content of the register is $(1 \quad 0 \quad 0 \quad 0)$. The contents at each time are

$$\begin{pmatrix} 1 & 0 & 0 & 0 \\ 1 & 1 & 0 & 0 \\ 0 & 1 & 1 & 0 \\ 1 & 0 & 1 & 1 \\ 0 & 1 & 0 & 1 \\ 0 & 0 & 1 & 0 \\ 0 & 0 & 0 & 1 \\ 1 & 0 & 0 & 0 \end{pmatrix}$$

The resulting output is given by the first column. It has period 7. ∎

## 7.4.2 Generator polynomial of a linear feedback shift register

Let us consider a linear feedback shift register with characteristic polynomial $h(q) = 1 + \sum_{i=1}^{k} h_i q^i$, $h_k \neq 0$, implementing Eq. (7.40) in $GF(p)$. Each output signal has a strictly proper rational series of the form $y(q) = \tilde{y}(q)/h(q)$, with $\deg \tilde{y}(q) < \deg h(q) = k$. The roots of $h(q)$ lie in the extension field $GF(p^k)$. By Lemma 1 (see also Remark 2) of Appendix II, $h(q)$ divides $1 - q^{p^k - 1}$. Let $g(q)$ be the quotient polynomial

$$h(q)g(q) = 1 - q^{p^k - 1} \tag{7.41}$$

Then

$$y(n) = \frac{\tilde{y}(q)}{h(q)} = \frac{\tilde{y}(q)g(q)}{1 - q^{p^k - 1}} \tag{7.42}$$

According to Example 7.3, $y(n)$ is periodic of period $p^k - 1$, although $p^k - 1$ may not be the fundamental period. Therefore the first $p^k - 1$ values of $y(n)$ are given by the coefficients of the polynomial $\tilde{y}(q)g(q)$. We conclude that the sequences produced by the LFSR are given within one period by multiples of $g(q)$. The polynomial $g(q)$ is determined from the characteristic polynomial of the register via Eq. (7.41) and is called *generator polynomial*. The outputs of the LFSR within one period can alternatively be computed by an FIR filter whose coefficients are the coefficients of $g(q)$ and whose input is a suitable polynomial.

***Example 7.20    Generator polynomial of a linear feedback shift register***
Consider the LFSR of Fig. 7.3. The characteristic polynomial is $h(q) = q^4 + q^3 + 1$. This is irreducible over $GF(2)$. The smallest $N$ such that $h(q)$ divides $1 - q^N$ is $N = 15$. Division of $1 - q^{15}$ by $h(q)$ gives

$$g(q) = q^{11} + q^{10} + q^9 + q^8 + q^6 + q^4 + q^3 + 1$$

Let $\hat{g}(n)$ denote the periodic extension of $g(n)$. Example 7.3 states that the corresponding series are related by

$$\hat{g}(q) = \frac{g(q)}{1 - q^N}$$

Comparison with Eq. (7.42) shows that $\hat{g}(q)$ is produced at the output of the LFSR if the initial condition is $\tilde{y}(q) = 1$. This also follows from Fig. 7.3. The contents of the register are listed below. The first column consists of the coefficients of $g(q)$.

$$
\begin{pmatrix}
q^0 : & 1 & 0 & 0 & 0 \\
q^1 : & 0 & 1 & 0 & 0 \\
q^2 : & 0 & 0 & 1 & 0 \\
q^3 : & 1 & 0 & 0 & 1 \\
q^4 : & 1 & 1 & 0 & 0 \\
q^5 : & 0 & 1 & 1 & 0 \\
q^6 : & 1 & 0 & 1 & 1 \\
q^7 : & 0 & 1 & 0 & 1 \\
q^8 : & 1 & 0 & 1 & 0 \\
q^9 : & 1 & 1 & 0 & 1 \\
q^{10} : & 1 & 1 & 1 & 0 \\
q^{11} : & 1 & 1 & 1 & 1 \\
q^{12} : & 0 & 1 & 1 & 1 \\
q^{13} : & 0 & 0 & 1 & 1 \\
q^{14} : & 0 & 0 & 0 & 1
\end{pmatrix}
$$

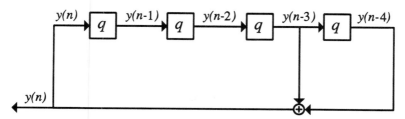

**Figure 7.3.** Linear feedback shift register in Example 7.20.

### 7.4.3 Linear feedback shift registers as pseudorandom generators

By proper selection of the characteristic polynomial of a linear feedback shift register, pseudorandom sequences whose second-order properties resemble those of white noise are produced. The precise meaning of this statement is as follows:

**Theorem 7.3** Suppose that the characteristic polynomial $h(q) = 1 + \sum_{i=1}^{k} h_i q^i$, $h_k \neq 0$, of a linear feedback shift register is primitive (see Appendix II) over $GF(2)$. Let $y(n)$ denote an arbitrary nonzero output sequence. The following statements hold:

1. $y(n)$ has maximal period $N = 2^k - 1$.
2. The Lie weight $w(y(n))$, that is, the number of ones of $y(n)$ within one period is approximately one-half: $w(y(n)) = 2^{k-1}$.
3. Any other nonzero output sequence $\tilde{y}(n)$ is obtained by a shift of $y(n)$; that is, there is $1 \leq l \leq N$ such that $\tilde{y}(n) = y(n - l)$.
4. The mean value is

$$m_y = \frac{1}{N} \sum_{n=0}^{N-1} y(n) = \frac{1}{2} + \frac{1}{2N}$$

and the covariance sequence is

$$r(i) = \frac{1}{N} \sum_{n=0}^{N-1} [y(n + i) - m][y(n) - m] = \begin{cases} \dfrac{1}{4} - \dfrac{1}{4N^2}, & i = 0 \\[2mm] \dfrac{-1}{4N} - \dfrac{1}{4N^2}, & i \neq 0 \end{cases}$$

**PROOF**

**1.** The output sequences produced by a LFSR are strictly proper rational. Therefore they are periodic with fundamental periods highlighted by Theorem 7.2. A primitive polynomial is always irreducible. Therefore all output sequences have period $N$, where $N$ is the order of a root of $h(q)$ in the extension field $GF(2^k)$. Because $h(q)$ is primitive, $N$ equals the maximum order $2^k - 1$.

**2.** All $k$-dimensional binary nonzero vectors are included in the maximum period sequence $y(n)$. It follows that $y(n)$ contains $2^{k-1}$ ones and $2^{k-1} - 1$ zeros. Maximal period sequences have a good balance of zeros and ones.

**3.** Due to linearity of Eq. (7.40), the set of output sequences forms a linear subspace. Uniqueness of solutions with respect to initial conditions implies that this vector space has dimension $k$. Hence there are precisely $2^k$ possible output sequences. Shift invariance warrants that $y(n - l)$ is also an output sequence for any $1 \le l \le N - 1$, and the conclusion follows.

**4.** The mean is $m_y = w(y(n))/N$, and the claim is a consequence of property 2. To compute the covariance sequence, we make use of the property

$$xy = \frac{1}{2}(x + y - x \oplus y)$$

where $+$, $-$ denote ordinary addition and subtraction and $\oplus$ denotes modulo 2 addition. The proof is easily inferred from the table

$$\begin{pmatrix} x & y & x+y & x \oplus y & (x+y-x \oplus y)/2 & xy \\ 0 & 0 & 0 & 0 & 0 & 0 \\ 0 & 1 & 1 & 1 & 0 & 0 \\ 1 & 0 & 1 & 1 & 0 & 0 \\ 1 & 1 & 2 & 0 & 1 & 1 \end{pmatrix}$$

Hence the Lie weight of the pointwise product $y(n + i)y(n)$ is

$$w(y(n+i)y(n)) = \frac{w(y(n+i)) + w(y(n)) - w(y(n+i) \oplus y(n))}{2}$$

The exclusive OR in the above expression is also an output sequence of the LFSR by property 2. It is the zero sequence if $i = 0$. Hence

$$w(y(n+i)y(n)) = \begin{cases} \dfrac{N+1}{2}, & i = 0 \\ \dfrac{N+1}{4}, & i \ne 0 \end{cases}$$

and the claim follows easily.                                                      ∎

The transformation $z(n) = -1 + 2y(n)$ converts the values 0 and 1 to $-1$ and 1, respectively. The statistics of $z(n)$ become

$$m_z = \frac{1}{N} \approx 0, \quad r_z(0) = 1 - \frac{1}{N^2} \approx 1, \quad r_z(i) = \frac{-1}{N} - \frac{1}{N^2} \approx \frac{-1}{N} \approx 0$$

We conclude that if $N$ is sufficiently large, $z(n)$ has the characteristics of a white noise sequence.

## 7.5 SERIES REPRESENTATION OF STATE SPACE MODELS

### 7.5.1 State space model in series form and the transfer function

Let us consider the state space model

$$x(n + 1) = Ax(n) + Bu(n) \tag{7.43}$$

$$y(n) = Cx(n) + Du(n) \tag{7.44}$$

Using the right shift property, we obtain $x(q) - x_0 = qAx(q) + qBu(q)$ or $(I - qA)x(q) = x_0 + qBu(q)$. Thus

$$x(q) = (I - qA)^{-1}x_0 + (I - qA)^{-1}qBu(q) \tag{7.45}$$

Substituting into the output equation, we find

$$y(q) = C(I - qA)^{-1}x_0 + \left[ C(I - qA)^{-1}qB + D \right]u(q) \tag{7.46}$$

where $y(q)$ is obtained from the superposition of two terms. The first term is written as $h^x(q)x_0$, where

$$h^x(q) = C(I - qA)^{-1} \tag{7.47}$$

It depends only on the initial state of the system and specifies the output when the input is set to zero. The second term has the form $h(q)u(q)$, where

$$h(q) = C(I - qA)^{-1}qB + D \tag{7.48}$$

It involves the input only and provides the output when the system is initially at rest. $h(q)$ is referred to as the *transfer function* of the state space representation. Since $(I - qA)^{-1}$ is rational, both $h(q)$ and $h^x(q)$ are rational matrices. We arrive at the above conclusions directly if we translate the variation of constants formula (Section 3.10.3) in series format.

Since $h^x(q)$ and $h(q)$ are rational, they are formed by linear combinations of combinatorial exponential signals whose parameters are characterized by the eigenvalues of $A$ and their multiplicities.

### 7.5.2 Equivalence of state space models

Consider the state space model (7.43)–(7.44) Let us see what happens when we change the coordinates by the linear invertible transformation $z = Px$. We have

$$z(n+1) = Px(n+1) = PAx(n) + PBu(n) = PAP^{-1}z(n) + PBu(n)$$

$$y(n) = CP^{-1}z(n) + Du(n)$$

Thus the state space parametrization

$$Q = \begin{pmatrix} A & B \\ C & D \end{pmatrix} \text{ converts to } \tilde{Q} = \begin{pmatrix} PAP^{-1} & PB \\ CP^{-1} & D \end{pmatrix}.$$

A remarkable property is that the transfer functions $h(q)$ and $\tilde{h}(q)$ of the two parametrizations coincide. Indeed,

$$\tilde{h}(q) = CP^{-1}(I - qPAP^{-1})^{-1}qPB + D = C(P - qPA)^{-1}qPB + D$$

$$= C(I - qA)^{-1}qP^{-1}PB + D = h(q)$$

The same transfer function $h(q)$ can be realized by an infinite number of different state space architectures of the form $\tilde{Q}$ as $P$ varies. Certain canonical types offering special features are discussed in Chapter 10.

**Example 7.21    Transfer function of a state space model**
Let

$$A = \begin{bmatrix} \frac{1}{2} & 2 \\ 0 & \frac{1}{4} \end{bmatrix}, \quad B = \begin{bmatrix} 0 \\ 1 \end{bmatrix}, \quad C = [1 \quad 1], \quad D = 0$$

Then

$$(I - qA)^{-1} = \begin{bmatrix} \dfrac{1}{1 - \frac{1}{2}q} & \dfrac{2q}{\left(1 - \frac{1}{2}q\right)\left(1 - \frac{1}{4}q\right)} \\ 0 & \dfrac{1}{1 - \frac{1}{4}q} \end{bmatrix}$$

Hence

$$h(q) = \frac{2q^2}{\left(1 - \frac{1}{2}q\right)\left(1 - \frac{1}{4}q\right)} + \frac{q}{1 - \frac{1}{4}q}$$

Likewise we have

$$h^x(q) = \begin{bmatrix} \dfrac{1}{1 - q/2} & \dfrac{(1 + 3q/2)}{(1 - q/2)(1 - q/4)} \end{bmatrix}$$

Calculation of the basic responses is easily accomplished with the control toolbox of MATLAB. The impulse response $h(n)$ from the $m$ input is determined by the command

```
h = dimpulse(A,B,C,D,m,n)
```

where $n$ specifies the time window. The response $h^x(n)$ is calculated by the command

```
hx = dlsim(A,B,C,D,zeros(1,length(n))
```

■

## 7.6  FAST CONVOLUTION ALGORITHMS

A fast convolution algorithm based on the fast Fourier transform was introduced in Section 5.9. An alternate approach based on the Chinese remainder theorem (see Appendix II) is developed in this section.

### 7.6.1  Polynomial representation of cyclic convolution

The series representation of linear convolution amounts to series multiplication. Cyclic convolution admits an algebraic representation as well. Consider the finite support signals $h(n)$ and $u(n)$ of duration $N$ and the associated polynomials $h(q)$ and $u(q)$. Let $y(q)$ denote the polynomial representing the cyclic convolution of $h(n)$ and $u(n)$. Then

$$y(q) = h(q)u(q)(\mathrm{mod}(q^N - 1)) \qquad (7.49)$$

Recall that given polynomials $a(q)$, $b(q)$, and $m(q)$, the relation $a(q) = b(q)(\mathrm{mod}\ m(q))$ means that there is polynomial $k(q)$ such that $a(q) = b(q) + k(q)m(q)$. If $\deg b(q) \leq \deg m(q) - 1$, the latter equation states that $b(q)$ is the remainder of the division of $a(q)$ by $m(q)$. Thus cyclic convolution is the remainder polynomial that results when the product $h(q)u(q)$ (linear convolution) is divided by $q^N - 1$.

To prove the statement, let $y'(q) = h(q)u(q)$. $y'(q)$ has at most degree $2N - 2$. Notice that

$$y'(q) = \sum_{n=0}^{2N-2} y'(n)q^n = \sum_{n=0}^{N-1} y'(n)q^n + \sum_{n=N}^{2N-2} y'(n)q^n$$

or $y'(q) = y_1(q) + q^N y_2(q)$, where

$$y_1(q) = \sum_{n=0}^{N-1} y'(n)q^n, \quad y_2(q) = \sum_{n=0}^{N-2} y'(N+n)q^n$$

Both the above polynomials have degree at most $N - 1$. We saw in Section 5.6.3 that $y(n) = y'(n) + y'(N + n)$, $0 \leq n \leq N - 1$. Hence

$$y(q) = y_1(q) + y_2(q)$$

Thus

$$y'(q) = y_1(q) + q^N y_2(q) = y_1(q) + q^N y_2(q) - y_2(q) + y_2(q)$$
$$= y_1(q) + y_2(q)(q^N - 1) + y_2(q)$$

or $y'(q) = y(q) + y_2(q)(q^N - 1)$.

If $N$ exceeds the sum of the degrees of $u(q)$ and $h(q)$, the multiplication modulus any polynomial of degree $N$ reduces to the product itself; in other words, the cyclic convolution reduces to the linear convolution.

***Example 7.22   Linear and cyclic convolution for signals of length 2***
Suppose that $h(n)$ and $u(n)$ have length 2 (see Example 5.14). Then

$$y(q) = h(0)u(0) + h(1)u(1) + (h(1)u(0) + h(0)u(1))q$$

$$y'(q) = h(0)u(0) + (h(0)u(1) + h(1)u(0))q + h(1)u(1)q^2$$

Thus

$$y'(q) - y(q) = -h(1)u(1) + h(1)u(1)q^2 = h(1)u(1)(q^2 - 1)$$

Therefore Eq. (7.49) holds.   ∎

### 7.6.2   Winograd short convolution algorithm

The Winograd short convolution algorithm offers a computationally attractive procedure for the determination of polynomial multiplication modulo a given polynomial. The main feature is the significant reduction of multiplications at the expense of additions. It is an elegant and direct application of the Chinese remainder theorem for polynomials. Say we wish to determine the product

$$y(q) = h(q)u(q) \ (\text{mod} \ (m(q)))$$

We consider the prime factorization of $m(q)$

$$m(q) = m_1(q)m_2(q) \ldots m_k(q)$$

and we assume that the coefficients of the prime factors $m_i(q)$ are small integers such as 0, ±1, and ±2. In the case of cyclic convolution, $m(q) = q^N - 1$, and the prime factors have small integer coefficients for values of $N$ less than 105.

We first compute the remainders of $u(q)$ and $h(q)$ by $m_i(q)$:

$$u_i(q) = u(q) \pmod{m_i(q)}$$

$$h_i(q) = h(q) \pmod{m_i(q)} \tag{7.50}$$

Then we form the products

$$y_i(q) = u_i(q)h_i(q) \pmod{m_i(q)} \tag{7.51}$$

As noted in Appendix II (see Section II.1.4), it holds that

$$y_i(q) = y(q) \pmod{m_i(q)}$$

Once the remainders of $y(q)$ by $m_i(q)$ are determined, $y(q)$ is recovered by the Chinese remainder theorem and the formula

$$y(q) = \sum_{i=1}^{k} y_i(q)a_i(q) \pmod{m(q)} \tag{7.52}$$

Equations (7.50), (7.51), and (7.52) constitute the main steps of the Winograd short convolution algorithm. The analogy with the DFT based computation of cyclic convolution is noticeable. The role of the DFT is now played by the Chinese remainder transform CR, which maps a polynomial $x(q)$ into the "lower resolution" remainders $x_i(q)$ resulting from the division of $x(q)$ by $m_i(q)$. This is a linear invertible transformation. Much like the DFT case the Chinese remainder transform and its inverse involve essentially multiplications with small integers. Such multiplications are replaced by additions. Indeed, inspection of the division algorithm for polynomials shows that the procedure involves small integers if the divider has small integer coefficients. The inverse Chinese remainder transform is essentially described by the polynomials $a_i(q)$ in (7.52). The computation of these polynomials is explained in Section II.1.4 of Appendix II. Again it is noticed that it involves small integers only, except some divisions with small integers in the Bezout identity. These can be absorbed by one of the given polynomials, say, $h(q)$. The mechanics of the algorithm are best explained by an example.

***Example 7.23  Illustration of the Winograd short conclusion algorithm***
We illustrate the algorithm with the computation of a $3 \times 2$ linear convolution. Let

$$h(q) = h_0 + h_1 q, \quad u(q) = u_0 + u_1 q + u_2 q^2$$

The linear convolution $y(q) = h(q)u(q)$ consists of 4 points. Direct computation requires 6 mps and 2 adds. The Winograd algorithm reduces the number of multiplications to 5. The product $u(q)h(q)$ modulo any polynomial of degree 4 will give $y(q)$. We pick $m(q)$ with coefficients, 0, $\pm 1$. One such choice is

$$m(q) = q(q - 1)(q^2 + 1) = m_1(q)m_2(q)m_3(q)$$

Division of $u(q)$ by $m_1(q)$, $m_2(q)$, and $m_3(q)$ gives the following remainders:

$$u_1(q) = u_0, \quad u_2(q) = u_2 + u_1 + u_0, \quad u_3(q) = u_1 q + u_0 - u_2$$

Apparently the CR transform of $u(q)$ involves additions only. In a similar fashion the CR transform of $h(q)$ gives

$$h_1(q) = h_0, \quad h_2(q) = h_0 + h_1, \quad h_3(q) = h_1 q + h_0$$

The lower resolution products are then formed:

$$y_1(q) = u_0 h_0$$
$$y_2(q) = (h_1 + h_0)(u_2 + u_1 + u_0)$$
$$y_3(q) = (h_1 q + h_0)(u_1 q + (u_0 - u_2))(\mathrm{mod}(q^2 + 1))$$

The compuation of $y_1(1)$ and $y_2(q)$ requires one multiplication each. $y_3(q)$ is

$$y_3(q) = (h_1(u_0 - u_2) + h_0 u_1)q + h_0(u_0 - u_2) - h_1 u_1$$

and seems to need four multiplications. It can be computed with only three multiplications if we notice that the two polynomial coefficients are written as

$$h_1(u_0 - u_2) + h_0 u_1 = (h_1 - h_0)(u_0 - u_2) + h_0(u_0 - u_2 + u_1)$$
$$h_0(u_0 - u_2) - h_1 u_1 = h_0(u_0 - u_2 + u_1) - (h_0 + h_1)u_1$$

The gain of one multiplication is secured if the inverse CR transform requires additions only. Using Eq. (II.6) of Appendix II, we find that

$$M_1(q) = q^3 - q^2 + q - 1, \quad M_2(q) = q^3 + q, \quad M_3(q) = q^2 - q$$

Application of the Euclidean algorithm three times gives

$$n_1(q) = q^2 - q + 1, \quad n_2(q) = -\frac{1}{2}(q^2 + q + 2), \quad n_3(q) = -\frac{1}{2}(q - 2)$$
$$N_1(q) = -1, \quad N_2(q) = \frac{1}{2}, \quad N_3(q) = \frac{1}{2}(q - 1)$$

The products $a_k(q) = M_k(q)N_k(q)$ are

$$a_1(q) = -(q^3 - q^2 + q - 1), \quad a_2(q) = \frac{1}{2}(q^3 + q), \quad a_3(q) = \frac{1}{2}(q^3 - 2q^2 + q)$$

The inverse CR converts the polynomials $y_1(q) = y_{10}$, $y_2(q) = y_{20}$, and $y_3(q) = y_{30} + y_{31}q$ into the polynomial $y(q)$ by

$$y(q) = -(q^3 - q^2 + q - 1)y_1(q) + \frac{1}{2}(q^3 + q)y_2(q)$$

$$+ \frac{1}{2}(q^3 - 2q^2 + q)y_3(q) \pmod{m(q)}$$

The first two terms have degrees smaller than 4, the degree of $m(q)$. The third term is divided by $m(q)$ and yields the remainder

$$\frac{1}{2}[(y_{30} - y_{31})q^3 - 2y_{30}q^2 + (y_{30} + y_{31})q]$$

Thus the inverse Chinese remainder transform is given by

$$y_0 = y_{10}$$

$$y_1 = -y_{10} + \frac{1}{2}(y_{20} + y_{30} + y_{31})$$

$$y_2 = y_{10} - y_{30}$$

$$y_3 = -y_{10} + \frac{1}{2}(y_{20} + y_{30} - y_{31})$$

Divisions by $1/2$ can be absorbed into $h$ and thus precomputed. ∎

### 7.6.3 Computation of the DFT using convolution algorithms

We show that the DFT computation can be reduced to convolution computation. As a consequence fast convolution algorithms relying on the Chineese remainder theorem can be used for the computation of the DFT.

***Rader prime algorithm.*** Consider a signal $x(n)$ of blocklength $N$ with DFT

$$X(k) = \sum_{n=0}^{N-1} x(n)w^{kn}$$

We assume that $N$ is a prime integer. Let $a$ be a primitive element in the Galois field $GF(N)$. All nonzero elements of the field are represented by powers of $a$. In particular,

$$a^{N-1} = 1 \tag{7.53}$$

For each nonzero element $n$ in $GF(N)$, let $r(n)$ denote the logarithm of $n$ with base $a$, that is, the unique integer such that $a^{r(n)} = n$. The mapping $r$ is a permutation of the nonzero elements of $GF(N)$. We will replace the indexes $n$ and $k$ in the DFT expression with their logarithms $r(n)$ and $r(k)$. Our intention is to change exponent products into sums and thus prepare the ground for the convolution formula. Since 0 is not obtained as a power of $a$, we treat it separately. Thus

$$X(0) = \sum_{n=0}^{N-1} x(n)$$

$$X(k) = x(0) + \sum_{n=1}^{N-1} w^{kn} x(n) = X(0) + \sum_{n=1}^{N-1} (w^{kn} - 1) x(n)$$

Using logarithms, we obtain

$$X(a^{r(k)}) = X(0) + \sum_{n=1}^{N-1} (w^{a^{r(k)+r(n)}} - 1) x(a^{r(n)}), \qquad 1 \le k \le N - 1 \qquad (7.54)$$

To create the index difference required in the convolutional formula, we set $m = N - 1 - r(n)$. Since $r$ is a permutation its range covers all integers $1, 2, \ldots, N - 1$. Hence the smallest value of $m$ is 0, obtained for $r(n) = N - 1$ and the largest value is $N - 2$, obtained for $r(n) = 1$. Thus

$$X(a^{r(k)}) = X(0) + \sum_{m=0}^{N-2} (w^{a^{r(k)-m}} - 1) x(a^{N-1-m}) \qquad (7.55)$$

Let us now define the scrambled signal of length $N - 1$,

$$\tilde{x}(n) = x(a^{N-1-n}), \qquad n = 0, 1, 2, \cdots, N - 2 \qquad (7.56)$$

$\tilde{x}(n)$ is obtained from $x(n)$ by an indexing permutation. We also consider the signal of length $N - 1$,

$$h(n) = w^{a^n} - 1, \qquad n = 0, 1, 2, \cdots, N - 2 \qquad (7.57)$$

The sum in the right-hand side of Eq. (7.55) is recognized as the cyclic convolution of $\tilde{x}(n)$ and $h(n)$ evaluated at times $r(k)$, $1 \le k \le N - 1$. Defining

$$\tilde{X}(n) = \tilde{x}(n) \odot h(n) \qquad (7.58)$$

Eq. (7.54) becomes

$$X(k) = \tilde{X}(r(k)) + X(0), \qquad 1 \le k \le N - 1 \qquad (7.59)$$

In summary, the computation of the DFT via the Rader scheme proceeds as follows: The signal $x(n)$ is shuffled in accordance with (7.56) to produce $\tilde{x}(n)$. This is subsequently convolved with $h(n)$, defined by (7.57). The resulting signal $\tilde{X}(n)$ is unscrambled by the indexing procedure (7.59) to give the desired DFT of $x(n)$. The Rader scheme can be viewed as a filtering operation with impulse response $h(n)$, preceded by a preprocessing scrambling permutation and followed by an unscrambling permutation. When the impulse response $h(n)$ is written in polynomial form it is referred to as *Rader polynomial*.

***Example 7.24  Rader prime algorithm for 7–point DFT***
Consider the Galois field $GF(7)$; pick 3 as a primitive element. Then $3^0 = 1$, $3^1 = 3$, $3^2 = 2$, $3^3 = 6$, $3^4 = 4$, $3^5 = 5$, and $3^6 = 1$. The permutation $r$ is

$$\begin{pmatrix} 1 & 2 & 3 & 4 & 5 & 6 \\ 6 & 2 & 1 & 4 & 5 & 3 \end{pmatrix}$$

The Rader polynomial is

$$h(q) = (w - 1) + (w^3 - 1)q + (w^2 - 1)q^2$$
$$+ (w^6 - 1)q^3 + (w^4 - 1)q^4 + (w^5 - 1)q^5$$

The scrambled input is

$$\tilde{x}(q) = x(1) + x(5)q + x(4)q^2 + x(6)q^3 + x(2)q^4 + x(3)q^5$$

The cyclic convolution is obtained as

$$\tilde{X}(q) = \tilde{x}(q)h(q) \quad (\mathrm{mod}(q^6 - 1))$$

Therefore the DFT is determined from the following expressions:

$$X(0) = x(0) + x(1) + x(2) + x(3) + x(4) + x(5) + x(6)$$
$$X(1) = \tilde{X}(6) + X(0)$$
$$X(2) = \tilde{X}(2) + X(0)$$
$$X(3) = \tilde{X}(1) + X(0)$$
$$X(4) = \tilde{X}(4) + X(0)$$
$$X(5) = \tilde{X}(5) + X(0)$$
$$X(6) = \tilde{X}(3) + X(0)$$

■

***The Winograd small FFT algorithm and the prime factor algorithm.*** The Winograd small FFT algorithm is obtained when the DFT computation is turned into a cyclic convolution computation by the Rader prime scheme and the cyclic convolution is determined by the Winograd short convolution algorithm. To control the number of additions from not getting too large, only short convolutions are allowed. Short convolutions are important because they are used by nesting schemes. Although the presentation is limited to prime blocklengths, extensions of the Rader algorithm, and consequently of the Winograd small FFT algorithm to other blocklengths, are possible yet more complicated.

The prime factor algorithm is a variant of the above scheme that applies to large blocklengths as well. Suppose that $N = N_1 N_2$, $N_1$, $N_2$ prime. First the Good-Thomas algorithm is employed to convert the DFT computation into 2-D DFTs. These in turn are expressed by several 1-D DFTs, as in Section 5.8.4. These smaller DFT's are converted by the Rader scheme into convolutions, and the latter are determined by the Winograd short convolution algorithm.

## BIBLIOGRAPHICAL NOTES

The series representation of signals and systems is commonly studied in the context of $z$ transform, which in this book is treated in Chapter 8. Series of digital signals are discussed in Peterson and Weldon (1972) and Gallagher (1968). Feedback shift registers are studied in Golomb (1967) and Stone (1973). Fast convolution algorithms are treated in Blahut (1984), McClellan and Rader (1979), and Lim (1990).

## PROBLEMS

**7.1.** Use partial fraction expansion to establish

$$x(q) = \begin{bmatrix} \dfrac{2}{1-q} & 0 \\ \dfrac{1-5q}{1-q^2} & 1+3q \end{bmatrix} \leftrightarrow x(n) = \begin{bmatrix} 2 & 0 \\ -2+3(-1)^n & (-3)^n \end{bmatrix}$$

**7.2.** Consider a multichannel system with impulse response and input given by

$$u(n) = \begin{bmatrix} \delta(n) \\ u_s(n) \end{bmatrix}, \quad h(n) = A^n, \quad A = \begin{bmatrix} \tfrac{1}{2} & 0 \\ 0 & \tfrac{1}{4} \end{bmatrix}$$

Compute the output.

**7.3.** Show that the impulse response of the feedback connection of two causal convolutional systems with sample responses $h_1$ and $h_2$ has the form

$$h(q) = [1 - h_1(q)h_2(q)]^{-1}h_1(q)$$

provided that $h_1(0)h_2(0) \neq 1$.

**7.4.** Derive the transfer function of the cascade, parallel and feedback connections of linear finite difference models.

**7.5.** Consider the system with state space parameters

$$A = \begin{bmatrix} -1/2 & 0 \\ 0 & -1/3 \end{bmatrix}, \quad B = \begin{bmatrix} 1 \\ 4 \end{bmatrix}, \quad C = [2 \quad 3], \quad D = 0$$

Show that the impulse response is given by

$$h(n) = 2\left(-\frac{1}{2}\right)^{n-1} + 12\left(-\frac{1}{3}\right)^{n-1}, \quad n \geq 0$$

**7.6.** Show that if the matrix $A$ in a linear time invariant single-input single-output state space model is diagonal, the impulse response is given by

$$h(n) = \sum_{i=1}^{k} r_i(\lambda_i)^{n-1}u_s(n-1) + D\delta(n), \quad r_i = b_ic_i$$

where $\lambda_i$ are the eigenvalues of $A$, $b_i$ are the entries of the column $B$, $c_i$ are the entries of the row $C$, and $D$ is a scalar.

**7.7.** Determine the periods of the binary sequences produced by the linear feedback shift registers whose characteristic polynomials are given below:
(a) $x^3 + x^2 + x + 1$
(b) $x^4 + x^3 + x^2 + x + 1$
(c) $x^5 + x^2 + 1$
(d) $x^6 + x^4 + x^2 + x + 1$
(e) $(x^2 + x + 1)^2(x^3 + x + 1)^2$

**7.8.** Determine the output of the systems below:
(a) $y(n) + 3y(n-1) = u(n)$, $u(n) = (\frac{1}{3})^n u_s(n)$, $y(-1) = 2$
(b) $y(n) - \frac{1}{2}y(n-1) = u(n) - \frac{1}{2}u(n-1)$, $u(n) = u_s(n)$, $y(-1) = 1$

Verify your answers with **MATLAB** simulation.

**7.9.** Construct a block diagram for a linear feedback shift register with characteristic polynomial $q^4 + q + 1$. Find the generator polynomial. Draw the FIR filter architecture that produces the outputs of the linear feedback shift register.

**7.10.** Working as in Example 7.23, compute the four-point by three-point linear convolution using

$$m(q) = q^2(q+1)(q-1)(q^2+1)$$

**7.11.** Use the Rader sheme to convert a five-point DFT to a linear convolution.

**7.12.** Work out the detailed computation of the five-point Winograd FFT algorithm.

## APPENDIX

***Irreducible factorization, poles, and zeros.*** Let $x(q)$ be a polynomial of degree $k$ with coefficients in the field $F$. There exist unique irreducible polynomials $f_1$, $f_2, \ldots, f_m$ of degrees $d_i$, respectively, and integers $k_1, k_2,\ldots, k_m$ such that

$$x(q) = f_1^{k_1}(q)f_2^{k_2}(q)\cdots f_m^{k_m}(q), \quad \sum_{i=1}^{m} d_i k_i = k \qquad (7.60)$$

Indeed, if $x(q)$ is irreducible over $F$, there is nothing to prove. If it is reducible, it can be factored as $x(q) = g_1(q)g_2(q)$. Repeating the same argument for $g_1$ and $g_2$, we eventually arrive at (7.60). The above result is known as *irreducible* or *prime factorization*.

If $F$ is the field of complex numbers, each irreducible factor $f_i(q)$ has degree 1 (such fields are called algebraically closed). In this case Eq. (7.60) takes the form

$$x(q) = (q - a_1)^{k_1}(q - a_2)^{k_2} \cdots (q - a_m)^{k_m} \qquad (7.61)$$

The field of real numbers can be embeded into $C$, where (7.61) holds. If $x(q)$ has real coefficients, each complex root $a$ is accompanied by its complex conjugate $a^*$. The corresponding real irreducible factor has degree 2 and is given by $q^2 - 2\Re aq + |a|^2$.

Next suppose that $F = GF(p^m)$ is a finite field. Consider the extension field $F^k = GF(p^{mk})$, where $k$ is the degree of $x(q)$. Let $a_1$ be a root of $f_1(q)$ in this extension field. Such a root exists. One way to see this is to construct the polynomial field $F/[f_1(q)]$ with coefficients in $F$. Operations in this field are carried out modulo the irreducible polynomial $f_1(q)$. As we recall from Appendix II, the element $a_1 = q$ is a root of $f_1(q)$. To obtain the desired conclusion, we embed the above field into $F^k$. Since $f_1(q)$ is irreducible over $F$, it coincides with the minimal polynomial of $a_1$. The remaining roots are the conjugate elements

$$a_1^p, \ a_1^{p^2}, \ldots, a_1^{p^{d_1-1}}$$

Application of the above argument to the remaining factors leads to the factorization

$$x(q) = (q - a_1)^{k_1}(q - a_1^p)^{k_1} \cdots (q - a_1^{p^{d_1-1}})^{k_1}$$

$$(q - a_2)^{k_2} \cdots (q - a_m)^{k_m} \cdots (q - a_m^{p^{d_m-1}})^{k_m} \qquad (7.62)$$

Suppose now that $x(q)$ is a rational series. The *poles* of $x(q)$ are given by the roots of the denominator (factored in an extension field) that remain after the cancellation of the common roots between denominator and numerator. In the same way *zeros* correspond to the roots of the numerator that remain after cancellations. Factorization of the numerator and the denominator leads to the *pole-zero* form

$$x(q) = K \frac{\prod(q - \rho_i)}{\prod(q - \lambda_i)}$$

**PROOF OF THEOREM 7.1**    Division of $b(q)$ by $a(q)$ gives quotient $b_0(q)$ and remainder $y(q)$ so that

$$x(q) = b_0(q) + \frac{y(q)}{a(q)}$$

We show that there are polynomials $b_1(q), b_2(q), \ldots, b_m(q)$ such that

$$\frac{y(q)}{a(q)} = \sum_{i=1}^{m} \frac{b_i(q)}{(1 - \lambda_i q)^{p_i}}, \quad \deg b_i(q) < p_i \qquad (7.63)$$

Indeed, we set

$$a_i(q) = \prod_{\substack{j=1 \\ j \neq i}}^{m} (1 - \lambda_j q)^{p_j} = \frac{a(q)}{(1 - \lambda_i q)^{p_i}} \qquad (7.64)$$

These polynomials are relatively prime because $\lambda_i \neq \lambda_j$ for $i \neq j$. Therefore $\gcd(a_1(q), a_2(q), \ldots, a_m(q)) = 1$. Consequently there are polynomials $k_i(q)$ such that

$$\sum_{i=1}^{m} k_i(q) a_i(q) = 1$$

If we multiply the latter expression by $y(q)$, we obtain

$$y(q) = \sum_{i=1}^{m} \hat{b}_i(q) a_i(q) = \sum_{i=1}^{m} \hat{b}_i(q) \frac{a(q)}{(1 - \lambda_i q)^{p_i}}, \qquad \hat{b}_i(q) = y(q) k_i(q) \qquad (7.65)$$

Each $\hat{b}_i(q)$ may not be what we are looking for because its degree may exceed $p_i$. Thus we divide it by $(1 - \lambda_i q)^{p_i}$ to obtain

$$\frac{\hat{b}_i(q)}{(1 - \lambda_i q)^{p_i}} = s_i(q) + \frac{b_i(q)}{(1 - \lambda_i q)^{p_i}}, \qquad \deg b_i(q) < p_i$$

Equation (7.65) now becomes

$$y(q) = \left( \sum_{i=1}^{m} s_i(q) \right) a(q) + \sum_{i=1}^{m} b_i(q) a_i(q) \tag{7.66}$$

Since $\deg b_i(q) < p_i$ and $\deg a_i(q) = \deg a(q) - p_i$, we have

$$\deg \left[ y(q) - \sum_{i=1}^{m} b_i(q) a_i(q) \right] < \deg a(q)$$

Thus Eq. (7.66) can hold only if $\sum_{i=1}^{m} s_i(q) = 0$, and then it becomes

$$y(q) = \sum_{i=1}^{m} b_i(q) a_i(q) = \sum_{i=1}^{m} \frac{b_i(q)}{(1 - \lambda_i q)^{p_i}} a(q)$$

which leads to Eq. (7.63). For each $1 \leq i \leq m$, we expand $b_i(q)$ in a Taylor series around $1/\lambda_i$; that is, we write

$$b_i(q) = r_{ip_i} + r_{ip_i-1}(1 - \lambda_i q) + r_{ip_i-2}(1 - \lambda_i q)^2 + \cdots + r_{i1}(1 - \lambda_i q)^{p_i-1} \tag{7.67}$$

This is always done in a unique way. The coefficients are obtained by

$$r_{ij} = \frac{(-1)^{p_i-j}}{(p_i - j)! \lambda_i^{p_i-j}} \frac{d^{p_i-j}}{dq^{p_i-j}} b_i(q) \Bigg|_{q=1/\lambda_i} \tag{7.68}$$

Then

$$\frac{b_i(q)}{(1 - \lambda_i q)^{p_i}} = \frac{1}{(1 - \lambda_i q)} r_{i1} + \frac{1}{(1 - \lambda_i q)^2} r_{i2} + \cdots + \frac{1}{(1 - \lambda_i q)^{p_i}} r_{ip_i}$$

and the expansion formula (7.11) has been derived. It remains to establish (7.12). Indeed, notice that

$$(1 - \lambda_i q)^{p_i} x(q) = b_i(q) + b_0(q)(1 - \lambda_i q)^{p_i} + \sum_{\substack{j=1 \\ j \neq i}}^{m} \frac{b_j(q)}{(1 - \lambda_j q)^{p_j}} (1 - \lambda_i q)^{p_i}$$

Hence the claim follows from (7.68), and the observation that the last term in the right-hand side of the latter expression has the form

$$M(q)(1 - \lambda_i q)^{p_i}, \quad M\left(\frac{1}{\lambda_i}\right) \neq 0$$

and therefore its first $p_i - 1$ derivatives are zero at $1/\lambda_i$.     ∎

PROOF OF THEOREM 7.2.   Let $1/\lambda_1, \ldots, 1/\lambda_l$ denote the poles of $x(n)$. If the poles are simple, $x(n)$ is written as

$$x(n) = \sum_{i=1}^{l} \lambda_i^n r_i \tag{7.69}$$

Each exponential $\lambda_i^n$ is periodic with fundamental period $N_i$. If $\lambda_j$ is conjugate to $\lambda_i$, it has order $N_i$ too. Indeed, $\lambda_j$ has the form $\lambda_j = \lambda_i^{p^k}$. According to Lemma 4 of Appendix II, the order of $\lambda_j$ is $N_i/\gcd(N_i, p^k)$. Since $N_i$ is a divisor of $p^m - 1$, $\gcd(N_i, p^k) = 1$. Suppose that $\lambda_i$ and $\lambda_j$ are not conjugate. Let $N$ be the order of the sum $z(n) = \lambda_i^n + \lambda_j^n$. We want to show that $N$ is given by the least common multiple of $N_i$ and $N_j$. It is clear that $\text{lcm}(N_i, N_j)$ is a period of $z(n)$ and hence a multiple of $N$. If $N_i = N_j$, the claim holds. Suppose that $N_i \neq N_j$ and $\lambda_j^N \neq 1$; that is, $N$ is not a period of $\lambda_j$. The equation $z(n + N) = z(n)$ becomes, for $n = N_i$, $\lambda_i^{N_i} + \lambda_j^{N_i+N} = 1 + \lambda_j^{N_i}$ or $\lambda_j^{N_i}(\lambda_j^N - 1) = 1 - \lambda_i^N$. In a similar fashion the equation $z(n + N) = z(n)$ becomes, for $n = N_j$, $\lambda_i^{N_j+N} + \lambda_j^N = 1 + \lambda_i^{N_j}$ or $\lambda_i^{N_j}(\lambda_i^N - 1) = 1 - \lambda_j^N$. Therefore $\lambda_i^{N_j}\lambda_j^{N_i}(\lambda_j^N - 1) = \lambda_j^N - 1$. By hypothesis, $\lambda_j^N \neq 1$. Therefore $\lambda_i^{N_j} = \lambda_j^{-N_i}$. Since $\lambda_i$ has order $N_i$, $\lambda_i^{N_j}$ has order $N_i/\gcd(N_i, N_j)$, by Lemma 4 of Appendix II. $\lambda_j$ and $1/\lambda_j$ have the same order. Hence $\lambda_j^{-N_i}$ has order $N_j/\gcd(N_i, N_j)$. Therefore $N_i = N_j$, a contradiction.
We finally consider the combinatorial exponential

$$x(n) = \binom{n+j}{j}\lambda^n$$

with rational series $1/(1 - q\lambda)^{j+1}$. Example 7.3 implies that $x(n)$ is periodic of period $N$ if $x(q)$ has the form $x(q) = x_0(q)/(1 - q^N)$ or $(1 - q\lambda)^{j+1}x_0(q) = 1 - q^N$. Equivalently, such an $N$ exists if $(1 - q\lambda)^{j+1}$ divides $1 - q^N$. The fundamental period $N$ is the smallest integer such that $(1 - q\lambda)^{j+1}$ divides $1 - q^N$. Let $N$ be of the form $Mp^r$ where $M$ is the order of $\lambda$ and $r$ is such that $j + 1 \leq p^r$. Then $1/\lambda$ has order $M$, and $1 - q\lambda$ divides $1 - q^M$. Hence $(1 - q\lambda)^{j+1}$ divides $(1 - q^M)^{j+1}$. By assumption, $j + 1 \leq p^r$. Thus $(1 - q\lambda)^{j+1}$ divides $(1 - q^M)^{p^r}$. By Lemma 5 of Appendix II, $(1 - q^M)^{p^r} = 1 - q^{Mp^r} = 1 - q^N$, and the claim has been established.     ∎

# 8

# REPRESENTATIONS OF SIGNALS AND SYSTEMS IN THE COMPLEX DOMAIN: z AND LAPLACE TRANSFORMS

The $z$ transform and the Laplace transform play an important role in the representation of signals and systems. They convert signals into analytic functions of a complex variable whose rich structure enables the extraction of important information. The $z$ transform concerns discrete signals, and the Laplace transform deals with analog signals. In each case the one-sided transform is first introduced to handle signals defined on positive times. Then the two-sided transforms are studied. Multichannel and multidimensional signals are also discussed. The chapter concludes with the characterization of BIBO stability in the complex domain and with an introduction to filter design.

## 8.1  z TRANSFORM

### 8.1.1  Definition

Let $x(n)$ be a real or complex-valued signal and $x(q)$ its series representation. If $q$ is replaced by a complex number $z^{-1}$, the function $z \to x(z)$:

$$x(z) = \sum_{n=0}^{\infty} x(n)z^{-n} \tag{8.1}$$

is called (one-sided) $z$ transform of $x(n)$. The $z$ transform, $x(z)$, is no longer a mere formal expression. Thus some care is needed to specify the domain of $x(z)$, that is the set of complex numbers $z$ over which $x(z)$ is well defined. It may happen that the infinite sum diverges everywhere in the complex plane, in which case the signal does not have a $z$ transform, although it always has a series representation. If, however, there is at least one $z_0 \neq 0$ such that $x(z_0)$ exists, then

**415**

there is $R_x > 0$ such that $x(z)$ converges for all $z$ outside the circle with center at the origin and radius $R_x$, and diverges for all $z$ inside the circle. For this reason the exterior of the circle

$$\{z \in C : |z| > R_x\}$$

is called the *region of convergence* of $x(z)$, and $R_x$ the *radius of convergence* of $x(z)$. The assignment

$$x(n) \xrightarrow{z} x(z) = \mathcal{Z}[x(n)]$$

defines the $z$ transform. If two signals $x(n)$ and $y(n)$ are equal, then $x(q) = y(q)$. If in addition the $z$ transform $x(z)$ exists, then $y(z)$ also exists and $x(z) = y(z)$. The converse is also true as we will shortly see in connection with the inverse $z$ transform.

### Example 8.1   Causal exponential signal

Let us consider the exponential signal

$$x(n) = a^n, \qquad n = 0, 1, 2, \ldots$$

where $a$ is a nonzero complex number. The $z$ transform is

$$x(z) = \sum_{n=0}^{\infty} a^n z^{-n} = \sum_{n=0}^{\infty} \left(\frac{a}{z}\right)^n = \frac{1}{1 - az^{-1}} = \frac{z}{z - a} \tag{8.2}$$

as long as $|az^{-1}| < 1$, and it diverges if $|az^{-1}| > 1$. We conclude that the $z$ transform is defined over the exterior of the circle of radius $|a|$. The unit step signal $x(n) = u_s(n) = 1$ can be viewed as an exponential signal with $a = 1$. Therefore

$$x(z) = \mathcal{Z}[u_s(n)] = \frac{1}{1 - z^{-1}} = \frac{z}{z - 1}, \qquad |z| > 1 \tag{8.3}$$

The region of convergence is the exterior of the unit circle.     ∎

The $z$ transform of a $m \times k$ multichannel signal $x(n)$ is given by Eq. (8.1), albeit now it has a matrix format. It is a matrix-valued function whose $ij$ entry is the $z$ transform of the $ij$ entry signal $x_{ij}(n)$. The region of convergence is the exterior of the circle with center the origin and radius $R = \max\{R_{ij}, 1 \le i \le m, 1 \le j \le k\}$, where $R_{ij}$ is the radius of convergence of $x_{ij}(z)$.

### Example 8.2   Multichannel signals

Let $x(n) = (\delta(n) \quad \delta(n-1) \quad \ldots \quad \delta(n-k))$, $k \ge 0$. Then $x(z)$ is given by $(1 \quad z^{-1} \quad \ldots \quad z^{-k})$ and is defined over the entire complex plane minus the origin. Likewise we easily deduce

$$x(n) = \begin{bmatrix} u_s(n) & \left(\dfrac{1}{2}\right)^n \\ 0 & 3^n \end{bmatrix}, \quad x(z) = \begin{bmatrix} \dfrac{z}{z-1} & \dfrac{z}{z-1/2} \\ 0 & \dfrac{z}{z-3} \end{bmatrix}$$

The radius of convergence is $R = \max\{1, 1/2, 3\} = 3$.  ∎

### 8.1.2  Existence of z transform

The series in the definition of the $z$ transform is interpreted in the sense of absolute convergence, $\sum |x(n)z^{-n}| < \infty$. A simple criterion to check whether the $z$ transform of a signal exists is given below.

**Theorem 8.1**  The $z$ transform of a complex multichannel signal $x(n)$ exists if and only if there are positive numbers $M$ and $r$ such that

$$|x(n)| \le Mr^n, \qquad 0 \le n < \infty \tag{8.4}$$

Equation (8.4) states that $x(n)$ grows no faster than an exponential signal. $|x(n)|$ denotes any induced norm of the matrix $x(n)$. The region of convergence contains all $z$ such that $|z| > r$.

**PROOF**  If $x(n)$ satisfies (8.4), it holds that $|x(n)z^{-n}| = |x(n)||z^{-1}|^n \le M|r/z|^n$. Therefore the Weierstrass test asserts that $x(z)$ converges absolutely for all $|z| > r$.

Conversely, if $\sum_{n=0}^{\infty} |x(n)z_0^{-n}| < \infty$, then (8.4) holds with $r = |z_0|$.  ∎

**Remark.**  The above existence condition supplies information about the radius of convergence. Indeed, (8.4) equivalently states that the sequence $\sqrt[n]{|x(n)|}$ is bounded. The radius of convergence is given by

$$R_x = \limsup_n \sqrt[n]{|x(n)|} \tag{8.5}$$

namely, the infimum of the upper bounds of the sequence $\sqrt[n]{|x(n)|}$.

### *Example 8.3    Radius of convergence*
Let $x(n) = (n+1)\lambda^n$, $n \ge 0$. Then $\sqrt[n]{|x(n)|} = \sqrt[n]{(n+1)|\lambda|^n} = |\lambda|\sqrt[n]{(n+1)}$. Since $\sqrt[n]{n} \to 1$ as $n \to \infty$, we have $\sqrt[n]{|x(n)|} \to |\lambda|$. Thus $R_x = \lim_n \sqrt[n]{|x(n)|} = |\lambda|$. In a similar fashion we can show that the combinatorial exponential signal $\binom{n+j}{j}\lambda^n$ has $z$ transform (see Example 7.4) $1/(1 - \lambda z^{-1})^{j+1} = z^{j+1}/(z - \lambda)^{j+1}$.  ∎

### *Example 8.4    Finite energy signals*

Consider a finite energy signal $x(n) \in \ell_2$. The $z$ transform exists and is an analytic function in the exterior of the unit disc: $|z| > 1$. Indeed, let $|z_0| > 1$. The Cauchy-Schwarz inequality gives

$$\sum_{n=0}^{\infty} |x(n)| |z_0^{-1}|^n \leq \sqrt{\sum |x(n)|^2} \sqrt{\sum |z_0^{-1}|^{2n}} = \frac{|x|_2}{\sqrt{1 - |z_0^{-2}|}} < \infty$$

In a similar fashion we can show that if $x \in \ell_1$, $x(z)$ exists and the region of convergence contains the unit circle as well: $|z| \geq 1$.    ∎

### 8.1.3    The inverse z transform

Suppose that the $z$ transform $x(z)$ of a signal $x(n)$ exists for $|z| > R_x$. Then $x(n)$ is uniquely recovered from $x(z)$ by

$$\mathcal{Z}^{-1}[x(z)] = x(n) = \frac{1}{2\pi j} \oint x(z) z^{n-1} dz, \qquad n \geq 0 \tag{8.6}$$

Integration is carried over a circle centered at the origin and contained in the region of convergence, or more generally, over any simple closed curve inside the region of convergence having the origin in its interior.

Formula (8.6) is a well-known result in the theory of complex numbers. It shows how the coefficients $x(n)$ of the Laurent series

$$x(z) = \sum_{n=0}^{\infty} x(n) z^{-n}, \qquad |z| > R_x \tag{8.7}$$

are determined from $x(z)$.

PROOF    Let us view Eq. (8.7) as a decomposition of $x(z)$ in terms of the exponential functions $z^{-n}$, $n \in Z^+$. We multiply both sides of Eq. (8.7) by $z^m$ and integrate over a circle of radius $r > R_x$ centered at the origin. Then

$$\oint x(z) z^m dz = \oint \sum_{n=0}^{\infty} x(n) z^{m-n} dz = \sum_{n=0}^{\infty} x(n) \oint z^{m-n} dz \tag{8.8}$$

Now it holds that

$$\oint z^{m-n} dz = \begin{cases} 2\pi j, & \text{If } m - n = -1 \\ 0, & \text{otherwise} \end{cases} \tag{8.9}$$

To see this, parametrize the circle as $z(t) = r e^{jt}$, $0 \leq t < 2\pi$. Then $z^{m-n}(t) =$

$r^{m-n}e^{j(m-n)t}$ and $dz(t) = rje^{jt}dt$. Hence

$$\oint z^{m-n}dz = \int_0^{2\pi} r^{m-n}e^{j(m-n)t}rje^{jt}\,dt = r^{m-n+1}j\int_0^{2\pi} e^{j(m-n+1)t}\,dt$$

The latter expression easily reduces to (8.9). The right-hand side of Eq. (8.8) becomes $2\pi jx(m+1)$, and the inversion formula (8.6) is established.  ∎

**Rational signals.** Suppose that $x(n)$ is a complex-valued signal such that the series $x(q)$ is rational. Suppose further that the constant term of the denominator polynomial is nonzero. Then the $z$ transform $x(z)$ exists, and the radius of convergence is given by the maximum of the moduli of the roots of the denominator which remain after pole-zero cancellation. Indeed, partial fraction expansion leads to Eq. (7.14). Each term of the form $\delta(n-j)$ is a finite extent signal. Hence the $z$ transform exists over the entire plane minus the origin. We showed in Example 8.3 that each combinatorial exponential with parameters $j$ and $\lambda_i$, has $z$ transform for all $|z| > |\lambda_i|$. Hence the claim is proved.

Entirely analogous conclusions are valid in the multichannel case. In particular, the $z$ transform of the matrix exponential $A^n$, $n \geq 0$ exists and is given by $(I - z^{-1}A)^{-1} = z(zI - A)^{-1}$. The radius of convergence is the maximum magnitude of the eigenvalues of $A$.

The $z$ transform of $x(n)$,

$$x(z) = \frac{\sum_{i=0}^m b_i z^{-i}}{\sum_{i=0}^k a_i z^{-i}}$$

can be written as a rational function of $z$ rather than $z^{-1}$ if we factor out the highest powers of $z^{-1}$:

$$x(z) = \frac{z^{-m}(b_0 z^m + b_1 z^{m-1} + \cdots + b_m)}{z^{-k}(a_0 z^k + a_1 z^{k-1} + \cdots + a_k)} = \frac{z^k(b_0 z^m + b_1 z^{m-1} + \cdots + b_m)}{z^m(a_0 z^k + a_1 z^{k-1} + \cdots + a_k)}$$

The denominator has degree $k+m$ because $a_0 \neq 0$. The numerator has degree at most $k+m$. Hence $x(z)$ is always proper rational in $z$ provided that $a_0 \neq 0$.

### 8.1.4   Convolution

Consider the multichannel signals $u(n)$ and $h(n)$ whose $z$ transforms $u(z)$ and $h(z)$ exist and have radii of convergence $R_u$ and $R_h$, respectively. Then the $z$ transform of the convolution $y(n) = h(n) * u(n)$ also exists; it is given by $y(z) = h(z)u(z)$, and the radius of convergence is smaller than $\max(R_h, R_u)$. Indeed, suppose that $R_u \geq R_h$. We show that $y(z)$ exists for all $|z| > R_u$. Let $|z_0| = r_1 > R_u$ and $R_h < r_2 < r_1$. Theorem 8.1 implies that $|u(n)| \leq M_1 r_1^n$ and $|h(n)| \leq M_2 r_2^n$. Then

$$|y(n)| \leq \sum_{k=0}^n |h(k)||u(n-k)| \leq M_1 M_2 \sum_{k=0}^n r_2^k r_1^{n-k} = M_1 M_2 r_1^n \sum_{k=0}^n \left(\frac{r_2}{r_1}\right)^k \leq C r_1^n$$

Thus $y(n)$ satisfies Eq.(8.4).

***Example 8.5***   ***Convolution of two signals***
Let

$$h(n) = Ma^n u_s(n), \quad u(n) = a\delta(n) + (1-a)u_s(n)$$

$M$ is a real constant. Then

$$h(z) = M\frac{z}{z-a}, \quad u(z) = a + (1-a)\frac{z}{z-1} = \frac{z-a}{z-1}$$

Hence

$$y(z) = M\frac{z}{z-a}\frac{z-a}{z-1} = M\frac{z}{z-1}, \quad y(n) = Mu_s(n)$$

If $|a| > 1$, $R_y = 1 < \max(R_u, R_h) = |a|$. This occurs because the pole of one signal is canceled by the zero of the second signal.   ∎

## 8.2 TWO-SIDED $z$ TRANSFORM

### 8.2.1   Definition

The two-sided $z$ transform concerns signals defined over all integers. It is given by the Laurent series

$$\mathcal{Z}[x(n)] = x(z) = \sum_{n=-\infty}^{\infty} x(n)z^{-n} \tag{8.10}$$

It requires that the one-sided series

$$x_1(z) = \sum_{n=-\infty}^{-1} x(n)z^{-n}, \quad x_2(z) = \sum_{n=0}^{\infty} x(n)z^{-n} \tag{8.11}$$

converge absolutely and $x(z) = x_1(z) + x_2(z)$. If $x(n)$ is causal, the two-sided $z$ transform coincides with the one sided $z$ transform. Consider the positive and negative part of $x(n)$:

$$x^+(n) = x(n)u_s(n), \quad x^-(n) = x(-n)u_s(n-1), \quad n \geq 0 \tag{8.12}$$

It holds that $x(n) = x^+(n) + x^-(-n)$. Let $x^+(z)$ and $x^-(z)$ be the $z$ transforms of $x^+(n), x^-(n)$ and $R^+, R^-$ the corresponding radii of convergence. It is easy to see that

$$x_1(z) = x^-(z^{-1}), \quad x_2(z) = x^+(z), \quad x(z) = x^-(z^{-1}) + x^+(z) \tag{8.13}$$

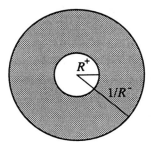

**Figure 8.1.** Region of convergence of the two sided z transform.

The $z$ transform exists provided that both one-sided transforms $x^+(z)$, $x^-(z^{-1})$ exist for $|z| > R^+$ and $|z| < 1/R^-$. Thus provided that $R^+ < (R^-)^{-1}$, the region of convergence is the annulus (see Fig. 8.1)

$$R^+ < |z| < \frac{1}{R^-} \tag{8.14}$$

**Example 8.6   Finite support signals**
The two-sided $z$ transform of a finite support signal has the form

$$X(z) = \sum_{n=M}^{N} x(n)z^{-n}$$

The region of convergence extends over the entire complex plane except the origin. ∎

**Example 8.7   Strictly noncausal exponential signal**
Let

$$x(n) = -a^n u_s(-n-1)$$

Straightforward computations lead to

$$x(z) = x^-(z^{-1}) = \frac{z}{z-a}, \qquad |z| < |a|$$

Note that the $z$ transform of the causal exponential signal and the $z$ transform of the strictly noncausal exponential signal have the same functional form but totally different regions of convergence. It is therefore absolutely necessary to specify the region of convergence when we deal with the two-sided $z$ transform. ∎

### 8.2.2  Inverse z transform and partial fraction expansion

In complete analogy with the one-sided transform, the inverse two-sided $z$ transform is specified by

$$x(n) = \frac{1}{2\pi j} \oint x(z) z^{n-1} dz \tag{8.15}$$

In the case of rational signals, inversion is carried out by partial fraction expansion. In fact, now general rational (not necessarily proper) functions are allowed. Let

$$x(z) = \frac{\sum_{i=0}^{k+l} d_i z^i}{z^k + \sum_{i=0}^{k-1} a_i z^i}$$

If $l \geq 0$, division gives

$$x(z) = \sum_{i=0}^{l} c_i z^i + \frac{\sum_{i=0}^{k-1} b_i z^i}{z^k + \sum_{i=0}^{k-1} a_i z^i}$$

The first sum is the quotient and corresponds to $\sum_{i=0}^{l} c_i \delta(n+i)$. The second term is strictly proper rational and is expanded to partial fractions. Caution is needed with the region of convergence. Each fraction $z/(z - \lambda_i)^j$ is associated with two signals. The right choice is inferred from the region of convergence of the given signal and is unique. Indeed, let $|\lambda_1| < |\lambda_2| < \cdots < |\lambda_m|$ be the distinct roots of the denominator. The region of convergence of $x(z)$ is necessarily one of the $m + 1$ annuli:

$$\{z \in C : |z| < |\lambda_1|\}, \ \{z \in C : |\lambda_1| < |z| < |\lambda_2|\}, \ \ldots,$$
$$\{z \in C : |\lambda_{m-1}| < |z| < |\lambda_m|\}, \{z \in C : |z| > |\lambda_m|\} \tag{8.16}$$

In each annulus every fraction $z/(z - \lambda_i)$ corresponds uniquely either to the strictly noncausal exponential $-\lambda_i^n u_s(-n-1)$ or to the causal exponential $\lambda_i^n u_s(n)$. If the region of convergence of $x(z)$ is not specified, there are exactly $m + 1$ signals whose $z$ transform has the same functional form. We illustrate the procedure with the following example:

***Example 8.8    Time domain reconstruction of a rational signal***
Let

$$x(z) = \frac{z}{(z+2)(z-1)}$$

Partial fraction expansion gives

$$x(z) = -\frac{1}{3}\frac{z}{z+2} + \frac{1}{3}\frac{z}{z-1}$$

The possible regions of convergence are

$$A = \{z \in C : |z| < 1\}, \quad B = \{z \in C : 1 < |z| < 2\}, \quad D = \{z \in C : |z| > 2\}$$

If the region of convergence of $x(z)$ is $A$, it holds that

$$\mathcal{Z}^{-1}\left[\frac{z}{z+2}\right] = -(-2)^n u_s(-n-1), \quad \mathcal{Z}^{-1}\left[\frac{z}{z-1}\right] = -u_s(-n-1)$$

Consequently

$$x(n) = \frac{1}{3}(-2)^n u_s(-n-1) - \frac{1}{3}u_s(-n-1)$$

If the region of convergence is $B$, it holds that

$$\mathcal{Z}^{-1}\left[\frac{z}{z+2}\right] = -(-2)^n u_s(-n-1), \quad \mathcal{Z}^{-1}\left[\frac{z}{z-1}\right] = u_s(n)$$

and the signal is

$$x(n) = \frac{1}{3}(-2)^n u_s(-n-1) + \frac{1}{3}u_s(n),$$

Likewise in the case of region $D$, we find that

$$x(n) = -\frac{1}{3}(-2)^n u_s(n) + \frac{1}{3}u_s(n) \qquad \blacksquare$$

Some of the basic properties of the $z$ transform are discussed in Exercise 8.8.

### 8.2.3  Relation with the discrete time Fourier transform

The similarity between the Fourier transform and the $z$ transform

$$X(\omega) = \sum_{n=-\infty}^{\infty} x(n)e^{-j\omega n}, \quad x(z) = \sum_{n=-\infty}^{\infty} x(n)z^{-n}$$

is evident. $X(\omega)$ is obtained from $x(z)$ if the latter is evaluated on the unit circle

$$\mathcal{F}(x)(\omega) = \mathcal{Z}(x)(z)|_{z=e^{j\omega}} \tag{8.17}$$

If $x(z)$ exists and the region of convergence of $x(z)$ contains the unit circle, the discrete time Fourier transform exists.

*Example 8.9    Computing the Fourier transform from the $z$ transform*
The $z$ transform of the causal exponential signal $x(n) = a^n u_s(n)$ is

$$x(z) = \frac{z}{z - a}, \qquad |z| > |a|$$

If the region of convergence contains the unit circle so that $x(z)$ is analytic for all $|z| \geq 1$, the discrete time Fourier transform exists and can be evaluated by Eq. (8.17). This occurs if $|a| < 1$. Then

$$X(\omega) = \frac{e^{j\omega}}{e^{j\omega} - a} = \frac{1}{1 - ae^{-j\omega}}$$

We obtain the same result as in Example 6.15.                            ■

### 8.2.4    Polyphase decomposition in the z domain

Consider the single channel signal $x(n), n \in Z$ and its polyphase components (see Section 6.2.2) $x_k(n) = x(Mn + k), 0 \leq k \leq M - 1$. The polyphase decomposition in the $z$ domain expresses the signal $z$ transform $x(z)$ in terms of the $z$ transforms of the signals $x_k(n)$. We proceed exactly as in the frequency domain. The time index $n$ is replaced by the pair $(l, k)$ obtained by dividing $n$ by $M$: $n = lM + k, 0 \leq k \leq M - 1$:

$$x(z) = \sum_{k=0}^{M-1} \sum_{l=-\infty}^{\infty} x(lM + k)z^{-(lM+k)} = \sum_{k=0}^{M-1} z^{-k} \sum_{l=-\infty}^{\infty} x_k(l)z^{-lM}$$

or

$$x(z) = \sum_{k=0}^{M-1} z^{-k} x_k(z^M) \tag{8.18}$$

The latter formula expresses the polyphase decomposition of $x(n)$ in the $z$ domain. To extract the polyphase components $x_k(z)$ from the signal $x(z)$, we evaluate (8.18) at $\omega^i z^{1/M}, 0 \leq i \leq M - 1, \omega = e^{-j2\pi/M}$ and add up. Then

$$\sum_{i=0}^{M-1} x(\omega^i z^{1/M}) = \sum_{i=0}^{M-1} \sum_{k=0}^{M-1} \omega^{-ki} z^{-k/M} x_k(\omega^{Mi} z)$$

$$= \sum_{k=0}^{M-1} z^{-k/M} x_k(z) \sum_{i=0}^{M-1} \omega^{-ki} = M x_0(z)$$

Therefore

$$x_0(z) = \frac{1}{M} \sum_{i=0}^{M-1} x(\omega^i z^{1/M}) \tag{8.19}$$

The above expression yields the decimated signal $x_M(n) = x(Mn) = x_0(n)$ in the complex domain. The remaining polyphase components are easily derived from (8.19).

### 8.2.5   z transform representation of systems

In this section the $z$ transform representation of systems is discussed.

**Convolutional systems.** A convolutional system is described by the input-output expression $y(n) = h(n) * u(n)$. If both the impulse response and the input are causal and grow no faster than an exponential, the output shares the same property. The $z$ transform is applicable, and the convolution theorem asserts that

$$y(z) = h(z)u(z)$$

$h(z)$ defines the *transfer function* of the system. The above result carries to noncausal convolutions via the two-sided $z$ transform provided that the regions of convergence of the two signals intersect.

**Linear finite difference models.** We demonstrated in Example 7.18 that the output series of a linear time-invariant finite difference model is uniquely defined in terms of the initial condition and the input and extends to all integers, if the input is defined over all integers. This might lead us to believe that all output signals of such models can be transformed in the $z$ domain. This is not true except in few cases. For instance, a glance at Eq. (7.30) shows that the $z$ transform cannot be obtained by simply adding the $z$ transforms of the three terms because the $z$ transform of the first term simply does not exist unless $y_0 = 0$. In some exceptional cases the first term is properly combined with the remaining terms so that the overall signal has a $z$ transform. To better understand this behavior, let us consider an input signal $u(n)$ defined on $Z$ with two-sided $z$ transform, $u(z)$ defined on $r_u^- < |z| < r_u^+$. Suppose that $y(n)$ is a solution of (7.27) whose $z$ transform exists. The shift property gives

$$y(z) = h(z)u(z), \quad h(z) = \frac{\sum b_i z^{-i}}{1 + \sum a_i z^{-i}}$$

$h(z)$ can be defined in any of the regions of convergence that has a nonempty intersection with the region of convergence of $u(z)$. For each such case a corresponding output signal is obtained. All such outputs are special cases of those obtained by the time domain analysis illustrated in Example 7.18 for particular choices of the initial condition.

***Example 8.10   Outputs of a finite difference model***
Consider the system of Example 7.18 with $b = 1$ and $a = \frac{1}{2}$. The input is the causal signal $u(n) = (\frac{1}{4})^n u_s(n)$. The transfer function is $h(z) = 1/(1 - z^{-1}/2)$. Both regions of convergence $|z| < 1/2$ and $|z| > 1/2$, intersect the domain of $u(z)$, $|z| > 1/4$. The resulting outputs are determined by partial fraction expansion. They are

$$\left[ 2\left(\frac{1}{2}\right)^n - \left(\frac{1}{4}\right)^n \right] u_s(n), \quad -2\left(\frac{1}{2}\right)^n u_s(-n-1) - \left(\frac{1}{4}\right)^n u_s(n)$$

Both are obtained from Eq. (7.39) for $y_0 = 0$ and $y_0 = -4$, respectively.   ∎

**Linear time invariant state space models.** We saw in Section 7.5 that the behavior of a linear time invariant state space model is governed by the rational series $h^\times(q)$ and $h(q)$. The one-sided $z$ transforms of these signals always exist. The region of convergence covers the exterior of the circle with radius the largest of the magnitudes of the eigenvalues of $A$. If the input is causal and has $z$ transform, the $z$ transform of the output exists and is given by Eq. (7.46) where $q$ is replaced by $z^{-1}$.

The form of the output produced by noncausal inputs is elucidated with the aid of the two-sided $z$ transform. If we apply the shift property, then arguing as with finite difference models, we find that

$$y(z) = h(z)u(z), \quad h(z) = C(zI - A)^{-1}B + D \tag{8.20}$$

$h(z)$ is proper rational. We have exactly the same situation with the finite difference model. If $m$ is the number of distinct eigenvalues of $A$, there are $m + 1$ disjoint regions of convergence giving rise to $m + 1$ different impulse responses. The region of convergence of the input will intersect more than one of these regions and thus will generate more than one possible outputs. Exactly one of these regions will contain the unit circle. In the latter case a frequency domain interpretation is possible. We simply evaluate Eq. (8.20) on the unit circle. The frequency response is

$$H(\omega) = C(e^{j\omega}I - A)^{-1}B + D \tag{8.21}$$

***Example 8.11   Outputs of a state space model***
Consider the state space system with parameters

$$A = \begin{pmatrix} \frac{1}{2} & 0 \\ 0 & \frac{1}{4} \end{pmatrix}, \quad B = \begin{pmatrix} 1 \\ 3 \end{pmatrix}, \quad C = (2 \quad 4), \quad D = 0$$

Let $u(n) = (\frac{1}{3})^n u_s(n)$. The transfer function is

$$h(z) = \frac{2}{z - 1/2} + \frac{12}{z - 1/4}$$

The region of convergence of the input intersects both domains $1/4 < |z| < 1/2$ and $1/2 < |z|$. The corresponding outputs have regions of convergence $1/3 < |z| < 1/2$ and $|z| > 1/2$. They are given by

$$-144 \left(\frac{1}{4}\right)^n u_s(n) - 12 \left(\frac{1}{2}\right)^n u_s(-n-1) + 132 \left(\frac{1}{3}\right)^n u_s(n)$$

$$-144 \left(\frac{1}{4}\right)^n u_s(n) + 12 \left(\frac{1}{2}\right)^n u_s(n) + 132 \left(\frac{1}{3}\right)^n u_s(n)$$

The region of convergence containing the unit circle is $1/2 < |z|$. The frequency response of the system is

$$H(\omega) = \frac{2}{e^{j\omega} - \frac{1}{2}} + \frac{12}{e^{j\omega} - \frac{1}{4}} \qquad \blacksquare$$

## 8.3  2-D z TRANSFORM

In close analogy with the one-dimensional case, the 2-D $z$ transform of the signal $x(n_1, n_2)$ is defined as the complex function of two complex variables

$$x(z_1, z_2) = \sum_{n_1=-\infty}^{\infty} \sum_{n_2=-\infty}^{\infty} x(n_1, n_2) z_1^{-n_1} z_2^{-n_2}$$

The set of complex pairs $(z_1, z_2)$ for which the double series converges absolutely is called *region of convergence*. It is visualized by means of its boundary which has a two-dimensional shape.

*Example 8.12   Separable signals*
If

$$x(n_1, n_2) = x_1(n_1) x_2(n_2),$$

then

$$x(z_1, z_2) = x_1(z_1) x_2(z_2)$$

where $x_1(z_1)$ is the 1-D $z$ transform of $x_1(n_1)$ and $x_2(z_2)$ the 1-D $z$ transform of $x_2(n_2)$. Moreover $(z_1, z_2)$ belongs to the region of convergence of $x(z_1, z_2)$ if and only if $z_1$ belongs to the region of convergence of $x(z_1)$ and $z_2$ belongs to the region of convergence of $x(z_2)$. Indeed,

$$x(z_1, z_2) = \sum_{n_1=-\infty}^{\infty} \sum_{n_2=-\infty}^{\infty} x_1(n_1)x_2(n_2)z_1^{-n_1}z_2^{-n_2}$$

$$= \sum_{n_1=-\infty}^{\infty} x(n_1)z_1^{-n_1} \sum_{n_2=-\infty}^{\infty} x(n_2)z_2^{-n_2} = x_1(z_1)x_2(z_2)$$

The $z$ transform of $x(n_1, n_2) = a_1^{n_1} a_2^{n_2} u_s(n_1, n_2)$ is

$$x(z_1, z_2) = \frac{z_1 z_2}{(z_1 - a_1)(z_2 - a_2)}, \qquad |z_1| > |a_1|, \ |z_2| > |a_2| \qquad \blacksquare$$

***Example 8.13    Finite support signals***
If $x(n_1, n_2)$ has finite support, the $z$ transform reduces to a finite sum

$$x(z_1, z_2) = \sum_{n_1=M_1}^{N_1} \sum_{n_2=M_2}^{N_2} x(n_1, n_2)z_1^{-n_1}z_2^{-n_2}$$

and the region of convergence extends everywhere except on the points of the lines $z_1 = 0$ and $z_2 = 0$. Thus the signal $x(n_1, n_2) = \delta(n_1 - k_1, n_2 - k_2)$ has $z$ transform

$$x(z_1, z_2) = z_1^{-k_1}z_2^{-k_2} \qquad \blacksquare$$

The region of convergence is more complicated if the signal is not separable. This is illustrated next.

***Example 8.14    A nonseparable signal***
Let

$$x(n_1, n_2) = a_1^{n_1} a_2^{n_2} \delta(n_1 - n_2)u_s(n_1, n_2)$$

Then

$$x(z_1, z_2) = \sum_{n_1=0}^{\infty}(a_1 a_2)^{n_1}(z_1 z_2)^{-n_1} = \frac{z_1 z_2}{z_1 z_2 - a_1 a_2}$$

The region of convergence is $|z_1 z_2| > |a_1 a_2|$. Taking logarithms, we obtain

$$\log|z_1| + \log|z_2| > \log|a_1 a_2|$$

In logarithmic scale, the region of convergence appears in Fig. 8.2. $\qquad \blacksquare$

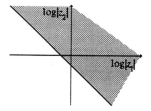

**Figure 8.2.** Region of convergence of 2-D z transform in Example 8.14.

The multidimensional $z$ transform is defined as

$$x(z_1, \ldots, z_m) = \sum_{n_1=-\infty}^{\infty} \cdots \sum_{n_2=-\infty}^{\infty} x(n_1, \ldots, n_m) z_1^{-n_1} \cdots z_m^{-n_m}$$

It can be shown that as in the 1-D case, the m-D $z$ transform is linear. Moreover the m-D convolution $x_1(n_1, \ldots, n_m) * x_2(n_1, \ldots, n_m)$ translates into $x_1(z_1, \ldots, z_m) x_2(z_1, \ldots, z_m)$. Additional properties are discussed in Exercise 8.36.

As in the one-dimensional case, the 2-D $z$ transform is inverted via the integral

$$x(n_1, n_2) = \left( \frac{1}{2\pi j} \right)^2 \oint_{C_2} \oint_{C_1} x(z_1, z_2) z_1^{n_1-1} z_2^{n_2-1} dz_1 dz_2$$

Integration is carried out on simple closed curves that include the origin in their interior and are contained in the region of convergence. If $x(z_1, z_2)$ does not have a simple form, inversion by the above formula is extremely complex and is rarely used. Unfortunately, and in contrast to the one-dimensional case, rational functions cannot be effectively handled. In the 2-D case partial fraction expansion is blocked by the fact that 2-D polynomials are rarely factorable into simple first-degree terms, and the roots are now described by surfaces. For instance, the signal of Example 8.14 has rational transform. The roots of the denominator are determined by the surface $z_1 z_2 = a_1 a_2$. Because of these difficulties the 2-D inverse $z$ transform is rarely computed analytically.

## 8.4 LAPLACE TRANSFORM

The Laplace transform constitutes the continuous time counterpart of the $z$ transform. Proceeding as in the case of the $z$ transform, we first discuss the (one-sided) Laplace transform for signals specified on the positive real axis $0 \le t < \infty$. The two-sided or bilateral Laplace transform is described later and refers to signals defined on the entire real line.

### 8.4.1  Definition

Let $x(t)$ be a continuous time complex signal defined on $0 \leq t < \infty$. The Laplace transform of $x(t)$ is the complex function of a complex variable

$$\mathcal{L}[x(t)] = x(s) = \int_0^\infty x(t)e^{-st}dt$$

The set of complex numbers $s$ over which $x(s)$ exists is called the *region of convergence* of $x(s)$. Existence of $x(s)$ is ensured if

$$\int_0^\infty |x(t)e^{-st}|dt < \infty \tag{8.22}$$

***Example 8.15    The polynomial exponential signal***
Let us first compute the Laplace transform of the exponential

$$x(t) = e^{at}, \qquad t \geq 0$$

where $a$ is an arbitrary complex number. Let $T > 0$. Then

$$x_T(s) = \int_0^T e^{at}e^{-st}dt = \frac{1}{a-s}\left[e^{(a-s)T} - 1\right], \qquad s \neq a$$

We know that $\lim_{T \to \infty} e^{(a-s)T}$ exists and is zero if and only if $\Re(a - s) < 0$. Therefore

$$x(s) = \int_0^\infty e^{at}e^{-st}dt = \frac{1}{s-a}, \qquad \Re s > \Re a \tag{8.23}$$

The region of convergence is the right half plane with boundary the line

$$\{s \in C : \quad \Re s = \Re a\}$$

It is illustrated in Fig. 8.3. The unit step signal equals the exponential signal with parameter $a = 0$, on the positive real axis. Thus

$$\mathcal{L}[u_s(t)] = \frac{1}{s}, \qquad \Re s > 0$$

The region of convergence is the half plane extending to the right of the imaginary axis. Differentiation of (8.23) with respect to $a$ gives

$$\int_0^\infty te^{at}e^{-st}dt = \mathcal{L}[te^{at}] = \frac{1}{(s-a)^2}, \qquad \Re s > \Re a \tag{8.24}$$

Repeated differentiation with respect to $a$ gives

$$\mathcal{L}\left[\frac{t^m}{m!}e^{at}\right] = \frac{1}{(s-a)^{m+1}}, \qquad \Re s > \Re a \tag{8.25}$$

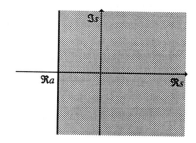

**Figure 8.3.** Region of convergence of the causal exponential signal.

The signal $t^m e^{at}/m!$ is the *polynomial exponential signal of order m,* and it plays a role similar to the combinatorial exponential signal. It gives rise to a multiple pole of order $m + 1$ at location $a$. ∎

We demonstrate in the Appendix at the end of the chapter that the Laplace transform of a signal satisfying (8.22) is an analytic function and that the region of convergence is a right half plane of the form $\Re s > \sigma$, as in Fig. 8.4.

### 8.4.2 Linearity and differentiation

Let $x(t)$ and $y(t)$ be matrix-valued signals of the same dimensions and $x(s)$ and $y(s)$ their Laplace transforms. Let $a$ and $b$ be complex numbers. Then

$$\mathcal{L}[ax(t) + by(t)] = ax(s) + by(s)$$

The region of convergence of $ax(s) + by(s)$ contains the intersection of the regions of convergence of $x(s)$ and $y(s)$. As a consequence, if $z(t) = Ax(t)$ and $A$ is a matrix of compatible size, then $z(s) = Ax(s)$.

The differentiation property relates the Laplace transforms of a signal $x(t)$ and its derivative $x^{(1)}(t)$. It is effectively employed in the study of differential equations. Suppose that $x(t)$ is continuous and has exponential order (see Appendix at the end of the chapter). Suppose further that $x^{(1)}(t)$ is piecewise

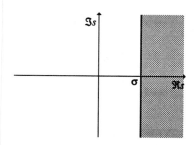

**Figure 8.4.** The region of convergence of the Laplace transform of a signal is a right half plane.

continuous and has exponential order. Then the Laplace transform of the derivative of $x(t)$ exists on the same region of convergence and is given by

$$\mathcal{L}\left[x^{(1)}(t)\right] = sx(s) - x(0)$$

Indeed, integration by parts gives

$$\mathcal{L}\left[x^{(1)}(t)\right] = \int_0^\infty e^{-st}\frac{d}{dt}x(t)dt = e^{-st}x(t)|_0^\infty$$

$$+ s\int_0^\infty e^{-st}x(t)dt = sx(s) - x(0)$$

Induction leads to the following expression for higher-order derivatives:

$$\mathcal{L}\left[x^{(p)}(t)\right] = s^p x(s) - \sum_{i=0}^{p-1} s^{p-1-i}x^{(i)}(0)$$

### 8.4.3   Signals with strictly proper rational Laplace transform

Let $x(t)$ be a signal with strictly proper rational Laplace transform

$$x(s) = \frac{b(s)}{a(s)} = \frac{\sum_{i=0}^{p-1} b_i s^i}{s^p + \sum_{i=0}^{p-1} a_i s^i}, \qquad \Re s > \sigma$$

$x(s)$ admits a partial fraction expansion. Motivated by Example 8.15, we consider a slight variant of the partial fraction decomposition studied in the context of the $q$ series and the $z$ transform. Entirely analogous arguments to those of Section 7.2 lead to

$$x(s) = \sum_{i=1}^m \sum_{j=1}^{p_i} \frac{1}{(s-\lambda_i)^j}r_{ij}$$

$\lambda_1, \lambda_2, \ldots, \lambda_m$ denote the distinct roots of the denominator polynomial $a(s)$ and $p_1, p_2, \ldots, p_m$ their multiplicities. Each partial fraction $1/(s-\lambda_i)^j$ is the Laplace transform of the polynomial exponential signal $t^{j-1}e^{\lambda_i t}/(j-1)!$ on the right half plane $\Re s > \Re\lambda_i$. Taking into account the superposition property of the Laplace transform, we conclude that

$$x(t) = \sum_{i=1}^m \sum_{j=1}^{p_i} \frac{t^{j-1}}{(j-1)!}e^{\lambda_i t}r_{ij}, \qquad \Re s > \sigma, \sigma = \max\{\Re\lambda_1, \ldots, \Re\lambda_m\} \quad (8.26)$$

where

$$r_{ij} = \frac{1}{(p_i-j)!}\frac{d^{p_i-j}}{ds^{p_i-j}}[(s-\lambda_i)^{p_i}x(s)]|_{s=\lambda_i} \quad (8.27)$$

If the Laplace transform of a multichannel signal is a strictly proper rational matrix, the signal takes the form (8.26) in the time domain except that the coefficients determined from (8.27) are matrices and $\lambda_1, \lambda_2, \ldots, \lambda_m$ are the distinct roots of the least common multiple of the denominators.

***Example 8.16*** ***Partial fraction expansion of a multichannel signal***
Let

$$x(s) = \begin{bmatrix} \dfrac{1}{s+1} & 0 \\[2mm] \dfrac{1}{s} & \dfrac{1}{s+2} \end{bmatrix}, \qquad \Re s > 0$$

Working entrywise we can easily calculate the inverse Laplace transform of each entry:

$$x(t) = \begin{bmatrix} e^{-t} & 0 \\ u_s(t) & e^{-2t} \end{bmatrix}, \qquad t \geq 0$$

Alternatively, we may apply the previous formulas. The least common multiple of the denominators is $s(s+1)(s+2)$ with roots $\lambda_1 = 0$, $\lambda_2 = -1$, $\lambda_3 = -2$. Therefore

$$x(s) = \frac{1}{s} R_1 + \frac{1}{s+1} R_2 + \frac{1}{s+2} R_3$$

$$x(t) = u_s(t) R_1 + e^{-t} u_s(t) R_2 + e^{-2t} u_s(t) R_3$$

$$R_1 = sx(s)|_{s=0} = \begin{bmatrix} \dfrac{s}{s+1} & 0 \\[2mm] 1 & \dfrac{s}{s+2} \end{bmatrix}_{s=0} = \begin{bmatrix} 0 & 0 \\ 1 & 0 \end{bmatrix}$$

$$R_2 = (s+1)x(s)|_{s=-1} = \begin{bmatrix} 1 & 0 \\[2mm] \dfrac{s+1}{s} & \dfrac{s+1}{s+2} \end{bmatrix}_{s=-1} = \begin{bmatrix} 1 & 0 \\ 0 & 0 \end{bmatrix}$$

$$R_3 = (s+2)x(s)|_{s=-2} = \begin{bmatrix} \dfrac{s+2}{s+1} & 0 \\[2mm] \dfrac{s+2}{s} & 1 \end{bmatrix}_{s=-2} = \begin{bmatrix} 0 & 0 \\ 0 & 1 \end{bmatrix}$$

The result agrees with the former calculation. ∎

### 8.4.4 The continuous time matrix exponential signal

We show that the Laplace transform of the matrix exponential signal exists and forms a strictly proper rational signal. Consider any matrix-induced norm. Then $|A^n| \leq |A|^n$, $n \geq 0$. Thus $|e^{At}| \leq \sum_{n=0}^{\infty} \frac{t^n}{n!} |A|^n = e^{|A|t}$, and the matrix exponential has exponential order (see the Appendix at the end of the chapter). Furthermore it is continuous. Thus the Laplace transform exists. We saw in Section 2.5.1 that the derivative of $x(t) = e^{At}$ satisfies $\dot{x}(t) = Ax(t)$. Therefore $\dot{x}(t)$ is also continuous and has exponential order. The differentiation property gives $sx(s) - x(0) = Ax(s)$. Since $x(0) = e^{A0} = I$, it follows that $(sI - A)x(s) = I$, or

$$\mathcal{L}(e^A t) = (sI - A)^{-1} = \frac{1}{\det(sI - A)} \text{adj}(sI - A) \qquad (8.28)$$

Hence $x(s)$ is a strictly proper rational function. Let $\lambda_1, \lambda_2, \ldots, \lambda_m$ be the roots of $\det(sI - A)$, that is, the eigenvalues of the matrix $A$ and $p_1, p_2, \ldots, p_m$ the corresponding multiplicities. Formula (8.28) gives

$$e^{At} = \sum_{i=1}^{m} \sum_{j=1}^{p_i} \frac{t^{j-1}}{(j-1)!} e^{\lambda_i t} R_{ij}, \qquad 0 \leq t < \infty \qquad (8.29)$$

$$R_{ij} = \frac{1}{(p_i - j)!} \frac{d^{p_i - j}}{ds^{p_i - j}} \left[ (s - \lambda_i)^{p_i} (sI - A)^{-1} \right]\Big|_{s=\lambda_i} \qquad (8.30)$$

Several important conclusions can be drawn from these expressions.

1. Each entry of the matrix exponential signal is formed by a linear combination of polynomial exponential signals whose parameters are specified by the eigenvalues of $A$.

2. If the eigenvalues of $A$ are pairwise distinct, we get a pure exponential expression

$$e^{At} = \sum_{i=1}^{m} e^{\lambda_i t} R_i, \qquad R_i = (s - \lambda_i)(sI - A)^{-1}\Big|_{s=\lambda_i} \qquad (8.31)$$

3. If $A$ is real, $e^{At}$ is real as well. The right-hand side of (8.29) reduces to a real matrix because the eigenvalues appear in conjugate pairs. The entries of $e^{At}$ include real polynomial exponentials as well as signals of the form $t^l \cos wt$, $t^l \sin wt$.

4. Since both sides of Eq. (8.29) are defined for negative values, Eq. (8.29) remains true on the entire real line if we observe that $e^{At} = e^{-A(-t)}$ for $t < 0$.

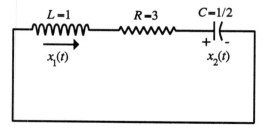

**Figure 8.5.** RLC circuit in Example 8.17.

We saw in Chapter 2 that the solution of

$$\dot{x}(t) = Ax(t), \quad x(0) = x_0 \tag{8.32}$$

is given by $x(t) = e^{At}x_0$. $A$ is a $k \times k$ matrix with real or complex entries, $x_0$ is an arbitrary initial $k \times l$ matrix. If Eq. (8.29) is substituted into the latter expression, a closed form formula results. All signals generated by a linear analog source are linear combinations of polynomial exponential signals. The decisive role of the eigenvalues of $A$ is evident.

### Example 8.17   RLC circuit

Consider the RLC circuit of Fig. 8.5. Let $x_1(t)$ be the current passing through the circuit elements and $x_2(t)$ the voltage across the capacitor. The equations describing the circuit are

$$\dot{x}_1 = -3x_1 - x_2, \quad \dot{x}_2 = 2x_1$$

The first equation follows upon applying Kirchhoff's voltage law on the loop formed by the three elements, while the second follows from the definition of the capacitor. The above equations are written in matrix form as

$$\begin{bmatrix} \dot{x}_1(t) \\ \dot{x}_2(t) \end{bmatrix} = \begin{bmatrix} -3 & -1 \\ 2 & 0 \end{bmatrix} \begin{bmatrix} x_1(t) \\ x_2(t) \end{bmatrix}$$

Let $A$ denote the matrix appeared above. Then

$$(sI - A)^{-1} = \frac{1}{(s+1)(s+2)} \begin{bmatrix} s & -1 \\ 2 & s+3 \end{bmatrix}$$

The eigenvalues of $A$ are $\lambda_1 = -1$ and $\lambda_2 = -2$. Therefore

$$e^{At} = e^{-t}R_1 + e^{-2t}R_2$$

The coefficients are found as follows:

$$R_1 = (s+1)(sI - A)^{-1}|_{s=-1} = \begin{bmatrix} -1 & -1 \\ 2 & 2 \end{bmatrix}$$

$$R_2 = (s+2)(sI - A)^{-1}|_{s=-2} = \begin{bmatrix} 2 & 1 \\ -2 & -1 \end{bmatrix}$$

Hence

$$e^{At} = \begin{bmatrix} -e^{-t} + 2e^{-2t} & -e^{-t} + e^{-2t} \\ 2e^{-t} - 2e^{-2t} & 2e^{-t} - e^{-2t} \end{bmatrix}$$

Suppose that initial current and voltage are $x_1(0) = I_0$ and $x_2(0) = V_0$. Then

$$\begin{bmatrix} x_1(t) \\ x_2(t) \end{bmatrix} = e^{At} \begin{bmatrix} x_1(0) \\ x_2(0) \end{bmatrix} = \begin{bmatrix} (-e^{-t} + 2e^{-2t})I_0 + (-e^{-t} + e^{-2t})V_0 \\ (2e^{-t} - 2e^{-2t})I_0 + (2e^{-t} - e^{-2t})V_0 \end{bmatrix} \qquad \blacksquare$$

### 8.4.5  The convolution property

Let $h(t)$ be an $m \times k$ matrix-valued signal and $u(t)$ a $k \times l$ matrix-valued signal. Suppose that both are piecewise continuous and have exponential order. Then the causal convolution

$$y(t) = (h * u)(t) = \int_0^t h(t - \tau)u(\tau)d\tau \qquad (8.33)$$

is piecewise continuous and has exponential order too. Moreover the Laplace transform of $y(t)$ is the product

$$y(s) = h(s)u(s) \qquad (8.34)$$

It is defined for all $s$ such that $s : \Re s > \max(\sigma_h, \sigma_u)$. As in the discrete case, it may exist on a larger region of convergence. The proof parallels the discrete counterpart and is left to the reader.

An immediate consequence is that if two signals have strictly proper rational Laplace transform, their convolution also has strictly proper rational Laplace transform. Therefore it is recovered by partial fraction expansion.

***Example 8.18   Computing the convolution by partial fraction expansion***
Compute the convolution $y(t) = h(t) * u(t)$ of the signals

$$u(t) = \begin{bmatrix} 3u_s(t) \\ 5u_s(t) \end{bmatrix}, \quad h(t) = e^{At}, \quad A = \begin{bmatrix} -1 & 0 \\ 0 & -2 \end{bmatrix}$$

We have

$$\mathcal{L}(e^{At}) = h(s) = \begin{bmatrix} \dfrac{1}{s+1} & 0 \\ 0 & \dfrac{1}{s+2} \end{bmatrix}, \quad u(s) = \begin{bmatrix} \dfrac{3}{s} \\ \dfrac{5}{s} \end{bmatrix}, \quad y(s) = \begin{bmatrix} \dfrac{3}{s(s+1)} \\ \dfrac{5}{s(s+2)} \end{bmatrix}$$

Partial fraction expansion gives

$$y(s) = \frac{1}{s} R_1 + \frac{1}{s+1} R_2 + \frac{1}{s+2} R_3$$

$$R_1 = \begin{bmatrix} \dfrac{3}{s+1} \\ \dfrac{5}{s+2} \end{bmatrix}_{s=0} = \begin{bmatrix} 3 \\ \dfrac{5}{2} \end{bmatrix} \quad R_2 = \begin{bmatrix} -3 \\ 0 \end{bmatrix} \quad R_3 = \begin{bmatrix} 0 \\ -\dfrac{5}{2} \end{bmatrix}$$

Thus

$$y(t) = \begin{bmatrix} 3u_s(t) - 3e^{-t} \\ \dfrac{5}{2} u_s(t) - \dfrac{5}{2} e^{-2t} \end{bmatrix}, \quad t \geq 0 \qquad \blacksquare$$

### 8.4.6 Finite derivative models in the s domain

Consider the linear time invariant finite derivative model

$$y^{(k)}(t) + \sum_{i=0}^{k-1} A_i y^{(i)}(t) = \sum_{i=0}^{r} B_i u^{(i)}(t) \tag{8.35}$$

The initial vectors specify successive input-output derivatives at time 0:

$$\begin{aligned} \mathbf{y}_0 &= \begin{bmatrix} y^T(0) & y^{(1)T}(0) & \cdots & y^{(k-1)T}(0) \end{bmatrix}^T \\ &= \begin{bmatrix} y_0^T & y_1^T & \cdots & y_{k-1}^T \end{bmatrix}^T \\ \mathbf{u}_0 &= \begin{bmatrix} u^T(0) & u^{(1)T}(0) & \cdots & u^{(r-1)T}(0) \end{bmatrix}^T \\ &= \begin{bmatrix} u_0^T & u_1^T & \cdots & u_{r-1}^T \end{bmatrix}^T \end{aligned} \tag{8.36}$$

The input $u(t)$ is sufficiently differentiable and the derivative of order $r$ has exponential order, so the Laplace transform can be used. The coefficients $A_0, A_1, \ldots, A_{k-1}$ are square real matrices, and the output $y(t)$ is a matrix-valued signal. Linearity and the differentiation property give

$$s^k y(s) - \sum_{i=0}^{k-1} s^{k-1-i} y_i = - \sum_{l=1}^{k-1} A_l \left[ s^l y(s) - \sum_{j=0}^{l-1} s^{l-1-j} y_j \right] - A_0 y(s)$$

$$+ \sum_{l=1}^{r} B_l \left[ s^l u(s) - \sum_{j=0}^{l-1} s^{l-1-j} u_j \right] + B_0 u(s)$$

Rearrangement of terms gives

$$\left( I s^k + \sum_{l=0}^{k-1} A_l s^l \right) y(s) = \sum_{i=0}^{k-1} s^{k-1-i} y_i + \sum_{l=1}^{k-1} A_l \sum_{j=0}^{l-1} s^{l-1-j} y_j$$

$$+ \sum_{l=0}^{r} B_l s^l u(s) - \sum_{l=1}^{r} B_l \sum_{j=0}^{l-1} s^{l-1-j} u_j$$

The double sums in the right-hand side are written

$$\sum_{j=0}^{k-2} \left( \sum_{l=j+1}^{k-1} A_l s^{l-1-j} \right) y_j, \qquad \sum_{j=0}^{r-1} \left( \sum_{l=j+1}^{r} B_l s^{l-1-j} \right) u_j$$

Hence

$$y(s) = h^y(s) \mathbf{y}_0 + h^u(s) \mathbf{u}_0 + h(s) u(s) \tag{8.37}$$

where $h(s)$, $h^y(s)$ and $h^u(s)$ are the proper rational matrices

$$h(s) = A^{-1}(s) B(s) \tag{8.38}$$

$$h^y(s) = A^{-1}(s) ( A_0(s) \quad A_1(s) \quad \cdots \quad A_{k-1}(s) ) \tag{8.39}$$

$$h^u(s) = - A^{-1}(s) ( B_0(s) \quad B_1(s) \quad \cdots \quad B_{r-1}(s) ) \tag{8.40}$$

The pertinent variables are the polynomial matrices

$$A(s) = \sum_{i=0}^{k} A_i s^i, \quad A_k = I, \quad B(s) = \sum_{i=0}^{r} B_i s^i \tag{8.41}$$

$$A_j(s) = \sum_{i=j+1}^{k} A_i s^{i-1-j}, \quad B_j(s) = \sum_{i=j+1}^{r} B_i s^{i-1-j} \tag{8.42}$$

The first term $h^y(s)\mathbf{y}_0$ in (8.37) depends only on the initial output values. It determines the output when the input is zero. The second term $h^u(s)\mathbf{u}_0$ indicates the contribution of initial input values. The third term $h(s)u(s)$ expresses the main effect of the input. The function $h(s)$ is called *transfer function* of the finite derivative model and characterizes its behavior when initial conditions are zero. It becomes strictly proper if $r \leq k - 1$. In this case the impulse response is a linear combination of polynomial exponential signals. In a similar fashion, $h^y(s)$ and $h^u(s)$ are rational. Inspection of (8.39)–(8.42) reveals that $h^y(s)$ is always strictly proper, while $h^u(s)$ has the same property when $r \leq k$. Thus the terms $h^y(s)\mathbf{y}_0$ and $h^u(s)\mathbf{u}_0$ are represented at the output by rational signals for any initial conditions. The parameters of the polynomial exponentials appearing in all three responses are characterized by the roots and multiplicities of the polynomial $\det A(s)$.

**Example 8.19    First-order system**
Consider the single-input single-output system

$$y^{(1)}(t) + ay(t) = b_0 u(t) + b_1 u^{(1)}(t)$$

We compute the pertinent polynomials

$$A(s) = s + a, \quad B(s) = b_1 s + b_0, \quad A_0(s) = 1, \quad B_0(s) = b_1$$

Thus

$$h(s) = \frac{b_1 s + b_0}{s + a}, \quad h^y(s) = \frac{1}{s + a}, \quad h^u(s) = \frac{-b_1}{s + a} \qquad \blacksquare$$

**Example 8.20    RLC circuit**
Consider the circuit of Fig. 3.24. The input-output relation is

$$y^{(2)}(t) + \frac{R}{L} y^{(1)}(t) + \frac{1}{LC} y(t) = \frac{1}{LC} u(t) \qquad (8.43)$$

The transfer function is

$$h(s) = \frac{1/LC}{s^2 + Rs/L + 1/LC}$$

If the independent source supplies the constant voltage $V$, that is $u(t) = V u_s(t)$, and the circuit is initially idle, the output is

$$y(s) = \frac{V}{s(LCs^2 + RCs + 1)} \qquad \blacksquare$$

**Example 8.21** *Mass-spring-dashpot system*

Consider the mass-spring-dashpot system of Fig. 3.25. The input-output relation is

$$y^{(2)}(t) + \frac{\lambda}{m} y^{(1)}(t) + \frac{k}{m} y(t) = \frac{1}{m} u(t) \tag{8.44}$$

The transfer function is

$$h(s) = \frac{1}{ms^2 + \lambda s + k} \qquad \blacksquare$$

**Example 8.22** *Multichannel linear source*

We want to compute the 2-channel signal generated by the source

$$y^{(2)}(t) = -A_1 y^{(1)}(t) - A_0 y(t)$$

where

$$A_1 = \begin{bmatrix} 3 & 2 \\ 0 & 3 \end{bmatrix}, \quad A_0 = \begin{bmatrix} 2 & 0 \\ 0 & 0 \end{bmatrix}$$

and the initial information

$$y(0) = \begin{bmatrix} 1 \\ 1 \end{bmatrix}, \quad y^{(1)}(0) = \begin{bmatrix} 2 \\ 0 \end{bmatrix}$$

The characteristic polynomial matrix is

$$A(s) = Is^2 + A_1 s + A_0 = \begin{bmatrix} s^2 + 3s + 2 & 2s \\ 0 & s^2 + 3s \end{bmatrix}$$

The determinant is

$$\det A(s) = (s^2 + 3s)(s^2 + 3s + 2) = s(s+1)(s+2)(s+3)$$

Therefore the solution $y(t)$ has the form

$$y(t) = u_s(t)R_1 + e^{-t}R_2 + e^{-2t}R_3 + e^{-3t}R_4, \quad t \geq 0$$

Now

$$A^{-1}(s) = \frac{1}{\det A(s)} \operatorname{adj} A(s) = \begin{bmatrix} \dfrac{1}{(s+1)(s+2)} & -\dfrac{2}{(s+1)(s+2)(s+3)} \\ 0 & \dfrac{1}{s(s+3)} \end{bmatrix}$$

$$A_1(s) = I, \quad A_0(s) = Is + A_1 = \begin{bmatrix} s+3 & 2 \\ 0 & s+3 \end{bmatrix}$$

and

$$A^{-1}(s)A_0(s) = \begin{bmatrix} \dfrac{s+3}{(s+1)(s+2)} & 0 \\ 0 & \dfrac{1}{s} \end{bmatrix}, \quad A^{-1}(s)A_1(s) = A^{-1}(s)$$

Therefore

$$h^y(s) = \begin{bmatrix} \dfrac{s+3}{(s+1)(s+2)} & 0 & \dfrac{1}{(s+1)(s+2)} & \dfrac{-2}{(s+1)(s+2)(s+3)} \\ 0 & \dfrac{1}{s} & 0 & \dfrac{1}{s(s+3)} \end{bmatrix}$$

The initial vector is

$$\mathbf{y}_0 = \begin{bmatrix} y_0^T & y_1^T \end{bmatrix}^T = \begin{bmatrix} 1 & 1 & 2 & 0 \end{bmatrix}^T$$

Hence

$$h^y(s)\mathbf{y}_0 = \begin{bmatrix} \dfrac{s+5}{(s+1)(s+2)} & \dfrac{1}{s} \end{bmatrix}^T$$

The partial fraction coefficients are

$$R_1 = sh^y(s)\mathbf{y}_0|_{s=0} = \begin{bmatrix} 0 \\ 1 \end{bmatrix}, \quad R_2 = (s+1)h^y(s)\mathbf{y}_0|_{s=-1} = \begin{bmatrix} 4 \\ 0 \end{bmatrix}$$

$$R_3 = (s+2)h^y(s)\mathbf{y}_0|_{s=-2} = \begin{bmatrix} -3 \\ 0 \end{bmatrix}, \quad R_4 = (s+3)h^y(s)\mathbf{y}_0|_{s=-3} = \begin{bmatrix} 0 \\ 0 \end{bmatrix}$$

and finally

$$y(t) = \begin{bmatrix} 4e^{-t} - 3e^{-2t} \\ 1 \end{bmatrix}, \quad t \geq 0$$

Notice that the pole $-3$ does not influence the solution. ∎

### 8.4.7   Linear time invariant state space models

Consider the state space model

$$\dot{x}(t) = Ax(t) + Bu(t)$$
$$y(t) = Cx(t) + Du(t)$$

The differentiation property enables us to write the state equation in the $s$ domain as $sx(s) - x_0 = Ax(s) + Bu(s)$, or

$$x(s) = (sI - A)^{-1}x_0 + (sI - A)^{-1}Bu(s) \tag{8.45}$$

Substitution in the output equation gives

$$y(s) = C(sI - A)^{-1}x_0 + [C(sI - A)^{-1}B + D]u(s) \tag{8.46}$$

The *transfer function* is

$$h(s) = C(sI - A)^{-1}B + D \tag{8.47}$$

It demonstrates the impact of the input and determines the output when the initial state is zero. It is given by the same expression as its discrete time counterpart; see Eq. (8.20). Therefore it is a proper rational function. It becomes strictly proper if $D = 0$. For the same reason $h^0(s) = C(sI - A)^{-1}$ is strictly proper and in the time domain becomes

$$h^0(t) = \mathcal{L}^{-1}[h^0(s)] = Ce^{At} \tag{8.48}$$

Note that $h(t)$ and $h^0(t)$ are linear combinations of polynomial exponential signals.

## 8.5   TWO-SIDED LAPLACE TRANSFORM

### 8.5.1   Definition and examples

The two-sided Laplace transform applies to signals defined over the entire real line. It is defined by the integral

$$x(s) = \int_{-\infty}^{\infty} x(t)e^{-st}dt \tag{8.49}$$

This presumes that the one-sided integrals

$$x_1(s) = \int_{-\infty}^{0} x(t)e^{-st}dt, \quad x_2(s) = \int_{0}^{\infty} x(t)e^{-st}dt$$

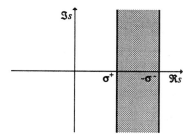

**Figure 8.6.**  Region of convergence of two-sided Laplace transform.

converge absolutely, and it holds that $x(s) = x_1(s) + x_2(s)$. Much like the $z$ transform the two-sided Laplace transform is obtained from the superposition $x(s) = x^+(s) + x^-(-s)$ of the one-sided Laplace transforms of the signals $x^+(t) = x(t)u_s(t)$ and $x^-(t) = x(-t)u_s(t)$. $x(s)$ is defined on the strip $\sigma^+ < \Re s < -\sigma^-$ provided that $\sigma^+ < -\sigma^-$. The form of the region of convergence is illustrated in Fig. 8.6. It follows from the above discussion that if the two-sided Laplace transform converges at two points $s_1$ and $s_2$ and $\Re s_1 < \Re s_2$, then it converges everywhere in the strip $\Re s_1 \leq \Re s \leq \Re s_2$.

### *Example 8.23  Exponential signals*

Consider the causal exponential signal $x(t) = e^{at}u_s(t)$. The two-sided Laplace transform coincides with the one-sided transform. Thus $x(s) = 1/(s - a)$, $\Re s > \Re a$. The region of convergence corresponds to the half plane $\sigma^- = -\infty$, $\sigma^+ = \Re a$. Consider next the strictly noncausal exponential $x(t) = -e^{at}u_s(-t)$. Then

$$x(s) = -\int_{-\infty}^{0} e^{at}e^{-st}dt = -\frac{1}{s-a}e^{(s-a)t}\Big|_{0}^{\infty} = \frac{1}{s-a}, \qquad \Re s < \Re a$$

We observe that the causal exponential and the strictly noncausal exponential are transformed to the same functional form but have totally distinct regions of convergence. Let us finally consider the signal $x(t) = e^{a|t|}$, $\Re a < 0$. We have $x^+(t) = e^{at}u_s(t)$, $x^-(t) = e^{at}u_s(t)$. Thus

$$x^+(s) = \frac{1}{s-a}, \quad x^-(s) = \frac{1}{s-a}, \qquad \Re s > \Re a$$

Since $\Re a < 0$, it holds that $\Re a < -\Re a$. Hence

$$x(s) = \frac{2a}{s^2 - a^2}, \qquad \Re a < \Re s < -\Re a. \qquad \blacksquare$$

## 8.5.2 Partial fraction expansion

The only distinction from the one-sided case regards the region of convergence. Consider the proper rational function $x(s)$, and let $\lambda_1, \lambda_2, \ldots, \lambda_m$ be the distinct roots of the denominator, ordered as $\Re\lambda_1 < \Re\lambda_2 < \cdots < \Re\lambda_m$. The complex plane is split into $m+1$ strips

$$\{s \in C : \Re s < \Re\lambda_1\}, \; \{s \in C : \Re\lambda_1 < \Re s < \Re\lambda_2\}, \ldots, \{s \in C : \Re s > \Re\lambda_m\}$$

In each strip every fraction $1/(s - \lambda_i)$ corresponds either to the strictly noncausal exponential $-e^{\lambda_i t}u_s(-t)$ or to the causal exponential $e^{\lambda_i t}u_s(t)$. The $m+1$ signals obtained this way constitute all possible signals whose Laplace transform have the same functional form. Among them the one having the given Laplace transform is determined from the given region of convergence. We illustrate the procedure with the following example:

***Example 8.24    Partial fraction expansion***
Let

$$x(s) = \frac{s-1}{(s+1)(s+2)}$$

The partial fraction expansion gives

$$x(s) = -\frac{2}{s+1} + \frac{3}{s+2}$$

All possible regions of convergence are

$$\Re s < -2, \quad -2 < \Re s < -1 \quad -1 < \Re s$$

The signals associated with each region are $x_1(t) = (2e^{-t} - 3e^{-2t})u_s(-t)$, $x_2(t) = 3e^{-2t}u_s(t) + 2e^{-t}u_s(-t)$, and $x_3(t) = (3e^{-2t} - 2e^{-t})u_s(t)$.    ∎

Further properties of the two-sided Laplace transform are given in the exercises.

## 8.5.3 Relationship of Laplace and Fourier transforms

The similarity between the Laplace and Fourier transform is evident. The Fourier transform $X(\omega)$ is obtained from the Laplace transform if the latter is evaluated on the imaginary axis

$$\mathcal{F}(x)(\omega) = \mathcal{L}(x)(s)|_{s=j\omega} \tag{8.50}$$

Equation (8.50) presumes that the region of convergence of $x(s)$ contains the imaginary axis. Consequently, if the Laplace transform of a signal exists, that is, it converges absolutely inside a strip and the region of convergence contains the imaginary axis, the signal is absolutely integrable and the Fourier transform exists. This fact can be used for the calculation of the Fourier transform of rational signals.

**Example 8.25  Computing the Fourier transform from the Laplace transform**
The region of convergence of the causal exponential $x(t) = e^{at}u_s(t)$ contains the imaginary axis if $\Re a < 0$. In this case Eq. (8.50) gives

$$X(\omega) = \frac{1}{j\omega - a}$$

We obtain the same result with that of Example 6.2. Note that if $\Re a > 0$, the Laplace transform of the exponential signal exists with region of convergence $\Re s > \Re a$, while the Fourier transform is not defined.  ■

### 8.5.4  Laplace transform representation of systems

**Convolutional systems.**  In complete analogy with the one-sided case, the two-sided Laplace transform converts the convolution into multiplication in the $s$ domain. Consequently the input-output representation of a noncausal convolutional system

$$y(t) = h(t) * u(t) = \int_{-\infty}^{\infty} h(t - \tau)u(\tau)d\tau$$

converts into

$$y(s) = h(s)u(s)$$

provided that the Laplace transforms of $h$ and $u$ exist and their regions of convergence have nonempty intersection. The transfer function $h(s)$ specifies the behavior of the transformed system.

**Example 8.26  Output of a convolutional system**
Let

$$u(t) = 3e^{-t}u_s(t), \quad h(t) = 2e^{-2t}u_s(t) + 4e^{3t}u_s(-t)$$

Passing to the $s$ domain, we obtain

$$u(s) = \frac{3}{s + 1}, \qquad \Re s > -1$$

$$h(s) = \frac{2}{s + 2} - \frac{4}{s - 3}, \qquad -2 < \Re s < 3$$

Therefore

$$y(s) = \frac{6}{(s+1)(s+2)} - \frac{12}{(s+1)(s-3)}, \qquad -1 < \Re s < 3$$

Thus

$$y(t) = 9e^{-t}u_s(t) - 6e^{-2t}u_s(t) + 3e^{3t}u_s(-t) \qquad \blacksquare$$

**Linear time invariant finite derivative models.** The causal case was considered in Section 8.4.6. In the noncausal case the differentiation property of the two-sided Laplace transform leads to

$$y(s) = h(s)u(s), \quad h(s) = A^{-1}(s)B(s) \qquad (8.51)$$

where the transfer function is rational. We will see in Section 10.1 that finite derivative models are converted to state space forms if $r \le k$. The variation of constants formula then shows that given any initial vector and any sufficiently differentiable input defined over all reals, the resulting output is uniquely defined over all reals. Among these output trajectories very few can be visualized in the $s$ domain, since very few have two-sided Laplace transforms. Much like the analysis of discrete signals in the $z$ domain, they are obtained from (8.51) if $h(s)$ is evaluated on each of the $m+1$ possible strips that have a nonempty intersection with the region of convergence of $u(s)$.

**Example 8.27  First-order system**
Consider the causal first order system

$$\dot{y}(t) + ay(t) = bu(t), \qquad a > 0 \qquad (8.52)$$

The transfer function is

$$h(s) = \frac{b}{s+a}, \qquad \Re s > -a$$

The frequency response and the impulse response are

$$H(\omega) = \frac{b}{j\omega + a}, \quad h(t) = be^{-at}u_s(t)$$

Likewise the step response is given by

$$y(t) = \left(\frac{b}{a} - \frac{b}{a}e^{-at}\right)u_s(t)$$

The steady state of the step response is $b/a$. The parameter $\tau = 1/a$ is called *time constant*, and it indicates the rate by which the system responses. At time $t = \tau$, the value of the impulse response is decreased by $1/e$ times its initial value while the step response reaches $1 - 1/e$ of its final value. The frequency response magnitude in dB is

$$20 \log_{10} |H(\omega)| = 20 \log_{10} \frac{\tau |b|}{\sqrt{1 + \tau^2 \omega^2}} = 20 \log_{10} \tau |b| - 10 \log_{10}(1 + \tau^2 \omega^2)$$

(8.53)

If $\omega\tau \ll 1$, it holds that $\log_{10}(1 + \tau^2 \omega^2) \approx \log_{10} 1 = 0$. Thus at low frequencies the magnitude is approximately constant, that is,

$$20 \log_{10} |H(\omega)| \approx 20 \log_{10} \tau |b|$$

(8.54)

At high frequencies $\omega\tau \gg 1$, we have

$$20 \log_{10} |H(\omega)| \approx 20 \log_{10} \tau |b| - 20 \log_{10} \tau - 20 \log_{10} \omega$$

(8.55)

If frequencies are represented in logarithmic scale, the magnitude is approximated by the line of slope $-20$. Bode plots are depicted in Fig. 8.7.

The low- and high-frequency asymptotes intersect at frequency $\omega = 1/\tau = a$. The difference between the actual value and the approximate value is $-10 \log_{10}(2) \approx -3$dB. For this reason the frequency $\omega = a$ is called the *3dB point*. Similar analysis applies for the phase. ∎

***Example 8.28*** ***Analysis of second-order systems***
Consider the causal system

$$y^{(2)}(t) + 2\zeta\omega_n y^{(1)}(t) + \omega_n^2 y(t) = \omega_n^2 u(t)$$

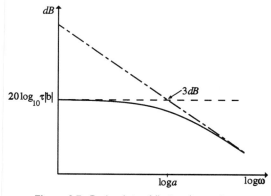

**Figure 8.7.** Bode plots of first-order system.

The transfer function is

$$h(s) = \frac{\omega_n^2}{s^2 + 2\zeta\omega_n s + \omega_n^2}$$

and the frequency response is

$$H(\omega) = \frac{1}{(j\omega/\omega_n)^2 + 2\zeta(j\omega/\omega_n) + 1}$$

$\zeta$ is referred to as the damping ratio and $\omega_n$ as the undamped natural frequency. If $0 < \zeta < 1$, the denominator of the frequency response has complex roots. Reconstruction in the time domain by partial fraction shows that the impulse response has a damped oscillatory behavior. If $\zeta > 1$, the poles of the transfer function are real, and the impulse response is expressed as a superposition of real exponential signals that asymptotically vanish; see Fig. 8.8(a).

If $0 < \zeta < 1$, the step response exhibits overshoot and ripple. If $\zeta \geq 1$, no ripples occur (see Fig. 8.8(b)). If $\zeta = 1$, the rising time is minimum, and the system has the fastest response. As $\zeta$ moves away from the critical value $\zeta = 1$, the response gets slower. If $\zeta < 1$ and $\omega_n$ increases, the ripples get more crowded. Bode plots of the second-order system are shown in Fig. 8.9. The magnitude of the frequency response is

$$20 \log_{10} |H(\omega)| = -10 \log_{10} \left( \left[ 1 - \left( \frac{\omega}{\omega_n} \right)^2 \right]^2 + 4\zeta^2 \left( \frac{\omega}{\omega_n} \right)^2 \right)$$

At low frequencies $\omega \ll \omega_n$, the Bode plot is approximated by the low-frequency asymptote line

$$20 \log_{10} |H(\omega)| \approx 0$$

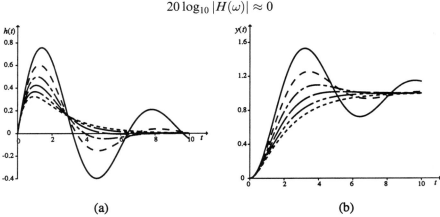

(a)                                                    (b)

**Figure 8.8.** (a) Impulse response of second-order system. (b) Step response of second-order system.

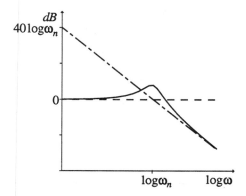

**Figure 8.9.** Bode plots for the second-order system: Frequency response magnitude, and low- and high-frequency asymptotes.

At high frequencies $\omega \gg \omega_n$, the Bode plot is approximated by the asymptote

$$-10\log_{10}\left(\frac{\omega^4}{\omega_n^4}\right) = -40\log_{10}(\omega) + 40\log_{10}(\omega_n)$$

The two lines intersect at $\omega = \omega_n$. Analogous plots are derived for the phase. ∎

**Bode plots for rational systems.** Bode plots of general rational transfer function models are developed via mere superposition of Bode plots of first- and second-order systems. Indeed, let $h(s) = b(s)/a(s)$. We can factor $b(s)$ and $a(s)$ into first- or second-order factors, and we can write $h(s)$ as a product of first- and second-order transfer functions as well as inverses of such systems. For example,

$$h(s) = \frac{3s+5}{s^2+7s+6} = (3s+5)\frac{1}{s+1}\frac{1}{s+6}$$

The systems $1/(s+1)$ and $1/(s+6)$ have order 1, while $3s+5$ is the inverse of the first-order system $1/(3s+5)$. If $G(\omega)$ is the frequency response of a first- or second-order system, the inverse system has magnitude

$$20\log_{10}\left|\frac{1}{G(\omega)}\right| = -20\log_{10}|G(\omega)|$$

The Bode plots of $h(s)$ are obtained from the superposition of Bode plots of order 1 or 2. The procedure is illustrated in Fig. 8.10.

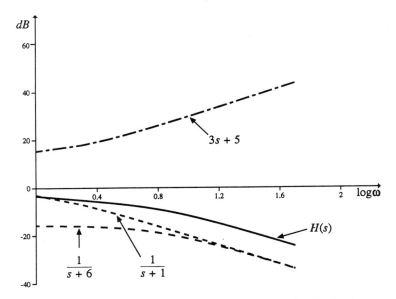

**Figure 8.10.** Bode plots for rational systems from the corresponding Bode plots of first and second order.

## 8.6  STABILITY

In Chapter 4 stability of convolutional systems was introduced. We saw that BIBO stability of a discrete system is ensured if the system impulse response $h(n)$ is absolutely summable. It follows from Theorem 8.1 that the $z$ transform of an absolutely summable signal exists and that the region of convergence contains the unit circle. Conversely, if the $z$ transform of $h(n)$ exists (in the sense of absolute convergence) and the region of convergence contains the unit circle, $h(n)$ is absolutely summable. Suppose that the convolutional model derives from a finite difference model. In this case $h(z)$ is proper rational. The region of convergence is one of the annuli described by Eq. (8.16). Among the systems with the same transfer function $h(z)$, there is a unique BIBO stable system. It corresponds to the annulus that contains the unit circle. If moreover $h(n)$ is causal, the region of convergence is necessarily the exterior of a disc. In this case $h(z)$ contains the unit circle if and only if all poles lie inside the unit disc. We have proved the following:

**Theorem 8.2**  The following statements hold:
   1. The convolutional system

$$y(n) = \sum_{k=-\infty}^{\infty} h(n-k)u(k)$$

is BIBO stable if and only if the region of convergence of the transfer function $h(z)$ contains the unit circle.

2. The causal finite difference model with zero initial conditions

$$y(n) + \sum_{i=1}^{k} a_i y(n-i) = \sum_{i=1}^{r} b_i u(n-i)$$

is BIBO stable if the poles of the transfer function $h(z) = b(z)/a(z)$ satisfy $|\lambda| < 1$. ∎

Entirely analogous statements are valid for continuous time systems.

**Theorem 8.3**
1. The convolutional system

$$y(t) = \int_{-\infty}^{\infty} h(t-\tau)u(\tau)d\tau$$

is BIBO stable if and only if the region of convergence of the transfer function $h(s)$ contains the imaginary axis.

2. The causal finite derivative model with zero initial conditions

$$y^{(k)}(t) + \sum_{i=0}^{k-1} a_i y^{(i)}(t) = \sum_{i=0}^{r} b_i u^{(i)}(t)$$

is BIBO stable if the poles of the transfer function $h(s) = b(s)/a(s)$ satisfy $\Re \lambda < 0$. ∎

*Example 8.29* **Stability of a finite difference model**
Consider the system

$$y(n) + \frac{5}{2}y(n-1) + y(n-2) = u(n) + 5u(n-1) + 6u(n-2)$$

The transfer function is

$$h(z) = \frac{z^2 + 5z + 6}{z^2 + \frac{5}{2}z + 1}$$

The roots of the denominator are $\lambda_1 = -2$, $\lambda_2 = -1/2$. The inequality $|\lambda_1| > 1$ may lead us to the wrong conclusion that the system is unstable. However, $\lambda_1$ is not a pole, since it is also a root of the numerator and cancels out. Thus $h(z)$ has a unique pole at $-1/2$ and the system is stable. ∎

*Example 8.30   Stability of backward and forward simulators*
As we saw in Chapter 4, the backward simulator of the finite derivative model
derives from the approximation of the differentiator with the backward difference

$$\frac{d}{dt}y(t) \approx \nabla y(n) = \frac{1}{T_s}(y(n) - y(n-1))$$

The Laplace transform of the differentiator $d/dt$ is $s$, while the $z$ transform of the
difference $\nabla$ is $(1 - z^{-1})/T_s$. Hence simulation is described by the following map
between the $z$ and the $s$ domain:

$$s = \frac{1}{T_s}(1 - z^{-1}) \tag{8.56}$$

The transfer function of the backward simulator $h_d(z)$ results from the transfer
function of the finite derivative model $h(s)$ via

$$h_d(z) = h(s)|_{s=(1-z^{-1})/T_s} \tag{8.57}$$

As $s$ varies on the imaginary axis $s = j\omega$, $z$ varies on a circle of center $(1/2, 0)$ and
radius $1/2$. It is illustrated in Fig. 8.11. Indeed,

$$\left(\frac{1}{1 - j\omega T_s} - \frac{1}{2}\right)^* \left(\frac{1}{1 - j\omega T_s} - \frac{1}{2}\right) = \frac{1}{4}$$

The circle is inside the unit disc. Moreover the left half plane is mapped in the
interior of the circle. Thus, if the analog system is stable for zero initial
conditions, the backward simulator is also stable for any sampling period.

In contrast, the forward simulator is not stable for every value of the sampling
period. The transformation has now the form $s = (z - 1)/T_s$. The line $s = j\omega$ is
mapped into the line $z = 1 + T_s\omega j$. For given $T_s$, we can easily construct a stable
analog system with unstable forward simulator. ∎

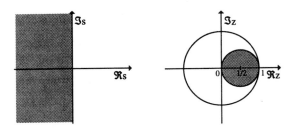

**Figure 8.11.** The map $s = (1 - z^{-1})/T_s$ carries the left half plane in the circle of radius $\frac{1}{2}$ and center
$(\frac{1}{2}, 0)$.

## 8.7 MULTIDIMENSIONAL LAPLACE TRANSFORM

In close analogy with the multidimensional $z$ transform, the $m$-$D$ Laplace transform is defined by

$$x(s_1, \ldots, s_m) = \int_{-\infty}^{\infty} \cdots \int_{-\infty}^{\infty} x(t_1, \ldots, t_m) \exp(-\sum_{i=1}^{m} s_i t_i) dt_1 \cdots dt_m \qquad (8.58)$$

The multidimensional Laplace transform has similar properties with the $z$ transform. Among others it is linear and converts multidimensional convolution to multiplication. If the signal $x(t_1, \ldots, t_m)$ is absolutely integrable the Fourier transform exists and is recovered from the Laplace transform by

$$X(\omega_1, \ldots, \omega_m) = x(s_1, \ldots, s_m)|_{(s_1, \ldots, s_m) = (j\omega_1, \ldots, j\omega_m)}$$

Inversion, on the other hand, is complicated. As in the discrete case partial fraction expansion is not applicable.

### 8.7.1 State affine and bilinear systems

Multidimensional Laplace and $z$ transforms are useful in the analysis of nonlinear state space models and Volterra expansions. Consider the single-input single-output system

$$x(n + 1) = f(x(n), u(n))$$
$$y(n) = g(x(n), u(n))$$

Suppose that $f(0, 0) = g(0, 0) = 0$ and that $f, g$ are sufficiently differentiable. If $f$ and $g$ are approximated by a finite Taylor expansion, the above equations give rise to approximate expressions where the right-hand side maps are polynomials. It can be shown that such polynomial state space models are approximately described by state affine systems. These have the form

$$x(n + 1) = \sum_{i=0}^{k-1} A_i x(n) u^i(n) + \sum_{i=1}^{k} b_i u^i(n)$$

$$y(n) = \sum_{i=0}^{k-1} c_i x(n) u^i(n) + \sum_{i=1}^{k} d_i u^i(n)$$

$A_i$ are square matrices, $b_i$ are column vectors, $c_i$ are row vectors, and $d_i$ are numbers. The state enters in an affine way, while the input enters polynomially. It can be shown that state affine systems are represented by Volterra systems having the property that the multidimensional $z$ transforms of the regular Volterra kernels have the form

$$h_{r,m}(z_1, z_2, \ldots, z_m) = \frac{P(z_1, z_2, \ldots, z_m)}{Q_1(z_1) \cdots Q_m(z_m)} \qquad (8.59)$$

and $P$, $Q_i$ are polynomial functions.

The corresponding continuous time result somewhat differs. The nice approximating systems are the bilinear systems. Thus each system of the form

$$\dot{x}(t) = f(x(t), u(t))$$

$$y(t) = g(x(t), u(t))$$

satisfying $f(0,0) = g(0,0) = 0$ is approximated by similar expressions involving polynomial maps $f$ and $g$. These polynomial systems are in turn approximately represented by bilinear systems of the form

$$\dot{x}(t) = Ax(t) + Ex(t)u(t) + bu(t)$$

$$y(t) = cx(t) + du(t)$$

Furthermore a bilinear system as above admits a Volterra representation with the property that the multidimensional Laplace transforms of the regular Volterra kernels have the form (8.59). No proofs are attempted. To get a feeling of the above results, let us look at the following example.

**Example 8.31   Volterra kernels of a bilinear system**
Consider the zero initialized bilinear system on $R^2$:

$$\dot{x}(t) = Ax(t)u(t) + bu(t)$$

$$y(t) = cx(t)$$

where $A$ is a nilpotent matrix satisfying $A^2 = 0$, $b$ is $2 \times 1$ column, and $c$ is $1 \times 2$ row. Let $A(t) = Au(t)$. The state equation has the form

$$\dot{x}(t) = A(t)x(t) + bu(t)$$

The variation of constants applies. It is easy to see that the transition matrix associated with $A(t)$ is given by

$$\Phi(t, t_0) = e^{A \int_{t_0}^{t} u(\tau)d\tau}$$

Expanding the exponential in power series and taking into account that $A$ is nilpotent gives

$$\Phi(t, t_0) = I + A \int_{t_0}^{t} u(\tau)d\tau$$

Thus

$$y(t) = \int_{0}^{t} cbu(\tau)d\tau + \int_{0}^{t}\int_{\tau_1}^{t} cAbu(\tau_1)u(\tau_2)d\tau_2 d\tau_1$$

The bilinear system is described by a Volterra system of degree 2. The Volterra kernel of degree 1 is $h_1(t) = cbu_s(t)$. The triangular kernel of degree 2 is $h_{t,2}(t_1, t_2) = cAbu_s(t_2)u_s(t_1 - t_2)$. The Laplace transform of $h_1(t)$ is $h_1(s) = cb/s$. The regular kernel of degree 2 is obtained from the triangular kernel using the formula $h_{r,2}(t_1, t_2) = h_{t,2}(t_1 + t_2, t_2) = cAbu_s(t_1)u_s(t_2)$. The latter has 2-D Laplace transform

$$h_{r,2}(s_1, s_2) = cAb \int_{0}^{\infty}\int_{0}^{\infty} e^{-s_1 t_1 - s_2 t_2} u_s(t_1)u_s(t_2)dt_1 dt_2$$

$$= cAb \int_{0}^{\infty} e^{-s_1 t_1} dt_1 \int_{0}^{\infty} e^{-s_2 t_2} dt_2 = \frac{cAb}{s_1 s_2}.$$

Both kernels have the form (8.59). ∎

### 8.7.2  Cumulant generating function

We will make no further use of $m$-$D$ Laplace transforms except their capacity to represent cumulant generating functions described next.

***Example 8.32    Cumulant generating function***
Let $X$ be a random variable with probability density function $f(x)$. The Fourier transform of $f(-x)$ defines the characteristic function of $X$. Likewise the Laplace transform of $f(-x)$ specifies the *moment generating function $M(s)$*. Thus

$$M(s) = \int_{-\infty}^{\infty} f(x)e^{sx}dx = E[e^{sX}] \qquad (8.60)$$

Expansion of the exponential in power series shows that higher-order moments are generated from the derivatives of $M(s)$ by

$$E[X^n] = \frac{d^n M(s)}{ds^n}\bigg|_{s=0}$$

and this explains the name of $M(s)$. The moment generating function of a

Gaussian random variable $N(m, \sigma^2)$ is

$$M(s) = \exp\left(sm + \frac{1}{2}s^2\sigma^2\right) \tag{8.61}$$

In many cases it is much simpler to work with the exponent of (8.61) which is a quadratic function of s, rather than $M(s)$ itself. This leads to the notion of the cumulant generating function defined by

$$K(s) = \log M(s) = \log E[\exp(sX)] \tag{8.62}$$

The coefficients $k_n$ in the power series expansion of $K(s)$ provide the cumulants of $X$

$$K(s) = \sum_{n=0}^{\infty} \frac{k_n}{n!} s^n$$

Thus $k_0 = \log E[e^{0X}] = \log 1 = 0$. Furthermore

$$\frac{dK(s)}{ds} = \frac{1}{M(s)} E[e^{sX} X] \tag{8.63}$$

Evaluation at $s = 0$ gives $k_1 = E[X]$. We write Eq. (8.63) as $M(s)[dK(s)/ds] = E[e^{sX} X]$ and differentiate to obtain

$$E[e^{sX} X]\frac{dK(s)}{ds} + M(s)\frac{d^2 K(s)}{ds^2} = E[e^{sX} X^2]$$

Evaluation at zero gives $k_2 = E[X^2] - E[X]^2 = \text{cov}[X, X]$. Continuing this way, we find

$$k_3 = E[(X - m_X)^3], \quad k_4 = E[(X - m_X)^4] - 3\sigma^4$$

Note that if $X$ is Gaussian, the cumulant generating function is the quadratic polynomial

$$K(s) = ms + \frac{1}{2}\sigma^2 s^2$$

and thus all cumulants of order higher than 2 are zero. This property is not shared by higher-order moments (see Exercise 8.35). As a consequence cumulants enable us to detect how much a random variable digresses from Gaussianity.

The cumulant generating function of a random vector is a direct extension of the previous concept. Suppose that $X = [x_1, x_2, \ldots, x_L]^T$ is a random vector with probability density $f(x)$. The $L$ dimensional Laplace transform of $f(-x)$ defines the moment generating function

$$M(s) = M(s_1, s_2, \ldots, s_L) = \int_{-\infty}^{\infty} \cdots \int_{-\infty}^{\infty} e^{\sum_{i=1}^{L} s_i x_i} f(x_1, \ldots, x_L) dx_1 \cdots dx_L$$

Clearly

$$M(s) = \int_{R^L} e^{s^T x} f(x) dx = E[e^{s^T X}] \tag{8.64}$$

The logarithm of $M(s)$ defines the cumulant generating function

$$K(s) = \log M(s) = \log E[e^{s^T X}] \tag{8.65}$$

As in the scalar case the cumulant generating function is expanded in a power series of $L$ variables:

$$K(s) = K(s_1, s_2, \ldots, s_L) = \sum_{i_1 i_2 \cdots i_L} k_{i_1 i_2 \cdots i_L} \frac{1}{i_1! i_2! \ldots i_L!} s_1^{i_1} s_2^{i_2} \cdots s_L^{i_L}$$

We will be interested only in the coefficient $k_{11\cdots1}$ associated with the term $s_1 s_2 \cdots s_L$, and we will call it the *L-order cumulant*. Thus

$$\text{cum}[x_1, x_2, \ldots, x_L] = k_{11\cdots1} = \frac{\partial K(s_1, s_2, \ldots, s_L)}{\partial s_1 \partial s_2 \cdots \partial s_L} \bigg|_{s=0} \tag{8.66}$$

Let us take, for instance, $L = 2$. Differentiation of (8.65) with respect to $s_1$ gives

$$\frac{\partial K(s_1, s_2)}{\partial s_1} = \frac{\partial \log E[e^{s_1 x_1 + s_2 x_2}]}{\partial s_1} = \frac{1}{M(s_1, s_2)} E[e^{s_1 x_1 + s_2 x_2} x_1] \tag{8.67}$$

or

$$M(s_1, s_2) \frac{\partial K(s_1, s_2)}{\partial s_1} = E[e^{s_1 x_1 + s_2 x_2} x_1]$$

Differentiation with respect to $s_2$ gives

$$\frac{\partial M(s_1, s_2)}{\partial s_2} \frac{\partial K(s_1, s_2)}{\partial s_1} + M(s_1, s_2) \frac{\partial^2 K(s_1, s_2)}{\partial s_1 \partial s_2} = E[e^{s_1 x_1 + s_2 x_2} x_1 x_2]$$

Setting $s_1 = s_2 = 0$, we find from Eqs. (8.64), (8.66), and (8.67) that

$$\text{cum}[x_1, x_2] = E[x_1 x_2] - E[x_1] E[x_2] = \text{cov}[x_1, x_2]$$

If the random vector $X$ is Gaussian and $L > 2$, the cumulant of order $L$ is zero.

Indeed, $K(s)$ has the form $K(s) = s^T m + \frac{1}{2} s^T C s$. This is a polynomial function of degree 2. If $L > 2$, the term $s_1 s_2 \cdots s_L$ does not appear in $K(s)$.   ∎

## 8.8   FILTER DESIGN

Frequency selective filters were introduced in Section 6.5 as convolutional systems whose frequency response magnitude has a piecewise characteristic. It was pointed out that these systems are noncausal and unstable. Filter design is concerned with the development of realizable systems that approximate the above ideal behavior. System architectures that have attracted most interest include linear time invariant finite derivative models (analog filters), finite difference models (IIR digital filters), and FIR systems (FIR digital filters). Conforming to standard terminology in filter design, discrete time filters are called *digital filters*. The first case involves analog design, while the second and third cases involve digital design.

In many applications phase information is equally important as magnitude information. A filter modifies both the magnitude and the phase spectrum of a signal. To avoid phase distortion, we employ filters with the linear phase property in which the phase response depends linearly on the frequency. More generally, we say that $H(\omega)$ has *linear phase* if it is written as

$$H(\omega) = c e^{-jK\omega} H_R(\omega)$$

where $c$ is a complex constant, $K$ is real (integer in the discrete case) and $H_R(\omega)$ is a real-valued function called *amplitude response* or *zero phase response*. The shift property of the Fourier transform (Section 6.1.4) implies that the output is a (scaled) and delayed version of the input in the passband. The extent to which phase distortion is tolerated is application dependent. The negative impact of nonlinear phase is critical in image processing.

The linear phase property is easy to establish within the class of FIR systems. An FIR filter with transfer function $h(z) = \sum_{n=0}^{L} h(n) z^{-n}$ has linear phase if and only if $h(n)$ is symmetric: $h(n) = h(L - n)$, $0 \leq n \leq L$, or antisymmetric $h(n) = -h(L - n)$ (see Exercise 8.37).

The ideal characteristic of a lowpass filter has an abrupt transition from the passband to the stopband. The admissible transfer function models we will consider have continuous frequency responses and thus employ a transition band in which a gradual transition from the passband to the stopband takes place. Design specifications are usually stated as tolerance limits on the frequency response magnitude and the width of the transition band. As Fig. 8.12 indicates, the frequency response magnitude of the lowpass filter in the passband is approximately 1 with error $\pm \delta_1$:

$$||H(\omega)| - 1| \leq \delta_1, \qquad |\omega| \leq \omega_p$$

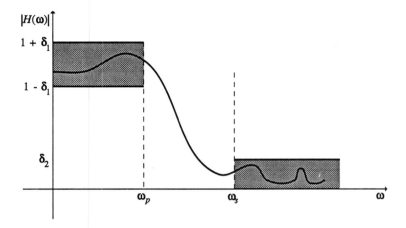

**Figure 8.12.** Design specifications for lowpass filters.

In the stopband it is zero with error $\delta_2$

$$|H(\omega)| \le \delta_2, \qquad |\omega| \ge \omega_s$$

Design specifications are expressed in terms of $\delta_1$, $\delta_2$ and the transition bandwidth $\omega_s - \omega_p$. Alternatively, the *attenuation* or *loss function* $A(\omega) = -20 \log_{10} |H(\omega)|$ is used. The tolerance limits convert to $A_s = -20 \log_{10} \delta_2$ (minimum stopband attenuation) and $A_p = -20 \log_{10}(1 - \delta_1)$ (peak passband value) in dB. Requirements become more stringent as the tolerance parameters get smaller.

### 8.8.1  Butterworth filters

A Butterworth filter of order $m$ provides the maximum possible flat behavior in the passband, in the sense that the magnitude response satisfies $|H(0)|^2 = 1$ and its first $2m - 1$ derivatives at $\omega = 0$ are zero. This flatness requirement is met by the function

$$|H(\omega)|^2 = \frac{1}{1 + (\omega/\omega_c)^{2m}} \tag{8.68}$$

$|H(\omega)|$ is monotonically decreasing and approaches 0 as $\omega \to \infty$. We also have

$$20 \log_{10} |H(\omega)| = -10 \log_{10} \left( 1 + \left( \frac{\omega}{\omega_c} \right)^{2m} \right)$$

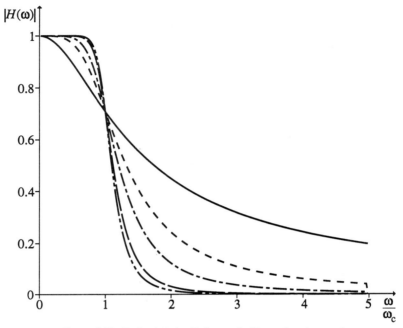

**Figure 8.13.** Bode plots for Butterworth filters of various orders.

The high-frequency asymptote corresponds to $\omega \gg \omega_c$ and is given by

$$-10\log_{10}\left(\frac{\omega}{\omega_c}\right)^{2m} = -20m\log_{10}\omega + 20m\log_{10}\omega_c$$

The plots of Fig. 8.13 show that the transition bandwidth decreases as the system order $m$ increases. The determination of $H(\omega)$ from its magnitude $|H(\omega)|^2$ is referred to as *spectral factorization*, and is discussed in detail in Chapter 12. Here we note that Eq. (8.68) is induced by the analytic function

$$h(s)h(-s) = \frac{1}{1 + (s/j\omega_c)^{2m}} \tag{8.69}$$

The poles of $h(s)h(-s)$ are

$$s_k = \omega_c e^{j[\frac{(2k+1)\pi}{2m} + \frac{\pi}{2}]}, \quad k = 0, 1, 2, \ldots, 2m - 1 \tag{8.70}$$

and are located on the circle of radius $\omega_c$, at equal distance of $\pi/m$ rads. None of the poles lie on the imaginary axis. Furthermore they do not lie on the real axis unless $m$ is odd.

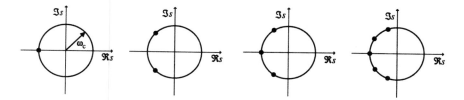

**Figure 8.14.** Poles for Butterworth filters of orders 1, 2, 3, 5.

To determine a causal and stable factor $h(s)$, we observe that if $s$ is a pole of $h(s)h(-s)$, then $-s$ is also a pole. Thus among the $2m$ poles (8.70), we choose those $m$ that are located in the left half plane. The cases $m = 1, 2, 3, 5$ are shown in Fig. 8.14. The resulting all pole transfer function $h(s)$ yields the $m$th-order Butterworth filter.

### 8.8.2  Chebyshev filters

The magnitude square of the frequency response of Butterworth filters is decreasing in the stopband and the passband. The approximation error in the passband is initially zero and grows as we approach the transition band. A better distribution of the error is accomplished by Chebyshev filters, whose frequency response exhibits ripples in the passband and is decreasing in the stopband.* The magnitude square of the frequency response of a Chebyshev filter of order $m$ has the form

$$|H(\omega)|^2 = \frac{1}{1 + \epsilon^2 T_m^2(\omega/\omega_c)} \tag{8.71}$$

where $T_m(x)$ is the Chebyshev polynomial of order $m$

$$T_m(x) = \cos(m\cos^{-1} x) = \cos m\theta, \qquad \theta = \cos^{-1} x \tag{8.72}$$

Thus

$$T_0(x) = \cos(0) = 1$$

$$T_1(x) = \cos(\cos^{-1} x) = x$$

$$T_2(x) = \cos(2\theta) = 2\cos^2 \theta - 1 = 2x^2 - 1$$

The trigonometric identity

$$\cos(m + 1)\theta + \cos(m - 1)\theta = 2\cos m\theta \cos \theta$$

* Structures with monotonically decreasing behavior in the passband and ripples in the stopband can be constructed in much the same way. These filters are referred to as *type 2* Chebyshev filters.

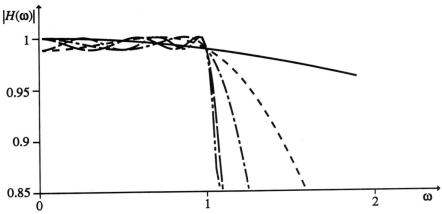

**Figure 8.15.** Chebyshev filters of orders 2, 3, 4, 5.

implies that Chebyshev polynomials obey the recursion

$$T_{m+1}(x) = 2xT_m(x) - T_{m-1}(x), \qquad m = 1, 2, \ldots$$

The Chebyshev filters for $m = 2, 3, 4, 5$ and $\omega_c = 1$ are plotted in Fig 8.15. An oscillatory behavior in the passband and a monotonically decreasing characteristic in the stopband are observed for $m \geq 3$.

If $|\omega| \leq 1$, the function $\cos^{-1} \omega$ is real. Otherwise, $\cos^{-1} \omega = j\theta$ and

$$\omega = \cos j\theta = \frac{1}{2}\{e^{j(j\theta)} + e^{-j(j\theta)}\} = \frac{1}{2}\{e^{-\theta} + e^{\theta}\} = \cosh \theta$$

or

$$\cos^{-1} \omega = j \cosh^{-1} \omega, \qquad \theta = \cosh^{-1} \omega$$

Hence

$$\cos\left(m \cos^{-1}\frac{\omega}{\omega_c}\right) = \cos\left(mj \cosh^{-1}\frac{\omega}{\omega_c}\right) = \cosh\left(m \cosh^{-1}\frac{\omega}{\omega_c}\right)$$

In summary, the Chebyshev filter is given by Eq. (8.71), where

$$T_m\left(\frac{\omega}{\omega_c}\right) = \begin{cases} \cos\left(m \cos^{-1}\frac{\omega}{\omega_c}\right), & |\omega| \leq \omega_c \\ \\ \cosh\left(m \cosh^{-1}\frac{\omega}{\omega_c}\right), & |\omega| > \omega_c \end{cases} \tag{8.73}$$

The structure of Chebyshev filters is analyzed in Exercise 8.39. Equation (8.72) implies that for $\omega = 0$,

$$|H(0)|^2 = \begin{cases} 1, & m \text{ odd} \\ \dfrac{1}{1+\epsilon^2}, & m \text{ even} \end{cases} \tag{8.74}$$

The parameter $1/(1 + \epsilon^2)$ specifies the tolerance limit $\delta_1$ of the corresponding error.

It can be shown that Chebyshev filters minimize the maximum error $\max |1 - H(\omega)|^2$ in the passband $|\omega| < \omega_c$. A trully optimal with respect to the sup norm procedure is provided by the *elliptic or Cauer filters*. Elliptic filters minimize the sup norm

$$\sup_{H(\omega)} |H_d(\omega) - H(\omega)| W(\omega)$$

over the entire frequency range. $H_d(\omega)$ is the ideal response, and $W(\omega)$ is a window reflecting the tolerance specifications. Elliptic filters admit close form expressions and lead to well-documented design procedures. The interested reader may consult the bibliographical notes at the end of the chapter.

Filter design (analog as well as discrete), can be approached by more general $\ell_p$ or $L_p$ optimization:

$$\min_{H(\omega)} \int W(\omega)|H_d(\omega) - H(\omega)|^p d\omega \tag{8.75}$$

The window function $W(\omega)$ indicates the special significance assessed to certain frequencies. Admissible systems are rational and hence admit a finite parametrization. Therefore minimization is performed over a finite-dimensional space. In general, no closed form expressions for the optimum exist, and iterative algorithms, such as the steepest descent and the Newton-Raphson method, must be employed. An exception is the minimization of energy ($p = 2$) over the class of FIR filters, discussed shortly.

### 8.8.3  IIR digital filters and the bilinear transformation

Discrete frequency selective filters are derived by simulating analog filters, such as Butterworth and Chebyshev. Depending on the simulation method several options exist. The most popular approach relies on the bilinear transformation and is discussed next. Other alternatives are discussed in Exercise 8.42.

The simulation which is based on the bilinear transform, expresses the transfer function model as a finite derivative model, realizes it with adders, scalors and integrators, and simulates each integrator with the trapezoid rule (see Section 3.12.2). We will see in Chapter 10 that every finite derivative model is realized in the above manner. The integrator with transfer function $h(s) = 1/s$ is simulated by the finite difference model

$$y(n) = y(n-1) + \frac{T_s}{2}(u(n) + u(n-1))$$

having transfer function

$$h(z) = \frac{T_s(1 + z^{-1})}{2(1 - z^{-1})}$$

The discrete filter results from the substitution

$$\frac{1}{s} = \frac{T_s}{2}\frac{1 + z^{-1}}{1 - z^{-1}} \tag{8.76}$$

and the relation

$$h(z) = h(s)|_{s=(2/T_s)(1-z^{-1})/(1+z^{-1})} \tag{8.77}$$

$h(z)$ is proper rational. The bilinear transform

$$s = \frac{2}{T_s}\frac{1 - z^{-1}}{1 + z^{-1}}, \quad z = \frac{1 + (T_s s/2)}{1 - (T_s s/2)} \tag{8.78}$$

carries the imaginary axis $s = j\omega$, on the unit circle $|z| = 1$, as we readily verify. Conversely, if $z = e^{j\Omega}$, then $s = j\omega$, and the two frequencies are related by

$$\frac{T_s}{2}\omega = \tan\left(\frac{\Omega}{2}\right) \tag{8.79}$$

This relation is plotted in Fig. 8.16.

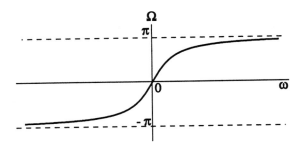

**Figure 8.16.** Mapping analog to discrete frequencies by the bilinear transform.

We easily deduce from (8.78) that if $\Re s > 0$, then $|z| > 1$, while if $\Re s < 0$, then $|z| < 1$. We conclude that the frequency response of the digital filter $H(\Omega)$ and the frequency response of the analog filter $H_a(\omega)$ are related by

$$H(\Omega) = H_a(\omega)|_{\omega=(2/T_s)\tan(\Omega/2)}$$

The nice characteristics offered by the analog filter are maintained by the discrete filter. If design specifications are given in terms of discrete frequencies $\Omega_s$, $\Omega_p$, the analog filter is chosen with analog frequencies

$$\omega_s = \frac{2}{T_s}\tan\left(\frac{\Omega_s}{2}\right), \quad \omega_p = \frac{2}{T_s}\tan\left(\frac{\Omega_p}{2}\right)$$

This is known as *prewarping*.

### 8.8.4  FIR filter design

Let us consider the filter design minimization problem (8.75) over the class of FIR systems with $p = 2$ and $W(\omega) = 1$. The performance criterion becomes

$$\int_{-\pi}^{\pi}|H_d(\omega) - \sum_{n=-N}^{N} h(n)e^{-j\omega n}|^2 d\omega \tag{8.80}$$

$H_d(\omega)$ is the frequency response of the ideal lowpass filter. This problem was solved in Exercise 5.9. For given length $N$, the optimal solution is given by

$$h(n) = \frac{\sin \omega_c n}{\pi n}, \qquad |n| \le N \tag{8.81}$$

The shifted sequence $h(n - N)$, $0 \le n \le 2N$ yields a causal solution.

The most serious weakness of the above method is the classical *Gibbs phenomenon*. Every partial sum of the Fourier expansion of the periodic function $H_d(\omega)$ exhibits ripples. The ripples closer to the discontinuity have a constant maximum independent of $N$. Of course, for every $\omega$ close to the cutoff frequency, the frequency response of the FIR filter gets arbitrarily close to one as $N$ increases, while the last ripples are pushed toward $\omega_c$. Partial control of Gibbs phenomenon is achieved by windows. The FIR filter to be designed is written as

$$h(n) = h_d(n)w(n) \tag{8.82}$$

where $w(n)$ is a time window. In particular, the rectangular window

$$w(n) = p_N(n) = \begin{cases} 1, & |n| \le N \\ 0, & \text{otherwise} \end{cases}$$

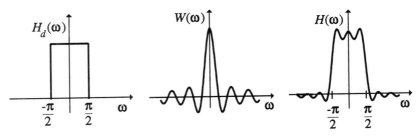

**Figure 8.17.** Visualization of the FIR frequency response for the case of the rectangular window.

leads to the optimal FIR filter discussed previously. Equation (8.82) converts in the frequency domain into

$$H(\omega) = \frac{1}{2\pi} \int_{-\pi}^{\pi} H_d(v) W(\omega - v) dv \tag{8.83}$$

The response $H(\omega)$ is given by the convolution of $H_d(\omega)$ and $W(\omega)$. The Fourier transform of the rectangular window is

$$W(\omega) = \frac{1 - e^{-j\omega N}}{1 - e^{-j\omega}} = e^{-j\omega(N-1)/2} \frac{\sin(\omega N/2)}{\sin \omega/2} \tag{8.84}$$

Equation (8.83) is illustrated in Fig. 8.17. The oscillatory behavior of the rectangular window causes ripples in the frequency response. We conclude from Eq. (8.83) that $H(\omega)$ tends to the ideal response $H_d(\omega)$, as the window approaches a Dirac function. Windows of short length are required to ensure low computational complexity and storage. On the other hand, the window must be long enough to achieve an impulsive behavior in the frequency domain. To meet these conflicting requirements, we fix $N$, and we try to determine $W(\omega)$ so that the main lobe width and the peak side lobe level are reduced. The main lobe width controls the filter transition bandwidth. The peak side lobe level controls the peak passband and stopband ripples. Some of the most popular windows are listed below and shown in Fig. 8.18.

- **Bartlett** (triangular window)

$$w(n) = \begin{cases} 1 + \dfrac{n}{N}, & -N \leq n \leq 0 \\ 1 - \dfrac{n}{N}, & 0 \leq n \leq N \end{cases} \tag{8.85}$$

- **Hanning and Hamming** They are given by

$$w(n) = a + (1 - a)\cos\frac{\pi n}{N}, \quad N/2 \leq n \leq N/2 \tag{8.86}$$

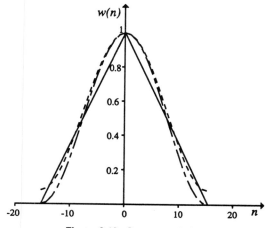

**Figure 8.18.** Common windows.

The Hanning window results for $a = 1/2$; the Hamming window uses the value $a = 0.54$.

- **Blackman** It is obtained from the previous windows with the addition of one term for the further reduction of side ripples. It has the form

$$w(n) = 0.42 + \frac{1}{2}\cos\frac{\pi n}{N} + 0.08\cos\frac{2\pi n}{N}, \quad -\frac{N}{2} \leq n \leq \frac{N}{2} \tag{8.87}$$

In all cases $w(n) = 0$ for $|n| > N/2$.

The most rational compromise between reduction of side lobe amplitude and main lobe width is achieved by Kaiser windows, which will not be discussed here. They constitute good approximations of the optimal windows discussed next.

*Window optimization.* Window design can be rationalized in several ways. One such efficient approach seeks for the finite duration window $w(n)$ that has unit energy and minimizes the energy over the stopband

$$\int_\sigma^\pi |W(\omega)|^2 d\omega \tag{8.88}$$

$\sigma$ is a design parameter indicating the stopband edge. Under the energy constraint

$$\frac{1}{\pi}\int_0^\pi |W(\omega)|^2 d\omega = 1 \quad \text{or} \quad \sum_{n=0}^N w^2(n) = 1$$

minimization of (8.88) is equivalent to maximization of

$$\int_0^\sigma |W(\omega)|^2 d\omega$$

Note that

$$\int_0^\sigma |W(\omega)|^2 d\omega = \int_0^\sigma \sum_{n=0}^N w(n) e^{-j\omega n} \sum_{k=0}^N w(k) e^{j\omega k} d\omega$$

$$= \sum_{n=0}^N w(n) \sum_{k=0}^N w(k) \int_0^\sigma e^{-j\omega(n-k)} d\omega = w^T P w$$

where

$$w = (\,w(0) \quad w(1) \quad \cdots \quad w(N)\,)^T$$

and $P$ is the real symmetric matrix with entries

$$P_{kn} = \frac{\sin((n-k)\sigma)}{n-k}$$

Thus the problem becomes

$$\max w^T P w \quad \text{subject to} \quad w^T w = 1$$

The solution to the above problem is given by an eigenvector corresponding to the largest eigenvalue of $P$. The windows $w(n)$ obtained above are called *prolate spheroidal wave* sequences.

### Example 8.33    MATLAB implementation of window optimization
The following MATLAB code designs FIR filters with window optimization based on prolate sequences:

```
n = [-N/2:N/2];                    % N is the filter order
h = (s/pi)*sinc((s/pi)*n);         % s is the cutoff frequency, h the
                                   % truncated ideal lowpass filter

p = [s,sin([1:N]*s)./[1:N]];
P = toeplitz(p);                   % generates the matrix P
[V,D] = eig(P);                    % computes the eigenvectors
                                   % and eigenvalues of P

d = diag(D);
lambda = max(d);
i = find(d = = lambda);
w = V(:,i);                        % finds eigevector of maximum eigenvalue
```

### 8.8.5  Computing the Fourier transform by the DFT and windowing

Various signal processing applications including speech processing, geophysical signal processing, and radar processing require the computation of the Fourier transform of analog waveforms. The role of the DFT in this task became evident in Sections 6.1.3 and 6.2.1. The sampling theorem helps us to recast the computational procedure as in Fig. 8.19. The lowpass filter creates a bandlimited version of the original signal minimizing the effect of aliasing. Multiplication of the sampled sequence $x_s(n)$ by $w(n)$ is referred to as *windowing*. Windowing reflects the finite length constraint of the DFT. The most obvious option is the rectangular window. For reasons similar to those exposed in the previous section, other windows like the Bartlett, Hamming, Hanning, and Kaiser windows are preferred due to their smoother rise and decay times. The use of such windows helps to smooth out sharp peaks and discontinuities in the signal spectrum. A final issue that needs attention is the frequency domain sampling imposed by the DFT. Evidently the sequence $X(k)$ is an estimate of the analog spectrum at discrete frequencies.

### 8.8.6  2-D filter design

Two-dimensional filter design methods include direct extensions of 1-D techniques, and approaches that have no 1-D counterpart. FIR filter design based on windows admits a straightforward extension to the 2-D case. If $H_d(\omega_1, \omega_2)$ denotes the ideal lowpass characteristic, the FIR system that minimizes the error energy

$$\int \int |H_d(\omega_1, \omega_2) - H(\omega_1, \omega_2)|^2 d\omega_1 d\omega_2$$

is given by the truncated Fourier series of $H_d(\omega_1, \omega_2)$. Windows are employed to alleviate the Gibbs phenomenon. A common choice is separable windows $w(n_1, n_2) = w_1(n_1)w_2(n_2)$ where $w_1(n_1)$, $w_2(n_2)$ are popular 1-D windows. Two-dimensional optimal windows can also be developed. It can be shown that the 2-D rectangular optimal window follows directly from two 1-D optimal windows (see Exercise 8.50). If the window is symmetric, the resulting FIR filter is symmetric and hence possesses the linear phase property. As we

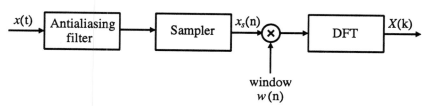

**Figure 8.19.** Computing the Fourier transform of an analog signal.

pointed out previously, the linear phase property is highly desirable in image processing.

Minimization of general $\ell_p$ norms over infinite impulse response 2-D structures is also employed. The dependence on the parameters is nonlinear, and iterative techniques are required. The interested reader may consult the bibliographical notes for further reading.

## BIBLIOGRAPHICAL NOTES

The $z$ and Laplace transform representation of signals and systems are discussed in most books devoted to signals and systems. They are also treated in engineering mathematics textbooks such as Kreyzig (1979). Multidimensional discrete systems in the $z$ domain and stability issues not pursued here, are discussed in Lim (1990). Filter design methods are covered in most signal processing books. The book of Antoniou (1979) is devoted to analog and digital filter design. Filter design methods geared toward subband coding and multirate filtering are systematically treated in Crochiere and Rabiner (1983). Volterra representations in the complex domain are developed in Rugh (1981). For a rigorous exposition of the Laplace transform the reader may consult Doetsch (1974).

## PROBLEMS

**8.1.** Determine the $z$ transform and the region of convergence for the signals:
(a) $(\frac{1}{3})^n [u_s(n) - u_s(n-6)]$
(b) $2(\frac{1}{2})^n \cos[2\pi n/6 + \pi/4] u_s(n)$

**8.2.** Reconstruct the signal from its transform for each of the following cases:

(a) $x(z) = \dfrac{1 - \frac{1}{2}z^{-1}}{1 + \frac{3}{4}z^{-1} + \frac{1}{8}z^{-2}}, \quad |z| > \dfrac{1}{2}$

(b) $x(z) = e^{1/z}$
(c) $x(z) = \log(1 - \frac{1}{2}z^{-1}), \quad |z| > \frac{1}{2}$

**8.3.** Let $x(n)$ be a real signal with $z$ transform, $x(z)$. Prove the relation $x(z) = x^*(z^*)$. Conclude that if $z = z_0$ is a pole (zero), then $z = z_0^*$ is also a pole (zero).

**8.4.** Represent the following systems in the $z$ domain:
(a) Difference $\Delta u(n) = u(n) - u(n-1)$
(b) Summer $y(n) = \sum_{k=0}^{n} u(k)$

**8.5.** Determine the $z$ transform of the signals produced by the following sources:
(a) $y(n) + 3y(n-1) = 0, \qquad y(-1) = 1$
(b) $y(n) - \frac{1}{2}y(n-1) = 0, \qquad y(-1) = 1$

Verify your answers with MATLAB.

**8.6.** *Initial and final value theorems.* (i) The initial value theorem for the one-sided $z$ transform states that if $x(n)$ has transform $x(z)$, then $x(0) = \lim_{z \to \infty} x(z)$. The final value theorem states that

$$\lim_{n \to \infty} x(n) = \lim_{z \to 1}(1 - z^{-1})x(z)$$

provided that the region of convergence of $(1 - z^{-1})x(z)$ includes the region $|z| \geq 1$. Prove the above theorems. *Hint*: To prove the final value theorem consider the sum

$$\sum_{n=0}^{\infty}[x(n) - x(n-1)]z^{-n}$$

(ii) Verify the initial and final value theorems for the signals:
(a) $x(n) = 2^{n+1}u_s(n)$
(b) $x(n) = (\frac{1}{3})^{|n|}$

(iii) Explain why the final value theorem does not apply to the signal

$$x(n) = \begin{cases} 1, & n \text{ even} \\ 0, & \text{otherwise} \end{cases}$$

**8.7.** Show that the exponential signal $a^n$, $n \in Z$ does not have a two-sided $z$ transform.

**8.8.** Establish the following properties of the two-sided $z$ transform:
(a) Linearity:

$$\mathcal{Z}[ax(n) + by(n)] = ax(z) + by(z)$$

for all $z$ in the intersection of the regions of convergence of $x(z)$ and $y(z)$ and any complex numbers $a$, $b$.
(b) Shift: For each $k \in Z$

$$x(n+k) \xrightarrow{\ z\ } z^k x(z), \quad R^+ < |z| < \frac{1}{R^-}$$

In contrast to the one-sided case, no initial conditions enter in the expression.
(c) Modulation property: If $c$ is a complex nonzero number, then

$$y(n) = c^n x(n), \quad y(z) = x\left(\frac{z}{c}\right), \quad |z| > |c|R_x$$

(d) Complex conjugate: If $x(n)$ is complex valued with $z$ transform $x(z)$, show that the signals $x^*(n)$, $\Re x(n)$, and $\Im x(n)$ have $z$ transforms $x^*(z^*)$, $\frac{1}{2}(x(z) + x^*(z^*))$, and $\frac{1}{2j}(x(z) - x^*(z^*))$, respectively.
(e) Differentiation: $nx(n) \xrightarrow{\ z\ } -z\frac{d}{dz}x(z)$, $|z| > R_x$

**8.9.** Compute the two-sided $z$ transform of the signals:
(a) $x(n) = \sin 2n\, u_s(n)$
(b) $x(n) = 2^{-|n|} \cos n$
(c) $x(n) = n^2 a^{-n}\, u_s(n)$

**8.10.** Let $h(z)$ be a proper rational function. The *complex cepstrum* $c_h(n)$ is defined as the inverse $z$ transform of $\log h(z)$. In analogy with the differentiation property of the $q$ series, show that the derivative of the logarithm of a proper rational function is also rational. Suppose that

$$h(z) = |A| \frac{\prod_{i=1}^{L_1}(1 - \rho_i z^{-1})}{\prod_{i=1}^{L_3}(1 - \lambda_i z^{-1})} \prod_{i=1}^{L_2}(1 - \mu_i z), \qquad |\lambda_i| < 1, |\rho_i| < 1, |\mu_i| < 1$$

Prove that the cepstrum coefficients are computed by $c_h(0) = \log|A|$, $c_h(n) = -A^{(n)}/n$ for $n > 0$ and $c_h(n) = B^{(-n)}/n$ for $n < 0$, where

$$A^{(n)} = \sum_{i=1}^{L_1} \rho_i^n - \sum_{i=1}^{L_3} \lambda_i^n, \qquad B^{(n)} = \sum_{i=1}^{L_2} \mu_i^n$$

Show that if $y(n) = h(n) * u(n)$, then $c_y(n) = c_h(n) + c_u(n)$.

**8.11.** Consider the causal finite difference model

$$y(n) + \frac{1}{2}y(n-1) = u(n)$$

Compute the transfer function, the impulse response, the frequency response, and the output produced by the following inputs.
(a) $u(n) = (\frac{1}{4})^n u_s(n)$
(b) $u(n) = \delta(n) + \frac{1}{2}\delta(n-1)$

Likewise find the outputs produced by the following inputs described in the frequency domain.
(a) $U(\omega) = 1/[(1 - \frac{1}{4}e^{-j\omega})(1 + \frac{1}{2}e^{-j\omega})]$
(b) $U(\omega) = 1 + 2e^{-4j\omega}$

Construct the input, output, and frequency response Bode plots.

**8.12.** Compute the transfer function of the system

$$y(n) - \frac{1}{2}y(n-1) + \frac{1}{4}y(n-2) = u(n)$$

and all possible impulse responses. Which one corresponds to a stable system?

**8.13.**  Let

$$y(n) = u(n) - e^{-8a}u(n-8), \qquad 0 < a < 1$$

Find the transfer function, the poles, the zeros, and the region of convergence. Determine the transfer function of the inverse system. For which region of convergence is the inverse system causal? stable?

**8.14.**  Find the transfer function, the poles, zeros, and the region of convergence of the digital filter of Fig. 8.20. For which values of $k$ is the system stable?

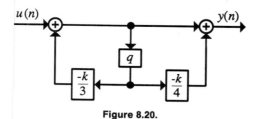

**Figure 8.20.**

**8.15.**  This exercise summarizes further properties of the one-sided Laplace transform. Let

$$x(t) \xrightarrow{\;\mathcal{L}\;} x(s), \qquad \Re s > \sigma$$

Prove the following statements:
(a) If $x(t)$ is periodic of period $T$ and absolutely integrable over a period, the Laplace transform exists and is given by

$$x(s) = \frac{1}{1 - e^{-sT}} \int_0^T x(t)e^{-st}\,dt, \qquad \Re s > 0$$

Show that the periodic square wave has Laplace transform

$$x(s) = \frac{M(1 - e^{sT/2})}{s(1 + e^{-sT/2})} = \frac{M}{s}\tanh\frac{T}{4}s$$

(b) For any complex $s_0$, the signal $y(t) = e^{s_0 t}x(t)$ has Laplace transform $y(s) = x(s - s_0)$, $\Re s > \sigma + \Re s_0$
(c) For all $t_0 > 0$, $x(t - t_0) \xrightarrow{\;\mathcal{L}\;} e^{-st_0}x(s)$, $\Re s > \sigma$, provided that $x(t) = 0$ for $t < 0$.
(d) $x(at) \xrightarrow{\;\mathcal{L}\;} (1/a)x(s/a)$, $\Re s > \sigma/a$, $\alpha > 0$
(e) $-tx(t) \xrightarrow{\;\mathcal{L}\;} dx(s)/ds$, $\Re s > \sigma$
(f) $x^*(t) \xrightarrow{\;\mathcal{L}\;} x^*(s^*)$, $\Re s > \sigma$

$$\Re x(t) \xrightarrow{\;\mathcal{L}\;} \frac{1}{2}\left(x(s) + x^*(s^*)\right), \qquad \Im x(t) \xrightarrow{\;\mathcal{L}\;} \frac{1}{2j}\left(x(s) - x^*(s^*)\right), \qquad \Re s > \sigma$$

(g) Final value: $\lim_{t \to \infty} x(t) = \lim_{s \to 0} sx(s)$. *Hint:* Compute $\lim_{s \to 0} sx(s) - x(0)$ and $\int_0^\infty \dot{x}(t)dt$. State the precise assumptions that must be met in analogy with the discrete case (see Exercise 8.6).

(h) Initial value: If $x(t)$ is analytic, it holds that $x(0) = \lim_{s \to \infty} sx(s)$. *Hint:* Expand $x(t)$ in Taylor series at the origin.

**8.16.** Find the Laplace transform of the following causal signals:
(a) Periodic triangular pulse of Fig. 8.21
(b) halfwave rectifier of Fig. 8.22
(c) sawtooth wave of Fig. 8.23
(d) stairwise function of Fig. 8.24
(e) $|\cos \omega t| \, u_s(t)$.

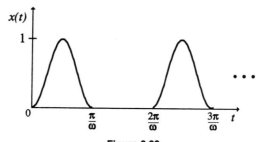

**Figure 8.21.**

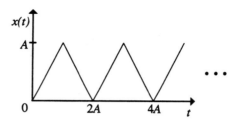

**Figure 8.22.**

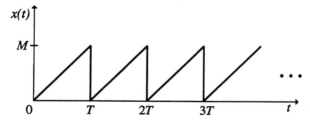

**Figure 8.23.**

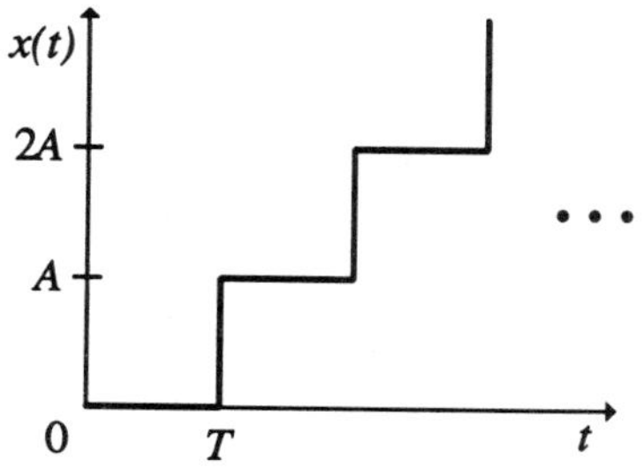

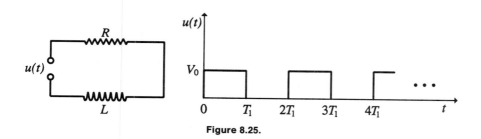

Figure 8.24.

**8.17.** Calculate the steady state value of the current in the circuit of Fig. 8.25 when the source supplies the rectangular periodic pulse shown in the same figure.

Figure 8.25.

**8.18.** Show the following more general statement of the initial value theorem. Let

$$x(0) = x^{(1)}(0) = x^{(2)}(0) = \cdots = x^{(n-1)}(0) = 0$$

and $x^{(n)}(0) \neq 0$. Then $\lim_{s \to \infty} s^{n+1} x(s) = x^{(n)}(0)$.

**8.19.** Verify the initial and final value theorem for the signals below:
(a) $x(t) = e^{-at} u_s(t)$
(b) $x(t) = t^2 e^{-at} u_s(t)$

**8.20.** Let $u(t)$ be the signal of Fig. 8.26 and $h(t) = e^{-3t} u_s(t)$. Using the final value theorem, compute the steady state value of the convolution $\lim_{t \to \infty} h(t) * u(t)$.

**8.21.** Compute the one-sided Laplace transform of the signals
(a) $x(t) = \sinh at$
(b) $x(t) = \sin t \cosh t - \cos t \sinh t$
(c) $x(t) = e^{-at}/t$
(d) $x(t) = 1/(1 + e^t)$

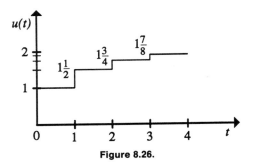

**Figure 8.26.**

**8.22.** Assuming the one-sided Laplace transform of $x(t)$ is $1/(s^2 + 4s + 8)$, compute the one-sided Laplace transform of the following signals
(a) $\dot{x}(t)/t$
(b) $x(t) \sin t$
(c) $x(5t + 3)$

**8.23.** Let
$$x(t) = (x_1(t) \quad x_2(t)) = (e^{-|t|} \quad (e^{-2t} + 4e^{-3t})u_s(t))$$

Compute the Fourier transform using the Laplace transform.

**8.24.** (a) Show that the exponential of the matrix
$$A = \begin{bmatrix} -1 & 0 \\ 2 & -3 \end{bmatrix}$$
is given by
$$e^{At} = \begin{bmatrix} e^{-t} & 0 \\ e^{-t} - e^{-3t} & e^{-3t} \end{bmatrix}$$

(b) Suppose that $A$ is a diagonalizable matrix with nonsimple eigenvalues (some are equal). Prove that $e^{At}$ involves pure exponentials only and is given by an expression of the form (8.31). Prove a similar statement for the discrete exponential $A^n$, $n \geq 0$.

**8.25.** Express the transfer function of the basic analog system interconnections (cascade, parallel, and feedback) in terms of the transfer functions of the constituent systems. Determine the impulse response of the interconnection depicted in Fig. 8.27 in terms of the impulse responses of the subsystems.

**8.26.** An analog single-channel source is specified by the equation
$$x^{(2)}(t) = -3x^{(1)}(t) - 2x(t)$$

Determine the signal emitted by the above source given that $x(0) = 0$, $x^{(1)}(0) = 1$. Provide a SIMULINK validation.

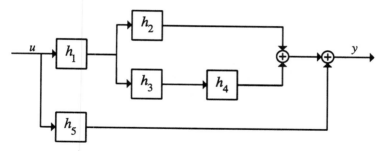

**Figure 8.27.**

**8.27.** Consider the causal finite derivative model

$$\dot{y}(t) + 3y(t) = 2u(t)$$

Compute the frequency response and construct Bode plots. Find the output spectrum produced by the input $u(t) = e^{-t}u_s(t)$. Repeat for the input with spectrum

$$U(\omega) = \frac{1 + j\omega}{2 + j\omega}$$

Verify your calculations with MATLAB simulations. The frequency response is determined from the transfer function by the command freqs. Bode plots are developed by the command bode. The output is estimated by the command lsim.

**8.28.** Determine the frequency response and the impulse response of the causal system

$$\ddot{y} + 6\dot{y} + 8y = 2u$$

Find the output produced by the input $u(t) = te^{-2t}u_s(t)$. Is the system BIBO stable? Validate your results with MATLAB.

**8.29.** Find the step response for the systems with the following transfer functions. Initial conditions are zero. Check BIBO stability.
(a) $h(s) = 1/(s^2 + 4s + 4)$
(b) $h(s) = (10s + 1)/(s^2 + 3s + 2)$
(c) $h(s) = (s + 2)/[(s + 4)(s + 1)]$

**8.30.** Find the transfer function of the systems interconnection depicted in Fig. 8.28.

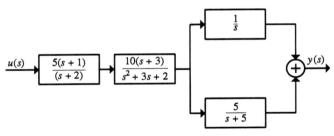

**Figure 8.28.**

**8.31.** The transfer function of a causal system is

$$h(s) = \frac{s+1}{s^2 + 2s + 2}$$

Compute and plot the output produced by the input $u(t) = e^{-|t|}$.

**8.32.** Find the Laplace transforms of the signals

$$\phi_n(t) = e^{-t/2} L_n(t) u_s(t) \quad L_n(t) = \frac{e^t}{n!} \frac{d^n}{dt^n} (t^n e^{-t}), \qquad n = 0, \ 1, \ 2, \ldots$$

Show that $L_n(t)$ are polynomials, called *Laguerre polynomials*. Prove that $\phi_n(t)$ are produced by the cascade connection of Fig. 8.29 for suitable transfer functions $h_1(s)$ and $h_2(s)$; $\delta(t)$ is the Dirac function.

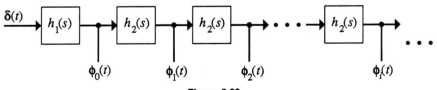

**Figure 8.29.**

**8.33.** If the input to a convolutional system is a causal signal with Laplace transform $(s + 2)/(s - 2)$ and the resulting output is $-\frac{2}{3} e^{2t} u_s(-t) + \frac{1}{3} e^{-t} u_s(t)$, find the transfer function and the region of convergence, the impulse response, and the output produced by the input $u(t) = e^{-4t} u_s(t)$.

**8.34.** Consider the satellite motion of Example 3.26. Prove that the coefficients in the partial fraction expansion of the impulse response are nonzero.

**8.35.** Show that the odd moments of a zero mean Gaussian random variable of variance $\sigma$ are zero and that the even moments are given by

$$E[x^n] = 1 \cdot 3 \cdots (n - 1) \sigma^n$$

**8.36.** Prove that the two-dimensional $z$ transform satisfies the following properties:

(a) $x(-n_1, -n_2) \xrightarrow{z} x(z_1^{-1}, z_2^{-1})$

(b) $x^*(n_1, n_2) \xrightarrow{z} x^*(z_1^*, z_2^*)$

(c) $x(n_1 \pm n_{10}, n_2 \pm n_{20}) \xrightarrow{z} z_1^{\pm n_{10}} z_2^{\pm n_{20}} x(z_1, z_2)$

(d) $a_1^{n_1} a_2^{n_2} x(n_1, n_2) \xrightarrow{z} x(z_1/a_1, z_2/a_2)$

Develop analogous properties for the 2-D Laplace transform. Extend the above to the general m-D case.

**8.37.** Prove that the real FIR system with transfer function $h(z) = \sum_{n=0}^{L} h(n) z^{-n}$ has linear phase if and only if $h(n) = \pm h(L - n)$. Show that the zeros of a linear phase filter $h(n)$ occur in reciprocal conjugate pairs, that is, $h(z) = 0$ if and only if $h(1/z^*) = 0$. Prove that if a causal transfer function model $h(z) = b(z)/a(z)$ with $b(z)$, $a(z)$ relatively prime, has linear phase, then it is an FIR system.

**8.38.** Consider the $RC$ circuit of Fig. 8.30. The source voltage $v_s(t)$ is the input and the capacitor voltage $v_c(t)$ is the output. Derive Bode plots and show that the circuit approximates the ideal lowpass characteristic. Repeat with the resistor voltage $v_r(t)$ as output. In the latter case show that the system approximates a highpass behavior.

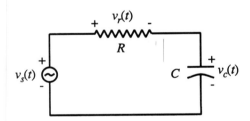

**Figure 8.30.**

**8.39.** Show that the roots of the Chebyshev polynomial $T_m(x)$ are given by

$$x_k = \cos \frac{(2k + 1)\pi}{2N}, \qquad k = 0, 1, \ldots, N - 1$$

and hence the Chebyshev filter exhibits ripples in the passband with values in the interval $[1/(1 + \epsilon^2), 1]$ and a monotonically decreasing behavior in the stopband.

**8.40.** Let $s_k = \sigma_k + j\omega_k$ denote the poles of $h(s)h(-s)$, where $h(s)$ is the transfer function of a Chebyshev filter of order $m$. Let

$$u + jv = \cosh^{-1}\left(\frac{s}{j\omega_c}\right)$$

Prove that

$$\cosh Nu \cos Nv = 0, \quad \sinh Nu \sin Nv = \pm \frac{1}{\epsilon}$$

and that

$$v = \frac{(2k+1)\pi}{2N}, \quad u = \pm \frac{1}{N} \sinh^{-1} \frac{1}{\epsilon}$$

Deduce that the poles are given by

$$\sigma_k = \pm \omega_c \sinh \left( \frac{1}{N} \sinh^{-1} \frac{1}{\epsilon} \right) \sin \frac{(2k+1)\pi}{2N}$$

$$\omega_k = \omega_c \cosh \left( \frac{1}{N} \sinh^{-1} \frac{1}{\epsilon} \right) \cos \frac{(2k+1)\pi}{2N}$$

and hence are located on an ellipse. Conclude that the Chebyshev filter is characterized by the $m$ poles with negative real part:

$$h(s) = \frac{h_0}{\prod_{k=0}^{N-1} (s - s_k)}, \quad s_k = \sigma_k + j\omega_k$$

where

$$h_0 = \begin{cases} \prod_{k=0}^{N-1} (-s_k), & N \text{ odd} \\ \dfrac{1}{1+\epsilon^2} \displaystyle\prod_{k=0}^{N-1} (-s_k), & N \text{ even} \end{cases}$$

**8.41.** Design of frequency selective filters other than lowpass can be accomplished with the aid of lowpass filter design and the use of suitable transformations of the form

$$h_d(\tilde{s}) = h_l(s)\big|_{s=g(\tilde{s})}$$

The transfer function $h_l(s)$ is built by lowpass filter design, $h_d(s)$ denotes the desired transfer function (highpass, bandpass, etc.), and $g(s)$ is a proper rational invertible transformation. Prove that highpass and bandpass filters are constructed from lowpass filters by the transformations $s = \lambda/\tilde{s}$ and $s = (\tilde{s} + \omega_0^2/\tilde{s})/B$, respectively.

**8.42.** This exercise summarizes the impulse invariance and the backward simulation method for digital filter design. Let $h_a(t)$ be the impulse response of an analog filter obtained by an analog filter design method.

The discrete simulation method of Section 3.12 leads to the discrete filter $h(n) = T_s h_a(nT_s)$. Prove that

$$H(\omega) = \sum_{k=-\infty}^{\infty} H_a\left(\frac{\omega}{T_s} - \frac{2\pi}{T_s}k\right)$$

If $h_a(t)$ is bandlimited and the Nyquist criterion is met, the sampling theorem asserts that both phase and magnitude are preserved. Show that $h_a(t)$ cannot be bandlimited. Hence aliasing effects are unavoidable. If the poles of $h_a(s)$ are simple as in the case of Chebyshev and Butterworth filters, prove that

$$h(n) = \sum \rho_i^n T_s r_i, \qquad \rho_i = e^{\lambda_i T_s}$$

Another alternative is the backward simulation of Section 3.12.2. Analyze the performance of the resulting discrete structures by MATLAB simulations.

**8.43.** Determine the transfer function and the frequency response of the lowpass Butterworth filter of order 4 and cutoff frequency 1. Verify your answers with MATLAB. The relevant commands are butter, buttord, and buttap.

**8.44.** Determine the transfer function of the lowpass Chebyshev filter of fifth order with cutoff frequency 1 and $A_p = 0.1$, and plot the frequency response and the attenuation.

**8.45.** Determine the order of the Chebyshev filter that satisfies the specifications $\omega_p = 1$ rad/s, $\omega_s = 3$ rad/s, $A_p = 0.3$, $A_s \geq 45$ dB, and find its transfer function. Determine the Butterworth filter for the same specifications. Verify your answers with MATLAB. The relevant commands are cheby1, cheb1ap, and cheb1ord.

**8.46.** Use the bilinear transform to design a digital filter from the filters of Exercise 8.45.

**8.47.** Design a FIR filter of order 11 that approximates the ideal lowpass filter with cutoff frequency 2 rad/s and sampling frequency 10 rad/s. Repeat the design with the Hanning, Hamming, and Blackman windows.

**8.48.** Design an FIR filter of order 11 for the highpass filter with passband $2.5 \leq |\omega| \leq 5$ rad/s and sampling frequency 10 rad/s. Repeat the design using the Hanning, Hamming, and Blackman windows.

**8.49.** In analogy with Examples 8.27 and 8.28 carry out an analysis of first- and second-order discrete systems.

**8.50.** Show that the $N_1 \times N_2$ unit energy window $W$ that minimizes

$$\int_{\sigma_1}^{\pi} \int_{\sigma_2}^{\pi} |W(\omega_1, \omega_2)|^2 d\omega_1 d\omega_2$$

is given by $W = w_1 w_2^T$, where $w_1, w_2$ are 1-D optimal windows of lengths $N_1, N_2$ respectively.


## APPENDIX

***Existence and inversion of Laplace transform.*** The Laplace transform can be interpreted as a Riemann integral, or more generally, as a Lebesgue integral. In the former case it is understood as the improper integral

$$\int_0^{\infty} e^{-st} x(t) dt = \lim_{\tau \to \infty} \int_0^{\tau} e^{-st} x(t) dt$$

This requires that $x(t)$ be integrable over any finite interval and in addition satisfy a certain growth condition. A useful class of functions that meets the first requirement is the class of piecewise continuous functions. $x(t)$ is piecewise continuous if it is continuous except possibly on a countable set, and at each discontinuity point $\tau$, both $\lim_{t \to \tau^+} x(t)$ and $\lim_{t \to \tau^-} x(t)$ exist and are finite. The second requirement is satisfied if $x(t)$ has *exponential order*; that is, there are positive numbers $M$, $\lambda$ and $t_0 \geq 0$ such that

$$|x(t)| \leq M e^{\lambda t}, \qquad t \geq t_0 \tag{8.89}$$

It is easy to show that if there is $\lambda$ such that $\lim_{t \to \infty} e^{-\lambda t} |x(t)|$ exists, then $x(t)$ has exponential order. On the other hand, if $\lim_{t \to \infty} e^{-\lambda t} |x(t)| = \infty$ for all $\lambda$, $x(t)$ is not of exponential order.

All polynomials are of exponential order. Indeed,

$$\lim_{t \to \infty} e^{-\lambda t} |t^m| = \lim_{t \to \infty} \frac{|t^m|}{e^{\lambda t}}$$

De l'Hôspital's rule shows that if $\lambda > 0$, the above limit goes to zero. On the other hand, the function $\exp(t^2)$ is not of exponential order because

$$\lim_{t \to \infty} \frac{e^{t^2}}{e^{\lambda t}} = \lim_{t \to \infty} e^{t(t-\lambda)} = \infty$$

for every $\lambda$.

**Theorem 8.4** If $x(t)$ is piecewise continuous and has exponential order, the Laplace transform $x(s)$ exists.

PROOF   Piecewise continuity implies that $x(t)$ is bounded over the range $0 \leq t \leq t_0$. Thus $|x(t)| \leq Me^{\lambda t}$, for all $t \geq 0$. Hence $|x(t)e^{-st}| = |x(t)|\,|e^{-st}| \leq Me^{(\lambda - \Re s)t}$. Moreover the integral $\int_0^\infty e^{(\lambda - \Re s)t}\,dt$ exists for $\Re s > \lambda$. Therefore $x(s)$ exists for every s: $\Re s > \lambda$.                                                                            ∎

The above proof shows that if $x(s)$ exists at a point $s = s_0$, it exists for all $\Re s > \Re s_0$. Hence we may take the infimum of all $\sigma > 0$ such that $x(s)$ exists for $\Re s > \sigma$ and diverges for $\Re s < \sigma$. The half plane $\{s \in C : \ \Re s > \sigma\}$ (see Fig. 8.4) constitutes the region of convergence of $x(s)$.

In close analogy with the development of the Fourier transform, the Laplace transform can be viewed as a Lebesgue integral. If (8.22) holds for $s = s_0$, it also holds for all $s$ with $\Re s \geq \Re s_0$. Indeed,

$$\int_0^\infty |e^{-st}x(t)|dt \leq \int_0^\infty |e^{-\Re s_0 t}|\,|x(t)|dt < \infty$$

Here we used the fact that if $|x(t)| \leq |y(t)|$ and $y(t)$ is Lebesgue integrable, then $x(t)$ is Lebesgue integrable as well and $\int |x| \leq \int |y|$.

**Theorem 8.5**   Let $\sigma$ be the infimum of positive numbers $\lambda$ such that (8.89) converges for $\Re s > \sigma$. Then $x(s)$ is analytic for any such $s$.

PROOF   The proof is analogous with the proof of the differentiation property of the Fourier transform. Let $h$ be a complex number such that $|h| < \zeta$ and $\Re s + \Re h > \sigma$, with $\zeta$ a small number. Then

$$\frac{x(s+h) - x(s)}{h} = \int_0^\infty \frac{1}{h}(e^{-ht} - 1)e^{-st}x(t)dt$$

If we take limits as $h \to 0$, we find that

$$\frac{dx(s)}{ds} = -\int_0^\infty te^{-st}x(t)dt = -\mathcal{L}[tx(t)]$$

The limit can be inserted into the integral provided that the requirements of the dominated convergence theorem are met. Thus we must show that the integrand is bounded with respect to $h$ by an integrable function and that the limit of the integrand exists almost everywhere. The second is immediate because

$$\frac{1}{h}(e^{-ht} - 1)e^{-st}x(t) \to -te^{-st}x(t) \quad \text{as } h \to O$$

Regarding boundedness, we observe that

$$e^{-ht} - 1 = \sum_{n=1}^\infty \frac{(-ht)^n}{n!} = -th\sum_{n=1}^\infty \frac{1}{n}\frac{(-ht)^{n-1}}{(n-1)!}$$

and

$$\left| \frac{1}{h}(e^{-ht} - 1)e^{-st}x(t) \right| \leq |t||e^{-ht}e^{-st}||x(t)| \leq |t|e^{-\sigma t}|x(t)|$$

The right-hand side is integrable by assumption.                          ∎

***Inverse Laplace transform.*** Let $x(t)$ be a signal with Laplace transform $x(s)$, $\Re s > \sigma$. Take any real $a > \sigma$. If $x(s)$ is evaluated on the line $\{a + j\omega, \omega \in R\}$, it coincides with the Fourier transform of the signal $\chi_a(t) = x(t)e^{-at}u_s(t)$. Due to Eq. (8.22), $\chi_a(t)$ is absolutely integrable, and its Fourier transform exists. Thus for any $s$ of the form $s = a + j\omega$, we have

$$x(s) = x(a + j\omega) = \int_0^\infty x(t)e^{-at-j\omega t}dt = \int_0^\infty \chi_a(t)e^{-j\omega t}dt$$

Application of the inverse Fourier transform leads to

$$x(t)e^{-at} = \frac{1}{2\pi}\int_{-\infty}^\infty x(a + j\omega)e^{j\omega t}d\omega$$

or

$$x(t) = \frac{1}{2\pi}\int_{-\infty}^\infty x(a + j\omega)e^{(a+j\omega)t}d\omega = \frac{1}{2\pi j}\int_{a-j\infty}^{a+j\infty} x(s)e^{st}ds \qquad (8.90)$$

The above expression is legitimate provided that the inverse Fourier transform exists. This is the case if $\chi_a(t)$ has finite energy, that is,

$$\int_{-\infty}^\infty e^{-2at}|x(t)|^2dt < \infty$$

and Eq. (8.90) is interpreted in the $L_2$ sense.

# 9

# SPECTRAL ANALYSIS OF STATIONARY PROCESSES

The harmonic analysis of deterministic signals discussed in Chapter 6 is extended to stochastic stationary signals in this chapter. First, wide sense stationary (WSS) processes are considered and the notion of power spectral density is introduced as the Fourier transform of the autocorrelation sequence. The integral of the power spectral density forms the spectral distribution of the signal. Unlike the power spectral density, the spectral distribution is always defined, and in addition characterizes the autocorrelation. Starting from the spectral distribution, the spectral process of the signal itself is constructed, and the Fourier analysis and synthesis equations are derived, leading to the so-called spectral representation of wide sense stationary processes. The final main theme of this chapter is devoted to higher-order spectral analysis. Higher-order cumulants are described in the time and frequency domain.

## 9.1 SPECTRAL DENSITY AND DISTRIBUTION

### 9.1.1 Power spectral density

Let $x(n)$ be a real discrete wide sense stationary signal. The Fourier transform of the autocorrelation sequence $r_x(n) = E[x(n+k)x(k)]$,

$$S_x(\omega) = \sum_{n=-\infty}^{\infty} r_x(n)e^{-j\omega n} \qquad (9.1)$$

is called *power spectral density* of the signal. The subscript $x$ is omitted when no confusion arises. If $r(n)$ is absolutely summable, the results of Section 6.2 assert that $S(\omega)$ is a well-defined continuous periodic function of period $2\pi$. Moreover the autocorrelation is recovered by the inverse Fourier transform

$$r(n) = \frac{1}{2\pi} \int_0^{2\pi} S(\omega) e^{j\omega n} d\omega \tag{9.2}$$

The situation is analogous if $r(n)$ is a finite energy signal. In this case the power spectral density exists in the $L_2(0, 2\pi)$ sense. The autocorrelation of a real signal is real and even, $r(n) = r(-n)$. It is not causal, unless it is zero. The symmetry properties of the Fourier transform imply that the power spectral density is real and even as well. We show next that the power spectral density is nonnegative definite.

**Proposition 9.1** Suppose that $r(n)$ belongs to $\ell_2$. Then $S(\omega) \geq 0$ for almost all $\omega$.

PROOF It suffices to show that $\int_I S(\omega) d\omega \geq 0$ for any interval $I$. Consider the indicator function of $I$:

$$\chi(\omega) = \begin{cases} 1, & \omega \in I \\ 0, & \text{otherwise} \end{cases}$$

$\chi(\omega)$ has finite support and is square integrable. Hence its periodic extension, also denoted for simplicity by $\chi$, is expanded in Fourier series. Let $c_n$ denote the Fourier series coefficients and $\chi_N(\omega) = \sum_{n=-N}^{N} c_n e^{j\omega n}$. Then $\lim_N \chi_N = \chi$ and $\lim_N |\chi_N|^2 = \chi$. Therefore

$$\int_I S(\omega) d\omega = \int S(\omega) \chi(\omega) d\omega = \langle S(\omega), \chi(\omega) \rangle = \lim_N \langle S(\omega), \chi_N(\omega) \chi_N^*(\omega) \rangle$$

$$= \lim_N \langle S(\omega), \sum_{n=-N}^{N} \sum_{k=-N}^{N} c_n c_k^* e^{j(n-k)\omega} \rangle$$

$$= \lim_N \sum_{n=-N}^{N} \sum_{k=-N}^{N} c_n^* c_k \langle S(\omega), e^{j(n-k)\omega} \rangle$$

$$= 2\pi \lim_N \sum_{n=-N}^{N} \sum_{k=-N}^{N} c_n^* c_k r(k - n) \geq 0$$

where we used the Fourier inversion formula $r(n) = \langle S, e^{-jn\omega} \rangle / 2\pi$ and the positive definiteness of $r(n)$ (see Section 2.2.2). ∎

In the spirit of Section 2.3, we identify the periodic function $S(\omega)$ with its truncation over $[-\pi, \pi]$. The truncation, designated by $S(\omega)$ as well, has all properties of a probability density function except that its integral is not equal to

1. Indeed,

$$\int_{-\infty}^{\infty} S(\omega)d\omega = \int_{-\pi}^{\pi} S(\omega)d\omega = 2\pi r(0)$$

A probability density function is obtained if the *normalized power spectral density*

$$S_n(\omega) = \frac{1}{2\pi r(0)} S(\omega)$$

is introduced.

If $x(n)$ is an $l$-channel WSS stochastic signal, the autocorrelation is an $l \times l$ matrix-valued sequence, and the power spectral density is an $l \times l$ matrix-valued function.

### Example 9.1   White noise

Let $x(n)$ be a white noise signal of unit variance. According to Section 2.2.2 the autocorrelation sequence is $r(n) = \delta(n)$. Invoking Example 6.17, we conclude that the power spectral density is

$$S(\omega) = 1, \qquad -\pi < \omega < \pi$$

In close analogy with white light, the spectrum extends equally over the entire range of frequencies. For this reason it is called white noise. ∎

### Example 9.2   Computation of the power spectral density from data

The power spectral density is the Fourier transform of the autocorrelation sequence. An obvious approach is to estimate the autocorrelation $r(n)$ from the signal data and then use the DFT to determine the Fourier transform $S(\omega)$. The autocorrelation sequence is estimated by

$$\hat{r}(n) = \frac{1}{N} \sum_{k=0}^{N-|n|-1} x(k)x(k+|n|), \qquad |n| \leq N-1 \tag{9.3}$$

It is easy to see that

$$E[\hat{r}(n)] = \left(\frac{N-|n|}{N}\right)r(n), \qquad |n| \leq N-1$$

Hence $\hat{r}(n)$ is a biased estimate of $r(n)$. The bias increases as $|n|$ approaches $N$. An unbiased estimate is

$$\hat{r}_u(n) = \frac{1}{N-|n|} \sum_{k=0}^{N-|n|-1} x(k)x(k+|n|), \qquad |n| \leq N-1 \tag{9.4}$$

The computation of the variance of the estimator $\hat{r}(n)$ is difficult. It is intuitively clear that the variance increases with increasing $|n|$ since fewer and fewer signal samples are considered in (9.3).

The sum in the right-hand side of (9.3) or (9.4) is the convolution of $x(n)$ and $x(-n)$. Hence it can be efficiently computed by the methods of Section 7.6, and in particular, with the FFT algorithm. The length of the FFT, $L$ is chosen to permit an efficient implementation. Once the autocorrelation sequence is determined the power spectral density is estimated by the FFT as in Section 6.2.1. A window $w(n)$ is incorporated to ameliorate the effects of the finite data record. To avoid the excessive increase of the variance of the autocorrelation estimators, only $M \ll N$ autocorrelation values are used in the computation. In summary, the FFT is invoked to compute the estimate

$$\hat{S}(\omega) = \sum_{n=-(M-1)}^{M-1} \hat{r}(n)w(n)e^{-j\omega n}, \qquad M \ll N$$

The window is real and even to ensure that the above estimate is a real function. The estimate $\hat{S}(\omega)$ remains nonnegative if the Fourier transform of $w(n)$ satisfies $W(\omega) \geq 0$. This condition is satisfied by the triangular window but is not satisfied by the rectangular, Hanning, Hamming, and Kaiser windows. Finally, the length $L$ of the FFT employed in the computation of the correlation sequence must satisfy $L \geq N + M - 1$. To avoid FFT's of very long length, we segment the data into $K$ records of $M$ samples each, $N = KM$. It is easy to see that

$$\hat{r}(n) = \frac{1}{N}\sum_{i=0}^{K-1} r^i(n)$$

$$r^i(n) = \sum_{k=0}^{M-|n|-1} x(k+iM)x(k+iM+|n|), \qquad |n| \leq M - 1$$

Each $r^i(n)$ corresponds to the correlation sequence of the $i$th data segment with values $x(k+iM)$, $0 \leq k \leq M - 1$.

The previous approach is sometimes called indirect because the computation of the power spectrum is preceded by the computation of the autocorrelation sequence. The direct approach determines the power spectral density directly from the data via the expression

$$\hat{S}(\omega) = \frac{1}{N}|X(\omega)|^2, \quad X(\omega) = \sum_{n=0}^{N-1} x(n)e^{-j\omega n}$$

The above estimator is called *periodogram*. The so-called modified periodogram

$$\hat{S}_m(\omega) = \frac{1}{N}|X_m(\omega)|^2, \quad X_m(\omega) = \sum_{n} x(n)w(n)e^{-j\omega n}$$

employs windowing. The MATLAB command **psd** implements the modified periodogram method using averaging over segments in a manner analogous with the indirect method. ∎

### *Example 9.3   Moving average signal*
Let

$$x(n) = \sum_{i=0}^{m} c_i e(n - i) \tag{9.5}$$

The *l*-channel signal $x(n)$ is obtained by filtering the *k*-channel signal $e(n)$ by an FIR filter with $l \times k$ matrix coefficients $c_i$. The autocorrelation of $x(n)$ is

$$r_x(n) = E[x(j)x^T(j - n)] = E\left[\sum_{i=0}^{m} c_i e(j - i)\left(\sum_{p=0}^{m} c_p e(j - n - p)\right)^T\right]$$

$$= \sum_i \sum_p c_i r_e(n + p - i)c_p^T$$

Applying the Fourier transform, we find that

$$S_x(\omega) = \sum_{n=-\infty}^{\infty} \sum_i \sum_p c_i r_e(n + p - i)c_p^T e^{-j(n+p-i)\omega} e^{j(p-i)\omega} = H(\omega)S_e(\omega)H^T(-\omega) \tag{9.6}$$

where $H(\omega) = \sum_{i=0}^{m} c_i e^{-ji\omega}$ is the filter frequency response. $x(n)$ is called a *moving average* process if it is obtained by (9.5), and $e(n)$ is white noise. In this case (9.6) becomes

$$S_x(\omega) = H(\omega)H^T(-\omega)$$

In the single-channel case the above equation reduces to

$$S_x(\omega) = |H(\omega)|^2 \tag{9.7}$$

Consider the MATLAB example:

```
e = randn(1,1014);
c = [-3, 1, 4, 1, -3];
x = filter(c,1,e);
[s, f] = spectrum (x, 64, 14, 64, 2*pi);     % computes the spectral density
[H, w] = freqz(c, 1, 512);                    % computes the frequency response
S1 = (abs(H)).^2;                             % computes the spectral density by (9.7)
subplot(121), plot(f, 10*log10(s(:,1)))       % compares plots
subplot(122), plot(w, 10*log10(S1))
```

∎

There are wide sense stationary processes whose autocorrelation is neither absolutely summable nor a finite energy signal. The harmonic process is such an example.

**Example 9.4   Harmonic process**
Consider the harmonic process

$$x(n) = \sum_{k=1}^{N} A_k \cos(\omega_k n + \theta_k)$$

Under the assumptions listed in Section 5.7, the autocorrelation is given by

$$r(n) = \frac{1}{2} \sum_{i=1}^{N} E[A_i^2] \cos(\omega_i n)$$

$r(n)$ does not belong to $\ell_2$. It is, however, bounded and can be viewed as a tempered distribution. It follows from Example 6.20 that the power spectral density is the tempered distribution

$$S(\omega) = \frac{\pi}{2} \sum_{i=1}^{N} E[A_i^2][\delta(\omega - \omega_i) + \delta(\omega + \omega_i)] \qquad \blacksquare$$

### 9.1.2   Spectral distribution

The autocorrelation sequence of a wide sense stationary process is always bounded. This follows from the Cauchy-Schwarz inequality:

$$|r(n)| = |E[x(m)x(m - n)]| = | < x(m), x(m - n) > |$$

$$\leq \sqrt{E[x^2(m)]}\sqrt{E[x^2(m - n)]} = r(0)$$

Since $r(n)$ belongs to $\ell_\infty$, it is interpreted as a tempered distribution. Then the power spectral density is defined as a tempered distribution. Generalized functions are avoided if the integral of the spectral density

$$F(\omega) = \int_{-\pi}^{\omega} S(v)dv \qquad (9.8)$$

is used instead. The function $F(\omega)$ is called *spectral distribution* and is always an ordinary function. The reason for the term is that the normalized function $F_n(\omega) = F(\omega)/2\pi r(0) = F(\omega)/F(\pi)$ is the probability distribution function associated with the probability density $S_n(\omega)$. The spectral distribution is nonnegative, bounded, and monotonically increasing. We recall from probability theory that the probability distribution exists even when the density does not. A similar situation occurs here, and the harmonic process discussed in Example 9.4 highlights it. The power spectral density consists of Dirac functions. In contrast, the spectral distribution is an ordinary piecewise constant function.

The spectral distribution $F(\omega)$ is explicitly written in terms of the autocorrelation $r(n)$ if Eq. (9.1) is substituted into Eq. (9.8):

$$F(\omega) = \int_{-\pi}^{\omega} \sum_{n=-\infty}^{\infty} r(n)e^{-jvn}dv = \sum_{n=-\infty}^{\infty} r(n) \int_{-\pi}^{\omega} e^{-jvn}dv$$

$$= 2\pi \sum_{n=-\infty}^{\infty} r(n)c_n(\omega) \tag{9.9}$$

where $c_n(\omega) = \int_{-\pi}^{\omega} e^{-jvn}dv/2\pi$ are the Fourier coefficients of the indicator function of the interval $[-\pi, \omega]$. The precise role of the spectral distribution and its ability to represent the autocorrelation sequence is elucidated by the Herglotz theorem. This in turn requires the notion of the Lebesgue-Stieltjes integral, briefly introduced next.

***Lebesgue-Stieltjes integral.*** Given a nonnegative bounded monotonically increasing function $F(\omega)$ the integral

$$\int h(\omega)dF(\omega)$$

is defined stepwise as follows:

- If $h$ is the indicator $h(\omega) = \chi_I$, $I = [\omega_1, \omega_2]$, we set

$$\int \chi_I dF(\omega) = F(\omega_2) - F(\omega_1)$$

- If $h$ is a simple function (i.e., a linear combination $h(\omega) = \sum_i c_i \chi_{I_i}$ of indicator functions $\chi_{I_i}$, $I_i = [\omega_i, \omega_{i+1}]$), we set

$$\int h(\omega)dF(\omega) = \sum_i c_i \int \chi_{I_i} dF(\omega) = \sum_i c_i[F(\omega_{i+1}) - F(\omega_i)]$$

- If $h$ is a measurable function (i.e., a limit of simple functions $h_N$), we set

$$\int h(\omega)dF(\omega) = \lim_{N \to \infty} \int h_N dF(\omega)$$

Measurable functions are extremely general.

**Theorem 9.1 Herglotz's theorem**    Let $x(n)$ be a real zero-mean wide sense stationary process with autocorrelation sequence $r(n)$. There exists an essentially unique real-valued function $F$ defined on $[-\pi, \pi]$ that is bounded, monotonically increasing, and continuous from the right and that satisfies $F(-\pi) = 0$ and

$$r(n) = \frac{1}{2\pi} \int_{-\pi}^{\pi} e^{jn\omega} dF(\omega) \qquad (9.10)$$

$F(\omega)$ is specified by Eq. (9.9) at continuity points. The proof will be omitted. ∎

Extension to the multichannel case is possible. Equation (9.10) is now a matrix equation. The spectral distribution is a matrix-valued function, continuous from the right and such that for any $\omega_1 < \omega_2$, $F(\omega_2) - F(\omega_1)$ is a nonnegative definite matrix.

### 9.1.3    Lebesgue decomposition of spectral distributions

Since $F(\omega)$ is monotonically increasing it is differentiable almost everywhere, and the derivative $S(\omega) = dF(\omega)/d\omega$ is integrable on $[-\pi, \pi]$. In general, $F$ is neither continuous nor even more so absolutely continuous. An absolutely continuous function is characterized by the property that the integral of the derivative identifies the function:

$$F(\omega) = \int_{-\pi}^{\omega} \frac{dF(v)}{dv} dv = \int_{-\pi}^{\omega} S(v) dv$$

Functions of the form

$$F(\omega) = \int_{-\pi}^{\omega} G(v) dv \qquad (9.11)$$

where $G(\omega)$ is any absolutely integrable function, provide important examples of absolutely continuous functions.

The structure of the spectral distribution $F$ is illuminated by Lebesgue decomposition which states that $F$ can be decomposed as

$$F = F_1 + F_2 + F_3$$

$F_1$ is an absolutely continuous distribution, $F_2$ is continuous having zero derivative almost everywhere, and $F_3$ is piecewise constant, formed by the jumps of $F$ at the discontinuity points. The construction of these distributions proceeds as follows: First, it is shown that the number of discontinuities of $F$ in $[-\pi, \pi]$ is at most countable. If $\omega_1, \omega_2, \ldots$ are those points, $F_3$ is defined by

$$F_3(\omega) = \sum_{i:\omega_i \leq \omega} \Delta F_i, \qquad \Delta F_i = F(\omega_i + 0) - F(\omega_i - 0)$$

where $F(\omega_i + 0) = \lim F(\omega)$, $\omega \to \omega_i$, $\omega > \omega_i$, and similarly for $F(\omega_i - 0)$. The function $F - F_3$ is now continuous and monotonically increasing. The absolutely continuous part is

$$F_1(\omega) = \int_{-\pi}^{\omega} \frac{d(F(v) - F_3(v))}{dv} \, dv$$

The difference $F - F_3 - F_1$ is the singular part with derivative equal to zero almost everywhere. The singular part is highly pathological and in practice is usually ignored. Notice that although it has zero derivative, it is not constant.

### 9.1.4 Spectral density of convolutional system outputs

Consider the causal multivariable convolutional system

$$y(n) = \sum_{k=-\infty}^{n} h(n - k)x(k) = \sum_{k=0}^{\infty} h(k)x(n - k) \tag{9.12}$$

The input $x(n)$ is a WSS process with zero mean, autocorrelation sequence $r_x(n)$, and spectral distribution $F_x(\omega)$. For simplicity $x(n)$ is a single-channel signal. Let $L_2(F_x)$ denote the set of (measurable) functions $H(\omega)$ on $[-\pi, \ \pi]$ that are square integrable with respect to $F_x$, that is,

$$\int_{-\pi}^{\pi} |H(\omega)|^2 dF_x(\omega) < \infty$$

$L_2(F_x)$ shares the basic features of the $L_2$ space of square integrable functions introduced in Section 2.8. In particular, it constitutes a Hilbert space with inner product

$$< H(\omega), G(\omega) >= \int_{-\pi}^{\pi} H(\omega)G^*(\omega)dF_x(\omega) \tag{9.13}$$

It is easy to see that $e^{jk\omega} \in L_2(F_x)$ for all $k \in Z$. It can be shown that $L_2(F_x)$ is spanned by the closure of $A = \{e^{jk\omega}, k \in Z\}$. Note, however, that $A$ is not in general orthogonal in $L_2(F_x)$. Let us form the partial sum of the impulse response coefficients

$$H_N(\omega) = \sum_{k=0}^{N} h(k)e^{-jk\omega}$$

Clearly $H_N(\omega) \in L_2(F_x)$. We assume that the trigonometric series

$$H(\omega) = \lim_{N} H_N(\omega) = \sum_{k=0}^{\infty} h(k)e^{-jk\omega}$$

exists in $L_2(F_x)$, that is,

$$\lim_{N \to \infty} \int_{-\pi}^{\pi} |H_N(\omega) - H(\omega)|^2 dF_x(\omega) = 0$$

Under the above assumptions we will show that the convolution representation yields a well-defined second-order output. Before we embark to prove the statement above we describe a practical condition that ensures the existence of $H(\omega)$ in $L_2(F_x)$. Suppose that the input autocorrelation sequence is absolutely summable. Then the power spectral density $S_x(\omega)$ exists and is bounded. Suppose that the impulse response has finite energy. Then $H(\omega)$ exists as a Fourier series in $L_2$. We need to show that $H(\omega)$ resides in $L_2(F_x)$. The Hölder inequality gives

$$\int_{-\pi}^{\pi} |H_N(\omega) - H(\omega)|^2 S_x(\omega) d\omega \le \left( \int_{-\pi}^{\pi} |H_N(\omega) - H(\omega)|^2 d\omega \right) \sup_{\omega \in [-\pi, \pi]} S_x(\omega)$$

The right-hand side tends to zero as $N \to \infty$.

Let us next analyze (9.12) in the general case where $\lim_N H_N(\omega)$ exists with respect to $L_2(F_x)$. Let $y_N(n) = \sum_{k=0}^{N} h(k)x(n-k)$. Then

$$E[|y_N(n) - y_M(n)|^2] = E\left[ \left| \sum_{k=M+1}^{N} h(k)x(n-k) \right|^2 \right]$$

$$= \sum_{k=M+1}^{N} \sum_{i=M+1}^{N} h(k)h(i)E[x(n-k)x(n-i)]$$

$$= \sum_{k=M+1}^{N} \sum_{i=M+1}^{N} h(k)h(i)r_x(i-k)$$

The Herglotz's theorem implies that

$$E[|y_N(n) - y_M(n)|^2] = \frac{1}{2\pi} \sum_{k=M+1}^{N} \sum_{i=M+1}^{N} h(k)h(i) \int_{-\pi}^{\pi} e^{j(i-k)\omega} dF_x(\omega)$$

$$= \frac{1}{2\pi} \int_{-\pi}^{\pi} \left| \sum_{k=M+1}^{N} h(k)e^{-jk\omega} \right|^2 dF_x(\omega)$$

The last term goes to zero as $N$, $M$ tend to infinity, because $H_N(\omega)$ is a Cauchy sequence in $L_2(F_x)$. Hence $y_N(n)$ is a Cauchy sequence living in the Hilbert space of second-order random variables, and it converges. We established that the output is well-defined. With analogous calculations we obtain the spectral distribution of the output. Continuity of the inner product and the Herglotz's theorem give

$$E[y(m)y(m-n)] =< y(m), y(m-n) >$$

$$=< \lim_N y_N(m), \lim_K y_K(m-n) >$$

$$= \lim_N \lim_K \sum_{k,i} h(k)h(i) < x(m-k), x(m-n-i) >$$

$$= \lim_N \lim_K \sum_{k,i} h(k)h(i)r_x(n+i-k)$$

$$= \frac{1}{2\pi} \lim_N \lim_K \sum_{k,i} h(k)h(i) \int_{-\pi}^{\pi} e^{j(n+i-k)\omega} dF_x(\omega)$$

$$= \frac{1}{2\pi} \lim_N \lim_K \int_{-\pi}^{\pi} e^{jn\omega} \sum_k h(k)e^{-jk\omega} \sum_i h(i)e^{ji\omega} dF_x(\omega)$$

Inserting limits inside the integral, we obtain

$$r_y(n) = \frac{1}{2\pi} \int_{-\pi}^{\pi} e^{jn\omega} |H(\omega)|^2 dF_x(\omega) \tag{9.14}$$

Uniqueness in the Herglotz's theorem implies that the output spectral distribution satisfies

$$dF_y(\omega) = |H(\omega)|^2 dF_x(\omega) \quad \text{or} \quad F_y(\omega) = \int_{-\pi}^{\omega} |H(v)|^2 dF_x(v) \tag{9.15}$$

If the input spectral distribution is absolutely continuous so that $dF_x(\omega) = S_x(\omega)d\omega$, Eq. (9.15) becomes

$$F_y(\omega) = \int_{-\pi}^{\omega} |H(v)|^2 S_x(v) dv$$

In the latter case the remark related to Eq. (9.11) implies that $F_y(\omega)$ is absolutely continuous as well and that

$$S_y(\omega) = |H(\omega)|^2 S_x(\omega)$$

### Example 9.5   White noise input
Suppose that $x(n)$ is a scalar white noise with $r_x(n) = \gamma_2\delta(n)$. Then $S_x(\omega) = \gamma_2$ and $F_x(\omega) = \gamma_2(\omega + \pi)$. The space $L_2(F_x)$ coincides with $L_2(-\pi, \pi)$, the space of square integrable functions on $[-\pi, \pi]$:

$$\int_{-\pi}^{\pi} |H(\omega)|^2 d\omega < \infty$$

It follows from our discussion of Fourier series that the above Hilbert space is isomorphic to $\ell_2(Z^+)$, the space of finite energy causal signals. Thus, if the input is white noise and the impulse response belongs to $\ell_2$, the output is defined in the quadratic mean sense, it is a zero mean WSS process, and its spectral distribution is given by $dF_y(\omega) = \gamma_2|H(\omega)|^2 d\omega$. In this case $F_y(\omega)$ is absolutely continuous, and the spectral density is given by

$$S_y(\omega) = |H(\omega)|^2\gamma_2 \tag{9.16}$$

∎

## 9.2   SPECTRAL REPRESENTATION OF WSS DISCRETE PROCESSES

The spectral representation of a wide sense stationary process extends the Fourier analysis and synthesis equations from the deterministic to the stochastic case. The stochastic case cannot be dealt with a direct extension of the methods developed for deterministic signals. Instead, the analysis builds upon the representation of the autocorrelation in terms of the spectral distribution.

### 9.2.1   Motivation

The discussion that follows serves to indicate the subtleties encountered in the Fourier analysis of the WSS process $x(n)$. Mimicking the deterministic case, let us define the Fourier transform via the equation

$$X(\omega) = \sum_{n=-\infty}^{\infty} x(n)e^{-j\omega n} \tag{9.17}$$

$X(\omega)$ is a complex random process whose realizations are the Fourier transforms of the corresponding realizations of $x(n)$. It is highly desirable that $X(\omega)$ be a second-order process. Unfortunately, this is not the case. The autocorrelation function of $X(\omega)$ is not an ordinary function; instead, it has an impulsive character. This is seen by the following computation:

$$R_X(\omega_1, \omega_2) = E[X(\omega_1)X^*(\omega_2)] = \sum_{n=-\infty}^{\infty}\sum_{k=-\infty}^{\infty} E[x(n)x(k)]e^{-j\omega_1 n + j\omega_2 k}$$

$$= \sum_{n=-\infty}^{\infty}\sum_{k=-\infty}^{\infty} r(n-k)e^{-j\omega_1 n + j\omega_2 k} = \sum_{n=-\infty}^{\infty}\sum_{i=-\infty}^{\infty} r(i)e^{-j\omega_2 i}e^{-j(\omega_1-\omega_2)n}$$

We recall from Section 6.2.3 that $\sum_{n=-\infty}^{\infty} e^{-j\omega n} = 2\pi\delta(\omega)$ for $|\omega| < \pi$, $\delta(\omega)$ is the Dirac function. Thus

$$R_X(\omega_1, \omega_2) = \sum_i r(i)e^{-j\omega_2 i}\sum_n e^{-j(\omega_1-\omega_2)n} = 2\pi S(\omega_2)\delta(\omega_1 - \omega_2) \tag{9.18}$$

The Dirac function $\delta(\omega)$ gives rise to an ordinary function when "integrated." In close analogy with the spectral analysis of the autocorrelation function, we consider the integral

$$x(\omega) = \int_{-\pi}^{\omega} X(v)dv \tag{9.19}$$

or

$$dx(\omega) = X(\omega)d\omega \tag{9.20}$$

If we insert Eq. (9.17) into (9.19) and integrate term by term, we obtain

$$x(\omega) = \sum_{n=-\infty}^{\infty} x(n) \int_{-\pi}^{\omega} e^{-jvn}dv = 2\pi \sum_{n=-\infty}^{\infty} x(n)c_n(\omega) \tag{9.21}$$

where $c_n(\omega) = (1/2\pi) \int_{-\pi}^{\omega} e^{-jvn}dv$ are the Fourier coefficients of the indicator function $\chi_\omega(v)$ of the interval $[-\pi, \omega]$. The latter equation demonstrates how the so-called *spectral process* $x(\omega)$ is obtained from the original signal $x(n)$ directly and not through $X(\omega)$. Taking into account Eq. (9.20), we infer that the signal $x(n)$ is reconstructed from the spectral process $x(\omega)$ via the Stieltjes integral

$$x(n) = \frac{1}{2\pi} \int_{-\pi}^{\pi} e^{j\omega n} X(\omega)d\omega = \frac{1}{2\pi} \int_{-\pi}^{\pi} e^{j\omega n} dx(\omega) \tag{9.22}$$

The above stochastic integral needs further elaboration. Leaving that aside for the moment, we proceed to demonstrate that $x(\omega)$ is a second-order process with autocorrelation function $R_x(\omega_1, \omega_2) = 2\pi F(\min(\omega_1, \omega_2))$. Let $\omega_1 \leq \omega_2$. Then

$$R_x(\omega_1, \omega_2) = E[x(\omega_1)x^*(\omega_2)] = \int_{-\pi}^{\omega_1} \int_{-\pi}^{\omega_2} E[X(v_1)X^*(v_2)]dv_2dv_1$$

$$= 2\pi \int_{-\pi}^{\omega_1} \int_{-\pi}^{\omega_2} S(v_2)\delta(v_1 - v_2)dv_2dv_1 = 2\pi \int_{-\pi}^{\omega_1} S(v_1)dv_1 = 2\pi F(\omega_1)$$

Therefore $R_x(\omega_1, \omega_2)$ is an ordinary function, and in particular, $E[|x(\omega)|^2] = 2\pi F(\omega)$. Note also that $x(\omega)$ is not a stationary process in general. Entirely analogous computations show that for any $\omega_1 \leq \omega_2$, it holds that

$$E[|x(\omega_2) - x(\omega_1)|^2] = 2\pi(F(\omega_2) - F(\omega_1)) \tag{9.23}$$

The latter equation is symbolically written as

$$E[|dx(\omega)|^2] = 2\pi dF(\omega) \tag{9.24}$$

It states that the power of the increments of $x(\omega)$ are given by the increments of the spectral distribution.

Another important observation stemming from Eq. (9.18) is that the values of the spectrum at two different frequencies are uncorrelated. How is this orthogonality property embedded into the process $x(\omega)$? It turns out that the latter is a stochastic process with orthogonal increments; that is, for any $\omega_1 < \omega_2 \leq \omega_3 < \omega_4$ in $[-\pi, \pi]$, it holds that

$$E[(x(\omega_2) - x(\omega_1))(x^*(\omega_4) - x^*(\omega_3))] = 0$$

Indeed, substituting Eq. (9.19), we obtain

$$E[(x(\omega_2) - x(\omega_1))(x^*(\omega_4) - x^*(\omega_3))] = \int_{\omega_1}^{\omega_2} \int_{\omega_3}^{\omega_4} E[X(v_1)X^*(v_2)]dv_2 dv_1$$

$$= 2\pi \int_{\omega_3}^{\omega_4} \left[ \int_{\omega_1}^{\omega_2} S(v_1)\delta(v_1 - v_2)dv_1 \right] dv_2$$

The inner integral is zero because $v_2$ does not reside in the interval $[\omega_1, \omega_2]$.

### 9.2.2  Stochastic Stieltjes integrals

Equation (9.22) requires a rigorous definition of the integral

$$y = \int h(\omega)dx(\omega)$$

Motivated by the preceding analysis, we assume that $x(\omega)$ is a stochastic process with orthogonal increments and that $h(\omega)$ is a deterministic (nonrandom) function. The integral $y$ will be a random variable. The construction of the stochastic integral proceeds along the same lines as the deterministic Stieltjes integral. If $h$ is the indicator function of an interval $I = [\omega_1, \omega_2]$, we set

$$y = \int h(\omega)dx(\omega) = x(\omega_2) - x(\omega_1)$$

If $h(\omega) = \sum_i c_i \chi_{I_i}$, $I_i = [\omega_i, \omega_{i+1}]$, we set

$$y = \sum_i c_i \int \chi_{I_i} dx(\omega) = \sum_i c_i[x(\omega_{i+1}) - x(\omega_i)]$$

For any $\omega_1 \leq \omega_2$ in $[-\pi, \pi]$, let $F$ be defined by

$$F(\omega_2) - F(\omega_1) = \frac{1}{2\pi} E[|x(\omega_2) - x(\omega_1)|^2] \tag{9.25}$$

$F$ is defined uniquely up to an additive constant and is monotonically increasing. Later $x(\omega)$ will be replaced by the spectral process, and $F$ will become the spectral distribution of $x(n)$. The most general class of integrands, we will consider consists of functions $h$ in $L_2(F)$. This means that $h$ is square integrable with respect to $F$:

$$\int_{-\pi}^{\pi} |h(\omega)|^2 dF(\omega) < \infty$$

It turns out that such an $h$ is obtained as the quadratic mean limit of a sequence of simple functions $h_N$. Thus we define

$$y = \lim_{N \to \infty} \int h_N dx(\omega)$$

It can be shown that the limit is well defined and does not depend on the particular sequence $h_N$.

### 9.2.3 Signal Hilbert spaces and $L_2(F)$ representations

Let $x(n)$ be a real zero mean WSS process with spectral distribution $F$. Given a set of integers $I$, the smallest closed linear subspace of second-order random variables generated by $x(n), n \in I$, is denoted $\mathcal{H}_x(I)$. It was originally introduced in Section 2.9.5. Each element of $\mathcal{H}_x(I)$ is obtained as the quadratic mean limit of a sequence of linear combinations of $x(n), n \in I$. Let $\mathcal{H}_x(\infty)$ and $\mathcal{H}_x(n)$ designate $\mathcal{H}_x(I)$ when $I = Z$ and $I = \{k \in Z : k \leq n\}$, respectively. The sequence $\mathcal{H}_x(n)$ is nested, namely

$$\mathcal{H}_x(n) \subset \mathcal{H}_x(n+1) \subset \mathcal{H}_x(\infty) = \overline{\cup_{n \in Z} \mathcal{H}_x(n)}$$

Since $\mathcal{H}_x(n)$ is spanned by the values $x(n), x(n-1), x(n-2), \ldots$, its elements are either finite sums of the form

$$y_N = \sum_{k=0}^{N} h_N(k) x(n-k) \tag{9.26}$$

or quadratic mean limits of such terms.

Let $L_2^+(F)$ denote the closed linear subspace of $L_2(F)$ generated by the exponentials $e^{-jk\omega}$, $k \geq 0$. Each element of $L_2^+(F)$ is either a finite linear combination

$$H_N(\omega) = \sum_{k=0}^{N} h_N(k) e^{-jk\omega}$$

or a limit of such linear combinations, $H(\omega) = \lim_N H_N(\omega)$. The limit is with respect to the distance induced by the inner product. We caution the reader that the elements of $L_2^+(F)$ are not trigonometric series because the coefficients $h_N(k)$ depend on $N$. We prove below that for each $n$, $\mathcal{H}_x(n)$ is isomorphic to $L_2^+(F)$. The isomorphism is explicitly described in the frequency domain. Indeed, let $y$ be an element of $\mathcal{H}_x(n)$. Then $y = \lim_N y_N$, $y_N = \sum_{k=0}^N h_N(k)x(n-k)$. Using the spectral representation, we obtain

$$y = \frac{1}{2\pi} \lim_N \sum_{k=0}^N h_N(k) \int_{-\pi}^{\pi} e^{j(n-k)\omega} dx(\omega) = \frac{1}{2\pi} \int_{-\pi}^{\pi} e^{jn\omega} H(\omega) dx(\omega) \qquad (9.27)$$

where

$$H(\omega) = \lim_N \sum_{k=0}^N h_N(k) e^{-jk\omega}$$

It turns out that $H(\omega) \in L_2^+(F)$. The map $y \to H(\omega)$ is the desired isomorphism. Equation (9.27) provides a frequency domain representation of the space $\mathcal{H}_x(n)$. This representation does not convert to a convolution representation in the time domain, unless we make the additional assumption that $H(\omega)$ is a trigonometric series in $L_2^+(F)$. In the latter case Eq. (9.27) takes the form $y = \sum_k h(k)x(n-k)$. Thus the convolution representation of the elements of $\mathcal{H}_x(n)$ is not always valid. Notice finally that $\mathcal{H}_x(n)$ and $\mathcal{H}_x(k)$ are isomorphic, being copies of $L_2^+(F)$. In a similar fashion $\mathcal{H}_x(\infty)$ can be identified with $L_2(F)$. The theorems that follow are proved in the Appendix at the end of the chapter.

**Theorem 9.2** Consider a real zero mean WSS process $x(n)$ with spectral distribution $F$. There exists a process with orthogonal increments $x(\omega)$ such that for any $y$ in $\mathcal{H}_x(\infty)$ there is a unique $H(\omega)$ in $L_2(F)$ such that

$$y = \frac{1}{2\pi} \int_{-\pi}^{\pi} H(\omega) dx(\omega) \qquad (9.28)$$

The Hilbert spaces $\mathcal{H}_x(\infty)$ and $L_2(F)$ are isomorphic. Preservation of inner products translates to the important identity

$$E\left[ \int_{-\pi}^{\pi} H(\omega) dx(\omega) \int_{-\pi}^{\pi} G(\omega) dx(\omega) \right] = \int_{-\pi}^{\pi} H(\omega) G^*(\omega) dF(\omega) \qquad (9.29)$$

∎

A direct consequence of the above theorem is the spectral representation of wide sense stationary processes stated next.

**Theorem 9.3** Let $x(n)$ be a wide sense stationary process with spectral distribution $F$. There exists a unique (aside a set of zero probability) process with orthogonal increments $x(\omega)$ such that

$$x(n) = \frac{1}{2\pi} \int_{-\pi}^{\pi} e^{j\omega n} dx(\omega), \quad E[|dx(\omega)|^2] = 2\pi dF(\omega) \tag{9.30}$$

The process $x(\omega)$ constitutes the *spectral process* of $x(n)$.  ∎

Suppose next that $y(n)$ is a wide sense stationary process belonging to $\mathcal{H}_x(\infty)$, the Hilbert space generated by the signal $x(n)$. Roughly speaking, $y(n)$ is obtained at the output of a linear filter acting on $x(n)$. This filter is expected to be time invariant if $y(n)$ and $x(n)$ are jointly wide sense stationary. The precise meaning of this statement is given next.

**Proposition 9.2**   Suppose that $x(n)$ and $y(n)$ are jointly wide sense stationary, that is, the multichannel signal $\left( x^T(n) \quad y^T(n) \right)^T$ is wide sense stationary, and that $y(n) \in \mathcal{H}_x(\infty)$. Then there exists $H(\omega)$ in $L_2(F_x)$ such that

$$y(n) = \frac{1}{2\pi} \int_{-\pi}^{\pi} e^{j\omega n} H(\omega) dx(\omega) \tag{9.31}$$

∎

The representation of the Hilbert space $\mathcal{H}_x(n)$ is given next.

**Proposition 9.3**   Every element $y$ of $\mathcal{H}_x(n)$ is expressed as

$$y = \frac{1}{2\pi} \int_{-\pi}^{\pi} e^{jn\omega} H(\omega) dx(\omega) \qquad H(\omega) \in L_2^+(F) \tag{9.32}$$

∎

As was pointed out above, (9.32) does not in general convert into convolution. The reason is that $H(\omega)$ is a limit of  trigonometric polynomials but not necessarily a trigonometric series.

### 9.2.4   WSS signals with absolutely continuous spectral distributions

A WSS process with absolutely continuous spectral distribution can be obtained by filtering white noise via a linear time invariant noncausal system. We often refer to such a signal as *colored noise*.

Suppose that $u(n)$ is a white noise signal of variance 1. Let

$$x(n) = \sum_{k=-\infty}^{\infty} h(k)u(n-k), \quad \sum_{k=-\infty}^{\infty} |h(k)|^2 < \infty \tag{9.33}$$

We saw in Example 9.5 that $x(n)$ is wide sense stationary with absolutely continuous spectral distribution and spectral density given by

$$S(\omega) = \frac{dF(\omega)}{d\omega} = |H(\omega)|^2 \tag{9.34}$$

Next we prove the converse statement. Suppose that $x(n)$ is wide sense stationary with absolutely continuous spectral distribution $F$. Let $dF(\omega) = S(\omega)d\omega$, where $S$ is the derivative of $F$. Consider a function $H(\omega)$ such that $|H(\omega)|^2 = S(\omega)$. Such a function always exists because $S(\omega) \geq 0$. Since $F(\omega)$ is absolutely continuous, $S(\omega)$ is integrable. Thus $H(\omega)$ belongs to $L_2$ and admits a Fourier series expansion. The Fourier coefficients of $H(\omega)$ determine the desired filter coefficients $h(n)$. To specify the white noise input signal, we assume for the moment that Eq. (9.33) is true. We plug into this equation the spectral representation of $u(n)$ to obtain

$$x(n) = \frac{1}{2\pi} \sum_k h(k) \int e^{j\omega(n-k)} du(\omega) = \frac{1}{2\pi} \int e^{j\omega n} H(\omega) du(\omega)$$

Uniqueness of the spectral process leads to

$$dx(\omega) = H(\omega)du(\omega), \quad \text{or} \quad du(\omega) = \frac{1}{H(\omega)} dx(\omega) \tag{9.35}$$

The above equation determines the spectral process of $u(n)$ and hence the noise itself provided that $H(\omega) \neq 0$.

Guided by the above analysis, we define the white noise by Eq. (9.35) or after integration by

$$u(\omega) = \int_{-\pi}^{\omega} \frac{1}{H(v)} dx(v) \tag{9.36}$$

The stochastic integral is well-defined provided that $1/H(\omega)$ belongs to $L_2(F)$. Indeed,

$$\int_{-\pi}^{\pi} \frac{1}{|H(\omega)|^2} dF(\omega) = \int_{-\pi}^{\pi} \frac{1}{|H(\omega)|^2} S(\omega)d(\omega) = \int_{-\pi}^{\pi} d\omega = 2\pi$$

Next we show that $u(n)$ is white noise. Taking into account Example 9.5, it suffices to prove that $E[|du(\omega)|^2] = 2\pi d\omega$. Note that

$$E[|du(\omega)|^2] = E[(u(\omega_2) - u(\omega_1))^2] = E\left[\left|\int_{\omega_1}^{\omega_2} \frac{1}{H(\omega)} dx(\omega)\right|^2\right]$$

$$= 2\pi \int_{\omega_1}^{\omega_2} \frac{1}{|H(\omega)|^2} dF(\omega) = 2\pi \int_{\omega_1}^{\omega_2} d\omega = 2\pi(\omega_2 - \omega_1) = 2\pi d\omega$$

Finally, we verify (9.33). Equation (9.35) and the spectral representation give

$$x(n) = \frac{1}{2\pi} \int e^{jn\omega} dx(\omega) = \frac{1}{2\pi} \int e^{jn\omega} H(\omega) du(\omega)$$

$$= \frac{1}{2\pi} \int e^{jn\omega} \sum_{k=-\infty}^{\infty} h(k) e^{-jk\omega} du(\omega)$$

$$= \frac{1}{2\pi} \sum_{k=-\infty}^{\infty} h(k) \int e^{j(n-k)\omega} du(\omega) = \sum_{k=-\infty}^{\infty} h(k) u(n-k)$$

In the above analysis we tacitly assumed that $S(\omega) > 0$. It can be shown that the conclusion is valid in the general case $S(\omega) \geq 0$.

### Example 9.6   ARMA processes
A WSS signal $x(n)$ is an ARMA process if it satisfies

$$x(n) + \sum_{i=1}^{k} a_i x(n-i) = \sum_{i=0}^{m} b_i e(n-i) \tag{9.37}$$

where $e(n)$ is a white noise signal of variance 1. The spectral representation theorem implies that

$$\int_{-\pi}^{\pi} e^{jn\omega} dx(\omega) + \sum_{i=1}^{k} a_i \int_{-\pi}^{\pi} e^{j(n-i)\omega} dx(\omega) = \sum_{i=0}^{m} b_i \int_{-\pi}^{\pi} e^{j(n-i)\omega} de(\omega) \tag{9.38}$$

Let

$$a(\omega) = 1 + \sum_{i=1}^{k} a_i e^{-ji\omega}, \quad b(\omega) = \sum_{i=0}^{m} b_i e^{-ji\omega}$$

Then Eq. (9.38) becomes

$$\int_{-\pi}^{\pi} e^{jn\omega} a(\omega) dx(\omega) = \int_{-\pi}^{\pi} e^{jn\omega} b(\omega) de(\omega)$$

Uniqueness of the spectral process leads to

$$dx(\omega) = \frac{b(\omega)}{a(\omega)} de(\omega)$$

Since $e(n)$ is a white noise signal with variance 1, it holds that $E[|de(\omega)|^2] = 2\pi d\omega$. The spectral distribution of $x(n)$, $F(\omega)$, satisfies

$$dF(\omega) = \frac{1}{2\pi} E[|dx(\omega)|^2] = \frac{1}{2\pi} \frac{|b(\omega)|^2}{|a(\omega)|^2} E[|de(\omega)|^2] = \frac{|b(\omega)|^2}{|a(\omega)|^2} d\omega$$

Therefore

$$F(\omega) = \int_{-\pi}^{\omega} \frac{|b(v)|^2}{|a(v)|^2} \, dv$$

The integral is well-defined provided that $|a(v)|^2$ does not vanish. Let

$$a(z) = 1 + \sum_{i=1}^{k} a_i z^{-i}, \quad b(z) = \sum_{i=0}^{m} b_i z^{-i}$$

It follows that $F(\omega)$ is well-defined if no roots of $a(z)$ lie on the unit circle. In this case $F(\omega)$ is absolutely continuous (see Eq. (9.11)), and the spectral density $S(\omega)$ is given by

$$S(\omega) = \frac{dF(\omega)}{d\omega} = \frac{|b(\omega)|^2}{|a(\omega)|^2} = \frac{b(z)b(z^{-1})}{a(z)a(z^{-1})}\bigg|_{z=e^{j\omega}} \tag{9.39}$$

If $b(z) = 1$, the process $x(n)$ is generated by

$$x(n) + \sum_{i=1}^{k} a_i x(n-i) = e(n) \tag{9.40}$$

and is called an *autoregressive process*. Besides $x(n)$, there are several other autoregressive signals that have the same power spectral density $S(\omega)$. Indeed, suppose for simplicity that the roots $\lambda_i$ of $a(z)$ are real and distinct, then the roots of $|a(z)|^2$ are $\lambda_1, \lambda_2, \ldots, \lambda_k$, and their reciprocals $1/\lambda_1, 1/\lambda_2, \ldots, 1/\lambda_k$. We modify the polynomial $a(z)$ by removing one root, say, $\lambda_1$, and replacing it with its reciprocal. The resulting autoregressive model will have the same power spectral density $S(\omega)$. There are $k$ such different polynomials. Next we replace two roots with their reciprocals. This can be done in $\binom{k}{2}$ ways, and all lead to the same power spectral density. Continuing this way we construct $\sum_{i=1}^{k} \binom{k}{i} = (1+1)^k = 2^k$ autoregressive filters with the same power spectral density.

As an example, consider the autoregressive model

$$x(n) - \frac{5}{6}x(n-1) + \frac{1}{6}x(n-2) = e(n)$$

The transfer function is

$$H_1(z) = \frac{1}{1 - \frac{5}{6}z^{-1} + \frac{1}{6}z^{-2}} = \frac{z^2}{\frac{1}{6} - \frac{5}{6}z + z^2}$$

The poles are $\lambda_1 = 1/2$, $\lambda_2 = 1/3$. Consider the pairs $(2, 1/3)$, $(1/2, 3)$, $(2, 3)$. The corresponding transfer functions are

$$H_2(z) = \frac{z^2}{\frac{2}{3} - \frac{7}{3}z + z^2} = \frac{1}{1 - \frac{7}{3}z^{-1} + \frac{2}{3}z^{-2}}$$

$$H_3(z) = \frac{z^2}{\frac{3}{2} - \frac{7}{2}z + z^2} = \frac{1}{1 - \frac{7}{2}z^{-1} + \frac{3}{2}z^{-2}}$$

$$H_4(z) = \frac{z^2}{6 - 5z + z^2} = \frac{1}{1 - 5z^{-1} + 6z^{-2}}$$

The resulting AR filters are

$$x(n) - \frac{7}{3}x(n-1) + \frac{2}{3}x(n-2) = e(n)$$

$$x(n) - \frac{7}{2}x(n-1) + \frac{3}{2}x(n-2) = e(n)$$

$$x(n) - 5x(n-1) + 6x(n-2) = e(n)$$

All four AR filters yield the same power spectral density

$$S(\omega) = \frac{1}{(1 - \frac{5}{6}e^{-j\omega} + \frac{1}{6}e^{-j2\omega})(1 - \frac{5}{6}e^{j\omega} + \frac{1}{6}e^{j2\omega})}$$

The autocorrelation function can be determined by partial fraction expansion. Let $\lambda_{\max}$ be the pole with the maximum magnitude among the poles lying inside the unit disc. Since the autocorrelation sequence $r(n)$ is real and even, the region of convergence is the annulus $|\lambda_{\max}| < |z| < 1/|\lambda_{\max}|$. In the present example $1/2 < |z| < 2$. Thus the autocorrelation has the form

$$r(n) = R_1 \left(\frac{1}{3}\right)^n u_s(n) + R_2 \left(\frac{1}{2}\right)^n u_s(n) - R_3 2^n u_s(-n-1) - R_4 3^n u_s(-n-1)$$

If $a(z) = 1$, the ARMA process reduces to a moving average process. If $\rho_1, \rho_2,$ ..., $\rho_m$ are the real distinct roots of the polynomial $b(z)$, then

$$S(\omega) = |b(\omega)|^2 = b(z)b(z^{-1})\big|_{z=e^{j\omega}}$$

As in the AR case there are $2^m$ different moving average models that produce the same power spectral density. ∎

## 9.3 SPECTRAL ANALYSIS OF WSS ANALOG SIGNALS

Harmonic analysis of continuous time signals proceeds along the same lines as the discrete case. We sketch the main results. We will consider zero mean wide sense stationary signals $x(t)$, $t \in R$, that are continuous in quadratic mean; that is, for any $t \in R$ they satisfy

$$\lim_{\tau \to 0} E[|x(t + \tau) - x(t)|^2] = 0$$

Quadratic mean continuity is equivalent to the continuity of the correlation function $r(t)$ (Exercise 9.5). The counterpart of the Herglotz's theorem is stated next.

**Theorem 9.4   Bochner's theorem**   If $x(t)$ is a quadratic mean continuous WSS process with correlation function $r(t)$, there exists a unique, bounded, nonnegative, monotone increasing, continuous from the right function $F(\omega)$ such that

$$r(t) = \frac{1}{2\pi} \int_{-\infty}^{\infty} e^{j\omega t} dF(\omega) \tag{9.41}$$

∎

The function $F(\omega)$ defines the *spectral distribution* of the process. If the correlation function $r(t)$ is absolutely integrable, the Fourier transform of $r(t)$

$$S(\omega) = \int_{-\infty}^{\infty} r(t) e^{-j\omega t} dt \tag{9.42}$$

is a well-defined bounded continuous function. It defines the *power spectral density* of the process. Since $r(t)$ is continuous (due to quadratic mean continuity of the process), bounded (consequence of the Cauchy-Schwarz inequality) and absolutely integrable, it follows that $S(\omega)$ is also absolutely integrable. The discussion of Section 6.1.5 asserts that the inversion formula

$$r(t) = \frac{1}{2\pi} \int_{-\infty}^{\infty} e^{jt\omega} S(\omega) d\omega \tag{9.43}$$

holds for all $t \in R$. If $r(t)$ does not belong to $L_1$ but has finite energy, we can still work with the power spectral density and Eqs. (9.42)–(9.43) provided that the latter are interpreted in the $L_2$ sense.

The continuous time counterpart of Theorem 9.3 reads as follows:

**Theorem 9.5**   Every quadratic mean continuous wide sense stationary process $x(t)$, $t \in R$, with spectral distribution $F$ admits the spectral representation

$$x(t) = \frac{1}{2\pi} \int_{-\infty}^{\infty} e^{j\omega t} dx(\omega)$$

The spectral process $x(\omega)$ has orthogonal increments and satisfies

$$E[|dx(\omega)|^2] = 2\pi dF(\omega)$$ ∎

**Example 9.7   Analog white noise**
Analog white noise is defined as a continuous stochastic process with flat power spectral density

$$S(\omega) = 1, \qquad \text{for all } \omega \in R$$

The autocorrelation function does not exist as an ordinary signal. It is given by the Dirac function $r(t) = \delta(t)$. ∎

## 9.4  SAMPLING THEOREM FOR WSS SIGNALS

In Section 6.7 the sampling theorem for deterministic bandlimited signals was developed. The stochastic version for wide sense stationary signals is described next. A quadratic mean continuous WSS process $x(t)$ with power spectral density $S(\omega)$ is *bandlimited* if $S(\omega) = 0$ for $|\omega| > W$, namely there is no average power for frequencies less than $-W$ and higher than $W$.

**Theorem 9.6   Sampling theorem**
Given a bandlimited signal $x(t)$ as above it holds

$$x(t) = \sum_{n=-\infty}^{\infty} x(nT_s)\mathrm{sinc}\left(\frac{t - nT_s}{T_s}\right) \qquad T_s = \frac{\pi}{W}$$

in the quadratic mean sense.

**PROOF**   Using the analog counterpart of Eq. (9.29) and the spectral representation theorem we obtain

$$E\left[\left|x(t) - \sum_n x(nT_s)\mathrm{sinc}\left(\frac{t - nT_s}{T_s}\right)\right|^2\right]$$

$$= E\left[\left|\frac{1}{2\pi}\int_{-\infty}^{\infty} e^{j\omega t} dx(\omega) - \frac{1}{2\pi}\sum_n \int_{-\infty}^{\infty} e^{j\omega nT_s} dx(\omega)\mathrm{sinc}\left(\frac{t - nT_s}{T_s}\right)\right|^2\right]$$

$$= \frac{1}{2\pi}\int_{-\infty}^{\infty}\left|e^{j\omega t} - \sum_n e^{j\omega nT_s}\mathrm{sinc}\left(\frac{t - nT_s}{T_s}\right)\right|^2 S(\omega)d\omega$$

$$= \frac{1}{2\pi}\int_{-W}^{W}\left|e^{j\omega t} - \sum_n e^{j\omega nT_s}\mathrm{sinc}\left(\frac{t - nT_s}{T_s}\right)\right|^2 S(\omega)d\omega \qquad (9.44)$$

For a fixed $t$, consider the deterministic signal which is zero outside $[-W, W]$ and equal to $e^{j\omega t}$ for $-W \le \omega \le W$. This is a finite support square integrable signal and its periodic extension admits a Fourier series. Therefore

$$e^{j\omega t} = \sum_n c_n e^{jnT_s\omega}, \quad -W \le \omega \le W$$

The Fourier coefficients are

$$c_n = \frac{1}{2W} \int_{-W}^{W} e^{j\omega t} e^{-jnT_s\omega} d\omega = \operatorname{sinc}\left(\frac{t - nT_s}{T_s}\right)$$

Therefore the integrand in (9.44) is zero and the proof is complete.  ∎

## 9.5  CUMULANTS AND POLYSPECTRA

Cumulants and polyspectra of a signal provide higher-order information in the time and frequency domain, respectively. They are useful in the analysis and processing of non-Gaussian signals. Such signals are observed as outputs of linear systems driven by non-Gaussian inputs. Alternatively, they are encountered as outputs of nonlinear systems. Although moments and cumulants convey equivalent information, the latter are preferred due to a number of attractive features discussed next.

### 9.5.1  Properties of cumulants

***Cumulants to moments conversion.*** Cumulants can be restored from moments and reversely. Let $x = (x_1, x_2, \ldots, x_k)^T$ be a random vector and $I_x = \{1, 2, \ldots, k\}$. For each subset of $I_x$, $I = \{i_1, i_2, \ldots, i_m\}$, let $x_I = (x_{i_1}, \ldots, x_{i_m})^T$ and

$$m_x(I) = E[x_{i_1} x_{i_2} \ldots x_{i_m}], \quad \operatorname{cum}_x(I) = \operatorname{cum}[x_{i_1}, \ldots, x_{i_m}]$$

A partition of $I_x$ is a collection of nonempty disjoint subsets of $I_x$ whose union covers $I_x$. For instance, if $k = 3$, the sets $\{1\}$, $\{2, 3\}$ define a partition. The number of sets in a given partition defines its order. The above partition has order 2. For each $q = 1, 2, \ldots, k$, let $\mathcal{P}_q$ be the set of partitions of $I_x$ of order $q$. Take, for example, $k = 3$. Then $\mathcal{P}_1$ consists of a single partition $\{1, 2, 3\}$. $\mathcal{P}_2$ consists of three partitions:

$$P_1 : \{1\}, \{2, 3\}, \quad P_2 : \{2\}, \{1, 3\}, \quad P_3 : \{3\}, \{1, 2\}$$

while $\mathcal{P}_3$ consists of the single partition $\{1\}, \{2\}, \{3\}$. The cumulant of order $k$ is recovered from the moments of orders less than $k$ according to the formula

$$\text{cum}[x_1, x_2, \ldots, x_k] = \sum_{q=1}^{k} (-1)^{q-1} (q-1)! \sum_{P \in \mathcal{P}_q} \prod_{I \in P} m_x(I) \qquad (9.45)$$

To illustrate the formula, we take $k = 3$. The sets $\mathcal{P}_q$ were identified above. For $q = 1$, the inner sum in (9.45) becomes $E[x_1 x_2 x_3]$. For $q = 2$, $\mathcal{P}_2$ has three partitions $P_1$, $P_2$, and $P_3$. The inner sum now becomes $E[x_1]E[x_2 x_3] + E[x_2]E[x_1 x_3] + E[x_3]E[x_1 x_2]$. Finally, for $q = 3$, $\mathcal{P}_3$ has a single partition leading to $E[x_1]E[x_2]E[x_3]$. Thus

$$\begin{aligned}
\text{cum}[x_1, x_2, x_3] = {} & E[x_1 x_2 x_3] - E[x_1]E[x_2 x_3] - E[x_2]E[x_1 x_3] \\
& - E[x_3]E[x_1 x_2] + 2E[x_1]E[x_2]E[x_3]
\end{aligned}$$

Moments can be reconstructed from cumulants by the formula

$$E[x_1 x_2 \cdots x_k] = \sum_{q=1}^{k} \sum_{P \in \mathcal{P}_q} \prod_{I \in P} \text{cum}_x(I) \qquad (9.46)$$

Formulas (9.45) and (9.46) can be proved by induction. The proof is omitted.

**Symmetry.** Cumulants are symmetric in their arguments:

$$\text{cum}[x_1, x_2, \ldots, x_k] = \text{cum}[x_{i_1}, x_{i_2}, \ldots, x_{i_k}]$$

where $(i_1, \ldots, i_k)$ is any permutation of $I_x$. Indeed, by definition, the cumulant is the coefficient of $s_1 s_2 \cdots s_k$ in the power series expansion of the cumulant generating function $K(s)$ and is given by Eq. (8.66). Thus it is not affected by a permutation of the indexes.

**Multilinearity.** Cumulants are multilinear mappings. For any constants $a$, $b$ it holds that

$$\text{cum}[ax_1 + by_1, x_2, \ldots, x_k] = a\,\text{cum}[x_1, x_2, \ldots, x_k] + b\,\text{cum}[y_1, x_2, \ldots, x_k]$$

Consider a partition $P$ containing the sets $I_1$, ..., $I_q$. Suppose that the index 1 belongs to $I_i$. Let

$$z = (ax_1 + by_1, x_2, \ldots, x_k), \quad x = (x_1, x_2, \ldots, x_k), \quad y = (y_1, x_2, \ldots, x_k)$$

Then

$$m_z(I_i) = E[(ax_1 + by_1)x_{i_1} \cdots x_{i_l}] = aE[x_1 x_{i_1} \cdots x_{i_l}] + bE[y_1 x_{i_1} \cdots x_{i_l}]$$

Hence $m_z(I_i) = a\,m_x(I_i) + b\,m_y(I_i)$ and

$$m_z(I_1)\cdots m_z(I_q) = a m_x(I_1)\cdots m_x(I_q) + b m_y(I_1)\cdots m_y(I_q)$$

The claim follows from Eq. (9.45). Cumulants owe their name to the above additive property.

**Independence.** Suppose that the random vectors $x$ and $y$ are statistically independent. Then

$$\text{cum}[x_1 + y_1, \ldots, x_k + y_k] = \text{cum}[x_1, \ldots, x_k] + \text{cum}[y_1, \ldots, y_k]$$

Let $z = x + y$. Then $K_z(s) = \log E[e^{s^T z}] = \log E[e^{s^T x}e^{s^T y}]$. Using the independence assumption, we obtain

$$K_z(s) = \log(E[e^{s^T x}]E[e^{s^T y}]) = \log E[e^{s^T x}] + \log E[e^{s^T y}] = K_x(s) + K_y(s)$$

The claim is proved. If some components of $x$ are independent from the rest, it holds that

$$\text{cum}[x_1, x_2, \ldots, x_k] = 0$$

Indeed, due to symmetry we may assume that $x_1, x_2, \ldots, x_r$ are independent from $x_{r+1}, \ldots, x_k$. Then

$$K_x(s) = \log E\left[\exp \sum_{i=1}^{r} s_i x_i\right] + \log E\left[\exp \sum_{i=r+1}^{k} s_i x_i\right] \tag{9.47}$$

The cumulant of order $k$ is obtained by differentiating $K_x(s)$ $k$ times with respect to $s_1, s_2, \ldots, s_k$. The first term in (9.47) does not depend on $s_i$, $i > r$; therefore the above partial derivative is zero. Likewise the derivative of the second term is zero.

A direct consequence of the above is that if $a$ is constant, it holds that

$$\text{cum}[a + x_1, x_2, \ldots, x_k] = \text{cum}[x_1, x_2, \ldots, x_k]$$

***Example 9.8   Suppression of Gaussian noise***
Let $x$ and $e$ be independent random vectors and $y = x + e$. Think of $y$ as a measurement of $x$ and $e$ as the measurement noise. Then

$$\text{cum}[y_1, y_2, \ldots, y_k] = \text{cum}[x_1, x_2, \ldots, x_k] + \text{cum}[e_1, e_2, \ldots, e_k]$$

If $e$ is Gaussian and $k \geq 3$, then $\text{cum}[e_1, e_2, \ldots, e_k] = 0$. Thus higher-order cumulants of the measurements completely suppress Gaussian additive noise and provide direct information on the cumulants of the true signal.     ∎

### 9.5.2 Polyspectra

Let $x(n)$ be a $k$-order stationary random process. Given integers $n_1, n_2, \ldots, n_{k-1}$, let

$$m(n_1, n_2, \ldots, n_{k-1}) = E[x(n + n_1)x(n + n_2) \cdots x(n + n_{k-1})x(n)]$$

Because of stationarity the $k$th-order moment depends only on the time lags $n_1$, $n_2, \ldots, n_{k-1}$. The multidimensional sequence $m(n_1, n_2, \ldots, n_{k-1})$ defines the $k$th-order moment sequence of the stochastic signal $x(n)$. Likewise the $k$th-order cumulant sequence is defined by

$$c(n_1, n_2, \ldots, n_{k-1}) = \text{cum}[x(n + n_1), x(n + n_2), \cdots, x(n + n_{k-1}), x(n)]$$

In several applications we work with slices of cumulants. A slice is restricted to a proper subset of $Z^{k-1}$. In particular, a 1-D slice results if $k - 2$ variables are kept constant. The one-dimensional signal $c(n, n, \ldots, n)$ forms a diagonal slice.

The Fourier transform of the $k$th-order cumulant defines the $k$th-order cumulant spectrum (or polyspectrum)

$$C(\omega_1, \ldots, \omega_{k-1}) = \sum_{n_1=-\infty}^{\infty} \cdots \sum_{n_{k-1}=-\infty}^{\infty} c(n_1, \ldots, n_{k-1}) e^{-j\sum_{i=1}^{k-1} \omega_i n_i} \qquad (9.48)$$

The polyspectrum exists and is a bounded continuous function provided that the cumulant sequence is absolutely summable. Crosscumulants are defined in a similar fashion. The third-order cumulant spectrum is known as the *bispectrum* and the fourth-order spectrum as the *trispectrum*.

### Example 9.9 Higher-order white noise

The covariance of a white noise signal is given by a scaled version of the unit sample signal. Hence signal values are uncorrelated at distinct instants. Higher-order white noise ensures a higher degree of independence between signal values via the requirement

$$c(n_1, n_2, \ldots, n_{k-1}) = \gamma_k \delta(n_1, n_2, \ldots, n_{k-1})$$

where $\delta(n_1, n_2, \ldots, n_{k-1})$ is the $(k - 1)$th-dimensional unit sample. ∎

## 9.6 HIGHER-ORDER SPECTRAL ANALYSIS OF CONVOLUTIONAL SYSTEMS

Consider the single-channel causal convolutional system

$$y(n) = h(n) * x(n) \qquad (9.49)$$

$x(n)$ is stationary of order $k$, and the impulse response is absolutely summable. It then follows that $y(n)$ is stationary of order $k$. Let $c_y(n_1, n_2, \ldots, n_{k-1})$ denote the $k$th-order cumulant of $y(n)$. We replace Eq. (9.49) inside the cumulant and invoke multilinearity to obtain

$$c_y(n_1, n_2, \ldots, n_{k-1}) = \sum_{i_1} h(i_1) \cdots \sum_{i_{k-1}} h(i_{k-1})$$

$$\sum_{i_0} h(i_0) \text{cum}[x(n + n_1 - i_1), x(n + n_2 - i_2), \ldots, x(n + n_{k-1} - i_{k-1}), x(n - i_0)]$$

$$= \sum_{i_0} \sum_{i_1} \cdots \sum_{i_{k-1}} h(i_0) h(i_1) \cdots h(i_{k-1}) c_x(n_1 + i_0 - i_1, \ldots, n_{k-1} + i_0 - i_{k-1})$$

The last term reminds us of multidimensional convolution. Indeed, the change of variables $m_1 = i_1 - i_0, m_2 = i_2 - i_0, \ldots, m_{k-1} = i_{k-1} - i_0$, gives

$$c_y(n_1, n_2, \ldots, n_{k-1}) = \sum_{i_0} \sum_{m_1} \cdots \sum_{m_{k-1}} h(i_0) h(i_0 + m_1) \cdots h(i_0 + m_{k-1})$$

$$c_x(n_1 - m_1, \ldots, n_{k-1} - m_{k-1})$$

$$= c_h(n_1, \ldots, n_{k-1}) * * \cdots * c_x(n_1, \ldots, n_{k-1}) \tag{9.50}$$

where

$$c_h(n_1, \ldots, n_{k-1}) = \sum_{i_0} h(i_0) h(i_0 + n_1) \cdots h(i_0 + n_{k-1}) \tag{9.51}$$

Next we convert the above expression to the frequency domain. The multidimensional Fourier transform of $c_h(n_1, \ldots, n_{k-1})$ is computed as follows: for each fixed $i_0$ the multidimensional signal $h(i_0 + n_1) h(i_0 + n_2) \cdots h(i_0 + n_{k-1})$ is separable. Hence its multidimensional Fourier transform is the product of the 1-D Fourier transforms of $h(i_0 + n_l)$, $l = 1, \ldots, k - 1$. Each $h(i_0 + n_l)$ is the $i_0$ shift of $h(n_l)$. The Fourier transform is $e^{j i_0 \omega_l} H(\omega_l)$. Putting these together, we find that

$$C_h(\omega_1, \ldots, \omega_{k-1}) = \sum_{i_0} h(i_0) e^{j i_0 \omega_1} e^{j i_0 \omega_2} \cdots e^{j i_0 \omega_{k-1}} H(\omega_1) \cdots H(\omega_{k-1})$$

$$= H\left(-\sum_{i=1}^{k-1} \omega_i\right) H(\omega_1) H(\omega_2) \cdots H(\omega_{k-1}) \tag{9.52}$$

Conversion of the multidimensional convolution (9.50) to the frequency domain leads to

$$C_y(\omega_1, \ldots, \omega_{k-1}) = C_x(\omega_1, \ldots, \omega_{k-1}) H\left(-\sum_{i=1}^{k-1} \omega_i\right) H(\omega_1) H(\omega_2) \cdots H(\omega_{k-1})$$

$$\tag{9.53}$$

If $x(n)$ is Gaussian and $k > 2$, the input polyspectrum of order $k$ is zero. Therefore the output polyspectrum is zero. Equation (9.53) is useful if the input is non-Gaussian. We will later see that higher-order output information enables the identification of the system frequency response. If $x$ is white noise of order $k$ with intensity $\gamma_k$, then $C_x(\omega_1, \omega_2, \ldots, \omega_{k-1}) = \gamma_k$, and Eq. (9.53) takes the simpler form

$$C_y(\omega_1, \omega_2, \ldots, \omega_{k-1}) = \gamma_k H\left( -\sum_{i=1}^{k-1} \omega_i \right) H(\omega_1) H(\omega_2) \cdots H(\omega_{k-1}) \qquad (9.54)$$

## BIBLIOGRAPHICAL NOTES

A tutorial presentation of the spectral representation of WSS processes is given in Papoulis (1991) and Priestley (1981). A rigorous exposition is offered in Doob (1960) and Rozanov (1960). Higher-order spectral properties are studied in Rosenblatt (1985). The signal processing viewpoint is stressed in Nikias and Petropoulou (1993). Other recommended sources for the material of this chapter are Caines (1988), Porat (1994), Wong and Hajek (1985), and Gray and Davisson (1986).

## PROBLEMS

**9.1.**  Let

$$x(n) = \frac{1}{2}e(n-1) + \frac{1}{4}e(n-2)$$

$e(n)$ is white noise of variance 1. Compute the autocorrelation sequence, the power spectral density, and the spectral distribution. Derive plots using MATLAB.

**9.2.**  Consider the first-order system

$$y(n) + \frac{1}{2}y(n-1) = u(n)$$

Suppose that $u(n)$ is white noise with power spectral density $S_u(\omega) = 1/4$. Compute the autocorrelation, the power spectal density and the spectral distribution of the output. Verify your results with MATLAB plots, as in Example 9.3.

**9.3.**  Consider the second order system

$$y(n) + \frac{4}{15}y(n-1) - \frac{1}{5}y(n-2) = u(n)$$

$u(n)$ is white noise with power spectral density $S_u(\omega) = 0.2$. The roots of

$z^2 + a_1 z + a_2$ are inside the unit disc. Compute the output spectral density and the autocorrelation. Construct **MATLAB** plots to validate your theoretical calculations.

**9.4.**  We saw in Exercise 6.7 that the Hilbert transform can be realized as a quadrature filter, namely as a convolutional system with frequency response $H(\omega) = -j\text{sgn}(\omega)$. Let $\hat{x}(t)$ denote the Hilbert transform of $x(t)$ and $z(t) = x(t) + j\hat{x}(t)$. Prove that

$$S_{\hat{x}}(\omega) = S_x(\omega), \quad S_z(\omega) = 4S_x(\omega)U_s(\omega)$$

where $U_s(\omega)$ is the Fourier transform of the unit step.

**9.5.**  Show that a WSS analog signal is quadratic mean continuous if and only if the autocorrelation function is continuous.

**9.6.**  The quadratic mean derivative of a second-order process $x(t)$, $\dot{x}(t)$ is defined by

$$\lim_{h \to 0} E\left[\left|\frac{1}{h}(x(t+h) - x(t)) - \dot{x}(t)\right|^2\right] = 0$$

Show that $\dot{x}(t)$ exists if and only if

$$\frac{\partial^2 R(t_1, t_2)}{\partial t_1 \partial t_2}$$

exists; $R(t_1, t_2)$ denotes the correlation function of $x(t)$. Show that if $x(t)$ is WSS, $\dot{x}(t)$ exists if and only if the autocorrelation function $r(t)$ is twice differentiable at $t = 0$. Moreover $\dot{x}(t) \in \mathcal{H}_x(\infty)$ and

$$\dot{x}(t) = \frac{1}{2\pi} \int_{-\infty}^{\infty} j\omega e^{j\omega t} dx(\omega)$$

**9.7.**  Suppose that the signal $y(t)$ is differentiable in the sense of Exercise 9.6 and satisfies

$$\dot{y}(t) + 3y(t) = u(t),$$

where $u(t)$ is white noise with power spectral density $S_u(\omega) = 1/2$. Compute the autocorrelation, the power spectal density, and the spectral distribution of the ouput. Verify your results with **MATLAB** plots.

**9.8.**  Suppose that $y(t)$ is twice differentiable and satisfies

$$\ddot{y}(t) + 5\dot{y}(t) + 6y(t) = u(t)$$

where $u(t)$ is white noise with power spectral density $S_u(\omega) = 1/3$. Compute the output spectral density and the autocorrelation.

**9.9.** Suppose that $x(t)$ is a cyclostationary process. The autocorrelation function $r_x(t + \tau, t)$ is periodic with period $T$ satisfying

$$\int_0^T |r_x(t + \tau, t)| dt < \infty$$

Let

$$\bar{r}_x(\tau) = \frac{1}{T} \int_{-T/2}^{T/2} r_x(t + \tau, t) dt$$

with Fourier transform $\bar{S}_x(\omega)$. The signal $x(t)$ is applied as input to a convolutional system with absolutely integrable impulse response $h(t)$. Show that the output is cyclostationary of period $T$ and that

$$\bar{r}_y(t) = h(t) * \bar{r}_x(t) * h(-t)$$

so that in the frequency domain it holds:

$$\bar{S}_y(\omega) = |H(\omega)|^2 \bar{S}_x(\omega)$$

**9.10.** Let $a(n)$ be a discrete WSS signal with autocorrelation sequence $r_a(n)$. Let $g(t)$ be an analog deterministic signal with Fourier transform $G(\omega)$. Show that the analog stochastic signal

$$v(t) = \sum_{n=-\infty}^{\infty} a(n) g(t - nT)$$

is cyclostationary with period $T$. Show that

$$\bar{r}_v(t) = \frac{1}{T} \sum_{n=-\infty}^{\infty} r_a(n) r_g(t - nT), \quad r_g(t) = g(t) * g(-t)$$

and that

$$\bar{S}_v(\omega) = \frac{1}{T} S_a(T\omega) |G(\omega)|^2$$

where $\bar{S}_v(\omega)$ is defined in Exercise 9.9.

**9.11.** Suppose that $x(t)$ is a cyclostationary process with period $T$ and that $\theta$ is a random variable independent of $x(t)$ and uniformly distributed on $(0, T)$. Show that $y(t) = x(t + \theta)$ is WSS having autocorrelation function

$$r_y(t) = \frac{1}{T} \int_{-T/2}^{T/2} r_x(\tau + t, \tau) d\tau$$

**9.12.** Suppose that $u(t)$ is a zero mean WSS process with autocorrelation $r_u(t)$. Prove that the AM modulated signal $y(t) = u(t)\cos \omega_0 t$ is cyclostationary with

$$\bar{S}_y(\omega) = \frac{1}{4}[S_u(\omega + \omega_0) + S_u(\omega - \omega_0)]$$

**9.13.** *Bandpass processes.* A bandpass process with center frequency $\omega_0$ and bandwidth $2W$, $(\omega_0 > W)$, is a quadratic mean continuous WSS process with $F(\omega) = 0$ for $|\omega - \omega_0| \geq W$.

The average power is concentrated in the frequency range $(-\omega_0 - W, -\omega_0 + W)$ and $(\omega_0 - W, \omega_0 + W)$. A bandpass signal $x(t)$ is bandlimited with bandwidth $\omega_0 + W$. Application of the sampling theorem requires a sampling rate of $2\omega_0 + 2W$. Since $x(t)$ has bandwidth $2W$, a sampling rate of $2W$ is expected rather than $2\omega_0 + 2W$. Commonly $\omega_0 \gg W$, so the difference is substantial. Efficient sampling rates are accomplished with the aid of the Hilbert transform and Rice representation. Using the spectral representation of the bandpass signal,

$$x(t) = \frac{1}{2\pi} \int_{-\omega_0 - W}^{-\omega_0 + W} e^{j\omega t} dx(\omega) + \frac{1}{2\pi} \int_{\omega_0 - W}^{\omega_0 + W} e^{j\omega t} dx(\omega)$$

and the Hilbert transform $\hat{x}(t)$ (see Exercise 9.4), show that

$$x(t) + j\hat{x}(t) = 2e^{j\omega_0 t} w(t) \tag{9.55}$$

$$x(t) - j\hat{x}(t) = 2e^{-j\omega_0 t} v(t)$$

$$x(t) = e^{j\omega_0 t} w(t) + e^{-j\omega_0 t} v(t) \tag{9.56}$$

where $w(t)$ and $v(t)$ are bandlimited to $W$:

$$w(t) = \frac{1}{2\pi} \int_{-W}^{W} e^{j\omega t} dx(\omega + \omega_0), \quad v(t) = \frac{1}{2\pi} \int_{-W}^{W} e^{j\omega t} dx(\omega - \omega_0)$$

$w(t)$ is the *complex envelope*. The real part of $w(t)$, $i(t)$, is the *in-phase* component of $x(t)$ and the imaginary part $q(t)$ is the *quadrature* component of $x(t)$. Let $r(t)$ and $\theta(t)$ be the magnitude and phase of $w(t)$. Show that they are bandlimited and that

$$x(t) = 2i(t)\cos \omega_0 t - 2q(t)\sin \omega_0 t = 2r(t)\cos(\omega_0 t + \theta(t))$$

Using the sampling theorem show that

$$x(t) = \sum_{n=-\infty}^{\infty} \mathrm{sinc}\left(\frac{t - nT_s}{T_s}\right) [x(nT_s)\cos \omega_0(t - nT_s)$$
$$- \hat{x}(nT_s)\sin \omega_0(t - nT_s)]$$

**9.14.** Using the cyclostationarity concepts of Exercises 9.9–9.11 compute the spectral density of the output of the periodically time varying linear system

$$y(t) = \int h(t,\tau)u(\tau)d\tau, \quad h(t+T,\tau+T) = h(t,\tau)$$

assuming that $u(t)$ is WSS (see also Section 6.4.3).

**9.15.** Let

$$z(n) = x(n)\cos\omega_c n + y(n)\sin\omega_c n$$

$x(n)$, $y(n)$ are zero mean independent stationary processes with identical autocorrelation sequences and equal third-order moments. Show that $z(n)$ is a wide sense stationary process but fails to be stationary of order 3.

**9.16.** Prove that the bispectrum satisfies

$$S(\omega_1,\omega_2) = S^*(-\omega_1,-\omega_2) = S(\omega_2,\omega_1)$$
$$= S(\omega_1,-\omega_1-\omega_2) = S(-\omega_1-\omega_2,\omega_2)$$

**9.17.** (a) Let $x(n)$ be a zero mean stationary process, and let $y(n) = x(n-m)$. Prove that $x(n)$ and $y(n)$ have the same cumulant spectra. Conclude that cumulant spectra suppress linear phase shifts.
(b) Suppose that $y_1(n)$ and $y_2(n)$ are two spatially separated sensor measurements:

$$y_1(n) = x(n) + w_1(n)$$
$$y_2(n) = x(n-D) + w_2(n)$$

Show that if $x(n)$, $w_1(n)$ and $w_2(n)$ are mutually uncorrelated and $w_1(n)$, $w_2(n)$ are zero mean, it holds that

$$r_{y_2 y_1}(n) = E[y_2(k)y_1(k-n)] = r_x(n-D)$$

Hence the time delay $D$ can be estimated as the value where $r_{y_2 y_1}(n)$ forms its peak.
(c) One important application of time delay estimation is for source bearing and range estimation. Figure 9.1 shows a source $S$ emitting energy and three receivers A, B and C, located at known distances $d_{AB}$, $d_{BC}$, and $d_{AC}$. Assuming that $r_B \gg d_{AB}$, prove that the *bearing angle* $\theta$ and the range $r_B$ are given by

$$\theta \approx \cos^{-1}\left(\frac{D_{AB}\upsilon}{d_{AB}}\right),$$

$$r_B \approx \frac{d_{AB}[1-(\Delta_{AB}^2/d_{AB}^2)] - d_{BC}[1-(\Delta_{BC}^2/d_{BC}^2)]\cos\phi}{2[(\Delta_{AB}/d_{AB}) + (\Delta_{BC}/d_{BC})\cos\phi + \sin\phi(1-(\Delta_{BC}^2/d_{BC}^2))]}$$

where $v$ is the signal speed of propagation, $D_{AB}$ the time delay between the receiver output signals, $\Delta_{AB} = D_{AB}v$, $\Delta_{BC} = D_{BC}v$, and $\phi$ is the angle $ABC$. Derive a simpler expression for the special case of a linear array: $\phi = 180^0$.

(d) If the noise signals $w_i(n)$ are correlated the approach outlined in (b) fails. Higher-order statistics can be utilized in this case. Prove that if $w_i(n)$ are Gaussian and $x(n)$ is non-Gaussian, it holds that

$$M_{3y_2y_1y_1}(\omega_1,\omega_2) = M_{3y_1}(\omega_1,\omega_2)e^{-j\omega_1 D}$$

Let $\Psi_{3y_2y_1y_1}(\omega_1,\omega_2)$ denote the phase spectrum of $M_{3y_2y_1y_1}(\omega_1,\omega_2)$ and

$$F(\omega_1,\omega_2) = \frac{M_{3y_2y_1y_1}(\omega_1,\omega_2)}{M_{3y_1}(\omega_1,\omega_2)} = \exp[j\Psi_{3y_2y_1y_1}(\omega_1,\omega_2) - j\Psi_{3y_1}(\omega_1,\omega_2)]$$

$$= e^{-j\omega_1 D}$$

Prove that $D$ can be estimated as the value where the inverse Fourier transform of $F$, $f(n_1, 0)$ peaks with respect to $n_1$:

$$D : \max_{n_1} f(n_1, 0) = f(D, 0)$$

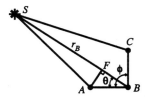

**Figure 9.1.**

**9.18.** Consider the moving average system

$$y(n) = u(n) - au(n-1)$$

$u(n)$ is zero mean non-Gaussian white noise with cumulants at the origin $\gamma_{2u}$, $\gamma_{3u}$ and $\gamma_{4u}$ (see Example 9.9). Calculate the power spectrum, the bispectrum, and the trispectrum of $y(n)$. Verify your results with MATLAB, and plot the corresponding magnitude and phase polyspectra.

**9.19.** Consider the discrete harmonic process

$$x(n) = \sum_{i=1}^{N} A_i \cos(\omega_i n + \phi_i)$$

$A_i$ are independent random variables and $\phi_i$ are independent and uniformly distributed over $[-\pi, \pi]$. Moreover $A_i$ and $\phi_j$ are independent for any $i$ and $j$. Show that the third-order cumulant of $x(n)$ is zero, while the fourth-order cumulant is given by

$$c_x(n_1, n_2, n_3) = \frac{1}{8} \sum_{i=1}^{N} E[A_i]^4 [\cos \omega_i(n_1 - n_2 - n_3) + \cos \omega_i(n_2 - n_3 - n_1)$$

$$+ \cos \omega_i(n_3 - n_1 - n_2)]$$

## APPENDIX

**PROOF OF THEOREM 9.2**   We have pointed out that $e^{jn\omega}$ belongs to $L_2(F)$. We define the isomorphism $\varphi$ as follows: Each value $x(n)$ is mapped by $\varphi$ to $e^{jn\omega}$. Each finite linear combination $\sum h(k)x(k)$ is mapped to $\sum h(k)e^{jk\omega}$. Finally, if $y$ is the limit of $y_N$ and each $y_N$ is a linear combination of signal values as above, we set $\varphi(y) = \lim \varphi(y_N)$. The limit operations are induced by the inner products of the Hilbert spaces $H_x(\infty)$ and $L_2(F)$. Recall that the inner product in $L_2(F)$ is given by (9.13) and that the inner product in $\mathcal{H}_x(\infty)$ is given by $< x, y > = E[xy]$. It is easy to see that $\varphi : \mathcal{H}_x(\infty) \to L_2(F)$ is a well-defined linear and continuous map. Moreover it preserves inner products modulo the constant $2\pi$; that is, for any $y$, $z$ in $\mathcal{H}_x(\infty)$ it holds that

$$< \varphi(y), \varphi(z) > = 2\pi < y, z >$$

Indeed, if $y$, $z$ are the finite sums $y = \sum_n h(n)x(n)$, $z = \sum_k g(k)x(k)$, we have

$$< \varphi(y), \varphi(z) > = < \sum_n h(n)e^{jn\omega}, \sum_k g(k)e^{jk\omega} > = \sum_{n,k} h(n)g(k) < e^{jn\omega}, e^{jk\omega} >$$

$$= \sum_{n,k} h(n)g(k) \int e^{j(n-k)\omega} dF(\omega) = 2\pi \sum_{n,k} h(n)g(k)r(n-k)$$

$$= 2\pi E\left[ \sum_{n,k} h(n)x(n)g(k)x(k) \right]$$

$$= 2\pi < \sum_n h(n)x(n), \sum_k g(k)x(k) > = 2\pi < y, z >$$

If $y = \lim y_N$ and $z = \lim z_N$, then

$$< \varphi(y), \varphi(z) > = < \lim \varphi(y_N), \lim \varphi(z_N) > = \lim < \varphi(y_N), \varphi(z_N) >$$

$$= 2\pi \lim < y_N, z_N > = 2\pi < y, z >$$

Since $\varphi$ preserves inner products, it is 1-1 and continuous. To complete the proof, it remains to show that $\varphi$ is onto. For this purpose we introduce the spectral process in accordance with the discussion of Section 9.2.1 and show that it is rigorously defined. Fix an arbitrary $\omega$ in $[-\pi, \pi]$. The indicator function $\chi_{[-\pi,\omega]}$ is square integrable and can be expanded in Fourier series

$$\chi_{[-\pi,\omega]}(v) = \sum_{n=-\infty}^{\infty} c_n(\omega) e^{jvn} \tag{9.57}$$

The notation $c_n(\omega)$ emphasizes the dependence on the fixed frequency $\omega$. We define

$$x(\omega) = 2\pi \sum_{n=-\infty}^{\infty} c_n(\omega) x(n) \tag{9.58}$$

For each $\omega$, $x(\omega)$ is a well-defined random variable. Indeed, according to the analysis of Section 9.1.4, it suffices to establish that $\sum_{n=-\infty}^{\infty} c_n(\omega) e^{-jvn} = \chi_{[-\pi,\omega]}(-v)$ belongs to $L_2(F)$. This is immediate because

$$\int_{-\pi}^{\pi} |\chi_{[-\pi,\omega]}(-v)|^2 dF(v) = \int_{-\omega}^{\pi} dF(v) = F(\pi) - F(-\omega)$$

Since $x(\omega)$ is a limit of linear combinations of the signal values $x(n)$, it belongs to $\mathcal{H}_x(\infty)$. Moreover

$$\varphi(x(\omega)) = 2\pi \sum c_n(\omega) \varphi(x(n)) = 2\pi \sum c_n(\omega) e^{jvn} = 2\pi \chi_{[-\pi,\omega]}(v)$$

Thus the spectral process at frequency $\omega$ is visualized in the space $L_2(F)$ by the indicator function of the interval $[-\pi, \omega]$.

Next we show that $x(\omega)$ has orthogonal increments. Indeed, let $\omega_1 < \omega_2 \le \omega_3 < \omega_4$. Then

$$E[(x(\omega_2) - x(\omega_1))(x^*(\omega_4) - x^*(\omega_3))] = <x(\omega_2) - x(\omega_1), x(\omega_4) - x(\omega_3)>$$

Since $\varphi$ preserves inner products, the latter is equal to

$$2\pi < \chi_{[-\pi,\omega_2]} - \chi_{[-\pi,\omega_1]}, \chi_{[-\pi,\omega_4]} - \chi_{[-\pi,\omega_3]} >= 2\pi < \chi_{[\omega_1,\omega_2]}, \chi_{[\omega_3,\omega_4]} >= 0$$

Preservation of distances gives

$$E(|dx(\omega)|^2) = E[|x(\omega_2) - x(\omega_1)|^2] = 2\pi \int |\chi_{[-\pi,\omega_2]} - \chi_{[-\pi,\omega_1]}|^2 dF(\omega)$$

$$= 2\pi \int |\chi_{[\omega_1,\omega_2]}|^2 dF(\omega) = 2\pi(F(\omega_2) - F(\omega_1)) \tag{9.59}$$

We finally prove that $\varphi$ is onto. Let $H(\omega) \in L_2(F)$ and

$$y = \int_{-\pi}^{\pi} H(\omega)dx(\omega)$$

Since $x(\omega)$ belongs to $\mathcal{H}_x(\infty)$, the definition of the stochastic integral outlined in Section 9.2.2 implies $y \in \mathcal{H}_x(\infty)$. It holds that

$$\varphi(y) = \varphi\left( \int_{-\pi}^{\pi} H(\omega)dx(\omega) \right) = 2\pi H(\omega) \tag{9.60}$$

Indeed, we write $H(\omega)$ as the quadratic mean limit of simple functions, $H = \lim s_N$, where $s_N = \sum_i a_N(i)\chi_{[\omega_{N,i}, \omega_{N,i+1}]}$. Then

$$\varphi\left( \int_{-\pi}^{\pi} H(\omega)dx(\omega) \right) = \lim_N \sum_i a_N(i)\varphi\left( \int \chi_{[\omega_{N,i}, \omega_{N,i+1}]}dx(\omega) \right)$$

$$= \lim_N \sum_i a_N(i)\varphi(x(\omega_{N,i+1}) - x(\omega_{N,i}))$$

$$= 2\pi \lim_N \sum_i a_N(i)(\chi_{[-\pi, \omega_{N,i+1}]} - \chi_{[-\pi, \omega_{N,i}]})$$

$$= 2\pi \lim_N \sum_i a_N(i)\chi_{[\omega_{N,i}, \omega_{N,i+1}]} = 2\pi H(\omega) \qquad \blacksquare$$

PROOF OF THEOREM 9.3    The spectral process $x(\omega)$ was constructed in the proof of the previous theorem as the element of $\mathcal{H}_x(\infty)$ that is mapped by $\varphi$ to the indicator function of the interval $[-\pi, \omega]$. We showed that $x(\omega)$ has the orthogonal increments property and that satisfies Eq. (9.59), which coincides with Eq. (9.30). Let $H(\omega) = e^{j\omega n}/2\pi$. Since $\varphi$ is one-to-one and $x(n)$ is mapped to $e^{j\omega n}$, Eq. (9.30) follows from Eq. (9.60). Uniqueness of the spectral process follows from Eq. (9.58) provided that $x(\omega)$ is zero at frequency $-\pi$ and continuous from the right. These properties can be imposed on the spectral process without harming generality.    $\blacksquare$

PROOF OF PROPOSITION 9.2    Since for each $n$, $y(n) \in \mathcal{H}_x(\infty)$, there is $\tilde{H}_n(\omega)$ in $L_2(F_x)$ such that

$$y(n) = \frac{1}{2\pi} \int_{-\pi}^{\pi} \tilde{H}_n(\omega)dx(\omega)$$

$\varphi$ maps $y(n)$ into $\tilde{H}_n(\omega)$ and $x(k)$ into $e^{jk\omega}$. Therefore preservation of inner products gives

$$E[y(n)x(k)] = <y(n), x(k)> = \frac{1}{2\pi} \int \tilde{H}_n(\omega)e^{-jk\omega}dF(\omega)$$

Likewise

$$E[y(0)x(k-n)] = \frac{1}{2\pi} \int \tilde{H}_0(\omega)e^{-j(k-n)\omega}dF(\omega)$$

Due to joint wide sense stationarity the left-hand side members of the latter equations are equal. Hence

$$\int e^{-jk\omega}[\tilde{H}_n(\omega) - e^{jn\omega}\tilde{H}_0(\omega)]dF(\omega) = <\tilde{H}_n(\omega) - e^{jn\omega}\tilde{H}_0(\omega), e^{jk\omega}> = 0$$

The above orthogonality equation is translated back to $\mathcal{H}_x(\infty)$. $e^{jk\omega}$ corresponds to $x(k)$, $\tilde{H}_n(\omega)$ corresponds to $y(n)$, and $e^{jn\omega}\tilde{H}_0(\omega)$ corresponds to

$$\tilde{y}(n) = \frac{1}{2\pi} \int e^{jn\omega}\tilde{H}_0(\omega)dx(\omega)$$

Thus $E[x(k)(y(n) - \tilde{y}(n))] = 0$. Since $y(n) - \tilde{y}(n)$ is orthogonal to $x(k)$ for every $k$, it is orthogonal to $\mathcal{H}_x(\infty)$. Moreover, it is an element of $\mathcal{H}_x(\infty)$. Therefore it must be the zero vector and Eq. (9.31) follows by setting $H(\omega) = \tilde{H}_0(\omega)$. ■

**PROOF OF PROPOSITION 9.3**    The elements $x(n-k)$, $k = 0, 1, 2, ...$, of $\mathcal{H}_x(n)$ are identified with $e^{jn\omega}e^{-jk\omega}$, via the isomorphism $\varphi$. Finite linear combinations $\sum_{k\geq 0} h(k)x(n-k)$ are translated to $e^{jn\omega}\sum_{k\geq 0} h(k)e^{-jk\omega}$, and limits of such sums are translated to $e^{jn\omega}H(\omega)$, where $H(\omega)$ is a limit of trigonometric polynomials. ■

# 10

## REALIZATION OF LINEAR
## RECURRENT STRUCTURES

The term realization was originally proposed to embrace methods devoted to the identification of a deterministic system from input-output information. The usage of the term adopted here is slightly different, since it refers to issues pertained to the passage from one system representation to another. Realization theory studies methods for the transfer between system representations. It searches for canonical forms with special characteristics and explores the uniqueness of representations and their bearing on minimum memory requirements.

The discussion of Chapter 8 showed that a convolutional system with transfer function $h(s)$, or impulse response $h(t)$, can be realized by a finite derivative model if and only if $h(s)$ is proper rational. Analogous results hold for discrete systems. Thus we will focus attention on the equivalence between finite derivative/difference models and state space forms. The analysis of Section 7.5 indicates that state space models always lead to finite derivative/difference models. Procedures for the reverse route will be of primary interest in this chapter. Given a rational transfer function model $h(s)$, there are infinitely many state space representations $(A, B, C, D)$ that realize the same transfer function $h(s)$. Some canonical forms are singled out and presented in detail, due to their importance in theory and applications.

Minimal realizations are of special interest, since they require minimum storage. The dimension of the state space of such realizations is minimum. Minimal realizations offer adequate coupling between input and state and between state and output. Good input-state coupling ensures that the state space cannot be decomposed into a smaller space where the input has no effect on the state. Thus it naturally leads to the notion of controllability. The state-output coupling property implies that the state space cannot be decomposed into a part that cannot be observed at the output. It naturally leads to the notion of observability. Controllability is a typical control property. It allows maneuvering in every direction on the state space by proper inputs. Observability is a typical estimation property. It is essential in estimating the state evolution from output measurements. Both properties are studied for linear time invariant systems, and equivalent algebraic tests are derived. Their strong duality is demonstrated and later used to translate control properties to estimation properties, and vice versa. Finally, they are employed to show that any two minimal realizations of the same transfer function model are equivalent.

## 10.1  CONVERSION OF FINITE DERIVATIVE MODELS TO STATE SPACE FORMS

Finite derivative models are converted in state space form in various ways. An instructive method mimicks the classical technique describing a particle's motion on the phase space. The method was illustrated in Example 3.24 where the equation of motion was expressed in terms of displacements and velocities in the state space. We apply this approach in the sequel starting with autoregressive systems.

### 10.1.1  Autoregressive systems

Consider the system

$$y^{(k)}(t) + \sum_{i=0}^{k-1} a_i y^{(i)}(t) = u(t) \tag{10.1}$$

Define the state variables

$$x_1(t) = y(t), x_2(t) = y^{(1)}(t), \ldots, x_k(t) = y^{(k-1)}(t) \tag{10.2}$$

If $k = 2$, and Eq. (10.1) represents the motion of a particle on a line, the state vector is specified by the displacement and velocity. We differentiate the above equations and express the derivatives of $y(t)$ in terms of the state variables. Then

$$\dot{x}_1(t) = y^{(1)}(t) = x_2(t)$$

$$\dot{x}_2(t) = y^{(2)}(t) = x_3(t)$$

$$\vdots \tag{10.3}$$

$$\dot{x}_{k-1}(t) = y^{(k-1)}(t) = x_k(t)$$

$$\dot{x}_k(t) = y^{(k)}(t) = -\sum_{i=0}^{k-1} a_i x_{i+1}(t) + u(t)$$

In the latter equation we made use of Eq. (10.1). If we write (10.3) in matrix form, we derive the state space representation

$$\dot{x}(t) = Ax(t) + Bu(t) \tag{10.4}$$

where

$$A = \begin{pmatrix} 0 & 1 & 0 & \cdots & 0 \\ 0 & 0 & 1 & \cdots & 0 \\ \vdots & \vdots & \vdots & \ddots & \vdots \\ 0 & 0 & 0 & \cdots & 1 \\ -a_0 & -a_1 & -a_2 & \cdots & -a_{k-1} \end{pmatrix}, \quad B = \begin{pmatrix} 0 \\ 0 \\ \vdots \\ 0 \\ 1 \end{pmatrix} \tag{10.5}$$

Note that $y(t) = x_1(t)$. Thus the output equation is

$$y(t) = Cx(t), \quad C = (1 \quad 0 \quad 0 \quad \cdots \quad 0) \tag{10.6}$$

The recursive model (10.1) and the state space representation (10.5)–(10.6) are equivalent in the sense that they produce the same output when they are excited by the same input and they are initialized by the same vector. The matrix $A$ in (10.5) is referred to as a *companion* matrix. Often the reversed order setup is used, where the state assignment (10.2) is replaced by

$$z_1(t) = y^{(k-1)}(t), \ldots, z_k(t) = y(t)$$

In the above coordinates the system is represented by the triple

$$A = \begin{pmatrix} -a_{k-1} & -a_{k-2} & \cdots & -a_1 & -a_0 \\ 1 & 0 & \cdots & 0 & 0 \\ \vdots & \vdots & \ddots & \vdots & \vdots \\ 0 & 0 & \cdots & 1 & 0 \end{pmatrix}, \quad B = \begin{pmatrix} 1 \\ 0 \\ \vdots \\ 0 \end{pmatrix}$$

$$C = (0 \quad 0 \quad \cdots \quad 0 \quad 1)$$

## 10.1.2   The infinite impulse response (IIR) case

Consider the infinite impulse response filter

$$y^{(k)}(t) + \sum_{i=0}^{k-1} a_i y^{(i)}(t) = \sum_{j=0}^{r} b_j u^{(j)}(t), \quad r \le k \tag{10.7}$$

The assumption $r \le k$ is crucial for the conversion to state space. Let $x(t)$ denote the output of the autoregressive filter

$$x^{(k)}(t) + \sum_{i=0}^{k-1} a_i x^{(i)}(t) = u(t) \tag{10.8}$$

A state space representation of (10.8) is given by (10.5) and (10.6). Next we express $y(t)$ and its derivatives in terms of $x(t)$ and its derivatives. We write (10.8) in the form

$$\sum_{i=0}^{k} a_i x^{(i)}(t) = u(t), \qquad a_k = 1 \tag{10.9}$$

Differentiation of (10.9) $l$ times gives

$$\sum_{i=0}^{k} a_i x^{(i+l)}(t) = u^{(l)}(t) \tag{10.10}$$

Multiplying (10.10) by $b_l$ and summing over $l$ gives

$$\sum_{l=0}^{r} b_l \sum_{i=0}^{k} a_i x^{(i+l)}(t) = \sum_{l=0}^{r} b_l u^{(l)}(t) \tag{10.11}$$

If we interchange the summations in the left-hand side and take into account (10.7), we obtain

$$\sum_{i=0}^{k} a_i \left( \sum_{l=0}^{r} b_l x^{(i+l)}(t) - y^{(i)}(t) \right) = 0$$

The latter equation is valid if we set $y(t) = \sum_{l=0}^{r} b_l x^{(l)}(t)$. Thus the recursive model (10.7) can be expressed in the state space form (10.4), where $A$ and $B$ are given by (10.5) and,

$$C = ( b_0 - b_k a_0 \quad b_1 - b_k a_1 \quad \cdots \quad b_{k-1} - b_k a_{k-1} ), \qquad D = b_k \tag{10.12}$$

provided that $r = k$. The two models have identical input-output characteristics if they are initially at rest. If $r < k$, $b_k = 0$, $D = 0$, and $C$ has the form

$$C = ( b_0 \quad b_1 \quad \cdots \quad b_r \quad 0 \quad \cdots \quad 0 )$$

### Example 10.1  Suspension system
Consider the suspension system of Example 3.21:

$$\ddot{y} + \lambda \dot{y} + ky = ku + \lambda \dot{u}$$

Equations (10.5)–(10.12) give

$$A = \begin{pmatrix} 0 & 1 \\ -k & -\lambda \end{pmatrix}, \quad B = \begin{pmatrix} 0 \\ 1 \end{pmatrix}, \quad C = ( k \quad \lambda ) \qquad \blacksquare$$

## 10.2  CONVERSION OF FINITE DIFFERENCE MODELS TO STATE SPACE FORMS

The previous methodology is easily carried over to the discrete case. We consider first the AR system

$$y(n) + \sum_{i=1}^{k} a_i y(n-i) = u(n) \tag{10.13}$$

We define the state space variables

$$x_1(n) = y(n-1), \; x_2(n) = y(n-2), \; \ldots, \; x_k(n) = y(n-k)$$

Shifting gives

$$x_1(n+1) = -\sum_{i=1}^{k} a_i x_i(n) + u(n)$$

$$x_2(n+1) = x_1(n)$$

$$\vdots$$

$$x_k(n+1) = x_{k-1}(n)$$

These equations have the form

$$x(n+1) = Ax(n) + Bu(n)$$

where

$$A = \begin{pmatrix} -a_1 & -a_2 & \cdots & -a_{k-1} & -a_k \\ 1 & 0 & \cdots & 0 & 0 \\ \vdots & \vdots & \ddots & \vdots & \vdots \\ 0 & 0 & \cdots & 1 & 0 \end{pmatrix}, \quad B = \begin{pmatrix} 1 \\ 0 \\ \vdots \\ 0 \end{pmatrix} \tag{10.14}$$

The output is given by $y(n) = x_1(n+1)$. Therefore $y(n) = Cx(n) + Du(n)$, where

$$C = (-a_1 \quad -a_2 \quad \cdots \quad -a_k), \quad D = 1 \tag{10.15}$$

Let us next consider the general finite difference model

$$y(n) + \sum_{i=1}^{k} a_i y(n-i) = \sum_{i=0}^{r} b_i u(n-i) \tag{10.16}$$

We set

$$L = \max(k, r), \quad a_i = 0, i > k, \quad b_j = 0, j > r$$

Let $x(n)$ be the output of the autoregressive model

$$\sum_{i=0}^{L} a_i x(n - i) = u(n), \qquad a_0 = 1 \tag{10.17}$$

Shifting (10.17) by $j$, multiplying both sides by $b_j$, and summing over $j$ gives

$$\sum_{i=0}^{L} a_i \left( \sum_{j=0}^{L} b_j x(n - i - j) - y(n - i) \right) = 0$$

The latter equation is satisfied if

$$y(n) = \sum_{j=0}^{L} b_j x(n - j) = b_0 x(n) + \sum_{j=1}^{L} b_j x(n - j)$$

$$= -\sum_{i=1}^{L} b_0 a_i x_i(n) + b_0 u(n) + \sum_{j=1}^{L} b_j x_j(n)$$

where the state assignment $x_i(n) = x(n - i)$, $1 \leq i \leq L$, was used. The resulting state space model is represented by the parameters

$$A = \begin{pmatrix} -a_1 & -a_2 & \cdots & -a_{L-1} & -a_L \\ 1 & 0 & \cdots & 0 & 0 \\ \vdots & \vdots & \ddots & \vdots & \vdots \\ 0 & 0 & \cdots & 1 & 0 \end{pmatrix}, \quad B = \begin{pmatrix} 1 \\ 0 \\ \vdots \\ 0 \end{pmatrix} \tag{10.18}$$

$$C = (-b_0 a_1 + b_1 \quad \cdots \quad -b_0 a_L + b_L), \quad D = b_0 \tag{10.19}$$

and is referred to as *controller form* because it is convenient in the design of feedback controllers. The parameters $A$, $B$, and $D$ are directly obtained from the coefficients of the rational model. $C$ becomes operations free if the output uses past values of the input only, that is, $b_0 = 0$; otherwise, it requires $L$ additions and multiplications. The state space dimension of the controller form is $L = \max(k, r)$.

The above analysis applies to digital systems without changes.

### Example 10.2    Linear feedback shift register
A linear feedback shift register is described by an autoregressive filter over $GF(2)$. The controller form is given by Eqs. (10.18)–(10.19). Take, for instance,

the register with characteristic polynomial $h(q) = q^4 + q + 1$. The state parameters are

$$A = \begin{pmatrix} 1 & 0 & 0 & 1 \\ 1 & 0 & 0 & 0 \\ 0 & 1 & 0 & 0 \\ 0 & 0 & 1 & 0 \end{pmatrix}, \quad B = \begin{pmatrix} 1 \\ 0 \\ 0 \\ 0 \end{pmatrix}, \quad C = (1 \quad 0 \quad 0 \quad 1), \quad D = 1$$

■

### *Example 10.3   Euclid's algorithm*
In this example we show that the greatest common divisor of two polynomials $a(z)$ and $b(z)$ is written as a polynomial combination of $a(z)$ and $b(z)$. For this purpose we convert the division algorithm (a nonlinear finite difference system) into state space form. Let $\deg b(z) \le \deg a(z)$. For notational simplicity we omit $z$. The greatest common divisor, $\gcd(a, b)$, is determined exactly as in the case of integers (see Appendix II), via the algorithm

$$y(n) = -q(y(n-2), y(n-1))y(n-1) + y(n-2) \tag{10.20}$$

After a finite number of steps $N$, we find that $y(N+1) = 0$, $y(N) = \gcd(a, b)$. Using the state variables $x_1(n) = y(n-2)$, $x_2(n) = y(n-1)$, we obtain

$$x(n+1) = \begin{pmatrix} x_1(n+1) \\ x_2(n+1) \end{pmatrix} = \begin{pmatrix} 0 & 1 \\ 1 & -q(x(n)) \end{pmatrix} x(n)$$

and in particular,

$$x(N+2) = \begin{pmatrix} \gcd(a, b) \\ 0 \end{pmatrix} = \begin{pmatrix} 0 & 1 \\ 1 & -q(x(N+1)) \end{pmatrix} x(N+1)$$

Successive substitutions give

$$x(N+2) = \Phi(N+1)x_0 = \Phi(N+1) \begin{pmatrix} a \\ b \end{pmatrix}$$

where

$$\Phi(N+1) = \begin{pmatrix} 0 & 1 \\ 1 & -q(x(N+1)) \end{pmatrix} \begin{pmatrix} 0 & 1 \\ 1 & -q(x(N)) \end{pmatrix} \cdots \begin{pmatrix} 0 & 1 \\ 1 & -q(x(0)) \end{pmatrix}$$

Taking the first coordinate of $x(N+2)$, we have

$$\gcd(a, b) = xa + yb \tag{10.21}$$

where $x$, $y$ are the polynomials located on the first row of $\Phi(N+1)$. Equation (10.21) expresses the greatest common divisor of $a$ and $b$ as a polynomial combination of $a$ and $b$.

■

## 10.3  TRANSFER FUNCTION MODELS AND STATE SPACE FORMS

Let us translate the preceding results to the $s$ and $z$ domains. A finite derivative model of the form (10.7) converts to a proper rational transfer function model

$$h(s) = \frac{b(s)}{a(s)}, \quad \deg a(s) \geq \deg b(s)$$

and conversely. A proper rational transfer function is always represented by a state space model. One such representation is given by the controller form. A quadruple $(A, B, C, D)$ such that

$$C(sI - A)^{-1}B + D = h(s) = \frac{b(s)}{a(s)} \tag{10.22}$$

is called a *(state space) realization* of the transfer function model $h(s)$ because the associated state space system

$$\dot{x}(t) = Ax(t) + Bu(t)$$

$$y(t) = Cx(t) + Du(t)$$

has transfer function equal to $h(s)$.

The situation is entirely analogous in the discrete case. A finite difference model is represented by the rational transfer function model

$$h(q) = \frac{b(q)}{a(q)}, \quad a_0 \neq 0$$

The corresponding $z$ transform $h(z)$ is then proper rational. A state space realization always exists, one example being the controller form. There are infinitely many realizations of $h(q)$.

We saw in Section 7.5 that a linear change of coordinates in the state space, $z = Px$, leads to a new realization $(PAP^{-1}, PB, CP^{-1}, D)$ of $h(q)$. A similar statement holds for the continuous time transfer function model. Important insight into realization theory is gained by the notions of controllability and observability. These are discussed in the next two sections.

## 10.4  CONTROLLABILITY

Controllability concerns the coupling between the state and the input and describes the ability to drive the system from state to state by suitable inputs. In this respect controllability is a typical control property and is expected to appear in various aspects of control theory.

Consider the system

$$x(n + 1) = f(x(n), u(n))$$

There are $k$ states and $m$ inputs. The output in this case is $y = x$. Let $\phi(n, x_0, u)$ denote the solution that emanates from $x_0$ at time 0 under the action of the input signal $u$. Inputs take values in $R^m$. The system is *controllable* if for any states $x_0$ and $x_1$ there is input $u$ that drives the system from $x_0$ to $x_1$, in finite time $N$, that is, $\phi(N, x_0, u) = x_1$. Controllability is nicely expressed in terms of the reachable sets. Let $x$ be a state and $N$ be a positive integer. The set

$$\mathcal{R}(N, x) = \{z \in R^k : \text{there is input } u : \phi(N, x, u) = z\}$$

is the *reachable set at time $N$ from $x$*. It consists of all states that can be reached at time $N$ from the initial state $x$ by all admissible inputs. The *reachable set from $x$*

$$\mathcal{R}(x) = \bigcup_{N > 0} \mathcal{R}(N, x)$$

consists of all states that are attained from $x$ at some finite time. The bigger the reachable set is, the more control power the system possesses. In particular, the system is *controllable* if $\mathcal{R}(x) = R^k$ for all $x \in R^k$.

Characterization of the reachable sets of nonlinear systems is hard. In the linear case, however, complete answers can be obtained. The discrete case is easier to handle and is treated first.

### 10.4.1 Controllability of discrete linear systems

Consider the system

$$x(n + 1) = Ax(n) + Bu(n), \quad x \in R^k, u \in R^m \tag{10.23}$$

Suppose that it is initially at rest. For the moment we assume that $m = 1$. The states that can be accessed in one unit of time have the form $x(1) = A0 + Bu(0) = Bu(0)$; $u(0)$ runs freely in $R$. Hence the reachable set $\mathcal{R}(1, 0)$ is the line passing from $B$ and the origin, $\mathcal{R}(1, 0) = < B >$; the symbol $< \cdot >$ denotes linear span. Let $x(2)$ be a state reached from the origin in two time units. Then $x(2) = Ax(1) + Bu(1) = ABu(0) + Bu(1)$. This says that the reachable set at time 2 is the plane defined by the origin and the vectors $AB$ and $B$. If $AB$ and $B$ are linearly dependent, the plane collapses to the line spanned by $B$. Continuing this way, we arrive at the following:

**Theorem 10.1** Consider the system (10.23). Define the matrix

$$C = (B \quad AB \quad \cdots \quad A^{k-1}B) \tag{10.24}$$

Let $C_{co} = < B, AB, \ldots, A^{k-1}B >$ denote the linear space spanned by the columns of $C$. The following statements hold:

1. For all $n > 0$, $\mathcal{R}(n, 0) = < B, AB, \ldots, A^{n-1}B >$ and
   $\mathcal{R}(n, x) = A^n x + \mathcal{R}(n, 0)$.
2. For any $n_1 \leq n_2$, $\mathcal{R}(n_1, 0)$ is contained in $\mathcal{R}(n_2, 0)$, and for $n \geq k$,
   $\mathcal{R}(n, 0) = R(k, 0) = \mathcal{C}_{co}$.
3. The system (10.23) is controllable if and only if it is controllable from 0, or
   equivalently

$$\text{rank } C = \text{rank}(B \quad AB \quad \cdots \quad A^{k-1}B) = k \qquad (10.25)$$

PROOF Statement 1 formalizes the discussion preceding the theorem. It states
that a state is reachable from the origin at time $n$ if it is a linear combination of the
vectors $A^i b_j$, $i = 0, 1, 2, \ldots, n - 1$, $j = 1, 2, \ldots, m$; $b_j$ denote the columns of $B$.
The reachable set from $x$ at time $n$ is a linear manifold passing from the point $A^n x$
and parallel to the subspace $\mathcal{R}(n, 0)$. This is a direct consequence of the variation
of constants formula.

The second statement asserts that if a state is reached from the origin at a
certain time, it can also be attained at any later time. Furthermore the increasing
sequence of the reachable sets $R(n, 0)$ stabilizes after $k$ steps, the dimension of the
state space. The first part of the assertion follows from statement 1. The second
part is a consequence of the Cayley-Hamilton theorem (see Exercise 10.1).
Indeed, the matrix $A$ nullifies its characteristic polynomial. Hence we can write

$$A^k = -\sum_{j=1}^{k} a_j A^{k-j}, \quad A^k B = -\sum_{j=1}^{k} a_j A^{k-j} B$$

Thus $A^k B$ belongs to $\mathcal{C}_{co}$. Finally, to prove statement 3, we note that a
controllable system is controllable from any $x$. In particular, it is controllable
from the origin. Then the reachable set $\mathcal{R}(k, 0)$ covers the entire space and
rank $C = k$. Conversely, if the rank condition holds, the linear manifold $\mathcal{R}(k, x)$
is the entire space, and hence the system is controllable from $x$. The proof is
complete. ∎

For obvious reasons the matrix $C$ is called *controllability matrix*, and the
column space of $C$ *controllability subspace*.

The controllability matrix $C$ has size $k \times km$. It is square if and only if $m = 1$,
namely the system has a single input. In the single-input case the rank condition
is satisfied if and only if $C$ is nonsingular, or the determinant of $C$ is nonzero.

***Example 10.4 Planar discrete system depending on a parameter***
Let

$$\begin{pmatrix} x_1(n+1) \\ x_2(n+1) \end{pmatrix} = \begin{pmatrix} 0 & \mu \\ -\mu & 0 \end{pmatrix} \begin{pmatrix} x_1(n) \\ x_2(n) \end{pmatrix} + \begin{pmatrix} 0 \\ 1 \end{pmatrix} u(n), \qquad \mu \neq 0$$

Then

$$B = \begin{pmatrix} 0 \\ 1 \end{pmatrix}, \qquad AB = \begin{pmatrix} \mu \\ 0 \end{pmatrix}$$

These are linearly independent for any $\mu \neq 0$; thus $C$ has rank 2, and the system is controllable. ∎

### 10.4.2 Controllability of digital systems

Digital systems are conveniently represented by state diagrams or state graphs. A *graph* $G = (N, B)$ consists of a set of points $N$, called *nodes* or *vertices*, and a set of *arcs* (also called *branches* or *edges*) $B$, connecting the nodes. Each arc connects a pair of nodes $(a, b)$, so it is identified by the set $\{a, b\}$. In a *directed graph* each arc is assigned a direction, schematically represented by an arrow as in Fig. 10.1. The arc connecting two nodes $a$, $b$ in a directed graph is identified with the pair $(a, b)$; that is, $B$ becomes a subset of $N \times N$. The node $a$ is then the *initial endpoint*, and $b$ is the *terminal endpoint* of the edge $(a, b)$. If $(a, b)$ is an arc of the directed graph $G$, we say that $b$ is *adjacent* to $a$. A graph $G_1 = (N_1, B_1)$ is a *subgraph* of $G$ if $N_1 \subset N$ and $B_1 \subset B$. A sequence of branches of the form $\{a_1, a_2\}, \{a_2, a_3\}, \ldots, \{a_{n-1}, a_n\}$ is called a *chain*, while a sequence of the form $(a_1, a_2), (a_2, a_3), \ldots, (a_{n-1}, a_n)$ is called a *path*. We say that the path is from $a_1$ to $a_n$ and has length $n - 1$. A chain is *simple* if all edges and all nodes, except possibly the first and last, are distinct. A simple path is similarly defined. A *loop* or *cycle* is a simple path of length at least 1 whose endpoints coincide (occasionally the above definition is reserved for a cycle, and a loop is a branch of the form $(a, a)$). A graph is *connected* if there is a chain between any two nodes. It is *strongly connected* if there is a path between any two nodes. The notions introduced so far are illustrated in Fig. 10.2.

A *sequential machine*, namely a digital state space representation, is readily mapped onto a so-called *state graph*. Imagine inputs, outputs, and states taking values over a binary field. Each node of the graph represents a point in the state space. Each directed arc $(x_1, x_2)$ signifies that $x_2$ is reached from $x_1$ in one unit of time, that is, $x_2 = f(x_1, u)$ for some $u$. The labels of the directed arcs designate the

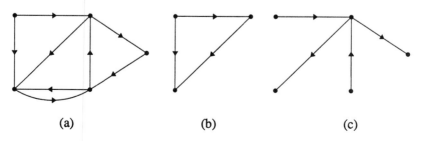

(a)  (b)  (c)

**Figure 10.1.** Examples of graphs; (b) and (c) are subgraphs of (a).

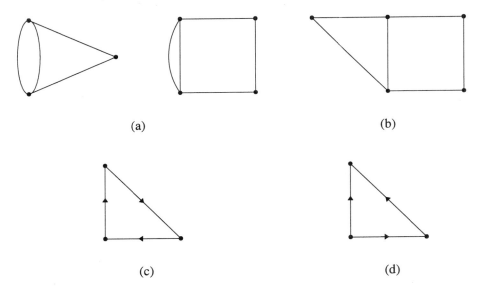

(a)                                         (b)

(c)                                         (d)

**Figure 10.2.** (a) Nonconnected graph, (b) connected graph, (c) strongly connected graph, and (d) nonstrongly connected graph.

corresponding input-output values. Alternatively, the information regarding next state transition can be summarized in a table as illustrated in Examples 10.5 and 10.6 below.

Controllability for digital systems is defined likewise. Theorem 10.1 is valid without changes. Controllability admits a nice interpretation in terms of the state diagram. A state $y$ is reached from state $x$ if there is a path emanating from $x$ and terminating at $y$. Therefore the system is controllable if the state diagram is strongly connected. In fact the reachable set from $x$ is the maximal connected subgraph containing $x$.

**Example 10.5**    *State diagram and controllability of a feedback shift register*
Consider the linear feedback shift register over $GF(2)$:

$$y(n) = y(n-1) + y(n-2) + u(n)$$

Conversion to state space gives

$$\begin{pmatrix} x_1(n+1) \\ x_2(n+1) \end{pmatrix} = \begin{pmatrix} 1 & 1 \\ 1 & 0 \end{pmatrix} \begin{pmatrix} x_1(n) \\ x_2(n) \end{pmatrix} + \begin{pmatrix} 1 \\ 0 \end{pmatrix} u(n)$$
$$y(n) = x_1(n) + x_2(n) + u(n)$$

Let 0,1,2,3 denote the four possible states 00, 01, 10, 11, respectively. One-step state transition is illustrated in the following table.

| Present state | Input | Next state | Output |
|---|---|---|---|
| 0 | 0 | 0 | 0 |
| 0 | 1 | 2 | 1 |
| 1 | 0 | 2 | 1 |
| 1 | 1 | 0 | 0 |
| 2 | 0 | 3 | 1 |
| 2 | 1 | 1 | 0 |
| 3 | 0 | 1 | 0 |
| 3 | 1 | 3 | 1 |

The state diagram is depicted in Fig. 10.3. The system is controllable because the matrix

$$(B \quad AB) = \begin{pmatrix} 1 & 1 \\ 0 & 1 \end{pmatrix}$$

has rank 2. The graph is strongly connected. Given any two states, there is a path that joins them.  ∎

**Example 10.6** *A linear machine with nonconnected state diagram*
Consider the linear machine

$$\begin{pmatrix} x_1(n+1) \\ x_2(n+1) \end{pmatrix} = \begin{pmatrix} 1 & 1 \\ 0 & 1 \end{pmatrix} \begin{pmatrix} x_1(n) \\ x_2(n) \end{pmatrix} + \begin{pmatrix} 1 \\ 0 \end{pmatrix} u(n)$$

$$y(n) = x_2(n) + u(n)$$

State transitions are described by the following table:

| Present state | Input | Next state | Output |
|---|---|---|---|
| 0 | 0 | 0 | 0 |
| 0 | 1 | 2 | 1 |
| 1 | 0 | 3 | 1 |
| 1 | 1 | 1 | 0 |
| 2 | 0 | 2 | 0 |
| 2 | 1 | 0 | 1 |
| 3 | 0 | 1 | 1 |
| 3 | 1 | 3 | 0 |

The state diagram is depicted in Fig. 10.4. The system is not controllable because the matrix

$$(B \quad AB) = \begin{pmatrix} 1 & 1 \\ 0 & 0 \end{pmatrix}$$

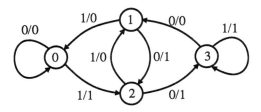

**Figure 10.3.** State diagram of the linear feedback shift register defined in Example 10.5.

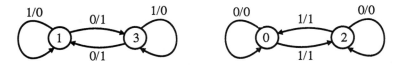

**Figure 10.4.** State diagram of the linear machine defined in Example 10.6

has rank 1. The graph is not connected. The states are decomposed into the connected subgraphs $\{0, 2\}$ and $\{1, 3\}$.  ∎

### 10.4.3  Controllability of continuous time linear systems

Consider the analog linear time invariant model

$$\dot{x}(t) = Ax(t) + Bu(t), \qquad x \in R^k, u \in R^m \tag{10.26}$$

For any positive real $t$ and $x \in R^k$ the reachable set $\mathcal{R}(t, x)$ consists of all states that can be reached from $x$ at time $t$ by piecewise continuous inputs with values in $R^m$. Let $\mathcal{C}$ be the controllability matrix (10.24) and $\mathcal{C}_{co}$ the controllability subspace formed by the linear span of the columns of $\mathcal{C}$. In analogy with the discrete case, we have

**Theorem 10.2**
1. $\mathcal{R}(t, 0) = \mathcal{C}_{co}$ for any $t > 0$.
2. $\mathcal{R}(t, x) = e^{At}x + \mathcal{C}_{co}$ for any $x$ in $R^k$ and $t > 0$.
3. The system is controllable if and only if it is controllable from the origin, or equivalently, rank $\mathcal{C} = k$.

**PROOF**  Let $x \in \mathcal{R}(t, 0)$. The variation of constants formula gives

$$x = \int_0^t e^{A(t-\tau)} Bu(\tau)d\tau \tag{10.27}$$

Using the series expansion of $e^{At}$, we obtain

$$x = \sum_{i=0}^{\infty} A^i B \frac{1}{i!} \int_0^t (t-\tau)^i u(\tau) d\tau$$

This is a series of vectors in $\mathcal{C}_{co}$. Since $\mathcal{C}_{co}$ is a subspace, the partial sum is inside $\mathcal{C}_{co}$. $\mathcal{C}_{co}$ is finite dimensional; therefore it is closed, and the limit is also in $\mathcal{C}_{co}$. Next we fix $t$, and we suppose that $\mathcal{R}(t,0) \neq \mathcal{C}_{co}$. The projection theorem implies that there is a nonzero vector $v \in \mathcal{C}_{co}$ which is orthogonal to all vectors in $\mathcal{R}(t,0)$, that is, $v^T x = 0$ for all $x \in \mathcal{R}(t,0)$. Therefore

$$\int_0^t v^T e^{A(t-\tau)} B u(\tau) d\tau = 0$$

for any admissible input $u(t)$. We choose $u(\tau) = B^T e^{A^T(t-\tau)} v$. Then

$$\int_0^t v^T e^{A(t-\tau)} B B^T e^{A^T(t-\tau)} v d\tau = \int_0^t |B^T e^{A^T(t-\tau)} v|^2 d\tau = 0$$

This occurs if and only if

$$v^T e^{As} B = 0 \qquad \text{for all } s \geq 0 \tag{10.28}$$

In particular, for $s = 0$ we obtain $v^T B = 0$. If we differentiate (10.28) and evaluate at zero, we find that $v^T AB = 0$. Repeated differentiation and evaluation at zero gives $v^T A^i B = 0$. Consequently $v$ is orthogonal to $\mathcal{C}_{co}$. Thus $v$ is zero, a contradiction. The proof of the remaining statements is straightforward. ∎

Theorems 10.1 and 10.2 show that the controllability matrix $\mathcal{C}$ is the same regardless of whether the parameters $A$ and $B$ specify a continuous or discrete time system. For this reason we often deal with both cases simultaneously. Moreover we are entitled to call a pair $(A, B)$ *controllable* if the controllability matrix $\mathcal{C}$ has maximum rank $k$.

It follows from statement 1 that for any $t_1$, $t_2$, $\mathcal{R}(t_1, 0) = \mathcal{R}(t_2, 0) = \mathcal{C}_{co}$. Therefore a state that can be reached from the origin at some time can also be reached at any other time, arbitrarily small or large. Rapid state transfers are accomplished at the expense of excessive control amplitudes as Exercise 10.13 demonstrates.

### Example 10.7 *Controllability of a constant resistance circuit*
Consider the circuit of Example 3.25. The controllability matrix is

$$\mathcal{C} = (B \quad AB) = \begin{pmatrix} \dfrac{1}{R_1 C} & \dfrac{-1}{R_1^2 C^2} \\[2mm] \dfrac{1}{L} & \dfrac{-R_2}{L^2} \end{pmatrix}$$

The system is uncontrollable if and only if $\det C = 0$, or

$$R_1 R_2 C = L \tag{10.29}$$

∎

Next we demonstrate that controllability is not destroyed by coordinate changes.

**Theorem 10.3**    Controllability remains invariant under a similarity transformation.

**PROOF**    Consider the pairs $(A, B)$ and $(\tilde{A}, \tilde{B})$, where $\tilde{A} = PAP^{-1}$, $\tilde{B} = PB$. Let $C$ and $\tilde{C}$ denote the corresponding controllability matrices. Then

$$\tilde{C} = \begin{pmatrix} \tilde{B} & \tilde{A}\tilde{B} & \cdots & \tilde{A}^{k-1}\tilde{B} \end{pmatrix} = \begin{pmatrix} PB & PAB & \cdots & PA^{k-1}B \end{pmatrix} = PC$$

Since $P$ is invertible, rank $C =$ rank $\tilde{C}$. ∎

**Example 10.8    Satellite control**
Let us consider the linearized state equation of the satellite motion in polar coordinates $r$ and $\theta$ (see Example 3.26). We scale the two inputs by $1/m$ and $1/mr_0$, respectively, so that the parameter $B$ has nonzero elements equal to 1. We remind the reader that the value $\omega^2 = k/r_0^3$ leads to periodic orbits with zero input. Changing coordinates to $r$, $\dot{r}$, $r_0\theta$, $r_0\dot{\theta}$, we obtain the state space representation

$$A = \begin{pmatrix} 0 & 1 & 0 & 0 \\ 3\omega^2 & 0 & 0 & 2\omega \\ 0 & 0 & 0 & 1 \\ 0 & -2\omega & 0 & 0 \end{pmatrix}, \quad B = \begin{pmatrix} 0 & 0 \\ 1 & 0 \\ 0 & 0 \\ 0 & 1 \end{pmatrix} \tag{10.30}$$

The controllability matrix is

$$C = \begin{pmatrix} 0 & 0 & 1 & 0 & 0 & 2\omega & -\omega^2 & 0 \\ 1 & 0 & 0 & 2\omega & -\omega^2 & 0 & 0 & -2\omega^3 \\ 0 & 0 & 0 & 1 & -2\omega & 0 & 0 & -4\omega^2 \\ 0 & 1 & -2\omega & 0 & 0 & -4\omega^2 & 2\omega^3 & 0 \end{pmatrix} \tag{10.31}$$

The dimensions of $C$ are $4 \times 8$. The first four columns are linearly independent. Thus the system is controllable. Controllability is verified by the MATLAB statement co = ctrb(A,B).

Suppose that control is exerted in only one direction; the second direction is ceased because of gas jet failure. If the tangential control is inactive, the system dynamics is governed by the pair $(A, b_1)$, where $b_1$ is the first column of $B$. The controllability matrix is easily inferred from (10.31). It is formed by the first, third, fifth, and seventh columns of $C$. The rank of the above matrix is 3, since the second and fourth columns are linearly dependent. Therefore the system is not controllable. We conclude that loss of tangential control weakens the control power of the system.

Using MATLAB we check the above as follows:

```
b1 = B(:,1);
co1 = ctrb(A,b1);
```

If the radial control is inactive, the system dynamics is governed by $(A, b_2)$ where $b_2$ is the second column of $B$. The controllability matrix now consists of the remaining columns of (10.31). We easily verify that the above matrix has rank 4, and hence the system is controllable. Thus loss of radial control does not limit the control power of the system. ∎

### 10.4.4  Controllability and controller canonical forms

Consider the single-input single-output system on $R^k$

$$x(n + 1) = Ax(n) + bu(n)$$

$$y(n) = cx(n) + du(n)$$

The parameters $B$ and $C$ of the general state space representation are column and row vectors, respectively, and are denoted by lowercase letters for emphasis. Likewise $d$ is scalar. Suppose that the pair $(A, b)$ is controllable. The controllability matrix

$$C = (e_1 \quad e_2 \quad \cdots \quad e_k)$$

is invertible. Let $Q = C^{-1}$. The similarity transformation $z = Qx$ converts the original parameters $A$, $b$, and $c$ into $\tilde{A} = QAQ^{-1}$, $\tilde{b} = Qb$, and $\tilde{c} = cQ^{-1}$. Let $\tilde{a}_i$ represent the columns of $\tilde{A}$. The expression $\tilde{A} = QAQ^{-1}$ is written as $C\tilde{A} = AC$, or columnwise, $C\tilde{a}_i = Ae_i$. Likewise $C\tilde{b} = b$. Recall that every vector $x$ is uniquely written as a linear combination of the basis vectors $e_i$:

$$x = \sum_{i=1}^{k} \mu_i e_i = C\mu, \qquad \mu = (\mu_1 \, \mu_2 \, \ldots \, \mu_k)^T$$

Therefore the expression $Ae_i = C\tilde{a}_i$ states that $\tilde{a}_i$ is the vector of coefficients of the representation of $Ae_i$ in terms of the basis $\{e_i\}$. Similarly $\tilde{b}$ is determined by the coefficients of the decomposition of $b$ in terms of $\{e_i\}$. Thus to determine $\tilde{b}$ and $\tilde{a}_i$, we must find the decompositions of $b$ and $Ae_i$ in the basis $\{e_i\}$. Note that $b = e_1$,

$Ae_1 = Ab = e_2$, $Ae_2 = A^2b = e_3$, ..., $Ae_{k-1} = e_k$, and

$$Ae_k = A^k b = -\sum_{i=1}^{k} a_i A^{k-i} b = -\sum_{i=1}^{k} a_i e_{k-i+1}$$

where $a_i$ are the coefficients of the characteristic polynomial. The last expression is a consequence of the Cayley-Hamilton theorem. Thus

$$\tilde{A} = \begin{pmatrix} 0 & 0 & \cdots & 0 & -a_k \\ 1 & 0 & \cdots & 0 & -a_{k-1} \\ \vdots & \vdots & \ddots & \vdots & \vdots \\ 0 & 0 & \cdots & 1 & -a_1 \end{pmatrix}, \quad \tilde{b} = \begin{pmatrix} 1 \\ 0 \\ \vdots \\ 0 \end{pmatrix} \qquad (10.32)$$

According to the results of Section 7.5, the system impulse response is given by $h(n) = cA^{n-1}bu_s(n-1) + d\delta(n)$. The parameter $\tilde{c}$ is determined by

$$\tilde{c} = cC = \begin{pmatrix} cb & cAb & \cdots & cA^{k-1}b \end{pmatrix} = \begin{pmatrix} h(1) & h(2) & \cdots & h(k) \end{pmatrix}$$

The realization $(\tilde{A}, \tilde{b}, \tilde{c})$ is known as *controllability form* because it relies on the controllability matrix and the basis formed by its columns.

Let us next consider a single-input controllable system

$$x(n+1) = Ax(n) + bu(n)$$

We will show that there exists a unique linear change of coordinates $P$ that converts the pair $(A, b)$ into $(\tilde{A}, \tilde{b}) = (PAP^{-1}, Pb)$ where the latter are in the controller form (10.14). Indeed, let

$$Q = P^{-1} = \begin{pmatrix} f_1 & f_2 & \cdots & f_k \end{pmatrix}$$

Then $Q\tilde{A} = AQ$, $Q\tilde{b} = b$. Let $\{e_i\}$ denote the standard Euclidean basis; $e_i$ is one at the $i$th position and zero everywhere else. Then $\tilde{b} = e_1$. Since $\tilde{A}$ is given by Eq. (10.14), its columns $\tilde{a}_i$ are expressed in the basis $\{e_i\}$ as follows:

$$\tilde{a}_1 = -a_1 e_1 + e_2, \quad \tilde{a}_2 = -a_2 e_1 + e_3, \quad \ldots, \quad \tilde{a}_{k-1} = -a_{k-1} e_1 + e_k, \quad \tilde{a}_k = -a_k e_1$$

Notice also that $Qe_i = f_i$. Since $Q\tilde{b} = Qe_1 = b$, we have $f_1 = b$. Equating the first columns in $Q\tilde{A} = AQ$, we obtain

$$Q\tilde{a}_1 = -a_1 Qe_1 + Qe_2 = -a_1 b + f_2 = Af_1 = Ab$$

Thus $f_2 = Ab + a_1 b$. Likewise equating the $i$th columns gives

$$Q\tilde{a}_i = Q(-a_i e_1 + e_{i+1}) = -a_i f_1 + f_{i+1} = Af_i$$

or

$$f_{i+1} = Af_i + a_i b, \qquad i = 1, 2, \ldots, k - 1$$

Using the variation of constants formula we write the above relation as

$$f_{i+1} = A^i b + \sum_{m=1}^{i} a_m A^{i-m} b \qquad (10.33)$$

The above equations specify the matrix $Q$ uniquely. There are two requirements that must be further met. $Q$ must satisfy the equality imposed by equating the last columns in $Q\tilde{A} = AQ$. Moreover $Q$ must be invertible. The first is a consequence of the Cayley-Hamilton theorem. Indeed, $Q\tilde{a}_k = -a_k f_1 = -a_k b$ and $Af_k = A^k b + \sum_{m=1}^{k-1} a_m A^{k-m} b$. The Cayley-Hamilton theorem implies that $A^k b = -\sum_{m=1}^{k} a_m A^{k-m} b$. Thus $Af_k = -a_k b$. Using Eq. (10.33), we write $Q$ as

$$Q = \mathcal{CM} = \begin{pmatrix} b & Ab & \cdots & A^{k-1}b \end{pmatrix} \begin{pmatrix} 1 & a_1 & a_2 & \cdots & a_{k-1} \\ 0 & 1 & a_1 & \cdots & a_{k-2} \\ 0 & 0 & 1 & \cdots & a_{k-3} \\ \vdots & \vdots & \vdots & & \vdots \\ 0 & 0 & 0 & \cdots & 1 \end{pmatrix}$$

The first factor $\mathcal{C}$ is the controllability matrix and is nonsingular by assumption. The second factor $\mathcal{M}$ is nonsingular because it is triangular with diagonal elements equal to 1. Thus $Q$ is invertible. We established that any controllable single–input system is converted to the controller form (10.14). This property is useful in the design of state feedback controllers as well as the design of state estimators, as we will see in Chapters 12 and 14.

It is easy to check that the controller form pair $(\tilde{A}, \tilde{b})$ is controllable. Indeed, the controllability matrix has the form

$$\tilde{C} = \begin{pmatrix} 1 & * & * & \cdots & * \\ 0 & 1 & * & \cdots & * \\ \vdots & \vdots & \vdots & & \vdots \\ 0 & 0 & 0 & \cdots & 1 \end{pmatrix}$$

The * represents "don't care" entries. $\tilde{C}$ is triangular with diagonal elements equal to 1. Hence it is nonsingular.

### 10.4.5 Input state connectivity and controllable reductions

The main result of this section asserts that an uncontrollable system is reduced to a controllable subsystem provided that the dynamics is restricted to a suitable lower-dimensional state space.

**Theorem 10.4**   Let

$$\dot{x}(t) = Ax(t) + Bu(t)$$

$A$ is a $k \times k$ matrix and $B$ is $k \times m$. Let $C$ be the controllability matrix, and let rank $C = k_1 < k$. Under a suitable change of coordinates the system is described by

$$\begin{pmatrix} \dot{x}_1(t) \\ \dot{x}_2(t) \end{pmatrix} = \begin{pmatrix} \hat{A}_1 & \hat{A}_2 \\ 0 & \hat{A}_3 \end{pmatrix} \begin{pmatrix} x_1(t) \\ x_2(t) \end{pmatrix} + \begin{pmatrix} \hat{B}_1 \\ 0 \end{pmatrix} u(t) \qquad (10.34)$$

where the $k_1$-dimensional subsystem

$$\dot{x}_1 = \hat{A}_1 x_1 + \hat{B}_1 u \qquad (10.35)$$

is controllable.

**PROOF**   We pick a basis $e_1, e_2, \ldots, e_{k_1}$ in the range of $C$, and we extend it to a basis $e_1, \ldots, e_{k_1}, e_{k_1+1}, \ldots, e_k$ in $R^k$. Let

$$P = \begin{pmatrix} e_1 & \cdots & e_{k_1} & e_{k_1+1} & \cdots & e_k \end{pmatrix} = (P_1 \quad P_2), \quad Q = P^{-1} = \begin{pmatrix} Q_1 \\ Q_2 \end{pmatrix}$$

The identity $QP = I$ implies that

$$Q_2 P_1 = 0 \qquad (10.36)$$

The columns of $P_1$ span the controllability subspace $C_{co}$. Therefore the latter equation implies that $Q_2 x = 0$ for all $x$ in $C_{co}$. Under the change of coordinates $z = Qx$, the system dynamics become $\hat{A} = QAP$, $\hat{B} = QB$. Hence

$$\hat{A} = \begin{pmatrix} Q_1 AP_1 & Q_1 AP_2 \\ Q_2 AP_1 & Q_2 AP_2 \end{pmatrix}, \quad \hat{B} = \begin{pmatrix} Q_1 B \\ Q_2 B \end{pmatrix}$$

From (10.36) we obtain $Q_2 B = 0$. Moreover it is easy to see that $C_{co}$ is $A$ invariant (see Exercise 10.2). Therefore Eq. (10.36) implies that $Q_2 AP_1 = 0$, and Eq. (10.34) is established.

Let $\hat{C}_1$ denote the controllability matrix associated with the pair $(\hat{A}_1, \hat{B}_1)$ and $\hat{C}$ the controllability matrix of (10.34). We have seen that rank $C = \dim C_{co} = \dim \hat{C}_{co} = k_1$. It is easy to show that

$$\hat{C} = \begin{pmatrix} \hat{B}_1 & \hat{A}_1 \hat{B}_1 & \cdots & \hat{A}_1^{k_1-1} \hat{B}_1 & \cdots & \hat{A}_1^{k-1} \hat{B}_1 \\ 0 & 0 & \cdots & 0 & \cdots & 0 \end{pmatrix} = \begin{pmatrix} \hat{C}_1 & M \\ 0 & 0 \end{pmatrix}$$

Since $\hat{C}_{1co}$ is the controllability subspace of $(\hat{A}_1, \hat{B}_1)$, all columns of $M$ are linearly dependent upon the columns of $\hat{C}_1$. Therefore rank $\hat{C} = $ rank $\hat{C}_1 = k_1$, and (10.35) is controllable. ∎

Inspection of (10.34) shows that if a system is not controllable, the part of the state designated by the $x_2$ coordinates is not affected by the input. Indeed, (10.34) is written

$$\dot{x}_1 = \hat{A}_1 x_1 + \hat{A}_2 x_2 + \hat{B}_1 u$$

$$\dot{x}_2 = \hat{A}_3 x_2$$

We observe that there is a lack of connectivity between inputs and states in the absence of controllability. On the other hand, if the state space is restricted to the $x_1$-space, namely to points of the form $(x_1, 0)$, $x_1 \in R^{k_1}$, the state trajectories of the original system coincide with the state trajectories of (10.35) initialized at $x_1$, and the dynamics of the reduced system are controllable. The reduced state space is the reachable set from the origin. If we set $z_1 = x_2$, $z_2 = x_1$, the system is described by

$$A_c = \begin{pmatrix} \hat{A}_3 & 0 \\ \hat{A}_2 & \hat{A}_1 \end{pmatrix}, \quad B_c = \begin{pmatrix} 0 \\ \hat{B}_1 \end{pmatrix} \tag{10.37}$$

The above form is referred to as *controllability staircase form*. The previous theorem is also valid for discrete systems and digital systems.

### Example 10.9   Satellite

In Example 10.8 we saw that the satellite motion is uncontrollable when only radial control is exerted. The first, third, and fifth column of the controllability matrix (10.31) are linearly independent. These appear as the first three columns of the similarity matrix

$$P = \begin{pmatrix} 0 & 1 & 0 & 0 \\ 1 & 0 & -\omega^2 & 0 \\ 0 & 0 & -2\omega & 0 \\ 0 & -2\omega & 0 & -\dfrac{1}{2\omega} \end{pmatrix}$$

The last column is chosen so that a basis in $R^4$ is obtained. The inverse of $P$ is

$$Q = \begin{pmatrix} 0 & 1 & -\dfrac{\omega}{2} & 0 \\ 1 & 0 & 0 & 0 \\ 0 & 0 & -\dfrac{1}{2\omega} & 0 \\ -4\omega^2 & 0 & 0 & -2\omega \end{pmatrix}$$

It is easy to check that the system in the new coordinates $z = Qx$ is in the controllability staircase form. It is validated in MATLAB by the commands $A = Q*A*inv(Q)$, $B = Q*B$. Reduction to the form (10.37) is accomplished in MATLAB by the statement

[Ac,Bc,Cc,P,m] = ctrbf(A,B,C).

$P$ designates the similarity transformation and $m$ is a vector of length $k$ whose elements are the ranks of the diagonal blocks. The direct application of the above command to the original parameters for geostatic orbits leads to inaccuracies due to numerical errors. ∎

## 10.5  OBSERVABILITY

Observability is concerned with the estimation of the state under the assumption that the input and output signals as well as the system parameters are known. Say a known input signal $u$ excites the state space model

$$x(n + 1) = f(x(n), u(n))$$

$$y(n) = g(x(n), u(n))$$

The initial state of the system is unknown. We measure the corresponding output signal $y(n)$ that obeys the relationship

$$y(n) = g(\phi(n, x_0, u), u(n)) \tag{10.38}$$

and we seek to determine the state signal $\phi(n, x_0, u)$. To this end it suffices to estimate $x_0$. Once $x_0$ is found, the entire state trajectory is uniquely specified by $\phi(n, x_0, u)$. Determination of $x_0$ amounts to solving Eq. (10.38) with respect to $x_0$. The latter equation always has one solution because $y(n)$ has been generated by the state equations and some initial state. We want to make sure that there will be no other solution because if there were, the two initial states would produce the same input-output behavior. In the latter case we cannot decide which of the two initial states generated the given measurements in response to the given excitation. If no ambiguity occurs, we say the system is observable or reconstructible in the sense that its state evolution can be observed or reconstructed from given input-output data.

Let $S_{x_0}$ denote the input-output transformation specified by the state space model and the initial state $x_0$. Thus

$$S_{x_0}(u)(n) = y(n) = g(\phi(n, x_0, u), u(n))$$

We say that states $x_0, z_0 \in R^k$ are *indistinguishable* if they give rise to the same input-output map, $S_{x_0} = S_{z_0}$. A system is called *observable* if any two indistinguishable states are equal. Thus an observable system does not allow different indistinguishable states.

Indistinguishability is an equivalence relation. Let $x$ be a state and $I(x)$ be the equivalence class of $x$, that is, the set of states indistinguishable from $x$. We can describe $I(x)$ in a somewhat more detailed way as follows: For each $n$ let

$$I(n, x) = \{z \in R^k : g(\phi(i, z, u), u(i)) = g(\phi(i, x, u), u(i)), \text{ for all } u \text{ and } i \leq n\}$$

$I(n, x)$ consists of states that are indistinguishable from $x$ with input-output information up to time $n$. We easily verify

$$I(x) = \bigcap_{n=0}^{\infty} I(n, x)$$

The above concepts are easily translated in the continuous case.

Consider two indistinguishable states $x$ and $z$ for the linear time invariant system

$$x(n + 1) = Ax(n) + Bu(n)$$
$$y(n) = Cx(n) + Du(n) \tag{10.39}$$

Let $v = x - z$. The variation of constants formula gives

$$CA^n x + C \sum_{i=0}^{n-1} A^{n-i-1} Bu(i) = CA^n z + C \sum_{i=0}^{n-1} A^{n-i-1} Bu(i)$$

Equivalently we have $CA^n v = 0$, $n = 0, 1, 2, \ldots$. By the Cayley-Hamilton theorem the latter holds if and only if $\mathcal{O}v = 0$, where

$$\mathcal{O} = \begin{pmatrix} C \\ CA \\ \vdots \\ CA^{k-1} \end{pmatrix} \tag{10.40}$$

We have proved the following:

**Theorem 10.5**  Consider the system (10.39). The following statements hold:

1. $I(n, 0) = \text{Ker} \begin{pmatrix} C \\ CA \\ \vdots \\ CA^{n-1} \end{pmatrix}$,  $I(n, x) = x + I(n, 0)$, $x \in R^k$.

2. For all $n \geq k$, $I(n, 0) = I(k, 0) = \text{Ker } \mathcal{O}$.
3. The system is observable if and only if $\text{Ker } \mathcal{O} = \{0\}$, or rank $\mathcal{O} = k$.  ∎

The matrix $\mathcal{O}$ is called the *observability matrix* and the subspace $\text{Ker } \mathcal{O}$ is called the *unobservable subspace*. The above result remains valid in the case of digital systems. An observable sequential machine is also called *diagnosable*, since it permits the identification of the state by observing the machine response. Observability of continuous time systems is handled in a similar way. The main result states that a linear time invariant system is observable if and only if the observability matrix $\mathcal{O}$ has rank $k$, exactly as in the discrete case.

### Example 10.10   Satellite motion

Consider the linearized equations of the satellite motion discussed in Example 10.8. The measured outputs are $r$ and $r_0\theta$. Thus

$$y = \begin{pmatrix} y_1 \\ y_2 \end{pmatrix} = \begin{pmatrix} 1 & 0 & 0 & 0 \\ 0 & 0 & 1 & 0 \end{pmatrix} x \tag{10.41}$$

The observability matrix is the $8 \times 4$ matrix

$$\mathcal{O} = \begin{pmatrix} 1 & 0 & 0 & 0 \\ 0 & 0 & 1 & 0 \\ 0 & 1 & 0 & 0 \\ 0 & 0 & 0 & 1 \\ 3\omega^2 & 0 & 0 & 2\omega \\ 0 & -2\omega & 0 & 0 \\ 0 & -\omega^2 & 0 & 0 \\ -6\omega^3 & 0 & 0 & -4\omega^2 \end{pmatrix}$$

The first four rows are linearly independent. Hence the system is observable. Is observability maintained if one of the measuring devices fails? Suppose that we measure the radial distance only. The matrix $C$ in this case consists of the first row of (10.41). The observability matrix $\mathcal{O}_1$ consists of the first, third, fifth, and seventh rows of $\mathcal{O}$ above. Thus

$$\mathcal{O}_1 = \begin{pmatrix} 1 & 0 & 0 & 0 \\ 0 & 1 & 0 & 0 \\ 3\omega^2 & 0 & 0 & 2\omega \\ 0 & -\omega^2 & 0 & 0 \end{pmatrix}$$

The second and fourth row are linearly dependent and the system is not observable. If the angle $\theta$ is the only available measurement, the observability matrix consists of the remaining rows of $\mathcal{O}$ and has nonzero determinant. Hence the system is observable.  ∎

### 10.5.1 State reconstruction and diagnosability

Observability ensures that state reconstruction from input-output data is feasible. One method for state recovery is described next. Consider the system (10.39). The output at time $t = 0$ is $y(0) = Cx_0 + Du(0)$. Hence $Cx_0 = y(0) - Du(0)$. Consecutive output samples provide additional linear equations for the initial state. Thus

$$y(1) = Cx(1) + Du(1) = CAx_0 + CBu(0) + Du(1)$$

and $CAx_0 = y(1) - Du(1) - CBu(0)$. Evaluation of the variation of constants formula for the first $k$ samples leads to the linear system of equations

$$\mathcal{O}x_0 = \mathbf{y}_0 - F\mathbf{u}_0 \tag{10.42}$$

where $\mathcal{O}$ is the observability matrix, and

$$\mathbf{y}_0 = \left( y^T(0) \quad y^T(1) \quad \ldots \quad y^T(k-1) \right)^T, \quad \mathbf{u}_0 = \left( u^T(0) \quad \ldots \quad u^T(k-1) \right)^T$$

Finally, $F$ is the triangular block Toeplitz matrix

$$F = \begin{pmatrix} D & 0 & 0 & \ldots & 0 & 0 \\ CB & D & 0 & \ldots & 0 & 0 \\ CAB & CB & D & \ldots & 0 & 0 \\ \vdots & \vdots & \vdots & \ddots & \vdots & \vdots \\ CA^{k-2}B & CA^{k-3}B & CA^{k-4}B & \ldots & CB & D \end{pmatrix}$$

The first block column of $F$ is the system's impulse response $h(n)$. The remaining columns are obtained by successive shifts of the impulse response. In the single-output case the observability matrix is square and invertible. The solution of (10.42) is

$$x_0 = \mathcal{O}^{-1}[\mathbf{y}_0 - F\mathbf{u}_0] \tag{10.43}$$

In the multichannel case we still have a unique solution because $\mathcal{O}$ has maximum rank $k$. An analogous method can be developed for continuous time systems provided that shifts are replaced by derivatives (see Exercise 10.14). The resulting scheme is not practical because it involves successive differentiation. Differentiators amplify noise and are avoided. An efficient approach relies on observers and is discussed in Section 12.4.5.

### Example 10.11 *Encryption using linear key generators*
Consider an encryption system employing a linear key generator. The key sequence $k(n)$ is produced at the output of an autonomous linear machine. The encrypted sequence $e(n)$ is produced by adding $k(n)$ to the message sequence

$m(n)$. The encryption operation is described by

$$
\begin{aligned}
x(n+1) &= Ax(n) \\
k(n) &= cx(n) \\
e(n) &= m(n) + k(n)
\end{aligned}
$$

(10.44)

All sequences take values in a binary field. The initial state of the generator $x_0 \in F^r$ represents the secret key, known only to the sender and the receiver. We will assume that the parameters $A$ and $c$ specifying the encryption system are publically known. We will further assume that $e(n)$ is known and that portions of the transmitted message $m(n)$ are known. As we pointed out in Section 4.8, this is a realistic assumption. It follows from the above equations that segments of the key sequence are available. The question we want to answer is how many successive digits of the key sequence are required in order to generate the entire key sequence. Without loss of generality the system may be assumed to be observable. In the opposite case it can be reduced to an equivalent observable system by the procedures described in subsequent sections. Collecting successive samples of $k(n)$, we find (see Eq. (10.43)) that

$$
( k(N) \quad k(N+1) \quad \cdots \quad k(N+r-1) )^T = \mathcal{O}x(N)
$$

Solution of the above linear system leads to the internal state of the machine $x(N)$. For any $n \geq N$ the value of the key sequence is determined from (10.44). If $A$ is invertible, the backward system $x(n) = A^{-1}x(n+1)$ enables the computation of the previous values of the key sequence all the way back to the initial key $x_0$. The performance of the above key generator is poor, unless the system memory $r$ is very large. Practical key generators are nonlinear (see Exercise 10.19). ∎

### 10.5.2  Duality

Comparison of Theorems 10.1 and 10.5, and in particular, the controllability and observability rank tests, shows a striking resemblance. This similarity is nicely expressed via the notion of duality. Consider the model

$$
\begin{aligned}
x(n+1) &= Ax(n) + Bu(n) \\
y(n) &= Cx(n) + Du(n)
\end{aligned}
$$

(10.45)

We refer to the above as the *primary system*, and we define its *dual system* by

$$
\begin{aligned}
x(n+1) &= A^T x(n) + C^T u(n) \\
y(n) &= B^T x(n) + D^T u(n)
\end{aligned}
$$

(10.46)

If the primary system has $k$ states, $r$ inputs, and $m$ outputs, its dual has $k$ states, $m$ inputs, and $r$ outputs. The above definition is the same in the continuous case. If the parameters of the primary system are specified by the matrix

$$P = \begin{pmatrix} A & B \\ C & D \end{pmatrix}$$

the dual system is represented by the transpose of $P$:

$$P^T = \begin{pmatrix} A^T & C^T \\ B^T & D^T \end{pmatrix}$$

It follows that the dual of the dual is the primary system. If $h(n)$ is the impulse response of the primary system, the impulse response $h^d(n)$ of the dual is $h^T(n)$. Indeed,

$$h^d(n) = B^T(A^{n-1})^T C^T u_s(n-1) + D^T \delta(n)$$

$$= (CA^{n-1}Bu_s(n-1) + D\delta(n))^T = h^T(n)$$

In particular, a single-input single-output system, and its dual have the same impulse response. Generally speaking, two properties are called dual if one is valid for one system each time the other is valid for its dual system. In this sense, controllability and observability are dual properties.

**Theorem 10.6 Duality**   If the primary system is controllable (observable), its dual is observable (controllable).

PROOF   Let $\mathcal{C}$ be the controllability matrix of the primary system and $\mathcal{O}$ the observability matrix of the dual. We have

$$\mathcal{O} = \begin{pmatrix} B^T \\ B^T A^T \\ \vdots \\ B^T(A^{k-1})^T \end{pmatrix}$$

Hence $\mathcal{O}^T = \mathcal{C}$. A matrix and its transpose always have the same rank. Therefore rank $\mathcal{C}$ = rank $\mathcal{O}$, and the claim follows.    ∎

The duality between controllability and observability is a manifestation of a rather deep duality between control and estimation. We will see more of that duality in later chapters. Using the duality theorem, the controllability results of the previous section are converted into statements concerning observability. It follows from Theorem 10.3 that observability is invariant under similarity transformations. The dual of Theorem 10.4 ensures the existence of a similarity transformation that converts the system (10.39) into the observability staircase form

$$\begin{pmatrix} x_1(n+1) \\ x_2(n+1) \end{pmatrix} = \begin{pmatrix} A_{uo} & A_{12} \\ 0 & A_o \end{pmatrix} \begin{pmatrix} x_1(n) \\ x_2(n) \end{pmatrix} + \begin{pmatrix} B_1 \\ B_2 \end{pmatrix} u(n), \quad y(n) = (0 \quad C_o)x(n)$$

where the pair $(A_o, C_o)$ is observable. Clearly only the $x_2$ portion of the state is observed at the output.

### 10.5.3  Observability and observer forms

The controller form was introduced in Section 10.2. It provides a natural means for the conversion of finite derivative and finite difference models to state space forms. The dual of the controller form is called *observer form* and is specified by

$$A = \begin{pmatrix} -a_1 & 1 & 0 & \cdots & 0 & 0 \\ -a_2 & 0 & 1 & \cdots & 0 & 0 \\ \vdots & \vdots & \vdots & & \vdots & \vdots \\ -a_{L-1} & 0 & 0 & \cdots & 0 & 1 \\ -a_L & 0 & 0 & \cdots & 0 & 0 \end{pmatrix}, \quad B = \begin{pmatrix} -b_0 a_1 + b_1 \\ \vdots \\ -b_0 a_L + b_L \end{pmatrix}$$

$$C = (1 \quad 0 \quad \cdots \quad 0), \quad D = b_0$$

The observer form is well suited for observer design, as we will later see. The *observability form* is the dual of the controllability form.

## 10.6  MINIMAL REALIZATIONS

Let $h(q)$ be the series of the signal $h(n)$. We have seen that if $h(q)$ has the form

$$h(q) = \frac{b(q)}{a(q)}, \quad a_0 \neq 0 \tag{10.47}$$

it is realizable, namely there is a quadruple $(A, B, C, D)$ such that the $q$ series of the state space model

$$x(n+1) = Ax(n) + Bu(n) \tag{10.48}$$

$$y(n) = Cx(n) + Du(n) \tag{10.49}$$

satisfies

$$h(q) = C(I - qA)^{-1}qB + D \tag{10.50}$$

Conversely, if $h(q)$ satisfies (10.50), it can be written as in (10.47). From now on and for the rest of the section, $h(q)$ will designate a realizable function.

If $(A, B, C, D)$ is a realization of $h(q)$, the size of $A$ defines the *order* of the realization. The order of a realization specifies the dimension of the state space of (10.50) and thus provides the memory requirements of the system. Suppose that (10.48)–(10.49) form a realization of $h(q)$, of order $L$. It is easy to construct a realization of $h(q)$, of order $L + k$, for any $k > 0$. We simply multiply the numerator and denominator of $h(q)$ by an arbitrary polynomial of degree $k$ and realize the resulting function by one of the canonical forms of the previous sections. Therefore, if $h(q)$ is realizable with $L$ memory elements, it is also realizable with any number of memory stages exceeding $L$. Although there is no upper bound on the order of realizations of $h(q)$, there is always a lower bound in memory requirements. A realization $(A, B, C, D)$ of $h(q)$ is called *minimal* if there is no realization of $h(q)$ with lower order. Minimal realizations necessarily have the same order called *minimal order* or *degree* of $h(q)$. Consider the state space representation associated with $(A, B, C, D)$. If we change coordinates, $w = Px$, the resulting quadruple $(PAP^{-1}, PB, CP^{-1}, D)$ is another realization of $h(q)$ having the same order (see Section 7.5). Therefore minimal realizations are not unique. We will shortly establish that minimal realizations are related by such a change of coordinates. This remarkable property is known as isomorphism of minimal realizations.

### 10.6.1 Controllability, observability, and minimal realizations

Insofar as minimality is concerned, the parameter $D$ is irrelevant, since it does not depend on the state. Put differently, $(A, B, C, D)$ is a minimal realization of $h(q)$ if and only if $(A, B, C)$ is a minimal realization of $h(q) - D$. Thus in the sequel we will assume that $h(q)$ is a rational series of the form (10.47), satisfying $b_0 = 0$ and thus is realizable by a triple $(A, B, C)$.

**Theorem 10.7** A realization $(A, B, C)$ of $h(q)$ is minimal if and only if the pair $(A, B)$ is controllable and the pair $(A, C)$ is observable.

**PROOF** Suppose that the given realization is minimal and has order $L$. Suppose also that the pair $(A, B)$ is not controllable. Using Theorem 10.4, we pass, via a suitable change of coordinates, to a state space realization of $h(q)$, $(\hat{A}, \hat{B}, \hat{C})$ whose parameters $\hat{A}$ and $\hat{B}$ have the form

$$\hat{A} = \begin{pmatrix} \hat{A}_1 & \hat{A}_2 \\ 0 & \hat{A}_3 \end{pmatrix}, \quad \hat{B} = \begin{pmatrix} \hat{B}_1 \\ 0 \end{pmatrix}, \quad \hat{C} = ( \hat{C}_1 \quad \hat{C}_2 )$$

We easily verify that the triple $(\hat{A}_1, \hat{B}_1, \hat{C}_1)$ is a realization of $h(q)$ of order smaller than $L$, contradicting minimality. Likewise the assumption that $(A, C)$ is not observable leads to a contradiction.

Next we assume the given realization is controllable and observable, and we prove that its order $L$ is minimal. Suppose that on the contrary this is not the case. Let $(A_1, B_1, C_1)$ be another realization of $h(q)$ having order $L_1$ and $L_1 < L$. The impulse response of the system, $h(n)$ is the same for both realizations. Hence

$$h(n) = CA^{n-1}B = C_1 A_1^{n-1} B_1, \qquad n \geq 1.$$

The Hankel matrix

$$H = \begin{pmatrix} h(1) & h(2) & \cdots & h(L) \\ \vdots & \vdots & \vdots & \vdots \\ h(L) & h(L+1) & \cdots & h(2L-1) \end{pmatrix}$$

is then factored as

$$H = \mathcal{O}\mathcal{C} = \begin{pmatrix} C \\ CA \\ \vdots \\ CA^{L-1} \end{pmatrix} \begin{pmatrix} B & AB & \cdots & A^{L-1}B \end{pmatrix} = \begin{pmatrix} \mathcal{O}_1 \\ * \end{pmatrix} \begin{pmatrix} \mathcal{C}_1 & * \end{pmatrix} \qquad (10.51)$$

$\mathcal{O}_1$ and $\mathcal{C}_1$ stand for the observability and controllability matrices of $(A_1, B_1, C_1)$. Since $(A, B, C)$ is controllable and observable, $\mathcal{O}$ and $\mathcal{C}$ have rank $L$. Therefore rank $H =$ rank $\mathcal{O}\mathcal{C} = L$. On the other hand, the matrices $\mathcal{O}_1, \mathcal{C}_1$ have dimensions $mL_1 \times L_1$ and $L_1 \times kL_1$, respectively. Therefore the rank of their product $H$ cannot exceed $L_1 < L$, and a contradiction results. ∎

The above theorem provides the means for developing a minimal realization from a given realization $(A, B, C)$. First, Theorem 10.4 is applied on $(A, B, C)$, and a realization $(A_1, B_1, C_1)$ is obtained so that $(A_1, B_1)$ is controllable. Subsequently the dual of Theorem 10.4 is invoked to obtain another realization $(A_2, B_2, C_2)$ so that $(A_2, B_2)$ remains controllable, and in addition $(A_2, C_2)$ is observable. We could as well start with the observable reduction, and then apply the controllable reduction. The above idea is crystallized in the next theorem.

**Theorem 10.8**    There exists a linear change of coordinates such that the system

$$x(n+1) = Ax(n) + Bu(n)$$
$$y(n) = Cx(n) + Du(n) \qquad (10.52)$$

is transformed to the canonical form

$$\begin{pmatrix} x_1(n+1) \\ x_2(n+1) \\ x_3(n+1) \end{pmatrix} = \begin{pmatrix} A_{uc} & 0 & 0 \\ A_{21} & A_{uo} & A_{c1} \\ A_{31} & 0 & A_{co} \end{pmatrix} \begin{pmatrix} x_1(n) \\ x_2(n) \\ x_3(n) \end{pmatrix} + \begin{pmatrix} 0 \\ B_{c1} \\ B_{co} \end{pmatrix} u(n) \qquad (10.53)$$

$$y(n) = \begin{pmatrix} C_1 & 0 & C_{co} \end{pmatrix} x(n) + Du(n) \qquad (10.54)$$

The triple $(A_{co}, B_{co}, C_{co})$ is controllable and observable. Furthermore the transfer function of (10.52) is expressed as

$$h(q) = C_{co}(I - qA_{co})^{-1}qB_{co} + D \qquad (10.55)$$

PROOF We use Theorem 10.4 to convert (10.52) into the controllability staircase form

$$\tilde{A} = \begin{pmatrix} A_{uc} & 0 \\ A_2 & A_c \end{pmatrix}, \quad \tilde{B} = \begin{pmatrix} 0 \\ B_c \end{pmatrix}, \quad \tilde{C} = (\, C_1 \quad C_2 \,) \qquad (10.56)$$

such that the pair $(A_c, B_c)$ is controllable. The transfer function is not affected by the change of coordinates. Thus

$$h(q) = (\, C_1 \quad C_2 \,) \begin{pmatrix} (I - qA_{uc})^{-1} & 0 \\ * & (I - qA_c)^{-1} \end{pmatrix} \begin{pmatrix} 0 \\ B_c \end{pmatrix} q + D$$

$$= C_2(I - qA_c)^{-1}qB_c + D \qquad (10.57)$$

Next we convert the pair $(A_c, C_2)$, into the observability staircase form

$$\overline{A}_c = \begin{pmatrix} A_{uo} & A_{c1} \\ 0 & A_{co} \end{pmatrix}, \quad \overline{C}_2 = (\, 0 \quad C_{co} \,), \quad \overline{B}_c = \begin{pmatrix} B_{c1} \\ B_{co} \end{pmatrix} \qquad (10.58)$$

If we substitute the above into Eq. (10.57), we obtain Eq. (10.55). Moreover Eq. (10.58) in conjunction with (10.56) yield the canonical representation. The reader is invited to show that controllability of the pair $(A_c, B_c)$ is passed to the controllable subsystem $(A_{co}, B_{co})$. ∎

### Example 10.12 *Direct canonical form*
Consider the finite difference model

$$y(n) + \sum_{i=1}^{k} a_i y(n - i) = \sum_{i=0}^{r} b_i u(n - i) \qquad (10.59)$$

The block diagram of Fig. 3.19 suggests the following definition of state

$$x(n) = (\, y(n - 1) \quad y(n - 2) \quad \cdots \quad y(n - k) \quad u(n - 1) \quad \cdots \quad u(n - r) \,)^T$$

Then

$$x(n + 1) = (\, y(n) \quad y(n - 1) \quad \cdots \quad y(n - k + 1) \quad u(n) \quad \cdots \quad u(n - r + 1) \,)^T$$

If the first entry is replaced by Eq. (10.59), we obtain

$$x(n+1) = \begin{pmatrix} -a_1 & \cdots & -a_{k-1} & -a_k & b_1 & \cdots & b_{r-1} & b_r \\ 1 & \cdots & 0 & 0 & 0 & \cdots & 0 & 0 \\ \vdots & \vdots & \vdots & \vdots & \vdots & \vdots & \vdots & \vdots \\ 0 & \cdots & 1 & 0 & 0 & \cdots & 0 & 0 \\ 0 & \cdots & 0 & 0 & 0 & \cdots & 0 & 0 \\ 0 & \cdots & 0 & 0 & 1 & \cdots & 0 & 0 \\ \vdots & \vdots & \vdots & \vdots & \vdots & \vdots & \vdots & \vdots \\ 0 & \cdots & 0 & 0 & 0 & \cdots & 1 & 0 \end{pmatrix} x(n) + \begin{pmatrix} b_0 \\ 0 \\ \vdots \\ 0 \\ 1 \\ 0 \\ \vdots \\ 0 \end{pmatrix} u(n)$$

$$y(n) = (-a_1 \quad -a_2 \quad \cdots \quad -a_k \quad b_1 \quad \cdots \quad b_r)x(n) + b_0 u(n)$$

The above realization is called direct. The parameters $A$, $B$, $C$, and $D$ are directly obtained from the coefficients of the numerator and denominator polynomials of $h(q)$ at no computational cost. The dimension of the state is $k + r$, and the representation is not minimal. As an illustration, take

$$y(n) + 2y(n-1) + 3y(n-2) = 4u(n-1) + 7u(n-2)$$

The direct form is

$$A = \begin{pmatrix} -2 & -3 & 4 & 7 \\ 1 & 0 & 0 & 0 \\ 0 & 0 & 0 & 0 \\ 0 & 0 & 1 & 0 \end{pmatrix}, \quad B = \begin{pmatrix} 0 \\ 0 \\ 1 \\ 0 \end{pmatrix}$$

$$C = (-2 \quad -3 \quad 4 \quad 7), \quad D = 0$$

Application of the MATLAB command [am,bm,cm,d] = minreal(A,B,C,D) yields a minimal realization via the controllable and observable reduction. Two states are removed. The command [b,a] = ss2tf(am,bm,cm,d) produces the transfer function of the original finite difference model:

$$b = (0 \quad 4 \quad 7), \quad a = (1 \quad 2 \quad 3) \qquad \blacksquare$$

### 10.6.2 Equivalence of minimal realizations

In this section we show that two minimal realizations of $h(q)$ are equivalent; that is, one is obtained from the other by a similarity transformation.

**Theorem 10.9** Let $(A, B, C)$ and $(A_1, B_1, C_1)$ be two minimal realizations of

$h(q)$ of order $L$. Then there exists an invertible matrix $P$ of order $L$, such that

$$(A_1, B_1, C_1) = (PAP^{-1}, PB, CP^{-1})$$

**PROOF** Let $C$ and $C_1$ denote the controllability matrices of $(A, B)$ and $(A_1, B_1)$, respectively. Likewise $\mathcal{O}$ and $\mathcal{O}_1$ denote the observability matrices of $(A, C)$ and $(A_1, C_1)$. Both realizations have the same impulse response, $h(n)$. Thus

$$h(n) = CA^{n-1}B = C_1 A_1^{n-1} B_1, \qquad n \geq 1$$

As in the proof of Theorem 10.7, the Hankel matrix

$$H = \begin{pmatrix} h(1) & h(2) & \cdots & h(L) \\ \vdots & \vdots & \vdots & \vdots \\ h(L) & h(L+1) & \cdots & h(2L-1) \end{pmatrix} \tag{10.60}$$

is factored as

$$H = \mathcal{O}C = \mathcal{O}_1 C_1 \tag{10.61}$$

In a similar fashion the shifted Hankel matrix

$$\tilde{H} = \begin{pmatrix} h(2) & h(3) & \cdots & h(L+1) \\ \vdots & \vdots & \vdots & \vdots \\ h(L+1) & h(L+2) & \cdots & h(2L) \end{pmatrix} \tag{10.62}$$

is factored as

$$\tilde{H} = \mathcal{O}AC = \mathcal{O}_1 A_1 C_1 \tag{10.63}$$

Observability ensures that $\mathcal{O}_1^T \mathcal{O}_1$ has rank $L$ and is invertible. If we multiply (10.61) by $\mathcal{O}_1^T$, we obtain

$$(\mathcal{O}_1^T \mathcal{O}_1)^{-1} \mathcal{O}_1^T \mathcal{O}C = C_1 \tag{10.64}$$

or

$$C_1 = PC, \quad P = (\mathcal{O}_1^T \mathcal{O}_1)^{-1} \mathcal{O}_1^T \mathcal{O} \tag{10.65}$$

We will show that $P$ is the desired isomorphism. Indeed, it follows from (10.65) that $A_1^i B_1 = PA^i B, i \geq 0$. In particular, $B_1 = PB$. From the same equation we obtain rank $C_1 \leq \min(\text{rank } P, \text{rank } C) = \min(\text{rank } P, L)$. Controllability implies $L \leq \min(\text{rank } P, L)$. Hence rank $P = L$, and $P$ is invertible. We post multiply

$C_1 = PC$ by $C^T$ to obtain $C_1 C^T = PCC^T$. By controllability, $CC^T$ is invertible. Hence

$$(C_1 C^T)(CC^T)^{-1} = P \tag{10.66}$$

Postmultiplying (10.61) by $C^T$ and taking into account (10.66), we obtain $\mathcal{O}CC^T = \mathcal{O}_1 C_1 C^T$, or $\mathcal{O} = \mathcal{O}_1 P$. This implies that $CA^i = C_1 A_1^i P$, $i \geq 0$, and in particular,

$$C = C_1 P \quad \text{or} \quad C_1 = CP^{-1}$$

It remains to establish the similarity of $A$ and $A_1$. We pre- and post-multiply (10.63) by $\mathcal{O}_1^T$ and $C^T$. Then taking into account (10.65) and (10.66), we obtain $(\mathcal{O}_1^T \mathcal{O}_1)^{-1} \mathcal{O}_1^T \mathcal{O} A = A_1 P$ or $PA = A_1 P$.    ∎

### 10.6.3   Irreducible transfer function models

Consider the single-input single-output state space model

$$x(n+1) = Ax(n) + Bu(n), \quad x \in F^k$$

$$y(n) = Cx(n) + Du(n) \tag{10.67}$$

$F$ is either the real field or a finite field. The associated transfer function

$$h(q) = C(I - qA)^{-1} qB + D$$

is rational and satisfies

$$h(q) = \frac{b(q)}{a(q)}, \quad a_0 \neq 0, \quad L_{a,b} = \max(\deg a(q), \deg b(q)) \leq k \tag{10.68}$$

The following holds:

**Theorem 10.10**   If (10.67) is controllable and observable, $a(q)$ and $b(q)$ are coprime and $L_{a,b} = k$. Conversely, if $h(q)$ is given by (10.68) and $b(q)$, $a(q)$ are coprime, the minimal order of $h(q)$ is $L_{a,b}$.

PROOF   Suppose that (10.67) is controllable and observable. Then (10.67) is a minimal realization of $h(q)$. It follows from (10.68) that $L_{a,b} \leq k$. If $L_{a,b} < k$, the controller form yields a realization of $h(q)$ of order smaller than $k$, contradicting minimality. Thus $L_{a,b} = k$. If $a(q)$, $b(q)$ are not coprime, there are polynomials $b_1(q)$, $a_1(q)$ with $h(q) = b_1(q)/a_1(q)$ and $L_{a_1,b_1} < L_{a,b}$. Employing the controller form, we obtain a realization of order less than $k$, a contradiction.

Next we prove the converse. Clearly there is a realization of order $L_{a,b}$. Suppose that there is a realization (10.67) of order $L_1 < L_{a,b}$. Then

$h(q) = b_1(q)/a_1(q)$ with $L_{a_1,b_1} \leq L_1$. Since $a(q)$, $b(q)$ are coprime, there is a polynomial $f(q)$ of at least degree 1 such that $b_1(q) = f(q)b(q)$ and $a_1(q) = f(q)a(q)$. Then $\deg b_1 > \deg b$ and $\deg a_1 > \deg a$. Thus $L_1 \geq L_{a,b}$, a contradiction. ∎

A transfer function model of the form (10.68) with coprime factors is called *irreducible*. If the system has real parameters and the $z$ transform is used, the transfer function $H(z)$ becomes proper rational:

$$H(z) = \frac{b(z)}{a(z)}, \qquad \deg b(z) \leq \deg a(z)$$

and $a(z)$, $b(z)$ are coprime. The minimal order is given by the degree of the denominator. The behavior of continuous time systems is analogous. The transfer function of a minimal realization is an irreducible proper rational function, and the minimal order is given by the degree of the denominator. Irreducible transfer function models have the following features:

**Theorem 10.11**  Suppose that $h(q)$ is a rational function

$$h(q) = \frac{b(q)}{a(q)}, \qquad a_0 \neq 0 \tag{10.69}$$

and

$$L = L_{a,b} = \max(\deg b(q), \deg a(q))$$

The following statements are equivalent:

1. $h(q)$ is irreducible.
2. There exist polynomials $x(q)$ and $y(q)$ such that the so-called Bezout identity holds:

$$a(q)x(q) + b(q)y(q) = 1 \tag{10.70}$$

3. The $2L \times 2L$ Sylvester matrix

$$M = \begin{pmatrix}
a_0 & 0 & \cdots & 0 & b_0 & 0 & \cdots & 0 \\
a_1 & a_0 & \cdots & 0 & b_1 & b_0 & \cdots & 0 \\
\vdots & \vdots & \ddots & \vdots & \vdots & \vdots & \ddots & \vdots \\
a_{L-1} & a_{L-2} & \cdots & a_0 & b_{L-1} & b_{L-2} & \cdots & b_0 \\
a_L & a_{L-1} & \cdots & a_1 & b_L & b_{L-1} & \cdots & b_1 \\
0 & a_L & \cdots & a_2 & 0 & b_L & \cdots & b_2 \\
\vdots & \vdots & \ddots & \vdots & \vdots & \vdots & \ddots & \vdots \\
0 & 0 & \cdots & a_L & 0 & 0 & \cdots & b_L
\end{pmatrix}$$

is nonsingular.

4. $L$ is the smallest integer satisfying

$$a_0 h(n) + \sum_{i=1}^{L} a_i h(n-i) = 0, \qquad n > L \qquad (10.71)$$

for some $a_i$.

5. For any integers $i, l \geq L$ the matrix

$$H_{i,l} = \begin{pmatrix} h(1) & h(2) & \cdots & h(l) \\ \vdots & \vdots & \ddots & \vdots \\ h(i) & h(i+1) & \cdots & h(i+l-1) \end{pmatrix}$$

has rank $L$.

**PROOF**   If $a(q)$, $b(q)$ are coprime, their greatest common divisor is 1. The Bezout identity (10.70) follows from the results of Section II.1.3 of Appendix II. The converse is also easy to prove. Next we establish the equivalence of 1 and 3. Uniqueness of coprime factorizations implies that there exist no polynomials $n(q)$ and $d(q)$ satisfying

$$\frac{b(q)}{a(q)} = \frac{n(q)}{d(q)}, \qquad \deg d(q) < \deg a(q)$$

The latter equation is also written as $-d(q)b(q) + n(q)a(q) = 0$. The representation of convolution as a matrix by vector multiplication (see Section 3.6.6) enables us to express the above as

$$M\begin{pmatrix} n \\ -d \end{pmatrix} = 0$$

where $n$ and $d$ stand for the vectors of coefficients of the corresponding polynomials. Coprimeness of $a(q)$ and $b(q)$ equivalently says that the null space of $M$ consists solely of the origin, which in turn is equivalent to the nonsingularity of $M$.

To establish the equivalence of irreducibility and statement 4, we write (10.69) as $a(q)h(q) = b(q)$, and we equate the first $L + 1$ coefficients to obtain

$$h_0 a_0 = b_0$$
$$h_1 a_0 + h_0 a_1 = b_1$$
$$\vdots$$
$$h_L a_0 + h_{L-1} a_1 + \cdots + h_0 a_L = b_L \qquad (10.72)$$

Equating coefficients of order higher than $L + 1$ gives (10.71). $L$ is the smallest positive integer such that (10.71) holds for some $a_i$; otherwise, an $L' < L$ can be found so that (10.71) holds with coefficients $a_i'$. If we insert $a_i'$ into (10.72), we determine coefficients $b_i'$ so that $h(q) = b'(q)/a'(q)$. The resulting ratio yields a realization of order $L'$, contradicting irreducibility of $h(q)$.

To prove the last assertion, we recall the decomposition $H_{L,L} = \mathcal{O}\mathcal{C}$, where $\mathcal{O}$ and $\mathcal{C}$ are the observability and controllability matrices of a minimal realization of $h(q)$ of order $L$. Both $\mathcal{O}$ and $\mathcal{C}$ have maximum rank $L$, by Theorem 10.10. Standard results in linear algebra assert that the rank of $H_{L,L}$ is also $L$. The claim follows from the observation that Eq. (10.71) implies that all columns and rows of the Hankel matrices $H_{L+i, L+l}$, $i, l \geq 0$ after the $L$th column and the $L$th row are linearly dependent upon the first $L$ columns and rows. ∎

### 10.6.4 Hankel realizations

The above theorem enables us to develop minimal realizations in one step. Consider a single-input single-output system with impulse response $h(n)$. We compute the maximum integer $L$ such that the Hankel matrix $H_{L,L}$ has rank $L$. From Theorem 10.11, $h(n)$ satisfies Eq. (10.71) with $a_0 = 1$. The parameters

$$A = \begin{pmatrix} 0 & 1 & 0 & \cdots & 0 \\ 0 & 0 & 1 & \cdots & 0 \\ \vdots & \vdots & \vdots & \ddots & \vdots \\ 0 & 0 & 0 & \cdots & 1 \\ -a_L & -a_{L-1} & -a_{L-2} & \cdots & -a_1 \end{pmatrix}, \quad B = \begin{pmatrix} h(1) \\ h(2) \\ \vdots \\ h(L) \end{pmatrix}$$

$$C = (1 \quad 0 \quad \cdots \quad 0), \quad D = h(0) \tag{10.73}$$

define a controllable and observable realization of $h(q)$. Indeed,

$$AB = \begin{pmatrix} h(2) \\ h(3) \\ \vdots \\ h(L+1) \end{pmatrix}, \quad A^2 B = \begin{pmatrix} h(3) \\ h(4) \\ \vdots \\ h(L+2) \end{pmatrix}$$

and so forth. Multiplying the first $L$ values of the impulse response by $A^i$ shifts these values by $i$. The controllability matrix coincides with $H_{L,L}$ and thus has rank $L$. The observability matrix is the identity, and the system is observable. Finally, the quadruple (10.73) forms a realization of $h(q)$ because $h(0) = D$ and

$$CA^i B = C(h(i+1) \quad h(i+2) \quad \cdots \quad h(L+i))^T = h(i+1)$$

Thus $CA^{n-1} Bu_s(n-1) + D\delta(n)$ coincides with $h(n)$.

### 10.6.5    Singular value decomposition and minimal realizations

The singular value decomposition is an important matrix representation finding many applications. Among its attractive features is that it allows efficient computation. For this reason it forms a basic block in several signal processing algorithms.

Every $m \times r$ real matrix $H$ of rank $L$ can be factored as

$$H = R \begin{pmatrix} \Sigma & 0 \\ 0 & 0 \end{pmatrix} Q^T \tag{10.74}$$

where $R^T R = RR^T = I_m$, $Q^T Q = QQ^T = I_r$, and $\Sigma = \text{diag}\{\sigma_1, \sigma_2, \ldots, \sigma_L\}$ with $\sigma_1 \geq \sigma_2 \geq \cdots \geq \sigma_L > 0$. The parameters $\sigma_i$ are called the *singular values* of $H$. It turns out that $\sigma_i^2$ are the eigenvalues of $H^T H$. $R$ and $Q$ are square orthogonal matrices, albeit not unique. Equation (10.74) forms the singular value decomposition (SVD) of $H$.

Minimal realizations of multivariable systems are obtained with the aid of the singular value decomposition. Consider the $m \times r$ rational series $h(q)$. All entries of $h(q)$ are irreducible. Let $a(q)$ be the least common multiple of all denominators of $h(q)$, and let $L$ denote the degree of $a(q)$. We consider the Hankel matrices $H$ and $\tilde{H}$ defined by Eq. (10.60) and (10.62), and we form the singular value decomposition of $H$, as given by Eq. (10.74). If $R_1$ consists of the first $L$ columns of $R$ and $Q_1$ consists of the first $L$ columns of $Q$, it follows that

$$H = R_1 \Sigma Q_1^T$$

Let $\mathcal{O} = R_1 \Sigma^{1/2}$, $\mathcal{C} = \Sigma^{1/2} Q_1^T$ so that $H = \mathcal{O}\mathcal{C}$. Motivated by the factorization (10.61), we are tempted to view $\mathcal{O}$ as the observability matrix and $\mathcal{C}$ as the controllability matrix. The question then becomes how to extract $A$, $B$, and $C$ from $\mathcal{O}$ and $\mathcal{C}$. Clearly $C$ is given by the first $m$ rows of $\mathcal{O}$. Likewise $B$ is given by the first $r$ columns of $\mathcal{C}$. It remains to determine $A$. This is where the factorization (10.63) and the shifted matrix $\tilde{H}$ take part. Thus

$$\tilde{H} = \mathcal{O}A\mathcal{C} = R_1 \Sigma^{1/2} A \Sigma^{1/2} Q_1^T$$

From $R^T R = I$, $Q^T Q = I$ it follows that $R_1^T R_1 = I$ and $Q_1^T Q_1 = I$. Hence

$$A = \Sigma^{-1/2} R_1^T \tilde{H} Q_1 \Sigma^{-1/2} = \mathcal{O}^+ \tilde{H} \mathcal{C}^+$$

$$\mathcal{O}^+ = \Sigma^{-1/2} R_1^T \quad \mathcal{C}^+ = Q_1 \Sigma^{-1/2} \tag{10.75}$$

In summary, the singular value decomposition of the Hankel matrix $H$ provides the controllability and observability matrices $\mathcal{O}$ and $\mathcal{C}$. Then $C$ and $B$ are automatically extracted by projection on the first rows and columns of $\mathcal{O}$ and $\mathcal{C}$. Finally, $A$ is obtained from the shifted Hankel matrix via (10.75).

*Technical remarks.* We prove next that the above parameters define a minimal realization of $h(q)$. Let $C_1 = [B, AB, \ldots, A^{L-1}B]$. Similarly $\mathcal{O}_1$ stands for the observability matrix of $(A, C)$. Then $H = \mathcal{O}_1 C_1$. Hence $L = \text{rank } H \leq \min(\text{rank } \mathcal{O}_1, \text{rank } C_1)$. Since $\mathcal{O}_1$ is an $mL \times L$ matrix, it holds rank $\mathcal{O}_1 \leq L$. Likewise rank $C_1 \leq L$. Therefore rank $\mathcal{O}_1 = \text{rank } C_1 = L$, and the system is controllable and observable. It remains to show that the impulse response is $h(n)$. Let

$$
M = \begin{pmatrix} 0 & I & \cdots & 0 \\ 0 & 0 & \cdots & 0 \\ \vdots & \vdots & & \vdots \\ -a_L I & -a_{L-1} I & \cdots & -a_1 I \end{pmatrix}, \quad N = \begin{pmatrix} 0 & \cdots & -a_{L-1} I & -a_L I \\ I & \cdots & -a_{L-2} I & -a_{L-1} I \\ \vdots & \vdots & & \vdots \\ 0 & \cdots & I & -a_1 I \end{pmatrix}
$$

It follows as in the derivation of (10.73) that $\tilde{H} = MH = HN$ and more generally

$$
M^i H = H N^i, \qquad i = 0, 1, \ldots \tag{10.76}
$$

Let

$$
H^+ = C^+ \mathcal{O}^+
$$

Then $HH^+H = \mathcal{O}CC^+\mathcal{O}^+\mathcal{O}C$. But $CC^+ = \Sigma^{1/2}Q_1^T Q_1 \Sigma^{-1/2} = I$. Likewise $\mathcal{O}^+\mathcal{O} = I$. Therefore

$$
HH^+H = H \tag{10.77}
$$

Now

$$
A^2 = (\mathcal{O}^+\tilde{H}C^+)(\mathcal{O}^+\tilde{H}C^+) = \mathcal{O}^+\tilde{H}H^+\tilde{H}C^+ = \mathcal{O}^+ MHH^+ HNC^+
$$

$$
= \mathcal{O}^+ MHNC^+ = \mathcal{O}^+ M^2 HC^+
$$

and more generally

$$
A^i = \mathcal{O}^+ M^i HC^+
$$

Let $I_m$ denote the projection of a block vector onto the first $m$ block rows and $I_r$ the projection onto the first $r$ block columns. Then $C = I_m \mathcal{O}$ and $B = CI_r$. Thus $CA^iB = I_m\mathcal{O}\mathcal{O}^+ M^i HC^+ CI_r$. We insert into the latter expression the identities $CC^+ = \mathcal{O}^+\mathcal{O} = I$ to obtain

$$
CA^iB = I_m\mathcal{O}CC^+\mathcal{O}^+ M^i HC^+\mathcal{O}^+\mathcal{O}CI_r
$$

We substitute $H = \mathcal{O}C$, $H^+ = C^+\mathcal{O}^+$ and Eqs. (10.76)–(10.77) into the above equation to obtain

$$CA^i B = I_m HH^+ M^i HH^+ HI_r = I_m HH^+ M^i HI_r = I_m HH^+ HN^i I_r$$

$$= I_m HN^i I_r = I_m M^i HI_r$$

The last expression is the left upper corner of $M^i H$. Due to the shifting structure of $M$, this element is $h(i + 1)$, and the proof is complete.

***Example 10.13    MATLAB commands related to realization***
In this example the MATLAB commands relevant to the material of this chapter are described.

| | |
|---|---|
| prony(h,nb,na) | % computes the numerator and denominator coefficients |
| | % (transfer function) of a system with impulse response h. |
| impulse(b,a,T) | % computes the impulse response from the transfer function |
| | % coefficients for a continuous time system over the time |
| | % interval T. |
| dimpulse(b,a,n) | % computes the impulse response from the transfer function |
| | % coefficients for a discrete system over the time |
| | % interval n. |
| tf2ss(b,a) | % computes state space parameters from transfer function |
| | % coefficients. |
| ss2tf(a,b,c,d) | % computes transfer function coefficients from |
| | % state space parameters . |
| tf2zp(b,a) | % computes zeros and poles from transfer function |
| | % coefficients. |
| ss2zp(a,b,c,d) | % computes zeros and poles from state space parameters.    ∎ |

## BIBLIOGRAPHICAL NOTES

There is an extensive literature on the topics discussed in this chapter. Our exposition is based on Kailath (1980) and Chen (1984). In the same references, realizations of multi-variable systems are treated. Stochastic realizations are considered in Caines (1988). Linear digital systems are analyzed in Peterson and Weldon (1972). The algebraic approach to realization is pursued in Kalman, Falb and Arbib (1969). Controllability, observability, and realizations of nonlinear systems are discussed in Isidori (1989) and Sontag (1990). State space realizations for Volterra systems are studied in Rugh (1981).

## PROBLEMS

**10.1.**    Let $\phi(\lambda)$ be the characteristic polynomial of a matrix $A$. The Cayley-Hamilton theorem states that $\phi(A) = 0$. Prove the statement for $2 \times 2$ matrices. Conduct MATLAB experiments to show the general validity of the theorem.

**10.2.** Let $A$ be a $k \times k$ real matrix and $M$ a linear subspace of $R^k$. $M$ is called $A$ *invariant* if $Ax \in M$ for any $x \in M$. Show the following:

1. $M$ is $A$ invariant if and only if $A^n x \in M$, $n \geq 0$, $x \in M$.
2. $M$ is $A$ invariant if and only if $e^{At}x \in M$, $t \geq 0$, $x \in M$. The above statements assert that the continuous and discrete flows generated by $A$ stay in $M$ if they start in $M$.
3. For any $b \in R^k$ the controllable subspace of the pair $(A, b)$ is the smallest $A$ invariant subspace containing $b$.

**10.3.** Consider the single-input single-output system

$$\dot{x}(t) = Ax(t) + bu(t), \quad y(t) = cx(t)$$

Suppose that $A$ is diagonal. Show that the system is controllable if and only if the eigenvalues of $A$ are pairwise distinct and all entries of $b$ are nonzero. Formulate and prove a similar property for observability.

**10.4.** Consider the single-input system $\dot{x}(t) = Ax(t) + bu(t)$ and the sampled system

$$x(n+1) = \Phi x(n) + \Gamma u(n), \quad \Phi = e^{AT_s}, \quad \Gamma = \int_0^{T_s} e^{A\tau} d\tau b$$

Suppose that $A$ is diagonal. Prove that the sampled system is controllable if the pair $(A, b)$ is controllable, and for any pair of eigenvalues of $A$ with $\Re\lambda_i = \Re\lambda_k$, it holds that

$$\Im(\lambda_i - \lambda_k) \neq \frac{2\pi m}{T_s}, \quad m \in Z$$

Formulate and prove a similar property for observability.

**10.5.** Using an argument similar to that employed in the proof of Theorem 10.2, show that the system (10.4) is controllable if and only if

$$W_c(t) = \int_0^t e^{A\tau} BB^T e^{A^T\tau} d\tau > 0$$

for any $t > 0$. Likewise prove that the discrete system (10.23) is controllable if and only if the following matrix is positive definite:

$$W_c(n) = \sum_{i=0}^{n} A^i BB^T (A^T)^i$$

for $n \geq k$. The matrices $W_c(t)$ and $W_c(n)$ are called *controllability grammians.* Establish similar statements for observability and the observability grammian

$$W_o(n) = \sum_{i=0}^{n} (A^T)^i C^T C A^i$$

**10.6.** Suppose that $A(t)$ is bounded and continuous. The continuous time system

$$\dot{x}(t) = A(t)x(t) + B(t)u(t) \tag{10.78}$$

is controllable at $t_0$, if there is $T > t_0$ such that for any pair of states $x_0$ and $x_1$, there is an input $u$ such that $\phi(T, x_0, u) = x_1$. Let $\Phi(t, t_0)$ be the transition matrix of (10.78). Show that (10.78) is controllable at $t_0$, if the *controllability grammian*

$$W(t_0, t_1) = \int_{t_0}^{t_1} \Phi(t_0, \tau)B(\tau)B^T(\tau)\Phi^T(t_0, \tau)d\tau$$

is positive definite for some $t_1 > t_0$. A similar result holds in the discrete case. Prove that $x(n + 1) = A(n)x(n) + B(n)u(n)$, $A(n)$ bounded, is controllable if and only if

$$W(n_0, n) = \sum_{i=n_0}^{n} \Phi(n, i)B(i)B^T(i)\Phi^T(n, i) > 0$$

for $n \geq n_0 + k$. State and prove similar results for observability.

**10.7.** Let

$$y(n) + \sum_{i=1}^{k} a_i y(n - i) = \sum_{i=1}^{k} b_i u(n - i)$$

The above is written

$$y(n) = -a_1 y(n - 1) + b_1 u(n - 1) + \xi_1(n - 1)$$
$$\xi_1(n) = -a_2 y(n - 1) + b_2 u(n - 1) + \xi_2(n - 1)$$

$$\vdots$$

$$\xi_{k-1}(n) = -a_k y(n - 1) + b_k u(n - 1)$$

Show that the state space model obtained by the state assignment

$$x(n) = ( y(n) \quad \xi_1(n) \quad \cdots \quad \xi_{k-1}(n) )^T$$

is the observer form.

**10.8.** Show that the rational model $h(q) = b(q)/a(q)$, $a_0 \neq 0$ admits a state space realization of the form $(A, B, C, 0)$ if and only if $b_0 = 0$. In the case of systems with real coefficients the above is equivalent to having a strictly proper rational transfer function $H(z)$.

**10.9.** If $(A, B, C, D)$ is a realization of $h(q)$ of order $L$ and $k$ a positive integer, show that

$$\hat{A} = \begin{pmatrix} A & * \\ 0 & * \end{pmatrix}, \quad \hat{B} = \begin{pmatrix} B \\ 0 \end{pmatrix}, \quad \hat{C} = (C \quad *), \quad \hat{D} = D$$

is a realization of $h(q)$ of order $L + k$.

**10.10.** Show that $(A, B)$ is controllable if and only if there exists no (left) eigenvector of $A$ that is orthogonal to all columns of $B$; that is, there exists no eigenvalue $\lambda$ of $A$ and no vector $a$ so that

$$a^T A = \lambda a^T \quad \text{and} \quad a^T b_j = 0$$

where $b_j$ are the columns of $B$. This is known as Popov-Belevitch-Hautus test. Derive a similar criterion for observability.

**10.11.** Show that the circuit of Example 10.7 is observable if and only if $R_1 R_2 L = C$.

**10.12.** Construct an example that shows that the controller form is not necessarily observable.

**10.13.** Consider the system on the real line

$$\dot{x}(t) = ax(t) + bu(t), \quad x \in R$$

Let $b \neq 0$. Find a control that drives the system from zero to $x$ at time $T$, and show that its amplitude grows inversely to $T$.

**10.14.** Consider an analog linear time invariant state space system initialized at $x_0$. Prove that $x_0$ can be determined by Eq. (10.43) where now $y_0$ and $\mathbf{u}_0$ consist of successive derivatives of $y$ and $u$ evaluated at 0.

**10.15.** Suppose that $h(q) = b(q)/a(q)$ is real rational with $a_0 \neq 0$. Using the prime factorization of $b(q)$ and $a(q)$, write $h(q)$ as a product of first-order and second-order systems. A first-order system has the form $(b_0 + b_1 q)/(a_0 + a_1 q)$. Second-order systems are similarly defined and correspond to complex roots. With the aid of the state space realizations of cascade connections discussed in Exercise 3.13, derive the so-called *cascade realization* of $h(q)$, using any canonical state space realization for the first- and second-order factors. In a similar way develop the *parallel realization* using the partial fraction expansion of $h(q)$. Each factor $1/(1 - \lambda q)^i$ is realized as a cascade of identical first-order systems.

Cascade and parallel architectures are preferred in an actual implementation due to their superior performance in finite arithmetic and robustness to round off errors.

**10.16.** Show that the controllability and observability grammians $W_c(L-1)$, $W_o(L-1)$ of the realization obtained by SVD are equal and diagonal.

**10.17.** Let

$$y(n) + \frac{5}{6}y(n-1) + \frac{1}{6}y(n-2) = 2u(n-1) + 3u(n-2)$$

Determine all canonical realizations introduced in this chapter. Validate your results with MATLAB. Repeat for the analog system

$$y^{(2)}(t) + 4y^{(1)}(t) + 3y(t) = u(t) + 4u^{(1)}(t)$$

**10.18.** Show that the pair $(A, C)$ is observable if $(A - KC, C)$ is observable for any matrix $K$ of compatible dimensions. This says that observability is maintained under output feedback.

**10.19.** The complexity performance of an encryption key generator (see Example 10.11) is improved if nonlinearities are introduced. Consider a linear feedback shift register with characteristic polynomial $h(q) = 1 + \sum_{i=1}^{r} h_i q^i$ and output $y(n)$; $h(q)$ is primitive. The key sequence is formed by the product

$$k(n) = y(n-i)y(n-j), \qquad 1 \le i,j \le r$$

Show that $k(n)$ can be realized by a linear feedback shift register of length $r(r+1)/2$. The roots of the corresponding characteristic polynomial are of the form $1/\lambda_i\lambda_j$, where $1/\lambda_i$, $1/\lambda_j$ are roots of $h(q)$.

# 11

## STABILITY OF RECURRENT STRUCTURES

In a broad sense, stability analysis of dynamical systems is concerned with the effects of input and initial state perturbations on the resulting output signals. Stability definitions for input-output operators were introduced in Chapter 4. In this chapter recurrent structures are considered and methods for studying stability issues of linear and nonlinear dynamical systems are supplied. First, a reference input signal is kept fixed, and the effects of initial state perturbations on the state and output signals are analyzed. Questions that arise include: Do nearby initial states generate state and output trajectories that remain close? If a deviation from a state trajectory occurs at a certain time, does it eventually die out and the system returns to its former path of evolution, or does it sustain and perhaps magnify over time? The classical Liapunov stability concepts provide a formal description to such questions. Liapunov functions and linearization are powerful tools that enable us to obtain concrete answers in certain important circumstances for continuous and discrete time systems. Subsequently input as well as initial state perturbations are explored, and sufficient conditions for total stability are developed.

Despite the unquestionable success of the tools described in this chapter, there is much more in the qualitative behavior of dynamical systems that goes beyond the scope of this book. Even innocent looking nonlinear systems on the real line may exhibit extremely complex behavioral patterns. The logistic dynamics is used to navigate the reader toward the world of bifurcation and chaos where these phenomena are observed.

## 11.1 BASIC STABILITY CONCEPTS

We will begin the study of stability issues with continuous time state space representations

$$\dot{x}(t) = f(x(t), u(t), t) \qquad (11.1)$$

$$y(t) = g(x(t), u(t), t) \qquad (11.2)$$

The output equation describes a static behavior. Hence it is not expected to present major difficulties once the behavior of the state is understood. So we will concentrate on the state equation. A reference signal $u^*(t)$ is picked and kept fixed. Thus Eq. (11.1) can be viewed as a time varying differential equation

$$\dot{x}(t) = f(x(t), t) \qquad (11.3)$$

We assume that $f$ is continuously differentiable and that the solutions are uniquely defined on $R^k$ given initial conditions. Following the notation of Chapter 3, $\phi(t, t_0, x_0)$ denotes the solution of (11.3) spanned by the initial state $x_0$ at time $t_0$. Let $x^*(t)$ be a specific solution generated by initial data $x_0^*$ and $t_0$, that is, $x^*(t) = \phi(t, t_0, x_0^*)$. $x^*(t)$ is referred to as the *nominal state trajectory*.

The solution $x^*(t)$ is called *stable (in the sense of Liapunov)* if, for any $t_0$ in $R$ and for all $\epsilon > 0$, there exists $\delta > 0$ such that, for all $x_0$ satisfying $|x_0 - x_0^*| < \delta$, $|\phi(t, t_0, x_0) - x^*(t)| < \epsilon$ for $t \geq t_0$. The notion is illustrated in Fig. 11.1. It roughly states that all state trajectories produced by the nominal input stay arbitrarily close to the nominal state trajectory provided that initial states remain close to the nominal initial state. The nominal solution $x^*(t)$ is *unstable* if it is not stable; that is, there is $t_0$ and $\epsilon > 0$ such that for all $\delta > 0$ an initial state $x_0$ can be found so that $|x_0 - x_0^*| < \delta$ and $|\phi(t, t_0, x_0) - x^*(t)| > \epsilon$ for some $t > t_0$. Instability of $x^*(t)$ means that there is $\epsilon > 0$ so that there are initial states arbitrarily close to the nominal initial state that eventually give rise to deviations greater than $\epsilon$.

An important special case occurs when the nominal state is an equilibrium point, $x^*(t) = x^*$ for all $t$. Recall from Section 3.10.6 that this holds if $f(x^*, t) = 0$ for all $t$. According to the former definition, $x^*$ is a *stable equilibrium* if for any

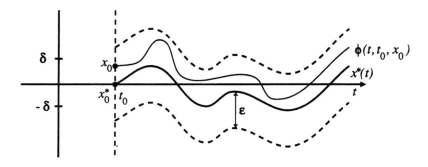

**Figure 11.1.** Illustration of stability of solutions.

initial time $t_0$ and any $\epsilon > 0$, there is $\delta > 0$ (depending on $\epsilon$ and $t_0$) such that for all $x_0$ satisfying $|x_0 - x^*| < \delta$ we have $|\phi(t, t_0, x_0) - x^*| < \epsilon$. $x^*$ is a *uniformly stable equilibrium* if $\delta$ does not depend on $t_0$. A uniformly stable equilibrium is also stable. If the system (11.3) is time invariant, that is, it has the form

$$\dot{x}(t) = f(x(t)) \tag{11.4}$$

the converse is also true due to shift invariance of solutions, namely

$$\phi(t, s_0, x_0) = \phi(t - s_0 + t_0, t_0, x_0)$$

Initial time is not important for time invariant systems. The flow in this case is characterized by the mapping $\phi(t, x) = \phi(t, 0, x)$. Thus $\phi(t, x)$ denotes the solution starting at time $t_0 = 0$ from $x$. In general, stability of an equilibrium point of a time varying system does not imply uniform stability.

Stability analysis of a solution $x^*(t)$ of (11.3) is reduced to stability of the zero equilibrium of the dynamical system defined by

$$\dot{z}(t) = h(z(t), t), \quad h(z, t) = f(z + x^*(t), t) - f(x^*(t), t) \tag{11.5}$$

Let $\psi(t, t_0, x_0)$ denote the flow of (11.5). The flows of (11.3) and (11.5) are related by

$$\psi(t, t_0, x_0) = \phi(t, t_0, x_0 + x_0^*) - x^*(t)$$

It is clear that 0 is a stable equilibrium state of (11.5) if and only if $x^*(t)$ is a stable trajectory of Eq. (11.3). Suppose, in particular, that Eq. (11.3) is affine in $x$:

$$\dot{x}(t) = A(t)x(t) + b(t) \tag{11.6}$$

Linear time varying state space representations are of this form. The dynamics specified by (11.5) become

$$h(z, t) = A(t)z(t) + A(t)x^*(t) + b(t) - A(t)x^*(t) - b(t) = A(t)z(t)$$

We conclude that if the zero solution of

$$\dot{x}(t) = A(t)x(t) \tag{11.7}$$

is stable, all solutions of Eq. (11.6) are stable. Conversely, if a particular solution of (11.6) is stable, then the zero solution of (11.7) is stable and hence all solutions of (11.6) are stable. Notice that conversion of (11.3) to (11.5) leads to time varying dynamics even if the original system is time invariant, unless the nominal trajectory is a critical point produced by a constant input.

A trajectory is asymptotically stable when small initial deviations produce small trajectory deviations which in addition eventually die out. Precise definitions follow.

The *region of attraction* of the equilibrium state $x^*$, also known as *basin of attraction*, consists of all states that are asymptotically attracted to $x^*$:

$$A(x^*) = \{x_0 \in R^k : \phi(t, t_0, x_0) \to x^*, t \to \infty, \text{ for all } t_0 \in R\}$$

Thus $x_0$ belongs to the region of attraction of $x^*$ if for all $t_0$ in $R$ and $\epsilon > 0$, there exists $t_1 > t_0$ such that $|\phi(t, t_0, x_0) - x^*| < \epsilon$ for all $t > t_1$. If $t_1$ does not depend on the initial time $t_0$ as in the time invariant case, $A(x^*)$ is called the *region of uniform attraction*. In general, the region of uniform attraction is properly contained in the region of attraction. $A(x^*)$ is nonempty since $x^* \in A(x^*)$. The size of $A(x^*)$ varies. There are systems for which $A(x^*)$ is just $x^*$. For some others the region of attraction extends to the entire space. This is illustrated next.

**Example 11.1   Region of attraction of certain planar linear systems**
Consider the linear system on the plane $\dot{x} = Ax$ with

$$A = \begin{pmatrix} \lambda & 0 \\ 0 & \mu \end{pmatrix} \tag{11.8}$$

The size of $A(0)$ depends on the sign of $\lambda$, $\mu$. In Fig. 11.2 (a) the region of attraction is just the origin, $A(0) = \{0\}$. In Fig. 11.2 (b) it is an one-dimensional subspace and coincides with the $x_1$ axis. In Fig. 11.2 (c), $A(0)$ covers the entire state space. In all cases the region of attraction is a linear space. This is a general characteristic of linear time varying systems and follows from the linearity of solutions with respect to initial states, $\phi(t, t_0, x_0) = \Phi(t, t_0)x_0$. $\Phi(t, t_0)$ designates the transition matrix (see Section 3.10.5). ∎

The equilibrium $x^*$ is an *attractor* if the region of attraction contains a neighborhood of $x^*$; that is, there is $\delta > 0$ such that each $x_0$ with $|x_0 - x^*| < \delta$ is contained in $A(x^*)$. The equilibrium state is a uniform attractor if the region of

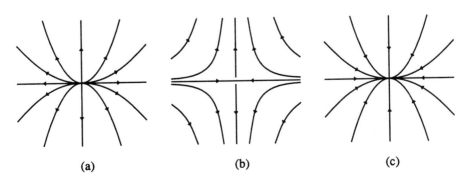

(a)                              (b)                              (c)

**Figure 11.2.** Solutions of the linear system in Example 11.1: (a) $\lambda > 0$, $\mu > 0$; (b) $\lambda < 0$, $\mu > 0$, (c) $\lambda < 0$, $\mu < 0$.

uniform attraction contains a neighborhood of $x^*$. Finally $x^*$ is (uniformly) asymptotically stable if it is (uniformly) stable and a (uniform) attractor.

The above definition is readily extended to an arbitrary state trajectory $x^*(t)$. We say that $x^*(t)$ is (uniformly) asymptotically (as.) stable if it is (uniformly) stable and there is $\delta > 0$ such that for any $t_0$ and $x_0$ satisfying $|x_0 - x_0^*| < \delta$ it holds that

$$\lim_{t \to \infty} |\phi(t, t_0, x_0) - x^*(t)| = 0$$

(uniformly in $t_0$).

*Remark.* Attraction and asymptotic stability ensure that the region of attraction is a thick set in the state space. In several practical situations, however, this may not be enough, and we may want to have an estimate of how big the region is. Let us consider the portrait of Fig. 11.3 (a). The system trajectories consist of (1) a critical point, (2) a circle of radius $R$, (3) spirals approaching the critical point in the interior of the circle, and (4) unbounded spirals moving away from the circle. The origin is as. stable. The region of attraction is the interior of the circle. Apparently, if $R$ is small, the interior of the circle becomes practically indistinguishable from its center. Thus, what we actually see is the diagram of Fig. 11.3 (b).

The case when the region of attraction covers the entire state space is important enough to deserve a special name. The equilibrium state $x^*$ is a *global attractor* if $\mathcal{A}(x^*) = R^k$. It is *globally (uniformly) asymptotically stable*, or *as. stable in the large*, if it is a (uniform) global attractor and (uniformly) stable.

The stability concepts introduced above are distinct. Fig. 11.4 (a) shows a

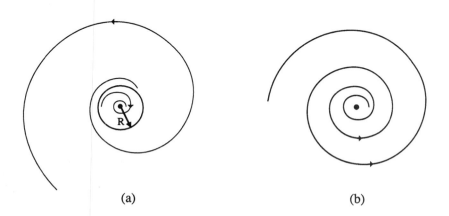

(a)                                      (b)

**Figure 11.3.** (a) As. stable system with the disc of radius $R$ as region of attraction. (b) Practically unstable system for small $R$.

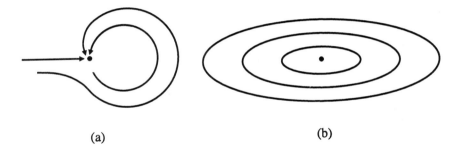

**Figure 11.4.** (a) Unstable global attractor; (b) uniformly stable equilibrium with $\mathcal{A}(x^*) = \{x^*\}$.

global attractor on the plane that is not stable. Fig. 11.4 (b) illustrates a system with periodic trajectories. The critical point is uniformly stable but not asymptotically stable.

## 11.2 STABILITY OF LINEAR TIME INVARIANT SYSTEMS

The stability behavior of linear time invariant systems

$$\dot{x}(t) = Ax(t), \qquad x \in R^k \tag{11.9}$$

is analyzed in this section. First, we complete the discussion on the asymptotic behavior of the matrix exponential signal started in Chapter 2. The representation of the matrix exponential in terms of polynomial exponential signals

$$e^{At} = \sum_{i=1}^{m} \sum_{j=1}^{p_i} \frac{t^{j-1}}{(j-1)!} e^{\lambda_i t} R_{ij} \tag{11.10}$$

derived in Section 8.4.4 forms the basis for subsequent developments. $\lambda_i$, $i = 1, 2, \ldots, m$, are the distinct eigenvalues of $A$ and $p_i$ the corresponding multiplicities. The asymptotic analysis of $e^{At}$ relies on the following lemma:

**Lemma 11.1** Given any eigenvalue $\lambda_i$ of $A$, at least one of the associated coefficients $R_{ij}, j = 1, 2, \ldots, p_i$, is nonzero.

PROOF    Suppose that, on the contrary, all coefficients $R_{ij}$ are zero. For simplicity we ignore the subscript $i$, and we write $\lambda$ for $\lambda_i$ and $R_j$ for $R_{ij}$. We consider the rational matrices

$$R_j(s) = \frac{1}{(p_i - j)!} \frac{d^{p_i - j}}{ds^{p_i - j}} \left[ (s - \lambda)^{p_i} (sI - A)^{-1} \right] = \frac{1}{(p_i - j)!} \frac{d^{p_i - j}}{ds^{p_i - j}} \frac{\text{adj}(sI - A)}{\chi(s)}$$

and $\chi(\lambda) \neq 0$. We set $M(s) = \text{adj}(sI - A)$. The partial fraction expansion formula of Section 8.4.3 implies that $R_j = R_j(\lambda)$. By assumption, $R_{p_i} = 0$. Therefore $\lambda$ is a root of $R_{p_i}(s)$ and hence a root of $M(s)$. For $j = p_i - 1$ we find that

$$R_{p_i-1}(s) = \frac{d}{ds}\frac{M(s)}{\chi(s)} = \frac{\dot{M}(s)\chi(s) - M(s)\dot{\chi}(s)}{\chi^2(s)}$$

Hence $R_{p_i-1} = 0 = R_{p_i-1}(\lambda) = \dot{M}(\lambda)/\chi(\lambda)$. It follows that $\dot{M}(\lambda) = 0$. Proceeding this way, we find that $M^{(j)}(\lambda) = 0, j = 0, \ldots, p_i - 1$. $\lambda$ is a root of every polynomial entry $M_{rl}(s)$ of $M(s)$ and of its derivatives of orders less than $p_i$. Thus $M_{rl}(s) = (s - \lambda)^{p_i}\tilde{M}_{rl}(s)$, and $M(s) = (s - \lambda)^{p_i}\tilde{M}(s)$. $\tilde{M}(s)$ is the polynomial matrix with entries $\tilde{M}_{rl}(s)$. Given a matrix $B$ of size $k$ and a constant $c$ it holds that $\det(cB) = c^k \det B$. Therefore $\det M(s) = (s - \lambda)^{p_i k} \det \tilde{M}(s)$, and $\lambda$ is a root of $\det M(s)$ of multiplicity at least $p_i k$. On the other hand, the identity $\text{adj}(sI - A) = \det(sI - A)(sI - A)^{-1}$ implies that $\det M(s) = (\det(sI - A))^{k-1}$. Thus the multiplicity of $\lambda$ is $(k - 1)p_i$, a contradiction.  ∎

**Theorem 11.1  Asymptotic behavior of $e^{At}$**  It holds that $\lim_{t\to\infty} e^{At} = 0$ if and only if all eigenvalues of $A$ have negative real part, $\Re\lambda(A) < 0$.

**PROOF**  Let $\lambda_i, i = 1, 2, \ldots, m$, be the distinct eigenvalues of $A$ and $p_i$ the corresponding multiplicities. If $\Re\lambda_i < 0$, $t^j e^{\lambda_i t}/j! \to 0$ as $t \to \infty$. The desired result follows from Eq. (11.10). To prove the converse, suppose that there is $\lambda_i$ such that $\Re\lambda_i > 0$. The previous lemma asserts that at least one of the coefficients $R_{ij}$ is nonzero. Therefore $e^{\lambda_i t}$ is present in (11.10) and prohibits $e^{At}$ from decaying to zero.  ∎

Next we state and prove the main stability result concerning linear autonomous systems.

**Theorem 11.2**  Consider the linear system (11.9). The following statements hold:

1. The origin is stable if and only if $e^{At}$ is bounded for all $t \geq 0$.
2. The origin is globally asymptotically stable if and only if each eigenvalue of $A$ has negative real part, $\Re\lambda(A) < 0$.
3. If there exists at least one eigenvalue of $A$ with positive real part, the origin is unstable.

**PROOF**

1. Assume that 0 is stable. Take $\epsilon = 1$. There is $\delta > 0$ such that for $|y| < \delta$, we have $|e^{At}y| < 1$ for all $t \geq 0$. A matrix-induced norm is employed. Let $x$ be a nonzero vector, $r > |x|/\delta$ and $x = ry$. Then $|y| < \delta$ and $|e^{At}x| = r|e^{At}y| < r, t \geq 0$. Since $e^{At}x$ is bounded for any $x$, $e^{At}$ is also

bounded. Conversely, let $M > 0$ such that $|e^{At}| < M$. Let $\epsilon > 0$ and $\delta = \epsilon/M$. Then, if $|x| < \delta$, $|e^{At}x| \le |e^{At}||x| \le M\delta = \epsilon$, and the origin is stable.

2. If all eigenvalues of $A$ have negative real part, Theorem 11.1 implies that the origin is a global attractor and the solutions are bounded. Therefore 0 is asymptotically stable by statement 1. The converse follows from Theorem 11.1.

3. The statement is established in a similar way.    ∎

The converse of statement 3 is not true. Instability of the origin does not necessarily imply that one eigenvalue has positive real part. Take, for instance, $A = 0$.

We illustrate the assertions of the previous theorem by looking at Eq. (11.9) on the line. In this case $A$ is a scalar. If $A < 0$, the origin is asymptotically stable, and if $A > 0$, it is unstable. Finally, if $A = 0$, all states are equilibria. In the latter case the origin is stable but not as. stable. The region of attraction of 0 is $\{0\}$. In higher dimensions more interesting possibilities occur.

***Example 11.2    Linear oscillator***
Consider the linear oscillator

$$\ddot{x} + \rho\dot{x} + kx = 0, \qquad k > 0, \; \rho \ge 0$$

We set $x_1 = x$, $x_2 = \dot{x}$. Then the above equation is written in state space form as

$$\begin{pmatrix} \dot{x}_1 \\ \dot{x}_2 \end{pmatrix} = \begin{pmatrix} 0 & 1 \\ -k & -\rho \end{pmatrix} \begin{pmatrix} x_1 \\ x_2 \end{pmatrix}$$

The characteristic polynomial is $s^2 + \rho s + k$. Suppose that the oscillator represents a mechanical system with displacement $x_1$, velocity $x_2$ and friction

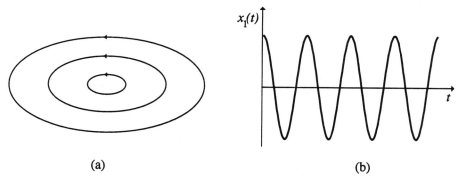

(a)    (b)

**Figure 11.5.** (a) Solution curves of undamped linear oscillator. (b) Displacement of undamped linear oscillator.

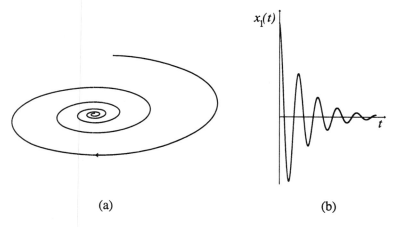

**Figure 11.6.** (a) Solution curves of linear oscillator for $0 < \rho < 2k^{1/2}$. (b) Displacement.

**Figure 11.7.** Rapid oscillations of linear oscillator for $\rho \rightarrow 2k^{1/2}$.

coefficient $\rho$. The following possibilities arise:

1. $\rho = 0$. $A$ has purely imaginary eigenvalues. The solution curves are illustrated in Fig. 11.5 (a). Displacement as a function of time is depicted in Fig. 11.5 (b).

2. $0 < \rho < 2\sqrt{k}$. The eigenvalues of $A$ are complex with negative real part. The state trajectories are sketched in Fig. 11.6 (a). The displacement is shown in Fig. 11.6 (b). In Fig. 11.6 the friction is small and the motion consists of slow oscillations. As $\rho$ increases toward $2\sqrt{k}$, oscillations decay faster and the state trajectories take the form of Fig. 11.7.

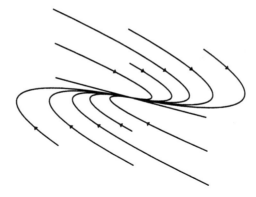

**Figure 11.8.** Solution curves of linear oscillator for $\rho = 2k^{1/2}$.

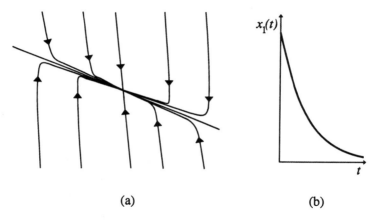

(a)                                        (b)

**Figure 11.9.** (a) Solution curves of linear oscillator with large friction; (b) displacement in the case of large friction.

3. $\rho = 2\sqrt{k}$. $A$ has equal eigenvalues and the solutions are given by

$$x(t) = e^{-\sqrt{k}t}r_1 + te^{-\sqrt{k}t}r_2$$

The state trajectories are sketched in Fig. 11.8.

4. $\rho > 2\sqrt{k}$. $A$ has two real distinct negative eigenvalues. The solutions behave essentially as in Fig. 11.9 (a) and the displacement is depicted in Fig. 11.9 (b). ∎

## 11.3 LIAPUNOV FUNCTIONS

Liapunov functions constitute a powerful tool in the qualitative theory of dynamical systems. They were introduced by Liapunov at the end of the nineteenth century as a means to study stability without prior knowledge of solutions. The basic idea of the method can be traced to a result stated by Lagrange around 1800 and later proved by Dirichlet. If the potential energy of a conservative mechanical system is minimum at the equilibrium state $x^*$, $x^*$ is stable. Liapunov functions are generalizations of the energy function. To gain insight into the nature of these functions and to their role in stability, we sketch the proof of Lagrange conjecture.

### Example 11.3   Scalar conservative system

Let us consider a particle of mass $m$ moving on a straight line under the action of a force $g(x)$. The motion is conservative; that is, the only acting force $g$ depends on only the displacement. The equation of motion is $m\ddot{y} = -g(y)$. The usual state space assignment $x_1 = y$, $x_2 = \dot{y}$ converts the latter equation into the state space format

$$\dot{x}_1 = x_2, \quad \dot{x}_2 = -\frac{1}{m}g(x_1) \tag{11.11}$$

The potential energy $U$ and the kinetic energy $T$ are

$$U(x_1) = \int_0^{x_1} g(z)dz, \quad T(x_2) = \frac{1}{2}mx_2^2$$

Hence the total energy is

$$V(x_1, x_2) = \int_0^{x_1} g(z)dz + \frac{1}{2}mx_2^2 \tag{11.12}$$

We assume that $g(0) = 0$. Then $(0,0)$ is an equilibrium point. We further assume that $x_1 g(x_1) > 0$ so that $U(x_1) > 0$ for $x_1 \neq 0$. Thus the total energy becomes minimum at the origin. We will demonstrate that $(0,0)$ is stable. The conservation of energy asserts that the total energy remains constant along a motion. To see this, let $x(t) = [x_1(t) \ x_2(t)]^T$ denote a solution of (11.11). Notice that

$$\nabla V(x_1, x_2) = \left( \frac{\partial V}{\partial x_1} \quad \frac{\partial V}{\partial x_2} \right) = (g(x_1) \quad mx_2)$$

and

$$\dot{x}(t) = \left( x_2(t) \quad -\frac{1}{m}g(x_1(t)) \right)^T$$

We compute the rate of change of energy along the motion

$$\frac{d}{dt}V(x(t)) = \nabla V(x(t))\frac{d}{dt}x(t) = 0$$

Therefore the total energy $V$ remains constant on $x(t)$. Let us now consider the family of curves $V(x_1, x_2) = c$, $c > 0$. Since $V$ is minimum at 0, these curves contain the origin in their interior. Given $\epsilon > 0$, we pick $c$ so that the corresponding curve and its interior lie inside the $\epsilon$ neighborhood of the origin. Then, if we start inside the interior of this curve, we stay there forever because the energy remains constant.    ∎

Motivated by the preceding discussion we shall next provide an intuitive demonstration of Liapunov method. Consider the time invariant system

$$\dot{x}(t) = f(x(t)), \qquad x \in R^k \tag{11.13}$$

and an equilibrium state $x^*$. The notion of Liapunov function extends the role of the energy function as follows: It is a real-valued function possessing the properties

1. $V(x^*) = 0$, and $V(x) > 0$ for $x \neq x^*$.
2. $\nabla V(x) f(x) = \sum_{i=1}^{k} (\partial V(x)/\partial x_i) f_i(x) \leq 0$ for any $x$.

The first property extends minimality of the potential energy at the rest position. The second property generalizes the conservation of energy. Indeed, if $\phi(t, x)$ denotes the flow of (11.13), we have

$$\dot{V} = \frac{d}{dt} V(\phi(t, x)) = \nabla V(\phi(t, x)) \frac{d}{dt} \phi(t, x) = \nabla V(\phi(t, x)) f(\phi(t, x)) \leq 0.$$

This means that $V$ is decreasing along the trajectories of the system. For stability purposes the condition $\dot{V} = 0$ can be relaxed to the weaker condition $\dot{V} \leq 0$. The theorem stated and proved by Liapunov asserts that if a function satisfying properties 1 and 2 can be found, then $x^*$ is stable. Indeed, consider the family of curves

$$\{x \in R^k : V(x) = c\}, \qquad c > 0$$

Because of property 2 these curves pictorially take the form shown in Fig. 11.10. Pick a point $x$ on the curve $V(x) = c$. The gradient $\nabla V(x)$ at $x$ is normal to this curve, that is, it is orthogonal to the tangent space of the curve at $x$. Property 2 implies that the angle between $f(x)$ and $\nabla V(x)$ is greater than 90 degrees. Hence $f(x)$ points inward the curve. On the other hand, it is tangent to the solution of (11.13) passing from $x$. Thus it forces this solution toward the interior of the curve. This is sufficient to guarantee stability. It turns out that if property 2 is replaced by strict inequality, trajectories are directed toward the critical point, which becomes asymptotically stable. The above statements are made precise next.

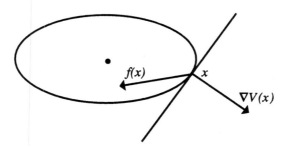

**Figure 11.10.** Illustration of Liapunov method.

### 11.3.1 Definition of Liapunov functions

Let $x^*$ be a point in $R^k$ and $V$ a smooth real valued function defined on an open neighborhood $S$ containing $x^*$. This means that the partial derivatives of $V$ exist and are continuous on $S$. We say $V$ is *nonnegative definite* or *positive semidefinite at* $x^*$ if $V(x^*) = 0$ and $V(x) \geq 0$ for any $x$ belonging in a sphere with center $x^*$ and contained in $S$. $V$ is called *positive definite* at $x^*$ if $V(x^*) = 0$ and $V(x) > 0$ for all $x \neq x^*$. Likewise $V$ is nonpositive definite at $x^*$ if $-V$ is nonnegative definite, that is, $V(x^*) = 0$ and $V(x) \leq 0$. Finally, $V$ is negative definite if $-V$ is positive definite.

### *Example 11.4  Nonnegative definite functions*
Consider the origin in $R^3$ and the function $V(x) = x_1^2 + x_3^2$. Clearly $V(0) = 0$ and $V(x) \geq 0$ for all $x$ in $R^3$. Hence $V$ is nonnegative definite at 0. It is not positive definite because it vanishes on the entire $x_2$ axis. The function

$$V(x) = x_1^2 + x_2^2 + x_3^2$$

is positive definite at the origin of $R^3$.

The above examples are special cases of quadratic forms on $R^k$, that is, functions of the form

$$V(x) = \sum_{i=1}^{k} \sum_{j=1}^{k} p_{ij} x_i x_j, \qquad V(x) = x^T P x \qquad (11.14)$$

It follows from the definitions that $V$ is nonnegative definite if and only if the square matrix $P$ is nonnegative definite. Likewise $V$ is positive definite if $P$ is a positive definite matrix. In the sequel, $P$ is always symmetric as well.  ∎

*Remark.* If $V$ is positive semidefinite at $x^*$, then $x^*$ is a local minimum of $V$. It becomes a strict minimum if $V$ is positive definite. Likewise, if $V$ is negative semidefinite at $x^*$, $x^*$ is a local maximum.

In the study of global asymptotic stability, we will use the concept of radial unboundedness. A positive definite function $V$ is called *radially unbounded* if $V(x) \to \infty$, $|x| \to \infty$. The function $V(x) = x^2/(x^2 + 3)$ is positive definite but not radially unbounded because $V(x) \to 1$ as $|x| \to \infty$. On the other hand, the positive definite quadratic form (11.14) is radially unbounded.

Let us next tie the above concepts to dynamical systems. Consider (11.13). Let $x^*$ denote a critical point and $V$ a smooth real-valued function defined on an open set $S$ containing $x^*$. The *derivative of $V$ along $f$* is the real-valued function defined on $S$:

$$\dot{V}(x) = \nabla Vf(x) = \sum_{i=1}^{k} \frac{\partial V}{\partial x_i} f_i(x)$$

***Example 11.5    Linear systems and quadratic forms***
Consider the linear system $\dot{x} = Ax$ and the quadratic form (11.14). The derivative of $V$ is

$$\dot{V}(x) = 2x^T PAx = x^T PAx + (x^T PAx)^T = x^T (PA + A^T P)x \qquad (11.15)$$

We conclude that $\dot{V}(x)$ is a quadratic form with associated matrix $PA + A^T P$.
∎

Now we are ready to state our main definitions.

**Definition.** A function $V$ as above is called a *Liapunov function* for the system (11.13) at $x^*$ if $V$ is positive definite at $x^*$, and its derivative $\dot{V}$ is negative semidefinite at $x^*$. If in addition $\dot{V}$ is negative definite, $V$ is called a *strict Liapunov function*. If $V$ is a strict Liapunov function over the entire space and is also radially unbounded, it is called *global Liapunov function*.

***Example 11.6    Scalar conservative system***
Consider the conservative system (11.11). The energy (11.12) is positive definite. The conservation of energy shows that $\dot{V} = 0$. Hence $V$ is a Liapunov function. Clearly it is not a strict Liapunov function.
∎

***Example 11.7    Constant resistance circuit***
Consider the constant resistance circuit described in Example 3.25. The circuit dynamics are

$$\dot{x} = \begin{pmatrix} \dfrac{-1}{R_1 C} & 0 \\ 0 & \dfrac{-R_2}{L} \end{pmatrix} x + \begin{pmatrix} \dfrac{1}{R_1 C} \\ \dfrac{1}{L} \end{pmatrix} u$$

The electric energy stored in the capacitor is $Cx_1^2/2$, and the magnetic energy stored in the inductor is $Lx_2^2/2$. The total energy stored is $V(x_1, x_2) = (Cx_1^2 + Lx_2^2)/2$. This is a quadratic form as in Eq. (11.14) where

$$P = \begin{pmatrix} C/2 & 0 \\ 0 & L/2 \end{pmatrix}$$

Since $L > 0$, $C > 0$, $P$ is positive definite. Suppose that the input source is switched off. The circuit equations take the form $\dot{x} = Ax$. According to Example 11.5, $\dot{V}$ is a quadratic form with associated matrix

$$PA + A^T P = \begin{pmatrix} \dfrac{-1}{R_1} & 0 \\ 0 & -R_2 \end{pmatrix}$$

This is a negative definite matrix. Hence $V$ is a strict Liapunov function over the entire space. Moreover it is radially unbounded. Hence it becomes a global Liapunov function. ∎

### 11.3.2  Stability and Liapunov functions

Consider the time invariant dynamical system

$$\dot{x}(t) = f(x(t)), \qquad x \in R^k \tag{11.16}$$

and an equilibrium state $x^*$. The following theorems are proved in the Appendix at the end of the chapter.

**Theorem 11.3  Stability**
 1. If there exists a Liapunov function at $x^*$, $x^*$ is stable.
 2. If $x^*$ has a strict Liapunov function, it is asymptotically stable.
 3. *LaSalle criterion.* Let $V$ be a Liapunov function at $x^*$ such that the set

$$\{x \in R^k : \dot{V}(x) = 0\}$$

contains no other positive trajectory $\phi(t, x)$, $t \geq 0$, besides $x^*$. Then $x^*$ is asymptotically stable.
 4. If there is a global Liapunov function, $x^*$ is globally asymptotically stable. Alternatively, if there is a radially unbounded Liapunov function defined on $R^k$ such that the set $\{x \in R^k : \dot{V}(x) = 0\}$ contains no positive trajectories besides $x^*$, $x^*$ is globally asymptotically stable.

Liapunov-type functions enable us to describe instability too.

**Theorem 11.4 Instability**   The critical point $x^*$ is unstable if there is a smooth real valued function $V$ defined on a neighborhood $Q$ of $x^*$, such that

1. $V(x^*) = 0$ and for all $\epsilon > 0$ there is $x$, $|x - x^*| \leq \epsilon$ and $V(x) < 0$ and either
2a. $\dot{V}(x)$ is negative definite or
2b. there is $a > 0$ and $V_1(x) \leq 0$ such that

$$\dot{V}(x) = aV(x) + V_1(x) \tag{11.17}$$

The following examples serve to illustrate the use of Liapunov functions.

***Example 11.8    Asserting stability by Liapunov functions***
Consider on $R^3$ the system

$$\dot{x}_1 = 2x_2(x_3 - 1)$$
$$\dot{x}_2 = - x_1(x_3 - 1)$$
$$\dot{x}_3 = - x_3^3$$

The origin $(0, 0, 0)$ is the only critical point. We take as Liapunov function candidate the quadratic form

$$V(x_1, x_2, x_3) = a_1 x_1^2 + a_2 x_2^2 + a_3 x_3^2, \qquad a_1, a_2, a_3 > 0.$$

$V$ is clearly positive definite. The derivative is

$$\dot{V} = 4a_1 x_1 x_2(x_3 - 1) - 2a_2 x_2 x_1(x_3 - 1) - 2a_3 x_3^4$$

If we pick $2a_1 = a_2$, we get $\dot{V} = -2a_3 x_3^4 \leq 0$. Hence $V$ is a Liapunov function and the origin is stable by Theorem 11.3.   ∎

***Example 11.9    Damped oscillator***
Consider the damped oscillator

$$\dot{x}_1 = x_2$$
$$\dot{x}_2 = - g(x_1) - x_2, \qquad x_1 g(x_1) > 0, \; x_1 \neq 0, \; g(0) = 0 \tag{11.18}$$

The origin is the only equilibrium and the total energy is positive definite. The derivative of the energy is

$$\dot{V} = g(x_1)x_2 + x_2\left(-g(x_1) - x_2\right) = -x_2^2 \tag{11.19}$$

Since $\dot{V} \leq 0$, stability is guaranteed. $\dot{V}$ is not negative definite and thus fails to be a strict Liapunov function. To deduce asymptotic stability, we will apply the

LaSalle criterion. Let

$$M = \{x \in R^2 : \dot{V}(x) = 0\} \tag{11.20}$$

$x \in M$, and let $x(t)$ be the solution of the damped oscillator emanating from $x$. Suppose that $x(t)$ is contained in $M$ for $t \geq 0$. Equation (11.19) yields $x_2(t) = 0$ for all $t \geq 0$. Likewise Eq. (11.18) gives $g(x_1(t)) = 0$. Since $x_1 g(x_1) > 0$ for all $x_1 \neq 0$, we get $x_1(t) = 0$. Therefore $x = (0, 0)$, and $M$ collapses to the origin. Hence the origin is asymptotically stable. ∎

### Example 11.10   Instability
Consider the system

$$\dot{x}_1 = -x_1 + x_2$$
$$\dot{x}_2 = x_1 + x_2 + x_2^3$$

The origin is a critical point. Let $V(x_1, x_2) = x_1^2 - x_2^2$. The derivative of $V$ is

$$\dot{V} = -2(x_1^2 + x_2^2 + x_2^4)$$

Thus $\dot{V}$ is negative definite. Moreover $V$ is negative on the sequence of points $(0, 1/n)$ which converges to zero. Invoking Theorem 11.4, we conclude that the origin is unstable. ∎

### Example 11.11   Global asymptotic stability
Consider the differential equation on the plane

$$\dot{x}_1 = x_2 - \lambda x_1(x_1^2 + x_2^2)$$
$$\dot{x}_2 = -x_1 - \lambda x_2(x_1^2 + x_2^2), \qquad \lambda > 0$$

It is easy to see that the only critical point is the origin. The function

$$V(x_1, x_2) = \alpha_1 x_1^2 + \alpha_2 x_2^2, \qquad \alpha_1, \alpha_2 > 0$$

is positive definite at the origin. We next compute the derivative of $V$

$$\dot{V} = 2\alpha_1 x_1\left(x_2 - \lambda x_1(x_1^2 + x_2^2)\right) + 2\alpha_2 x_2\left(-x_1 - \lambda x_2(x_1^2 + x_2^2)\right)$$

To get rid of the product $x_1 x_2$, we choose $\alpha_1 = \alpha_2$. Then

$$\dot{V} = -2\alpha_1 \lambda(x_1^2 + x_2^2)^2$$

Therefore $V$ is a strict Liapunov function. Moreover it is radially unbounded. Consequently the origin is globally asymptotically stable. ∎

### 11.3.3  Gradient Flows

Gradient flows form a class of dynamical systems for which much can be said about their qualitative behavior. A gradient flow is given by

$$\dot{x} = -[\nabla V(x)]^T \tag{11.21}$$

where $V$ is a smooth real-valued function. An equilibrium point $x^*$ is a stationary point of $V$, $\nabla V(x^*) = 0$. Let $x^*$ be an isolated stationary point of $V$. Then there is a neighborhood of $x^*$ where no other stationary point of $V$ is found. If $x^*$ is a local minimum of $V$, it is asymptotically stable. Indeed, the function $V_1(x) = V(x) - V(x^*)$ is positive definite at $x^*$. Moreover

$$\dot{V}_1(x) = -\nabla V(x)[\nabla V(x)]^T = -|\nabla V(x)|^2 \leq 0$$

Furthermore $\dot{V}_1(x)$ is negative definite; otherwise, $x^*$ would not be an isolated stationary point.

If $x^*$ is not a local minimum, $V_1(x)$ satisfies condition 1 of Theorem 11.4, and $\dot{V}_1(x)$ is negative definite. Hence $x^*$ is unstable. In conclusion, an isolated stationary point of $V$ is asymptotically stable if it is a local minimum of $V$ and an unstable point if it is not a local minimum.    ■

### 11.3.4  Analog Hopfield Neural Networks

Analog Hopfield neural networks are dynamical systems that were proposed by Hopfield as content addressable memory devices. In contrast to address addressable memory where recall of a pattern requires a complete correct address, associative memory recalls information on the basis of partial or noisy information of the item. The analog Hopfield neural network is a dynamical system with isolated asymptotically stable equilibria. Associative memory is achieved as follows: The stored items are encoded onto the asymptotically stable equilibria. They are referred to as fundamental memories or prototype states. A piece of information $x_0$ is submitted to the network as initial state. $x_0$ conveys incomplete or noisy information about a prototype state $x^*$. If $x_0$ is sufficiently close to $x^*$, it belongs to the region of attraction of $x^*$ and will be asymptotically attracted to $x^*$. Thus presenting the partial information item $x_0$ suffices to retrieve the correct memory item $x^*$.

Dynamical systems employed as content addressable memories must enjoy the following features: They must provide for sufficiently large number of isolated equilibria states at affordable computational complexity. Furthermore, they must control the extent of the region of attraction. Roughly speaking, the larger the region of attraction is, the smaller the amount of information required from the input pattern.

Hopfield networks are formed as interconnections of neurons whose physical structure is depicted in Fig. 11.11. The output of each neuron represents

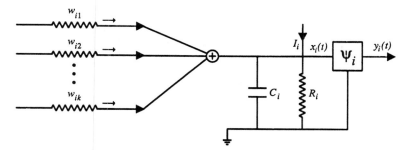

**Figure 11.11.** Physical description of the Hopfield neuron.

potential. It is formed at the output of a sigmoidal nonlinearity

$$y_i(t) = \psi_i(x_i(t)) \tag{11.22}$$

The current flowing toward the input of the nonlinear element is formed by the sum of the currents transmitted from the other neurons and the current of an external source representing a bias term $\sum_{j=1}^{k} w_{ij} y_j(t) + I_i$. The synaptic weights are conductances. The nonlinear device has an input resistance $R_i$ and a leakage capacitance $C_i$. The $RC$ characteristics are related by the Kirchhoff's current law:

$$C_i \frac{dx_i(t)}{dt} = -\frac{1}{R_i} x_i(t) + \sum_{j=1}^{k} w_{ij} y_j(t) + I_i \tag{11.23}$$

Equations (11.22) and (11.23) are written in matrix form as

$$C\dot{x} = Wy + I - R^{-1}x \tag{11.24}$$

$$y = \psi(x) \tag{11.25}$$

$C$ and $R$ are diagonal matrices with diagonal entries $C_i$ and $R_i$, respectively; $I$ is the vector of source currents, $W$ is the matrix of synaptic weights and is assumed to be symmetric; the vector $x$ is the input to all sigmoid nonlinearities and $y$ is the output vector. Finally, $\psi(x) = (\psi_1(x_1) \ \psi_2(x_2) \ \dots \ \psi_k(x_k))$. We will further assume that each real-valued function $\psi_j$ is monotone increasing so that both $\psi_j$ and its inverse function $\psi_j^{-1}$ are differentiable and $\psi_j(0) = 0$, $\psi_j(x) = -\psi_j(-x)$. Examples are the logistic function, the arctangent, and the hyperbolic function. Equations (11.24) and (11.25) define a gradient-like system. Indeed, let us introduce the function

$$V(y) = -\frac{1}{2} \sum_{i=1}^{k} \sum_{j=1}^{k} w_{ij} y_i y_j - \sum_{i=1}^{k} I_i y_i + \sum_{i=1}^{k} R_i^{-1} \int_0^{y_i} \psi_i^{-1}(z) dz \tag{11.26}$$

Then

$$\frac{\partial V(y)}{\partial y_i} = -\sum_{j=1}^{k} w_{ij}y_j - I_i + R_i^{-1}\psi_i^{-1}(y_i)$$

which, modulo the minus sign, coincides with the $i$th entry of the right-hand side of Eq. (11.24). To translate Eq. (11.26) into a matrix format, we set

$$r^T = \text{diag}(R^{-1}) = \left( R_1^{-1} \cdots R_k^{-1} \right)$$

$$g(y) = (g(y_1) \cdots g(y_k))^T, \quad g(y_i) = \int_0^{y_i} \psi_i^{-1}(z)dz$$

Then

$$V(y) = -\frac{1}{2}y^T Wy - I^T y + r^T g(y) \tag{11.27}$$

If we differentiate Eq.(11.25) and take into account Eq. (11.24), we obtain

$$\dot{y} = -C(y)[\nabla V(y)]^T, \quad C(y) = \left[\frac{\partial\psi}{\partial x}C^{-1}\right]\Big|_{x=\psi^{-1}(y)} \tag{11.28}$$

The matrix $C(y)$ is positive definite. Indeed, $\partial\psi/\partial x$ is diagonal with positive diagonal entries because each $\psi_i$ is increasing. Moreover $C$ is diagonal with positive diagonal entries.

The equilibria points of the Hopfield network are the stationary points of $V(y)$. Taking into account Eq. (11.27), these stationary points satisfy the equation

$$RWy + RI = \psi^{-1}(y), \quad \psi^{-1}(y) = (\psi^{-1}(y_1)\cdots\psi^{-1}(y_k)) \tag{11.29}$$

A stationary point $y^*$ is minimum if the matrix of second-order derivatives of $V$ is positive definite:

$$-W + R\frac{\partial\psi^{-1}(y)}{\partial y}\Big|_{y*} > 0 \tag{11.30}$$

Aside the factor $C(y)$, Eq. (11.28) is a gradient system. It shares all stability features of gradient systems because $C(y)$ is positive definite. The following theorem summarizes the properties of the Hopfield network:

**Theorem 11.5** Consider the Hopfield network as described by Eqs. (11.27) and (11.28). Each isolated local minimum of $V(y)$ is an asymptotically stable equilibrium.

**PROOF**   We proceed exactly as in the case of gradient flows. The function $V_1(y) = V(y) - V(y^*)$ is a strict Liapunov function at $y^*$ because $V_1(y) > 0$ and

$$\dot{V}_1 = \left[ -C(y)[\nabla V(y)]^T \right]^T [\nabla V(y)]^T = -\nabla V(y) C(y) [\nabla V(y)]^T < 0$$

We took into account that $C(y) > 0$ and $\nabla V(y) \neq 0$ for $y \neq y^*$. The latter holds because $y^*$ is an isolated local minimum.  ∎

Suppose that we want to design a Hopfield network with stored patterns the given vectors $v_1, \ldots, v_M$. We compute the vectors $u_i = \psi^{-1}(v_i)$. Each $v_i$ satisfies Eq. (11.29):

$$RWv_i + RI = u_i, \qquad 1 \le i \le M \tag{11.31}$$

This is a set of linear equations for the parameters $RW$ and $RI$. A solution is determined and subsequently factored into $RW$ and $RI$ accordingly, with $R$ diagonal positive and $W$ symmetric. The idea is illustrated by an example. Clearly at least $k + 1$ stored patterns are needed to cater for the unknown parameters $RW$ and $RI$ in (11.31).

***Example 11.12***
We wish to store the patterns $v_1 = 1/2$, $v_2 = 1/4$. We pick $k = 1$ and $\psi(x) = 1/(1 + e^{-x})$. Then $\psi^{-1}(y) = \log(y/(1 - y))$. Let $G = RW$ and $h = RI$. Equation (11.31) becomes

$$\frac{G}{2} + h = 0, \qquad \frac{G}{4} + h = -\log 3$$

Hence $G = 4\log 3$, $h = -2\log 3$. A Hopfield network with parameters $R = \log 3$, $W = 4$ and $I = -2$ will store the given patterns. Note that $3/4$ is an additional (spurious) equilibrium point. It is easy to check that the above equilibria points satisfy (11.30) and hence are asymptotically stable.  ∎

The above analysis indicates that the Hopfield network can operate as associative memory unit. Systematic design procedures that determine a Hopfield architecture from a given set of stored items have been developed. One efficient method included in the **MATLAB** neural networks toolbox relies on a detailed investigation of the system evolution inside a cube. In practice, several issues need thorough consideration. One major concern is that the Hopfield network and its variants give rise to additional equilibria, besides the desired stored patterns. Furthermore the number of stored patterns requires a significantly higher number of neurons.

### 11.3.5 Linear time invariant systems and Liapunov equations

In this subsection we apply the method of Liapunov functions to linear time invariant systems:

$$\dot{x} = Ax, \qquad x \in R^k \tag{11.32}$$

The theorem that follows complements Theorem 11.2.

**Theorem 11.6** Consider (11.32), and suppose that there exists a positive definite matrix $Q$ such that the so-called *Liapunov equation*

$$PA + A^T P = -Q \tag{11.33}$$

admits a positive definite solution $P$. Then the origin is globally asymptotically stable. Conversely, if the origin is asymptotically stable, for any positive definite matrix $Q$, Eq. (11.33) has a unique positive definite solution $P$, which is analytically expressed by the integral

$$P = \int_0^\infty e^{A^T t} Q e^{At} dt \tag{11.34}$$

The proof is given in the Appendix.                                     ■

*Example 11.13   Linear damped oscillator*
Consider the linear damped oscillator in state space format

$$\begin{bmatrix} \dot{x}_1 \\ \dot{x}_2 \end{bmatrix} = \begin{bmatrix} 0 & 1 \\ -1 & -k \end{bmatrix} \begin{bmatrix} x_1 \\ x_2 \end{bmatrix}, \qquad k > 0 \tag{11.35}$$

We explore the stability behavior of the origin by looking at the Liapunov equation. We set $Q = I$, for a lack of a better choice. Then we have

$$\begin{pmatrix} p_{11} & p_{12} \\ p_{12} & p_{22} \end{pmatrix} \begin{pmatrix} 0 & 1 \\ -1 & -k \end{pmatrix} + \begin{pmatrix} 0 & -1 \\ 1 & -k \end{pmatrix} \begin{pmatrix} p_{11} & p_{12} \\ p_{12} & p_{22} \end{pmatrix} = \begin{pmatrix} -1 & 0 \\ 0 & -1 \end{pmatrix}$$

Performing the matrix multiplications and equating corresponding entries, we obtain

$$-2p_{12} = -1$$
$$p_{11} - kp_{12} - p_{22} = 0$$
$$2(p_{12} - kp_{22}) = -1$$

This is a linear system of equations. The unique solution is given by

$p_{11} = k/2 + 1/k$, $p_{12} = 1/2$, $p_{22} = 1/k$. Thus

$$P = \begin{pmatrix} \dfrac{k}{2} + \dfrac{1}{k} & \dfrac{1}{2} \\ \dfrac{1}{2} & \dfrac{1}{k} \end{pmatrix}$$

Finally, $P > 0$ because the principal determinants $k/2 + 1/k$, $1/k^2 + 1/4$ are positive. Thus the damped oscillator is globally asymptotically stable. ∎

The above theorem relies on strict Liapunov functions. We deduce a weaker variant if we make use of LaSalle's test. In doing so, observability enters in.

**Theorem 11.7** Consider the linear system (11.32). Let $C$ be a row vector such that the pair $(A, C)$ is observable. If the Liapunov equation

$$PA + A^T P = -C^T C \tag{11.36}$$

has a positive definite solution $P$, the system (11.32) is globally as. stable.

**PROOF** Let $V(x) = x^T P x$. Then $\dot{V}(x) = -x^T C^T C x$. Let $z(t)$ be a solution of (11.32) starting from $z$. Suppose that $z(t)$ is contained in

$$\{x \in R^k : \dot{V}(x) = 0\} = \{x \in R^k : x^T C^T C x = 0\} = \{x \in R^k : Cx = 0\}$$

Then $Cz(t) = 0$ for all $t \geq 0$. Successive differentiation and evaluation at the origin give $CA^n z = 0$ for all $n \in Z^+$. Observability implies that $z = 0$, and LaSalle's criterion proves the assertion. ∎

## 11.4  LINEARIZATION

In Section 11.2 we derived some important conclusions regarding the qualitative behavior of linear time invariant systems. The closed form expresssion of the matrix exponential signal played a decisive role in this respect. Now let us consider the nonlinear system

$$\dot{x} = f(x) \tag{11.37}$$

and the equilibrium point $x^*$. Lack of closed form expressions makes the stability analysis of the above system much harder. In this section we infer useful statements about the behavior of (11.37) from the linear dynamical system associated with the first-order Taylor approximation of $f$ at $x^*$. Consider the linearized equation of (11.37) at $x^*$:

$$\dot{x} = Ax, \quad A = \left.\frac{\partial f}{\partial x}\right|_{x^*} \qquad (11.38)$$

The following theorem is commonly referred to as the *linearization theorem* or *small signal analysis* in circuit theory:

**Theorem 11.8 Linearization**    Consider the nonlinear system (11.37), the equilibrium $x^*$ and the linearized equation (11.38). The following statements hold:

1. If 0 is as. stable for (11.38), that is, all eigenvalues of $A$ are in the left half plane $\Re\lambda(A) < 0$, then $x^*$ is asymptotically stable for the original nonlinear system (11.37).
2. If there is at least one eigenvalue of $A$ in the right half plane $\Re\lambda(A) > 0$, then $x^*$ is unstable for the nonlinear system (11.37).

The proof is supplied in the Appendix.                                            ∎

***Example 11.14    Linearization of a gradient flow***
Consider the gradient flow (11.21) where

$$V(x_1, x_2) = (x_1 - 2)^2(x_1 - 1)^2 + 3x_2^2$$

Then

$$\dot{x}_1 = -\frac{\partial V}{\partial x_1} = -2(x_1 - 2)(x_1 - 1)(2x_1 - 3)$$

$$\dot{x}_2 = -\frac{\partial V}{\partial x_2} = -6x_2 \qquad (11.39)$$

The critical points are

$$a = (1, 0), \quad b = \left(\frac{3}{2}, 0\right), \quad c = (2, 0)$$

If the Jacobian of (11.39) is evaluated at each critical point, it gives

$$A(a) = \begin{pmatrix} -2 & 0 \\ 0 & -6 \end{pmatrix}, \quad A(b) = \begin{pmatrix} 1 & 0 \\ 0 & -6 \end{pmatrix}, \quad A(c) = \begin{pmatrix} -2 & 0 \\ 0 & -6 \end{pmatrix}$$

Thus the points $a$ and $c$ are as. stable, while $b$ is unstable. Notice that $a$ and $c$ are local minima of $V$, so our findings fit with those of Subsection 11.3.3. The flow is illustrated in Fig. 11.12.                                            ∎

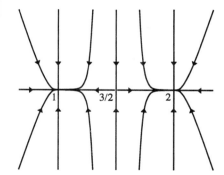

**Figure 11.12.** Phase portrait of the gradient flow in Example 11.14.

If the eigenvalues of $A$ satisfy $\Re\lambda(A) \leq 0$, and there is at least one eigenvalue with zero real part, then we can not decide on the stability of the nonlinear system on the basis of the linearized equation only. The following example demonstrates the inadequacy of the linear test:

**Example 11.15   Insufficiency of linearization**
Consider the system

$$\dot{x}_1 = x_2 - x_1(x_1^2 + x_2^2)$$

$$\dot{x}_2 = -x_1 - x_2(x_1^2 + x_2^2)$$

Let $r = |x|^2 = x_1^2 + x_2^2$. The equilibria points are the solutions of the system

$$x_2 - x_1 r = 0, \quad -x_1 - x_2 r = 0$$

The origin is the only equilibrium point. The linearized system at the origin corresponds to the matrix

$$A = \begin{pmatrix} 0 & 1 \\ -1 & 0 \end{pmatrix}$$

The eigenvalues are $\pm j$; they lie on the imaginary axis and are simple. The linear system is stable with trajectories forming concentric circles. To analyze the behavior of the nonlinear system, we consider $r(t) = x_1^2(t) + x_2^2(t)$. We easily compute

$$\dot{r} = 2x_1\dot{x}_1 + 2x_2\dot{x}_2 = 2x_1(x_2 - x_1 r) + 2x_2(-x_1 - x_2 r)$$

or $\dot{r} = -2r^2$. The solutions are calculated by direct integration

$$r(t) = \frac{r_0}{1 + 2r_0 t}$$

Thus, for every initial state $x_0 \neq 0$, $r_0 = |x_0|^2 \neq 0$ and $r(t) \to 0$ as $t \to \infty$. In effect, the above calculation establishes that $V(x) = r = x_1^2 + x_2^2$ is a global Liapunov function. Hence the system is globally asymptotically stable at the origin.

Let us next consider the system

$$\dot{x}_1 = x_2 + x_1(x_1^2 + x_2^2)$$

$$\dot{x}_2 = -x_1 + x_2(x_1^2 + x_2^2)$$

As before, the origin is the only critical point. Linearization leads to the same matrix as above. Let $r(t)$ denote the square of the length. Straightforward computations give this time $\dot{r} = 2r^2$ or $r(t) = r_0/(1 - 2r_0 t)$. We observe that for any initial state $x_0$ different from zero, the trajectory becomes infinite at time $t = 1/2r_0$, $r_0 = |x_0|^2$. Therefore the system is unstable. ∎

## 11.5 STABILITY OF DISCRETE STATE SPACE SYSTEMS

In this section we consider discrete state space representations and their stability properties. The presentation closely follows that of the previous sections, so only the main points are highlighted.

### 11.5.1 Basic definitions

Let us first consider the time invariant system

$$x(n + 1) = f(x(n)), \qquad x_0 \in R^k, n \geq 0 \tag{11.40}$$

Let $\phi(n, x_0)$ denote the solution of (11.40) emanating from $x_0$ at time 0

$$\phi(n, x_0) = \underbrace{f(f(\cdots f(x_0)))}_{n \text{ times}} = f^n(x_0)$$

Recall a point $x^*$ is an equilibrium or a stationary point if $\phi(n, x^*) = x^*$, or equivalently, $f(x^*) = x^*$.

The concepts of stability, attraction, and asymptotic stability are defined in complete analogy with the continuous time case. For instance, $x^*$ is *stable* if for all $\epsilon > 0$ there exists $\delta > 0$ such that if $| x_0 - x^* | < \delta$, $| \phi(n, x_0) - x^* | < \epsilon$ for all $n \geq 0$. Likewise $x^*$ is *asymptotically stable* if it is stable, and there is $\eta > 0$ such that for all $x_0$ satisfying $| x_0 - x^* | < \eta$, $\lim_{n \to \infty} \phi(n, x_0) = x^*$. Stability of trajectories is similarly defined.

The stability behavior of a stationary point of (11.40) is tightly related with the solution of nonlinear equations and the convergence properties of Picard algorithm (see Appendix III). A straightforward application of the fixed point theorem proves the following claim:

**Theorem 11.9**   If $f : K \to K$ is a contraction on a closed subset $K$ of $R^k$, (11.40) has a unique equilibrium $x^* \in K$, and the region of attaction of $x^*$ contains $K$.

∎

Discrete linear time invariant systems exhibit a similar behavior with continuous time systems. The main results are stated next. Proofs are left to the reader.

**Theorem 11.10**
1. $\lim_{n \to \infty} A^n = 0$ if and only if all eigenvalues of $A$ have magnitude less than 1, that is, $|\lambda(A)| < 1$.
2. The discrete linear system on $R^k$:

$$x(n+1) = Ax(n), \qquad n \geq 0 \tag{11.41}$$

is asymptotically stable at the origin if and only if all eigenvalues of $A$ lie inside the unit disc: $|\lambda(A)| < 1$. If there exists at least one eigenvalue $\lambda$, with $|\lambda| > 1$, the origin is unstable.

∎

### 11.5.2  Discrete Liapunov functions

The theory of Liapunov functions readily extends to the discrete case. A discrete Liapunov function is a positive real-valued function that is decreasing along trajectories. More precisely, we have the following:

*Definition.* Let $x^*$ be a stationary point of (11.40). A real-valued function $V$ defined on a neighborhood of $x^*$ is called a *Liapunov function* for (11.40) at $x^*$ if it is positive definite at $x^*$ and

$$V(f(x)) \leq V(x), \qquad x \neq x^* \tag{11.42}$$

A *strict Liapunov function* satisfies the above with strict inequality. If $V$ is a strict Liapunov function over the entire space and in addition is radially unbounded, it is called *global Liapunov function*.

*Remark.* Condition (11.42) implies that $V$ is decreasing along the system trajectories. Indeed,

$$V(x(n)) = V(f(x(n-1))) \leq V(x(n-1))$$

Let us introduce the difference function

$$\Delta V(x) = V(f(x)) - V(x) \tag{11.43}$$

Then a Liapunov function is determined by the specification $V > 0$, $\Delta V \le 0$.

Consider a positive definite matrix $P > 0$. The quadratic form $V(x) = x^T P x$ is a strict Liapunov function for (11.41) if $x^T A^T P A x < x^T P x$, or

$$A^T P A - P < 0$$

**Theorem 11.11**    Consider (11.40) and the equilibrium $x^*$. If $x^*$ has a Liapunov function, it is stable. If it has a strict Liapunov function, it is asymptotically stable. If it has a global Liapunov function, it is globally asymptotically stable. ∎

### Example 11.16    Square root evaluation
Square root evaluation via the Gauss-Newton algorithm is discussed in Appendix III. The algorithm dynamics is

$$f(x) = \frac{1}{2}\left(x + \frac{a}{x}\right), \qquad x > 0$$

The unique critical point over the positive reals is $\sqrt{a}$. The function $V(x) = (x - \sqrt{a})^2$ is a Liapunov function. Indeed, it is positive definite at $\sqrt{a}$. Straightforward calculations show that

$$\Delta V(x) = -\frac{(x - \sqrt{a})^2}{4x^2}(x + \sqrt{a})(3x - \sqrt{a})$$

$\Delta V(x)$ is zero at $\sqrt{a}$ and is negative for $x > \sqrt{a}/3$. Hence $\sqrt{a}$ is as. stable. ∎

**Theorem 11.12**    If for some matrix $Q > 0$ the matrix equation

$$A^T P A - P = -Q \tag{11.44}$$

has a positive definite solution $P$, the origin is globally asymptotically stable for (11.41). Conversely, if the origin is asymptotically stable, Eq. (11.44) has for any $Q > 0$ a unique solution $P > 0$ which is given by

$$P = \sum_{n=0}^{\infty} (A^T)^n Q A^n \tag{11.45}$$

∎

**Theorem 11.13 Linearization** Consider (11.40), a stationary point $x^*$, and the linearized equation

$$x(n+1) = Ax(n), \quad A = \left. \frac{\partial f}{\partial x} \right|_{x^*} \tag{11.46}$$

If all eigenvalues of $A$ are inside the unit disc, $x^*$ is as. stable. If at least one eigenvalue of $A$ has magnitude greater than 1, $x^*$ is unstable. If none of the eigenvalues of $A$ are outside the unit disc but, some are on the unit circle, no inference regarding stability can be made on the basis of the linear test only. ∎

*Example 11.17 Logistic dynamics*
The quadratic or logistic dynamics

$$x(n+1) = f(x(n), \mu) = \mu x(n)(1 - x(n)), \quad \mu > 0 \tag{11.47}$$

is encountered in several applications. In the context of population dynamics, it appears as follows: Let $y(n)$ be the population of a species evolving in accordance with the law

$$y(n+1) - y(n) = ky(n)[L - y(n)]$$

The model postulates that the species population grows with time provided that it does not exceed a certain value $L$, after which it tends to decrease. The latter equation converts to (11.47) if we set $x(n) = ky(n)/\mu$, $\mu = 1 + kL$.

The equilibria points satisfy $\mu x(1 - x) = x$, from which we obtain

$$x = 0, \quad x = p_\mu = \frac{\mu - 1}{\mu}, \quad \mu \neq 1$$

If $\mu = 1$, there is a single fixed point at the origin; if $\mu \neq 1$, an additional fixed point is borned that varies with $\mu$, becoming negative for $\mu < 1$ and positive for $\mu > 1$.

To check the local behavior of the fixed points, we perform linearization:

$$\frac{\partial f}{\partial x} = \mu - 2\mu x, \quad \left. \frac{\partial f}{\partial x} \right|_0 = \mu, \quad \left. \frac{\partial f}{\partial x} \right|_{p_\mu} = 2 - \mu$$

If $\mu = 1$, we cannot decide on the stability of the unique fixed point $x = 0$, on the basis of the linear test only. Graphical analysis, however, easily shows that 0 is unstable.

If $0 < \mu < 1$, 0 is as. stable and $p_\mu$ is unstable, whereas if $1 < \mu < 3$, 0 is unstable and $p_\mu$ is as. stable. We observe that a significant change in the dynamics occurs as the parameter varies. When $0 < \mu < 1$, there exist two fixed points, 0 and $p_\mu$. $p_\mu$ is repelling and 0 is attracting. When $\mu$ attains the value $\mu = 1$, the fixed point $p_\mu$ dies, and 0 becomes unstable. Once $\mu$ exceeds 1 and as long as it stays smaller than 3, a new fixed point $p_\mu$ is born again, which this time is attracting while 0 is repelling. The study of such changes is the subject of bifurcation theory. Bifurcation means a splitting apart, a division in two. We will continue the discussion on bifurcation in Section 11.8. ∎

## 11.6 TIME VARYING DYNAMICAL SYSTEMS

Stability of time varying systems is considered in this section. We discuss the qualitative behavior of linear systems, the characterization of stability by Liapunov functions, and linearization. The presentation covers the discrete case. Aspects of continuous time linear systems are studied in Exercises 11.5–11.6.

### 11.6.1 Time varying Liapunov functions

Consider the system

$$x(n+1) = f(x(n), n) \tag{11.48}$$

and the real-valued function $W(x, n)$. Let

$$\Delta W(x, n) = W(f(x, n), n+1) - W(x, n) \tag{11.49}$$

If $W(x, n)$ is evaluated along a trajectory $x(n)$ of (11.48), the difference function records the increments of $W$ between successive trajectory states:

$$\Delta W(x(n), n) = W(x(n+1), n+1) - W(x(n), n)$$

The function $W(x, n)$ is called *nonnegative definite* if (1) $W(x^*, n) = 0$ for all $n \geq 0$ and (2) $W(x, n) \geq 0$ for all $x$ near $x^*$ and $n \geq 0$. $W(x, n)$ is *positive definite* if in addition there exists a positive definite function $V(x)$ such that $W(x, n) \geq V(x)$ for all $x$ near $x^*$, and $n \geq 0$. If $V(x)$ is radially unbounded, $W(x, n)$ is called *radially unbounded*. Finally, $W(x, n)$ is called *decrescent* if there is a positive definite function $\hat{V}(x)$ such that $W(x, n) \leq \hat{V}(x)$ for all $x$ near $x^*$, and $n \geq 0$.

### Example 11.18 *Positive definite functions*
The function

$$W(x, n) = n^2 x^2$$

is nonnegative definite. It is not positive definite because $W(x, 0) = 0$. The function $W(x, n) = a^n x^2$, $|a| < 1$, satisfies $W(x, n) > 0$ for all $x \neq 0$. It is not positive definite because positivity is not uniform in $n$. Indeed, if there was a positive definite function $V(x)$ bounding $W$ from below, it would hold that $a^n \geq V(x)/x^2$, $x \neq 0$. This cannot occur because the exponential decays to zero. The function

$$W(x, n) = \frac{2n^2 + 1}{n^2 + 1} x^2$$

is positive definite, radially unbounded, and decrescent because $x^2 \leq W(x, n) < 2x^2$. The above examples are expressed as time varying quadratic forms, namely

$$W(x, n) = x^T P(n) x \qquad (11.50)$$

Given symmetric matrices $A$ and $B$, we write $A \leq B$ if $B - A \geq 0$, that is, $B - A$ is nonnegative definite. $A > B$ is similarly defined. Equation (11.50) defines a positive definite function if $P(n)$ is bounded below: $0 < Q \leq P(n)$. Likewise (11.50) is decrescent if $P(n)$ is bounded above: $P(n) \leq L$. Time varying Liapunov functions are defined in close analogy with the time invariant case.

**Theorem 11.14**  Suppose that $x^*$ is a stationary point of (11.48): $f(x^*, n) = x^*$ for all $n \geq 0$. If there is a positive definite decrescent function $W(x, n)$ at $x^*$ such that the difference function $\Delta W(x, n)$ is negative semidefinite, $x^*$ is uniformly stable. If in addition $\Delta W(x, n)$ is negative definite, $x^*$ is uniformly as. stable.

■

### 11.6.2  Linear systems

Let us consider the linear time varying system

$$x(n + 1) = A(n)x(n) \qquad (11.51)$$

$A(n)$ is bounded. We have already noted that stability of an arbitrary solution is equivalent to stability of the zero solution. Thus we will talk about stability of (11.51) without specific reference to trajectories. Concrete and detailed characterizations, as those of Theorem 11.2, are not available for general time varying matrices $A(n)$. The following theorem provides some general statements. We recall that the transition matrix $\Phi(n, n_0)$ satisfies Eq. (11.51) and the initial condition $\Phi(n_0, n_0) = I$ for any $n_0 \geq 0$.

**Theorem 11.15**  The following statements hold:

1. The system (11.51) is asymptotically stable if $\lim \Phi(n, n_0) = 0$ as $n \to \infty$. Uniform as. stability results if convergence is uniform in $n_0$.
2. If (11.51) is uniformly asymptotically stable, there are constants $M > 0$,

and $0 < \lambda < 1$ such that

$$|\Phi(n, n_0)| \leq M\lambda^{n-n_0}, \qquad n \geq n_0 \tag{11.52}$$

3. If $Q(n) > 0$ and the matrix difference equation

$$A^T(n)P(n+1)A(n) - P(n) = -Q(n) \tag{11.53}$$

has a bounded solution $0 < P_- < P(n) < P_+$, then (11.51) is uniformly as. stable.                                                                          ∎

Along the lines of Theorem 11.8, the following is proved:

**Theorem 11.16**   Suppose that 0 is an equilibrium for (11.48). Consider the first-order expansion

$$f(x, n) = A(n)x + h(x, n)$$

where

$$A(n) = \left.\frac{\partial f(x, n)}{\partial x}\right|_{x=0}, \qquad \lim_{|x| \to 0} \frac{|h(x, n)|}{|x|} = 0 \qquad \text{uniformly in } n,$$

If $A(n)$ is bounded and the system $x(n+1) = A(n)x(n)$ is uniformly as. stable, 0 is uniformly as. stable for the nonlinear system (11.48).                    ∎

A translation of Theorem 11.7 to discrete time varying systems is found useful in the convergence analysis of adaptive algorithms.

**Theorem 11.17**   Consider the system (11.51). Suppose that $A(n)$ is bounded and that there is a constant matrix $P > 0$ such that

$$A^T(n)PA(n) - P = -C^T(n)C(n) \tag{11.54}$$

for some matrix $C(n)$. If the pair $(A(n), C(n))$ is uniformly completely observable, namely there are positive constants $c_1$ and $c_2$ and a positive integer $N$ such that the observability grammian (see Exercise 10.6) satisfies the bound

$$c_1 I \leq \sum_{k=0}^{N-1} \Phi^T(n+k, n)C^T(n+k)C(n+k)\Phi(n+k, n) \leq c_2 I, \quad n \geq 0 \tag{11.55}$$

then (11.51) is uniformly as. stable.

**PROOF**   Let $V(x) = x^T P x$. Then $V(x) \leq |P||x|$. Furthermore

$$\Delta V(x,n) = x^T A^T(n) P A(n) x - x^T P x = -x^T C^T(n) C(n) x \leq 0 \qquad (11.56)$$

Let $x(n+k), k \geq 0$ be the solution of (11.51) emanating from $x$ at time $n$. Then the observability condition becomes

$$c_1 x^T x \leq \sum_{k=0}^{N-1} x^T \Phi^T(n+k,n) C^T(n+k) C(n+k) \Phi(n+k,n) x \leq c_2 x^T x$$

or

$$c_1 x^T x \leq \sum_{k=0}^{N-1} x^T(n+k) C^T(n+k) C(n+k) x(n+k) \leq c_2 x^T x \qquad (11.57)$$

Equation (11.56) gives

$$V(x(N+n)) - V(x(n)) = \sum_{k=n}^{N+n-1} \Delta V(x(k)) = - \sum_{k=n}^{N+n-1} x^T(k) C^T(k) C(k) x(k)$$

$$= - \sum_{k=0}^{N-1} x^T(n+k) C^T(n+k) C(n+k) x(n+k)$$

The latter equation in conjunction with (11.57) leads to

$$V(x(N+n)) - V(x(n)) \leq - \frac{c_1}{|P|} V(x(n))$$

Let $\lambda = 1 - c_1/|P|$. Then

$$V(x(N+n)) \leq \lambda V(x(n)) \qquad (11.58)$$

Moreover $0 < \lambda < 1$ because $V(x(n+N)) > 0$. Thus

$$V(x(k)) \leq \lambda^k, \qquad k \geq 2N$$

Since $P > 0$, it follows that $x(n)$ goes to zero as $n \to \infty$.   ∎

## 11.7  TOTAL STABILITY

In the previous sections the input was kept fixed and the effects of initial state perturbations on the state trajectories were analyzed. In this section both input and initial state deviations from nominal values are considered. We focus on linear state space models. Nonlinear systems are studied along the same lines by linearization. Let

$$\dot{x}(t) = Ax(t) + Bu(t)$$

$$y(t) = Cx(t) + Du(t) \tag{11.59}$$

The above system is called *totally stable* if for any initial state $x_0$ and any constant $L > 0$, there is $M > 0$ such that if $|u(t)| \leq L$, $t \geq 0$, it holds that $|y(t)| \leq M$, $t \geq 0$. The following theorem holds:

**Theorem 11.18**   The system (11.59) is totally stable if and only if the following hold:

1. Poles of the transfer function $h(s) = C(sI - A)^{-1}B$ belong to the left half plane.
2. Poles of

$$h_1(s) = C(sI - A)^{-1} \tag{11.60}$$

satisfy $\Re \lambda_i \leq 0$, while those with zero real part are simple.

PROOF   The variation of constants formula

$$y(t) = Ce^{At}x_0 + \int_0^t Ce^{A(t-\tau)}Bu(\tau)d\tau + Du(t)$$

enables us to study separately the contribution of the initial state

$$y_1(t) = Ce^{At}x_0 \tag{11.61}$$

and the contribution of the input

$$y_2(t) = \int_0^t Ce^{A(t-\tau)}Bu(\tau)d\tau + Du(t) \tag{11.62}$$

Let us start with (11.62). Assume that $|u(t)| \leq L$. Then $|Du(t)| \leq |D|L$. The convolutional integral in the right side was studied in Section 4.1 and Section 8.6. It is BIBO stable if and only if condition 1 holds.

The output (11.61) is written

$$y_1(t) = \sum_{i=1}^m \sum_{j=1}^{p_i} \frac{t^{j-1}}{(j-1)!} e^{\lambda_i t} CR_{ij}x_0 \tag{11.63}$$

$y_1(t)$ stays bounded for every $x_0 \in R^k$ because, by statement 2 of the theorem, all eigenvalues of $A$ with positive real part are excluded, and in addition, if an eigenvalue has zero real part, it is allowed only if it is simple. Such eigenvalues give rise to signals of the form $\cos \omega_i t$, $\sin \omega_i t$, which are bounded. Multiple eigenvalues with real part zero lead to signals of the form $t^l \cos \omega_i t$, $t^l \sin \omega_i t$, $l \geq 1$, which are unbounded. The converse is easy to establish; see also the proof of Theorem 11.2. ■

**Corollary 11.1**   If the eigenvalues of $A$ have negative real part, the system is totally stable.

Discrete systems exhibit entirely analogous behavior.

**Theorem 11.19**   The system

$$x(n+1) = Ax(n) + Bu(n) \tag{11.64}$$

$$y(n) = Cx(n) + Du(n) \tag{11.65}$$

is totally stable if the poles of the transfer function $h(z) = C(zI - A)^{-1}B + D$ lie inside the unit circle. In addition the poles of $h_1(z) = C(zI - A)^{-1}$ satisfy $|\lambda_i| \leq 1$, while those of magnitude one are simple. In particular, if the eigenvalues of $A$ are inside the unit circle, the system is totally stable. ■

Development of total stability results in a time varying or nonlinear setup is usually pursued along the direction of the previous Corollary. Consider the time varying system

$$x(n+1) = A(n)x(n) + B(n)u(n) \tag{11.66}$$

$$y(n) = C(n)x(n) + D(n)u(n) \tag{11.67}$$

Assume that all pertinent matrices are bounded. In particular, let $|B(n)| \leq K_B$, $|C(n)| \leq K_C$, $|D(n)| \leq K_D$. Moreover the autonomous system $x(n+1) = A(n)x(n)$ is uniformly asymptotically stable. Let $|u(n)| \leq L$, $n \geq n_0$. Using property 2 of Theorem 11.15 and the variation of constants formula, we obtain

$$|y(n)| \leq K_C M \lambda^{n-n_0}|x_0| + K_C LMK_B \frac{1 - \lambda^{n-n_0}}{1 - \lambda} + K_D L, \quad n \geq n_0$$

Thus the output rattles around $K_C LMK_B/(1 - \lambda) + K_D L$ after a transient period.

*Example 11.19   Analysis of the normalized LMS algorithm*
The LMS algorithm was introduced in Section 4.5.1. The technical subtleties appearing in the convergence analysis are next pointed out. Let us assume that

the data sequences are generated by the linear time varying model

$$y(n) = \varphi^T(n)\theta_0(n) + v(n) \tag{11.68}$$

The true parameter $\theta_0(n)$ obeys a model of the form

$$\theta_0(n+1) = \theta_0(n) + \xi(n) \tag{11.69}$$

The LMS recursion is given by (see Section 4.5.1)

$$\theta(n+1) = \theta(n) + \mu(n)\varphi(n)[y(n) - \varphi^T(n)\theta(n)]$$

Consider the adaptation error

$$\tilde{\theta}(n) = \theta_0(n) - \theta(n)$$

Then

$$\tilde{\theta}(n+1) = \theta_0(n) + \xi(n) - \theta(n) - \mu(n)\varphi(n)[y(n) - \varphi^T(n)\theta(n)]$$

Substituting (11.68) into the above equation, we obtain

$$\tilde{\theta}(n+1) = \tilde{\theta}(n) + \xi(n) - \mu(n)\varphi(n)[\varphi^T(n)\tilde{\theta}(n) + v(n)]$$

or

$$\tilde{\theta}(n+1) = [I - \mu(n)\varphi(n)\varphi^T(n)]\tilde{\theta}(n) + \xi(n) - \mu(n)\varphi(n)v(n) \tag{11.70}$$

This is a linear time varying system of the form (11.66). The eigenvalues of the associated matrix $A(n) = I - \mu(n)\varphi(n)\varphi^T(n)$ are equal to 1 except that given by $\rho = 1 - \mu(n)\varphi^T(n)\varphi(n)$ which has eigenvector $\varphi(n)$. The normalized LMS chooses the stepsize $\mu(n)$ so that $|\rho| < 1$. Then in the direction of the regressor vector $\varphi(n)$ the adaptation error is controlled. The best convergence rate is achieved when $\rho$ is zero:

$$\mu(n) = \frac{1}{\varphi^T(n)\varphi(n)}$$

Problems will arise with the above choice when the regressor is zero. A more safe selection is

$$\mu(n) = \frac{\mu}{1 + \mu\varphi^T(n)\varphi(n)} \tag{11.71}$$

where $\mu$ is small enough. In this manner division by zero is avoided, while $\rho$ remains smaller than one.

If the disturbances $v(n)$ and $\xi(n)$ are interpreted as deterministic signals of bounded amplitude, Eq. (11.70) becomes a linear time varying dynamical system. In accordance with the analysis of (11.66), the autonomous system

$$\tilde{\theta}(n+1) = A(n)\tilde{\theta}(n), \quad A(n) = I - \mu(n)\varphi(n)\varphi^T(n) \tag{11.72}$$

plays a dominant role. Let us assume that there exist constants $c_1 > 0$, $c_2 > 0$, and a positive integer $N$ such that for all $n$,

$$0 < c_1 I \leq \sum_{n=0}^{N-1} \varphi(n)\varphi^T(n) \leq c_2 I \tag{11.73}$$

We will show that (11.72) is uniformly asymptotically stable utilizing Theorem 11.17. Let $V(\tilde{\theta}) = \tilde{\theta}^T\tilde{\theta} = |\tilde{\theta}|^2$. Then $V > 0$, and the difference function is

$$\Delta V(\tilde{\theta}, n) = \tilde{\theta}^T A^T(n) A(n)\tilde{\theta} - \tilde{\theta}^T\tilde{\theta} = \tilde{\theta}^T Q(n)\tilde{\theta}$$

where

$$Q(n) = A^T(n)A(n) - I = [\mu^2(n)|\varphi(n)|^2 - 2\mu(n)]\varphi(n)\varphi^T(n)$$

$$= -\frac{\mu(2 + \mu|\varphi(n)|^2)}{(1 + \mu|\varphi(n)|^2)^2}\varphi(n)\varphi^T(n)$$

Therefore $P = I$ is the solution of Eq. (11.54), with

$$C(n) = \frac{\sqrt{\mu(2 + \mu|\varphi(n)|^2)}}{1 + \mu|\varphi(n)|^2}\varphi(n)$$

Next we use output feedback to simplify the observability condition (11.55). In analogy with Exercise 10.18 the pair $(A(n), C^T(n))$ is uniformly completely observable if and only if the pair $(A(n) - K(n)C^T(n), C^T(n))$ with bounded feedback gain $K(n)$ is uniformly completely observable. The special form of $A(n)$ and $C(n)$ enables the selection of a feedback gain that converts $A(n)$ into the identity matrix. Indeed, take

$$K(n) = -\sqrt{\frac{\mu}{2 + \mu|\varphi(n)|^2}}\,\varphi(n)$$

Then $A(n) - K(n)C^T(n) = I$. Clearly $K(n)$ is bounded. Condition (11.73) implies that $\varphi(n)$ is bounded. Hence there exist constants $k_1, k_2$ such that

$$k_1 \leq \frac{\sqrt{\mu(2 + \mu|\varphi(n)|^2)}}{1 + \mu|\varphi(n)|^2} \leq k_2$$

The transition matrix related to the identity matrix is the identity, and the observability condition required in Theorem 11.17 is a consequence of (11.73).

Next we consider the stochastic setup. Disturbances are now modeled as stochastic processes. The adaptation error becomes a random signal. The emphasis is shifted on the statistics of $\tilde{\theta}(n)$, such as the mean and covariance. Unfortunately, these parameters cannot be readily computed because the coefficients entering (11.70) are random rather than deterministic. To illustrate this further, let us take expected values on both sides of (11.70). Then

$$E[\tilde{\theta}(n+1)] = E\left[(I - \mu(n)\varphi(n)\varphi^T(n))\tilde{\theta}(n)\right]$$

It is highly tempting to factor the right-hand side as $E\left[I - \mu(n)\varphi(n)\varphi^T(n)\right]E[\tilde{\theta}(n)]$. This is not justifiable on formal grounds because $\varphi(n)$ and $\theta(n)$ include data up to time $n-1$ (an exception is the case of one-dimensional $\theta$) and therefore are correlated. Proceeding, however, as if the above was allowable, we obtain

$$E[\tilde{\theta}(n+1)] = (I - \mu(n)E[\varphi(n)\varphi^T(n)])E[\tilde{\theta}(n)]$$

If the regressor vector is stationary, $E[\varphi(n)\varphi^T(n)] = R$. The above equation becomes a linear time invariant system. Assuming a constant learning rate the eigenvalues of $I - \mu R$ are $1 - \mu\lambda$, $\lambda$ eigenvalue of $R$. Therefore, if $R > 0$ and

$$0 < \mu < \frac{2}{\lambda_{\max}}$$

it holds that $E[\tilde{\theta}(n)] \to 0$.

Despite the lack of mathematical rigor, the above analysis led to many useful developments both in theory and practice. It turns out the conclusions derived by the above reasoning are correct, yet a rigorous derivation requires more sophisticated machinery. ∎

*Example 11.20   Analysis of the RLS algorithm*
Suppose that the data are generated by the linear regression (11.68) and the true parameters are governed by Eq. (11.69). The RLS algorithm with exponential forgetting factor (see Section 4.5.2) obeys the recursion

$$\theta(n+1) = \theta(n) + w(n)e(n), \quad w(n) = P(n)\varphi(n)$$

Much like the analysis of the LMS algorithm, the adaptation error satisfies the linear time varying system

$$\tilde{\theta}(n+1) = [I - w(n)\varphi^T(n)]\tilde{\theta}(n) - w(n)v(n) + \xi(n)$$

Consider the linear time varying system

$$\tilde{\theta}(n+1) = A(n)\tilde{\theta}(n), \quad A(n) = I - w(n)\varphi^T(n) \tag{11.74}$$

Then

$$A(n) = P(n)P^{-1}(n) - P(n)\varphi(n)\varphi^T(n) = P(n)\lambda P^{-1}(n-1)$$

because

$$P^{-1}(n) = R(n) = \sum_{k=0}^{n} \lambda^{n-k}\varphi(k)\varphi^T(k)$$

The transition matrix of (11.74) is easily evaluated to be

$$\Phi(n, n_0) = \lambda^{n-n_0} P(n)P^{-1}(n-1)P(n-1)\cdots P^{-1}(n_0) = \lambda^{n-n_0} P(n)P^{-1}(n_0)$$

Thus, if

$$0 < c_1 I \leq \sum_{k=0}^{n} \lambda^{n-k}\varphi(k)\varphi^T(k) \leq c_2 I$$

$P(n)$ remains bounded, and uniform as. stability is guaranteed. Moreover $\varphi(n)$ is bounded, and hence $w(n)$ is bounded. It follows that the adaptation error is bounded if $\xi(n)$ and $v(n)$ are bounded.  ∎

## 11.8  OBSERVING BIFURCATION AND CHAOS

The study of the qualitative behavior of the logistic dynamics was initiated in Example 11.17. We saw that if $0 < \mu < 1$, the stationary points 0 and $p_\mu$ occur. 0 is attracting and $p_\mu$ is repelling. Once $\mu$ hits the value 1, a behavioral change occurs. The fixed point $p_\mu$ disappears, and the origin becomes unstable. In fact it is not difficult to unveil the entire structure of the flow. Since $f(x) = x - x^2$, $\phi(n, x)$ is a decreasing sequence. If $\phi(n, x) \to p$, $p$ must be an equilibrium. The only possibility is 0. Then $\phi(n, x) \geq 0$ for any $n$, which implies that $0 \leq x \leq 1$. Thus $\phi(n, x) \to 0$ for $0 \leq x \leq 1$, and $\phi(n, x) \to -\infty$ otherwise. The origin is unstable because it attracts no negative points.

Bifurcations such as the above occur in the vicinity of nonhyperbolic equilibria. An equilibrium $x^*$ is called *hyperbolic* if the Jacobian $A = \partial f/\partial x|_{x^*}$ has no eigenvalues on the unit circle. The linearization theorem asserts that each hyperbolic fixed point is either unstable (there exists $\lambda$ with $|\lambda| > 1$), or asymptotically stable (all eigenvalues of $A$ satisfy $|\lambda| < 1$). Now suppose that the value $\mu = \mu_0$ leads to a hyperbolic point $p_{\mu_0}$. If the dynamics depend smoothly on $\mu$, the eigenvalues of $A$ depend continuously on $\mu$. Therefore all values of $\mu$ near $\mu_0$ will give rise to hyperbolic points. This remark is made precise by the implicit function theorem, stated without proof.

**Theorem 11.20  Implicit function theorem**    Consider the differentiable function $F(x, y)$ defined on the open set $U \subset R^k \times R^m$, and the equation

$$F(x, y) = c$$

Let $(x_0, y_0) \in U$ such that $F(x_0, y_0) = c$. Suppose further that $\partial F(x_0, y_0)/\partial y$ is invertible. Then there are open sets $U_1 \subset R^k$ and $U_2 \subset R^m$ and a unique differentiable function $g : U_1 \to U_2$, $x \to y = g(x)$ such that $g(x_0) = y_0$ and

$$F(x, g(x)) = c, \qquad x \in U_1 \tag{11.75}$$

∎

Differentiation of (11.75) with respect to $x$ gives

$$\frac{\partial F}{\partial x} + \frac{\partial F}{\partial y}\frac{\partial g}{\partial x} = 0$$

or

$$\frac{\partial g}{\partial x} = -\left[\frac{\partial F(x, g(x))}{\partial y}\right]^{-1}\frac{\partial F(x, g(x))}{\partial x}$$

Consider, for example, the equation

$$x^2 + y^2 = 1$$

Let $(x_0, y_0)$ be a point on the circle. If $y_0 > 0$, we have $y = g(x) = \sqrt{1 - x^2}$, while if $y_0 < 0$, it holds that $y = g(x) = -\sqrt{1 - x^2}$.

The next theorem employs the implicit function theorem to show that bifurcations do not occur near hyperbolic fixed points.

**Theorem 11.21**    Consider the dynamical system

$$x(n + 1) = f(x(n), \mu), \qquad x \in R^k \tag{11.76}$$

Suppose that $f(x_0, \mu_0) = x_0$; that is, $x_0$ is a fixed point for $\mu = \mu_0$, and $x_0$ is hyperbolic. Then there is a neighborhood $S$ of $x_0$, a neighborhood $N$ of $\mu_0$, and a smooth map $p : N \to S$ such that

$$p(\mu_0) = x_0, \quad f(p(\mu), \mu) = p(\mu)$$

Moreover (11.76) has no other fixed points in $S$.

**PROOF** Consider the function $G(x, \mu) = f(x, \mu) - x$. By hypothesis, $G(x_0, \mu_0) = 0$. Moreover

$$\left. \frac{\partial G}{\partial x} \right|_{(x_0, \mu_0)} = \left. \frac{\partial f}{\partial x} \right|_{(x_0, \mu_0)} - I.$$

The latter matrix is invertible because $(x_0, \mu_0)$ defines a hyperbolic point. The claim follows from the implicit function theorem. ∎

We recall from Example 11.17 that as $\mu$ ranges in the interval $(1, 3)$, the origin is unstable and $p_\mu$ is as. stable. A complete analysis of the flow is described in Exercise 11.10.

Let us next set $\mu = 3$. Both fixed points are unstable. The nonzero equilibrium is $p_\mu = 2/3$, and the corresponding linearization is $-1$. Thus $p_\mu$ is a nonhyperbolic point and alarms for the possible occurrence of a bifurcation. This is indeed the case. The resulting bifurcation is called *period doubling bifurcation* and is typical. The logistic dynamics undergoes a sequence of periodic doubling bifurcations as the parameter increases and eventually enters a chaotic regime. The precise nature of period doubling bifurcations is elucidated in the following theorem:

**Theorem 11.22 Period doubling bifurcation** Consider the dynamical system (11.76) with $k = 1$. Suppose that $x_0$ is a nonhyperbolic point $f(x_0, \mu_0) = x_0$, $\partial f(x_0, \mu_0)/\partial x = -1$. Let $f^2(x, \mu) = f(f(x, \mu), \mu)$ and

$$\frac{\partial}{\partial \mu} \frac{\partial f^2}{\partial x}(x_0, \mu_0) \neq 0$$

There exist neighborhoods $N$ and $M$ about $x_0$ and $\mu_0$, respectively, and a smooth map $p : N \to M$ such that

$$p(x_0) = \mu_0, \quad p'(x_0) = 0, \quad f(x, p(x)) \neq x, \quad f^2(x, p(x)) = x$$

**PROOF** We set

$$G(x, \mu) = f^2(x, \mu) - x$$

and

$$H(x, \mu) = \begin{cases} \dfrac{G(x, \mu)}{x - x_0}, & \text{if } x \neq x_0 \\[2ex] \dfrac{\partial G}{\partial x}(x_0, \mu), & \text{if } x = x_0 \end{cases}$$

We will next apply the implicit function theorem on $H$. Note that

$$H(x_0, \mu_0) = \frac{\partial G}{\partial x}(x_0, \mu_0) = \frac{\partial f}{\partial x}(f(x, \mu), \mu)\frac{\partial f}{\partial x}(x, \mu)|_{(x_0, \mu_0)} - 1 = 0$$

Moreover

$$\frac{\partial H}{\partial \mu}(x_0, \mu_0) = \frac{\partial}{\partial \mu}\frac{\partial f^2}{\partial x}(x_0, \mu_0) \neq 0$$

Therefore $H$ satisfies the requirements of the implicit function theorem, and the equation $H(x, \mu) = 0$ can be solved with respect to $\mu$. More precisely, there are neighborhoods $N$ and $M$ of $x_0$ and $\mu_0$, respectively, and a smooth function $p : N \to M$ such that $p(x_0) = \mu_0$ and

$$H(x, p(x)) = 0 \tag{11.77}$$

Thus $G(x, p(x)) = 0$ for $x \neq x_0$, and $f^2(x, p(x)) = x$. The latter equation says that $x$ is a periodic point of period 2. ∎

The period doubling bifurcation describes the behavior of the quadratic family for $\mu = 3$. Once $\mu$ exceeds 3 yet stays close to 3, both fixed points are unstable and repelling, and a periodic point of period two is born. This is because the curve $p(x)$ anticipated by the bifurcation has a maximum at $x_0$ (see Exercise 11.12). Thus it will intersect the range $\mu > 3$. Let us denote this periodic point by $p_1$. Computer simulations show that if $\mu$ slightly exceeds 3, for instance, $\mu = 3.1$, the periodic orbit $\{p_1, f(p_1)\}$ becomes an attractor attracting all points in the open interval $(0, 1)$. Thus all trajectories emanating in $(0, 1)$ eventually oscillate between the values $p_1$ and $f(p_1)$. This behavior is observed for parameter values in the range $3 < \mu < M_1$. $M_1$ is approximately $M_1 = 3.14$. Once $\mu$ exceeds $M_1$, a new bifurcation occurs. The periodic point of period 2, $p_1$, becomes unstable and nonhyperbolic and gives rise to a period doubling bifurcation. As a consequence a new periodic point $p_2$ of period 4 is born to the right of $M_1$. All trajectories in the open unit interval oscillate among the values $\{p_2, f(p_2), f^2(p_2), f^3(p_2)\}$ as long as $\mu$ stays smaller than a value $M_2$. $M_2$ is approximately 3.4495. In this way an increasing sequence $M_j$ results such that if $\mu$ lies in the interval $M_j < \mu < M_{j+1}$, the previously born periodic point $p_j$ of period $2^j$ becomes repelling, and a new periodic point $p_{j+1}$ of period $2^{j+1}$ appears, attracting all trajectories starting in $(0, 1)$. It turns out that the sequence $M_j$ quickly converges to a particular critical value $\mu_\infty$, of size roughly 3.5699. Moreover the ratio of distances between successive $M_j$ decreases geometrically; that is, there is a constant $\delta$ called *Feigenbaum's number*, with value approximately $4.66920160910299097\cdots$ so that

$$\frac{M_j - M_{j-1}}{M_{j+1} - M_j} \to \delta$$

Once the parameter $\mu$ exceeds $\mu_\infty$, dynamics enter into the chaotic regime. A rigorous explanation of the above genealogy of periodic points and the route to chaos via period doubling bifurcations is accomplished by symbolic dynamics and the kneading theory. These topics are beyond the scope of this book. In the remaining of this section we give a precise definition of chaotic behavior, and we demonstrate its appearance in the logistic dynamics for $\mu = 4$.

When the parameter $\mu$ reaches the value $\mu = 4$, the trajectories of the logistic map show an erratic behavior regardless of initial conditions in the unit interval. This is easily illustrated by MATLAB simulation. Evolution appears totally unpredictable and the power spectral density shows a white-noise-like characteristic. There are several ways to formally define chaotic behavior. A topological definition relies on the notion of transitivity and sensitivity to initial conditions. The system

$$x(n+1) = f(x(n)) \qquad (11.78)$$

is called *transitive* if, given any pair of open sets $A$ and $B$, there is a point $x$ in $A$ such that the trajectory emanating from $x$ eventually enters $B$. Thus trajectories starting close to a given point $x$ will approach arbitrarily close to any point in the state space.

The system (11.78) has *sensitive dependence on initial conditions* if there is $M > 0$ such that, for any $x$ and any neighborhood $A$ of $x$, there is $y$ in $A$ and $n \geq 0$ such that

$$|\phi(n, x) - \phi(n, y)| > M$$

The above statement asserts that there is a point arbitrarily close to a given point $x$ that is eventually separated from $x$ by at least $M$.

We say the system (11.78) is *chaotic* on a set $A$ if it is transitive on $A$, it has sensitive dependence on initial conditions in $A$, and the set of periodic points is dense in $A$. Unpredictability of a chaotic system is a consequence of sensitive dependence on initial conditions. Transitivity implies that the state space cannot be partitioned in subspaces where dynamics are disassembled into simpler constituents. These two properties capitulate the randomness features of the system. The third property imposes additional regularity, since it requires that the periodic orbits abound; it states that every point in the state space is arbitrarily close to a periodic orbit.

**Example 11.21   A chaotic system**
Let $A$ be the interval $[0, 2\pi)$, and

$$x(n+1) = f(x(n)) = 2x(n)(\text{mod } 2\pi)$$

The system is visualized on the circle. Each point $\theta$ in $[0, 2\pi)$ is identified with the arc of angular distance $\theta$. If we scale by $2\pi$, we obtain the equivalent system on the unit interval

$$y(n + 1) = 2y(n)(\text{mod } 1) \tag{11.79}$$

The evolution of the above system is better understood if each initial state is represented in binary form

$$y = \sum_{n=1}^{\infty} y_n 2^{-n} = (y_1 \, y_2 \, y_3 \, y_4 \cdots)$$

The digits $y_n$ are 0 and 1. An example is

$$y = 0.11111000110011 \cdots \tag{11.80}$$

The integer part is zero because $0 < y < 1$. The effect of the dynamics $f$ on $y$ is to shift the fractional part to the left and remove the integer part. Indeed,

$$f(y) = \sum_{n=1}^{\infty} y_n 2^{-n+1} = \sum_{n=0}^{\infty} y_{n+1} 2^{-n} = y_1 + \sum_{n=1}^{\infty} y_{n+1} 2^{-n}$$

or

$$f(y) = \sum_{n=1}^{\infty} y_{n+1} 2^{-n} (\text{mod } 1) = (y_2 \, y_3 \, y_4 \cdots)$$

Thus (11.80) becomes

$$y = 0.1111000110011 \cdots$$

The system (11.79) possesses sensitive dependence on initial conditions. Indeed, take any two different initial states $y$ and $z$. Let $r$ be the most significant digit where they differ. Then

$$y = (y_1 \, y_2 \cdots y_r \, y_{r+1} \cdots), \quad z = (z_1 z_2 \cdots z_r z_{r+1} \cdots)$$

and

$$y - z = (0\,0 \cdots y_r - z_r \, y_{r+1} - z_{r+1} \cdots), \quad y_r \neq z_r$$

We let the system run for $r$ time instants. Then

$$\phi(r, y) - \phi(r, z) = (y_r - z_r \, y_{r+1} - z_{r+1} \cdots), \quad y_r \neq z_r$$

and

$$|\phi(r, y) - \phi(r, z)| > \frac{1}{2}$$

Thus any two different initial states are eventually separated by at least $1/2$.

Transitivity follows from the observation that there are orbits that together with their limit sets (see Appendix) cover the entire space. Indeed, consider the initial state whose binary form is

$$y = (0, 1|00, 01, 10, 11|000, 001, \cdots)$$

The first block lists all sequences of length 1, the second block lists all sequences of length 2, and so forth. Let $z$ be any state in the interval $(0, 1)$ and $z_k$ the segment of the first $k$ digits. Then $z_k$ resides somewhere in the $k$th block of $y$. If the system operates for some time, it will produce a state whose first segment coincides with $z_k$. The trajectory emanating from $y$ approaches arbitrarily close to $z$, and the claim is established.

Finally, we prove that the set of periodic points is dense in $(0, 1)$. Let $y$ be any point in $(0, 1)$. We truncate the binary representation of $y$ up to the first $N$ digits and take the resulting periodic extension:

$$\tilde{y}_N = (y_1 y_2 \ldots y_N y_1 \ldots y_N y_1 \ldots)$$

$\tilde{y}_N$ produces a periodic trajectory of period $N$. Indeed, after $N$ steps the point $\tilde{y}_N$ is shifted $N$ places to the right, while the first $N$ digits are dropped. The result is $\phi(N, \tilde{y}_N) = \tilde{y}_N$. Moreover

$$|\tilde{y}_N - y| = \sum_{n=N}^{\infty} y_n 2^{-n} \le \sum_{n=N}^{\infty} 2^{-n} = \frac{1}{2^{N-1}}$$

Hence $\tilde{y}_N \to y$ as $N \to \infty$. ∎

*Remark.* The above example is reminiscent of the shift system and the symbolic dynamics introduced in Example 3.6. Exactly the same arguments show that the shift system exhibits a chaotic behavior. The shift system dynamics are transparent. Because of this property, the shift system and symbolic dynamics provide a model of chaotic behavior in the sense that many chaotic systems can be mapped to the shift system. A very simple illustration of this idea is delineated next for the logistic dynamics.

**Example 11.22** *Chaotic behavior of the logistic dynamics for $\mu = 4$*
Consider the change of variables

$$y = 1 - 2x \quad \text{or} \quad x = \frac{1}{2}(1 - y)$$

In the new coordinates the logistic dynamics for $\mu = 4$ become

$$y(n + 1) = 1 - 2x(n + 1) = 1 - 8x(n)(1 - x(n)) = 1 - 2(1 - y(n))(1 + y(n))$$

or

$$y(n + 1) = 2y^2(n) - 1 \tag{11.81}$$

The logistic dynamics is chaotic if and only if the latter equation is chaotic. Next we consider the transformation $y = \cos z$. The dynamics of (11.81) are transformed into

$$y(n + 1) = \cos z(n + 1) = 2\cos^2 z(n) - 1 = \cos 2z(n)$$

or

$$z(n + 1) = 2z(n)(\bmod 2\pi) \tag{11.82}$$

Equation (11.82) defines a chaotic system in accordance with Example 11.21. Since the cosine function maps the interval $[0, 2\pi)$ onto $[-1, 1]$, we easily infer that (11.81) is chaotic. ∎

## BIBLIOGRAPHICAL NOTES

Classical tutorial expositions of stability theory for dynamical systems are provided by Kalman and Bertram (1960), Willems (1970), Hirch and Smale (1974), Vidyasagar (1978), and Arnold (1973). For further reading on the stability concepts, the reader is referred to Hahn (1967). The deterministic analysis of adaptive algorithms is pursued in Anderson et al. (1986). Convergence of adaptive algorithms in a stochastic setup is treated in Goodwin and Sin (1984), Ljung and Soderstrom (1983), and Kumar and Varaiya (1986). Several papers have been written on the synthesis of Hopfield networks. The method employed in the neural network MATLAB toolbox is based on Li, Mitchell, and Porod (1989). Additional references can be found in Haykin (1994). The brief section on bifurcation and chaos is intended to inform the reader about the complex behavioral patterns encountered in nonlinear dynamical systems. Structural changes of the qualitative behavior and the route to chaotic regimes can be systematically analyzed by an elegant theory. The interested reader may consult Devaney (1986), Tsonis (1992), and the references therein.

## PROBLEMS

**11.1.** Consider the matrix

$$A = \begin{pmatrix} 0 & 1 & 3 \\ -2 & -3 & 0 \\ 0 & 0 & -3 \end{pmatrix}$$

Show that $A$ is asymptotically stable. Compute the unique solution $P$ of the equation $PA + A^T P = -I$. Verify Eq. (11.34) by computing $e^{At}$. Validate all steps by MATLAB.

**11.2.** Consider $\dot{x} = Ax$ on the plane $R^2$. Let $\lambda_1, \lambda_2$ be the eigenvalues of $A$, with $\Re\lambda_1 \leq \Re\lambda_2$. Develop phase portraits using MATLAB for each possible category. Repeat for the discrete planar system $x(n+1) = Ax(n)$.
   *Hint:* Possible categories are (a) distinct real eigenvalues

$$\lambda_2 > \lambda_1 > 0, \quad \lambda_1 < 0 < \lambda_2, \quad 0 < \lambda_1 < \lambda_2,$$
$$\lambda_1 = 0 < \lambda_2, \quad \lambda_1 < \lambda_2 = 0$$

(b) complex eigenvalues $\Re\lambda < 0$, $\Re\lambda > 0$, and $\Re\lambda = 0$; and (c) equal eigenvalues (diagonalizable and nondiagonalizable case) $\lambda < 0$, $\lambda > 0$, and $\lambda = 0$.

**11.3.** The circuit of Fig. 11.13 consists of a linear capacitor with capacitance $C = 1$ F, a linear inductor with inductance $L = 1$ H, a linear resistor with resistance $R = 1$ $\Omega$, an independent voltage source supplying a constant voltage $u = 1$ V, and a nonlinear current-controlled resistance with voltage-current relationship

$$i = g(v) = v(v - 1)(v - 4)$$

Let $x_1$ be the current passing through the inductor and $x_2$ the voltage across the capacitor. Derive the state space equations

$$\dot{x}_1 = -x_1 + x_2$$
$$\dot{x}_2 = -x_1 + x_2(1 - x_2)(x_2 + 3)$$

Find the equilibria points. Use linearization to explore the stability behavior in the vicinity of each equilibrium.

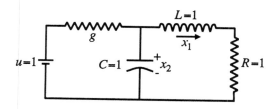

**Figure 11.13.** Nonlinear circuit of Exercise 11.3.

**11.4.** Consider the system

$$x(n+1) = Ax(n) + Bu(n)$$

where

$$A = \begin{pmatrix} 0 & 1 \\ -\frac{1}{8} & \frac{3}{4} \end{pmatrix}, \qquad B = \begin{pmatrix} 5 \\ 6 \end{pmatrix}$$

Show that $A$ is as. stable using the discrete Liapunov equation (11.44). Show that the system is BIBO stable.

**11.5.** Consider the system $\dot{x}(t) = A(t)x(t)$. $A(t)$ is locally integrable and bounded. Let $\Phi(t, t_0)$ be the transition matrix. Show that the origin is stable if and only if $|\Phi(t, t_0)| < M(t_0)$. It is uniformly stable if the bound does not depend on $t_0$. It is (uniformly) asymptotically stable if it is (uniformly) stable and $\lim_{t \to \infty} \Phi(t, t_0) = 0$ (convergence is uniform in $t_0$). Furthermore prove that if the system is uniformly asymptotically stable, there are $M > 0$ and $\lambda > 0$ such that

$$|\Phi(t, t_0)| \le M e^{-\lambda(t-t_0)} \qquad \text{for all } t_0, \ t \ge t_0$$

**11.6.** The system $\dot{x}(t) = A(t)x(t)$ is called *periodic* if $A(t + T) = A(t)$ for some $T > 0$ and for all $t$ in R. Let $\Phi(t, t_0)$ denote the transition matrix of the above periodic system. Prove the following statements:

(a) $\Phi(t + T, t_0) = \Phi(t, t_0)\Phi(t_0 + T, t_0)$.
(b) Let $C = \Phi^{-1}(t, 0)\Phi(t + T, 0)$. Show that $C = \Phi(T, 0)$ is a non-singular constant matrix.
(c) Given any nonsingular matrix $B$, there exists a matrix $A$ called a logarithm of $B$ such that $e^A = B$. Establish the claim in the special case of diagonalizable matrix $B$.
(d) Define $R$ such that $C = e^{TR}$ and $P(t) = \Phi(t, 0)e^{-Rt}$. Prove that $P(t)$ is nonsingular and periodic.
(e) Show that

$$\Phi(t, t_0) = P(t)e^{R(t-t_0)}P^{-1}(t_0)$$

(f) The eigenvalues of $R$, $\rho_i$, are called *the characteristic exponents*. The eigenvalues of $C$, $\mu_i$, are the *characteristic multipliers*. It follows from (c) that $\mu_i = e^{\rho_i T}$. Using Exercise 11.4, show that

$$\prod \mu_i = e^{\int_0^T \text{tr}(A(\tau))d\tau}$$

(g) 0 is uniformly asymptotically stable if and only if the characteristic exponents have negative real parts.

**11.7.** Consider the system

$$\dot{x}_1 = -x_1 + x_2, \qquad \dot{x}_2 = -x_1 - x_2 + x_2^3$$

Show that the origin is asymptotically stable by means of the function $V(x) = x_1^2 + x_2^2$. In a similar fashion show that the system

$$\dot{x}_1 = -x_1 + x_2, \quad \dot{x}_2 = -x_1 - x_2 - x_2^3$$

is globally asymptotically stable.

**11.8.** This exercise shows that Liapunov functions provide estimates of the region of attraction. Let $V(x)$ be a strict Liapunov function for $\dot{x} = f(x)$ at the equilibrium $x^*$ and $a > 0$. Suppose that the region $M = \{x : V(x) < a\}$ is bounded, $V(x)$ is positive definite in $M$, and $\dot{V}$ is negative definite in $M$. Following the argument used in the proof of Theorem 11.3 show that $M$ is contained in the region of attraction of $x^*$.

**11.9.** Consider the system

$$\dot{x}_1 = x_1^2 - x_2^2$$
$$\dot{x}_2 = -2x_1 x_2$$

Use $V(x) = 3x_1 x_2^2 - x_1^3$ to show that the origin is unstable.

**11.10.** Consider the logistic dynamics. Prove the following statements:
(a) If $\mu > 1$ and $x < 0$, or $x > 1$, $\phi(n, x) \to -\infty$ as $n \to \infty$.
(b) If $1 < \mu < 2$, and $0 < x < 1$, then $\phi(n, x) \to p_\mu = (\mu - 1)/\mu$. Verify by simulation that the same conclusion holds if $2 \leq \mu < 3$.

**11.11.** Prove that the origin of $\dot{x} = Ax$ is a stable equilibrium if and only if all eigenvalues of $A$ have nonpositive real part, while for those eigenvalues $\lambda$ with zero real part, it holds that

$$\frac{d^{p_i-j}}{ds^{p_i-j}} \left( (s - \lambda_i)^{p_i} (sI - A)^{-1} \right) \bigg|_{s=\lambda_i} = 0, \quad j = 2, 3, \cdots, p_i$$

**11.12.** Prove by differentiating (11.77) that $p'(x_0) = 0$ and $p''(x_0) < 0$; that is, the curve $p(x)$ has a maximum at $x_0$.

**11.13.** Let $\lambda_1, \ldots, \lambda_k$ denote the eigenvalues of $A$. Prove that the Liapunov equation

$$PA + A^T P = -Q, \quad Q > 0$$

has a unique solution $P$ if and only if

$$\lambda_i + \lambda_j \neq 0, \quad i, j = 1, 2, \ldots, k$$

In a similar fashion prove that the discrete Liapunov equation

$$A^T PA - P = -Q, \quad Q > 0$$

has a unique solution $P$ if and only if

$$\lambda_i \lambda_j \neq 1, \qquad i, j = 1, 2, \ldots, k$$

*Hint*: Use Kronecker products and Exercise 5.3.

**11.14.** Consider the spinning vehicle of Exercise 3.17. Investigate the stability behavior of the equilibria points.

**11.15.** Set up a complete SIMULINK experiment showing the route to chaos for the logistic dynamics as $\mu$ varies.

**11.16.** Simulate the pendulum motion in the absence of air resistance by SIMULINK. Find the equilibria and the linearized equations. Show by simulations that if the angular velocity exceeds a certain threshold, the pendulum executes full rotations, indefinitely.

**11.17.** Show that the scalar dynamical systems defined by the Chebyshev polynomials
(a) $f(x) = 4x^3 - 3x$
(b) $f(x) = 8x^4 - 8x^2 + 1$

are chaotic in $[-1, 1]$. Simulate the above dynamics.
*Hint*: Use the transformation $x = \cos y$ and Eq. (8.72) to convert (a) and (b) into the systems $y(n+1) = 3y(n) \pmod{2\pi}$, $y(n+1) = 4y(n) \pmod{2\pi}$, respectively.

**11.18.** Identify equilibria, stability behavior and bifurcations for the following dynamics:
(a) $f(x, \mu) = \mu x - x^3$, $\quad 0 < \mu < 2$
(b) $f(x, \mu) = x + \mu^2$, $\quad 0 < \mu \leq 1/4$

**11.19.** Let $I = [0, 1]$. A map $f : I \to I$ is called *unimodal* if $f(0) = f(1) = 0$ and $f$ has a unique stationary point $c$ in $0 < c < 1$, $df(c)/dx = 0$. The itinerary of a point $x$ in $I$ is a digital signal $s(n)$ with values 0, 1, and 2 defined by $s(n) = 0$ if $\phi(n, x) < c$, $s(n) = 1$ if $\phi(n, x) > c$, and $s(n) = 2$ if $\phi(n, x) = c$. The itinerary of $c$ is referred to as the *kneading sequence* $K(f)$ of $f$. Itineraries translate the dynamical properties of the system $f$ into a symbolic dynamics environment.
    Write a MATLAB function that computes the itinerary and the kneading sequence. Show that
(a) $\mu x(1 - x)$, $\quad 0 < \mu \leq 4$
(b) $\mu \sin(\pi x)$, $\quad 0 < \mu < 1$

are unimodal maps. List all itineraries of the logistic map for $1 < \mu < 2$. Compute the kneading sequence for $\mu = 2$ and $\mu = 4$.

**11.20.** Consider the planar dynamical system defined by the Henon map

$$x_1(n + 1) = 1 - ax_1^2(n) + x_2(n)$$

$$x_2(n + 1) = bx_1(n)$$

Determine the equilibria. Analyze the stability behavior by linearization, when $a = b = 1/4$. Verify your results with MATLAB simulations.

**11.21.** Consider the linear time varying system $\dot{x}(t) = A(t)x(t)$, where

$$A(t) = \begin{pmatrix} -1 & e^{2t} \\ 0 & -1 \end{pmatrix}$$

The eigenvalues of $A(t)$ are constant and equal to $-1$. Prove that the origin is unstable by showing that the transition matrix is unbounded. Construct a similar example for the discrete case.

**11.22.** Consider a controllable and observable state space model

$$\dot{x}(t) = Ax(t) + Bu(t)$$

$$y(t) = Cx(t) + Du(t)$$

(a) Show that the system is totally stable if and only if $A$ is stable; that is, the eigenvalues of $A$ are in the left half plane. Establish the discrete counterpart.

(b) Suppose that $A$ is as. stable. Show that the controllability and observability grammians

$$P = \int_0^\infty e^{At} BB^T e^{A^T t} dt, \quad Q = \int_0^\infty e^{A^T t} C^T C e^{At} dt$$

are well-defined positive definite matrices that satisfy the Liapunov equations

$$PA^T + AP + BB^T = 0, \quad A^T Q + QA + C^T C = 0$$

Show that if $(A, B, C)$ is converted to $(A_1, B_1, C_1) = (LAL^{-1}, LB, CL^{-1})$, the products $PQ$ and $P_1 Q_1$ are similar, $P_1 Q_1 = LPQL^{-1}$.

**11.23.** In analogy with Example 11.12, design an analog Hopfield network that will store the patterns $(\frac{1}{2}, \frac{1}{2})$, $(\frac{1}{2}, \frac{3}{4})$ and $(\frac{1}{4}, \frac{1}{2})$. Take $\psi^{-1}(y) = \log(y/(1 - y))$.

**11.24.** *Discrete Hopfield network.* The discrete Hopfield neural nework has the form

$$y_i(n+1) = \text{sgn}\left( \sum_{j=1}^{k} w_{ij}y_j(n) + I_i \right), \qquad 1 \le i \le k$$

The signum function returns 1 if the argument is positive and $-1$ if it is negative. The above expression is written in matrix form as

$$y(n+1) = \text{sgn}(Wy(n) + I)$$

$I$ is a $k \times 1$ vector and $W$ a $k \times k$ matrix. Apparently the coordinates of each state are either 1 or $-1$. For any real $a$ consider the function

$$V_a(y) = -\frac{1}{2}y^T Wy - I^T y + a$$

(a) Prove that if $W$ is nonnegative definite, $V_a(y(n))$ is decreasing and bounded.
(b) Show that the patterns $(-1, 1)$ and $(1, -1)$ can be stored by a Hopfield network with weights and biases given by

$$W = \begin{pmatrix} 1 & -1 \\ -1 & 1 \end{pmatrix}, \quad I = \begin{pmatrix} 0.5 \\ -0.5 \end{pmatrix}$$

Give a geometrical interpretation. Are there spurious stored patterns?

## APPENDIX

In this appendix the proofs of the main stability theorems are given.

**PROOF OF STATEMENT 1 OF THEOREM 11.3**   Let $\epsilon > 0$ and $0 < \eta < \epsilon$ such that the neighborhood $S(x^*, \eta)$ of $x^*$ is contained in the domain $S$ of the Liapunov function $V$. Let $H(x^*, \eta)$ be the boundary

$$H(x^*, \eta) = \{x \in S : | x - x^* | = \eta\}$$

and

$$m = \inf\{V(y) : y \in H(x^*, \eta)\}$$

Since $V$ is positive definite, $m$ is nonnegative. We will show that $m > 0$. Suppose that $m = 0$. There is a sequence $y_n$ with $|y_n - x^*| = \eta$ and $V(y_n) \to 0$. Since $H(x^*, \eta)$ is compact, there is a subsequence $y_{n_k}$ that converges to a point $y$ in

$H(x^*, \eta)$. By continuity of $V$, we have

$$\lim_{k \to \infty} V(y_{n_k}) = V(\lim_{k \to \infty} y_{n_k}) = V(y) = 0$$

This is a contradiction because $y \neq x^*$. Next we consider the set

$$Q = \{y \in S(x^*, \eta) : V(y) < m\}$$

$Q$ is an open neighborhood of $x^*$ because $V$ is continuous and $V(x^*) = 0$. Let $\delta > 0$ such that $S(x^*, \delta) \subset Q$. We claim that if $z \in S(x^*, \delta)$, $\phi(t, z) \in S(x^*, \epsilon)$ for all $t \geq 0$. Indeed, if the claim is not true, there is $z \in S(x^*, \delta)$ and $T > 0$, with $|\phi(T, z) - x^*| > \epsilon$. The trajectory $\phi(t, z)$ starts inside the sphere $S(x^*, \eta)$ and eventually gets out. Connectedness of the sphere ensures that the trajectory will eventually hit the boundary. Thus there is $s$, $0 \leq s \leq T$, with $\phi(s, z) \in H(x^*, \eta)$. Now $V(\phi(s, z)) \leq V(z)$ because $V$ is decreasing along trajectories. Furthermore $V(z) < m$ because $z \in Q$. Thus $V(\phi(s, z)) < m$. On the other hand, $\phi(s, z) \in H(x^*, \eta)$. Therefore $V(\phi(s, z)) \geq m$, by definition of $m$. Thus we arrived at a contradiction. ∎

**Limit set.** To prove statement 2 of Theorem 11.3, we need the concept of the limit set. The limit set of a point $x$ consists of all states that are asymptotically reached from $x$. In a sense the limit set of $x$ and the region of attraction of $x$, namely the set of states that are asymptotically driven to $x$, are dual concepts. The precise definition is as follows:

Consider the system (11.16) and a state $x$ in $R^k$. The *limit set of $x$* is denoted $\Lambda^+(x)$ and is defined as

$$\Lambda^+(x) = \{y \in R^k : \text{there is } t_n \to \infty : \phi(t_n, x) \to y\}$$

The next theorem summarizes some useful properties of the limit set. Given a point $x \in R^k$ and a closed subset $M$ of $R^k$ the distance of $x$ from $M$ is $d(x, M) = \inf\{d(x, m) : m \in M\}$; $d$ denotes a distance function on $R^k$. We say that a sequence $x_n$ in $R^k$ converges to $M$ if the sequence of real numbers $d(x_n, M)$ converges to 0. In a similar fashion $\phi(t, x) \to M$ if $d(\phi(t, x), M) \to 0$ as $t \to \infty$.

**Theorem 11.23** The limit set enjoys the following properties:

1. $\Lambda^+(x)$ is invariant; that is, for any $t \in R$ and any $y \in \Lambda^+(x)$, $\phi(t, y) \in \Lambda^+(x)$, as well.
2. $\Lambda^+(x)$ is a closed set.
3. If $\Lambda^+(x)$ is nonempty and compact, $\phi(t, x) \to \Lambda^+(x)$, $t \to \infty$.

**PROOF**

1. Let $y \in \Lambda^+(x)$ and $t \in R$. Let $t_n$ be a real sequence with $t_n \to \infty$ and $\phi(t_n, x) \to y$. Continuity of the flow and the semigroup property imply that

$$\phi(t + t_n, x) = \phi(t, \phi(t_n, x)) \to \phi(t, y)$$

Hence $\phi(t, y) \in \Lambda^+(x)$.

2. We wish to show that if $y = \lim y_n$ and $y_n \in \Lambda^+(x)$, then $y \in \Lambda^+(x)$. Since $y_n \in \Lambda^+(x)$ for each $n$, we can find a sequence $s_{n_k}$ such that $s_{n_k} \to \infty$, $\phi(s_{n_k}, x) \to y_n$ for all $n$. For each $n$ there exists $k(n)$ such that $s_{n_{k(n)}} \geq n$ and $d\big(\phi(s_{n_{k(n)}}, x), y_n\big) \leq 1/n$. Let $s_n = s_{n_{k(n)}}$. Then $s_n \to \infty$ and

$$d(\phi(s_n, x), y) \leq d(\phi(s_n, x), y_n) + d(y_n, y) \leq \frac{1}{n} + d(y_n, y) \to 0$$

Therefore $y \in \Lambda^+(x)$, and the limit set is closed.

3. Suppose that the conclusion is not true. Then there is $\epsilon > 0$ and a sequence $t_n \to \infty$ such that $d\big(\phi(t_n, x), \Lambda^+(x)\big) \geq \epsilon$. Since $\Lambda^+(x)$ is nonempty and compact, there exists $s_n \to \infty$ such that $d\big(\phi(s_n, x), \Lambda^+(x)\big) \leq \epsilon$. Furthermore we can choose $s_n \leq t_n$. The solution curve $\phi(t, x)$ emanates from $x$, is continuous, and on the interval $[s_n, t_n]$ has its endpoints inside and outside the neighborhood

$$S\big(\Lambda^+(x), \epsilon\big) = \{y \in R^k : d(y, \Lambda^+(x)) < \epsilon\}$$

Connectedness of the neighborhood ensures that the curve will hit the boundary. Thus there is $\tau_n : s_n \leq \tau_n \leq t_n$ with $\phi(\tau_n, x) \in H\big(\Lambda^+(x), \epsilon\big) = H$. Since $H$ is compact, there is a convergent subsequence $\phi(\tau_{n_k}, x) \to y \in H$, $\tau_{n_k} \to \infty$. Thus $d(y, \Lambda^+(x)) = \epsilon$. On the other hand, $y \in \Lambda^+(x)$, by definition, and we arrived at a contradiction. ∎

The next proposition relates limit sets to Liapunov functions.

**Theorem 11.24**  Consider a function $V$ defined on a closed neighborhood of $x^*$, $K$, such that $\dot{V} \leq 0$, and $\dot{V}(x^*) = 0$. Suppose that $\phi(t, x)$ is contained in $K$ for all $t \geq 0$. Then

$$\Lambda^+(x) \subset \{y : \dot{V}(y) = 0\}$$

If $V$ is a strict Liapunov function, the latter relation becomes

$$\Lambda^+(x) = \{x^*\}$$

**PROOF**  We show first that $V$ is constant on the limit set $\Lambda^+(x)$. Since the positive time trajectory of $x$ belongs to $K$, $K$ is closed and $V$ is decreasing along trajectories, $\Lambda^+(x)$ is also contained in $K$. Let $y, z$ in $\Lambda^+(x)$ with $V(y) < V(z)$. We consider sequences $t_n \to \infty, s_n \to \infty$ such that $\phi(t_n, x) \to y$ and $\phi(s_n, x) \to z$.

Without loss of generality, we may take $s_n \geq t_n$. Taking into account the semigroup property of the flow (see Exercise 3.9), we obtain

$$V(\phi(s_n, x)) = V(\phi(s_n - t_n, \phi(t_n, x))) \leq V(\phi(t_n, x))$$

Taking limits, we have $V(z) \leq V(y)$, a contradiction.

Let $y \in \Lambda^+(x)$. Since $\Lambda^+(x)$ is invariant, $\phi(t, y) \in \Lambda^+(x)$ for all $t \in R$. Thus the function $t \to V(\phi(t, y))$ is constant, and its derivative with respect to $t$ is zero. Hence

$$\dot{V}(y) = \frac{d}{dt} V(\phi(t, y)) \bigg|_{t=0} = \nabla V(y) \cdot f(y) = 0$$

If $V$ is a strict Liapunov function, $x^*$ is the only point where $\dot{V}$ vanishes. ∎

With the aid of the limit set and its properties we complete the proof of Theorem 11.3.

PROOF OF STATEMENT 2 OF THEOREM 11.3   Consider the set $Q$ introduced in the proof of statement 1. Let $x \in Q$. $\Lambda^+(x)$ is nonempty. Indeed, take any sequence $\phi(t_n, x)$ with $t_n \to \infty$. Then $\phi(t_n, x) \in Q$ because $Q$ is invariant. Moreover $Q$ is bounded; therefore there exists a subsequence $\phi(s_n, x) \to y$, and $y \in Q$. By definition, $y \in \Lambda^+(x)$. We have shown that $\Lambda^+(x)$ is nonempty. It follows from Theorem 11.24 that $\Lambda^+(x) = \{x^*\}, x \in Q$. Application of statement 3 of Theorem 11.23 yields $\phi(t, x) \to x^*$, as $t \to \infty$, for all $x \in Q$. Thus $Q$ is contained in the region of attraction of $x^*$ and contains a neighborhood of $x^*$. Therefore $x^*$ is asymptotically stable. ∎

PROOF OF STATEMENT 3 OF THEOREM 11.3   Tracing the proof of statement 2 we observe that the limit set of any point $x$ in the set $Q$ is nonempty and compact, is invariant by Theorem 11.23, and is contained in the set of zeros of $\dot{V}$ by Theorem 11.24. Let $z \in \Lambda^+(x)$. Since the limit set is invariant, the positive trajectory $\phi(t, z)$, $t \geq 0$, belongs to $\Lambda^+(x)$. By assumption, $z = x^*$ and $\Lambda^+(x) = x^*$. The rest of the proof follows as in statement 2. ∎

PROOF OF STATEMENT 4 OF THEOREM 11.3   Let $x \in R^k$. The solution $\phi(t, x)$ is bounded for $t \geq 0$. Indeed, if $\phi(t, x) \to \infty$, then $V(\phi(t, x)) \to \infty$ as well because $V$ is radially unbounded. This cannot occur because $V$ is a Liapunov function and $V(\phi(t, x)) \leq V(x)$. Since $\phi(t, x)$ is bounded, the limit set $\Lambda^+(x)$ is nonempty and compact. The rest of the proof proceeds along the lines of statements 2 and 3 of Theorem 11.3. ∎

PROOF OF THEOREM 11.4   Suppose that conditions 1 and 2a hold. Let $\epsilon > 0$ such that the closed neighborhood $K$ with radius $\epsilon$ and center $x^*$ is contained in the domain of $V$. Pick $x$ as in property 1. If the trajectory $\phi(t, x)$

escapes $K$, the system is unstable. If it remains in $K$ for all positive times, the corresponding limit set is a nonempty subset of $K$. Property 2a implies that $V$ is decreasing along trajectories. Theorem 11.24 implies that $\Lambda^+(x) = \{x^*\}$. Since $\dot{V}(x) < 0$, we have $V(x^*) = \lim V(\phi(t_n, x)) \leq V(x) < 0$, a contradiction. Thus $\phi(t, x)$ cannot stay in $K$ for ever and the system is unstable.

Next suppose that conditions 1 and 2b hold. Choose $\epsilon$ and $x$ as above. Let $c(t) = V(\phi(t, x))$, $d(t) = V_1(\phi(t, x))$. Notice that $c(t) < 0$ for all $t \geq 0$. Because of property 2b, $c(t)$ satisfies $\dot{c} = ac + d$. Since $V_1 \leq 0$, we have $\dot{c} \leq ac$. Taking into account that $c(t) < 0$, we find that $c(t) \geq V(x)e^{at}$. Thus $c(t) \to \infty$, a contradiction. ∎

**PROOF OF THEOREM 11.6**   Consider the quadratic form $V(x) = x^T P x$, where $P$ is the positive definite solution of Eq. (11.33). $V$ is positive definite because $P > 0$. Example 11.5 implies that $\dot{V}(x) = x^T(PA + A^T P)x = -x^T Q x$. Thus $V$ is a global Liapunov function, and the origin is globally asymptotically stable.

Next we prove the converse statement. Consider the matrix $P$ given by formula (11.34). Since $A$ is asymptotically stable, the eigenvalues of $A$, $\lambda_i$, have a negative real part. Then each polynomial exponential signal $t^j e^{\lambda_i t}$ is absolutely integrable on $[0, \infty]$. Equation (11.10) implies that $e^{At}$ is also absolutely integrable and that $P$ is well defined. $P$ is clearly symmetric and

$$x^T P x = \int_0^\infty (e^{At}x)^T Q(e^{At}x)dt \geq 0$$

because $Q > 0$. Moreover $P > 0$. Indeed, if there is $x$ such that $x^T P x = 0$, then $Qe^{At}x = 0$, and since $Q$ and $e^{At}$ are invertible, $x = 0$. Next we show that $P$ is a solution of (11.33). We have

$$PA + A^T P = \int_0^\infty (e^{A^T t}Qe^{At}A + A^T e^{A^T t}Qe^{At})dt$$

$$= \int_0^\infty \left( e^{A^T t}Q\frac{d}{dt}e^{At} + \frac{d}{dt}(e^{A^T t})Qe^{At} \right) dt$$

$$= \int_0^\infty \frac{d}{dt}(e^{A^T t}Qe^{At})dt = e^{A^T t}Qe^{At}|_0^\infty = 0 - IQI = -Q$$

The linear map $f : P \to f(P) = PA + A^T P$ is onto by the preceding discussion. It follows from standard results in linear algebra that it is one-to-one, as well. Thus $P$ is the unique solution of (11.33). ∎

**PROOF OF THEOREM 11.8**   1. Since the eigenvalues of $A$ have negative real part, the linear system (11.38) is as. stable. By Theorem 11.6, the Liapunov equation

$$PA + A^T P = -I \tag{11.83}$$

has a unique positive definite solution $P$. Consider the positive definite function

$$V(x) = (x - x^*)^T P(x - x^*) \tag{11.84}$$

We will show that if $V$ is properly confined on a neighborhood of $x^*$, it is a strict Liapunov function for the original system at $x^*$. Consider the Taylor series expansion of $f$ at $x^*$

$$f(x) = A(x - x^*) + h(x)$$

with

$$\lim_{|x-x^*| \to 0} \frac{|h(x)|}{|x - x^*|} = 0 \tag{11.85}$$

and for $x$ near $x^*$. Now the derivative of $V$ becomes

$$\dot{V}(x) = 2(x - x^*)^T P[A(x - x^*) + h(x)] = -|x - x^*|^2 + 2(x - x^*)^T P h(x)$$

Because of (11.85) we can find $\delta > 0$ such that

$$\frac{|h(x)|}{|x - x^*|} < \frac{1}{4|P|}, \qquad |x - x^*| < \delta$$

$|P|$ is an induced norm of $P$. The Cauchy-Schwarz inequality implies that

$$(x - x^*)^T P h(x) \le |x - x^*||P||h(x)| \le \frac{1}{4}|x - x^*|^2$$

Hence

$$\dot{V}(x) \le -\frac{1}{2}|x - x^*|^2 < 0, \qquad |x - x^*| < \delta$$

Therefore $\dot{V}$ is negative definite near $x^*$. Consequently $V$ is a strict Liapunov function, and $x^*$ is as. stable for the nonlinear system (11.37).

2. Let $\lambda_1, \ldots, \lambda_k$ denote the eigenvalues of $A$. It can be shown (see Exercise 11.13) that (11.83) has a unique solution $P$ if and only if

$$\lambda_i + \lambda_j \ne 0, \qquad i, j = 1, 2, \ldots, k \tag{11.86}$$

$A$ may violate the above condition. Thus we pick $\epsilon > 0$ so that the matrix $A - \epsilon I$ satisfies (11.86) and has at least one eigenvalue with positive real part. Let $P$ be the solution of

$$P(A - \epsilon I) + (A^T - \epsilon I)P = -I \quad \text{or} \quad PA + A^T P - 2\epsilon P = -I$$

and $V(x) = (x - x^*)^T P(x - x^*)$. $P$ cannot be positive definite because then $A - \epsilon I$ would be asymptotically stable. Therefore $V$ satisfies condition 1 of instability Theorem 11.4. Moreover

$$\dot{V}(x) = 2(x - x^*)^T P f(x) = 2(x - x^*)^T P[A(x - x^*) + h(x)] = 2\epsilon V(x) + V_1(x)$$

where

$$V_1(x) = -|x - x^*|^2 + 2(x - x^*)^T P h(x)$$

Using the Cauchy-Schwarz inequality, it follows exactly as in the proof of the preceding statement that $V_1(x)$ is negative definite. The claim then follows from Theorem 11.4. ∎

# 12

# PREDICTION, FILTERING, AND IDENTIFICATION

This chapter continues the discussion on prediction, filtering, and identification that was originated in Chapter 4. Infinite memory linear predictors are derived using the mean squared formulation. The basic elements of the Kolmogorov-Wiener theory are highlighted, and important insight into the structure of wide sense stationary processes is gained through the Wold decomposition. The design of digital pulse amplitude modulation systems for transmission over baseband channels is delineated. An alternative approach to filtering is offered by the Kalman filter. Kalman filters rely on state space representations, and they can cope with nonstationary signals and time varying characteristics. Time invariant architectures are also analyzed. The extended Kalman filter is intended for nonlinear state space models and relies on linearization. Identification is discussed in the context of convolutional and finite difference models. The crosscorrelation and the prediction error approach are treated. Finally, a glimpse of Volterra system identification is provided.

## 12.1  LINEAR PREDICTION

In this section we concentrate on infinite horizon linear prediction where the entire past history of the signal is taken into account in order to predict future signal values. We will assume that $x(n)$ is a zero mean wide sense stationary single-channel signal with spectral distribution $F(\omega)$. For each integer $n$, $\mathcal{H}_x(n)$ denotes the Hilbert space of second-order variables that can be obtained as quadratic mean limits of linear combinations of $x(n - k), k \geq 0$. The linear mean square predictor is defined as the system with input $x(n)$ and output $\hat{x}(n) = E^*[x(n)|\mathcal{H}_x(n - 1)]$, the projection of $x(n)$ onto the subspace $\mathcal{H}_x(n - 1)$. The projection theorem asserts that $\hat{x}(n)$ minimizes the mean squared error $E[\tilde{x}(n)^2]$ where

$$\tilde{x}(n) = x(n) - \hat{x}(n)$$

Moreover $\hat{x}(n)$ is characterized by the orthogonality principle

$$E[\tilde{x}(n)x(n-k)] = 0, \quad k = 1, 2, \ldots$$

A WSS process $x(n)$ yielding a nonzero minimum prediction error is called a *regular* process. Otherwise, it is a *singular* process. If $x(n)$ is a singular process, then $\hat{x}(n) = x(n)$ almost everywhere and $x(n) \in \mathcal{H}_x(n-1)$. The structure of the linear predictor is eluminated in the sequel using spectral domain techniques and the developments of Chapter 9. We will show that under broad conditions the predictor is described by a linear time invariant causal system whose frequency response is determined by the spectral distribution $F(\omega)$.

### 12.1.1 Wold decomposition

Wold decomposition provides important insight into the structure of WSS processes. In addition it paves the way for the solution of linear prediction and filtering.

***The innovation sequence.*** In this paragraph we show that the estimation error $\tilde{x}(n)$ is a wide sense stationary white noise signal. Let $W(n)$ be the orthogonal complement of $\mathcal{H}_x(n-1)$ in $\mathcal{H}_x(n)$:

$$\mathcal{H}_x(n) = \mathcal{H}_x(n-1) \oplus W(n) \tag{12.1}$$

$W(n)$ is spanned by the estimation error $\tilde{x}(n)$ at time $n$, and it carries the entirely "new" information in $\mathcal{H}_x(n)$, not already included in $\mathcal{H}_x(n-1)$. The orthogonality principle implies that $\tilde{x}(n)$ is a white noise signal. Indeed, let $n > k$. Since $x(k) \in \mathcal{H}_x(k)$ and $\hat{x}(k) \in \mathcal{H}_x(k-1)$, $\tilde{x}(k)$ belongs to $\mathcal{H}_x(k)$. Moreover $\mathcal{H}_x(k) \subset \mathcal{H}_x(n-1)$ and $\tilde{x}(n)$ is orthogonal to $\mathcal{H}_x(n-1)$. Therefore $E[\tilde{x}(n)\tilde{x}(k)] = 0$. We will shortly see that for $n = k$ the minimum mean squared error

$$\sigma_1^2 = E[\tilde{x}^2(n)]$$

is independent of $n$, and thus $\tilde{x}(n)$ is zero mean wide sense stationary.

***The regular and singular components.*** In the rest of this section we assume that $x(n)$ is a regular process. The normalized innovation sequence

$$e(n) = \frac{1}{\sigma_1}\tilde{x}(n) \tag{12.2}$$

is a WSS process forming an orthonormal set. Equation (12.1) is recursive in $\mathcal{H}_x(n)$. The subspaces $W(n)$, and in particular, the innovation sequence, will be capable to reproduce $\mathcal{H}_x(n)$ if the remote past of the signal is zero. We will show that

$$\mathcal{H}_x(n) = \sum_{j \leq n} W(j) \oplus \mathcal{H}_x(-\infty) \quad \text{where } \mathcal{H}_x(-\infty) = \cap_{n=-\infty}^{\infty} \mathcal{H}_x(n) \qquad (12.3)$$

Formula (12.3) states that $\mathcal{H}_x(n)$ is composed of the space spanned by the innovation sequence up to time $n$ and the remote past of the signal $\mathcal{H}_x(-\infty)$. Consider the space $\mathcal{H}_e(n)$, formed by the closure of the linear span of $e(n - k)$, $k = 0, 1, 2, \ldots$. Clearly $\mathcal{H}_e(n) = \sum_{j \leq n} W(j)$. Let $u(n)$ be the projection of $x(n)$ onto $\mathcal{H}_e(n)$. Then

$$x(n) = u(n) + v(n) \qquad (12.4)$$

and $v(n)$ is orthogonal to $\mathcal{H}_e(n)$. Since $e(n - k), k \geq 0$, form an orthonormal set in $\mathcal{H}_e(n)$, Theorem 2.5 asserts that $u(n)$ is uniquely expressed as

$$u(n) = \sum_{k=0}^{\infty} h(k)e(n - k), \quad \sum_{k=0}^{\infty} |h(k)|^2 < \infty \qquad (12.5)$$

The coefficients $h(k)$ are determined by Eq. (2.66) of Section 2.9.3:

$$h(k) = <u(n), e(n - k)> = E[u(n)e(n - k)] \qquad (12.6)$$

Thus (12.4) becomes

$$x(n) = \sum_{k=0}^{\infty} h(k)e(n - k) + v(n), \quad \sum_{k=0}^{\infty} |h(k)|^2 < \infty \qquad (12.7)$$

Clearly $u(n)$ and $v(n)$ and WSS processes.

**Lemma 12.1**   The following statements hold:
  1. $E[e(n)v(m)] = E[u(n)v(m)] = 0$ for all integers $n, m$.
  2. $\mathcal{H}_x(n) = \mathcal{H}_e(n) \oplus \mathcal{H}_x(-\infty)$.
  3. $\mathcal{H}_x(n) = \mathcal{H}_u(n) \oplus \mathcal{H}_v(n)$.

**PROOF**
1. If $n \leq m$, then $v(m)$ is orthogonal to $\mathcal{H}_e(n)$ and both $e(n), u(n)$ belong to $\mathcal{H}_e(n)$; hence the statement holds. Next suppose $n > m$. Then

$$E[e(n)v(m)] = E[e(n)(x(m) - u(m))] = E[e(n)x(m)] - E[e(n)u(m)]$$

The first term of the right-hand side is zero because $x(m)$ belongs to $\mathcal{H}_x(m) \subset \mathcal{H}_x(n-1)$ and $e(n)$ is orthogonal to $\mathcal{H}_x(n-1)$. Taking into account Eq. (12.5), the second term becomes

$$E[e(n)u(m)] = \sum_{k=0}^{\infty} h(k)E[e(n)e(m-k)] = 0$$

because $m - k \le m < n$ and $e(n)$ is white noise. Thus for each $m$, $v(m)$ is orthogonal to every subspace $\mathcal{H}_e(n)$. In particular $v(m)$ is orthogonal to $u(n)$.

2. Let $y \in \mathcal{H}_x(-\infty)$. Then $y \in \mathcal{H}_x(n)$ and $y \in \mathcal{H}_x(n-1)$. Since $e(n)$ is orthogonal to $\mathcal{H}_x(n-1)$, it is orthogonal to $y$. Therefore $y$ belongs to the orthogonal complement of $\mathcal{H}_e(n)$ for all $n$. Conversely, if $y$ belongs to $\mathcal{H}_x(n)$ and is orthogonal to $e(n)$, then it lies in $\mathcal{H}_x(n-1)$. By the same token, it lies in $\mathcal{H}_x(n-2)$, and so on. This proves that $y$ belongs to $\mathcal{H}_x(-\infty)$, and the claim is established.

3. Obvioulsy $\mathcal{H}_u(n) \subset \mathcal{H}_e(n)$ and $\mathcal{H}_v(n) \subset \mathcal{H}_x(-\infty)$. Since the pairs of subspaces sum up to $\mathcal{H}_x(n)$, it follows that

$$\mathcal{H}_u(n) = \mathcal{H}_e(n), \quad \mathcal{H}_v(n) = \mathcal{H}_x(-\infty) \tag{12.8}$$

∎

**Theorem 12.1 Wold decomposition**   A regular WSS process $x(n)$ with spectral distribution $F$ is written

$$x(n) = u(n) + v(n) = \sum_{k=0}^{\infty} h(k)e(n-k) + v(n) \tag{12.9}$$

with

$$\sum_{k=0}^{\infty} |h(k)|^2 < \infty, \qquad h(0) > 0$$

$e(n)$ is white noise of unit variance and $e(n) \in \mathcal{H}_x(n)$. $u(n)$ and $v(n)$ are uncorrelated signals; $v(n)$ is singular and belongs to $\mathcal{H}_x(-\infty)$. The process $u(n)$ is regular with absolutely continuous spectral distribution. Finally $h(n)$ and $e(n)$ are unique.

PROOF   Consider the decompositions (12.4)–(12.5). The process $v(n)$ is singular because, by Eq. (12.8), $\mathcal{H}_v(n) = \mathcal{H}_x(-\infty) = \mathcal{H}_v(n-1)$. On the other hand, $u(n)$ is regular. Indeed, from Eq. (12.8) we have

$$\hat{u}(n) = E^*[u(n)|\mathcal{H}_u(n-1)] = E^*[u(n)|\mathcal{H}_e(n-1)]$$

Equation (12.5) is written

$$u(n) = h(0)e(n) + \sum_{k=1}^{\infty} h(k)e(n-k)$$

The second term in the right-hand side belongs to $\mathcal{H}_e(n-1)$ and is orthogonal to the first term because $e(n)$ is white noise. Let $h(q) = \sum_{k=0}^{\infty} h(k)q^k$ be the series representation of $h(n)$. Using the notation of Section 7.1.1 we obtain

$$\hat{u}(n) = \sum_{k=1}^{\infty} h(k)e(n-k) = (h(q) - h(0))e(n), \quad \sum_{k=0}^{\infty} |h(k)|^2 < \infty \qquad (12.10)$$

The minimum prediction error is

$$E[(u(n) - \hat{u}(n))^2] = E[(h(0)e(n))^2] = h^2(0)$$

$h(0)$ is nonzero because, by (12.6) and statement 1 of Lemma 12.1,

$$h(0) = E[u(n)e(n)] = E[x(n)e(n)] = \frac{1}{\sigma_1} E[\tilde{x}^2(n)] = \sigma_1$$

Since $x(n)$ is regular, $\sigma_1 > 0$. Therefore $h(0) > 0$ and $u(n)$ is also regular. The rest of the proof is omitted. ∎

A consequence of $h(0) = \sigma_1$ is that $\tilde{x}(n)$ is wide sense stationary. A regular process whose singular component is zero is called *completely nondeterministic*.

### 12.1.2 Structure of the linear predictor

The Wold decomposition reveals the structure of the predictor filter. Suppose that $x(n)$ is a regular WSS process with spectral distribution function $F$. The derivative of $F(\omega)$, $S(\omega)$ always exists, even if $F$ is not absolutely continuous. Since the projection is a linear map and $v(n) \in \mathcal{H}_x(n-1)$, we have

$$\hat{x}(n) = E^*[u(n) + v(n)|\mathcal{H}_x(n-1)] = E^*[u(n)|\mathcal{H}_x(n-1)] + E^*[v(n)|\mathcal{H}_x(n-1)]$$

or

$$\hat{x}(n) = E^*[u(n)|\mathcal{H}_x(n-1)] + v(n)$$

Statement 3 of Lemma 12.1 implies that

$$E^*[u(n)|\mathcal{H}_x(n-1)] = E^*[u(n)|\mathcal{H}_u(n-1)] + E^*[u(n)|\mathcal{H}_v(n-1)]$$

$$= E^*[u(n)|\mathcal{H}_u(n-1)] = \hat{u}(n)$$

We used the fact that $u(n)$ is orthogonal to every $\mathcal{H}_v(n)$. The predictor $\hat{u}(n)$ was computed in Eq. (12.10). Thus

$$\hat{x}(n) = [h(q) - h(0)]e(n) + v(n) = \sum_{k=1}^{\infty} h(k)e(n-k) + v(n) \qquad (12.11)$$

The above equation expresses the predictor as a convolutional system driven by the innovation sequence $e(n)$ and the singular process $v(n)$. In particular, if $x(n)$ is completely nondeterministic, Eq. (12.11) reduces to

$$\hat{x}(n) = [h(q) - h(0)]e(n) \qquad (12.12)$$

Further information on the predictor structure is extracted if we pass to the frequency domain.

### 12.1.3  Predictor representation in the frequency domain

Equation (12.11) combined with the orthogonality of the WSS processes $e(n)$ and $v(n)$ implies that $\hat{x}(n)$ is a WSS process, jointly stationary with $x(n)$. Since $\hat{x}(n)$ belongs to $\mathcal{H}_x(n-1)$, Proposition 9.2 enables us to write

$$\hat{x}(n) = \frac{1}{2\pi} \int_{-\pi}^{\pi} e^{j(n-1)\omega} G(\omega) dx(\omega), \qquad G(\omega) \in L_2^+(F) \qquad (12.13)$$

Equation (12.11) becomes

$$\hat{x}(n) = \frac{1}{2\pi} \int_{-\pi}^{\pi} e^{jn\omega} [H(\omega) - h(0)] de(\omega) + \frac{1}{2\pi} \int_{-\pi}^{\pi} e^{jn\omega} dv(\omega)$$

Wold's decomposition in the frequency domain gives taking into account (9.35)

$$dx(\omega) = du(\omega) + dv(\omega) = H(\omega)de(\omega) + dv(\omega)$$

The spectral processes $de(\omega)$ and $dv(\omega)$ are uncorrelated. Thus Eq. (12.13) yields

$$\hat{x}(n) = \frac{1}{2\pi} \int_{-\pi}^{\pi} e^{jn\omega} e^{-j\omega} G(\omega) H(\omega) de(\omega) + \frac{1}{2\pi} \int_{-\pi}^{\pi} e^{jn\omega} e^{-j\omega} G(\omega) dv(\omega)$$

Uniqueness of decompositions in the direct sum described in statement 3 of Lemma 12.1 implies that

$$H(\omega) - h(0) = e^{-j\omega} G(\omega) H(\omega) \qquad (12.14)$$

for almost all $\omega$ and at the same time

$$e^{-j\omega} G(\omega) = 1$$

Equation (12.14) is valid almost everywhere because $e(n)$ is white noise. The last two equations can hold simultaneously only if the second holds on a subset $A$ of $[-\pi, \pi]$ where (12.14) does not hold. Hence $A$ is negligible (has Lebesgue measure zero), and the spectral distribution of $v(n)$ is totally concentrated in the set $A$:

$$\int_{-\pi}^{\pi} dF_v(\omega) = \int_A dF_v(\omega)$$

It follows that the predictor is characterized by the function

$$G(\omega) = \begin{cases} e^{j\omega} \dfrac{H(\omega) - h(0)}{H(\omega)}, & \omega \in [-\pi, \pi] - A \\ e^{j\omega}, & \omega \in A \end{cases} \tag{12.15}$$

If $x(n)$ is completely nondeterministic the predictor is directly expressed in terms of $x(n)$ in the frequency domain by

$$\hat{x}(n) = \frac{1}{2\pi} \int_{-\pi}^{\pi} e^{jn\omega} \frac{H(\omega) - h(0)}{H(\omega)} \, dx(\omega) \tag{12.16}$$

Let us consider the Lebesgue decomposition $F = F_1 + F_2 + F_3$. $F_1$ is the absolutely continuous part of $F$, $F_2$ is the singular part of $F$ and $F_3$ is the discontinuous part of $F$. Using (9.30) we find that

$$E[|dx(\omega)|^2] = E[|du(\omega)|^2] + E[|dv(\omega)|^2]$$

or

$$F(\omega) = F_u(\omega) + F_v(\omega)$$

where $F_u(\omega)$, $F_v(\omega)$ are the spectral distributions of $u(n)$ and $v(n)$. Clearly $F_u(\omega)$ is absolutely continuous and

$$dF_u(\omega) = |H(\omega)|^2 d\omega$$

Uniqueness of Lebesgue decomposition implies that

$$F_u(\omega) = F_1(\omega), \quad F_v(\omega) = F_2(\omega) + F_3(\omega)$$

Moreover $S(\omega) = dF_1(\omega)/d\omega = dF_u(\omega)/d\omega$. Thus the power spectral density of $u(n)$ is $S(\omega)$. In particular, if the spectral distribution of the regular process $x(n)$, $F(\omega)$ is absolutely continuous, $x(n)$ is completely nondeterministic. We stress that the above holds provided that $x(n)$ is regular. We will later see examples of processes that are not regular and have absolutely continuous spectral distribution.

### 12.1.4    Predictor spectral factor

Equations (12.13) and (12.15) describe the predictor in the frequency domain. The determination of the predictor function $G(\omega)$ requires knowledge of the function $H(\omega)$. The latter is uniquely specified by the Wold decomposition and Eq. (12.6). Equation (12.6) is not easy to compute directly. In the frequency domain we have

$$|H(\omega)|^2 = S(\omega) \qquad (12.17)$$

Therefore $H(\omega)$ is a spectral factor of $S(\omega)$. $S(\omega)$ has more than one spectral factors. The uniqueness of the coefficients $h(n)$ in Wold decomposition asserts that only one of them is assigned to the predictor. In the sequel we refer to this particular factor as *the predictor spectral factor*. Identification of the predictor spectral factors requires the extraction of additional features of $H(\omega)$. This is done in the next theorem, which is given without proof.

**Theorem 12.2**    The predictor spectral factor $H(\omega)$ of the regular process $x(n)$ with spectral distribution $F(\omega)$ and derivative of $F(\omega)$, $S(\omega)$, is uniquely determined by the following properties.

1. $|H(\omega)|^2 = S(\omega)$.
2. $H(\omega)$ has one-sided Fourier transform in the $L_2$ sense, that is,

$$H(\omega) = \sum_{k=0}^{\infty} h(k)e^{-j\omega k}, \quad \sum_{k=0}^{\infty} |h(k)|^2 < \infty, \qquad h(0) > 0$$

3. The $z$ transform $H(z) = \sum_{k=0}^{\infty} h(k)z^{-k}$ is well-defined outside the unit disc and has no zeros in there, that is,

$$H(z) \neq 0, \qquad |z| > 1 \qquad \blacksquare$$

### 12.1.5    Minimum phase and convolutional representation

The preceding analysis led to a complete characterization of the predictor in the frequency domain. Let us assume that $x(n)$ is completely nondeterministic. Much like the discussion of Section 9.2.3, formula (12.16) does not convert to a convolutional format because $G(\omega)$ may not admit a trigonometric expansion. This barrier is surpassed if we make the additional assumption that $1/H(\omega)$ is square integrable:

$$\int_{-\pi}^{\pi} \frac{1}{|H(\omega)|^2} d\omega < \infty$$

In this case $1/H(\omega)$ is written as a Fourier series

$$\frac{1}{H(\omega)} = \sum_{k=0}^{\infty} \hat{h}(k)e^{-jk\omega}, \quad \hat{h}(0) = \frac{1}{h(0)}$$

Therefore

$$1 - \frac{h(0)}{H(\omega)} = -\sum_{k=1}^{\infty} h(0)\hat{h}(k)e^{-jk\omega}$$

and

$$\hat{x}(n) = -\sum_{k=1}^{\infty} h(0)\hat{h}(k)x(n-k)$$

A linear time invariant filter $H(\omega)$ with causal square integrable frequency response whose inverse is also causal and square integrable is called a *minimum phase* filter.

### 12.1.6    Regularity and the Paley-Wiener condition

The preceding developments critically depend on the assumption of regularity. By definition, a regular process leads to a positive minimum mean squared error, a property hard to check a priori. An analytic test is provided by the Paley-Wiener condition. The Paley-Wiener condition states that $x(n)$ is regular if and only if the function $\log S(\omega)$ is absolutely integrable over $[-\pi, \pi]$. The above discussion is summarized in the following theorem given without proof.

**Theorem 12.3**  Suppose that $S(\omega) > 0$ almost everywhere and that the Paley-Wiener condition

$$\int_{-\pi}^{\pi} \log S(\omega)d\omega > -\infty \tag{12.18}$$

holds. Then there exists a real causal finite energy signal $h(n)$ with Fourier transform $H(\omega)$ and $z$ transform $h(z)$ such that $h(0) > 0$, $h(z) \neq 0$ for $|z| > 1$, and $|H(\omega)|^2 = S(\omega)$. Moreover $x(n)$ is regular. Conversely, if $x(n)$ is regular, $S(\omega) > 0$ almost everywhere, and the Paley-Wiener condition (12.18) holds. ∎

*Synopsis.* Let us summarize the main points of the preceding discussion. A wide sense stationary process $x(n)$ has always a differentiable spectral distribution $F(\omega)$. The derivative of $F(\omega)$, $S(\omega)$ is integrable on $[-\pi, \pi]$. If $S(\omega)$ satisfies the Paley-Wiener condition, $x(n)$ is regular; otherwise, it is singular. In the latter case $x(n)$ is a limit of linear combinations of past values and $\hat{x}(n) = x(n)$. In the regular case the mean square error is always positive. If $F(\omega)$ is absolutely continuous the Paley-Wiener condition may or may not hold. According to Section 9.2.4 a process with absolutely continuous spectral distribution can be generated at the output of a linear filter excited by white noise. This filter is not in general causal. The Paley-Wiener condition enforces causality and leads to regularity. The predictor is characterized by the function $G(\omega)$ specified by Eq. (12.15), which in turn requires $H(\omega)$. The predictor spectral factor $H(\omega)$ is the spectral factor of $S(\omega)$ that has one-sided Fourier series and $z$ transform featuring no zeros outside the unit circle. It turns out that these requirements determine a unique spectral factor of $S(\omega)$.

Under the stronger condition that $\log S(\omega)$ is square integrable (recall that the space of square integrable functions over a finite interval is contained in the space of absolutely integrable functions over the same interval), the construction of the predictor spectral factor is somewhat easier to understand. Indeed, let us assume that $S(\omega) > 0$ almost everywhere and that $\log S(\omega)$ belongs to $L_2$ over $[-\pi, \pi]$. Consider the Fourier series expansion of $\log S(\omega)$

$$\log S(\omega) = \sum_{k=-\infty}^{\infty} a_k e^{j\omega k}$$

Since $S(\omega)$ is real, $\log S(\omega)$ is real as well, and $a_{-k} = a_k^*$. In particular,

$$a_0 = \frac{1}{2\pi} \int_{-\pi}^{\pi} \log S(\omega) d\omega$$

Thus

$$\log S(\omega) = a_0 + \sum_{k=-\infty}^{-1} a_k e^{j\omega k} + \sum_{k=1}^{\infty} a_k e^{j\omega k}$$

$$= a_0 + \sum_{k=1}^{\infty} a_k^* e^{-j\omega k} + \sum_{k=1}^{\infty} a_k e^{j\omega k}$$

Let

$$H(\omega) = e^{a_0/2 + \sum_{k=1}^{\infty} a_k^* e^{-j\omega k}}$$

Then

$$S(\omega) = e^{\log S(\omega)} = |H(\omega)|^2$$

Since $H(\omega)$ is a spectral factor of the integrable function $S(\omega)$, it is square integrable. If the exponential is expanded in power series, the one-sided Fourier series of $H(\omega)$ is obtained. The $z$ transform involves no zeros outside the unit circle due to the exponential nature of $H(\omega)$. The minimum prediction error is given by

$$\sigma_1^2 = h^2(0) = e^{\int_{-\pi}^{\pi} \log S(\omega)d\omega/2\pi}$$

### 12.1.7 Predictors for ARMA signals

Suppose that $x(n)$ is generated by the ARMA model

$$x(n) + \sum_{i=1}^{k} a_i x(n-i) = \sum_{i=0}^{r} b_i w(n-i) \tag{12.19}$$

where $w(n)$ is a white noise signal of variance 1. As we saw in Section 9.2.4 the spectral distribution is absolutely continuous and the power spectral density is given by

$$S(\omega) = \frac{dF(\omega)}{d\omega} = \frac{|b(\omega)|^2}{|a(\omega)|^2} = \frac{b(z)b(z^{-1})}{a(z)a(z^{-1})}\bigg|_{z=e^{j\omega}}. \tag{12.20}$$

where

$$a(z) = 1 + \sum_{i=1}^{k} a_i z^{-i}, \quad b(z) = \sum_{i=0}^{r} b_i z^{-i}$$

provided that no poles are located on the unit circle. It can be shown that the Paley-Wiener condition holds, and therefore ARMA signals with no poles of unit magnitude are always regular and completely nondeterministic. The rational function $b(z)/a(z)$ constitutes a spectral factor of $S$ but may not correspond to the predictor, since it may fail to satisfy properties 2 and 3 of Theorem 12.2. For a rational spectral factor, property 2 requires that the poles be inside the unit disc. The poles of $b(z)/a(z)$ are the roots of the reversed polynomial $a_{\#}(z)$:

$$a(z) = z^{-k} a_{\#}(z), \qquad a_{\#}(z) = a_k + a_{k-1} z + \cdots + z^k$$

To determine the denominator $d(z)$ of the predictor spectral factor, we split the roots of $a_{\#}(z)$ into the roots $\lambda_s$ located inside the unit disc and the roots $\lambda_u$ lying outside the unit disc:

$$a_{\#}(z) = a_s(z)a_u(z), \quad a_s(z) = \prod_{|\lambda_s|<1} (z - \lambda_s), \quad a_u(z) = \prod_{|\lambda_u|>1} (z - \lambda_u)$$

Remember no pole is allowed on the unit circle. The predictor denominator is then formed from the stable poles $\lambda_s$ and the reciprocals of the unstable poles:

$$d_\#(z) = a_s(z) \prod_{|\lambda_u|>1} \left(z - \frac{1}{\lambda_u}\right)$$

Notice that

$$(z - \lambda_u)(z^{-1} - \lambda_u) = \lambda_u^2\left(z - \frac{1}{\lambda_u}\right)\left(z^{-1} - \frac{1}{\lambda_u}\right)$$

Therefore

$$a(z)a(z^{-1}) = a_\#(z)a_\#(z^{-1})$$

$$= \prod(z - \lambda_s)\prod(z - \lambda_u)\prod(z^{-1} - \lambda_s)\prod(z^{-1} - \lambda_u)$$

$$= a_s(z)\left(\prod(z - \lambda_u)(z^{-1} - \lambda_u)\right)a_s(z^{-1})$$

$$= \left(\prod_{\lambda_u}\lambda_u^2\right)a_s(z)a_s(z^{-1})\prod\left(z - \frac{1}{\lambda_u}\right)\left(z^{-1} - \frac{1}{\lambda_u}\right)$$

Consequently

$$a(z)a(z^{-1}) = d(z)d(z^{-1}), \quad d(z) = \prod_{\lambda_u}\lambda_u z^{-k}d_\#(z)$$

The numerator is treated likewise. Thus

$$b(z)b(z^{-1}) = c(z)c(z^{-1})$$

$$c(z) = \prod_{\rho_u}\rho_u z^{-r}c_\#(z), \quad c_\#(z) = b_s(z)\prod\left(z - \frac{1}{\rho_u}\right)$$

The predictor spectral factor is $c(z)/d(z)$. Next we determine the predictor. Since $F$ is absolutely continuous, the Wold decomposition asserts that the singular component of $x(n)$ is zero. Therefore

$$\hat{x}(n) = \int_{-\pi}^{\pi} e^{jn\omega}\frac{H(\omega) - h(0)}{H(\omega)}dx(\omega)$$

and in the frequency domain

$$d\hat{x}(\omega) = \left(1 - \frac{h(0)}{H(\omega)}\right)dx(\omega) = \frac{c(\omega) - h(0)d(\omega)}{c(\omega)}dx(\omega)$$

or

$$c(\omega)d\hat{x}(\omega) = [c(\omega) - h(0)d(\omega)]dx(\omega)$$

Hence

$$\sum_k c_k e^{-jk\omega} d\hat{x}(\omega) = \sum_k c_k e^{-jk\omega} dx(\omega) - h(0)\sum_k d_k e^{-jk\omega} dx(\omega)$$

Integration over $[-\pi, \pi]$ and the spectral representation lead to

$$\sum c_i \hat{x}(n - i) = \sum (c_i - h(0)d_i)x(n - i)$$

The predictor is modeled as an ARMA filter driven by the signal $x(n)$. It is written as a transfer function model with transfer function

$$\frac{c(q) - h(0)d(q)}{c(q)}$$

### Example 12.1 Prediction for an ARMA signal

Let

$$x(n) - \frac{7}{2}x(n - 1) + \frac{3}{2}x(n - 2) = w(n) - 4w(n - 1)$$

The $z$ transform of the autocorrelation function is

$$S(z) = \frac{(1 - 4z^{-1})(1 - 4z)}{(1 - \frac{7}{2}z^{-1} + \frac{3}{2}z^{-2})(1 - \frac{7}{2}z + \frac{3}{2}z^2)}$$

We have

$$a(z) = 1 - \frac{7}{2}z^{-1} + \frac{3}{2}z^{-2} = z^{-2}\left(z - \frac{1}{2}\right)(z - 3)$$

Hence $\lambda_s = 1/2$, $\lambda_u = 3$. The denominator of the predictor spectral factor is

$$d(z) = 3z^{-2}\left(z - \frac{1}{2}\right)\left(z - \frac{1}{3}\right) = 3 - \frac{5}{2}z^{-1} + \frac{1}{2}z^{-2}$$

In a similar fashion we have

$$b(z) = 1 - 4z^{-1} = z^{-1}(z - 4)$$

$$c(z) = 4z^{-1}\left(z - \frac{1}{4}\right) = 4 - z^{-1}$$

The predictor spectral factor is

$$h(z) = \frac{c(z)}{d(z)} = \frac{4 - z^{-1}}{3 - \frac{5}{2}z^{-1} + \frac{1}{2}z^{-2}}$$

from which it follows that $h(0) = 4/3$. The minimum prediction error is $\sigma_1^2 = 16/9$. The transfer function of the ARMA predictor is

$$\frac{c(z) - h(0)d(z)}{c(z)} = \frac{\frac{7}{12}z^{-1} - \frac{1}{6}z^{-2}}{1 - \frac{1}{4}z^{-1}}$$

The ARMA predictor is given in the time domain

$$\hat{x}(n) - \frac{1}{4}\hat{x}(n-1) = \frac{7}{12}x(n-1) - \frac{1}{6}x(n-2)$$    ∎

### Example 12.2    Bandpass signals

Suppose that $x(n)$ is obtained at the output of a bandpass filter excited by white noise

$$x(n) = \sum_{k=-\infty}^{\infty} h(k)w(n-k)$$

The frequency response $H(\omega)$ is 1 inside the band $[W_1, W_2]$ and zero elsewhere. According to Section 9.1 the power spectral density $S(\omega)$ of $x(n)$ is $S(\omega) = |H(\omega)|^2 = H(\omega)$. The Paley-Wiener condition fails because $S(\omega)$ is zero outside $[W_1, W_2]$, and the logarithm becomes infinite. Therefore $x(n)$ is singular.    ∎

### Example 12.3    Harmonic process

Suppose that $x(n)$ is a harmonic process. The power spectral density does not exist in the ordinary sense (see Example 9.4). The spectral distribution is a piecewise constant function. Lebesgue decomposition of the spectral distribution and the discussion of Section 12.1.3 imply that $x(n)$ is a singular process.

∎

## 12.2    WIENER FILTER

The Wiener filter is an infinite memory filter that processes the measurement signal $y(n)$ with the aim to reconstruct the desired signal $x(n)$ at the output. Let $\mathcal{H}_y(n)$ be the Hilbert space generated by the past measurements $y(n-k)$,

$k = 0, 1, 2, \ldots$, and $\hat{x}(n)$ the projection of $x(n)$ onto $\mathcal{H}_y(n)$:

$$\hat{x}(n) = E^*[x(n)|\mathcal{H}_y(n)]$$

where $\hat{x}(n)$ is the best linear mean square estimator of $x(n)$ with respect to measurements up to time $n$. The Wiener filter represents the system with input $y(n)$ and output $\hat{x}(n)$. Under broad conditions it is implemented by a linear time invariant causal system whose frequency response is determined by the power spectral density of $y$ and the cross spectrum of $y$ and $x$.

Suppose that $x(n)$ and $y(n)$ are zero mean jointly wide sense stationary processes and that the spectral distribution of the augmented vector $[x(n), y(n)]$ is absolutely continuous. Let $S_y(\omega)$ denote the power spectral density of $y(n)$, and let $S_{xy}(\omega)$ denote the cross spectral density of $x(n)$ and $y(n)$. Then

$$r_{xy}(n) = E[x(k)y(k-n)] = \frac{1}{2\pi} \int_{-\pi}^{\pi} e^{jn\omega} S_{xy}(\omega) d\omega \qquad (12.21)$$

Suppose that $y(n)$ is regular. It follows from the Wold decomposition that it is completely nondeterministic. Let $e(n) = \tilde{y}(n)/(E[\tilde{y}^2(n)])^{1/2}$ be the normalized innovation of $y(n)$, and let $H(\omega)$ be the predictor spectral factor having a well-defined square integrable inverse $1/H(\omega) = \sum_{k=0}^{\infty} \tilde{h}(k)e^{-jk\omega}$. Then the innovation sequence is obtained by the whitening filter

$$e(n) = \sum_{k=0}^{\infty} \tilde{h}(k)y(n-k) \qquad (12.22)$$

The Wiener filter is given by the expression

$$\hat{x}(n) = E^*[x(n)|\mathcal{H}_y(n)] = \sum_{k=0}^{\infty} g_f(k)e(n-k) \qquad (12.23)$$

Since $e(n-k)$, $k \geq 0$, form an orthonormal set, we have

$$g_f(k) = <x(n), e(n-k)> = r_{xe}(k)u_s(k) \qquad (12.24)$$

The crosscorrelation $r_{xe}(k)$ is obtained from (12.22):

$$r_{xe}(k) = <x(n), e(n-k)> = <x(n), \sum_{i=0}^{\infty} \tilde{h}(i)y(n-k-i)>$$

$$= \sum_{i=0}^{\infty} \tilde{h}(i) <x(n), y(n-k-i)> = \sum_{i=0}^{\infty} \tilde{h}(i)r_{xy}(k+i)$$

Passing to the frequency domain, we obtain

$$S_{xe}(\omega) = S_{xy}(\omega)\tilde{H}(-\omega) = \frac{S_{xy}(\omega)}{H(-\omega)} \qquad (12.25)$$

In summary the Wiener filter coefficients $g_f(k)$ are determined by the following procedure:

- Compute the predictor spectral factor of $y(n)$, $H(\omega)$
- Compute $S_{xe}(\omega)$ from Eq. (12.25)
- Compute $g_f(k)$ from Eq. (12.24).

One important point the reader should keep in mind is that Eq. (12.24) holds only for $k \geq 0$. If $k$ takes negative values, $g_f(k)$ is zero by causality, while $r_{xe}(k)$ is nonzero in general. As a consequence Eq. (12.24) does not convert in the frequency domain to $G_f(\omega) = S_{xy}(\omega)/H(-\omega)$.

If $S_{xe}(\omega)$ is rational, the $z$ transform of $r_{xe}(k)u_s(k)$, $g_f(z)$ is determined from the $z$ transform of $r_{xe}(k)$ directly using partial fraction expansion. Suppose, for simplicity, that the $z$ transform of $r_{xe}(k)$, $S_{xe}(z)$ has simple poles. Partial fraction expansion gives

$$S_{xe}(z) = \sum_i \frac{c_i}{1 - \lambda_i z^{-1}} + \sum_i \frac{d_i}{1 - \rho_i z^{-1}}, \qquad |\lambda_i| < 1, \qquad |\rho_i| > 1$$

It follows that the $z$ transform of $r_{xe}(k)u_s(k)$, $g_f(z)$ is rational and its poles are included in the poles of $S_{xe}(z)$. Since $r_{xe}(k)u_s(k)$ is causal, its poles are precisely $1/\lambda_i$. Hence

$$g_f(z) = \sum_i \frac{c_i}{1 - \lambda_i z^{-1}}$$

In the preceding discussion the Wiener filter is expressed in terms of the innovation sequence $e(n)$ via the impulse response $g_f(k)$. A direct representation in terms of the signal $y(n)$ is readily obtained from Eq. (12.22). It has the form

$$\hat{x}(n) = \sum_{k=0}^{\infty} g(k)y(n - k)$$

where

$$g(k) = \tilde{h}(k) * g_f(k)$$

or in the frequency domain

$$G(\omega) = \frac{G_f(\omega)}{H(\omega)}$$

*Example 12.4   ARMA signal with noisy measurement*
Suppose that the desired signal is generated by the ARMA model

$$x(n) + 3x(n-1) = w(n) + \frac{5}{6}w(n-1)$$

The process noise $w(n)$ is a white noise signal of variance 1. Let

$$y(n) = x(n) + v(n)$$

where $v(n)$ represents the measurement noise. It is white noise with variance 5/9 and is uncorrelated with $w(n)$. Since $x(n)$ is obtained by filtering $w(n)$ with a linear filter, it is also uncorrelated with $v(n)$. Hence

$$r_y(k) = E[y(n)y(n-k)] \, =<x(n)+v(n), x(n-k)+v(n-k)>$$

$$=<x(n), x(n-k)> + <v(n), v(n-k)> = r_x(k) + r_v(k)$$

Passing to the frequency domain, we obtain

$$S_y(\omega) = S_x(\omega) + \sigma_v^2 = S_x(\omega) + \frac{5}{9}$$

Likewise the crosscorrelation is found to be

$$r_{yx}(k) =<y(n), x(n-k)>=<x(n)+v(n), x(n-k)>= r_x(k)$$

and

$$S_{yx}(\omega) = S_x(\omega)$$

Proceeding as in Example 12.1, we have

$$S_{yx}(z) = \frac{(1+\frac{5}{6}z^{-1})(1+\frac{5}{6}z)}{(1+3z^{-1})(1+3z)} \tag{12.26}$$

and

$$S_y(z) = \frac{(1+\frac{5}{6}z^{-1})(1+\frac{5}{6}z)}{(1+3z^{-1})(1+3z)} + \frac{5}{9}$$

$S_y(z)$ is proper rational with numerator

$$\frac{261}{4} + \frac{45}{2}z + \frac{45}{2}z^{-1}$$

To determine the predictor spectral factor of $y(n)$ we need to factor the above in the form $K(1 + cz)(1 + cz^{-1})$. It is easy to verify that $c$ is a root of

$$c^2 - \frac{29}{10}c + 1 \tag{12.27}$$

and that $K$ is given by $K = 45/2c$. The roots of (12.27) are $c_1 = 2/5$, $c_2 = 5/2$. Notice that $c_1 = 1/c_2$. The two values of $K$ are $225/4$ and $9$. The power spectral density of $y$ is written

$$S_y(z) = \frac{(1 + \frac{5}{2}z)(1 + \frac{5}{2}z^{-1})}{(1 + 3z)(1 + 3z^{-1})}$$

The predictor spectral factor associated with the signal $y$ is

$$h(z) = \frac{\frac{5}{2} + z^{-1}}{3 + z^{-1}}$$

Thus

$$S_{xe}(z) = \frac{\frac{5}{6}(1 + \frac{6}{5}z^{-1})(1 + \frac{5}{6}z^{-1})}{3(1 + \frac{1}{3}z^{-1})(1 + \frac{5}{2}z^{-1})}$$

The $z$ transform of $g_f(k)$ is obtained from the partial fraction expansion of $S_{xe}(z)$ by maintaining only the pole outside the unit disc, which is the pole $-3$. Thus

$$g_f(k) = R_1\left(-\frac{1}{3}\right)^k u_s(k) \qquad \blacksquare$$

Continuous time prediction and filtering is developed along the same lines.

## 12.3  DIGITAL PULSE AMPLITUDE MODULATION AND DEMODULATION

The problem addressed by Wiener filters arises in signal demodulation. In fact, if the causality requirement of the filter is not explicitly accounted in the design, the optimum noncausal Wiener filter is much easier to develop. It is treated in Exercise 12.20. A closer look at digital pulse amplitude modulation and demodulation systems is taken in this section. The design of transmitting and receiving filters that cope with the bandwidth limitations of the channel is described.

### 12.3.1  Digital PAM transmission

We pointed out in Section 4.7.1 that transmission media like microwave links and optical fibers transmit signals in analog form. In a digital communication

**Figure 12.1.** PAM signal.

system the channel encoder delivers information in digital form. The role of the modulator is then to convert a digital stream to an analog waveform suitable for transmission. In binary pulse amplitude modulation (PAM), the information bit 1 is represented by a pulse of amplitude $a$ and duration $T$, and the information bit 0 is represented by a pulse of amplitude $-a$ and duration $T$ (see Fig. 12.1). In practical implementations pulses have smoother characteristics. Suppose, more generally, that the channel encoder output is a stream of binary vectors $v_i$ and that each vector $v_i$ consists of $n$ bits. The PAM signal generated by the modulator has the form

$$v(t) = \sum_i a(i) g_T(t - iT) \tag{12.28}$$

For each $i$, $a(i)$ is one of $M = 2^n$ pulse amplitude values. The function $g_T(t)$ is deterministic and has approximately the shape of a rectangular pulse. The sequence $a(i)$ represents the amplitude levels used to represent each value of the coded sequence $v_i$ and is taken to be a wide sense stationary discrete signal with autocorrelation $r_a(i)$ and power spectral density $S_a(\omega)$. The block diagram of a digital PAM system is illustrated in Fig. 12.2. The transmitting filter with impulse response $g_T(t)$ and frequency response $G_T(\omega)$ converts the digital stream $a(i)$ into an analog waveform $v(t)$ in accordance with Eq. (12.28) and transmits the signal through the channel. The channel is represented by the mathematical model described in Section 6.6. The filter $C(\omega)$ specifies the frequency characteristics of the physical medium and $w(t)$ the additive noise degrading the transmitted signal. We will assume that $w(t)$ has zero mean and square integrable power spectral density $S_w(\omega)$. The received analog waveform $y(t)$ is processed by the receiving filter with impulse response $g_R(t)$ and frequency response $G_R(\omega)$. The output of the receiving filter is sampled quantized and subsequently passed to the channel decoder. The aim is to design the transmitting filter $G_T(\omega)$ and the

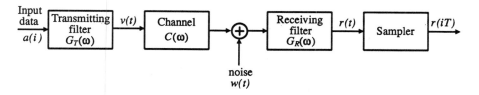

**Figure 12.2.** Digital PAM system.

receiving filter $G_R(\omega)$ so that the bandwidth limitations imposed by the channel frequency response $C(\omega)$ are satisfied. Note in this respect that the bandlimited nature of the channel generally precludes the use of the rectangular pulse as impulse response for the transmitting filter $g_T(t)$.

We will confine our analysis to bandlimited baseband channels. The frequency response of these channels satisfies $C(\omega) = 0$ for $|\omega| > W$. The methods described in the sequel can be combined with suitable modulation schemes and be applied to bandpass channels such as optical fibers and satellite links.

The block diagram of Fig. 12.2 readily translates into the following expression:

$$r(t) = g_R(t) * y(t) = g_R(t) * [c(t) * v(t) + w(t)]$$

or

$$r(t) = (g_R(t) * c(t)) * v(t) + g_R(t) * w(t) \qquad (12.29)$$

The convolution of a deterministic signal $z(t)$ with a PAM signal $v(t)$ of the form (12.28) is also a PAM signal with impulse response $z(t) * g_T(t)$. Indeed,

$$z(t) * v(t) = \int_{-\infty}^{\infty} z(\tau)v(t - \tau)d\tau = \sum_i a(i) \int_{-\infty}^{\infty} z(\tau)g_T(t - \tau - iT)d\tau$$

or

$$z(t) * v(t) = \sum_i a(i)(z * g_T)(t - iT)$$

Applying the above remark to (12.29), we obtain

$$r(t) = \sum_i a(i)h(t - iT) + \xi(t)$$

where

$$\xi(t) = g_R(t) * w(t) \qquad (12.30)$$

and

$$h(t) = g_R(t) * c(t) * g_T(t) \qquad (12.31)$$

In the sequel, $n$ stands for discrete time. The sampler's output is

$$r(nT) = \sum_i a(i)h(nT - iT) + \xi(nT)$$

or

$$r(n) = a(n) * h(n) + \xi(n), \qquad h(n) = h(nT) \qquad (12.32)$$

The latter expression indicates that reception of $a(n)$ is degraded by the filter $h(n)$ and the noise $\xi(n)$. The effect of $h(n)$ is somewhat more explicitly displayed if we write (12.32) as

$$r(n) = h(0)a(n) + \sum_{i \neq n} a(i)h(n - i) + \xi(n) \qquad (12.33)$$

The middle term of the right-hand side represents the intersymbol interference (ISI). It shows that the symbol $a(n)$ is affected by the symbols $a(i)$, $i \neq n$.

### 12.3.2 Cancellation of intersymbol interference

To remove the intersymbol interference, we search for transmitting and receiving filters that will ensure

$$a(n) * h(n) = a(n) \quad \text{or} \quad h(n) = \delta(n)$$

$\delta(n)$ is the unit sample signal with Fourier transform equal to 1. Note that $h(n)$ is obtained from $h(t)$ by sampling. Hence the Fourier transform of $h(n)$ relates to the Fourier transform of $h(t)$ by Eq. (6.97) of Section 6.7. Therefore ISI is removed if the following condition holds:

$$\sum_{k=-\infty}^{\infty} H\left(\frac{\omega + 2k\pi}{T}\right) = T \qquad (12.34)$$

Equation (12.31) converts in the frequency domain to

$$H(\omega) = G_R(\omega)C(\omega)G_T(\omega) \qquad (12.35)$$

Since $C(\omega)$ is bandlimited with bandwidth $W$, $H(\omega)$ is also bandlimited with bandwidth $W$. If the Nyquist criterion $W \leq 2\pi/T$ is satisfied, the spectrum of $h(n)$ consists of nonoverlapping shifts of the analog spectrum $H(\omega)$. Therefore Eq. (12.34) cannot be satisfied. If $W \geq 2\pi/T$, the shifts of the analog spectra overlap, and Eq. (12.34) is satisfied for numerous choices of $H(\omega)$. One choice that has been widely used is the *raised cosine*

$$h(t) = \text{sinc}\left(\frac{t}{T}\right) \frac{\cos(\pi a t/T)}{1 - 4a^2 t^2/T^2}$$

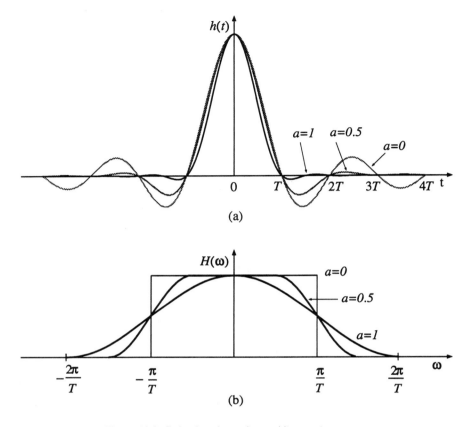

**Figure 12.3.** Raised cosine pulse and its spectrum.

with Fourier transform

$$H(\omega) = \begin{cases} T, & 0 \le |\omega| \le \pi\dfrac{1-a}{T} \\[2ex] \dfrac{T}{2}\left\{1 + \cos\left[\dfrac{\pi T}{a}\left(\dfrac{|\omega|}{2\pi} - \dfrac{1-a}{2T}\right)\right]\right\}, & \pi\dfrac{1-a}{T} < |\omega| \le \pi\dfrac{1+a}{T} \\[2ex] 0, & |\omega| > \pi\dfrac{1+a}{T} \end{cases}$$

The parameter $a$ is called the *rolloff factor*. It satisfies $0 \le a \le 1$. Figure 12.3 illustrates the raised cosine pulse and its spectrum for $a = 0$, $1/2$, $1$.

### 12.3.3 Optimum transmitting and receiving filters

For the remaining of this section, $H(\omega)$ will denote the raised cosine spectrum. We will further confine our search of transmitting and receiving filters to bandlimited filters: $G_T(\omega) = G_R(\omega) = 0$, for $|\omega| \geq W$. Equation (12.32) takes the form $r(n) = a(n) + \xi(n)$. The noise free receiving filter output has power $E[a^2(n)]$. Likewise the noisy component at the receiver output has power $E[\xi^2(n)]$. We seek to determine the receiver and transmitter filters so that the signal to noise ratio

$$\frac{E[a^2(n)]}{E[\xi^2(n)]}$$

is maximized for a given level of transmitted signal power. Equivalently we seek to minimize the noise to signal ratio $E[\xi^2(n)]/E[a^2(n)]$.

The power of the noise component is easy to find using the developments of Section 9.1.4. and Eq. (12.30). Thus

$$E[\xi^2(n)] = r_\xi(0)$$

$$= \frac{1}{2\pi} \int_{-\infty}^{\infty} S_\xi(\omega) d\omega = \frac{1}{2\pi} \int_{-W}^{W} |G_R(\omega)|^2 S_w(\omega) d\omega \qquad (12.36)$$

Let us next compute the average transmitted power $E[v^2(t)]$. Equation (12.28) yields

$$E[v^2(t)] = E\left[ \sum_k a(k) g_T(t - kT) \sum_i a(i) g_T(t - iT) \right]$$

$$= \sum_k \sum_i r_a(k - i) g_T(t - kT) g_T(t - iT)$$

$$= \sum_l r_a(l) \sum_k g_T(t - kT + lT) g_T(t - kT)$$

Assuming that $a(n)$ is zero mean white noise, we obtain

$$E[v^2(t)] = r_a(0) \sum_k g_T^2(t - kT) \qquad (12.37)$$

Note that $E[v^2(t)]$ generally depends on $t$. In fact it is periodic with period $T$. It turns out that the PAM signal $v(t)$ is not wide sense stationary but cyclostationary with period $T$ (see Exercise 9.10). The average transmitted power of $v(t)$ is defined as

$$P_{av} = \frac{1}{T} \int_0^T E[v^2(t)] dt \qquad (12.38)$$

Equation (12.37) gives

$$P_{av} = \frac{1}{T} r_a(0) \int_0^T \sum_k g_T^2(t - kT) dt = \frac{E[a^2(n)]}{T} \int_{-\infty}^{\infty} g_T^2(t) dt$$

Taking into account Parseval's identity, the above equation gives

$$E[a^2(n)] = \frac{2\pi T \; P_{av}}{\int_{-\infty}^{\infty} |G_T(\omega)|^2 d\omega} \tag{12.39}$$

The frequency response of the transmitting filter is expressed in terms of the frequency response of the receiving filter by means of Eq. (12.35) as

$$|G_T(\omega)| = \frac{|H(\omega)|}{|C(\omega)||G_R(\omega)|}, \qquad |\omega| \le W \tag{12.40}$$

and $G_T(\omega) = 0$ for $|\omega| \ge W$. Combining Eqs. (12.36), (12.39), and (12.40), we seek to determine $G_R(\omega)$ so that for a given average transmitted power level $P_{av}$ the noise to signal ratio

$$\frac{E[\xi^2(n)]}{E[a^2(n)]} = \frac{1}{4\pi^2 T \; P_{av}} \int_{-W}^{W} |G_R(\omega)|^2 S_w(\omega) d\omega \int_{-W}^{W} \frac{|H(\omega)|^2}{|C(\omega)|^2 |G_R(\omega)|^2} d\omega$$

is minimized. The Cauchy-Schwarz inequality yields the lower bound

$$\frac{E[\xi^2(n)]}{E[a^2(n)]} \ge \frac{1}{4\pi^2 T \; P_{av}} \left[ \int_{-W}^{W} \frac{|H(\omega)| \sqrt{S_w(\omega)}}{|C(\omega)|} d\omega \right]^2$$

which is independent of $|G_R(\omega)|^2$. We recall from Section 2.8 that the Cauchy-Schwarz inequality becomes equality if the relevant vectors are linearly dependent. In our case this leads to

$$|G_R(\omega)| \sqrt{S_w(\omega)} = K \frac{|H(\omega)|}{|C(\omega)||G_R(\omega)|}, \qquad |\omega| \le W$$

Thus the magnitude of the receiving filter is given by

$$|G_R(\omega)| = K_1 \frac{|H(\omega)|^{1/2}}{|C(\omega)|^{1/2} [S_w(\omega)]^{1/4}}, \qquad |\omega| \le W \tag{12.41}$$

In accordance with Eq. (12.40), the magnitude of the transmitting filter is

$$|G_T(\omega)| = \frac{1}{K_1} \frac{|H(\omega)|^{1/2} [S_w(\omega)]^{1/4}}{|C(\omega)|^{1/2}}, \qquad |\omega| \le W \tag{12.42}$$

If the additive noise is white Gaussian with variance $N_0/2$, we obtain

$$|G_R(\omega)| = M_1 \frac{|H(\omega)|^{1/2}}{|C(\omega)|^{1/2}}, \quad |G_T(\omega)| = M_2 \frac{|H(\omega)|^{1/2}}{|C(\omega)|^{1/2}} \qquad |\omega| \le W$$

In this case it holds that $|G_T(\omega)| = L|G_R(\omega)|$ for $|\omega| \leq W$. An obvious solution is $G_T(\omega) = M_2\sqrt{H(\omega)/C(\omega)}\, e^{-j\omega t_1}$ and $G_R(\omega) = M_1 G_T^*(\omega)e^{-j\omega t_2}$. The receiving filter is *matched* to the transmitting filter. The delays $t_1$ and $t_2$ are inserted to ensure approximate causality.

### 12.3.4  Channel equalization

In the preceding analysis the channel frequency response $C(\omega)$ was assumed known. In practice, we are often confronted with channels that are unknown or time varying. Mobile cellular radio channels are an example. To cope with this case, we return to Eq. (12.33), and we assume that intersymbol interference affects a finite number of symbols. The convolution can be approximated by an FIR filter

$$r(n) = \sum_{i=M}^{N} a(i)h(n-i) + \xi(n) \tag{12.43}$$

and the problem can be viewed as an FIR identification problem. A training data sequence $a(n)$ is transmitted, and the received sequence $r(n)$ is measured. The coefficients $h(k)$ are estimated by the methods and algorithms of Chapter 4. The resulting filters are called *channel equalizers*. For time varying channels, input data are transmitted periodically. Adaptive equalizers are adaptive filters that update their parameters using the LMS algorithm, the RLS algorithm, or their variants.

An alternative approach to channel equalization relies on the observation that the output of the FIR filter (12.43) is realized as the output of a finite state machine plus noise. This is a consequence of the fact that $a(n)$ is a digital signal with values in a finite alphabet. The Viterbi algorithm described in Section 14.8 can then be used to determine the maximum likelihood estimate of $a(n)$ based on the received sequence $r(n)$.

## 12.4  KALMAN FILTER

In Sections 12.1 and 12.2, filtering and prediction were studied in an infinite horizon environment involving stationary signals. Wiener filters are essentially infinite memory time invariant causal linear systems whose frequency response is determined by a suitable spectral factor of the input power spectral density and the cross spectrum of the input and desired response. In this section finite memory filtering is reapproached using state space representations. Finite windows whose size grows with the observation interval are considered. It is shown that if the data sequences are produced by a linear state space model, the least mean square estimator is realized by a time varying linear state space model, known as the *Kalman filter*. Analogous results are valid in the continuous time and lead to the Kalman-Bucy filter.

### 12.4.1 Recursiveness in linear least squares estimation

Let $\hat{X}_1 = E^*[X|Y_1]$ be the linear least squares estimator of $X$, given $Y_1$, and let $\tilde{X}_1 = X - \hat{X}_1$ be the resulting error. Suppose that a new measurement $Y_2$ becomes available. We want to compute $\hat{X}_2 = E^*[X|Y]$, $Y^T = \begin{pmatrix} Y_1^T & Y_2^T \end{pmatrix}$ in a way that utilizes the computed estimate $\hat{X}_1$. Suppose first that the newly arrived information $Y_2$ is uncorrelated with the past measurement $Y_1$. Then the covariance matrix $C_{Y_1 Y_2} = \text{cov}[Y_1, Y_2]$ is zero. In a sense $Y_2$ carries entirely new information in comparison to $Y_1$, since no information in $Y_2$ can be extracted from $Y_1$. Using Eq. (4.18) of Section 4.2.3, we have

$$\hat{X}_2 = E^*[X|Y] = C_{XY} C_{YY}^{-1}(Y - m_Y) + m_X \tag{12.44}$$

But

$$m_Y^T = \begin{pmatrix} m_{Y_1}^T & m_{Y_2}^T \end{pmatrix}$$

and

$$C_{XY} = E[(X - m_X)(Y - m_Y)^T] = \begin{pmatrix} C_{XY_1} & C_{XY_2} \end{pmatrix}$$

Likewise

$$C_{YY} = \begin{pmatrix} C_{Y_1 Y_1} & C_{Y_1 Y_2} \\ C_{Y_2 Y_1} & C_{Y_2 Y_2} \end{pmatrix} = \begin{pmatrix} C_{Y_1 Y_1} & 0 \\ 0 & C_{Y_2 Y_2} \end{pmatrix}$$

$C_{YY}$ is block diagonal because $Y_1$, $Y_2$ are uncorrelated. Thus

$$C_{YY}^{-1} = \begin{pmatrix} C_{Y_1 Y_1}^{-1} & 0 \\ 0 & C_{Y_2 Y_2}^{-1} \end{pmatrix}$$

and Eq. (12.44) becomes

$$\hat{X}_2 = C_{XY_1} C_{Y_1 Y_1}^{-1}(Y_1 - m_{Y_1}) + C_{XY_2} C_{Y_2 Y_2}^{-1}(Y_2 - m_{Y_2}) + m_X$$

or

$$\hat{X}_2 = \hat{X}_1 + E^*[X|Y_2] - m_X = E^*[X|Y_1] + E^*[X|Y_2] - m_X \tag{12.45}$$

A similar argument leads to the following update of the covariance error:

$$C_{\tilde{X}_2 \tilde{X}_2} = C_{XX} - C_{XY} C_{YY}^{-1} C_{YX}$$
$$= C_{XX} - C_{XY_1} C_{Y_1 Y_1}^{-1} C_{Y_1 X} - C_{XY_2} C_{Y_2 Y_2}^{-1} C_{Y_2 X}$$

The first two terms in the right-hand side yield $C_{\tilde{X}_1 \tilde{X}_1}$. Thus

$$C_{\tilde{X}_2 \tilde{X}_2} = C_{\tilde{X}_1 \tilde{X}_1} - C_{XY_2} C_{Y_2 Y_2}^{-1} C_{Y_2 X} \tag{12.46}$$

Equation (12.45) states that the estimate of $X$ based on the collective information $Y_1$, $Y_2$, is determined by simply adding the best estimate of $X$ formed by the recent measurement $Y_2$, to the previously computed estimate $\hat{X}_1$ and by subtracting $m_X$. Let $m_X = m_Y = 0$. The projection theorem provides a geometric interpretation of the above result. $\hat{X}_2$ is the projection of $X$ onto the space spanned by the measurements $Y_1$, $Y_2$. By assumption $Y_1$, $Y_2$ are uncorrelated, hence orthogonal. Therefore the above projection is the sum of the projections of $X$ on $Y_1$ and $Y_2$.

Let us next tackle the general case. Motivated by the above geometric interpretation it is reasonable to replace $Y_2$ with a vector $\tilde{Y}_2$ that together with $Y_1$ spans the space $\{Y_1, Y_2\}$ and is orthogonal to $Y_1$. The construction of $\tilde{Y}_2$ follows the Gram-Schmidt orthogonalization procedure. It produces $\tilde{Y}_2$ via the projection of $Y_2$ onto the orthogonal complement of $Y_1$: $\tilde{Y}_2 = Y_2 - \hat{Y}_1$. Since $\hat{Y}_1$ is the projection of $Y_2$ onto $Y_1$, it is given by the linear least squares estimator of $Y_2$ based on $Y_1$. Therefore $\tilde{Y}_2 = Y_2 - E^*[Y_2|Y_1]$. Orthogonality of $\tilde{Y}_2$ and $Y_1$ means that $\tilde{Y}_2$ represents the piece of information in measurement $Y_2$ that is entirely new and is not already conveyed in $Y_1$. If we replace $Y_1$, $Y_2$ by $Y_1$ and $\tilde{Y}_2$ and make use of orthogonality, Eq. (12.45) applies and gives

$$\hat{X}_2 = E^*[X|Y_1, Y_2] = E^*[X|Y_1] + E^*[X|\tilde{Y}_2] \tag{12.47}$$

The term $E^*[X|\tilde{Y}_2]$ admits an alternative and useful description. Since $\tilde{X}_1 = X - \hat{X}_1$, linearity of the estimator leads to

$$E^*[X|\tilde{Y}_2] = E^*[\hat{X}_1|\tilde{Y}_2] + E^*[\tilde{X}_1|\tilde{Y}_2]$$

Since $\hat{X}_1$ is the projection of $X$ on $Y_1$, $\tilde{Y}_2$ is orthogonal to $Y_1$, and $m_X = 0$, we have $C_{\hat{X}_1 \tilde{Y}_2} = 0$ and $E^*[\hat{X}_1|\tilde{Y}_2] = 0$. In summary, Eq. (12.47) is also written as

$$\hat{X}_2 = E^*[X|Y_1, Y_2] = E^*[X|Y_1] + E^*[\tilde{X}_1|\tilde{Y}_2] \tag{12.48}$$

Combining (12.47) and (12.46), the covariance of the error becomes

$$C_{\tilde{X}_2 \tilde{X}_2} = C_{\tilde{X}_1 \tilde{X}_1} - C_{X\tilde{Y}_2} C_{\tilde{Y}_2 \tilde{Y}_2}^{-1} C_{\tilde{Y}_2 X} \tag{12.49}$$

Since $X = \hat{X}_1 + \tilde{X}_1$, we have $C_{X\tilde{Y}_2} = C_{\hat{X}_1 \tilde{Y}_2} + C_{\tilde{X}_1 \tilde{Y}_2}$. We saw above that $C_{\hat{X}_1 \tilde{Y}_2} = 0$. Thus $C_{X\tilde{Y}_2} = C_{\tilde{X}_1 \tilde{Y}_2}$ and

$$C_{\tilde{X}_2 \tilde{X}_2} = C_{\tilde{X}_1 \tilde{X}_1} - C_{\tilde{X}_1 \tilde{Y}_2} C_{\tilde{Y}_2 \tilde{Y}_2}^{-1} C_{\tilde{Y}_2 \tilde{X}_1} \tag{12.50}$$

It is left to the readers to convince themselves that Eqs. (12.48), (12.49), and (12.50) remain valid if $X$ and $Y$ have nonzero means.

### 12.4.2   Stochastic state space models

The stochastic state space model aims to capture dynamics evolving in an uncertain environment. Both state and output signals are corrupted by stochastic disturbances. The linear time varying model is the most common. It has the form

$$x(n+1) = A(n)x(n) + B(n)u(n) + D(n)w(n) \qquad (12.51)$$

$$y(n) = C(n)x(n) + v(n) \qquad (12.52)$$

$u(n)$ is the measurable input signal. We assume that it is deterministic. $w(n)$ is the plant or process noise and $v(n)$ is the measurement noise. The state equation is initiated at the random vector $x_0$ with mean $E[x_0] = m_0$ and covariance $\text{cov}[x_0, x_0] = P_0$. Both disturbances are nonstationary white noise signals. More precisely, they are zero mean stochastic processes with covariances

$$\text{cov}[v(n), v(k)] = R_v(n)\delta(n-k), \quad \text{cov}[w(n), w(k)] = R_w(n)\delta(n-k) \quad (12.53)$$

Moreover we will assume that the two noise signals are uncorrelated to each other and to the initial state, namely for all $n, k \geq 0$,

$$\text{cov}[v(n), w(k)] = \text{cov}[v(n), x_0] = \text{cov}[w(n), x_0] = 0 \qquad (12.54)$$

Since $u(n)$ is deterministic, it is uncorrelated with $x_0$. This means that the system operates in an open loop. Indeed, if there is feedback, $u(n)$ depends on the output $y(n)$ and hence becomes correlated to $x_0$. The state and output trajectories $x(n)$ and $y(n)$ are expressed by the variation of constants formula

$$x(n) = \Phi(n, 0)x_0 + \sum_{k=0}^{n-1} \Phi(n, k+1)B(k)u(k)$$
$$+ \sum_{k=0}^{n-1} \Phi(n, k+1)D(k)w(k) \qquad (12.55)$$

$$y(n) = C(n)\Phi(n, 0)x_0 + \sum_{k=0}^{n-1} C(n)\Phi(n, k+1)B(k)u(k)$$
$$+ \sum_{k=0}^{n-1} C(n)\Phi(n, k+1)D(k)w(k) + v(n) \qquad (12.56)$$

The orthogonality assumptions lead to the next proposition.

**Proposition 12.1**  The following statements hold:

1. $\text{cov}[y(n), v(m)] = R_v(n)\delta(n - m)$, $\delta(n)$ is the unit sample signal.
2. $\text{cov}[x(n), v(m)] = 0$.
3. $\text{cov}[y(n), w(m)] = C(n)\Phi(n, m + 1)D(m)R_w(m)$ for $m \leq n - 1$, and zero otherwise.
4. $\text{cov}[x(n), w(m)] = \Phi(n, m + 1)D(m)R_w(m)$ for $m \leq n - 1$, and zero otherwise.

PROOF  To establish the first statement we invoke Eq. (12.56). Then $\text{cov}[y(n), v(m)]$ is given by

$$C(n)\Phi(n, 0)\text{cov}[x_0, v(m)] + \sum_{k=0}^{n-1} C(n)\Phi(n, k + 1)B(k)\text{cov}[u(k), v(m)]$$

$$+ \sum_{k=0}^{n-1} C(n)\Phi(n, k + 1)D(k)\text{cov}[w(k), v(m)] + \text{cov}[v(n), v(m)]$$

The first and third term are zero due to Eq. (12.54). The second term is also zero because $u(n)$ is deterministic. In a similar fashion we prove property 2. To prove property 3, notice that $\text{cov}[y(n), w(m)]$ is written

$$C(n)\Phi(n, 0)\text{cov}[x_0, w(m)] + \sum_{k=0}^{n-1} C(n)\Phi(n, k + 1)B(k)\text{cov}[u(k), w(m)]$$

$$+ \sum_{k=0}^{n-1} C(n)\Phi(n, k + 1)D(k)\text{cov}[w(k), w(m)] + \text{cov}[v(n), w(m)]$$

The first and fourth terms are zero because of (12.54). The second term is zero because $u(n)$ is deterministic. Since $w(n)$ is white noise, the third term is nonzero only if $m \leq n - 1$, and in this case it reduces to the desired expression. Property 4 is established in a similar way.  ∎

The state and output means are given by

$$m_x(n + 1) = A(n)m_x(n) + B(n)u(n)$$

$$m_y(n) = C(n)m_x(n)$$

The means are governed by the same state space form, albeit the disturbances are suppressed. As far as the second-order statistics are concerned, only the diagonal slice of the output covariance function $P_y(n) = \text{cov}[y(n), y(n)]$ will be considered. If $y(n)$ were wide sense stationary, $P_y(n)$ would be constant. This is not the case even if the original state space model is time invariant. Using property 2 of

Proposition 12.1, we obtain

$$P_y(n) = \text{cov}[C(n)x(n) + v(n), C(n)x(n) + v(n)]$$
$$= C(n)P_x(n)C^T(n) + R_v(n) \qquad (12.57)$$

Taking into account property 4 of Proposition 12.1 and the assumption that $u(n)$ is deterministic, we find that

$$P_x(n+1) = A(n)P_x(n)A^T(n) + D(n)R_w(n)D^T(n), \qquad P_x(0) = P_0 \qquad (12.58)$$

The above recursion shows that even if $A(n)$, $D(n)$, and $R_w(n)$ are constant, $P_x(n)$ is time varying in general.

### 12.4.3  Development of the Kalman predictor

Let us assume that the signals $u(n)$ and $y(n)$ are related by the state space Eqs. (12.51)–(12.52). The input-output information up to time $n - 1$ is collected into the vector

$$Z_{n-1} = (\, y(n-1) \quad \cdots \quad y(0) \quad u(n-1) \quad \cdots \quad u(0)\,)$$

Let $\hat{y}(n) = E^*[y(n)|Z_{n-1}]$ be the linear least squares estimator of $y(n)$ given prior input-output information. It is shown below that the above system is also realized in state space form. The relevant architecture constitutes the Kalman predictor.

**Theorem 12.4 Kalman predictor**    With the formalism and hypotheses of the previous section the estimator $\hat{y}(n) = E^*[y(n)|Z_{n-1}]$ is determined by the linear time varying state space model

$$\hat{x}(n+1) = [A(n) - K(n)C(n)]\hat{x}(n) + B(n)u(n) + K(n)y(n) \qquad (12.59)$$

$$\hat{y}(n) = C(n)\hat{x}(n) \qquad (12.60)$$

The state $\hat{x}(n)$ coincides with the best linear estimator of the system state given past measurements, $E^*[x(n)|Z_{n-1}]$. The parameter $K(n)$ is referred to as the *Kalman gain* and is given by

$$K(n) = A(n)P(n)C^T(n)\big(C(n)P(n)C^T(n) + R_v(n)\big)^{-1} \qquad (12.61)$$

Finally, the square matrix $P(n)$ is computed by the recursion

$$P(n+1) = (A(n) - K(n)C(n))P(n)(A(n) - K(n)C(n))^T$$
$$+ D(n)R_w(n)D^T(n) + K(n)R_v(n)K^T(n) \qquad (12.62)$$

and the initial condition

$$P(0) = \text{cov}[x_0, x_0] = P_0$$

The proof is supplied in the Appendix at the end of the chapter.  ∎

The parameters of the Kalman predictor have the following physical significance: Inputs are the data sequences $u(n)$ and $y(n)$. The state is the best state estimate given past data. The output is the best output estimate given past data. $u(n)$ enters the predictor state equation after it is multiplied by $B(n)$, exactly as in the original system format. $y(n)$ enters the predictor state equation after it is

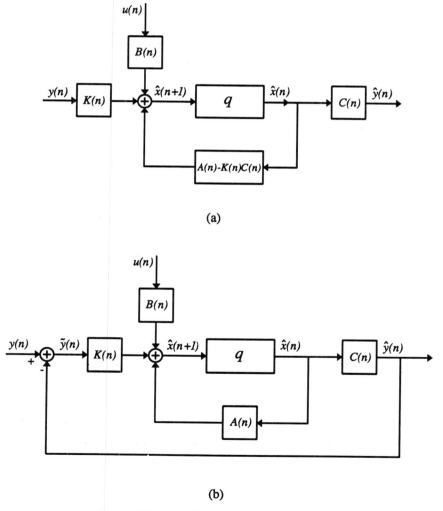

(a)

(b)

**Figure 12.4.** Two architectures for the Kalman predictor.

modified by the Kalman gain. The matrix $A(n)$ of the original system modifies to $A(n) - K(n)C(n)$ in the state estimator. The crucial parameter in the calculation of the Kalman gain is $P(n)$. $P(n)$ is a square matrix of size equal to the dimension of the state space, and it turns out to be the covariance of the state estimation error:

$$P(n) = \text{cov}[\tilde{x}(n), \tilde{x}(n)], \qquad \tilde{x}(n) = x(n) - \hat{x}(n) \qquad (12.63)$$

The architecture of the Kalman predictor is depicted in Fig. 12.4. Fig. 12.4 (a) depicts Eqs. (12.59)–(12.60). Fig. 12.4 (b) illustrates the equation

$$\hat{x}(n+1) = A(n)\hat{x}(n) + B(n)u(n) + K(n)[y(n) - C(n)\hat{x}(n)]$$

trivially derived from (12.59). The inputs are the sequences $u(n)$ and $y(n)$. Two feedback loops are utilized in Fig. 12.4 (b). The gain in the first loop is $A(n)$; the second loop is unity gain feedback.

We observe that even if the system parameters and the noises variances are constant, the Kalman gain is time varying due to the time varying nature of the error covariance $P(n)$. Hence the Kalman predictor is a linear time varying system. Constant solutions of Eq. (12.62) and time invariant structures are discussed in a later section.

*Remark 1. The Gaussian case.* If the noise signals $w(n)$ and $v(n)$ are Gaussian processes and the initial state is a Gaussian vector, the Kalman predictor provides a state space representation of the best (nonlinear) least squares estimator given by the conditional mean $E[y(n)|Z_{n-1}]$.

*Remark 2. Alternate expressions for P(n).* An alternative expression for the determination of $P(n)$ is obtained if we make use of the covariance error update (12.50). Let us use the more precise notation $\hat{x}(n|n-1)$ to designate the estimate of $x(n)$ based on the information up to time $n-1$. In a similar way $\tilde{x}(n|n-1)$ stands for $x(n) - \hat{x}(n|n-1)$. Then the following formula can be shown (see Exercise 12.4):

$$P(n+1) = A(n)\left[P(n) - P(n)C^T(n)\left(C(n)P(n)C^T(n) + R_v(n)\right)^{-1}C(n)P(n)\right]A^T(n)$$

$$+ D(n)R_w(n)D^T(n) \qquad (12.64)$$

## 12.4.4 Kalman filter

The state estimator $\hat{x}(n)$ discussed in the previous section has a predictive nature, since it is based on past measurements $y(n-1), y(n-2), \ldots, y(0)$. We explicitly indicate this property by writing $\hat{x}(n) = \hat{x}(n|n-1) = E^*[x(n)|Z_{n-1}]$. Two other quantities of interest are $\hat{x}(n|n) = E^*[x(n)|Z_n]$ and $\hat{x}(n+d|n) = E^*[x(n+d)|Z_n]$, $d > 1$. Estimation of $\hat{x}(n|n)$ is referred to as filtering, and estimation of $\hat{x}(n+d|n)$ as $d$-steps-ahead prediction.

The $d$ steps ahead prediction is discussed in Exercise 12.5. The Kalman filter estimator $\hat{x}(n|n)$ is described next.

**Theorem 12.5** The Kalman filter is realized by the following state space model:

$$\hat{x}(n|n) = \hat{x}(n|n-1) + M(n)[y(n) - C(n)\hat{x}(n|n-1)]$$

$$= (I - M(n)C(n))\hat{x}(n|n-1) + M(n)y(n) \qquad (12.65)$$

The gain is determined by

$$M(n) = P(n|n-1)C^T(n)[C(n)P(n|n-1)C^T(n) + R_v(n)]^{-1} \qquad (12.66)$$

$P(n|n-1)$ is the covariance of the one-step prediction estimation error and is computed by Eq. (12.62).

The proof is supplied in the Appendix at the end of the chapter. ∎

## 12.4.5 Observers

In Section 10.5.1 we saw how to compute the state from input-output information. A more effective approach relies on the notion of *observer* or *state estimator*. Consider the deterministic linear time invariant state space model on $R^k$

$$x(n+1) = Ax(n) + Bu(n)$$

$$y(n) = Cx(n) + D_1u(n) \qquad (12.67)$$

An observer is a linear time invariant system that processes the input-output data sequences $u(n)$ and $y(n)$ in order to reconstruct the state $x(n)$. It has the form

$$\hat{x}(n+1) = F\hat{x}(n) + L_1u(n) + Ly(n) \qquad (12.68)$$

Observer design seeks to determine the parameters $F$, $L_1$, and $L$ so that the estimation error

$$e_o(n) = x(n) - \hat{x}(n)$$

is small in an appropriate sense. The class of allowable observer structures can be narrowed down by requiring that for any initial state $x_0$ and input $u(n)$, the system and the observer have identical state trajectories if they are both initialized at $x_0$. In other words, if the observer succeeds in capturing the state of the system at a certain time, then it will reproduce the state perfectly at all future times. Let $x(n)$ and $\hat{x}(n)$ denote the system and observer state trajectories generated by $x(0) = \hat{x}(0) = x_0$ and the input $u(n)$. Perfect reconstruction at time $n = 1$ gives $\hat{x}(1) = x(1)$, or

$$F\hat{x}(0) + L_1 u(0) + Ly(0) = Fx_0 + L_1 u(0) + L(Cx_0 + D_1 u(0)) = Ax_0 + Bu(0)$$

Since the above holds for all $x_0$ and $u(0)$, we can conclude that

$$L_1 = B - LD_1, \quad F = A - LC$$

Thus the class of observers has the form

$$\hat{x}(n + 1) = (A - LC)\hat{x}(n) + (B - LD_1)u(n) + Ly(n) \tag{12.69}$$

The only parameter left is the $k \times m$ matrix $L$. The design reduces to finding $L$ so that the error $e_o(n) = x(n) - \hat{x}(n)$ becomes small. It is easy to see that the error satisfies the recursive equation

$$e_o(n + 1) = (A - LC)e_o(n) \tag{12.70}$$

We observe that the error dynamics does not depend on the parameters $B$, $D_1$, and the input signal $u$. A natural criterion is then to choose $L$ so that the error decays to zero with controlled rate of convergence.

**Example 12.5  Observer for a scalar system**
Consider the scalar system

$$x(n + 1) = ax(n) + bu(n)$$
$$y(n) = cx(n) + du(n)$$

The observer has the form

$$\hat{x}(n + 1) = (a - lc)\hat{x}(n) + (b - ld)u(n) + ly(n)$$

The estimation error satisfies the equation

$$e_o(n + 1) = (a - lc)e_o(n)$$

Thus $e_o(n) = (a - lc)^n e_o(0)$. It is clear that if $c \neq 0$, then for any $0 \leq |\lambda| < 1$ the choice $l = (a - \lambda)/c$ leads to errors that decay to zero like $\lambda^n$. Observability enables the extension of the above result to the general case. ∎

**Theorem 12.6**  The pair $(A, C)$ is observable if and only if for any polynomial $a_d(z)$ there exists a matrix $L$ such that the characteristic polynomial of $A - LC$ coincides with $a_d(z)$.

PROOF Assume first that $(A, C)$ is observable. For simplicity, we consider only the single-output case. Let $P$ be the similarity transformation that converts $(A, C)$ to the observer form $(\hat{A}, \hat{C})$ (see Section 10.5.3). Let $\hat{L} = PL$. Then

$$A - LC = P^{-1}\hat{A}P - L\hat{C}P = P^{-1}(\hat{A} - \hat{L}\hat{C})P$$

It follows that $A - LC$ and $\hat{A} - \hat{L}\hat{C}$ are similar. Therefore they have the same characteristic polynomial. Hence it suffices to find $\hat{L}$ so that $\hat{A} - \hat{L}\hat{C}$ has $a_d(z)$ as characteristic polynomial. It is easy to see that $\hat{A} - \hat{L}\hat{C}$ retains the companion form. Indeed,

$$\hat{A} - \hat{L}\hat{C} = \begin{pmatrix} -a_1 - \hat{l}_1 & 1 & 0 & \cdots & 0 & 0 \\ -a_2 - \hat{l}_2 & 0 & 1 & \cdots & 0 & 0 \\ \vdots & & \vdots & \vdots & & \vdots & \vdots \\ -a_{k-1} - \hat{l}_{k-1} & 0 & 0 & \cdots & 0 & 1 \\ -a_k - \hat{l}_k & 0 & 0 & \cdots & 0 & 0 \end{pmatrix}$$

The characteristic polynomial of $\hat{A} - \hat{L}\hat{C}$ is

$$s^k + (a_1 + \hat{l}_1)s^{k-1} + (a_2 + \hat{l}_2)s^{k-2} + \cdots + (a_k + \hat{l}_k)$$

Equating the coefficients of the above polynomial with the coefficients $a_{d_i}$ of the desired polynomial $a_d(z)$, we find that $\hat{l}_i = a_{d_i} - a_i$.

Conversely, suppose that $(A, C)$ is not observable. We pass to the observability staircase form under a suitable similarity transformation $P$ (see Section 10.5.2). Let $\hat{L} = PL$ and

$$\hat{L} = \begin{pmatrix} \hat{L}_1 \\ \hat{L}_2 \end{pmatrix}$$

In the new coordinates the observer error dynamics is given by

$$\hat{A} - \hat{L}\hat{C} = \begin{pmatrix} \hat{A}_1 & * \\ 0 & \hat{L}_2\hat{C}_2 \end{pmatrix}$$

and $\hat{A}_1$ does not depend on $\hat{L}$. The block triangular structure of the above matrix implies that its eigenvalues consist of the eigenvalues of $\hat{A}_1$ and the eigenvalues of $\hat{L}_2\hat{C}_2$. Clearly the eigenvalues of $\hat{A}_1$ are not influenced by the observer parameter $L$ and hence cannot be placed at arbitrary positions. ∎

The observer theory discussed above remains valid in the continuous time case. An analog observer has the form

$$\dot{\hat{x}}(t) = (A - LC)\hat{x}(t) + (B - LD)u(t) + Ly(t)$$

and the error dynamics is governed by the linear differential equation

$$\dot{e}_o(t) = (A - LC)e_o(t)$$

The counterpart of Theorem 12.6 asserts that analog observers can be designed with controlled estimation error if the pair $(A, C)$ is observable.

In the discrete case the highest convergence rate for the estimation error is achieved when the eigenvalues are placed at zero, namely $a_d(z) = z^k$. In this case $A - LC$ is nilpotent, and $(A - LC)^N = 0$ for some $N \leq k$. Observers of this form are referred to as *deadbeat observers*.

Let us consider the time invariant system

$$x(n + 1) = Ax(n) + Bu(n) + Dw(n)$$
$$y(n) = Cx(n) + v(n) \tag{12.71}$$

with stationary disturbances

$$\text{cov}[w(n), w(m)] = R_w \delta(n - m), \quad \text{cov}[v(n), v(m)] = R_v \delta(n - m)$$

The Kalman predictor can be viewed as an observer with time varying gain. These observers have the form (12.69), albeit the observer gain varies with time

$$\hat{x}(n + 1) = (A - L(n)C)\hat{x}(n) + Bu(n) + L(n)y(n) \tag{12.72}$$

The Kalman predictor is a member of this class with $L(n) = K(n)$, given by (12.61). It is easy to see that when the state dynamics includes disturbances and is given by (12.71), the estimation error committed by a time varying observer (12.72) satisfies

$$e_o(n + 1) = (A - L(n)C)e_o(n) + Dw(n) - L(n)v(n)$$

In accordance with the analysis of Section 12.4.2, the error covariance $P_L(n) = \text{cov}[e_o(n), e_o(n)]$ of the observer with gain $L(n)$, obeys the equation

$$P_L(n + 1) = [A - L(n)C]P_L(n)[A - L(n)C]^T + DR_w D^T + L(n)R_v L^T(n) \tag{12.73}$$

The next theorem demonstrates the optimum performance of the Kalman predictor as a time varying observer.

**Theorem 12.7**    Among all time varying observers of the form (12.72) initialized at a nonrandom vector $\hat{x}_0$, the Kalman predictor yields the minimum covariance error in the sense that $P_L(n) \geq P(n)$ or, equivalently, $a^T P_L(n)a \geq a^T P(n)a$ for any vector $a$.

PROOF   The proof proceeds by induction. The initial covariance error produced by an observer of the above type is

$$P_L(0) = \text{cov}[e_o(0), e_o(0)] = \text{cov}[x(0) - \hat{x}_0, x(0) - \hat{x}_0] = \text{cov}[x_0, x_0] = P(0)$$

Assume that the conclusion holds for $n$. We will show it remains true for $n + 1$. Equation (12.73) gives

$$\begin{aligned}
P_L(n+1) &= AP_L(n)A^T + DR_wD^T + L(n)CP_L(n)C^TL^T(n) - L(n)CP_L(n)A^T \\
&\quad - AP_L(n)C^TL^T(n) + L(n)R_vL^T(n) \\
&= AP_L(n)A^T + DR_wD^T + L(n)\big[R_v + CP_L(n)C^T\big]L^T(n) \\
&\quad - L(n)CP_L(n)A^T - AP_L(n)C^TL^T(n)
\end{aligned}$$

Completing the squares with respect to $L(n)$, we write $P_L(n + 1)$ as

$$\begin{aligned}
&AP_L(n)A^T + DR_wD^T - AP_L(n)C^T\big[R_v + CP_L(n)C^T\big]^{-1}CP_L(n)A^T \\
&+ \big[L(n) - AP_L(n)C^T(R_v + CP_L(n)C^T)^{-1}\big](R_v + CP_L(n)C^T) \\
&\times \big[L(n) - AP_L(n)C^T(R_v + CP_L(n)C^T)^{-1}\big]^T
\end{aligned}$$

The first three terms do not depend on $L(n)$, and by the induction hypothesis, $P_L(n)$ is minimized by $P(n)$. The fourth term is nonnegative definite and achieves its minimum value for

$$L(n) = AP_L(n)C^T\big(R_v + CP_L(n)C^T\big)^{-1} = K(n)$$

This is precisely the expression for the Kalman gain.   ■

### 12.4.6   Time invariant Kalman predictor

In this section we consider the time invariant system (12.71). We have seen that the Kalman predictor is a time varying system due to the time varying nature of the error covariance matrix $P(n)$. Let $P_0$ be a nonnegative matrix, and $P(n, n_0, P_0)$ denote the solution of

$$P(n+1) = (A - K(n)C)P(n)(A - K(n)C)^T + DR_wD^T + K(n)R_vK^T(n) \tag{12.74}$$

that satisfies the initial condition $P(n_0, n_0, P_0) = P_0$, for $n_0 \geq 0$. The Kalman gain is given by

$$K(n) = AP(n)C^T\big(CP(n)C^T + R_v\big)^{-1}$$

When no confusion arises, we set $P(n) = P(n, n_0, P_0)$. A sensible time invariant structure is obtained if $P(n)$ is replaced by its limit $P = \lim_{n \to \infty} P(n)$ provided that the latter exists. If we take limits on both sides of Eq. (12.74), we obtain the following discrete algebraic Ricatti equation:

$$P = (A - KC)P(A - KC)^T + DR_wD^T + KR_vK^T \tag{12.75}$$

where

$$K = APC^T (CPC^T + R_v)^{-1}$$

We will show that under mild assumptions the steady state Ricatti equation (12.75) has a unique positive definite solution that attracts all solutions of (12.74), starting from nonnegative definite initial matrices. Moreover the resulting time invariant Kalman predictor is totally stable.

**Theorem 12.8**    Consider the time invariant system (12.71) with noise variances $R_w$ and $R_v > 0$. Let $R_w = R_1 R_1^T$. Assume that the pair $(A, DR_1)$ is controllable and that the pair $(A, C)$ is observable. Then for an initial condition $P_0 \geq 0$, the solution of the Ricatti equation (12.74) converges to the unique positive definite solution of (12.75). The resulting time invariant Kalman predictor

$$\hat{x}(n + 1) = (A - KC)\hat{x}(n) + Bu(n) + Ky(n)$$

is totally stable.

The proof is supplied in the Appendix at the end of the chapter. ∎

***Example 12.6    Kalman filter for a scalar system***
Consider the time invariant scalar system

$$x(n + 1) = ax(n) + bu(n) + dw(n)$$

$$y(n) = cx(n) + v(n)$$

All parameters are scalar. The noise variances are $r_w$ and $r_v$. The time invariant Kalman gain is

$$k = \frac{apc}{c^2p + r_v}$$

$p$ is the solution of the algebraic Ricatti equation (12.75):

$$p = (a - kc)^2 p + d^2 r_w + k^2 r_v$$

Substituting the Kalman gain into the above equation, we have

$$p = \left( \frac{ar_v}{c^2p + r_v} \right)^2 p + d^2 r_w + \left( \frac{apc}{c^2p + r_v} \right)^2 r_v$$

or

$$p = \frac{a^2 r_v p (c^2 p + r_v)}{(c^2 p + r_v)^2} + d^2 r_w$$

After straightforward computations we arrive at the quadratic equation

$$c^2 p^2 + (r_v - a^2 r_v - d^2 r_w c^2) p - d^2 r_w r_v = 0$$

The roots are real because $-c^2 d^2 r_w r_v \le 0$. There is a unique positive root provided that $c \ne 0$ and $d \ne 0$. These are precisely the controllability and observability conditions of Theorem 12.8.    ∎

### Example 12.7  Satellite
A simplified model for the motion of a satellite is

$$\ddot{\theta}(t) = u(t) + w(t)$$

where $\theta(t)$ is the angle of the satellite with respect to an inertial reference system, $u$ is proportional to the control torque, and $w$ is proportional to the disturbance torque caused by wind gusts. The satellite is described in state space by the parameters

$$a = \begin{pmatrix} 0 & 1 \\ 0 & 0 \end{pmatrix}, \quad b = \begin{pmatrix} 0 \\ 1 \end{pmatrix}, \quad c = (1 \quad 0)$$

MATLAB simulation is carried out as follows:

```
a = [0 1; 0 0]; b = [0 ; 1]; c = [1 0];
t = (0:.1:100)';            % defines the time window
w = 0.1*randn(size(t));
u = sin(t);
plot(t,u);
pause
U = [u,w];                  % multiplexes the two inputs
B = [b,b];
D = [0,0];
[y,x] = lsim(a,B,c,D,U,t);

                            % A discrete simulator is built as follows:
Ts = .1;                    % defines the sampling period
[a1,b1] = c2d(a,b,Ts);      % computes the state parameters of the discrete simulator
t1 = 0:0.1:100;             % defines the time window
u1 = sin(t1)';
w1 = 0.1*randn(size(t1))';
U1 = [u1,w1];               % multiplexes the two inputs
B1 = [b1,b1];
[y1,x1] = dlsim(a1,B1,c,D,U1);  % computes state and output trajectories
                            % Computation of the Kalman predictor follows.
```

```
rw = .01;                        % process noise variance
rv = .1;                         % measurement noise variance
[K,m,P] = dlqe(a1,b1,c,rw,rv);   % computes the Kalman gain K and the
                                 % solution of the algebraic Ricatti equation P.
                                 % The  Kalman filter dynamics are determined by
ak = a1-K*c;
bk = [b1 K];
                                 % Simulation of the Kalman filter is done by
vk = [u1 y1];
[yk xk] = dlsim(ak,bk,c,D,vk);
abs(eig(ak))                     % checks total stability of the Kalman predictor
                                 % Tracking performance is analyzed by the plot commands
subplot(211), plot(t1',x1(:,1),t1',xk(:,1))
subplot(212),plot(t1',x1(:,2),t1',xk(:,2))
```

Performance of the Kalman predictor is illustrated in Fig. 12.5. ∎

### 12.4.7  Extended Kalman filter

The extended Kalman filter (EKF) provides an approximation of the least

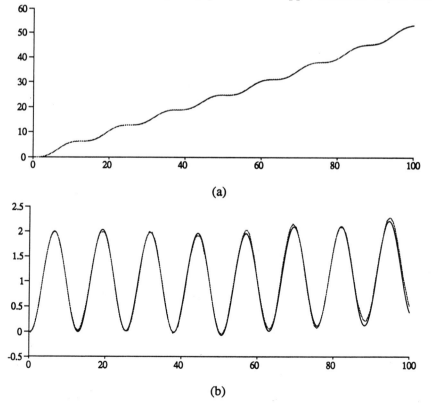

(a)

(b)

**Figure 12.5.** (a) Simulated and estimated angle; (b) simulated and estimated angular velocity.

squares estimator when the data sequences are produced by a nonlinear state space model. Let us consider the following affine in input system

$$x(n+1) = f(x(n)) + g_0(x(n))u(n) + g_1(x(n))w(n) \quad (12.76)$$

$$y(n) = h(x(n)) + v(n) \quad (12.77)$$

Both equations are nonlinear. The state dynamics depends linearly on $u$ and $w$. Suppose that an approximation of the one-step predictor at time $n$, $\hat{x}(n)$ has been obtained and the update $\hat{x}(n+1)$ is seeked. We consider the first-order Taylor approximations of $f$ and $h$ and the zero-order Taylor approximations of $g_0$ and $g_1$, at $\hat{x}(n)$. Then

$$x(n+1) \approx f(\hat{x}(n)) + A(n)[x(n) - \hat{x}(n)] + g_0(\hat{x}(n))u(n) + g_1(\hat{x}(n))w(n)$$

$$y(n) \approx h(\hat{x}(n)) + C(n)[x(n) - \hat{x}(n)] + v(n) \quad (12.78)$$

where

$$A(n) = \left.\frac{\partial f}{\partial x}\right|_{\hat{x}(n)}, \quad C(n) = \left.\frac{\partial h}{\partial x}\right|_{\hat{x}(n)} \quad (12.79)$$

Equations (12.78) have the form

$$x(n+1) = A(n)x(n) + \tilde{u}(n) + D(n)w(n)$$

$$y(n) = C(n)x(n) + \alpha(n) + v(n) \quad (12.80)$$

where

$$\tilde{u}(n) = f(\hat{x}(n)) - A(n)\hat{x}(n) + g_0(\hat{x}(n))u(n), \quad D(n) = g_1(\hat{x}(n))$$

$$\alpha(n) = h(\hat{x}(n)) - C(n)\hat{x}(n) \quad (12.81)$$

Under the assumption that $\hat{x}(n)$ is known, the above equation has the familiar form (12.51)–(12.52) except for the constant term $\alpha(n)$ in the measurement equation. Following the arguments used to derive the main Kalman filter recursions, we find that

$$\hat{x}(n+1) = A(n)\hat{x}(n) + \tilde{u}(n) + K(n)[y(n) - \hat{y}(n)] \quad (12.82)$$

and

$$\hat{y}(n) = h(\hat{x}(n)) + C(n)\hat{x}(n) - C(n)\hat{x}(n) = h(\hat{x}(n)) \quad (12.83)$$

If we substitute (12.83) and (12.81) into (12.82) we obtain

$$\hat{x}(n+1) = f(\hat{x}(n)) + g_0(\hat{x}(n))u(n) + K(n)[y(n) - h(\hat{x}(n))] \qquad (12.84)$$

The Kalman gain is computed by Eqs. (12.61)–(12.62), where $A(n)$ and $C(n)$ are defined by (12.79). Equation (12.84) in conjunction with Eqs. (12.61)–(12.62) defines the extended Kalman filter. The EKF is a linear filter associated with a linearized approximation of the original system and does not coincide with the best nonlinear estimator $E[x(n)|Z_{n-1}]$ or the best linear estimator $E^*[x(n)|Z_{n-1}]$.

**Example 12.8    EKF for a nonlinear time delay system**
Consider the nonlinear time delay system

$$y(n) = h(u(n-1), \ldots, u(n-k)) + v(n)$$

We convert the above in state space form. The state assignment

$$x(n) = (\, u(n-1) \quad u(n-2) \quad \ldots \quad u(n-k)\,)^T$$

leads to the model

$$x(n+1) = Ax(n) + Bu(n)$$

$$y(n) = h(x(n)) + v(n)$$

where

$$A = \begin{pmatrix} 0 & 0 & \cdots & 0 & 0 \\ 1 & 0 & \cdots & 0 & 0 \\ \vdots & & & & \\ 0 & 0 & \cdots & 1 & 0 \end{pmatrix}, \quad B = \begin{pmatrix} 1 \\ 0 \\ \vdots \\ 0 \end{pmatrix}$$

The extended Kalman filter becomes

$$\hat{x}(n+1) = A\hat{x}(n) + Bu(n) + K(n)[y(n) - h(\hat{x}(n))] \qquad \blacksquare$$

## 12.5  IDENTIFICATION OF CONVOLUTIONAL SYSTEMS

Identification of discrete systems is considered in this section, complementing the discussion initiated in Chapter 4. Both the correlation and the prediction error approach are treated.

The correlation approach uses the crosscorrelation between the measured input and output to suppress noise effects. The system's frequency response is then determined by dividing the input-output cross spectrum by the input power spectral density. When no input measurements are available, we are confronted with the blind identification problem. In this case the magnitude of the frequency response is determined from the autocorrelation of the output. Determination of the frequency response phase requires higher-order output statistics.

## 12.5.1  Correlation analysis

An unknown system is observed through a pair of sequences $u(n)$ and $y(n)$. We write

$$y(n) = z(n) + v(n) \tag{12.85}$$

$y(n)$ is viewed as the measurement of the actual system output $z(n)$; $v(n)$ represents measurement noise. The unknown system has a convolutional format and operates in an uncertain environment. Thus it has the form

$$z(n) = g(n) * u(n) + w(n) \tag{12.86}$$

The process noise $w(n)$ and the measurement noise $v(n)$ are zero mean wide sense stationary uncorrelated signals. Both are uncorrelated with the input $u(n)$. We form the crosscorrelation of $y(n)$ and $u(n)$. Then

$$r_{yu}(n) = r_{zu}(n)$$

It is easy to see that $r_{zu}(n) = g(n) * r_u(n)$. Hence

$$r_{yu}(n) = g(n) * r_u(n) \tag{12.87}$$

Since both $u(n)$ and $y(n)$ are measured, estimates of $r_{yu}(n)$ and $r_u(n)$ can be obtained. The solution of the above equation is immediately obtained if $u$ is second-order white noise. In that case we have

$$g(n) = \frac{1}{\gamma_2} r_{yu}(n) \tag{12.88}$$

where $\gamma_2$ is the input variance. On the basis of the above equation, a practical estimator is built as follows: Data measurements $u(n)$, $y(n)$ $1 \leq n \leq N$ are obtained from a data acquisition device. Second-order parameters $\gamma_2$ and $r_{yu}(n)$ are estimated as in Section 9.1. Then the impulse response $g(n)$ is calculated from (12.88).

For general inputs the solution of (12.87) can be obtained by passing to the frequency domain. If we assume that $r_u(n)$ and $g(n)$ are absolutely summable, we obtain

$$G(\omega) = S_{yu}(\omega) S_u^{-1}(\omega) \tag{12.89}$$

In the single-channel case the latter equation becomes

$$G(\omega) = \frac{S_{yu}(\omega)}{S_u(\omega)} \tag{12.90}$$

### Example 12.9    FIR identification

Consider the FIR system

$$y(n) = u(n-1) + 2u(n-2) + 5u(n-3) + w(n)$$

This has the form of Eq. (12.86), with $v(n) = 0, g(1) = 1, g(2) = 2, g(3) = 5$, and $g(n) = 0$ otherwise. Data points are generated by the following **MATLAB** simulation:

```
u = sqrt(gamma2)*randn(1,1014);   % generates 1014 input samples from
                                  % gaussian white noise with variance gammma2
w = randn(1,1014);                % generates 1014 noise samples
b = [0,1,2,5];
y = filter(b,1,u) + w;
[s,f] = spectrum(u,y,64,14,64,2*pi);
[N,D] = invfreqz (s(:,4),f,3,0);  % N is an estimate of b
```
■

Suppose next that the process noise $w(n)$ is colored and is obtained by filtering white noise

$$w(n) = h(n) * e(n) \tag{12.91}$$

According to the developments of Section 9.2.4, a wide sense stationary process $w(n)$ is obtained at the output of a linear filter excited by white noise if its spectral distribution is absolutely continuous. Determination of the impulse response $h(n)$ is often desired. For instance, knowledge of $h(n)$ enables us to come up with a better prediction estimate of the system output $z(n)$. Since $g(n)$ has already been determined, the signal $y_1(n) = y(n) - g(n) * u(n)$ is known. The question then is how to estimate $h(n)$ from the equation $y_1(n) = h(n) * e(n) + v(n)$, on the basis of the $y_1(n)$ statistics only. This is the blind identification problem. If $e(n)$ has unit variance and $v(n) = 0$, the frequency response $H(\omega)$ is a spectral factor of the power spectral density of $y_1(n)$. Since there are more than one spectral factors, second-order statistics do not suffice to identify $h(n)$. If we assume that $h(n)$ is minimum phase, then the method developed in connection with the linear prediction problem enables us to determine $h(n)$. In the next section higher-order statistics are employed to specify $h(n)$.

### 12.5.2    Blind identification

We employ the notation

$$y(n) = h(n) * e(n) + v(n) \tag{12.92}$$

where $v$ and $e$ have zero mean and are independent. We assume that $y$ is measured and that we have no access on $e(n)$. The correlation method is not applicable because $r_{ye}(n)$ can no longer be determined. Under these circumstances we are forced to consider the statistics of the output signal per se with the hope that they contain sufficient information to determine the impulse response $h(n)$. Clearly not much is gained from the output mean. Let us proceed to second-order output statistics. The analysis of Section 9.1.4 leads to

$$r_y(n) = h(n) * r_e(n) * h(-n) + r_v(n)$$

or in the frequency domain

$$S_y(\omega) = |H(\omega)|^2 S_e(\omega) + S_v(\omega)$$

Apparently second-order output information does not suffice to determine $h(n)$. Suppose, for instance, that both $e(n)$ and $v(n)$ are white noise signals of known variances. Then the above equation yields the magnitude spectrum but is incapable to reconstruct the phase spectrum. We say that second-order statistics are *phase blind*. Phase information is provided if we move up to higher-order statistics. We will be dealing with the single-channel case only. Let $c_{ky}(n_1, n_2, \ldots, n_{k-1})$ denote the $k$th-order cumulant of $y(n)$. Suppose that $e(n)$ is non-Gaussian white noise with nonzero intensity of order $k > 2$, $\gamma_k$:

$$C_{ke}(\omega_1, \omega_2, \ldots, \omega_{k-1}) = \gamma_k$$

and that the measurement noise $v(n)$ is Gaussian. Invoking the results of Section 9.6, we obtain

$$C_{ky}(\omega_1, \omega_2, \ldots, \omega_{k-1}) = \gamma_k H\left(-\sum_{i=1}^{k-1} \omega_i\right) H(\omega_1) H(\omega_2) \cdots H(\omega_{k-1}) \qquad (12.93)$$

The above equation is phase sensitive and enables the computation of $H(\omega)$. We will first illustrate the plausibility of this statement with the special case of moving average systems and then establish the claim in the general case.

**Example 12.10    Blind identification of moving average systems**
Consider the moving average system

$$y(n) = \sum_{i=0}^{k} h(i)e(n-i) + v(n), \qquad h(0) = 1$$

The signal $e(n)$ is non-Gaussian white noise such that $\gamma_{3e} = \gamma_3 \neq 0$. In contrast, the measurement noise $v(n)$ is Gaussian. Since $h(n)$ has finite support and

$c_{3e}(n_1, n_2) = \gamma_3 \delta(n_1, n_2)$, Eq. (9.50) becomes

$$c_{3y}(n_1, n_2) = \gamma_3 \sum_{i=0}^{k} h(i)h(i + n_1)h(i + n_2)$$

Evaluation at $n_1 = k$ and $n_2 = 0$ gives

$$c_{3y}(k, 0) = \gamma_3 h(k)$$

Likewise, if we set $n_1 = k$ and $n_2 = n$, $1 \leq n \leq k$, we have

$$c_{3y}(k, n) = \gamma_3 h(k)h(n) = c_{3y}(k, 0)h(n)$$

Thus

$$h(n) = \frac{c_{3y}(k, n)}{c_{3y}(k, 0)} \tag{12.94}$$

The above equation determines the impulse response with the aid of third-order output cumulants. It utilizes for this purpose only the 1-D slice $c_{3y}(k, .)$. If $\gamma_3 = 0$, then $c_{3y}(k, 0) = 0$, and the above formula is no longer applicable. This case occurs, for instance, if the probability density of the signal is symmetric. The difficulty is bypassed by using a similar formula involving higher-order intensities. For instance, if $\gamma_{4e} \neq 0$, it holds (see Exercise 12.10) that

$$h(n) = \frac{c_{4y}(k, 0, n)}{c_{4y}(k, 0, 0)}, \qquad n = 0, 1, \ldots, k \tag{12.95}$$

∎

Phase recovery of the frequency response is established next.

**Theorem 12.9** Consider Eq. (12.93) with $H(\omega) \neq 0$ for all $\omega$, and

$$\sum_{n=-\infty}^{\infty} |n| |h(n)| < \infty \tag{12.96}$$

Moreover $\gamma_k \neq 0$ for some $k \geq 3$. Then the phase response $\arg H(\omega)$ is uniquely determined from Eq. (12.93), except for a linear factor and the sign of $H(0) = \sum_{n=-\infty}^{\infty} h(n)$.

PROOF  Evaluation of Eq. (12.93) at zero gives

$$C_{ky}(0, 0, \ldots, 0) = \gamma_k H^k(0)$$

For $k = 2$ we obtain $C_{2y}(0) = |H(0)|^2$. Thus

$$\gamma_k = \left( \frac{|H(0)|}{H(0)} \right)^k C_{ky}(0, 0, \ldots, 0) C_{2y}^{-k/2}(0)$$

Let $\Psi(\omega) = \arg(H(0)H(\omega)/|H(0)|)$. Since $h(n)$ is real, $\Psi(\omega)$ is odd, and thus $\Psi(0) = 0$. Equation (12.93) implies that

$$\Psi(\omega_1) + \Psi(\omega_2) + \cdots + \Psi(\omega_{k-1}) - \Psi\left( \sum_{i=1}^{k-1} \omega_i \right)$$

$$= \arg\left( \left( \frac{H(0)}{|H(0)|} \right)^k \frac{1}{\gamma_k} C_{ky}(\omega_1, \ldots, \omega_{k-1}) \right)$$

$$= \arg(C_{2y}^{k/2}(0) C_{ky}^{-1}(0, \ldots, 0) C_{ky}(\omega_1, \ldots, \omega_{k-1})) = Q(\omega_1, \ldots, \omega_{k-1})$$

where $Q(\omega_1, \ldots, \omega_{k-1})$ is a known function. The differentiation property of the Fourier transform (see Section 6.1.4) ensures that the frequency response $H(\omega)$ is differentiable, due to (12.96). Let $\omega_{k-1} = \omega$ and $\omega_1 = \omega_2 = \cdots = \omega_{k-2} = v$. Then

$$\Psi(\omega) + (k - 2)\Psi(v) - \Psi(\omega + (k - 2)v) = Q(v, \ldots, v, \omega) \qquad (12.97)$$

Let $\Psi'(\omega) = d\Psi(\omega)/d\omega$. If we divide (12.97) by $(k - 2)v$ and take limits as $v \to 0$, we obtain

$$\Psi'(0) - \Psi'(\omega) = \lim_{v \to 0} \frac{1}{(k - 2)v} Q(v, \ldots, v, \omega) = Q_1(\omega)$$

The function $Q_1$ is uniquely defined by the cumulants of $y(n)$. Thus

$$\Psi(\omega) = \Psi'(0)\omega - Q_2(\omega), \quad Q_2(\omega) = \int_0^\omega Q_1(v)dv \qquad (12.98)$$

The above equation specifies $\Psi(\omega)$ except for the constant $\Psi'(0) = c$, which is determined as follows: Since $h(n)$ is real, $H(\pi) = \sum h(2n) - \sum h(2n + 1)$ is also real. Therefore $\Psi(\pi) = m\pi$ for some integer $m$. Equation (12.98) implies that

$$\Psi(\pi) = m\pi = c\pi - Q_2(\pi) \quad \text{or} \quad c = m + \frac{Q_2(\pi)}{\pi}$$

We infer that $c$ is known modulo an integer shift.  ∎

### 12.5.3  Predictors for convolutional models

The prediction error approach to identification was introduced in Chapter 4 as a system design formulation. It relies on the notion of predictor and the minimization of an error criterion between the measured output and the predicted output. Finite memory predictors based on the mean square error and the total square error were introduced in Section 4.3.2. More generally, a predictor is a system that at each time $n$ processes the given input-output sequences $u(n)$ and $y(n)$ up to time $n - 1$ and produces a predictive estimate of $y(n)$, $\hat{y}(n)$. Infinite memory predictors are defined as multichannel Wiener filters. Let $\mathcal{H}_{u,y}(n)$ denote the closed linear space spanned by the data samples $u(n - k)$, $y(n - k)$, $k = 0, 1, 2, \ldots$. The output of the infinite memory predictor is defined as

$$\hat{y}(n) = E^*[y(n)|\mathcal{H}_{u,y}(n - 1)]$$

The predictor is a model that produces a noise clean copy of the system output. The determination of the predictor requires an excursion to multichannel linear prediction and will not be further pursued. Some parts of the multichannel theory are developed by a direct extension of the single-channel case. The main difficulties arise in connection with the spectral factorization and the Paley-Wiener condition. These topics are now much harder to handle. Here we will compute the predictor under the assumption that the data sequences are generated by the convolutional model

$$y(n) = g(n) * u(n) + w(n) = \sum_{k=1}^{\infty} g(k)u(n - k) + w(n) = g(q)u(n) + w(n)$$

$$w(n) = h(n) * e(n) = \sum_{k=0}^{\infty} h(k)e(n - k) = h(q)e(n) \tag{12.99}$$

The signal $w(n)$ may model measurement noise. In this case the true system output $g(q)u(n)$ involves no further uncertainty; namely no process noise is present. Alternatively, $w(n)$ represents process noise, and measurements are error free. The noise $w(n)$ is colored. We will assume that $u(n)$ and $w(n)$ are mutually uncorrelated. Moreover both $h(n)$ and $g(n)$ are absolutely summable and $h(n)$ is minimum phase, so the frequency responses $G(\omega)$, $H(\omega)$, and $1/H(\omega)$ exist. Linearity of the projection operator gives

$$\hat{y}(n) = E^*[g(q)u(n)|\mathcal{H}_{u,y}(n - 1)] + E^*[w(n)|\mathcal{H}_{u,y}(n - 1)]$$

Since $g(q)u(n)$ belongs to $\mathcal{H}_{u,y}(n - 1)$, we have

$$\hat{y}(n) = g(q)u(n) + E^*[w(n)|\mathcal{H}_{u,y}(n - 1)] \tag{12.100}$$

We claim that

$$E^*[w(n)|\mathcal{H}_{u,y}(n - 1)] = E^*[w(n)|\mathcal{H}_w(n - 1)] = \hat{w}(n)$$

In other words, the projection of $w(n)$ onto $\mathcal{H}_{u,y}(n-1)$ coincides with its projection on $\mathcal{H}_w(n-1)$. Indeed, it suffices to show that the predictor $\hat{w}(n)$ belongs to $\mathcal{H}_{u,y}(n-1)$ and that the prediction error $\tilde{w}(n) = w(n) - \hat{w}(n)$ is orthogonal to all past data samples $u(n-k)$, $y(n-k)$, $k = 1, 2, \ldots$. It follows from (12.99) that $\mathcal{H}_w(n-1) \subset \mathcal{H}_{u,y}(n-1)$. Therefore $\hat{w}(n) \in \mathcal{H}_{u,y}(n-1)$. Since $\tilde{w}(n) \in \mathcal{H}_w(n)$ and $w(n)$ is uncorrelated with $u(n)$, $\tilde{w}(n)$ is orthogonal to $u(n-k)$. Notice finally that $\tilde{w}(n)$ is orthogonal to $w(n-k)$, $k = 1, 2, \ldots$. Moreover $w(n-k) = y(n-k) - g(q)u(n-k)$, and $\tilde{w}(n)$ is orthogonal to $u(n-k)$. Therefore $\tilde{w}(n)$ is orthogonal to $y(n-k)$, and the claim is proved. According to the linear prediction theory of Section 12.1, Eq. (12.100) becomes

$$\hat{y}(n) = g(q)u(n) + \left(1 - \frac{h(0)}{h(q)}\right)w(n) = g(q)u(n) + \left(1 - \frac{h(0)}{h(q)}\right)(y(n) - g(q)u(n))$$

or

$$\hat{y}(n) = \left(1 - \frac{h(0)}{h(q)}\right)y(n) + \frac{h(0)g(q)}{h(q)}u(n) \tag{12.101}$$

The prediction error dynamics is

$$y(n) - \hat{y}(n) = \frac{h(0)}{h(q)}y(n) - \frac{h(0)g(q)}{h(q)}u(n) = h(0)e(n) \tag{12.102}$$

**Example 12.11  Predictor dynamics**
Let

$$y(n) = b_1 u(n-1) + b_2 u(n-2) + c_0 e(n) + c_1 e(n-1)$$

Then $g(q) = b_1 q + b_2 q^2$ and $h(q) = c_0 + c_1 q$. Both impulse responses have finite support and thus are absolutely summable. $h(q)$ has minimum phase if $|c_1| < |c_0|$. In this case the predictor dynamics is

$$\hat{y}(n) = \frac{c_1 q}{c_0 + c_1 q}y(n) + \frac{c_0(b_1 q + b_2 q^2)}{c_0 + c_1 q}u(n)$$

or

$$c_0\hat{y}(n) + c_1\hat{y}(n-1) = c_1 y(n-1) + c_0 b_1 u(n-1) + c_0 b_2 u(n-2) \qquad \blacksquare$$

## 12.6  IDENTIFICATION OF LINEAR FINITE DIFFERENCE MODELS

In this section the identification problem is considered in the framework of finite difference models. As before $y(n)$ is a measurement of the actual system output and has the form

$$y(n) = z(n) + v(n) \tag{12.103}$$

where $v(n)$ is the measurement noise. The output $z(n)$ is obtained from the measurable input $u(n)$ by a linear time invariant finite difference model and is affected by the process noise $w(n)$. Thus

$$z(n) + \sum_{i=1}^{n_a} a_i z(n - i) = \sum_{i=1}^{n_b} b_i u(n - i) + w(n) \tag{12.104}$$

The signals $u(n)$, $w(n)$, and $v(n)$ are mutually uncorrelated and have zero mean.

### 12.6.1   Correlation approach

We proceed exactly as in Section 12.5.1. Since $u(n)$ and $v(n)$ are uncorrelated, Eq. (12.103) gives, upon taking crosscorrelations,

$$r_{yu}(n) = r_{zu}(n) \tag{12.105}$$

Next we multiply both sides of (12.104) with $u(n - l)$ and take expectations to obtain

$$E[z(n)u(n - l)] + \sum_{i=1}^{n_a} a_i E[z(n - i)u(n - l)]$$

$$= \sum_{i=1}^{n_b} b_i E[u(n - i)u(n - l)] + E[w(n)u(n - l)]$$

Since $u(n)$ and $w(n)$ are uncorrelated, the above equation together with (12.105) give

$$r_{yu}(n) + \sum_{i=1}^{n_a} a_i r_{yu}(n - i) = \sum_{i=1}^{n_b} b_i r_u(n - i) \tag{12.106}$$

Thus we first determine estimates of $r_u(n)$ and $r_{yu}(n)$, and then we apply the techniques of Section 7.3.1 to determine the coefficients $a_i$ and $b_i$. These are uniquely defined provided that the polynomials $a(q)$ and $b(q)$ are relatively prime.

Suppose next that the process noise $w(n)$ is modeled as an **ARMA** signal

$$w(n) = \frac{c(q)}{d(q)} e(n) \tag{12.107}$$

Exactly as in Section 12.5.1 the signal $y_1(n) = y(n) - [b(q)/a(q)]u(n)$ is known, and Eq. (12.103) takes the form

$$y_1(n) = \frac{c(q)}{a(q)d(q)} e(n) + v(n)$$

Techniques for the determination of $c(q)$ and $d(q)$ are discussed next.

## 12.6.2  Blind identification of ARMA models

Consider the ARMA model

$$y(n) = \frac{b(q)}{a(q)} e(n) + v(n), \qquad b_0 = 1, \quad a_0 = 1 \qquad (12.108)$$

or

$$y(n) + \sum_{i=1}^{n_a} a_i y(n-i) = \sum_{i=0}^{n_b} b_i e(n-i) + \sum_{i=0}^{n_a} a_i v(n-i) \qquad (12.109)$$

Both $e(n)$ and $v(n)$ are mutually uncorrelated white noise signals. Moreover $v(n)$ is Gaussian, and $e(n)$ is non-Gaussian with nonvanishing intensity $\gamma_3$. The AR coefficients $a_i$ can be recovered from the autocorrelation sequence of the output. Indeed, we can multiply the above equation by $y(n-k)$ and take expected values. Then

$$E[y(n)y(n-k)] + \sum_{i=1}^{n_a} a_i E[y(n-i)y(n-k)]$$

$$= \sum_{i=0}^{n_b} b_i E[e(n-i)y(n-k)] + \sum_{i=0}^{n_a} a_i E[v(n-i)y(n-k)]$$

Let $k > \max(n_a, n_b)$. The right-hand side terms are zero because $y(n-k)$ depends on past values $e(n-k-l)$, $v(n-k-l)$, $l = 0, 1, \ldots$, while $e(n-i)$ and $v(n-i)$ are not among them, and both are white noise signals. We conclude

$$r_y(k) + \sum_{i=1}^{n_a} a_i r_y(k-i) = 0, \qquad k > \max(n_a, n_b) = L \qquad (12.110)$$

Collecting the above equations for $n_a$ successive values of $k$ leads to the following linear systems of equations:

$$
\begin{pmatrix}
r_y(L) & r_y(L-1) & \cdots & r_y(L-n_a+1) \\
r_y(L+1) & r_y(L) & \cdots & r_y(L-n_a+2) \\
\vdots & \vdots & & \vdots \\
r_y(L+n_a-1) & r_y(L+n_a-2) & \cdots & r_y(L)
\end{pmatrix}
\begin{pmatrix}
a_1 \\ a_2 \\ \vdots \\ a_{n_a}
\end{pmatrix}
$$

$$
= -
\begin{pmatrix}
r_y(L+1) \\
r_y(L+2) \\
\vdots \\
r_y(L+n_a)
\end{pmatrix}
\tag{12.111}
$$

The matrix is nonsymmetric Toeplitz and the AR coefficients can be computed by a variant of the Levinson algorithm (see Exercise 4.15).

We calculated the denominator of $h(q) = b(q)/a(q)$. Next we compute the first $n_b$ values of the impulse response $h(n)$, $1 \le n \le n_b$ using higher-order statistics. The numerator polynomial is readily determined from (see Section 7.3)

$$
h(n) + \sum_{i=1}^{n_a} a_i h(n-i) = \sum_{i=0}^{n_b} b_i \delta(n-i) = b_n, \qquad 1 \le n \le n_b \tag{12.112}
$$

We assume, for simplicity, that $\gamma_{3e} = \gamma_3 \neq 0$. The general case is treated in a similar way. We set $a_0 = 1$ in (12.112) to obtain

$$
\sum_{i=0}^{n_a} a_i h(n-i) = \sum_{i=0}^{n_b} b_i \delta(n-i), \qquad n \ge 0 \tag{12.113}
$$

Next we evaluate Eq. (9.50) for $k = 3$:

$$
c_{3y}(n_1, n_2) = \gamma_3 \sum_{k=0}^{\infty} h(k)h(k+n_1)h(k+n_2)
$$

If we set $n_1 = n - i$, $0 \le i \le n_a$, multiply by $a_i$, and sum up with respect to $i$, we obtain

$$
\sum_{i=0}^{n_a} a_i c_{3y}(n-i, n_2) = \gamma_3 \sum_{k=0}^{\infty} h(k)h(k+n_2) \sum_{i=0}^{n_a} a_i h(k+n-i)
$$

$$
= \gamma_3 \sum_{k=0}^{-n+n_b} h(k)h(k+n_2)b_{n+k}
$$

The change of variables $n + k = l$ leads to

$$\sum_{i=0}^{n_a} a_i c_{3y}(n - i, n_2) = \gamma_3 \sum_{l=0}^{n_b} h(l - n)h(l - n + n_2)b_l \qquad (12.114)$$

If the latter expression is evaluated on the slice $n = n_b$, the sum in the right-hand side involves only the term corresponding to $l = n_b$, due to causality. Thus

$$\sum_{i=0}^{n_a} a_i c_{3y}(n_b - i, n_2) = \gamma_3 h(n_2)b_{n_b} \qquad (12.115)$$

In particular,

$$\sum_{i=0}^{n_a} a_i c_{3y}(n_b - i, 0) = \gamma_3 b_{n_b}$$

In conclusion,

$$h(j) = \frac{\sum_{i=0}^{n_a} a_i c_{3y}(n_b - i, j)}{\sum_{i=0}^{n_a} a_i c_{3y}(n_b - i, 0)} \qquad (12.116)$$

### 12.6.3  Prediction error approach

The prediction error approach minimizes a criterion of the error between the measured output sequence $y(n)$ and the predicted output $\hat{y}(n)$ obtained at the output of a predictor. Predictors can be designed in a variety of ways. The predictors considered here derive from the convolutional structures of Section 12.5.3. It turns out that if the data sequences are generated by a rational model, these predictors are also rational and hence finitely parametrizable.

We will mainly work with two important special cases of the setup delineated by Eqs. (12.103)–(12.104): the measurement noise free case and the process noise free case.

**Measurement noise free case, $v = 0$.** If $v = 0$ the system is described by

$$y(n) + \sum_{i=1}^{n_a} a_i y(n - i) = \sum_{i=1}^{n_b} b_i u(n - i) + w(n) \qquad (12.117)$$

It is compactly written as

$$a(q)y(n) = b(q)u(n) + w(n)$$

where

$$a(q) = 1 + \sum_{i=1}^{n_a} a_i q^i, \qquad b(q) = \sum_{i=1}^{n_b} b_i q^i$$

Several special cases are worth singling out.

***ARX models.*** If the signal $w(n)$ is white noise, Eq. (12.117) is referred to as autoregressive exogenous (ARX) model. The term $a(q)y(n)$ forms the autoregressive part and the term $b(q)u(n)$ forms the exogenous part. We assume that the roots of $a(q)$ are outside the unit circle, so absolute summability of the impulse response is ensured. The predictor associated with the ARX model is readily determined using the results of Section 12.5.3 (Eq. (12.101)) and the identification

$$g(q) = \frac{b(q)}{a(q)}, \quad h(q) = \frac{1}{a(q)}, \quad h(0) = 1$$

Thus

$$\hat{y}(n) = (1 - a(q))y(n) + b(q)u(n) \tag{12.118}$$

or

$$\hat{y}(n) = -\sum_{i=1}^{n_a} a_i y(n - i) + \sum_{i=1}^{n_b} b_i u(n - i)$$

***ARMAX models.*** The autoregressive moving average exogenous (ARMAX) model results when the noise is modeled as a moving average signal. We then have

$$y(n) + \sum_{i=1}^{n_a} a_i y(n - i) = \sum_{i=1}^{n_b} b_i u(n - i) + e(n) + \sum_{i=1}^{n_c} c_i e(n - i) \tag{12.119}$$

where $e(n)$ is white noise. The above equation is compactly rewritten as

$$a(q)y(n) = b(q)u(n) + c(q)e(n)$$

where $c(q) = 1 + \sum_{i=1}^{n_c} c_i q^i$. Assuming that the roots of $c(q)$ are outside the unit circle, we determine the predictor via the identification

$$g(q) = \frac{b(q)}{a(q)}, \quad h(q) = \frac{c(q)}{a(q)}$$

We thus have

$$\hat{y}(n) = \left(1 - \frac{a(q)}{c(q)}\right)y(n) + \frac{b(q)}{c(q)}u(n)$$

or

$$\hat{y}(n) + \sum_{i=1}^{n_c} c_i \hat{y}(n - i) = \sum_{i=1}^{n_c} c_i y(n - i) - \sum_{i=1}^{n_a} a_i y(n - i) + \sum_{i=1}^{n_b} b_i u(n - i)$$

**Process noise free case,** $w = 0$. If the process noise is absent, Eqs. (12.103)–(12.104) become

$$y(n) = \frac{b(q)}{a(q)} u(n) + v(n) \tag{12.120}$$

This is a typical representative of the so-called *output error structures*. A fairly general model results when the disturbance $v(n)$ is obtained at the output of a finite difference model excited by white noise

$$v(n) + \sum_{i=1}^{n_d} d_i v(n - i) = e(n) + \sum_{i=1}^{n_c} c_i e(n - i)$$

If we define the polynomials $c(q)$ and $d(q)$ in the obvious manner, we rewrite the above compactly as

$$y(n) = \frac{b(q)}{a(q)} u(n) + \frac{c(q)}{d(q)} e(n) \tag{12.121}$$

The above expression constitutes the *Box-Jenkins* model. The identification

$$g(q) = \frac{b(q)}{a(q)}, \qquad h(q) = \frac{c(q)}{d(q)}$$

leads to the predictor

$$\hat{y}(n) = \left(1 - \frac{d(q)}{c(q)}\right) y(n) + \frac{b(q)d(q)}{c(q)a(q)} u(n) \tag{12.122}$$

The roots of $a(q)$ and $c(q)$ must be outside the unit circle.

The preceding discussion demonstrates that if the data sequences are produced by a linear time invariant difference model, the above predictor also satisfies a linear time invariant difference model and admits a finite parametrization. Linear dependence of the predictor output on the unknown parameters is highly desirable. In this case quadratic in the error optimization criteria lead to unique global solutions obtained from the solution of linear systems of equations. Unfortunately, this is not always the case. Complications arise when the predictor is not described by an FIR filter, but at each time $n$, $\hat{y}(n)$ requires past predictor values $\hat{y}(n - i)$. Successive substitution of these values in the error $y(n) - \hat{y}(n)$ creates polynomial dependencies on the coefficients. The optimization problem leads to nonlinear equations that may have more than one local optima and which require iterative algorithms. Before we treat the general case, we will take up the special but important case of predictors with linear dependence on the parameters.

### 12.6.4  Linear Regression

Consider the ARX model (12.117) and the predictor (12.118). Let

$$\theta = \begin{pmatrix} a_1 & a_2 & \cdots & a_{n_a} & b_1 & b_2 & \cdots & b_{n_b} \end{pmatrix}^T$$

be the vector of unknown coefficients. As we saw in Section 4.4, Eq. (12.118) is written as a linear regression

$$\hat{y}(n) = \varphi^T(n)\theta$$

Minimization of the mean square error

$$J(\theta) = E[e^2(n)] = E[(y(n) - \varphi^T(n)\theta)^2]$$

leads to the normal equations

$$R\theta = d, \quad R = E[\varphi(n)\varphi^T(n)], \quad d = E[\varphi(n)y(n)] \tag{12.123}$$

Likewise, if the total square error

$$J(\theta) = \sum_{n=1}^{N}(y(n) - \varphi^T(n)\theta)^2$$

is employed, the following equations result

$$R(N)\theta(N) = d(N), \quad R(N) = \sum_{n=1}^{N}\varphi(n)\varphi^T(n), \quad d(N) = \sum_{n=1}^{N}\varphi(n)y(n)$$

$$\tag{12.124}$$

Suppose that the input-output sequences are produced by a model of the form (12.117). Let $\theta^*$ denote the true parameter. Exactly as in Example 4.6, we find that

$$\theta = \theta^* + R^{-1}s, \quad s = E[\varphi(n)w(n)]$$

The true parameter is recovered if $R$ is positive definite, so $R^{-1}$ exists, and $s$ is the zero vector. The vector $s$ is zero if $w(n)$ is uncorrelated with the regressor vector. The latter certainly holds if $n_a = 0$, namely the true system is an FIR system. This case was treated in Example 4.6. If $n_a \geq 1$, we have $s = 0$ if $w(n)$ is orthogonal to both $u(n - i)$ and $y(n - i)$. By assumption, $w(n)$ is orthogonal to $u(n - i)$. It will be also orthogonal to $y(n - i)$ if $w(n)$ is white noise.

In conclusion, the true parameter is recovered if $R > 0$, $u(n)$, $w(n)$ are uncorrelated, and moreover either $n_a = 0$, or $n_a \geq 1$ and $w(n)$ is white noise.

Positive definiteness of $R$ requires that the input signal be sufficiently informative. This is explained next.

**Persistent excitation.** A stochastic signal $x(n)$ is called *persistently exciting of order $m$* if the autocorrelation matrix $R_m = E[x_m(n)x_m^T(n)]$ of

$$x_m(n) = (\, x(n-1) \quad x(n-2) \quad \cdots \quad x(n-m) \,)^T$$

is positive definite. It is *persistently exciting* if it is persistently exciting of any order. Roughly speaking, persistent excitation ensures that the power spectral density of $x(n)$ is not overwhelmed by spectral nulls. Indeed, suppose that $x(n)$ is a scalar wide sense stationary process having power spectral density $S(\omega)$. Let $a$ be a real vector of length $m$. Then

$$a^T R a = \sum_{k=1}^{m} \sum_{l=1}^{m} a_k a_l r(k-l) = \frac{1}{2\pi} \sum_{k=1}^{m} \sum_{l=1}^{m} a_k a_l \int_{-\pi}^{\pi} e^{j(k-l)\omega} S(\omega) d\omega$$

or

$$a^T R a = \frac{1}{2\pi} \int_{-\pi}^{\pi} |A(\omega)|^2 S(\omega) d\omega \geq 0$$

where $A(\omega) = \sum_{k=1}^{m} a_k e^{-jk\omega}$. Hence $a^T R a = 0$, if $|A(\omega)|^2 S(\omega) = 0$ for all $\omega$. If $S(\omega)$ is nonzero at $m$ frequency points, the polynomial $A(q)$ will have more than $m$ roots. Hence it is the zero polynomial. Therefore $a = 0$ and $R > 0$. In particular, if $S(\omega) > 0$, almost everywhere the signal is persistently exciting of any order. The latter condition was encountered in linear prediction (see Theorem 12.3) and in the correlation analysis of Section 12.5.1.

### Example 12.12  White noise and ARMA signals

A white noise signal is persistently exciting because $S(\omega)$ is a positive constant. If $x(n)$ is an ARMA process, it is persistently exciting because the power spectral density is of the form $|b(\omega)|^2/|a(\omega)|^2$ and thus is positive.  ∎

### 12.6.5  Instrumental variables

The method of instrumental variables is a generalization of the correlation approach discussed in Section 12.6.1. Let us assume that the data sequences are generated by the linear regression model

$$y(n) = \varphi^T(n)\theta^* + w(n) \tag{12.125}$$

If $w(n)$ is white noise, the least squares formulation gives the true parameter provided that the identification experiment is carefully designed so that the input is sufficiently informative. If $w(n)$ is not white, the least squares estimate fails to reproduce $\theta^*$. In a sense there are additional predictive attributes in $w(n)$ that are not accounted by the predictor. One approach is to model the disturbance by a rational model and build a predictor along the lines of Section 12.6.6. In this way the predictor will extract the additional predictive structure of $w(n)$. The price paid is the nonlinear dependence on the unknown parameter. An alternative technique that maintains the linear system of equations for the determination of the parameters is described next. We multiply both sides of (12.125) by a vector $\zeta(n)$ called the *instrument* and take expected values. Then

$$E[\zeta(n)y(n)] = E[\zeta(n)\varphi^T(n)]\theta^* + E[\zeta(n)w(n)]$$

We choose $\zeta(n)$ so that it is uncorrelated with $w(n)$ and the matrix

$$R_{iv} = E[\zeta(n)\varphi^T(n)]$$

is nonsingular. Then $\theta^*$ is determined by the linear system

$$R_{iv}\theta = r_{iv}, \quad r_{iv} = E[\zeta(n)y(n)] \tag{12.126}$$

The correlation technique of Section 12.6.1 basically employs an instrument of the form

$$\zeta(n) = (\, u(n-1) \quad u(n-2) \quad \cdots \quad u(n-m)\,)^T$$

A common choice of the instrument is dictated by the form of the regressor vector (see Section 4.4). A popular choice is

$$\zeta(n) = (\, x(n-1) \quad x(n-2) \quad \cdots \quad x(n-k) \quad u(n-1) \quad \cdots \quad u(n-m)\,)^T$$

where $x(n)$ is the output of a linear filter excited by $u(n)$:

$$x(n) = \frac{f(q)}{g(q)} u(n)$$

The above choice warrants that $\zeta(n)$ be always uncorrelated to $w(n)$. It can be shown that the matrix $R_{iv}$ is invertible for almost all choices of polynomials $f(q)$ and $g(q)$.

If the total square error formulation is employed, the instrumental variables estimate is obtained from the linear system

$$R_{iv}(N)\theta(N) = r_{iv}(N), R_{iv}(N) = \sum_{i=1}^{N} \zeta(n)\varphi^T(n), r_{iv}(N) = \sum_{i=1}^{N} \zeta(n)y(n) \tag{12.127}$$

*Example 12.13   MATLAB simulation and identification*
The purpose of this example is to illustrate the previous identification methods. Consider the system

$$y(n) + \frac{3}{4}y(n-1) + \frac{1}{8}y(n-2) = 2u(n-1) + w(n)$$

Input-output data are generated by MATLAB simulation. The commands arx and iv4 from the identification toolbox are used to identify the parameters for various noise scenaria:

```
u = randn(1,300);
w = randn(1,300);
a = [1 3/4 1/8];
b = [0 2];
y = filter(b,a,u) + w;
z = [y',u'];
theta = arx(z,[2,1,1]);          % The arguments 2,1,1 declare the orders of the
                                 % polynomials a(q), b(q) and the input delay.
present(theta)                   % displays theta
c = [12 -7 1];
w = filter(c,1,w);               % repeat with coloured moving average noise
y = filter(b,a,u) + w;
theta = arx(z,[2,1,1]);
theta1 = iv4(z,[2,1,1]);         %  estimates parameters with the instrumental variables
present(theta)
present(theta1)
```
∎

## 12.6.6   Nonlinear regression and the prediction error algorithm

With the exception of the ARX model, predictors of the form (12.122) exhibit a nonlinear dependence on the unknown parameter vector. Minimization of the total square error is a nonlinear optimization problem. An iterative algorithm based on steepest descent involves the recursion

$$\theta(n+1) = \theta(n) - \mu(n)\nabla J(\theta(n))$$

where

$$J(\theta) = \frac{1}{2}\sum_{k=1}^{n} e^2(k,\theta), \quad e(k,\theta) = y(k) - \hat{y}(k,\theta)$$

The notation $\hat{y}(n,\theta)$ emphasizes the dependence of the predictor output on $\theta$. Now

$$\nabla J(\theta) = -\sum_{k=1}^{n} e(k,\theta)\psi(k,\theta), \quad \psi(k,\theta) = \frac{\partial \hat{y}(k,\theta)}{\partial \theta}$$

In analogy with the LMS algorithm we replace the above average with the instantaneous error to obtain

$$\theta(n+1) = \theta(n) + \mu(n)e(n, \theta(n))\psi(n, \theta(n)) \tag{12.128}$$

where $e(n, \theta(n))$ is the prediction error at time $n$, corresponding to the parameter $\theta(n)$. The stepsize $\mu(n)$ is usually set to a small positive number. Alternatively, it is chosen to normalize the predictor gradient much like the normalized LMS algorithm

$$\mu(n) = \frac{\mu}{1 + \mu\psi^T(n)\psi(n)}$$

where $\psi(u) = \psi(n, \theta(n))$. Eq. (12.128) requires computation of the predictor output $\hat{y}(n, \theta(n))$ and its gradient $\psi(n)$. This is best illustrated by the following example.

**Example 12.14    Stochastic gradient algorithm for ARMAX identification**
Suppose that the data sequences are produced by the ARMAX model

$$a(q)y(n) = b(q)u(n) + c(q)w(n), \qquad c_0 = 1$$

The zeros of $c(q)$ (as a polynomial in $q$) are outside the unit disc. $w(n)$ is white noise. The predictor is chosen so that the prediction error $e(n) = y(n) - \hat{y}(n)$ equals $w(n)$. We rewrite the above equation as

$$c(q)e(n) = c(q)[y(n) - \hat{y}(n)] = a(q)y(n) - b(q)u(n) \tag{12.129}$$

Solving with respect to $\hat{y}(n)$, we find that

$$c(q)\hat{y}(n) = (c(q) - a(q))y(n) + b(q)u(n) \tag{12.130}$$

We obtained the same structure with the model of Section 12.6.3. We rewrite Eq. (12.129) in the time domain as

$$e(n) + \sum_{i=1}^{n_c} c_i e(n-i) = y(n) + \sum_{i=1}^{n_a} a_i y(n-i) - \sum_{i=1}^{n_b} b_i u(n-i)$$

Therefore

$$e(n) = y(n) - \varphi^T(n, \theta)\theta$$

where $\theta = \begin{pmatrix} a^T & b^T & c^T \end{pmatrix}^T$ and

$$\varphi(n, \theta) = (-y(n-1) \cdots - y(n-n_a)\ u(n-1) \cdots u(n-n_b)\ e(n-1) \cdots e(n-n_c))^T \tag{12.131}$$

Hence the predictor is written as $\hat{y}(n) = \varphi^T(n, \theta)\theta$.

The above expression is referred to as *pseudolinear regression*. The dependence on $\theta$ is nonlinear because the regressor depends on the error $e(n)$, which in turn depends on $\theta$. For simplicity of notation, we write $\varphi(n) = \varphi(n, \theta(n))$ and similarly for the gradient. Storage and complexity are controlled by setting $e(n - i, \theta(n - i))$ in place of $e(n - i, \theta(n))$ in Eq. (12.131). To determine the predictor gradient, we differentiate (12.130) with respect to $\theta$:

$$c(q) \frac{\partial \hat{y}}{\partial a_i} = - q^i y(n), \qquad 1 \leq i \leq n_a$$

$$c(q) \frac{\partial \hat{y}}{\partial b_i} = q^i u(n), \qquad 1 \leq i \leq n_b$$

$$c(q) \frac{\partial \hat{y}}{\partial c_i} + q^i \hat{y}(n) = q^i y(n) \quad \text{or} \quad c(q) \frac{\partial \hat{y}}{\partial c_i} = q^i e(n), \qquad 1 \leq i \leq n_c$$

Putting the above together, we find that

$$\psi(n, \theta) = \frac{1}{c(q)} \varphi(n, \theta)$$

The predictor gradient dynamics is stable because the roots of $c(q)$ are outside the unit disc.

In summary the stochastic gradient algorithm for **ARMAX** models becomes

$$\theta(n + 1) = \theta(n) + \mu(n)e(n)\psi(n)$$

$$e(n) = y(n) - \hat{y}(n)$$

$$\hat{y}(n) = \varphi^T(n)\theta(n)$$

$$\varphi(n) = (-y(n - 1) \cdots - y(n - n_a) \, u(n - 1) \cdots u(n - n_b) \, e(n - 1) \cdots e(n - n_c))^T$$

$$\psi(n) + \sum_{k=1}^{n_c} c_k(n)\psi(n - k) = \varphi(n) \qquad \blacksquare$$

The more general Box-Jenkins model is handled in a similar fashion. The stochastic gradient algorithm relies on the steepest descent direction and the minimization of the total squared error. Extensions to more general differentiable cost functions and to other descent directions (see Appendix III) is possible. All these schemes are unified under the term *prediction error algorithms*.

## 12.7 IDENTIFICATION OF SECOND-ORDER VOLTERRA SYSTEMS

This section is concerned with the identification of Volterra systems of degree 2:

$$y(n) = \sum_{i=0}^{\infty} h_1(i)u(n-i) + \sum_{i_1=0}^{\infty}\sum_{i_2=0}^{\infty} h_2(i_1, i_2)u(n-i_1)u(n-i_2) + v(n) \quad (12.132)$$

The noise $v(n)$ is independent of the measurable input $u(n)$. Both $u(n)$ and $v(n)$ have zero mean. In addition $u(n)$ is a Gaussian process with aurocorrelation sequence $r_u(n)$ and power spectral density $S_u(\omega)$ satisfying $S_u(\omega) > 0$, everywhere. The kernels $h_1$ and $h_2$ are causal absolutely summable sequences. Moreover $h_2(k_1, k_2)$ is symmetric. Symmetry does not harm generality, according to Section 3.7.2. Note that the special case of finite support Volterra systems leads to a linear regression and that the identification problem can be handled by the methods of Section 4.4. Here we are interested in infinite extent kernels and in correlation analysis. In analogy with the linear case, we will generate sufficient input-output crosscumulant information, that will likely suppress the noise and provide the means for the estimation of the kernels.

We first calculate the cumulant of $y$ with one copy of $u$. Multilinearity of the cumulants and the independence of $u(n)$ and $v(n)$ give

$$c_{uy}(-k_1) = \text{cum}[u(n-k_1), y(n)] = \sum_{i=0}^{\infty} h_1(i)\text{cum}[u(n-k_1), u(n-i)]$$

$$+ \sum_{i_1=0}^{\infty}\sum_{i_2=0}^{\infty} h_2(i_1, i_2)\text{cum}[u(n-k_1), u(n-i_1)u(n-i_2)] \quad (12.133)$$

Cumulants can be computed from input-output data. The above expression provides a linear type of equation for the determination of the kernels. An additional equation is obtained if the crosscumulant of $y$ with two copies of $u$ is formed. Thus

$$c_{uuy}(-k_1, -k_2) = \text{cum}[u(n-k_1), u(n-k_2), y(n)]$$

$$= \sum_{i=0}^{\infty} h_1(i)\text{cum}[u(n-k_1), u(n-k_2), u(n-i)]$$

$$+ \sum_{i_1=0}^{\infty}\sum_{i_2=0}^{\infty} h_2(i_1, i_2)\text{cum}[u(n-k_1), u(n-k_2), u(n-i_1)u(n-i_2)]$$

$$(12.134)$$

To come up with a better understanding of the above equations, we must analyze the respective input cumulants. Consider Eq. (12.133). The associated cumulants are covariances. Hence

$$\text{cum}[u(n - k_1), u(n - i)] = r_u(i - k_1) = r_u(k_1 - i)$$

and

$$\text{cum}[u(n - k_1), u(n - i_1)u(n - i_2)] = E[u(n - k_1)u(n - i_1)u(n - i_2)]$$

$$- E[u(n - k_1)]E[u(n - i_1)u(n - i_2)] = 0$$

The two terms in the right-hand side are zero because $u(n)$ is zero mean Gaussian (see Exercise 8.35). Thus Eq. (12.133) becomes

$$c_{uy}(-k_1) = \sum_{i=0}^{\infty} h_1(i)r_u(k_1 - i)$$

Exactly as in Section 12.5.1, we pass to the frequency domain to deduce that

$$H_1(\omega) = \frac{C_{uy}(-\omega)}{S_u(\omega)}$$

Next we consider the input cumulants appearing in Eq. (12.134). Since $u(n)$ is Gaussian, it holds that

$$\text{cum}[u(n - k_1), u(n - k_2), u(n - i)] = 0$$

Moreover it follows from the moments to cumulants conversion formula (Eq. (9.45), of Section 9.5.1) that

$$\text{cum}[u(n - k_1), u(n - k_2), u(n - i_1)(u - i_2)]$$
$$= E[u(n - k_1)u(n - k_2)u(n - i_1)u(n - i_2)] -$$
$$E[u(n - k_1)u(n - k_2)]E[u(n - i_1)u(n - i_2)]$$
$$= E[u(n - k_1)u(n - k_2)u(n - i_1)u(n - i_2)] - r_u(k_2 - k_1)r_u(i_2 - i_1) \quad (12.135)$$

Now we use the cumulants to moments conversion formula to express the fourth-order cumulant as

$$E[u(n - k_1)u(n - k_2)u(n - i_1)u(n - i_2)] = r_u(k_2 - k_1)r_u(i_2 - i_1) +$$
$$r_u(k_1 - i_1)r_u(k_2 - i_2) + r_u(k_2 - i_1)r_u(k_1 - i_2)$$

Substituting the above into Eq. (12.135), we obtain

$$\text{cum}[u(n - k_1), u(n - k_2), u(n - i_1)u(n - i_2)]$$
$$= r_u(k_1 - i_1)r_u(k_2 - i_2) + r_u(k_2 - i_1)r_u(k_1 - i_2)$$

If we plug the above into Eq. (12.134) and use the symmetry of the kernel, we have

$$c_{uuy}(-k_1, -k_2) = 2 \sum_{i_1=0}^{\infty} \sum_{i_2=0}^{\infty} h_2(i_1, i_2)r_u(k_1 - i_1)r_u(k_2 - i_2)$$

Conversion to the frequency domain via the 2-D Fourier transform leads to

$$H_2(\omega_1, \omega_2) = \frac{1}{2} \frac{C_{uuy}(-\omega_1, -\omega_2)}{S_u(\omega_1)S_u(\omega_2)}$$

## BIBLIOGRAPHICAL NOTES

Classical sources for the Kolmogorov-Wiener theory of linear prediction and filtering are Doob (1960) and Rozanov (1967). A detailed exposition is given in Caines (1989). The reader may also consult Wong and Hajek (1985) and Porat (1994). The exposition on Kalman filtering is based on Anderson and Moore (1979) and Rhodes (1971). System identification is treated in Goodwin and Sin (1984), Ljung and Soderstrom (1983), Ljung (1987), Soderstrom and Stoica (1989), Norton (1986), and Hannan and Deistler (1988). The use of higher order statistics in system identification is studied in Nikias and Petropoulou (1993). Volterra identification is studied in Rugh (1981) and Schetzen (1980). Other major approaches to system identification not included in the book are subspace identification (De Moor and Overschee 1995), set membership identification (Milanese and Vicino 1991; Walter and Piet-Lahanier 1990) and the Frisch scheme and errors-in-variables identification (Frisch 1934; Kalman 1982; Guidorzi 1993).

## PROBLEMS

**12.1.**   Consider the MA signal

$$x(n) = w(n) - \frac{3}{5}w(n - 1)$$

and the measurement $y(n) = x(n) + v(n)$. The noise sequences $w(n)$ and $v(n)$ are uncorrelated with variances $R_w = 1/2$, $R_v = 8/25$. Compute the Wiener filter $E^*[x(n)|\mathcal{H}_y(n)]$.

**12.2.** Consider the AR signal

$$x(n) - \frac{1}{2}x(n-1) = w(n)$$

and the measurement $y(n) = x(n) + v(n)$. The noise sequences $w(n)$ and $v(n)$ are uncorrelated with variances $R_w = 1/4$, $R_v = 3/5$. Compute the Wiener filter $E^*[x(n)|\mathcal{H}_y(n)]$.

**12.3.** Suppose that the noise signals $w(n)$ and $v(n)$ are correlated. Let $\text{cov}[w(n), v(m)] = Q(n)\delta(n-m)$. Show that the state equations of the Kalman filter remain unaltered but that the Kalman gain is computed by the equations

$$K(n) = \left(A(n)P(n)C^T(n) + D(n)Q(n)\right)\left(C(n)P(n)C^T(n) + R_v(n)\right)^{-1}$$

$$P(n+1) = (A(n) - K(n)C(n))P(n)(A(n) - K(n)C(n))^T$$
$$+ D(n)R_w(n)D^T(n) + K(n)R_v(n)K^T(n)$$
$$- D(n)Q(n)K^T(n) - K(n)Q^T(n)D^T(n)$$

**12.4.** Prove Eq. (12.64).

**12.5.** Show that the $d$-steps-ahead prediction $(d \geq 2)$ is determined by the recursion

$$\hat{x}(n+d|n) = \Phi(n+d, n+1)\hat{x}(n+1|n) + \sum_{i=n+1}^{n+d-1} \Phi(n+d, i+1)B(i)u(i)$$

**12.6.** Show that the Kalman filter and the Kalman predictor are related by

$$\hat{x}(n+1|n) = A(n)\hat{x}(n|n)$$

Likewise, show that the error covariances satisfy

$$P(n|n) = P(n|n-1)$$
$$- P(n|n-1)C^T(n)[C(n)P(n|n-1)C^T(n) + R_v(n)]^{-1}$$
$$C(n)P(n|n-1)$$
$$P(n+1|n) = A(n)P(n|n)A^T(n) + D(n)R_w(n)D^T(n)$$

**12.7.** The discussion of Section 12.4.6 is next extended to the case of correlated noises (see Exercise 12.3)
(a) Prove that the steady state error covariance and Kalman gain

become

$$K = (APC^T + DS)(CPC^T + R_v)^{-1}$$

$$P = (A - KC)P(A - KC)^T + DR_wD^T + KR_vK^T - DSK^T - KSD^T$$

(b) Define

$$A_1 = A - DSR_v^{-1}C \quad K_1 = K - DSR_v^{-1}$$

Prove

$$K_1 = A_1PC^T(CPC^T + R_v)^{-1}, \quad A - KC = A_1 - K_1C$$

and

$$P = (A_1 - K_1C)P(A_1 - K_1C)^T + K_1R_vK_1^T + D(R_w - SR_v^{-1}S^T)D^T$$

(c) Conclude that $P$ is the limiting error covariance associated with a filtering problem defined by $\{A_1, D, R_{1w} = R_w - SR_v^{-1}S^T, R_v\}$ and uncorrelated noises. Using Theorem 12.8 deduce that the Kalman predictor with correlated noises is totally stable provided that $(A - DSR_v^{-1}C, DR_2)$, where $R_2R_2^T = R_{1w}$, is controllable or $A - DSR_v^{-1}C$ is as. stable. The remaining claims of Theorem 12.8 are proved by the same analysis.

**12.8.** Suppose that the data sequences $y(n)$ and $u(n)$ are generated by the linear regression $y(n) = \phi^T(n)\theta^* + e(n)$. The input is persistently exciting and the disturbance is a finite support signal

$$e(n) = \sum_{k=1}^{m} b_k\delta(n - k)$$

Prove that the least squares solution (12.124) asymptotically recovers $\theta^*$.

**12.9.** Prove (12.95). Derive the following expressions for the intensities:

$$\gamma_{3e} = \frac{c_{3y}(0,0)}{\sum_{n=0}^{k}[c_{3y}(k,n)/c_{3y}(k,0)]^3}$$

$$\gamma_{4e} = \frac{c_{4y}(0,0,0)}{\sum_{n=0}^{k}[c_{4y}(k,0,n)/c_{4y}(k,0,0)]^3}$$

**12.10.** Show that the parameters $a_i$ and the impulse response coefficients $h(i)$

of the ARMA model (12.113) can be computed by a linear system involving output cumulants of order $m$.

**12.11.** Consider a finite Volterra system of degree 2 containing a constant term. Using the technique of Section 12.7 compute the kernels $h_0$, $h_1$, and $h_2$.

**12.12.** The motion of a rotating antenna is described by

$$\ddot{\theta}(t) + \frac{B}{J}\dot{\theta}(t) = \frac{k}{J}u(t)$$

where $B$ is the coefficient of viscous friction, $J$ the moment of inertia, $u(t)$ the input voltage to the motor.
(a) Write the above in state space form.
(b) Set $J = 10kgm^2$, $B/J = a = 4.6s^{-1}$, $k/J = 0.787rad/(Vs^2)$; construct a continuous time observer with estimation poles located at $-50 \pm j50s^{-1}$. Simulate the system and the observer in MATLAB. Derive plots of the angular position and the reconstructed angular position for a constant input voltage $u(t) = -10V$, $t \geq 0$.
(c) Use MATLAB to develop a discrete simulator and a corresponding discrete observer with sampling period $T_s = 0.1s$ and estimation poles $e^{-5\pm j5}$. Plot the resulting discretized angular position and the reconstructed version.

**12.13.** Write a MATLAB program for the computation of the Kalman filter in the general time varying case.

**12.14.** Consider the motion of a rotating antenna under the presence of a disturbing torque acting upon the shaft of the motor

$$\ddot{\theta}(t) + \frac{B}{J}\dot{\theta}(t) = \frac{k}{J}u(t) + \frac{1}{J}w(t)$$

Let $y(t) = \theta(t) + v(t)$. Let $R_w = 10N^2m^2s$, $R_v = 10^{-2}rad^2s$.
(a) Perform discrete simulation in MATLAB, using the numerical values of Exercise 12.12 and the c2d command.
(b) Determine the optimal time invariant Kalman predictor and the Kalman predictor itself. What are the poles of the time invariant predictor?
(c) Design an observer with estimation poles equal to the poles of the Kalman predictor. Plot the angular position and the reconstructed versions obtained by the observer, the Kalman predictor, and the time invariant Kalman predictor.

**12.15.** *Information filter.* The information filter and predictor update the inverse of the covariance matrices $R(n|n-1) = P^{-1}(n|n-1)$ and $R(n|n) = P^{-1}(n|n)$, and the vectors $a(n|n-1) = R(n|n-1)\hat{x}(n|n-1)$, $a(n|n) = R(n|n)\hat{x}(n|n)$. Establish the following recursions, taking $B = 0$

in Eq. (12.51).

$$R(n|n) = R(n|n-1) + C^T(n)R_v^{-1}(n)C(n)$$

$$R(n+1|n) = [I - E(n)D^T(n)]H(n)$$

$$H(n) = (A^{-1}(n))^T R(n|n)A^{-1}(n)$$

$$a(n|n) = a(n|n-1) + C^T(n)R_v^{-1}(n)y(n)$$

$$a(n+1|n) = [I - E(n)D^T(n)](A^{-1}(n))^T a(n|n)$$

*Hint.* Apply the matrix inversion lemma of the Appendix in Chapter 4 on the error covariance formulas of Exercise 12.6.

**12.16.** *Square root filtering and the Potter algorithm.* Finite precision arithmetic may lead to updates of the Riccati equation that fail to be nonnegative definite matrices. The Potter algorithm alleviates the problem by updating a square root of the covariance error. Let $P(n|n-1) = S(n|n-1)S^T(n|n-1)$ and $P(n|n) = S(n|n)S^T(n|n)$.

(a) Using the updates of Exercise 12.6 prove that

$$P(n|n-1) = S(n|n-1)[I - a(n)G(n)G^T(n)]S^T(n|n-1)$$

where

$$G(n) = S^T(n)C^T(n) \quad a^{-1}(n) = G^T(n)G(n) + R_v(n)$$

$a(n)$ is a scalar.

(b) A recursion for $S(n)$ is possible if

$$I - a(n)G(n)G^T(n) = [I - \gamma(n)G(n)G^T(n)][I - \gamma(n)G(n)G^T(n)]$$

Deduce a suitable value for $\gamma(n)$.

(c) Establish that

$$S(n|n) = S(n|n-1)[I - \gamma(n)G(n)G^T(n)]$$

$$S(n+1|n) = A(n)S(n|n-1) + D(n)R_w^{1/2}(n)$$

**12.17.** Consider the extraction of a WSS signal $x(n)$ from corrupted measurements $y(n)$:

$$y(n) = g(n) * x(n) + v(n)$$

Extraction of $x(n)$ is to be accomplished by a noncausal filter $\hat{x}(n) = h(n) * y(n)$.

(a) Show using the projection theorem, that the optimum filter $h(n)$ that minimizes the error $E[(x(n) - \hat{x}(n)]^2$ is given by

$$H(\omega) = \frac{S_{xy}(\omega)}{S_y(\omega)} = \frac{G^*(\omega)S_x(\omega)}{|G(\omega)|^2 S_x(\omega) + S_v(\omega)}$$

or

$$H(\omega) = \frac{|\rho(\omega)|^2}{G(\omega)} = \frac{1/G(\omega)}{1 + r^{-1}(\omega)}$$

where $|\rho(\omega)|$ is the coherence magnitude and $r(\omega)$ is the ratio of spectral densities of distorted signal and noise:

$$|\rho(\omega)| = \frac{|S_{xy}(\omega)|}{(S_y(\omega)S_x(\omega))^{1/2}}, \quad r(\omega) = \frac{|G(\omega)|^2 S_x(\omega)}{S_v(\omega)}$$

Deduce that if the SNR $r(\omega)$ is large, the filter acts as an equalizer $(1/G)$, while if the SNR is low, the filter acts as an attenuator.

(b) Show that the minimum mean squared error is given by

$$J = \frac{1}{2\pi} \int_{-\pi}^{\pi} S_x(\omega)(1 - |\rho(\omega)|^2)d\omega = \frac{1}{2\pi} \int_{-\pi}^{\pi} \frac{S_x(\omega)}{1 + r(\omega)} d\omega$$

**12.18.**  Consider the AR signal

$$x(n) - 0.8x(n - 1) = w(n)$$

$$y(n) = x(n) + v(n)$$

where $R_w = 0.36$, $R_v = 1$. Prove the following statements:

(a) The optimum noncausal filter (see Exercise 12.20) has impulse response

$$h(n) = 0.3\left(\frac{1}{2}\right)^{|n|}$$

while the mean squared error is 0.3.

(b) The (causal) Wiener filter is given by

$$h(n) = 0.375\left(\frac{1}{2}\right)^n u_s(n)$$

and it is realized by the finite difference model

$$\hat{x}(n) - 0.5\hat{x}(n - 1) = 0.375y(n), \quad n \geq 0$$

The mean squared error is 0.375.

(c) Realize the system in state space form, and compute the time invariant Kalman predictor and the Kalman predictor. Compare the mean square error with the preceding cases.

**12.19.** Consider the linear regression $y(n) = \varphi^T(n)\theta^* + w(n)$ where

$$\varphi(n) = (y(n-1) \cdots y(n-n_a) \quad u(n-1) \cdots u(n-n_b))$$

$$\theta^* = (a_1\ a_2 \cdots a_{n_a}\ \ b_1\ b_2 \cdots b_{n_b})^T$$

Show that if $w(n)$ is white noise of nonzero variance and $u(n)$ is persistently exciting, $R = E[\varphi(n)\varphi^T(n)]$ has the form

$$R = \begin{pmatrix} A_1 + A_2 & B \\ B^T & C \end{pmatrix}$$

where $A_2 > 0$ and $C > 0$. Conclude that $R > 0$. *Hint:* $A_1 = E[\mathbf{yy}^T]$, $A_2 = E[\mathbf{yv}^T]$, $C = E[\mathbf{uu}^T]$, where

$$a(q)v(q) = w(n)$$

$$\mathbf{y} = (y(n-1) \cdots y(n-n_a))^T$$

Likewise $\mathbf{u}$ and $\mathbf{v}$ are shifted versions of $u(n)$ and $v(n)$.

**12.20.** Consider the state space model

$$x(n+1) = Ax(n) + w(n)$$

$$y(n) = Cx(n) + v(n)$$

with the standard assumptions discussed in Section 12.4.6. We further assume that $A$ is asymptotically stable. Prove the following statements:
(a) $y(n)$ is equivalently realized by the innovation model

$$\hat{x}(n+1) = A\hat{x}(n) + K\tilde{y}(n)$$

$$y(n) = C\hat{x}(n) + \tilde{y}(n) \tag{12.136}$$

where $K$ is the time invariant Kalman gain.
(b) The transfer function of the above model is

$$h(q) = C(I - qA)^{-1}qK + I, \qquad h(0) = I$$

(c) The inverse of the above system is realized as

$$\hat{x}(n+1) = (A - KC)\hat{x}(n) + Ky(n)$$

$$\tilde{y}(n) = -C\hat{x}(n) + y(n)$$

and the transfer function is

$$h^{-1}(q) = -C(I - qA + qKC)^{-1}qK + I$$

(d) The power spectral density is

$$S_y(\omega) = H(\omega)(R_v + CPC^T)H^T(-\omega) \qquad (12.137)$$

and both $h(q)$ and its inverse $h^{-1}(q)$ are stable.

(e) Conclude that (12.137) yields the spectral factorization of the power spectral density of $y(n)$ and specify the predictor spectral factor.

(f) Show that (a)–(e) remain true if the noises $w(n)$ and $v(n)$ are correlated, provided that either $A - DSR_v^{-1}$ is as. stable or the pair $(A - DSR_v^{-1}, DR_2)$ is controllable (see Exercise 12.7).

**12.21.** Consider the ARMA process

$$y(n) + ay(n - 1) = w(n) + cw(n - 1), \qquad |a| < 1, \; c \neq 0$$

$w(n)$ is white noise of variance 1. Prove the following:

(a) The predictor spectral factor is

$$h(q) = \frac{1 + cq}{1 + aq}, \qquad |c| \leq 1$$

$$h(q) = -c\frac{1 + c^{-1}q}{1 + aq}, \qquad |c| \geq 1$$

(b) Write the given system in state space form as

$$x(n + 1) = \begin{pmatrix} -a & c \\ 0 & 0 \end{pmatrix} x(n) + \begin{pmatrix} 1 \\ 1 \end{pmatrix} w(n)$$

$$y(n) = (-a \quad c)x(n) + w(n)$$

and show that the spectral factorization deduced by means of the Kalman filter as discussed in Exercise 12.20 leads to the same spectral factor.

(c) Repeat for the minimal realization

$$x(n + 1) = -ax(n) + (c - a)w(n)$$

$$y(n) = x(n) + w(n)$$

**12.22.** Simulate the system

$$y(n) - 0.9y(n - 1) = u(n - 1) + 0.5u(n - 2) + w(n)$$

where $u(n)$ is a sufficiently exciting signal, such as a moving average signal, and $w(n)$ is white Gaussian noise. Determine the least squares estimate for various system orders. Determine the instrumental variable estimates.

**12.23.** (a) Develop a MATLAB function that implements Eq. (12.132) under the assumption that the kernels have compact support.
(b) Construct MATLAB functions that estimate the crossmoments, the crosscumulants and their spectra.
(c) Write a MATLAB program for the identification of second-order Volterra system. Apply your program to the output sequence produced in (a) and compare the estimated kernels with those used in (a).

**12.24.** (a) Using Theorem 12.5 and Exercise 12.6 prove that

$$\hat{x}(n+1|n+1) = [I - M(n+1)C(n+1)]A(n)\hat{x}(n|n)$$
$$+ [I - M(n+1)C(n+1)]B(n)u(n) + M(n+1)y(n)$$

(b) The corresponding time invariant Kalman filter is

$$\hat{x}(n+1|n+1) = (I - MC)A\hat{x}(n|n) + (I - MC)Bu(n) + My(n)$$

where

$$M = PC^T[CPC^T + R_v]^{-1}$$

Show that the matrix $(I - MC)A$ and the matrix $A - KC$ associated with the Kalman predictor have the same eigenvalues. Deduce that the Kalman filter is totally stable under the assumptions of Theorem 12.8. *Hint*: Use the fact that for square matrices $A, B$, the products $AB$ and $BA$ have the same eigenvalues.

**12.25.** Determine the optimum transmitting and receiving filters for a binary communication system that transmitts data over a channel with frequency response magnitude

$$|C(\omega)| = 1/\sqrt{1 + \left(\frac{\omega}{W}\right)^2}, \qquad |\omega| \le W$$

The additive noise is zero mean white noise with variance $N_0/2$. Use a raised cosine with rolloff factor $a = 1$.

## APPENDIX

PROOF OF THEOREM 12.4   Using the system output equation and linearity
of the estimator, we obtain

$$\hat{y}(n) = C(n)\hat{x}(n) + E^*[v(n)|Z_{n-1}]$$

The second term in the right-hand side is zero due to property 1 of Proposition
12.1. The output equation (12.60) has been established. We will next deal with
$\hat{x}(n)$. Since $u(n)$ is deterministic, it holds that

$$\hat{x}(n) = E^*[x(n)|Z_{n-1}] = E^*[x(n)|Y_{n-1}] \qquad (12.138)$$

where

$$Y_{n-1} = (y(n-1) \quad y(n-2) \quad \dots \quad y(1))$$

Using Eq. (12.48) we have

$$\hat{x}(n+1) = E^*[x(n+1)|Y_n] = E^*[x(n+1)|Y_{n-1}, y(n)]$$
$$= \hat{x}(n+1|n-1) + E^*[\tilde{x}(n+1|n-1)|\tilde{y}(n)] \qquad (12.139)$$

where

$$\hat{x}(n+1|n-1) = E^*[x(n+1)|Y_{n-1}], \quad \tilde{x}(n+1|n-1) = x(n+1) - \hat{x}(n+1|n-1).$$

Since the variables $\tilde{x}(n+1|n-1)$, $\tilde{y}(n)$ have zero mean and the linear least
squares estimator is involved, the above equation is rewritten as

$$\hat{x}(n+1) = \hat{x}(n+1|n-1) + K(n)\tilde{y}(n) \qquad (12.140)$$

$$K(n) = \text{cov}[\tilde{x}(n+1|n-1), \tilde{y}(n)](\text{cov}[\tilde{y}(n), \tilde{y}(n)])^{-1} \qquad (12.141)$$

The linearity of the estimator and the state equation (12.51) give

$$\hat{x}(n+1|n-1) = E^*[A(n)x(n) + B(n)u(n) + D(n)w(n)|Y_{n-1}]$$
$$= A(n)\hat{x}(n) + B(n)u(n) + D(n)E^*[w(n)|Y_{n-1}]$$

$E^*[w(n)|Y_{n-1}]$ is zero due to property 3 of Proposition 12.1. Thus

$$\hat{x}(n+1|n-1) = A(n)\hat{x}(n) + B(n)u(n) \qquad (12.142)$$

It follows then that

$$\tilde{x}(n+1|n-1) = A(n)\tilde{x}(n) + D(n)w(n) \qquad (12.143)$$

Using (12.60) established above, we obtain

$$\tilde{y}(n) = y(n) - \hat{y}(n) = y(n) - C(n)\hat{x}(n) \qquad (12.144)$$

Equations (12.142)–(12.144) in conjunction with Eq. (12.140) establish (12.59).

Next we consider the Kalman gain as defined by Eq. (12.141) and show that it can be computed by Eqs. (12.61) and (12.62). Substitution of the output equation (12.52) in (12.144) gives

$$\tilde{y}(n) = C(n)x(n) + v(n) - C(n)\hat{x}(n) = C(n)\tilde{x}(n) + v(n) \qquad (12.145)$$

where $v(n)$ and $\tilde{x}(n)$ are uncorrelated. Indeed, $v(n)$ and $x(n)$ are uncorrelated by property 2 of Proposition 12.1. Likewise $v(n)$ and $\hat{x}(n)$ are uncorrelated because $\hat{x}(n)$ depends linearly on $Y_{n-1}$, and $v(n)$ is uncorrelated with $y(n-i)$, $i \geq 1$, by property 1 of Proposition 12.1. Consequently (12.145) gives

$$\text{cov}[\tilde{y}(n), \tilde{y}(n)] = C(n)\text{cov}[\tilde{x}(n), \tilde{x}(n)]C^T(n) + R_v(n)$$

$$= C(n)P(n)C^T(n) + R_v(n) \qquad (12.146)$$

Now we are ready to compute $\text{cov}[\tilde{x}(n+1|n-1), \tilde{y}(n)]$. Equation (12.145) implies that

$$\text{cov}[\tilde{x}(n+1|n-1), \tilde{y}(n)]$$
$$= \text{cov}[\tilde{x}(n+1|n-1), C(n)\tilde{x}(n) + v(n)]$$
$$= \text{cov}[\tilde{x}(n+1|n-1), \tilde{x}(n)]C^T(n) + \text{cov}[\tilde{x}(n+1|n-1), v(n)] \qquad (12.147)$$

Note that

$$\text{cov}[\tilde{x}(n+1|n-1), v(n)] = \text{cov}[x(n+1), v(n)] - \text{cov}[\hat{x}(n+1|n-1), v(n)] = 0$$

The first term in the right-hand side is zero by property 2 of Proposition 12.1. The second term is zero due to property 1 of Proposition 12.1 and the linear dependence of $\hat{x}(n+1|n-1)$ on $y(n-i)$, $i \geq 1$. Returning to Eq. (12.147) and using (12.143) we have

$$\text{cov}[\tilde{x}(n+1|n-1), \tilde{x}(n)] = \text{cov}[A(n)\tilde{x}(n) + D(n)w(n), \tilde{x}(n)]$$

$$= A(n)P(n) + D(n)\text{cov}[w(n), \tilde{x}(n)]$$

We have already pointed out that $\text{cov}[w(n), \tilde{x}(n)]$ is zero, by properties 3 and 4 of Proposition 12.1. Equation (12.61) has been established. To complete the proof, we must verify Eq. (12.62). Recall that $P(n) = \text{cov}[\tilde{x}(n), \tilde{x}(n)]$. Using (12.59) and

(12.51), the following state space model for the state estimation error is obtained:

$$\tilde{x}(n+1) = x(n+1) - \hat{x}(n+1) = A(n)x(n) + B(n)u(n) + D(n)w(n)$$
$$- A(n)\hat{x}(n) + K(n)C(n)\hat{x}(n) - B(n)u(n) - K(n)y(n)$$

or

$$\tilde{x}(n+1) = (A(n) - K(n)C(n))\tilde{x}(n) + (D(n) \quad - K(n))\begin{pmatrix} w(n) \\ v(n) \end{pmatrix} \qquad (12.148)$$

Thus Eq. (12.58) gives

$$P(n+1) = [A(n) - K(n)C(n)]P(n)[A(n) - K(n)C(n)]^T$$
$$+ (D(n) \quad -K(n))\begin{pmatrix} R_w(n) & 0 \\ 0 & R_v(n) \end{pmatrix}\begin{pmatrix} D^T(n) \\ -K^T(n) \end{pmatrix}$$

which coincides with Eq. (12.62). The derivation is complete.     ∎

**PROOF OF THEOREM 12.5**   The main recursive formula (12.48) gives

$$\hat{x}(n|n) = E^*[x(n)|Z_{n-1}, y(n), u(n)] = E^*[x(n)|Z_{n-1}] + E^*[\tilde{x}(n)|\tilde{y}(n)]$$

or

$$\hat{x}(n|n) = \hat{x}(n|n-1) + M(n)[y(n) - C(n)\hat{x}(n|n-1)]$$

Equation (12.65) is a direct consequence. The gain is determined from

$$M(n) = \text{cov}[\tilde{x}(n), \tilde{y}(n)](\text{cov}[\tilde{y}(n), \tilde{y}(n)])^{-1}$$

Using Eq. (12.145) and Eq (12.146), we obtain (12.66).     ∎

The lemmas that follow are needed to prove Theorem 12.8. The assumptions and notation of Theorem 12.8 are valid.

**Lemma 12.2.**   The sequence $P(n)$ is bounded.

**PROOF**   According to Theorem 12.6 an observer with constant gain $L$ can be designed such that the matrix $A - LC$ has eigenvalues in the unit circle. The covariance error sequence of this observer, $P_L(n)$, is a convergent sequence. Indeed, $P_L(n)$ satisfies (12.73). A simple induction argument shows that $P_L(n)$ is

given by

$$P_L(n) = \sum_{i=0}^{n-1} F^i Q (F^T)^i + F^n P_0 (F^T)^n$$

where

$$F = A - LC, \quad Q = DR_w D^T + LR_v L^T$$

Since $F$ is asymptotically stable, the results of Section 11.5.2 assert that the limit

$$\lim_{n \to \infty} P_L(n) = \sum_{i=0}^{\infty} F^i Q (F^T)^i = P_L$$

exists and is the unique solution of the discrete Lyapunov equation

$$P_L = FP_L F^T + Q$$

Therefore $P_L(n)$ is a bounded sequence. Invoking Theorem 12.7, we conclude that $P(n) \le P_L(n)$, and thus $P(n)$ is bounded. ∎

Within the family of solutions $P(n) = P(n, n_0, P_0)$, those solutions that emanate from the zero matrix are increasing and hence converge. This is the content of the next lemma.

**Lemma 12.3** For any $n_0$, $P(n, n_0) = P(n, n_0, 0)$ is a bounded increasing sequence with respect to $n$, and $\lim_{n \to \infty} P(n, n_0)$ does not depend on $n_0$.

**PROOF** Equation (12.74) is time invariant. Shift invariance of solutions implies that $P(n + 1, n_0) = P(n, n_0 - 1)$. Thus it suffices to show

$$P(n, n_0 - 1) \ge P(n, n_0), \qquad n \ge n_0 \tag{12.149}$$

We use induction. For $n = n_0$, (12.149) becomes

$$P(n_0, n_0 - 1) \ge P(n_0, n_0) = 0$$

which is true because (12.74) generates nonnegative matrices. Next assume that (12.149) holds for all integers between $n_0$ and $n - 1$. The induction hypothesis yields

$$P(n, n_0 - 1) = (A - K(n - 1, n_0 - 1)C)P(n - 1, n_0 - 1)(A - K(n - 1, n_0 - 1)C)^T$$
$$+ DR_w D^T + K(n - 1, n_0 - 1)R_v K^T(n - 1, n_0 - 1)$$
$$\ge (A - K(n - 1, n_0 - 1)C)P(n - 1, n_0)(A - K(n - 1, n_0 - 1)C)^T$$
$$+ DR_w D^T + K(n - 1, n_0 - 1)R_v K^T(n - 1, n_0 - 1)$$

The right-hand side is certainly greater than

$$\geq \min_{L(n)}\{(A - L(n)C)P(n - 1, n_0)(A - L(n)C)^T + DR_wD^T + L(n)R_vL^T(n)\}$$

because $K(n - 1, n_0 - 1)$ is simply one of the possible gains $L(n)$. According to Theorem 12.7, the above minimum is achieved by $L(n) = K(n, n_0)$ and then becomes equal to $P(n, n_0)$.

Since $P(n, n_0)$ is a bounded increasing sequence it converges to a nonnegative definite matrix. Time invariance of Eq. (12.74) implies that

$$\lim_{n\to\infty} P(n, n_0) = \lim_{n\to\infty} P(n - n_0, 0) = \lim_{k\to\infty} P(k, 0) = P \qquad \blacksquare$$

The above lemma proves that the algebraic Ricatti equation always has a nonnegative definite solution. The same argument proves the following:

**Lemma 12.4.**   If $P_0 \leq Q_0$ then

$$P(n, n_0, P_0) \leq P(n, n_0, Q_0)$$

Next we establish total stability.                                    ∎

**Lemma 12.5.**   The matrix $(A - KC)$ is asymptotically stable.

**PROOF**   Suppose that there is an eigenvalue $\lambda$ of $A - KC$ with $|\lambda| \geq 1$. Then $\lambda$ is also an eigenvalue of $(A - KC)^T$. Let $a$ be an eigenvector, $(A - KC)^T a = \lambda a$. Since $\lambda$ is complex in general, $a$ has complex entries. Let $a^*$ denote the complex conjugate of $a$. Then

$$a^{*T} Pa = a^{*T}(A - KC)P(A - KC)^T a + a^{*T} DR_1 R_1^T D^T a + a^{*T} KR_vK^T a$$

or

$$a^{*T} Pa = |\lambda|^2 a^{*T} Pa + a^{*T} DR_1 R_1^T D^T a + a^{*T} KR_vK^T a$$

Therefore

$$(1 - |\lambda|^2)a^{*T} Pa = a^{*T} DR_1 R_1^T D^T a + a^{*T} KR_vK^T a$$

The left-hand side is negative, while the right-hand side is positive. Therefore all terms are zero. In particular, $R_1^T D^T a = 0$ and $K^T a = 0$. The relation $K^T a = 0$ in conjunction with $(A - KC)^T a = \lambda a$, gives $A^T a = \lambda a$. Thus $R_1^T D^T A^T a = \lambda R_1^T D^T a = 0$. More generally, it holds that $R_1^T D^T A^{kT} a = 0$. Thus the nonzero vector $a$ is orthogonal to the space spanned by $A^k DR_1$, $k = 0, 1, 2, ...$, contradicting controllability of $(A, DR_1)$.          ∎

**Lemma 12.6.** The solution $P(n)$ of (12.74) can alternatively be computed by the following recursions:

$$P(n + 1) = [A - K(n)C]P(n)A^T + DR_wD^T \qquad (12.150)$$

$$P(n + 1) = AP(n)[A - K(n)C]^T + DR_wD^T \qquad (12.151)$$

In a similar fashion the solution of the algebraic Ricatti equation (12.75) satisfies

$$P = [A - KC]PA^T + DR_wD^T, \quad P = AP[A - KC]^T + DR_wD^T \qquad (12.152)$$

**PROOF**    Equation (12.74) gives

$$\begin{aligned}
P(n + 1) &= [A - K(n)C]P(n)A^T - [A - K(n)C]P(n)C^TK^T(n) \\
&\quad + DR_wD^T + K(n)R_vK^T(n) \\
&= [A - K(n)C]P(n)A^T + DR_wD^T \\
&\quad - \{AP(n)C^T - K(n)[CP(n)C^T + R_v]\}K^T(n)
\end{aligned}$$

The term in braces is zero by Eq. (12.61), and the statement is proved. The remaining expressions are established in a similar way.    ■

**Lemma 12.7**    The error $\tilde{P}(n) = P(n) - P$ obeys the recursion

$$\tilde{P}(n + 1) = [A - KC]\tilde{P}(n)[A - K(n)C]^T \qquad (12.153)$$

Notice that one of the factors is time constant. This fascilitates subsequent convergent analysis.

**PROOF**    Subtracting Eq. (12.152) from Eq. (12.151), we obtain

$$\begin{aligned}
\tilde{P}(n + 1) &= AP(n)[A - K(n)C]^T - [A - KC]PA^T \\
&= [A - KC]\tilde{P}(n)[A - K(n)C]^T \\
&\quad + KCP(n)[A - K(n)C]^T - [A - KC]PC^TK^T(n)
\end{aligned}$$

We will show that the sum of the second and third terms at the right-hand side is zero. Indeed,

$$KCP(n)[A - K(n)C]^T - [A - KC]PC^TK^T(n)$$

$$= K\{CP(n)A^T - CP(n)C^TK^T(n) + CPC^TK^T(n)\} - APC^TK^T(n) \qquad (12.154)$$

Now recall from Eq. (12.61) that

$$K(n)[R_v + CP(n)C^T] = AP(n)C^T, \quad K[R_v + CPC^T] = APC^T$$

Thus Eq. (12.154) becomes

$$K\{R_v K^T(n) + CP(n)C^T K^T(n) - CP(n)C^T K^T(n)$$
$$+ CPC^T K^T(n)\} - APC^T K^T(n)$$

$$= K(R_v + CPC^T)K^T(n) - APC^T K^T(n) = APC^T K^T(n) - APC^T K^T(n) = 0$$

and the proof is complete. ∎

PROOF OF THEOREM 12.8    Consider Eq. (12.153). Successive substitution gives

$$\tilde{P}(n) = (A - KC)^n \tilde{P}_0 \Phi^T(n, 0)$$

where $\tilde{P}_0 = P_0 - P$ and

$$\Phi(n, 0) = [A - K(n - 1)C] \cdots [A - K(0)C]$$

Lemma 12.5 implies that $(A - KC)^n$ decays to zero. We next demonstrate that $\Phi(n, 0)$ is a bounded sequence. Then $\tilde{P}(n)$ converges to zero. We first assume that $P_0$ is of the form $P_0 = \rho I$, $\rho > 0$. It is easy to see that the corresponding solution $P(n)$ of (12.74) satisfies an expression of the form

$$P(n) = \rho \Phi(n, 0) \Phi^T(n, 0) + \mathcal{M} \qquad \mathcal{M} \geq 0$$

Thus

$$\Phi(n, 0) \Phi^T(n, 0) \leq \frac{1}{\rho} P(n)$$

By Lemma 12.2, $P(n)$ is bounded. Hence $\Phi(n, 0)$ is bounded. Next suppose that $P_0$ is an arbitrary nonnegative definite matrix. Let $\rho$ such that $P_0 \leq \rho I$. Lemma 12.4 asserts that

$$P(n, n_0, 0) \leq P(n, n_0, P_0) \leq P(n, n_0, \rho I)$$

Taking limits and using Lemma 12.3, we obtain

$$P = \lim_{n \to \infty} P(n, n_0, 0) \leq \lim_{n \to \infty} P(n, n_0, P_0) \leq P$$

and the claim follows.

We finally note that Eq. (12.74) has a unique positive definite solution. Indeed, if there is another nonnegative definite solution $P_1$, the solution starting from $P_1$ is constant, namely $P(n, 0, P_1) = P_1$, and hence $P = \lim_{n \to \infty} P(n, 0, P_1) = P_1$. ∎

# 13

## CONTROL SYSTEMS

Control system design was motivated in Chapter 4. The purpose of this chapter is to introduce the reader to specific design methods. The first design approach is carried out exlusively in the $s$ or $z$ domain and relies on the transfer function description of systems. Hence it is referred to as *transform design*. Stabilization, tracking of reference signals, and rejection of deterministic disturbances are performed. The design formulation reduces to the solution of a Diophantine equation.

The second design approach is referred to as *pole placement* and uses state space ideas. It determines the feedback gain that places the poles of the given plant at desired locations under full state information. It also computes the observer that reconstructs the state from output measurements. Finally, it combines the two procedures into the controller structure. Controllability and observability are critical properties in the above design.

The third design approach is referred to as *regulator design*. Just like pole placement, it combines a control phase for the determination of the feedback gain under full state information and an estimation phase for the determination of the state from output measurements. Estimation is performed by the Kalman filter. The control phase computes the feedback gain by minimizing a quadratic function. The two phases are dual and hence are handled by the same algorithm.

## 13.1 TRANSFORM DESIGN

Let us consider the feedback configuration of Fig. 4.27. All systems are single-variable, causal, linear, time invariant with proper rational transfer functions. Let

$$h_u(s) = \frac{b(s)}{a(s)}, \qquad \deg b(s) \le \deg a(s) \tag{13.1}$$

be the transfer function from $u$ to $y$. Likewise

$$h_w(s) = \frac{b_w(s)}{a(s)}, \qquad \deg b_w(s) \le \deg a(s) \tag{13.2}$$

is the transfer function from the disturbance $w$ to $y$. Without loss of generality the above fractions have the same denominator. The measurement $z$ is

$$z(s) = y(s) + v(s) = \frac{b(s)}{a(s)} u(s) + \frac{b_w(s)}{a(s)} w(s) + v(s) \tag{13.3}$$

where $v(s)$ designates the additive measurement noise. The controller processes the reference signal $r$ and the measurement signal $z$, and produces at the output the signal

$$u(s) = \frac{n_r(s)}{d(s)} r(s) + \frac{n(s)}{d(s)} z(s) \tag{13.4}$$

We require

$$\deg n_r(s) \le \deg d(s), \quad \deg n(s) \le \deg d(s)$$

so that the controller is causal and proper rational. Inspection of diagram 4.27 gives ($s$ is omitted for simplicity)

$$y = \frac{b}{a} u + \frac{b_w}{a} w = \frac{b}{a} \left( \frac{n_r}{d} r + \frac{n}{d} (y + v) \right) + \frac{b_w}{a} w$$

or

$$\left( 1 - \frac{bn}{ad} \right) y = \frac{bn_r}{ad} r + \frac{b_w}{a} w + \frac{bn}{ad} v$$

or

$$y = \frac{bn_r}{ad - bn} r + \frac{db_w}{ad - bn} w + \frac{bn}{ad - bn} v \tag{13.5}$$

In a similar way the controller output is expressed as

$$u = \frac{an_r}{ad - bn} r + \frac{nb_w}{ad - bn} w + \frac{an}{ad - bn} v \tag{13.6}$$

It is clear from Eqs. (13.5) and (13.6) that the dynamics of the signals $u$ and $y$ is governed by the common denominator $c = ad - bn$. A bounded behavior is secured if the roots of $c$ lie in the left half plane. Thus it is reasonable to look for polynomials $n$ and $d$ so that the polynomial $c$ has roots in the left half plane. It turns out that it is more convenient to analyze the problem in the following more demanding form. Given polynomials $a(s)$, $b(s)$, and $c(s)$, find polynomials $n(s)$ and $d(s)$ such that

$$a(s)d(s) - b(s)n(s) = c(s) \tag{13.7}$$

The roots of $c(s)$ are placed in the left half plane to enforce stability. If, in particular, the above equation has solutions for every $c(s)$, disturbance rejection and asymptotic tracking are achieved by enhanced control capability. Equation (13.7) is the polynomial version of the classical Diophantine equation. The following theorem discusses existence of solutions. It constitutes a generalization of Theorem II.4 of Appendix II.

**Theorem 13.1 Diophantine equation** Let $a(s)$, $b(s)$ and $c(s)$ be polynomials with coefficients in a field. There exist polynomials $n(s)$, $d(s)$ satisfying (13.7) if and only if

$$\gcd(a(s), b(s)) | c(s) \tag{13.8}$$

In particular, if $a(s)$, $b(s)$ are relatively prime, Eq. (13.7) has a solution for every $c(s)$.

PROOF Suppose that Eq. (13.8) holds. Let $k(s)$ be a polynomial such that $c(s) = k(s) \gcd(a(s), b(s))$. Using Euclid's algorithm (see Section II.1.7), we construct polynomials $\tilde{n}$ and $\tilde{d}$ such that

$$\gcd(a(s), b(s)) = a(s)\tilde{d}(s) + b(s)\tilde{n}(s) \tag{13.9}$$

Hence $c(s) = k(s)a(s)\tilde{d}(s) + k(s)b(s)\tilde{n}(s)$. The polynomials $d(s) = k(s)\tilde{d}(s)$, $n(s) = -k(s)\tilde{n}(s)$, are solutions of (13.7). Conversely, if $n(s)$, $d(s)$ solve (13.7), then $\gcd(a(s), b(s))$ divides both $d(s)a(s)$ and $n(s)b(s)$, and hence divides $c(s)$. ∎

Note that if $d_0(s)$, $n_0(s)$ form a particular solution of the Diophantine equation, for any polynomial $q(s)$ the polynomials $d(s) = d_0(s) + q(s)b(s)$, $n(s) = n_0(s) + q(s)a(s)$ are also solutions.

It follows from the above theorem that if the given system is irreducible, namely $a(s)$ and $b(s)$ are coprime, the poles of the closed loop system can be arbitrarily assigned by feedback. The controller's parameters $n(s)$ and $d(s)$ are determined via the solution of a Diophantine equation.

To come up with causal controller architectures, particular solutions of the Diophantine equation that meet the degree constraints must be selected.

**Theorem 13.2  Causality**  Let $n_b \leq n_a$ denote the degrees of the coprime polynomials $a$, $b$. Let

$$n_c = 2n_a \tag{13.10}$$

For any polynomial $c$ of degree $n_c$, there exist polynomials $n, d$ satisfying Eq. (13.7) and $\deg n < \deg d$.

**PROOF**  Let $\tilde{n}, \tilde{d}$ be solutions of (13.7). We divide $\tilde{n}$ with $a$ to obtain $\tilde{n} = ga + r$. Then $(\tilde{d} - bg)a - br = c$. The polynomials $d = \tilde{d} - bg$, $n = r$ are solutions of (13.7). Note that $n_b \leq n_a$, by assumption, and that $\deg r \leq n_a - 1$, by the division. Hence $\deg br \leq 2n_a - 1$, and $\deg(c + br) = \deg c = 2n_a$. Therefore $\deg d = \deg(ad) - \deg a = \deg(c + br) - n_a = n_a$. We have proved that $\deg n \leq n_a - 1 = \deg d - 1$.  ■

Suppose next that $r(t)$, $w(t)$, and $v(t)$, asymptotically tend to zero. Disturbance rejection and tracking are established if the closed loop system is stable. In the general case additional information must be supplied. Here we will assume that the above signals are proper rational

$$r(s) = \frac{b_r(s)}{d_r(s)}, \quad w(s) = \frac{n_w(s)}{d_w(s)}, \quad v(s) = \frac{n_v(s)}{d_v(s)} \tag{13.11}$$

Let $\chi(s)$ be the least common multiple of the unstable parts of the signals $r(s)$ and $w(s)$. $\chi(s)$ consists of those roots of $d_r(s)$ and $d_w(s)$ that lie in the right half plane $\Re s \geq 0$. Let $\psi(s)$ be the unstable part of the measurement noise. We assume that $\chi(s)$ and $\psi(s)$ are known polynomials and that $\deg \psi \leq \deg \chi$.

It is clear that $\chi(s)$ and $\psi(s)$ are the main obstacles to tracking and disturbance rejection. Their influence is removed as the next theorem demonstrates.

**Theorem 13.3  Disturbance rejection and tracking**  Consider the unity feedback system of Fig. 4.28. The plant is described by Eqs. (13.1)–(13.2). The controller has the form (13.4). The reference signal and the disturbances are given by the rational models (13.11). Let $c(s)$ be any polynomial with roots in the left half plane and of given degree

$$n_c = 2(n_a + n_\chi) \tag{13.12}$$

$n_a$ and $n_\chi$ are the degrees of $a(s)$ and $\chi(s)$. Suppose that $\deg \psi(s) \leq \deg \chi(s)$ and the following coprimeness conditions hold:

$$\gcd(a(s), b(s)) = \gcd(a(s), \psi(s)) = \gcd(b(s), \chi(s)) = \gcd(\chi(s), \psi(s)) = 1$$

Then there exist solutions $n^*(s)$ and $d^*(s)$ of the Diophantine equation

$$(a(s)\chi(s))d^*(s) - (b(s)\psi(s))n^*(s) = c(s) \tag{13.13}$$

such that the controller

$$n(s) = n^*(s)\psi(s), \quad d(s) = d^*(s)\chi(s), \quad n_r(s) = -n(s) \tag{13.14}$$

is strictly proper rational and achieves disturbance rejection and asymptotic tracking.

PROOF  The coprimeness conditions and the causality constraint (13.12) warrant that Eq. (13.13) has a solution satisfying $\deg n^* < \deg d^*$. The hypothesis $\deg \psi \leq \deg \chi$ implies that $n$ and $d$ specified by (13.14) satisfy $\deg n < \deg d$. Thus the controller is strictly proper rational. To establish the disturbance rejection capability, we first examine the effect of $w$ on $y$. As Eq. (13.5) indicates, this effect is described by the term

$$y_w = \frac{db_w}{ad - bn} w = \frac{d^* b_w n_w}{c} \frac{\chi}{d_w}$$

The unstable roots of $d_w$ are canceled by the zeros of $\chi$. Partial fraction expansion and conversion to time domain gives $y_w(t) \to 0$ when $t \to \infty$. Therefore the controller asymptotically rejects the effect of disturbance $w$. In a similar way the effect of measurement noise at the output is given by

$$y_v = \frac{bn}{ad - bn} v = \frac{bn^* n_v}{c} \frac{\psi}{d_v}$$

The unstable roots of $d_v$ are canceled by the corresponding roots of $\psi$. Returning to the time domain, we find that $y_v(t) \to 0$ when $t \to \infty$. Next we establish asymptotic tracking. The unity feedback structure of Fig. 4.28 implies that the controller is steered by $r - z$. Since $n_r(s) = -n(s)$, we have in the transform domain

$$e = r - y = r - \frac{bn_r}{c} r - y_w - y_v$$

or

$$e = \frac{ad - bn + bn}{c} r - y_w - y_v = \frac{ad^* b_r}{c} \frac{\chi}{d_r} - y_w - y_v$$

The unstable roots of $d_r$ are canceled by the roots of $\chi$. The asymptotic rejection of disturbances implies that $y_w(t)$, $y_v(t)$ go to zero. Hence the tracking error also goes to zero.  ∎

Figure 13.1 illustrates the controller architecture resulting from the above theorem. The controller design is based on the *internal model principle* because

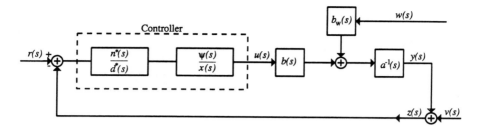

**Figure 13.1.** Unity feedback and the internal model principle.

the controller builds upon the unstable portions of the reference and the disturbance signals. The internal model principle utilizes a controller in the form of a cascade connection of $n^*(s)/d^*(s)$ and $\psi(s)/\chi(s)$. The first subsystem $n^*(s)/d^*(s)$ carries the poles of the given system at desired prespecified locations. The design parameters $n^*(s)$ and $d^*(s)$ are determined from the solution of the Diophantine equation. The second subsystem $\psi(s)/\chi(s)$ captures the unstable dynamics of the disturbances and the reference signal. The system $\psi(s)/\chi(s)$ is prespecified and reflects a priori available information regarding the uncertainties and the tracking target.

The previous theory remains exactly the same in the discrete time case. Extension of the transform design method to multivariable systems requires the solution of the Diophantine equation for polynomial matrices. The interested reader can consult the recommended references at the end of the chapter.

### *Example 13.1   Tracking antenna*
Figure 13.2 illustrates an antenna aiming to track a satellite or an airplane. This is a typical servomechanism position control problem. The antenna is rotated by an electric motor. The error signal is proportional to the difference between the

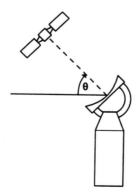

**Figure 13.2.** Antenna tracking a satellite.

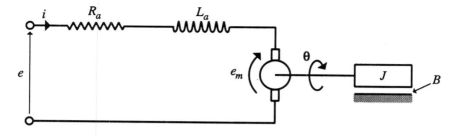

**Figure 13.3.** The dc motor.

pointing direction of the antenna and the line of sight to the satellite. It is amplified and used to steer the dc motor so that the error is reduced. The motor is illustrated in Fig. 13.3. We have

$$e_m = K_b \frac{d\theta}{dt} \tag{13.15}$$

where $\theta$ is the motor shaft position, and $K_b$ a constant depending on the motor. The rotation of the antenna is governed by $J\ddot{\theta} + B\dot{\theta} = T_c + T_w$, where $J$ is the moment of inertia, $B$ is the total viscous friction, $T_c$ the torque developed by the motor and $T_w$ the wind gust. The motor torque is $T_c = K_T i$, where $i$ is the armature-controlled current. We assume that the inductance $L_a$ is neglible. Then $e = iR_a + e_m$. The latter equation together with Eq. (13.15) implies that $i = e/R_a - K_b\dot{\theta}/R_a$, and that

$$T_c = \frac{K_T}{R_a} e - \frac{K_T K_b}{R_a} \dot{\theta} = J\ddot{\theta} + B\dot{\theta} - T_w$$

We thus obtain the finite derivative model

$$\ddot{\theta} + \alpha\dot{\theta} = u + w \tag{13.16}$$

where

$$\alpha = \frac{BR_a + K_T K_b}{JR_a}, \quad u = \frac{K_T}{JR_a} e, \quad w = \frac{T_w}{J}$$

The equivalent transfer function model is

$$y(s) = \frac{1}{s(s + \alpha)} u(s) + \frac{1}{s(s + \alpha)} w(s)$$

The reference signal is $r(t) = kt$. It is transformed into $r(s) = k/s^2$. Suppose that the measurement noise is negligible and that the disturbance $w$ has the characteristic of a step plus a dirac function:

$$w(s) = K_1 + \frac{K_2}{s} = \frac{K_1 s + K_2}{s}$$

Hence $\chi(s) = s^2$, $\psi(s) = 1$. The Diophantine equation becomes

$$d^*(s)s(s + \alpha)s^2 - n^*(s) = c(s)$$

The coprimeness assumptions are clearly met. The causality constraint (13.12) requires that $c(s)$ have degree 8. Let us take, for example,

$$c(s) = \prod_{i=1}^{8}(s + i)$$

Since $b(s)\psi(s) = 1$, a solution of (13.13) is obtained by the quotient and the remainder from the division of $c(s)$ by $a(s)\chi(s)$. Let $\alpha = 1$. Division of $c(s)$ by $s^3(s + 1)$ gives

$$d^*(s) = 18424 + 4025s + 511s^2 + 35s^3 + s^4$$

$$n^*(s) = 40320 + 109584s + 118124s^2 + 48860s^3$$

The above computations are easily carried out in MATLAB as follows:

```
t=-1:-1:-8;                % time interval
c=poly(t);                 % desired polynomial c(s)
a1=[1 1 0 0 0 ];           % coefficients of  s^4+s^3
[dstar,nstar]=deconv(c,a1);  % desired polynomials are found by division   ■
```

## 13.2   POLE PLACEMENT AND STABILIZATION

The pole placement design uses state space ideas. Consider a plant described by the state space equation

$$x(n + 1) = Ax(n) + Bu(n) \tag{13.17}$$

Typically (13.17) is a discrete simulator of an analog system (see Section 3.12.3). We first assume that the entire state $x(n)$ is measured and is accessible for control. Let $r(n)$ denote the reference signal. We seek to determine a static controller of the form

$$u(n) = -Fx(n) + r(n) \tag{13.18}$$

so that the resulting closed loop system

$$x(n+1) = (A - BF)x(n) + Br(n) \qquad (13.19)$$

is BIBO stable. The dynamics of the closed loop system are governed by the eigenvalues of the matrix $A - BF$. So we look for a matrix $F$ that places these eigenvalues at desired locations in the unit disc. Then the system will be totally stable. The problem presents a striking resemblance with the observer design treated in Section 12.4.5. This is demonstrated in the next theorem.

**Theorem 13.4** The pair $(A, B)$ is controllable if and only if, for any polynomial $a_d(z)$, there exists a matrix $F$ such that the characteristic polynomial of $A - BF$ coincides with $a_d(z)$.

**PROOF** By Theorem 10.6, the pair $(A, B)$ is controllable if and only if the pair $(A^T, B^T)$ is observable. Taking into account Theorem 12.6 of Section 12.4.5, the latter is true if and only if the eigenvalues of $A^T - LB^T$ can be arbitrarily assigned. The claim is proved if we set $F = L^T$. ∎

Comparison of Theorems 12.6 and 13.4 shows the duality between observer design and feedback control. It also shows that both cases are handled with the same algorithms.

In reality, measuring the entire state vector is either impossible or prohibitively costly. Let us consider the stabilization problem in the more realistic situation of a plant governed by

$$x(n+1) = Ax(n) + Bu(n)$$
$$y(n) = Cx(n) + Du(n) \qquad (13.20)$$

The idea is to maintain the feedback gain $F$ found in the case of full state information and to use the feedback law (13.18) with $x(n)$ replaced by an observer estimate. More precisely we consider the observer

$$\hat{x}(n+1) = (A - LC)\hat{x}(n) + (B - LD)u(n) + Ly(n)$$
$$= (A - LC)\hat{x}(n) + Bu(n) + LCx(n) \qquad (13.21)$$

and Eq. (13.18) with $\hat{x}(n)$ replacing $x(n)$,

$$u(n) = -F\hat{x}(n) + r(n) \qquad (13.22)$$

The controller is now a dynamic system with state $\hat{x}(n)$. The resulting closed loop system is given by

$$\begin{pmatrix} x(n+1) \\ \hat{x}(n+1) \end{pmatrix} = \begin{pmatrix} A & -BF \\ LC & (A - LC) - BF \end{pmatrix} \begin{pmatrix} x(n) \\ \hat{x}(n) \end{pmatrix} + \begin{pmatrix} B \\ B \end{pmatrix} r(n)$$

$$y(n) = ( C \quad -DF ) \begin{pmatrix} x(n) \\ \hat{x}(n) \end{pmatrix} + Dr(n) \tag{13.23}$$

A more convenient form is obtained if we apply the coordinate change $(x, \hat{x}) \rightarrow (x, e_o)$, where $e_o$ is the observer estimation error. It is easy to see that the corresponding similarity transformation is given by

$$T = \begin{pmatrix} I & 0 \\ I & -I \end{pmatrix}$$

The closed loop system is expressed in the new coordinates as

$$\begin{pmatrix} x(n+1) \\ e_o(n+1) \end{pmatrix} = \begin{pmatrix} A - BF & BF \\ 0 & A - LC \end{pmatrix} \begin{pmatrix} x(n) \\ e_o(n) \end{pmatrix} + \begin{pmatrix} B \\ 0 \end{pmatrix} r(n)$$

$$y(n) = ( C - DF \quad DF ) \begin{pmatrix} x(n) \\ e_o(n) \end{pmatrix} + Dr(n)$$

The eigenvalues of the closed loop system consist of the eigenvalues of $A - BF$ and the eigenvalues of $A - LC$, due to the block triangular structure of the associated matrix. If the pair $(A, B)$ is controllable, the eigenvalues of $A - BF$ are arbitrarily assigned by proper choice of $F$. Likewise, if the pair $(A, C)$ is observable, the eigenvalues of $A - LC$ are arbitrarily assigned by proper choice of $L$. We have proved the following:

**Theorem 13.5**  Consider the controllable and observable system (13.20). An observer-based controller of the form (13.21)–(13.22) places the poles of the closed loop system at arbitrary locations by suitable choice of the observer gain $L$ and the feedback gain $F$.                                        ■

The above design approach incorporates two phases: a state estimation phase and a control phase under full state information. The state estimation phase determines the observer gain $L$ so that the state estimation error decays to zero in a controlled rate. The control under complete state information phase determines a feedback gain so that the system poles are placed at desired locations. Both phases are implemented by the same method, thanks to duality. Moreover they are executed in parallel, independently from each other. The decomposition of the design into the above independent phases is known as *separation principle*.

The previous theory remains exactly the same in the continuous time case.

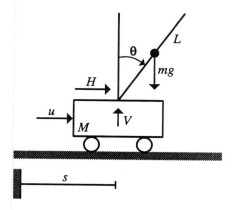

**Figure 13.4.** Inverted pendulum.

## *Example 13.2   Inverted pendulum*

The inverted pendulum consists of a metal rod mounted on a cart, as shown in Fig. 13.4. The cart is driven by an electric motor and aims to stabilize the pendulum at the vertical position. A space booster on takeoff or a robotic arm can be modeled this way. We assume the motion is planar. The pendulum has mass $m$ and length $2L$. $\theta(t)$ denotes the angle of the pendulum from the vertical position. $J$ is the moment of inertia. $s(t)$ denotes the displacement of the cart from a fixed reference point. $M$ is the mass of the cart. The force exerted by the cart to the rod consists of a horizontal component $H(t)$ and a vertical component $V(t)$. $u(t)$ denotes the control force produced by the motor. The motion of the cart is described by Newton's equation

$$M\ddot{s}(t) = -H(t) - k\dot{s}(t) + u(t)$$

where $k$ represents the friction coefficient. Assuming that $m << M$, we can ignore the influence of $H(t)$ on the motion of the cart. Hence we can rewrite the above equation in the simpler form

$$M\ddot{s}(t) = -k\dot{s}(t) + u(t)$$

If we view the rod as a particle located at the center of gravity, we obtain the following equations of motion:

$$m\frac{d^2}{dt^2}[s(t) + L\sin\theta(t)] = H(t) \tag{13.24}$$

$$m\frac{d^2}{dt^2}[L\cos\theta(t)] = V(t) - mg \tag{13.25}$$

The angular displacement satisfies

$$J\ddot{\theta}(t) = V(t)L\sin\theta(t) - H(t)L\cos\theta(t) \tag{13.26}$$

If we perform differentiation in (13.24) and (13.25), we obtain

$$m\ddot{s}(t) + mL\ddot{\theta}(t)\cos\theta(t) - mL\dot{\theta}^2(t)\sin\theta(t) = H(t)$$

$$-mL\ddot{\theta}(t)\sin\theta(t) - mL\dot{\theta}^2(t)\cos\theta(t) = V(t) - mg$$

If we substitute $H(t)$ and $V(t)$ as given by the above equations into (13.26), we obtain

$$\ddot{\theta}(t) - \frac{g}{l}\sin\theta(t) + \frac{1}{l}\ddot{s}(t)\cos\theta(t) = 0, \quad l = \frac{J + mL^2}{mL}$$

The state assignment $x_1 = s$, $x_2 = \dot{s}$, $x_3 = \theta$, $x_4 = \dot{\theta}$, leads to the nonlinear system

$$\dot{x}_1 = x_2$$

$$\dot{x}_2 = -\frac{k}{M}x_2 + \frac{1}{M}u$$

$$\dot{x}_3 = x_4$$

$$\dot{x}_4 = \frac{g}{l}\sin x_3 + \frac{k}{lM}x_2\cos x_3 - \frac{1}{lM}\cos x_3 u$$

A SIMULINK model is readily constructed as in Example 3.32. Simulations show that the zero state under the zero input is an unstable equilibrium. Linearization gives

$$\dot{x} = \begin{pmatrix} 0 & 1 & 0 & 0 \\ 0 & -\dfrac{k}{M} & 0 & 0 \\ 0 & 0 & 0 & 1 \\ 0 & \dfrac{k}{Ml} & \dfrac{g}{l} & 0 \end{pmatrix} x + \begin{pmatrix} 0 \\ \dfrac{1}{M} \\ 0 \\ -\dfrac{1}{Ml} \end{pmatrix} u$$

A more convenient coordinate representation is obtained by the assignment $x_1 = s$, $x_2 = \dot{s}$, $x_3 = s + l\theta$, $x_4 = \dot{s} + l\dot{\theta}$. The system parameters in the new coordinates are

$$A = \begin{pmatrix} 0 & 1 & 0 & 0 \\ 0 & -\dfrac{k}{M} & 0 & 0 \\ 0 & 0 & 0 & 1 \\ -\dfrac{g}{l} & 0 & \dfrac{g}{l} & 0 \end{pmatrix}, \quad B = \begin{pmatrix} 0 \\ \dfrac{1}{M} \\ 0 \\ 0 \end{pmatrix} \qquad (13.27)$$

$A$ is block lower triangular. The characteristic polynomial is factored as

$$s\left(s + \frac{k}{M}\right)\left(s^2 - \frac{g}{l}\right)$$

The eigenvalue $s = \sqrt{g/l}$ is positive. The linearization Theorem 11.8 implies that the original nonlinear system is unstable. The following numerical values will be used

$$\frac{g}{l} = 11.65 \text{ s}^{-2}, \quad \frac{k}{M} = 1 \text{ s}^{-1}, \quad l = 0.842 \text{ m}, \quad M = 1 \text{ kgr}$$

The controllability matrix for the above values has rank 4. Hence the system is controllable. The poles can be arbitrarily assigned by feedback. The controller gain $F$ can be found by the MATLAB command

```
F = place(A,B,p)
```

where $p$ is the vector of the desired poles. It is instructive to calculate the feedback analytically as well. The closed loop dynamics is governed by the characteristic polynomial of the matrix

$$A - BF = \begin{pmatrix} 0 & 1 & 0 & 0 \\ -\dfrac{f_1}{M} & -\dfrac{k+f_2}{M} & -\dfrac{f_3}{M} & -\dfrac{f_4}{M} \\ 0 & 0 & 0 & 1 \\ -\dfrac{g}{l} & 0 & \dfrac{g}{l} & 0 \end{pmatrix}$$

given by

$$s^4 + \left(\frac{k+f_2}{M}\right)s^3 + \left(\frac{f_1}{M} - \frac{g}{l}\right)s^2 - \frac{g(k+f_2+f_4)}{lM}s - \frac{g(f_1+f_3)}{lM}$$

Suppose that the desired poles are placed at the location $-a$. The polynomial with roots at $-a$ is

$$s^4 + 4as^3 + 6a^2s^2 + 4a^3s + a^4.$$

The feedback gain is determined by equating the coefficients of the above polynomials.

Consider next the output measurement $y(t) = \theta(t) = (x_3(t) - x_1(t))/l$, or in matrix form

$$y(t) = \left(-\frac{1}{l} \quad 0 \quad \frac{1}{l} \quad 0\right)x(t) \tag{13.28}$$

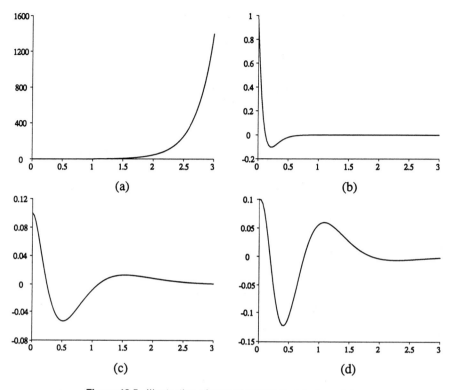

**Figure 13.5.** Illustration of control for the inverted pendulum.

We easily check that the system is not observable. The cart displacement $x_1(t)$ is employed as an additional measurement. The resulting output matrix is

$$C = \begin{pmatrix} 1 & 0 & 0 & 0 \\ -\dfrac{1}{l} & 0 & \dfrac{1}{l} & 0 \end{pmatrix}$$

Now the system is observable. Hence there exists a $4 \times 2$ matrix $L$ such that the observer

$$\frac{d}{dt}\hat{x}(t) = (A - LC)\hat{x}(t) + Ly(t) + Bu(t)$$

estimates the state at any desired accuracy. The **MATLAB** command

```
L = place(A',C',p)'
```

is employed again for the observer gain selection.

Figure 13.5 illustrates the behavior of $\theta(t)$ when (a) no control is applied, (c) state feedback is used, or (d) dynamic output feedback is employed. Finally, observer tracking performance is shown in (b). Stabilization is slower in panel (d) because of the time taken by the observer to estimate the state. The following MATLAB commands demonstrate the performance of pole placement design:

```
% Part 1: Simulation of linearized inverted pendulum
k = 1; M = 1; I = 0.0842; g = 9.81;
A = [0 1 0 0; 0 -k/M 0 0; 0 0 0 1; -g/I 0 g/I 0];   % linearized dynamics
B = [0; 1/M; 0; 0];
C = [1 0 0 0; -1/I 0 1/I 0];
D = [0 0]';
t = 0:0.01:3;                                       % defines time axis
u = zeros(length(t),1);
x0 = [0 0 I/10 0]';                                 % initial state of the open loop system
y = lsim(A, B, C, D, u, t, x0);                     % system simulation under zero input

% Part 2: Simulation of observer dynamics
ob = rank(obsv(A,C));                               % checks observability
p1 = [-9 + j -9-j -8 + j -8-j];                     % desired estimation poles
L = place(A',C', p1)';
xo0 = [0 0 0 0]';                                   % initial observer state
ao = A-L*C;                                         % observer dynamics
bo = [B , L];
co = [-1/I 0 1/I 0];
do = [0 0 0];
uo = [u, y];                                        % observer inputs
yo = lsim(ao,bo,co,do,uo,t,xo0);                    % observer simulation

% Part 3: Control under full state information
cont = rank(ctrb(A,B));                             % checks controllability
p = [-3 + j -3-j -4 + j -4-j];                      % desired poles
F = place(A,B,p);                                   % computes the feedback gain
a1 = A-B*F;
y1 = lsim(a1, B,C,D,u,t,xo);                        % simulation of closed loop system with state
                                                    % feedback

% Part 4: Control with output dynamic feedback
aa = [A, -B*F; L*C, A-L*C-B*F];
bb = [B; B];
cc = [C, -D*F];
x00 = [x0; zeros(length(x0),1)];                    % initial state of the closed loop system
yy = lsim(aa, bb, cc, D, u,t, x00);                 % simulation of closed loop system with output
                                                    % dynamic feedback

subplot(221); plot(t,y(:,2))
subplot(222); plot(t,(y(:,2)-yo)./y(:,2))
subplot(223); plot(t,y1(:,2))
subplot(224); plot(t,yy(:,2))
```

The above analysis is readily carried out in discrete time. The system is approximated by a digital simulator via the command

```
[ad,bd] = c2d(a,b,0.1);
```

and the above process is repeated with lsim replaced by dlsim.          ■

## 13.3 LINEAR QUADRATIC REGULATOR AND TRACKING

Let us consider the system

$$x(n + 1) = Ax(n) + Bu(n) + w(n), \qquad x(0) = x_0 \qquad (13.29)$$

$x_0$ is a random vector with mean $m_0$ and covariance $P_0$. $w(n)$ is a zero mean white noise process of variance $E[w(n)w^T(n)] = R_w$. $w(n)$ and $x_0$ are uncorrelated. $A$ and $B$ are matrices of compatible dimensions. Let $r(n)$ denote the reference signal and $y(n)$ the variable that we wish to control so that it tracks $r(n)$. The controlled variable is linearly related to the state

$$y(n) = C_1 x(n) \qquad (13.30)$$

Our primary objective is to design a static feedback controller that achieves a small tracking error

$$e(n) = r(n) - y(n)$$

at affordable input cost. We interpret these requirements in a least squares sense via the objective function

$$J = E\left[ \sum_{n=0}^{N-1} e^T(n + 1)e(n + 1) + u^T(n)Ru(n) \right] \qquad (13.31)$$

The first term penalizes the mean squared tracking error over an interval of system operation. The second term penalizes the input energy. $R$ is a positive definite matrix. The initial term $e(0) = r(0) - C_1 x_0$ is not affected by the input; hence there is no point to include it in the cost. Likewise the final value $u(N)$ affects the state $x(N + 1)$, which is beyond the final time $N$ and thus is excluded. We seek to determine a linear static time varying controller of the form

$$u(n) = -F(n)x(n) + F_1(n)r(n) \qquad (13.32)$$

that minimizes the cost (13.31) over all parameters $F(n)$ and $F_1(n)$. The need for using time varying gains will be made clear as we go along. It relates to a recursive Riccati type equation that has to be solved, much the same way Riccati equations pop up in the determination of the time varying Kalman gain. If the reference signal $r(n)$ is zero, the resulting problem is known as the *(stochastic) regulator problem*. In this case the cost takes the form

$$J = E\left[ \sum_{n=0}^{N-1} y^T(n + 1)y(n + 1) + u^T(n)Ru(n) \right] \qquad (13.33)$$

while the class of admissible controllers is given by

$$u(n) = -F(n)x(n) \tag{13.34}$$

The general case is known as *stochastic tracking problem*. We will first study the regulator problem. Subsequently we will show that under certain conditions the tracking problem can be viewed as a regulator problem.

### 13.3.1   Solution of the regulator problem

We start with a lemma that enables us to express the cost (13.33) in a more convenient form. We will deal with time varying representations because the control gain in (13.34) is time varying.

**Lemma 13.1**   Consider a system of the form

$$x(n+1) = A(n)x(n) + B(n)w(n), \qquad x(0) = x_0$$

$x_0$ is a random vector. $w(n)$ is a zero mean stochastic process, uncorrelated with $x_0$ and with variance $R_w$. Let $Q(n)$ be any nonnegative definite matrix-valued sequence and

$$J = E\left[\sum_{n=0}^{N} x^T(n)Q(n)x(n)\right]$$

The following expression holds:

$$J = \mathrm{tr}\left\{ E[x_0 x_0^T]P(0) + \sum_{n=0}^{N-1} B(n)R_w B^T(n)P(n+1) \right\} \tag{13.35}$$

where the nonnegative definite matrix sequence $P(n)$ satisfies the recursion

$$P(N) = Q(N)$$

$$P(n) = A^T(n)P(n+1)A(n) + Q(n), \qquad 0 \le n \le N-1 \tag{13.36}$$

The above matrix difference equation starts at time $N$ and runs backward. $\mathrm{tr}(A)$ denotes the trace of $A$.

PROOF   To bring into the picture the given covariance matrices, we use the trace operator and its familiar properties $\mathrm{tr}(A+B) = \mathrm{tr}(A) + \mathrm{tr}(B)$, $\mathrm{tr}(AB) = \mathrm{tr}(BA)$, $\mathrm{tr}(A) = \mathrm{tr}(A^T)$. Furthermore let $y$, $z$ be random column vectors, and let $A$ be a nonrandom matrix of compatible dimensions so that $y^T A z$ is a scalar. The trace of a scalar is the scalar itself. Moreover it is easy to see that the trace and the expectation commute. Therefore

$$E[y^T Az] = \text{tr}(E[y^T Az]) = E[\text{tr}(Azy^T)] = \text{tr}(E[Azy^T]) = \text{tr}(AE[zy^T])$$

Since $w(n)$ and $x_0$ are uncorrelated, the variation of constants formula and the properties of the trace operator yield

$$J = E\left[\sum_{n=0}^{N} x_0^T \Phi^T(n,0)Q(n)\Phi(n,0)x_0\right]$$

$$+ E\left[\sum_{n=0}^{N}\sum_{k=0}^{n-1} w^T(k)B^T(k)\Phi^T(n,k+1)Q(n)\Phi(n,k+1)B(k)w(k)\right]$$

Moreover

$$J = \text{tr}\left\{\sum_{n=0}^{N} \Phi^T(n,0)Q(n)\Phi(n,0)E[x_0 x_0^T]\right.$$

$$\left. + \sum_{n=0}^{N}\sum_{k=0}^{n-1} B^T(k)\Phi^T(n,k+1)Q(n)\Phi(n,k+1)B(k)R_w\right\}$$

We set

$$P(k) = \sum_{n=k}^{N} \Phi^T(n,k)Q(n)\Phi(n,k) \tag{13.37}$$

Interchanging summation, we write $J$ as

$$J = \text{tr}\left\{P(0)E[x_0 x_0^T] + \sum_{k=0}^{N-1} B(k)R_w B^T(k) \sum_{n=k+1}^{N} \Phi^T(n,k+1)Q(n)\Phi(n,k+1)\right\}$$

The latter equation takes the form (13.35) if Eq. (13.37) is taken into account. Next we establish (13.36), using the properties of the transition matrix

$$\Phi(n,k) = \Phi(n,k+1)A(k), \quad \Phi(n,n) = I$$

Indeed, $P(N) = \Phi^T(N,N)Q(N)\Phi(N,N) = Q(N)$, and

$$P(k) = Q(k) + A^T(k)\left[\sum_{n=k+1}^{N} \Phi^T(n,k+1)Q(n)\Phi(n,k+1)\right]A(k)$$

from which (13.36) follows.    ∎

We show in the sequel that the cost (13.33), associated with the regulator problem, can be written in the form (13.35). The closed loop system is

$$x(n+1) = [A - BF(n)]x(n) + w(n) \tag{13.38}$$

The controlled variable satisfies

$$y(n+1) = C_1 x(n+1) = C_1[A - BF(n)]x(n) + C_1 w(n)$$

Since $w(n)$ and $x_0$ are uncorrelated, the first term of (13.33) becomes

$$E\left[\sum_{n=0}^{N-1} x^T(n)[A - BF(n)]^T C_1^T C_1[A - BF(n)]x(n) + w^T(n)C_1^T C_1 w(n)\right]$$

If we incorporate the second term of (13.33) and take into account (13.34), we obtain

$$J = E\left[\sum_{n=0}^{N-1} x^T(n)Q(n)x(n)\right] + N\mathrm{tr}\{R_w R_1\}$$

where

$$R_1 = C_1^T C_1 \tag{13.39}$$

$$Q(n) = [A - BF(n)]^T R_1[A - BF(n)] + F^T(n)RF(n), \qquad 0 \le n \le N-1 \tag{13.40}$$

We set $Q(N) = 0$, and we apply Lemma 13.1, noting that $x(n)$ satisfies (13.38):

$$J = \mathrm{tr}\left\{E[x_0 x_0^T]P(0) + \sum_{n=0}^{N-1} R_w[P(n+1) + R_1]\right\} \tag{13.41}$$

where

$$P(N) = Q(N) = 0$$

$$P(n) = [A - BF(n)]^T P(n+1)[A - BF(n)] + Q(n), \; 0 \le n \le N-1 \tag{13.42}$$

Taking into account (13.40), Eq. (13.42) is rewritten as

$$P(n) = [A - BF(n)]^T[P(n+1) + R_1][A - BF(n)] + F^T(n)RF(n), \tag{13.43}$$

for $0 \le n \le N-1$.

We finally set

$$\Pi(n) = P(n) + R_1$$

Then

$$\Pi(N) = R_1$$

$$\Pi(n) = [A - BF(n)]^T \Pi(n+1)[A - BF(n)]$$
$$+ F^T(n)RF(n) + R_1, \qquad 0 \le n \le N - 1 \qquad (13.44)$$

The resulting expression for the cost is

$$J = \text{tr}\left\{ E[x_0 x_0^T](\Pi(0) - R_1) + \sum_{n=0}^{N-1} R_w \Pi(n+1) \right\} \qquad (13.45)$$

Equation (13.44) is initialized at time $n = N$ and runs backward in time.

The resemblance of (13.44) with the Riccati equation encountered in the context of Kalman filtering is evident and is a further manifestation of the duality between estimation and control. In the sequel we exploit this duality to solve the regulator problem, utilizing the estimation theory developed in Section 12.4. We recall from Section 12.4.5 that the estimation error of a time varying observer with gain $L(n)$ satisfies the recursion

$$P_L(n+1) = [A - L(n)C]P_L(n)[A - L(n)C]^T + DR_1R_1^TD^T + L(n)R_vL^T(n)$$

The following identifications make the resemblance of the above equation with (13.44) transparent. Recall that $R_w = R_1R_1^T$.

| | regulator | | | | | |
|---|---|---|---|---|---|---|
| regulator | $A$ | $B$ | $C_1$ | $R$ | $\Pi(n)$ | $F(n)$ |
| observer | $A^T$ | $C^T$ | $(DR_1)^T$ | $R_v$ | $P_L(N-n)$ | $L^T(N-n)$ |

The solution $\Pi(n)$ associated with the regulator problem corresponds to the solution $P_L(N-n)$ for $0 \le n \le N - 1$. The cost (13.45) is minimized if $\Pi(n)$ is minimized. According to Theorem 12.7, the observer covariance error is minimal if the observer gain coincides with the Kalman gain. Dualizing the expression for the Kalman gain (see Eq. (12.61)), we obtain the optimum feedback gain

$$F(n) = [R + B^T \Pi(n)B]^{-1}B^T \Pi(n)A \qquad (13.46)$$

The above findings are summarized below.

**Theorem 13.6**   Consider the plant (13.29). Among all linear time varying static controllers of the form (13.34), the one that minimizes the cost (13.33) is given by

$$u(n) = - F(n)x(n)$$

$$F(n) = [R + B^T \Pi(n)B]^{-1} B^T \Pi(n)A, \qquad 0 \leq n \leq N - 1$$

$$\Pi(n) = [A - BF(n)]^T \Pi(n + 1)[A - BF(n)]$$

$$+ F^T(n)RF(n) + R_1, \qquad 0 \leq n \leq N - 1$$

$$\Pi(N) = R_1 \qquad \blacksquare$$

Theorem 13.6 is extended to time varying systems. It can be shown that the above control law is optimum among all nonlinear static systems $u(n) = f(x(n), n)$.

### 13.3.2    Time invariant linear quadratic regulators (lqr)

The developments of Section 12.4.6 regarding the time invariant Kalman predictor can be used to design time invariant feedback controllers for the regulator problem. Suppose that the pair $(A, B)$ is controllable and the pair $(A, C_1)$ is observable.

Let $\Pi_N(n)$ denote the solution of (13.44) satisfying $\Pi_N(N) = R_1$. If we let $N$ go to infinity, the limit $\Pi = \lim_N \Pi_N(n)$ satisfies the algebraic equation

$$\Pi = [A - BF]^T \Pi [A - BF] + F^T RF + R_1 \qquad (13.47)$$

where

$$F = [R + B^T \Pi B]^{-1} B^T \Pi A$$

Equation (13.47) has a unique positive definite solution. The resulting time invariant feedback

$$u(n) = -Fx(n) \qquad (13.48)$$

transfers the poles of the closed loop matrix $A - BF$ inside the unit disc and thus achieves stabilization. Moreover it can be shown that (13.48) minimizes the error

$$\lim_{N \to \infty} \frac{1}{N} E \left[ \sum_{n=0}^{N-1} y^T(n + 1)y(n + 1) + u^T(n)Ru(n) \right]$$

### 13.3.3    Regulator design under partial state information

In our discussion of the regulator design, we  assumed that the full state is available for control. Let us next consider the general case of partial state information. In the light of the pole placement design and the separation principle, it is reasonable to look for a controller of the type postulated by the regulator design but with the state replaced by an estimate derived from output measurements. Kalman filter is an obvious estimation scheme. It then turns out that the above procedure is optimal in an appropriate mean square error sense. Consider the system (13.29) and the measurement equation

$$z(n) = Cx(n) + Du(n) + v(n) \tag{13.49}$$

The usual assumptions about the process noise $w(n)$ and the measurement noise $v(n)$, made in the context of Kalman filtering, are valid here as well. The controlled variable is given by (13.30). The performance criterion is (13.33). Among all controllers of the form

$$u(n) = \sum_{i=1}^{n} f_i(n)z(n-i), \qquad 0 \le n \le N-1$$

the one that minimizes the cost is expressed as

$$u(n) = -F(n)\hat{x}(n)$$

The gain $F(n)$ is the solution of the regulator problem and is determined as in Theorem 13.6, while $\hat{x}(n)$ is the state estimate derived from the Kalman predictor.

In analogy with the pole placement approach, the design procedure is composed of a control phase and an estimation phase. The control phase solves the regulator problem under full state information. The estimation phase computes a Kalman predictor. The two phases are dual and can be carried out by the same algorithms. Moreover they are processed independently of each other, a property referred to as *stochastic separation*.

A time invariant controller can be employed by combining the time invariant Kalman predictor and the time invariant linear quadratic regulator feedback gain.

### 13.3.4  Tracking

Tracking can be formulated as a regulator problem and solved by the methods described above. We model the reference signal as

$$r(n) = C_r x_r(n)$$
$$x_r(n+1) = A_r x_r(n) + w_r(n)$$

The matrix $A_r$ is asymptotically stable. We augment states and disturbances as

$$x_a(n) = \begin{pmatrix} x(n) \\ x_r(n) \end{pmatrix}, \qquad w_a(n) = \begin{pmatrix} w(n) \\ w_r(n) \end{pmatrix}$$

Then

$$x_a(n+1) = \begin{pmatrix} A & 0 \\ 0 & A_r \end{pmatrix} x_a(n) + \begin{pmatrix} B \\ 0 \end{pmatrix} u(n) + w_a(n)$$

The measurement equation (13.49) is written as

$$z(n) = C_a x_a(n) + D u(n) + v(n), \quad C_a = (C \quad 0)$$

The tracking error is

$$e(n) = r(n) - y(n) = C_{1a} x_a(n), \qquad C_{1a} = (-C_1 \quad C_r)$$

The performance criterion is written as

$$J = E\left[ \sum_{n=0}^{N-1} x_a^T(n+1) C_{1a}^T C_{1a} x_a(n+1) + u^T(n) R u(n) \right]$$

The tracking problem is now reduced to the regulator formalism, and Theorem 13.6 is applicable. The steady state case with the time invariant feedback requires some caution, since the controllability assumption is not satisfied; see Exercise 13.9. A detailed account is given in the recommended references.

### Example 13.3   Inverted pendulum
The following MATLAB code demonstrates the linear quadratic regulator design for the inverted pendulum. The system parameters are defined in Example 13.2.

```
% Part 1 : Discrete Simulation
[ad,bd] = c2d(A,B,0.001);
Bd = [bd, bd];
td = 1:3000;                        % defines the time axis
u = zeros(length(td),1);           % zero input
w = sqrt(0.1)*randn(length(td),1);
U = [u,w];
DD = [D' ; D'];
yd = dlsim(ad,Bd,C,DD,U,x0);       % simulation of the discrete system
v = sqrt(0.1)*randn(size(yd));     % in the presence of disturbances
yd = yd + v;


% Part 2 : Linear Quadratic Regulator Design
C1 = C(2,:);                       % the controlled variable is the angular displacement
R1 = C1'*C1;
R = 1;
flqr = dlqr(ad,bd,R1,R);           % computes the optimal feedback gain by Eq.(13.48)
ar = ad-bd*flqr;
yr = dlsim(ar,Bd,C1,D',U,x0);      % simulation of the closed loop system
```

```
% Part 3: Kalman Predictor
Rw = 0.1;                          % process noise covariance
Rv = 0.1*eye(2);                   % measurement noise covariance
klqe = dlqe(ad,bd,C,Rw,Rv);        % determines the Kalman gain
x0k = [0 0 0 0]';                  % initial state of Kalman predictor
ak = ad-klqe*C;                    % Kalman predictor dynamics
bk = [bd, klqe];
ck = C1;
dk = [D ;0]';
uk = [u, yd];                      % Kalman predictor inputs
yk = dlsim(ak,bk,ck,dk,uk,x0k);    % simulation of Kalman predictor

% Part 4: Regulator design under partial state information
% The equations below are of the form (13.23) plus the
% disturbances. The state of the closed loop system combines
% the state of the plant and the state of the Kalman predictor

ark = [ad, -bd*flqr; klqe*C, ad-klqe*C-(bd + klqe*D)*flqr];
brk = [bd , zeros(4,2); zeros(4,1) , klqe];
W = [w, v];
crk = [C, zeros(2,4)];
x00 = [x0; x0k];
drk = [0 1 1; 0 1 1];
yrk = dlsim(ark,brk,crk,drk,W,x00);                                 ■
```

## BIBLIOGRAPHICAL NOTES

The progress of control theory over the recent decades is impressive. Many crucial issues in analysis and design have been studied successfully, and a wide body of knowledge is consolidated in existing books. Standard references are D'Azzo and Houpis (1988), Franklin and Powell (1980), Kwakernaak and Sivan (1972), Kailath (1980), and Chen (1984). For further study of transform design, sensitivity, and robustness, the reader is referred to Vidyasagar (1985) and Doyle (1992). Adaptive control is concerned with the identification of partially known systems. Control schemes are linked to the identification algorithms of Chapter 12. The interested reader may consult Goodwin and Sin (1984). Significant insight has been gained in the theory of nonlinear control systems. Isidori (1989), Nijmeijer and Van der Schaft (1990), and Sontag (1990) provide an account of nonlinear control. Besides quadratic criteria, various other cost functions have been analyzed. An account of $H_\infty$ optimization is provided in Francis (1987). Minimum time optimal control is a distinguished optimization paradigm over nonconventional feasible inputs. The reader may consult Lee and Markus (1967) and Athans and Falb (1965).

## PROBLEMS

**13.1.** In this exercise the satellite control problem is approached by the transform design method. The reader is urged to review Example 10.8.
   (a) Compute the eigenvalues of the discrete simulator of the linearized state representation with output $y = r$, using MATLAB.

(b) Show that $z = 1$ is a common root of the numerator and denominator of the transfer function. Conclude that the Diophantine equation cannot be handled.

**13.2.** Consider the linearized equations of the inverted pendulum (13.27). Suppose that only the angular displacement is measured and that the static output feedback

$$u(t) = -f\theta(t) = \left( -\frac{f}{l} \quad 0 \quad \frac{f}{l} \quad 0 \right) x(t)$$

is tried, the rationale being that if the pendulum starts falling to the right the cart must also move to the right. This is accomplished if the motor force is proportional to the angle. Show that the above law cannot stabilize the system. Demonstrate your analysis with MATLAB simulation.

**13.3.** Let

$$H(s) = \frac{s^2 + 1}{s^2 + 2s - 2}$$

Design a unity feedback controller which stabilizes the system by placing the poles of the system at the roots of $c(s) = s^4 + 12s^3 + 49s^2 + 78s + 40$. Verify your calculations by MATLAB.

**13.4.** Let

$$H(s) = \frac{s^2 - 1}{s^2 - 3s + 1}$$

Design a unity feedback controller that carries the poles of the system at positions $-1, -2, \cdots, -8$, and tracks step reference signals in the presence of exponential disturbances of the form $e^{3t}u_s(t)$. Measurement noise is negligible. Verify your calculations by MATLAB.

**13.5.** (a) A system $\dot{x}(t) = Ax(t) + Bu(t)$ is called *stabilizable* if there is state feedback $u(t) = -Fx(t)$ such that the eigenvalues of $A - BF$ are located in the left half plane. Write the system in the controllability staircase form

$$\dot{x}(t) = \begin{pmatrix} A_1 & A_2 \\ 0 & A_3 \end{pmatrix} x(t) + \begin{pmatrix} B_1 \\ 0 \end{pmatrix} u(t)$$

and prove that it is stabilizable if and only if the eigenvalues of $A_3$ are in the left half plane.

(b) The system $\dot{x}(t) = Ax(t) + Bu(t), y(t) = Cx(t)$ is called *detectable* if there is a time invariant observer with observer gain $L$ such that

$A - LC$ is asymptotically stable. Show that detectability is dual to stabilizability. Deduce an analogous criterion with (a) using the observability staircase form.

**13.6.** Consider the single-input system $\dot{x}(t) = Ax(t) + Bu(t)$ and the feedback $u(t) = -Fx(t) + r(t)$. Let $a_F(s) = \det(sI - A + BF)$ be the characteristic polynomial of the closed loop system and $a(s)$ the characteristic polynomial of $A$. Using the determinant identity

$$\det(I - MN) = \det(I - NM)$$

for arbitrary matrices with compatible dimensions, prove that

$$a_F(s) - a(s) = a(s)F(sI - A)^{-1}B$$

Suppose that $A$ is diagonal. Use partial fraction expansion to deduce that

$$f_i = \frac{1}{b_i} \frac{\prod_j(\lambda_i - \mu_j)}{\prod_{j \neq i}(\lambda_i - \lambda_j)}$$

where $\mu_i$ are the desired poles, $\lambda_i$ are the eigenvalues of $A$, and $b_i, f_i$ are the components of $B$, $F$, respectively. Observe that large eigenvalue spread leads to high-feedback gain amplitudes. Validate the above analysis in the discrete case.

**13.7.** Consider the system

$$\dot{x}(t) = \begin{pmatrix} -1 & -2 & -2 \\ 0 & -1 & 1 \\ 1 & 0 & -1 \end{pmatrix} x(t) + \begin{pmatrix} 2 \\ 0 \\ 1 \end{pmatrix} u(t), \quad y(t) = (1 \quad 1 \quad 0)x(t)$$

Use state feedback to transfer the eigenvalues to $-1$, $-2$, and $-3$. Compute an observer with eigenvalues $-2$, $-3$, and $-4$. Simulate the original system and the closed loop system in the case of static feedback and in the case of dynamic feedback.

**13.8.** Consider the deterministic system

$$x(n + 1) = Ax(n) + Bu(n), \qquad x(0) = x_0$$

and the minimization of the quadratic cost

$$J = \sum_{n=0}^{N-1} y^T(n + 1)y(n + 1) + u^T(n)Ru(n)$$

Determine the optimal feedback $u(n) = -F(n)x(n)$. Compute the minimum error. What is the difference with the stochastic case?

**13.9.** Prove that the analysis of the time invariant Kalman predictor discussed in Section 12.4.6 can be carried out in the more general case where controllability is replaced by stabilizability and observability is replaced by detectability (see Exercise 13.5). The only exception is that the solution of the algebraic Riccati equation is nonnegative definite rather than positive definite. Use duality to translate the theory to the regulator problem. With the above weaker assumptions show that the tracking problem is directly translated into the regulator problem and that the conclusions of Section 13.3.4 are valid.

**13.10.** Consider the stochastic tracking problem discussed in Section 13.3.4. Let $u(n) = -F(n)x(n) - F_r(n)x_r(n)$ be the optimal controller. Prove that the gain $F(n)$ is independent of the reference signal $r(n)$. *Hint*: Partition $P(n)$ as

$$P(n) = \begin{pmatrix} P_1(n) & P_2(n) \\ P_2^T(n) & P_3(n) \end{pmatrix}$$

and show that the Riccati equation associated with $P_1(n)$ involves only $P_1(n)$.

**13.11.** The angular velocity of the shaft of a dc motor satisfies the scalar differential equation

$$\dot{w}(t) = -aw(t) + kV(t) + w(t)$$

where $V(t)$ is the input voltage and $w(t)$ is white noise. We seek to steer the angular velocity to a constant value $w_0$.
(a) Show that if

$$x(t) = w(t) - w_0, \quad u(t) = V(t) - V_0, \quad V_0 = \frac{a}{k}w_0$$

the problem becomes equivalent to the stabilization of

$$\dot{x}(t) = -ax(t) + ku(t) + w(t)$$

to zero.
(b) Let $a = 0.5 \text{ s}^{-1}$, $k = 150 \text{ rad}/(\text{Vs}^2)$ and $V(t) = 1V$. Compute the discrete simulator by the developments of Section 3.12.3. Determine the optimal time invariant feedback gain. For MATLAB simulations take $R = 1000 \text{ rad}^2/(\text{V}^2\text{s}^2)$, $R_w = 0.001$, $T_s = 0.1$. Plot the state trajectory of the resulting closed loop system.
(c) Suppose that the tachometer used to measure $x(t)$ introduces noise. Then $y(t) = x(t) + v(t)$. Determine the optimal time invariant

Kalman predictor. Design a controller using the stochastic separation principle. Plot the state trajectory.

**13.12.** Consider the position stabilization problem specified by the system

$$\ddot{\theta}(t) + a\dot{\theta}(t) = ku(t) + w(t)$$

(see also Example 13.1). The goal is to bring the angular position to a constant value $\theta_0$.
  (a) Show that shifting to $x_1(t) = \theta(t) - \theta_0$, $x_2(t) = \dot{\theta}(t)$ and analogously for the input converts the problem into a standard regulator problem. Using the numerical values given in Exercise 13.11 carry out an analysis similar to Exercise 13.11. More precisely:
  (b) Compute the discrete simulator by the developments of Section 3.12.3. Determine the optimal time invariant state feedback gain. Plot the state trajectory of the resulting closed loop system.
  (c) Consider the measurement $y(t) = x(t) + v(t)$. Design a dynamic output feedback controller using the stochastic separation principle. Plot the state trajectory. Verify your calculations with MATLAB simulations.
  (d) How do the closed loop poles behave as $R \to \infty$?

# 14

## COMPRESSION AND ERROR CONTROL CODING

The discussion on compression and error control coding that was originated in Chapter 4 is completed here. Two major compression methods are presented; transform coding via the discrete cosine transform (DCT) and subband coding. Transform coders quantize a transformed representation of the signal. The aim of the transform is to reduce redundancy by removing correlation in the signal samples. In addition the transform may lead to a representation where the significant information is concentrated in only a few of the transform coefficients. This is the *energy compaction* property. An example of the latter type is offered by the discrete Fourier transform. Indeed, suppose that the signal spectrum is limited to few spectral coefficients. Quantization in the frequency domain allocates more bits to information bearing frequency components and fewer bits to less significant spectral lines. It further appears that the human auditory and visual systems perceive superior quality with coding performed in the frequency domain. This explains the popularity of frequency domain techniques in speech and image processing.

Subband coding relies on the related observation that many signals of practical interest are split in two frequency bands with greater energy in the low-frequency range and less energy in the higher-frequency band. Therefore a quantizer can allocate more bits in the higher-energy band and fewer bits in the lower-energy band. The signal is partitioned into two lower resolution components by an analysis filter bank, the lower resolution signals are decimated by a factor of two, and bit allocation is performed on each subband depending on energy content.

Subband coding techniques can be extended to the multidimensional case. Significant gains in image compression have been achieved with acceptable quality.

The final subject of this chapter is error control coding. Two topics are discussed; BCH codes and convolutional decoding via the Viterbi algorithm. BCH codes are linear block codes that can be designed to correct an arbitrary number of channel errors. This important feature is complimented with efficient decoding algorithms and satisfactory rates. For this reason BCH codes form an appealing class of codes finding widespread applications such as compact disc players. A distinguished class of BCH codes are the Reed-Solomon codes which in a sense have a maximal correcting capability. The Viterbi algorithm is an efficient decoding method for convolutional codes. It determines optimal length paths in trellis diagrams. In particular, it performs maximum likelihood decoding of convolutional codes. It also provides the maximum likelihood estimate of the transmitted sequence over a bandlimited channel with intersymbol interference. The Viterbi algorithm and its variants have been successfully applied to numerous applications, including deep space and satellite communications.

## 14.1   TRANSFORM CODING

Consider a wide sense stationary signal $x(n)$. A practical transform coder slices the signal into blocks $\mathbf{x}(n) = (\, x(n) \quad x(n+1) \quad \cdots \quad x(n+N-1)\,)^T$ and maps each block $\mathbf{x}(n)$ linearly into $\mathbf{y}(n) = A\mathbf{x}(n)$. In addition, $A$ is orthogonal, that is, $A^{-1} = A^T$. In the complex case $A$ becomes unitary, namely the inverse of $A$ is given by its complex conjugate transpose. Orthogonality ensures that the introduction of the transform does not affect average distortion. In accordance with Fig. 14.1 we have

$$E[(x - \hat{x})^T(x - \hat{x})] = E[(y - \hat{y})^T(y - \hat{y})]$$

### 14.1.1   Karhunen-Loève transform

The Karhunen-Loève transform is an orthogonal transform $\mathbf{y} = A\mathbf{x}$ with the property that $\mathbf{y}$ has uncorrelated components. Since the autocorrelation matrix $R_y = E[\mathbf{y}\mathbf{y}^T]$ is diagonal, it holds that

$$R_y = E[\mathbf{y}\mathbf{y}^T] = AE[\mathbf{x}\mathbf{x}^T]A^T = AR_xA^T$$

It follows from Section 2.5 that $A$ consists of the eigenvectors of the autocorrelation matrix $R_x$ provided that $R_x$ is diagonalizable. Since $R_x$ is real and symmetric, well-known results from linear algebra (see Exercise 2.31) assert that $R_x$ is always diagonalizable by an orthogonal matrix and has real eigenvalues. Diagonalization of $R_x$ is a special case of the singular value decomposi-

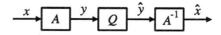

**Figure 14.1.** Transform coding.

tion. The columns of $A$ consist of eigenvectors of $R_x$. The variances of the transform coefficients are the eigenvalues of $R_x$. Compression gains are incurred when some of the transform coefficients have very small variance and thus can be quantized with zero bits. The above ideas are easily carried over to multi-dimensional signals.

Under certain assumptions it is shown that the Karhunen-Loève transform minimizes the mean square distortion among all orthogonal transforms. This optimal performance is counterbalanced by a costly implementation, as the coder $A$ depends on the autocorrelation matrix $R_x$ and the latter may vary with the input vector. Due to its increased complexity the Karhunen-Loève transform has limited use.

A transform with allegedly comparable performance to that of the Karhunen-Loève transform and yet signal independent is the discrete cosine transform (DCT).

### 14.1.2  Discrete cosine transform

Among the transform coders that perform frequency domain conversion the discrete cosine transform is the most popular. In fact most transform coding applications employ the DCT. The two-dimensional DCT is an essential component of CCITT standard for still image compression (JPEG) and video compression (MPEG). Consider a finite support signal $x(n)$ that is zero outside $0 \leq n \leq N - 1$. An example is shown in Fig. 14.2 (a). The periodic extension of $x(n)$ is shown in Fig. 14.2 (b). The discontinuity appearing at the edge $N$ causes the appearance of high-frequency spectral coefficients and thus deteriorates energy compaction. To avoid the edge discontinuity, we consider the symmetrized sequence of length $2N$ as in Fig. 14.3:

$$y(n) = x(n) + x(2N - 1 - n) = \begin{cases} x(n), & 0 \leq n \leq N - 1 \\ x(2N - 1 - n), & N \leq n \leq 2N - 1 \end{cases}$$

The artificial discontinuities of $x(n)$ are no longer present. The $2N$-point DFT of $y(n)$ is

$$Y(k) = \sum_{n=0}^{2N-1} y(n) w_{2N}^{kn} = \sum_{n=0}^{N-1} x(n) w_{2N}^{kn} + \sum_{n=N}^{2N-1} x(2N - n - 1) w_{2N}^{kn}$$

The change of variables $2N - n - 1 \rightarrow n$ yields

$$Y(k) = w_{2N}^{-k/2} \sum_{n=0}^{N-1} x(n) [w_{2N}^{k(n+1/2)} + w_{2N}^{-k(n+1/2)}]$$

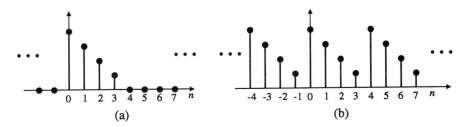

**Figure 14.2.** (a) Finite support signal; (b) periodic extension.

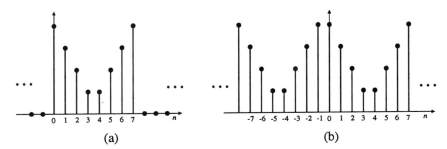

**Figure 14.3.** (a) Symmetrized signal; (b) periodic extension.

and finally

$$Y(k) = w_{2N}^{-k/2} \sum_{n=0}^{N-1} 2x(n) \cos \frac{\pi k}{2N} (2n+1), \qquad 0 \leq k \leq 2N - 1 \qquad (14.1)$$

The DCT of $x(n)$ is defined by

$$C(k) = \begin{cases} \sum_{n=0}^{N-1} 2x(n) \cos \dfrac{\pi k}{2N} (2n+1), & 0 \leq k \leq N - 1 \\ 0, & \text{otherwise} \end{cases}$$

Notice that the DCT of the vector $x = ( x(0) \quad x(1) \quad \cdots \quad x(N - 1))$ is also an $N$-dimensional vector. Moreover it is real if $x$ is real.

Let us next compute the inverse DCT. It follows from Eq. (14.1) that the discrete Fourier transform of $y(n)$, $Y(k)$, satisfies

$$Y(k) = w_{2N}^{-k/2} C(k), \qquad 0 \leq k \leq N - 1$$

Moreover $Y(N) = 0$ and $Y(k) = w_{2N}^{-k} Y(2N - k) = -w_{2N}^{-k/2} C(2N - k)$ for

$N + 1 \leq k \leq 2N - 1$. For $0 \leq n \leq N - 1$, it holds that $x(n) = y(n)$. Therefore

$$x(n) = \frac{1}{2N} \sum_{k=0}^{2N-1} Y(k)w_{2N}^{-kn} = \frac{1}{2N} \left[ \sum_{k=0}^{N-1} C(k)w_{2N}^{-k(2n+1)/2} + \sum_{k=1}^{N-1} C(k)w_{2N}^{k(2n+1)/2} \right]$$

or

$$x(n) = \frac{1}{N} \left[ \frac{C(0)}{2} + \sum_{k=1}^{N-1} C(k) \cos \frac{\pi}{2N} k(2n + 1) \right]$$

The computation of the DCT involves three steps: (1) computation of the $2N$-point sequence $y(n) = x(n) + x(2N - 1 - n)$, (2) computation of the $2N$-point DFT $Y(k)$, and (3) computation of the DCT by $C(k) = w_{2N}^{k/2} Y(k)$. The DFT computation is performed by an FFT algorithm. It turns out that an $N$-point FFT rather than $2N$-point FFT suffices (see Exercise 14.1). The two-dimensional DCT is derived along the same lines and is discussed in Exercise 14.2. Some properties of the DCT are given in Exercise 14.3.

A weakness of transform coders is the blocking effect caused by the slicing of the signal into blocks. Blocking in images shows discontinuities across block boundaries. In speech coding, blocking creates extraneous tones. Blocking effects are reduced by subband coding and lapped transforms, as we explain next.

## 14.2   SUBBAND CODING

Decimation of a discrete signal by $M$ is an obvious way to achieve compression. For instance, the signal $y(n) = x(2n)$ retains only half of the samples of $x(n)$. The frequency domain representation of the decimator indicates when and how the original signal is recovered from the decimated signal. We saw in Section 6.2.2 that the decimator $y(n) = x(Mn)$ is represented in the frequency domain by

$$Y(\omega) = \frac{1}{M} \sum_{k=0}^{M-1} X\left(\frac{\omega - 2\pi k}{M}\right) \tag{14.2}$$

The spectrum of the original signal, $X(\omega)$, stretched by a factor $M$, corresponds to $k = 0$. The remaining $M - 1$ terms are shifted versions of this stretched form and give rise to aliasing. Reconstruction of the original signal $x(n)$ is not possible if the terms causing aliasing are present. To get rid of the aliasing terms a lowpass filter with cutoff frequency $\omega_c = \pi/M$ is employed, as in Fig. 14.4. The cascade of the lowpass filter and the decimator is referred to as a *decimation filter*. The output spectrum of the decimation filter is $X(\omega/M)/M$ provided that $X(\omega)$ is bandlimited of bandwidth $2W < 2\pi/M$. Reconstruction of the signal $x(n)$ requires unstretching of the spectrum by a factor of $M$. The interpolation

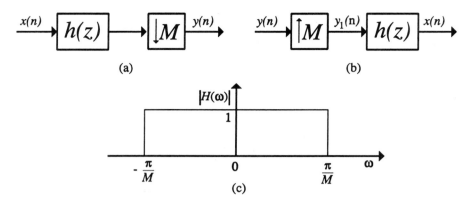

**Figure 14.4.** (a) Decimation filter, (b) interpolation filter, and (c) lowpass filter.

property of the Fourier transform states that unstretching is accomplished by upsampling. The signal

$$y_1(n) = \begin{cases} y(n/M), & \text{if } n \text{ is a multiple of } M \\ 0, & \text{otherwise} \end{cases}$$

has Fourier transform $Y_1(\omega) = Y(M\omega)$. As illustrated in Fig. 14.5, the spectrum

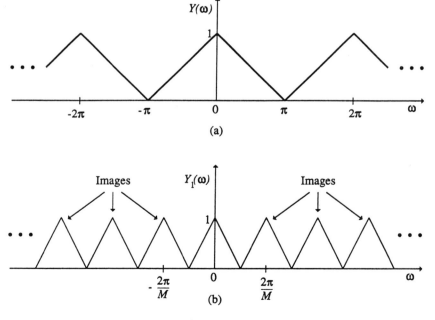

**Figure 14.5.** Imaging effect.

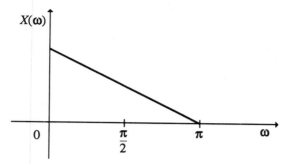

**Figure 14.6.** Illustration of subband partition.

$X(\omega)$ is now accompanied by $M - 1$ copies, called *images*. The imaging effect is removed by a lowpass filter. The resulting structure forms the interpolation filter shown in Fig. 14.4. Most signals do not have sharp bandlimited characteristics. As a result the decimator-interpolator structures either remove useful information content or cause unacceptable aliasing. In applications such as audio signal processing, aliasing can be very disturbing. The presence of exponentials that are not harmonically related to the input introduces nonharmonic distortion. On the other hand, the spectrum of many real signals like speech can be partitioned in two or more frequency bands, as illustrated in Fig. 14.6. The signal energy in the low-frequency range $0 < \omega < \pi/2$ is higher than the energy in the higher-frequency band $\pi/2 < \omega < \pi$. Therefore a quantizer can allocate more bits in the higher-energy band and less bits in the lower-energy band. It is therefore reasonable to split the signal into two lower resolution components by an analysis bank of two filters $h_0(z)$ and $h_1(z)$ having frequency responses as in Fig. 14.7, decimate each signal by a factor of two, and allocate bits on each subband depending on energy content. If $x(n)$ is a digitized signal at 10,000 samples per second, an 8-bit quantizer gives rise to a data rate of 80 kbits/s. If a subband coding system is employed using 8 bits for the high-energy subband and

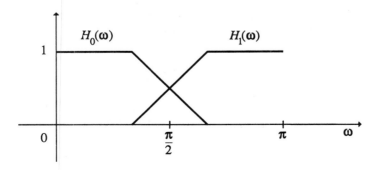

**Figure 14.7.** Analysis filter bank for the decomposition of the signal in Fig. 14.6 into 2 subbands.

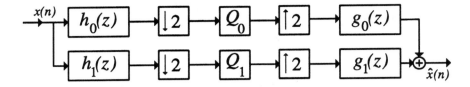

**Figure 14.8.** A two-channel subband coding system.

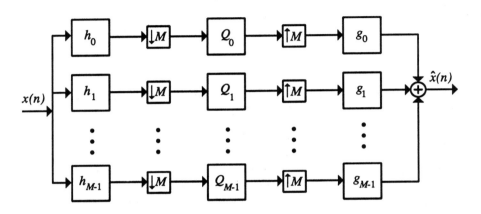

**Figure 14.9.** A general subband coding system.

4 bits for the low-energy subband, the sampling rate of the decimated signals is 5000 samples/s, and the data rate is $40 + 20 = 60$ kbits/s. A compression gain of 4/3 is achieved. The signal is reconstructed from its subband components by a synthesis bank of interpolation filters $g_0(z)$ and $g_1(z)$ as in Fig. 14.8. A general subband coding system is based on the analysis-synthesis filter bank of Fig. 14.9. Subband splitting and bit allocation require a spectral estimation phase that determines the input energy distribution. Since practical subband coding systems involve nonideal filters, aliasing occurs. Therefore filters must be carefully designed so that distortions are minimized. These issues are discussed next.

### 14.2.1 Perfect reconstruction filter banks

Figure 14.10 shows the filter bank of Fig. 14.9 without the quantizers. $y_i(n)$ denotes the $i$th subband component of $x(n)$. It is obtained by decimating $h_i(z)x(z)$, the output of the $i$th filter. The results of Section 8.2.4 give

$$y_i(z) = \frac{1}{M} \sum_{k=0}^{M-1} h_i(w^k z^{1/M}) x(w^k z^{1/M}), \quad w = e^{-j2\pi/M}$$

Let

$$y(z) = (\, y_0(z) \quad y_1(z) \quad \cdots \quad y_{M-1}(z) \,)^T$$

Then

$$y(z) = \frac{1}{M} H_m(z^{1/M}) x_m(z^{1/M}) \tag{14.3}$$

where

$$H_m(z) = \begin{pmatrix} h_0(z) & h_0(wz) & h_0(w^2z) & \cdots & h_0(w^{M-1}z) \\ h_1(z) & h_1(wz) & h_1(w^2z) & \cdots & h_1(w^{M-1}z) \\ \vdots & \vdots & \vdots & \ddots & \vdots \\ h_{M-1}(z) & h_{M-1}(wz) & h_{M-1}(w^2z) & \cdots & h_{M-1}(w^{M-1}z) \end{pmatrix}$$

and

$$x_m(z) = (\, x(z) \quad x(wz) \quad x(w^2z) \quad \cdots \quad x(w^{M-1}z) \,)^T$$

$H_m(z)$ defines the *modulated matrix*. Its first column contains the transfer functions of the analysis filters. The remaining columns are the modulated versions of the first column. Evaluation of $h_i(w^k z)$ on the unit circle yields the shifted frequency response $H_i(\omega - 2\pi k/M)$. The latter corresponds to the modulated signal $e^{j2\pi kn/M} h_i(n)$ in the time domain. Apparently $H_m(z)$ is specified by its first column. The analysis filter bank introduces distortion into $y(z)$, caused by the stretching factor $z^{1/M}$ in (14.3) and by the presence of the modulated signals $x(w^k z)$, $k = 1, 2, \ldots, M-1$. Let us see how the synthesis filters undo the stretch. As each signal $y_i(z)$ goes through the interpolator, it transforms into $y_i(z^M)$. Filtered by the synthesis filter $g_i(z)$, $y_i(z)$ becomes $g_i(z)y_i(z^M)$. The overall output is obtained as the superposition

$$\hat{x}(z) = \sum_{i=0}^{M-1} g_i(z) y_i(z^M)$$

or in compact form

$$\hat{x}(z) = G^T(z) y(z^M) \tag{14.4}$$

where $G^T(z) = (\, g_0(z) \quad \cdots \quad g_{M-1}(z) \,)$. If we substitute (14.3) into (14.4), we have

$$\hat{x}(z) = \frac{1}{M} G^T(z) H_m(z) x_m(z)$$

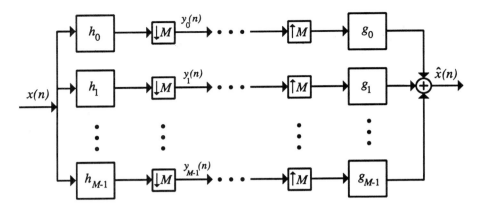

**Figure 14.10.** Analysis and synthesis multichannel filter bank.

Interpolation has removed the stretching effect. Aliasing is canceled if the modulated terms $x(w^k z)$, $k = 1, \ldots, M - 1$, are canceled. This is accomplished if the filter bank satisfies

$$G^T(z)H_m(z) = ( Mf(z) \quad 0 \quad 0 \quad \cdots \quad 0 ) \tag{14.5}$$

In this case the aliasing terms are canceled and

$$\hat{x}(z) = f(z)x(z) \tag{14.6}$$

The filter bank architecture of Fig. 14.10 is a linear time varying system due to the appearance of the decimators and interpolators (see Example 3.14). If the alias free condition (14.5) holds, the time varying system is equivalent to a linear time invariant system (14.6) with transfer function $f(z)$. If the alias free condition (14.5) is secured and in addition $f(z)$ has the form $f(z) = cz^{-m}$, we say that the system satisfies the *perfect reconstruction property*. In this case the filter bank output is a delayed and scaled version of the input: $\hat{x}(n) = cx(n - m)$. Alias free reconstruction is possible if the modulated matrix $H_m(z)$ is invertible, namely $\det H_m(z)$ is not identically zero. Then the synthesis filters are determined from the analysis bank by the linear system of equations (14.5).

If condition (14.5) is evaluated at points of the form $w^i z$, $0 \le i \le M - 1$, the synthesis and analysis modulated matrices interact evenly via the expression (see Exercise 14.5)

$$G_m^T(z)H_m(z) = M \; \text{diag}( f(z) \quad f(wz) \quad \cdots \quad f(w^{M-1}z) ) \tag{14.7}$$

The synthesis modulated matrix $G_m(z)$ has the same format as $H_m(z)$ except for the first column which consists of the synthesis filters $g_i(z)$. The modulated matrix includes redundancy because the first column produces the rest by

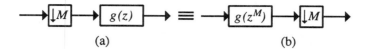

**Figure 14.11.** Interchanging decimation and filtering.

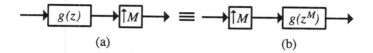

**Figure 14.12.** Interchanging interpolation and filtering.

modulation. An alternate and more compact test for alias free and perfect reconstruction based on the polyphase decomposition is developed in the sequel.

Let us first derive the useful identity illustrated in Fig. 14.11. The output of the cascade in the right-hand side is obtained by decimating the signal $g(z^M)x(z)$. Therefore it is given by

$$y(z) = \frac{1}{M} \sum_{k=0}^{M-1} g(w^{kM} z^{M/M}) x(w^k z^{1/M}) = g(z) \frac{1}{M} \sum_{k=0}^{M-1} x(w^k z^{1/M})$$

The right-hand side is the output of the filter $g(z)$ when the latter is excited by the $M$-decimation of $x(n)$. We write the above identity symbolically as

$$g(z)D_M = D_M g(z^M) \tag{14.8}$$

$D_M$ stands for decimation by $M$. In a similar way the cascades in Fig. 14.12 coincide. We write

$$I_M g(z) = g(z^M) I_M \tag{14.9}$$

$I_M$ denotes the $M$-interpolator. With reference to Fig. 14.11 we decompose each analysis filter into its polyphase components

$$h_k(z) = \sum_{i=0}^{M-1} z^{-i} h_{ki}(z^M), \qquad 0 \leq k \leq M-1 \tag{14.10}$$

In a similar way we consider the reverse polyphase decomposition of each synthesis filter

$$g_k(z) = \sum_{i=0}^{M-1} z^{-(M-1-i)} \tilde{g}_{ki}(z^M), \qquad 0 \leq k \leq M-1$$

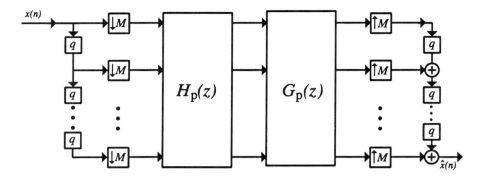

**Figure 14.13.** Equivalent realization of the filter bank using polyphase representation.

Each decimation filter $D_M h_k(z)$ is identical to $\sum_{i=0}^{M-1} z^{-i} D_M h_{ki}(z^M)$. Using identity (14.8), the latter filter becomes $\sum_{i=0}^{M-1} z^{-i} h_{ki}(z) D_M$. In a similar fashion invoking (14.9), we write each interpolation filter as

$$\sum_{i=0}^{M-1} z^{-(M-1-i)} \tilde{g}_{ki}(z^M) I_M = \sum_{i=0}^{M-1} z^{-(M-1-i)} I_M \tilde{g}_{ki}(z)$$

In this manner the diagram of Fig. 14.10 is equivalently drawn as in Fig. 14.13. The matrix $H_p(z)$ constitutes the *polyphase matrix* of the analysis bank. The $k$th row of $H_p(z)$ consists of the polyphase components of $h_{k-1}(z)$. $G_p(z)$ is the reverse polyphase matrix associated with the synthesis filters. Its $k$th column consists of the reverse polyphase components of $g_{k-1}(z)$. Setting

$$P(z) = G_p(z) H_p(z) \tag{14.11}$$

we obtain the diagram of Fig. 14.14, known as the *blocking system*. The blocking

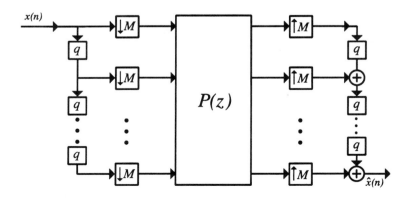

**Figure 14.14.** Equivalent realization of the filter bank by a blocking system.

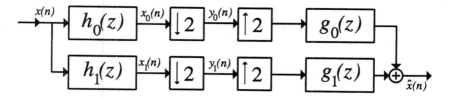

**Figure 14.15.**  Two channel filter bank.

operation is a demultiplexer that converts a signal $x(n)$ into the $M$-blocked form $(x(Mn) \quad x(Mn-1) \quad \cdots \quad x(Mn-M+1))$. The deblocking operation is a multiplexer that serializes the multichannel input. The output of the blocking-deblocking cascade is $\hat{x}(z) = z^{-M+1}x(z)$ or $\hat{x}(n) = x(n-M+1)$.

It is apparent from the above discussion that the perfect reconstruction property holds if

$$P(z) = G_p(z)H_p(z) = z^{-k}I, \qquad k \geq 0$$

This is the case if the analysis polyphase matrix $H_p(z)$ is invertible. Once the analysis filters satisfy the constraint $\det H_p(z) \neq 0$, the polyphase components of the synthesis filters and hence the synthesis filters themselves are determined from

$$G_p(z) = z^{-k}H_p^{-1}(z) \tag{14.12}$$

**Perfect reconstruction by FIR filters.**  Perfect reconstruction requires the inversion of the analysis polyphase matrix. If the analysis bank consists of FIR filters, the synthesis filters are IIR in general. FIR synthesis filters result if

$$\det H_p(z) = z^{-k} \tag{14.13}$$

In this case the synthesis bank is given by the FIR filters whose polyphase components are $G_p(z) = \mathrm{adj}(H_p(z))$.

**Orthogonal filter banks.**  Inversion of the polyphase matrix is straightforward if $H_p(z)$ is *paraunitary:*

$$H_p^T(z^{-1})H_p(z) = cI$$

We assume that filters have real coefficients. Evaluation on the unit circle gives

$$H_p^{*T}(\omega)H_p(\omega) = cI$$

This says that $H_p(\omega)$ is a unitary matrix for each $\omega$. A scalar transfer function is paraunitary if $|H(\omega)|^2 = 1$. The associated system is called *allpass*, since it lets all input frequencies pass through. A linear time invariant BIBO stable system whose transfer function is paraunitary is called *lossless*. A filter bank with paraunitary polyphase matrix is referred to as *orthogonal filter bank*. The above concepts are clarified next for two-channel filter banks.

### Example 14.1    Two-channel filter bank

Consider the two-channel filter bank illustrated in Fig. 14.15. Note that $w = e^{-j\pi} = -1$. The modulated matrix is

$$H_m(z) = \begin{pmatrix} h_0(z) & h_0(-z) \\ h_1(z) & h_1(-z) \end{pmatrix}$$

The alias free condition (14.5) becomes

$$g_0(z)h_0(z) + g_1(z)h_1(z) = 2f(z) \tag{14.14}$$

$$g_0(z)h_0(-z) + g_1(z)h_1(-z) = 0 \tag{14.15}$$

Aliasing is canceled if and only if (14.15) holds and $f(z)$ is nonzero. One of the earlier solutions of the above equation is known as the QMF (quadrature mirror filter). It employs a lowpass filter for $h_0(z)$ and then specifies the remaining filters by

$$h_1(z) = h_0(-z), \quad g_0(z) = h_0(z), \quad g_1(z) = -h_1(z) = -h_0(-z)$$

$h_1(z)$ is highpass, since it is obtained from $h_0(z)$ by modulation (frequency shifting). Likewise $g_0(z)$ is lowpass and $g_1(z)$ is highpass. Condition (14.15) is clearly satisfied. The QMF solution achieves perfect reconstruction if $f(z)$ in Eq. (14.14) has the form $f(z) = cz^{-m}$:

$$h_0^2(z) - h_0^2(-z) = 2cz^{-m} \tag{14.16}$$

The name QMF is due to the fact that on the unit circle $h_0(-z)$ is the mirror image of $h_0(z)$ above $\pi/2$, a quarter of the sampling frequency. Equation (14.16) does not admit an FIR solution apart from the Haar filters (see Exercise 14.6). We will discuss this condition later.

Let us next consider FIR perfect reconstruction. It follows from Eq. (14.13) as well as Exercise 14.4 that perfect reconstruction is accomplished if $\det H_m(z) = az^{-l}$. Thus

$$\det H_m(z) = h_0(z)h_1(-z) - h_0(-z)h_1(z) = az^{-l} \tag{14.17}$$

Let $A(z) = h_0(z)h_1(-z)$. Then $\det H_m(z) = A(z) - A(-z)$ and hence is an odd function of $z$. Therefore $l$ is odd, and (14.17) becomes

$$\det H_m(z) = A(z) - A(-z) = az^{-2k-1} \tag{14.18}$$

Solving Eqs. (14.14)–(14.15) with respect to $g_i(z)$, we find that

$$g_0(z) = \frac{2c}{a} z^{2k+1-m} h_1(-z)$$

$$g_1(z) = -\frac{2c}{a} z^{2k+1-m} h_0(-z) \tag{14.19}$$

If the analysis filters are causal, the synthesis filters will also be causal provided that $m > 2k + 1$. Let us next consider the polyphase matrix. Condition (14.13) becomes

$$h_{00}(z)h_{11}(z) - h_{01}(z)h_{10}(z) = z^{-k} \tag{14.20}$$

We fix $h_0(z)$ and hence its polyphase components $h_{00}(z)$ and $h_{01}(z)$. Equation (14.20) can be viewed as a Diophantine equation with unknowns the polyphase components of $h_1(z)$. Existence of solutions is guaranteed if the polyphase components of $h_0(z)$ are coprime. This occurs if and only if $h_0(z)$ has no pair of zeros at $z = \alpha$ and $z = -\alpha$. Indeed, suppose that $h_0(z)$ has such a pair of zeros. Then $B(z^2) = z^{-2} - \alpha^{-2}$ divides $h_0(z)$. The polyphase components are computed by

$$h_{00}(z^2) = \frac{1}{2}(h_0(z) + h_0(-z)), \quad h_{01}(z^2) = \frac{z}{2}(h_0(z) - h_0(-z))$$

$B(z^2)$ is a common divisor of $h_{00}(z)$ and $h_{01}(z)$, and coprimeness fails. The converse follows also easily. As an example, the binomial filter $h_0(z) = (1 + z^{-1})^m$ has coprime polyphase components.

Let us summarize our discussion on FIR perfect reconstruction. The analysis filter $h_0(z)$ is chosen so that it contains no pairs of roots at $z = \alpha$ and $z = -\alpha$. The second analysis filter is obtained from its polyphase components through the solution of the Diophantine equation (14.20). This also has coprime components. Finally, the synthesis filters are computed by Eq. (14.19).

***Orthogonal case.*** Consider a two-channel FIR filter bank with paraunitary polyphase matrix. Since $H_m(z)$ is also paraunitary, we have

$$h_0(z)h_0(z^{-1}) + h_0(-z)h_0(-z^{-1}) = c \tag{14.21}$$

$$h_1(z)h_1(z^{-1}) + h_1(-z)h_1(-z^{-1}) = c \tag{14.22}$$

$$h_0(z)h_1(z^{-1}) + h_0(-z)h_1(-z^{-1}) = 0 \tag{14.23}$$

The first two equations translate in the frequency domain into

$$|H_i(\omega)|^2 + |H_i(\omega + \pi)|^2 = c, \qquad i = 0, 1$$

Filters of the above form are known as *power complementary*. The third condition specifies one of the filters in terms of the other. Indeed, let $\alpha$ be a root of $h_0(z)$. Then $-\alpha$ cannot be a root of $h_0(z)$ because the perfect reconstruction property holds. Equation (14.23) implies that $h_1(-\alpha^{-1}) = 0$. Consequently $h_1(z)$ has the form

$$h_1(z) = -c'z^{-i}h_0(-z^{-1}) \tag{14.24}$$

It follows from (14.22) that $|c'| = 1$. Moreover $\alpha = c'c$ and $i = 2k + 1$. Hence

$$h_1(z) = -c'z^{-2k-1}h_0(-z^{-1}) \tag{14.25}$$

In summary, an orthogonal filter bank is specified by

$$|H_i(\omega)|^2 + |H_i(\omega + \pi)|^2 = c, \qquad i = 0, 1 \tag{14.26}$$

$$h_1(z) = -c'z^{-2k-1}h_0(-z^{-1}), \qquad c' = \pm 1 \tag{14.27}$$

$$g_0(z) = c'h_1(-z) = z^{-(2k+1)}h_0(z^{-1}) \tag{14.28}$$

$$g_1(z) = -c'h_0(-z) = z^{-(2k+1)}h_1(z^{-1}) \tag{14.29}$$

■

### 14.2.2 Lapped orthogonal transform (LOT)

Transform coders can be viewed as filter banks. Suppose that the signal $x(n)$ is sliced into blocks of the form

$$\mathbf{x}(n) = (x(Mn) \quad x(Mn-1) \quad \cdots \quad x(Mn-M+1))^T$$

Let $\mathbf{y}(n) = A\mathbf{x}(n)$ be the transformed block. The $i$th component of the transformed vector is

$$\mathbf{y}_i(n) = \sum_{j=0}^{M-1} a_{ij}x(Mn-j)$$

Comparison with Fig. 14.10 shows that the transformed vector $\mathbf{y}(n)$ can be obtained at the output of an analysis filter bank with analysis filters $h_i(z) = \sum_j a_{ij}z^{-j}$ and decimation by $M$. It is obvious that the $M$-polyphase components are $a_{ij}$. Hence $H_p(z) = A$. If the transform is unitary, the polyphase matrix is unitary and an orthogonal filter bank results.

As we mentioned earlier, transform coders suffer from blocking. Blocking is reduced by signal overlap. A transform with overlap converts a block of length $L$ into a block of length $M$ with $L \geq M$. The input stream is sliced into overlapping blocks

$$\mathbf{x}(n) = (\, x(Mn) \quad x(Mn-1) \quad \cdots \quad x(Mn-L+1)\, )^T$$

and each block is transformed into a block of length $M$, $\mathbf{y}(n) = A\mathbf{x}(n)$. $A$ is an $M \times L$ matrix. Proceeding as in the nonoverlapping case, we infer that the transform is described by an analysis bank of $M$ filters of length $L$ and decimation by $M$. The transfer function of the analysis filter bank is

$$H(z) = A e_{L-1}(z), \quad e_{L-1}(z) = \begin{pmatrix} 1 & z^{-1} & \cdots & z^{-(L-1)} \end{pmatrix}^T$$

Let $L = 2M$. If we partition $A$ as $A = (\, A_0 \ A_1 \,)$, we obtain

$$H(z) = A_0 e_{M-1}(z) + z^{-M} A_1 e_{M-1}(z) = (A_0 + z^{-M} A_1) e_{M-1}(z)$$

Comparison with (14.10) shows that the analysis polyphase matrix is

$$H_p(z) = A_0 + z^{-1} A_1$$

The transform with overlap $A$ defines a *lapped orthogonal transform* if

$$(\, A_0 \quad A_1 \,) \begin{pmatrix} A_0^{*T} \\ A_1^{*T} \end{pmatrix} = I \quad \text{or} \quad A_0 A_0^{*T} + A_1 A_1^{*T} = I$$

and $A_0 A_1^{*T} = 0$. It is straightforward to see that the corresponding polyphase matrix $H_p(z)$ is paraunitary. Hence the LOT transform corresponds to an orthogonal FIR filter bank with analysis filters of length $2M$. Perfect reconstruction is accomplished with the synthesis filters $g_i(n) = h_i(2M - 1 - n)$.

### 14.2.3  Orthogonal expansions

Two-channel orthogonal filter banks generate a significant class of orthonormal expansions of $\ell_2(Z)$. Let us consider the representation of an orthogonal filter bank by Eqs. (14.26)–(14.29), and let us assume that the output $\hat{x}(n)$ coincides with the input. Then

$$g_i(n) = h_i(-n) \tag{14.30}$$

Clearly each $g_i(z) = h_i(z^{-1})$ is power complementary, and Eqs. (14.21)–(14.22) take the form

$$g_i(z)g_i(z^{-1}) + g_i(-z)g_i(-z^{-1}) = 2 \tag{14.31}$$

Let $s_i(z) = g_i(z^{-1})g_i(z)$ be the $z$ transform of the deterministic autocorrelation sequence

$$r_i(n) = \sum_k g_i(n+k)g_i(k)$$

Then Eq. (14.31) is rewritten as

$$s_i(z) + s_i(-z) = 2 \quad \text{or} \quad 2\sum_{n \text{ even}} r_i z^{-n} = 2$$

Therefore $r_i(0) = 1$ and $r_i(n) = 0$ for $n$ even. Hence

$$< g_i(n), g_i(n-2k) >= \sum_n g_i(n)g_i(n-2k) = \delta(k) \tag{14.32}$$

Thus for any $k \neq 0$ the sequences $g_i(n)$, $g_i(n-2k)$ are orthogonal. It follows more generally that the even shifts

$$\{g_0(n-2k), k \in Z\}$$

form an orthonormal set in $l_2(Z)$. Likewise the even shifts of $g_1(n)$ constitute an orthonormal set. Equation (14.23) then states that the combined set formed by the even shifts of $g_0(n)$ and $g_1(n)$ is orthogonal. Indeed,

$$< h_0(n-2k), h_1(n-2l) >= \sum_n h_0(n-2k)h_1(n-2l) = r_{01}(2(l-k))$$

where $r_{01}(n)$ is the crosscorrelation of $h_0(n)$ and $h_1(n)$ with $z$ transform $s_{01}(z) = h_0(z^{-1})h_1(z)$. In a manner analogous to that used to derive (14.32), we verify that Eq. (14.23) is equivalently expressed as $r_{01}(n) = 0$ for $n$ even, establishing the claim.

The perfect reconstruction property states that the orthogonal set

$$\{g_0(n-2k), k \in Z\} \cup \{g_1(n-2k), k \in Z\}$$

spans $\ell_2(Z)$. Indeed, set

$$\psi_{2k}(n) = g_0(n-2k), \quad \psi_{2k+1}(n) = g_1(n-2k)$$

Let $x(n)$ be a finite energy signal. Let $y_0(n)$ and $y_1(n)$ be the lower resolution components formed at the outputs of the decimators in the orthogonal filter bank. Thanks to perfect reconstruction, $x(n)$ appears at the output of the synthesis bank. We assume no delay at the output of the filter bank. Hence

$$x(n) = \sum_k y_0(k)g_0(n-2k) + \sum_k y_1(k)g_1(n-2k)$$

$$= \sum_k y_0(k)\psi_{2k}(n) + \sum_k y_1(k)\psi_{2k+1}(n)$$

Therefore the lower resolution components of $x(n)$, $y_i(n)$ represent $x(n)$ in the above orthogonal expansion.

### Example 14.2   Haar basis
The discrete Haar basis in $\ell_2(Z)$ consists of the even shifts of the two sequences

$$g_0(n) = \begin{cases} \dfrac{1}{\sqrt{2}}, & n = 0, 1 \\ 0, & \text{otherwise} \end{cases} \qquad g_1(n) = \begin{cases} \dfrac{1}{\sqrt{2}}, & n = 0 \\ -\dfrac{1}{\sqrt{2}}, & n = 1 \\ 0, & \text{otherwise} \end{cases}$$

Thus

$$\psi_{2k}(n) = g_0(n - 2k), \quad \psi_{2k+1}(n) = g_1(n - 2k)$$

The sequences $g_0(n)$ and $g_1(n)$ are plotted in Fig. 14.16. In the $z$ domain we obtain

$$g_0(z) = \frac{1}{\sqrt{2}}(1 + z^{-1}), \quad g_1(z) = \frac{1}{\sqrt{2}}(1 - z^{-1})$$

We set

$$h_0(z) = g_0(z^{-1}) = \frac{1}{\sqrt{2}}(1 + z), \quad h_1(z) = g_1(z^{-1}) = \frac{1}{\sqrt{2}}(1 - z)$$

It is straightforward to verify that conditions (14.21)–(14.23) are met. Hence the Haar sequences form an orthonormal basis in $\ell_2(Z)$. ∎

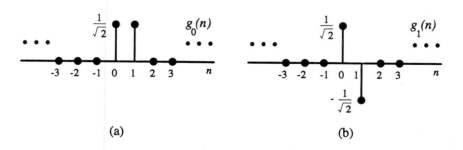

(a)  (b)

**Figure 14.16.** The Haar basis is formed by the even shifts of $g_0(n)$ shown in (a) and the even shifts of $g_1(n)$ shown in (b).

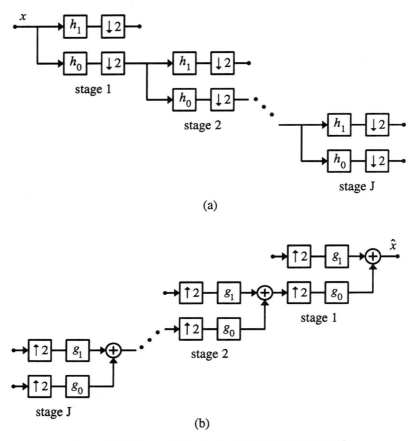

**Figure 14.17.** Iterated filter bank: (a) Analysis; (b) synthesis.

### 14.2.4   Discrete time wavelet series

A particularly useful type of multichannel filter bank is obtained by the iterated architecture of Fig. 14.17(a). A two-channel bank is employed to split the frequency band into a lowpass and a highpass part. The resulting lowpass is then divided into a lowpass and a highpass part, and the procedure is repeated. The synthesis bank is illustrated in Fig. 14.17(b). At each stage a two-channel synthesis filter bank combines the highpass component with the lowpass component formed by the two-channel synthesis filter at the previous stage.

Using identities (14.8) and (14.9), the filter bank of Fig. 14.17 is realized by the multichannel filter bank of Fig. 14.18 featuring different decimator/interpolator orders. Proceeding as in Example 14.1, we can show that the polyphase matrix associated with Fig. 14.18 is paraunitary, when $h_0$, $h_1$ are paraunitary. It then follows that the signals with transforms $g_1(z)$, $g_0(z)g_1(z^2)$, $g_0(z)g_0(z^2)g_1(z^4)$, . . . ,

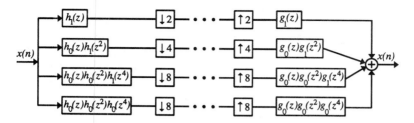

**Figure 14.18.** Multichannel realization of the iterative filter bank, $J = 3$.

$g_0(z)g_0(z^2)\cdots g_1(z^{2^{J-1}})$, and $g_0(z)g_0(z^2)\cdots g_0(z^{2^{J-1}})$ form an orthonormal basis in $\ell_2(Z)$. Hence every finite energy signal can be decomposed as

$$x(n) = \sum_{k=-\infty}^{\infty} y^{(J)}(2k)g_0^{(J)}(n - 2^J k) + \sum_{j=1}^{J}\sum_{k=-\infty}^{\infty} y^{(j)}(2k + 1)g_1^{(j)}(n - 2^j k) \quad (14.33)$$

The signals $g_1^{(j)}(n)$, $g_0^{(J)}(n)$ have $z$ transforms

$$g_0^{(J)}(z) = g_0^{(J-1)}(z)g_0(z^{2^{J-1}})$$

$$g_1^{(j)}(z) = g_0^{(j-1)}(z)g_1(z^{2^{j-1}}), \qquad 1 \leq j \leq J$$

The coefficients in (14.33) are given by the inner products

$$y^{(j)}(2k + 1) = \;< h_1^{(j)}(2^j k - n), x(n) >, \qquad 1 \leq j \leq J \qquad (14.34)$$

$$y^{(J)}(2k) = \;< h_0^{(J)}(2^J k - n), x(n) > \qquad (14.35)$$

Equation (14.33) provides a representation of the signal $x(n)$ over $J$ stages. At every stage $j$ the signal is split into two lower resolution components, a highpass or fine component and a lowpass or coarse component. The lowpass version is further divided into a lowpass and a highpass constituent. After $J$ stages the signal $x(n)$ is synthesized by the final coarsest component (first term of Eq. (14.33)) and the highpass components formed in the $J$ intermediate stages (the second term in Eq. (14.33)). As we pointed out, the signals $g_1^{(j)}(n - 2^j k)$, $g_0^{(J)}(n - 2^J l)$, $k, l$ integers form a complete orthonormal set in $\ell_2(Z)$. Thus

$$< g_0^{(J)}(n - 2^J k), g_0^{(J)}(n - 2^J l) >= \delta(k - l)$$

$$< g_1^{(j)}(n - 2^j k), g_1^{(i)}(n - 2^i l) >= \delta(j - i)\delta(k - l)$$

$$< g_1^{(j)}(n - 2^j k), g_0^{(J)}(n - 2^J l) >= 0$$

Equation (14.33) constitutes the discrete wavelet series representation of a finite energy signal. The coefficients (14.34)–(14.35) form the discrete wavelet coefficients. The discrete wavelet series involves a hierarchy of resolutions and thus is a *multiresolution representation*. The signal is composed of a very coarse constituent, the final lowpass version given by the first term of (14.33) and the various *details* given by the highpass components of the intermediate stages. Reconstruction of the signal by means of the multiresolution representation starts off with the very coarse approximation and adds details. As more details are superimposed, higher resolutions result along with better approximations of the signal. When all details have been considered, perfect reconstruction is achieved. The above progressive construction is quite natural in image compression and computer vision and somehow models the function of the human visual system. As a consequence compression techniques relying on multiresolution ideas are expected to enhance perceptual quality. A related popular technique is pyramid coding.

## 14.3  CYCLIC CODES

Cyclic codes are linear codes that are invariant under shifts. A cyclic code is generated by an FIR filter whose impulse response forms the generator polynomial of the code. BCH codes constitute the most prominent class of cyclic codes and are introduced later on. The generator polynomial of a BCH code is characterized by roots that are consecutive powers in an extension field. Thanks to this property, they can correct any number of channel errors. In the sequel we restrict to fields of characteristic 2.

### 14.3.1  Definitions

Let $v = (v_0 \quad v_1 \quad \cdots \quad v_{n-1})$, $v_i \in F$, be a codeword of a $(n, k)$ linear code $C$ over the field $F$. The one step *right shift* of $v$ is

$$v^{(1)} = (v_{n-1} \quad v_0 \quad v_1 \quad \cdots \quad v_{n-2})$$

The digits of $v$ are shifted to the right, and the first position is occupied by the last digit. The $m$ steps right shift is

$$v^{(m)} = (v_{n-m} \quad v_{n-m+1} \quad \cdots \quad v_{n-m-1})$$

Let $v(q)$ be the polynomial associated with the codeword $v$:

$$v(q) = \sum_{i=0}^{n-1} v_i q^i = v_0 + v_1 q + v_2 q^2 + \cdots + v_{n-1} q^{n-1}$$

The polynomial associated with the shift $v^{(1)}(q)$ is

$$v^{(1)}(q) = v_{n-1} + v_0 q + v_1 q^2 + \cdots + v_{n-2} q^{n-1}$$

To see the relationship between the two, we multiply $v(q)$ by $q$ and subtract

$$qv(q) - v^{(1)}(q) = v_{n-1} q^n - v_{n-1} = v_{n-1}(q^n - 1)$$

Thus $v^{(1)}(q)$ is the remainder of the division of $qv(q)$ by $q^n - 1$. Hence

$$qv(q) = v^{(1)}(q)(\mathrm{mod}(q^n - 1))$$

It follows $q^2 v(q) = v^{(2)}(q)(\mathrm{mod}(q^n - 1))$, or more generally that

$$q^m v(q) = v^{(m)}(q)(\mathrm{mod}(q^n - 1)) \qquad (14.36)$$

In words, the $m$ shift $v^{(m)}(q)$ is the remainder of the division of $q^m v(q)$ by $q^n - 1$.

The linear code $C$ is a *cyclic code* if the one step right shift of every codeword is also a codeword. Clearly the $m$ right shift of each codeword of a cyclic code is a codeword too.

***Generator polynomial.*** Let $g(q)$ denote the lowest-degree nonzero code polynomial in the cyclic $(n, k)$ code $C$, with leading coefficient 1. $g(q)$ exists because $C$ is a finite linear space. Moreover it is unique. Indeed, if there is another polynomial $g_1(q)$ of minimal degree and leading coefficient 1, then the difference $g(q) - g_1(q)$ is a code polynomial of smaller degree, contradicting minimality. Let $m$ denote the degree of $g(q)$. The polynomials $qg(q)$, $q^2 g(q)$, ..., $q^{n-1-m} g(q)$ have degrees less than $n$. According to (14.36) they coincide with the shifts $g^{(1)}(q)$, $g^{(2)}(q)$, ..., $g^{(n-1-m)}(q)$ and thus are codewords. For any $u_0, u_1, ..., u_{n-m-1}$ in $F$, the linear combination

$$\begin{aligned} v(q) &= u_0 g(q) + u_1 qg(q) + \cdots + u_{n-m-1} q^{n-m-1} g(q) \\ &= (u_0 + u_1 q + \cdots + u_{n-m-1} q^{n-m-1})g(q) = u(q)g(q) \qquad (14.37) \end{aligned}$$

is also a code polynomial. Conversely, every code polynomial $v(q)$ is written as $v(q) = u(q)g(q)$, $\deg u(q) \leq n - m - 1$. Indeed, division of $v(q)$ by $g(q)$ gives $v(q) = u(q)g(q) + f(q)$, $\deg f(q) < \deg g(q) = m$. The preceding discussion asserts that $u(q)g(q)$ is a code polynomial. Since $v(q)$ is a code polynomial and $C$ is a linear code, $f(q)$ is a code polynomial with degree less than $m$. Thus minimality of $g(q)$ is violated unless $f(q)$ is the zero polynomial. We have proved the following:

**Theorem 14.1**  Let $C$ be a $(n, k)$ cyclic code and $g(q)$ as above. A polynomial $v(q)$

of degree at most $n - 1$ is a code polynomial if and only if it is written

$$v(q) = u(q)g(q), \quad \deg u(q) \le n - 1 - \deg g(q) \qquad (14.38)$$

∎

The polynomial $g(q)$ characterizes the cyclic code and is called the *generator polynomial*. The following properties are direct consequences of Theorem 14.1:

1. $\deg g(q) = m = n - k$. Suppose that $F$ has $N$ elements. The code contains $N^k$ codewords and by Eq. (14.38), the number of code polynomials is $N^{n-m}$. Thus $n - m = k$.
2. $g(q)$ divides $q^n - 1$. Indeed, Eq. (14.36) implies that

$$q^k g(q) = a(q^n - 1) + g^{(k)}(q) = a(q^n - 1) + u(q)g(q)$$

3. The constant term of $g(q)$ is always nonzero. If $g_0 = 0$, $q$ divides $g(q)$. By property 2, it divides $q^n - 1$, a contradiction.

We recognize in Eq. (14.38) the input-output expression of an FIR filter. Each codeword is obtained as the output of an FIR filter with taps the coefficients of the generator polynomial $g(q)$ and input the message polynomial $u(q)$. Fast convolution algorithms based on FFT or the Chinese remainder theorem are applicable. Alternatively, the code can be implemented as a linear encoder by means of the generator matrix and a matrix by vector multiplication circuit. This is described next.

***Generator matrix.*** Equation (14.37) demonstrates that the product $v(q) = u(q)g(q)$ converts into the vector by matrix multiplication $v = uG$, where $G$ is formed by $g(q)$ and the first $k - 1$ shifts of $g(q)$:

$$G = \begin{pmatrix} g(q) \\ qg(q) \\ \vdots \\ q^{k-1}g(q) \end{pmatrix} = \begin{pmatrix} g_0 & g_1 & \cdots & g_{m-1} & 1 & 0 & \cdots & 0 \\ 0 & g_0 & \cdots & g_{m-2} & g_{m-1} & 1 & \cdots & 0 \\ \vdots & \vdots & \ddots & \vdots & \vdots & \vdots & \ddots & \vdots \\ 0 & 0 & \cdots & 0 & g_0 & \cdots & g_{m-1} & 1 \end{pmatrix} \qquad (14.39)$$

The rows of $G$, $q^i g(q)$, $0 \le i \le k - 1$, are code polynomials, since they are shifts of $g(q)$. Moreover they are linearly independent because the constant term $g_0$ is nonzero.

## 14.3.2 Systematic encoding

Encoding of a cyclic code by FIR filtering or by matrix by vector multiplication does not lead to systematic forms. Code polynomials of systematic encoders have the form

$$v(q) = v_0 + v_1 q + \cdots + v_{n-k-1} q^{n-k-1} + u_0 q^{n-k} + \cdots + u_{k-1} q^{n-1}$$
$$= \hat{v}(q) + q^{n-k} u(q)$$

Since the code is cyclic, there is a polynomial $\hat{u}(q)$ such that $v(q) = \hat{u}(q)g(q)$. Thus

$$q^{n-k} u(q) = \hat{u}(q)g(q) + \hat{v}(q) \qquad \deg \hat{v}(q) \le n - k - 1 \qquad (14.40)$$

The parity polynomial $\hat{v}(q)$ is the remainder of the division of $q^{n-k} u(q)$ by $g(q)$. Consequently systematic encoding is accomplished by a division circuit.

In Section 7.4 we saw that the output sequences of a linear feedback shift register are periodic and that the corresponding finite support truncations are realized by an FIR filter. This statement allows the implementation of systematic encoding by linear feedback shift registers. More precisely consider the polynomial $h(q)$ such that

$$g(q)h(q) = q^n - 1 \qquad (14.41)$$

$h(q)$ exists by property 2 above and has degree $k$. It is called the *parity check polynomial* of the code. It is a monic polynomial ($h_k = 1$) with nonzero constant term. Let $v(q)$ be a codeword. Then

$$v(q) = \hat{u}(q)g(q) = \hat{u}(q) \frac{q^n - 1}{h(q)}$$

Let $v_p(q)$ be the periodic extension of $v(q)$. In accordance with Example 7.3 we have

$$v_p(q) = \frac{v(q)}{1 - q^n} = \frac{\hat{u}(q)g(q)}{1 - q^n} = -\frac{\hat{u}(q)}{h(q)}$$

It follows that $v_p(q)$ is rational. The analysis of Section 7.4 confirms that $v_p(q)$ is produced by a linear feedback shift register whose taps are provided by the coefficients of the polynomial $h(q)$:

$$h_0 v_i + h_1 v_{i-1} + h_2 v_{i-2} + \cdots + h_k v_{i-k} = 0, \qquad i \ge k, \ h_0 \ne 0, h_k = 1$$

The above equation runs backward in time since the initial values of the code vector are placed at the last $k$ positions. Thus

$$v_j = -h_{k-1} v_{j+1} - h_{k-2} v_{j+2} - \cdots - h_1 v_{j+k-1} - h_0 v_{j+k}$$

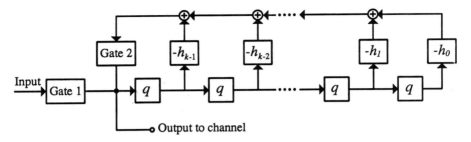

**Figure 14.19.** Encoding, using a linear feedback shift register and the parity polynomial.

for $j = n - k - 1, n - k - 2, \ldots, 0$. The encoder is realized by the linear feedback shift register of Fig. 14.19. Initially gate 1 is on, and the register is loaded with the message bits. Then gate 1 is switched off. Gate 2 is on, and the circuit produces the parity digits. We illustrate the procedure by the following example:

### Example 14.3 Systematic encoding

Consider the binary (7,4) cyclic code with generator polynomial $g(q) = 1 + q + q^3$. Suppose that the message sequence is $u = (1110)$, or in polynomial form $u(q) = 1 + q + q^2$. A nonsystematic encoder produces the codeword $v(q) = (1 + q + q^2)(1 + q + q^3) = 1 + q^4 + q^5$, or $v = (1000110)$. A systematic encoder forms the product $q^{n-k}u(q) = q^3(1 + q + q^2) = q^3 + q^4 + q^5$ and divides it by $1 + q + q^3$. The remainder is $\hat{v}(q) = q$. Therefore the codeword is $v = (0101110)$. The parity polynomial is obtained by dividing $q^7 - 1$ by $1 + q + q^3$. The result is $h(q) = q^4 + q^2 + q + 1$. The difference equation becomes

$$v_j = v_{j+2} + v_{j+3} + v_{j+4}, \qquad j = 2, 1, 0$$

The initial values are $(v_3, v_4, v_5, v_6) = u = (1110)$. Thus $v_2 = v_4 + v_5 + v_6 = 0$, $v_1 = v_3 + v_4 + v_5 = 1$, and $v_0 = v_2 + v_3 + v_4 = 0$. Therefore $\hat{v}(q) = q$ or $v = (0101110)$, verifying the previous calculation. ∎

### Example 14.4 Cyclic Hamming codes

The Hamming codes introduced in Example 4.10 are cyclic codes. Consider a primitive element $a$ in the field $GF(2^m)$. The columns of the parity matrix $H$ are all nonzero $m$ tuples and can be represented as successive powers of $a$ by the exponential representation of finite fields (see Appendix II). Therefore $H$ takes the form

$$H = \begin{pmatrix} 1 & a & a^2 & \cdots & a^{n-1} \end{pmatrix}, \qquad n = 2^m - 1$$

A codeword $v$ satisfies $vH^T = 0$ or

$$v_0 + v_1 a + v_2 a^2 + \cdots + v_{n-1} a^{n-1} = v(a) = 0$$

Thus a polynomial of degree at most $n - 1$ is a codeword if and only if it is zero at $a$. Property 2 of minimal polynomials (see Section II.2.3) implies that $v(q)$ is a multiple of the minimal polynomial of $a$. It follows that the code is cyclic and is generated by the minimal polynomial of $a$.

Consider the $(7, 4)$ Hamming code introduced in Example 14.3. Consider the field $GF(2^3)$. Since 3 is Mersenne prime, all elements of the field different from 0 and 1 are primitive. Let $a$ be a root of the irreducible polynomial $q^3 + q + 1$. Then $g(q) = q^3 + q + 1$ is the generator polynomial of the Hamming code. ∎

### 14.3.3 Interleaving

Consider an $(n, k)$ cyclic code with generator polynomial $g(q)$. An interleaver of order $L$ is an $(nL, kL)$ cyclic code with generator polynomial $g(q^L)$. Let $u(q)$ be an information sequence of length $Lk$ fed as input to the interleaver. Working as in Section 8.2.4, we obtain the following polyphase representation, using the $q$-series formalism:

$$u(q) = \sum_{i=0}^{L-1} q^i u_i(q^L)$$

$u_i(q)$ is the $q$-series of the polyphase component $u(Ln + i)$. The resulting code polynomial is

$$v(q) = g(q^L)u(q) = \sum_{i=0}^{L-1} q^i g(q^L) u_i(q^L)$$

This states that the polyphase components of $v(q)$ are

$$v_i(q) = g(q)u_i(q)$$

In other words, the interleaved code polynomial $v(q)$ is specified by interleaving the code polynomials of the original code produced by the polyphase components of the information sequence. It follows that if the original code can correct burst errors of length $t$ or less, the interleaved code can correct any burst error of length $Lt$, or less.

## 14.4 BCH CODES

The motivation that follows illustrates the definition of a BCH code as an extension of Hamming codes.

### 14.4.1 Motivation

We saw in Examples 4.10 and 14.4 that Hamming codes are single error correcting codes with blocklength $n = 2^m - 1$ and $m$ parity bits. The parity matrix $H$ is represented as

$$H = \begin{pmatrix} 1 & a & a^2 & \cdots & a^{n-1} \end{pmatrix}$$

Since $m$ bits are employed to correct one error, it is reasonable to inquire whether $m$ additional bits suffice to correct two errors. We thus examine matrices of the form

$$H = \begin{pmatrix} 1 & a & a^2 & \cdots & a^{n-1} \\ f(1) & f(a) & f(a^2) & \cdots & f(a^{n-1}) \end{pmatrix}$$

$f$ is the rule that specifies the additional $m$ bits residing at each column. Suppose that two errors occurred at positions $i$ and $j$. The error vector has the form $e = (0 \cdots 1 \cdots 1 \cdots 0)$. Evaluation of the syndrome $s = eH^T$ gives

$$s = (s_1 \quad s_2) = \begin{pmatrix} a^i + a^j & f(a^i) + f(a^j) \end{pmatrix}$$

We set $\beta_i = a^i$ and seek to specify $f$ so that the equations

$$\beta_i + \beta_j = s_1, \qquad f(\beta_i) + f(\beta_j) = s_2 \qquad (14.42)$$

are uniquely solvable for all $s_1, s_2$ in $GF(2^m)$. Clearly linear functions $f$ are ruled out because the second equation yields $s_2 = f(\beta_i + \beta_j) = f(s_1)$. Therefore the system of equations cannot be solved for all syndromes but only for those satisfying $f(s_1) = s_2$. We turn next to nonlinear functions, and we examine functions that preserve multiplication, that is, $f(a) = a^\rho$, $\rho \geq 2$. The simplest case is $\rho = 2$. Then $\beta_i^2 + \beta_j^2 = s_2$ or $(\beta_i + \beta_j)^2 = s_2$ or $s_1^2 = s_2$. Hence the system (14.42) cannot be solved for all $s_1, s_2$. We proceed to the next simplest test $\rho = 3$. Now

$$s_2 = \beta_i^3 + \beta_j^3 = (\beta_i + \beta_j)(\beta_i^2 + \beta_j^2 + \beta_i\beta_j) = s_1(\beta_i^2 + \beta_j^2 + \beta_i\beta_j)$$

If $s_1 = 0$, then $s_2 = 0$ and $s = 0$. Therefore no error occurs. If $s_1 \neq 0$, we have

$$\beta_i + \beta_j = s_1, \quad \frac{s_2}{s_1} = s_1^2 + \beta_i\beta_j \quad \text{or} \quad \beta_i\beta_j = \frac{s_2 + s_1^3}{s_1}$$

Thus $\beta_i, \beta_j$ are the roots of the quadratic polynomial

$$q^2 + s_1 q + \frac{s_2 + s_1^3}{s_1}$$

and can be determined by exhaustive search. This means that the above polynomial is evaluated at all elements of $GF(2^m)$ until the roots are found. A single error occurs if either $\beta_i$ or $\beta_j$ is zero, that is, $s_2 = s_1^3$. In this case the error is determined from $\beta_i = s_1$. If $s_2 \neq s_1^3$, two errors have occurred. The decoding procedure is complete.

With the above choice of $f$, the parity matrix takes the form

$$H = \begin{pmatrix} 1 & a & a^2 & \cdots & a^{n-1} \\ 1 & a^3 & a^{3\cdot 2} & \cdots & a^{3\cdot(n-1)} \end{pmatrix} \tag{14.43}$$

$H$ defines a cyclic code. Indeed, a vector $v$ is a codeword if $vH^T = 0$. Due to the special form of $H$ the latter equation is equivalent to

$$v(a) = 0, \qquad v(a^3) = 0$$

Thus a code polynomial has a root at $a$ and at $a^3$ and, by property 2 of Section II.2.3, is a multiple of the minimal polynomials $\phi_1(q)$, $\phi_3(q)$ of $a$ and $a^3$, respectively. It follows that the generator polynomial of the code is

$$g(q) = \text{lcm}(\phi_1(q), \phi_3(q))$$

The triple error correcting BCH code can be developed along the same lines. The parity matrix $H$ is formed by augmenting the parity matrix (14.43) of the double error correcting code by $m$ additional bits. The simplest possible rule for the selection of these additional bits turns out to be $f(a) = a^5$. The corresponding parity matrix takes the form

$$H = \begin{pmatrix} 1 & a & a^2 & \cdots & a^{n-1} \\ 1 & a^3 & a^{3\cdot 2} & \cdots & a^{3\cdot(n-1)} \\ 1 & a^5 & a^{5\cdot 2} & \cdots & a^{5\cdot(n-1)} \end{pmatrix} \tag{14.44}$$

$H$ defines a cyclic code, and the codewords are characterized by

$$v(a) = 0, \quad v(a^3) = 0, \quad v(a^5) = 0$$

In words, they are zero at the first three odd consecutive powers of $a$. The generator polynomial is

$$g(q) = \text{lcm}(\phi_1(q), \phi_3(q), \phi_5(q))$$

$\phi_5(q)$ is the minimal polynomial of $a^5$. Extension of the above methodology to general $t$-error-correcting codes is cumbersome. Thus we use the above findings to define BCH codes as cyclic codes whose generator polynomials vanish at consecutive odd powers of a primitive element and proceed from there to establish their good properties.

### 14.4.2 Definition of BCH codes

Let $m$ be given and $n = 2^m - 1$. Let $a$ be a primitive element of the field $GF(2^m)$. For any positive integer $t$, the $t$-error-correcting BCH code of blocklength $n$ is the cyclic code with generator polynomial

$$g(q) = \text{lcm}(\phi_1(q), \phi_3(q), \ldots, \phi_{2t-1}(q)) \tag{14.45}$$

where $\phi_i(q)$ is the minimal polynomial of $a^i$ with coefficients in $GF(2)$. The roots of $g(q)$ are the roots of $\phi_i(q)$ and reside in the extension field $GF(2^m)$. We deduce that $g$ vanishes at the first $t$ consecutive odd powers of $a$:

$$g(a) = g(a^3) = g(a^5) = \cdots = g(a^{2t-1}) = 0 \tag{14.46}$$

Since the code is cyclic, the number of parity check digits is given by the degree of $g$. Since $\deg \phi_i(q) \leq m$, this number never exceeds $mt$. It can be strictly smaller than $mt$, as we will show shortly. The number of parity checks and the number of information digits are determined after the degree of $g$ is computed. It follows from Eq. (14.46) that the first $t$ even powers of $a$ are also roots of $g(q)$. Indeed, let $1 \leq i \leq 2t$, $i$ even. Then $i$ is written as $i = j2^l$, $l$ odd, $j < i$, and $l \geq 1$. Thus $a^i = (a^j)^{2^l}$. Therefore $a^j$ and $a^i$ are conjugate elements and have the same minimal polynomial (see Appendix II). The generator polynomial is written

$$g(q) = \text{lcm}(\phi_1(q), \phi_2(q), \ldots, \phi_{2t}(q)) \tag{14.47}$$

Minimal polynomials are irreducible. The least common multiple is the product of those minimal polynomials that are pairwise distinct.

The BCH codes defined above are referred to as *binary primitive narrow sense BCH codes*. They are binary because the minimal polynomials are taken with coefficients in $GF(2)$. They are primitive because $a$ is a primitive element in $GF(2^m)$ and are narrow sense because the powers originate from $a$. The general definition is as follows:

Let $N$ and $m$ be given and $b$ a primitive element of $GF(N^m)$ of order $n$. The $t$-error-correcting BCH code is the cyclic code with generator polynomial

$$g(q) = \text{lcm}(\phi_{j_0}(q), \phi_{j_0+1}(q), \ldots, \phi_{j_0+2t-1}(q)) \tag{14.48}$$

where $\phi_j(q)$ is the minimal polynomial of $b^j$. The choice $j_0 = 1$ is the most common as it ensures low-degree generators and thus better code rates. We will prove below that the minimum distance of a $t$-error-correcting BCH code, $d_{\min}$ satisfies the *BCH bound*

$$d_{\min} \geq 2t + 1 \tag{14.49}$$

and thus can correct at least $t$ errors. First, we will discuss a few examples.

***Example 14.5   BCH codes over GF(2⁴)***

Consider the field $GF(2^4)$ generated by $a^4 + a + 1 = 0$, where $a$ is a primitive element. The single-error correcting BCH code of length 15 is spanned by

$$g_1(q) = \phi_1(q) = q^4 + q + 1$$

This is a (15, 11) cyclic Hamming code. The BCH bound gives $d_{min} \geq 3$. Since $g(q)$ is a codeword of weight 3, $d_{min} = 3$.

The double-error correcting BCH code is generated by

$$g_2(q) = \text{lcm}(\phi_1(q), \phi_3(q)) = g_1(q)\phi_3(q)$$

The minimal polynomial of $a^3$ is $\phi_3(q) = q^4 + q^3 + q^2 + q + 1$ (see Section II.2.5). Therefore

$$g_2(q) = q^8 + q^7 + q^6 + q^4 + 1$$

The BCH bound implies that $d_{min} \geq 5$. Since $g_2(q)$ is a codeword of weight 5, $d_{min} = 5$.

The minimal polynomial of $a^5$ is $\phi_5(q) = q^2 + q + 1$. Hence the triple-error correcting BCH code of blocklength 15 is generated by

$$g_3(q) = \text{lcm}(\phi_1(q), \phi_3(q), \phi_5(q)) = g_2(q)\phi_5(q)$$
$$= q^{10} + q^8 + q^5 + q^4 + q^2 + q + 1$$

It is a (15,5) cyclic code with $d_{min} \geq 7$. Since $g_3(q)$ has weight 7, the minimum distance is exactly 7. Note that the number of parity checks is 10 and is smaller than $mt = 12$.

Finally, the BCH code of blocklength 15 correcting 4 errors is generated by

$$g_4(q) = g_3(q)\phi_7(q)$$

Since $\phi_7(q) = q^4 + q^3 + 1$, we find that

$$g_4(q) = q^{14} + q^{13} + q^{12} + q^{11} + q^{10} + q^9 + q^8 + q^7 + \cdots + q^2 + q + 1$$

The resulting code is a (15,1) cyclic code. It is a simple repetition low rate code where each bit is repeated 14 times. ∎

The generator polynomial $g(q)$ of a $t$-error-correcting BCH code vanishes at $2t$ consecutive powers of $a$. Since each code polynomial $v(q)$ is a multiple of $g(q)$, it also vanishes at $2t$ consecutive powers of $a$:

$$v(a) = v(a^2) = \cdots = v(a^{2t}) = 0 \qquad (14.50)$$

Conversely, if a polynomial $v(q)$ of length $n - 1$ satisfies the above condition, it is a code polynomial. Indeed if $v(a^i) = 0$, property 2 of Section II.2.3 implies that $v(q)$ is a multiple of $\phi_i(q)$. Thus $v(q)$ is a multiple of $g(q)$. Equations (14.50) are written in matrix form as $vH^T = 0$, where

$$
H = \begin{pmatrix}
1 & a & a^2 & \cdots & a^{n-1} \\
1 & a^2 & a^{2 \cdot 2} & \cdots & a^{2 \cdot (n-1)} \\
\vdots & \vdots & \vdots & \ddots & \vdots \\
1 & a^{2t} & a^{2 \cdot 2t} & \cdots & a^{2t(n-1)}
\end{pmatrix}
\tag{14.51}
$$

$H$ represents the parity check matrix of the code.

**Theorem 14.2    BCH bound** Let $C$ be a cyclic code whose generator polynomial vanishes at $d - 1$ consecutive powers of an element $a$ in some extension field: $g(a^l) = g(a^{l+1}) = \cdots = g(a^{l+d-2}) = 0$. Then the minimum distance of the code is at least $d$.

PROOF    If the bound fails there exists a codeword

$$
c(q) = c_{j_1} q^{j_1} + c_{j_2} q^{j_2} + \cdots + c_{j_w} q^{j_w}
$$

with weight $w \le d - 1$. Then by assumption $c(a^i) = 0$, $l \le i \le l + w - 1$. These equations are written in matrix form as $c_1 H_1^T = 0$ where

$$
c_1 = \begin{pmatrix} c_{j_1} & \cdots & c_{j_w} \end{pmatrix}, \quad
H_1 = \begin{pmatrix}
a^{j_1 l} & a^{j_2 l} & \cdots & a^{j_w l} \\
a^{j_1 (l+1)} & a^{j_2 (l+1)} & \cdots & a^{j_w (l+1)} \\
\vdots & \vdots & \ddots & \vdots \\
a^{j_1 (l+w-1)} & a^{j_2 (l+w-1)} & \cdots & a^{j_w (l+w-1)}
\end{pmatrix}
$$

The determinant of the square matrix $H_1$ is zero because $c_1$ is a nonzero vector. Linearity of determinants with respect to columns shows that the determinant of $H_1$ is $a^{(j_1 + j_2 + \cdots + j_w)l}$ times a Vandermonde determinant spanned by $a^{j_1}, a^{j_2}, \ldots, a^{j_w}$. Hence the determinant is nonzero, and we arrive at a contradiction.    ∎

## 14.5    DECODING OF BCH CODES

### 14.5.1    Syndrome computation

Suppose that $v(q)$ is the transmitted code polynomial, $r(q)$ the received polynomial, and $e(q)$ the error. Then

$$
r(q) = v(q) + e(q)
\tag{14.52}
$$

Suppose that $\nu$ errors occurred during transmission. The error polynomial has the form

$$e(q) = \gamma_1 q^{j_1} + \gamma_2 q^{j_2} + \cdots + \gamma_\nu q^{j_\nu} \qquad (14.53)$$

$j_1, j_2, \ldots, j_\nu$ denote the error locations. $\gamma_k$ is the intensity of the error at location $j_k$. In the binary case (14.53) becomes

$$e(q) = q^{j_1} + q^{j_2} + \cdots + q^{j_\nu} \qquad (14.54)$$

The syndrome $s = (\, s_1 \quad s_2 \quad \cdots \quad s_{2t} \,)$ satisfies $s = rH^T = eH^T$. The parity check matrix is given by Eq. (14.51). Therefore

$$s_i = r(a^i) = r_0 + r_1 a^i + r_2 a^{2i} + \cdots + r_{n-1} a^{(n-1)i}$$

or

$$s_i = e(a^i) = \gamma_1 a^{j_1 i} + \gamma_2 a^{j_2 i} + \cdots + \gamma_\nu a^{j_\nu i}, \qquad 1 \leq i \leq 2t \qquad (14.55)$$

Each value $s_i$ is computed by evaluating the received polynomial at $a^i$. This is carried out by Horner's rule (see Example 4.4). A more efficient syndrome computation is the following: The received sequence is divided by each minimal polynomial

$$r(q) = a_i(q)\phi_i(q) + b_i(q), \qquad \deg b_i(q) < \deg \phi_i(q)$$

Since $\phi_i(a^i) = 0$, we obtain $s_i = r(a^i) = b_i(a^i)$. Therefore syndrome computation includes at most $t$ divisions by $\phi_i(q)$ circuits and $2t$ polynomial evaluation circuits. The evaluation of $b_i(q)$ is much simpler than the evaluation of $r(q)$ because the latter has degree of the order of $n \approx 2^m$, while the former has degree at most $m$. Much like the discussion of Section 14.4.1 it is more convenient to work with the powers

$$\beta_i = a^{j_i}, \qquad 1 \leq i \leq \nu \qquad (14.56)$$

rather than the error locations themselves. Of course $j_i$ are uniquely determined from $\beta_i$. In the sequel $\beta_i$ are referred to as error locations.

### 14.5.2 Estimation of the number of errors and error locations

Equation (14.55) gives

$$s_i = e(a^i) = \gamma_1 \beta_1^i + \gamma_2 \beta_2^i + \cdots + \gamma_\nu \beta_\nu^i, \qquad 1 \leq i \leq 2t$$

or in detail

$$s_1 = \gamma_1\beta_1 + \gamma_2\beta_2 + \cdots + \gamma_\nu\beta_\nu$$
$$s_2 = \gamma_1\beta_1^2 + \gamma_2\beta_2^2 + \cdots + \gamma_\nu\beta_\nu^2$$
$$\vdots = \qquad \vdots$$
$$s_{2t} = \gamma_1\beta_1^{2t} + \gamma_2\beta_2^{2t} + \cdots + \gamma_\nu\beta_\nu^{2t} \qquad (14.57)$$

This is a set of nonlinear equations with unknowns $\beta_i$, $\gamma_i$, and $\nu$.

To determine $\beta_i$ and $\nu$, we observe that the syndrome sequence is written as a linear combination of exponential signals with parameters $\beta_1, \beta_2, \ldots, \beta_\nu$. Let $\tilde{s}(i)$ denote the sequence appearing in the right side of (14.57). To be compatible with the series formalism of Chapter 7, we write

$$\tilde{s}(i) = \sum_{j=1}^{\nu} \gamma_j\beta_j^{i+1} = \sum_{j=1}^{\nu} \gamma_j\beta_j\beta_j^i, \qquad i \geq 0$$

The first $2t$ values of $\tilde{s}(i)$ coincide with the syndrome digits:

$$\tilde{s}(i) = s_{i+1}, \quad 0 \leq i \leq 2t - 1 \qquad (14.58)$$

The results of Section 7.2.1 assert that the series associated with $\tilde{s}$, $\tilde{s}(q)$ is strictly proper rational:

$$\tilde{s}(q) = \frac{\chi(q)}{\zeta(q)}, \quad \deg \chi(q) < \deg \zeta(q) \qquad (14.59)$$

$\chi(q)$ and $\zeta(q)$ form a coprime factorization. Moreover $\tilde{s}(i)$ can be delivered at the output of a linear feedback shift register with characteristic polynomial

$$\zeta(q) = (1 - \beta_1 q)(1 - \beta_2 q) \cdots (1 - \beta_\nu q)$$

The coefficients $\gamma_i\beta_i$ are recognized as the partial fraction coefficients. According to formula (7.13) they are given by

$$\gamma_i\beta_i = (1 - \beta_i q)\tilde{s}(q)|_{q=\beta_i^{-1}} \qquad (14.60)$$

The polynomial $\zeta(q)$ is called *error locator polynomial* because its roots are the inverses of the error locations $\beta_i$.

The coefficients of the polynomials $\chi(q)$ and $\zeta(q)$ can be determined by the linear systems of equations developed in Section 7.3.1 (Eqs. (7.35)–(7.36)). Indeed, let

$$\zeta(q) = 1 + \zeta_1 q + \zeta_2 q^2 + \cdots + \zeta_\nu q^\nu$$

The sequence $\tilde{s}(i)$ satisfies the difference equation

$$\tilde{s}_i + \zeta_1 \tilde{s}_{i-1} + \zeta_2 \tilde{s}_{i-2} + \cdots + \zeta_\nu \tilde{s}_{i-\nu} = 0, \qquad i \geq \nu \qquad (14.61)$$

If we evaluate (14.61) for $\nu \leq i \leq 2\nu - 1$, and take into account (14.58) and the hypothesis $\nu \leq t$, we obtain the linear system

$$S(\nu)\zeta = s(\nu + 1) \qquad (14.62)$$

where

$$S(\nu) = \begin{pmatrix} s_\nu & s_{\nu-1} & \cdots & s_1 \\ s_{\nu+1} & s_\nu & \cdots & s_2 \\ \vdots & \vdots & \ddots & \vdots \\ s_{2\nu-1} & s_{2\nu-2} & \cdots & s_\nu \end{pmatrix}, \quad \zeta = \begin{pmatrix} \zeta_1 \\ \zeta_2 \\ \vdots \\ \zeta_\nu \end{pmatrix}, \quad s(\nu+1) = \begin{pmatrix} -s_{\nu+1} \\ -s_{\nu+2} \\ \vdots \\ -s_{2\nu} \end{pmatrix}$$

We observe that the number of errors $\nu$ is the degree of the denominator $\zeta(q)$ in the coprime factorization (14.59) of $\tilde{s}(q)$. Indeed, if $\chi(q)$ and $\zeta(q)$ are not coprime, then $\tilde{s}(q) = \chi_1(q)/\zeta_1(q)$ with $\deg \zeta_1(q) \leq \deg \zeta(q)$, and the syndrome sequence would be written in the form (14.57) with $\nu$ replaced by $\nu_1$. Then the number of errors would be smaller than $\nu$.

Since the number of errors is given by the degree of the denominator of the coprime factorization of $\tilde{s}(q)$, it equals the rank of the associated Hankel matrix (see Theorem 10.11). Note that $S(\nu) = JH_{\nu,\nu}$, where $J$ is the reversing order matrix (see Section 4.3.4) and $H_{\nu,\nu}$ is the Hankel matrix discussed in Theorem 10.11.

The assumption $\nu \leq t$ and Theorem 10.11 suggest the following procedure for the determination of $\nu$. Starting from $i = t$, compute the determinant $\det S(i)$. If $\det S(i) \neq 0$, deduce that the number of errors is $t$. Otherwise, set $i = t - 1$, and compute $\det S(i)$. It holds $\nu = t - 1$ if $\det S(i) \neq 0$. Continue until the first integer $i$ such that $\det S(i) \neq 0$ is reached. The resulting integer $i$ gives the number of errors $\nu$. Once $\nu$ is available, the coefficients of the error locator polynomial are determined from Eq. (14.62). The matrix $S(\nu)$ is nonsymmetric Toeplitz. Hence the linear system of equations (14.62) and the associated determinant computations can be effected by a variant of the Levinson algorithm. Once the coefficients of the error locator polynomial are found, exhaustive search can be employed to determine the roots and hence the error locations. The above scheme is referred to as *Peterson-Zierler-Görenstein* algorithm. It is illustrated in the following example.

### Example 14.6   Decoding by the Peterson-Zierler-Görenstein algorithm
Consider the (15,5) triple-error correcting BCH code with generator polynomial

$$g(q) = q^{10} + q^8 + q^5 + q^4 + q^2 + q + 1$$

Suppose that the transmitted codeword is the all-zero word and the received sequence is

$$r = 010000000010000$$

Thus $r(q) = q + q^{10}$. Using the arithmetic of $GF(2^4)$, we compute the syndromes

$$s_1 = r(a) = a + a^{10} = a^2 + 1 = a^8$$

$$s_2 = r(a^2) = a^2 + a^5 = a$$

$$s_3 = r(a^3) = a^3 + 1 = a^{14}$$

$$s_4 = r(a^4) = a^4 + a^{10} = a^2$$

$$s_5 = r(a^5) = a^5 + a^5 = 0$$

$$s_6 = r(a^6) = a^6 + 1 = a^{13}$$

Alternatively the syndromes are determined as follows: The division of $r(q)$ by $\phi_1(q)$, $\phi_3(q)$, and $\phi_5(q)$ give

$$b_1(q) = q^2 + 1, \quad b_3(q) = q + 1, \quad b_5(q) = 0$$

Therefore $s_1 = b_1(a) = a^8$, $s_2 = b_1(a^2) = a^4 + 1 = a$, $s_3 = b_3(a^3) = a^3 + 1 = a^{14}$, $s_4 = b_1(a^4) = a^8 + 1 = a^2$, $s_5 = b_5(a^5) = 0$, $s_6 = b_3(a^6) = a^{13}$, validating the previous calculations.

Next we set $i$ to its maximum value $i = 3$. Then

$$\det S(3) = \det \begin{pmatrix} s_3 & s_2 & s_1 \\ s_4 & s_3 & s_2 \\ s_5 & s_4 & s_3 \end{pmatrix} = \det \begin{pmatrix} a^{14} & a & a^8 \\ a^2 & a^{14} & a \\ 0 & a^2 & a^{14} \end{pmatrix} = 0$$

For $i = 2$ we have

$$\det S(2) = \det \begin{pmatrix} s_2 & s_1 \\ s_3 & s_2 \end{pmatrix} = a^{12} \neq 0$$

Therefore $\nu = 2$, and two errors have occurred. The coefficients of the error locator polynomial are determined by

$$\begin{pmatrix} \zeta_1 \\ \zeta_2 \end{pmatrix} = \begin{pmatrix} s_2 & s_1 \\ s_3 & s_2 \end{pmatrix}^{-1} \begin{pmatrix} s_3 \\ s_4 \end{pmatrix} = \begin{pmatrix} a^4 & a^{11} \\ a^2 & a^4 \end{pmatrix} \begin{pmatrix} a^{14} \\ a^2 \end{pmatrix}$$

or $\zeta_1 = a^8$, $\zeta_2 = a^{11}$. Thus

$$\zeta(q) = 1 + a^8 q + a^{11} q^2$$

The roots of $\zeta(q)$ are found by exchaustive search over $GF(2^4)$. Thus

$$\zeta(1) = 1 + a^8 + a^{11} = a^3 + a = a^9 \neq 0$$

$$\zeta(a) = 1 + a^9 + a^{13} = a^2 + a = a^5 \neq 0$$

Continuing this way, we find that the roots of $\zeta(q)$ are $a^5$ and $a^{14}$. The reciprocals are $a^{10}$ and $a$. Therefore the errors are correctly identified.    ∎

***Computation of the error magnitudes.*** Once the number of errors and the error locations are found, Eqs. (14.57) become linear with respect to the error magnitudes. The associated matrix has a Vandermonde structure and is nonsingular because $\beta_i \neq \beta_j$ for $i \neq j$. Alternatively the partial fraction formulas can be invoked:

$$\gamma_i = \beta_i^{-1}(1 - \beta_i q)\tilde{s}(q)|_{q=\beta_i^{-1}} = \frac{\beta_i^{-1}\chi(\beta_i^{-1})}{\prod_{j\neq i}(1 - \beta_j\beta_i^{-1})} \qquad (14.63)$$

The denominator is directly obtained from $\zeta(q)$. The numerator is computed as follows: The syndrome polynomial

$$s(q) = \sum_{i=0}^{2t-1} s_{i+1}q^i \qquad (14.64)$$

constitutes the first $2t$ terms of the series $\tilde{s}(q)$. Therefore $\tilde{s}(q)$ is written as $\tilde{s}(q) = s(q) + q^{2t}s_1(q)$. Thus $\chi(q) = \tilde{s}(q)\zeta(q) = s(q)\zeta(q) + q^{2t}s_1(q)\zeta(q)$. We conclude that

$$\chi(q) = s(q)\zeta(q)(\text{mod } q^{2t}) \qquad (14.65)$$

The above equation determines $\chi(q)$. Equations (14.63) and (14.65) are known as *Forney's algorithm*.

### 14.5.3  The Berlekamp-Massey algorithm

We saw in the previous section that the syndromes are produced at the output of a linear feedback shift register with taps the coefficients of the error locator polynomial $\zeta(q)$. The computation of the number of errors $\nu$ and the error locator polynomial coefficients $\zeta_i$ becomes a linear feedback shift register synthesis problem: Given a sequence of length $2t$, find the smallest length linear feedback shift register that generates it. One register that produces the given sequence is the cyclic shift of length $2t$ with characteristic polynomial $q^{2t} - 1$. The content of the rightmost stage is fed back to the leftmost stage. For each $1 \leq r \leq 2t$, consider the minimum length linear feedback shift register that realizes the initial segment $s_1$, $s_2$, ..., $s_r$. This register is characterized by the memory size $d_r$ and the polynomial representing its taps $\zeta^{(r)}(q) = 1 + \zeta_1^{(r)}q + \zeta_2^{(r)}q^2 + \cdots + \zeta_{d_r}^{(r)}q^{d_r}$. $d_r$ is a positive integer and $\zeta^{(r)}(q)$ a vector of length $d_r$. The degree of $\zeta^{(r)}(q)$ may be smaller than its length $d_r$, in other words $\zeta_{d_r}^{(r)}$ may be zero. The Berlekamp–Massey algorithm is a recursive procedure for the sequential computation of $d_r$ and $\zeta^{(r)}(q)$ with respect to $r$. Suppose that $d_{r-1}$ and $\zeta^{(r-1)}(q)$ have been computed. The corresponding feedback shift register realizes the segment $s_1$, $s_2$, ...., $s_{r-1}$. To determine the minimum length shift register that realizes the augmented record $s_1, s_2, \ldots, s_{r-1}, s_r$ we check the error between $s_r$ and the output of the available register:

$$\Delta_r = s_r + \sum_{i=1}^{d_{r-1}} \zeta_i^{(r-1)} s_{r-i} = \sum_{i=0}^{d_{r-1}} \zeta_i^{(r-1)} s_{r-i}, \qquad \zeta_0^{(r-1)} = 1 \qquad (14.66)$$

The condition

$$r - d_{r-1} \geq 1 \qquad (14.67)$$

ensures that only the available data enter the computation. $\Delta_r$ is known as the $r$th *discrepancy*. If $\Delta_r = 0$, the current register realizes the augmented sequence, and no change is needed, $\zeta^{(r)}(q) = \zeta^{(r-1)}(q)$. Suppose that $\Delta_r \neq 0$. Then a new register must be constructed. Let us try the following updating law:

$$\zeta^{(r)}(q) = \zeta^{(r-1)}(q) + Ab(q) \qquad (14.68)$$

$A$ is a field element and $b(q)$ a polynomial aiming to correct the previous estimate $\zeta^{(r-1)}(q)$. Two requirements must be satisfied: First $\zeta^{(r)}(q)$ must produce the segment $s_1$, $s_2$, ...., $s_r$ and secondly it must have minimum length. The first requirement is met if the a posteriori error

$$\tilde{\Delta}_k^{(r)} = s_k + \sum_{i=1}^{d_r} \zeta_i^{(r)} s_{k-i} = \sum_{i=0}^{d_r} \zeta_i^{(r)} s_{k-i}, \qquad 2 \leq k \leq r$$

is zero. Notice that $\tilde{\Delta}_k^{(r)}$ coincides with the $k$th coefficient of the product $\zeta^{(r)}(q)s(q)$, $[\zeta^{(r)}(q)s(q)]_k$. For $2 \leq k \leq r - 1$, Eq. (14.68) gives

$$\tilde{\Delta}_k^{(r)} = [\zeta^{(r)}(q)s(q)]_k = [\zeta^{(r-1)}(q)s(q)]_k + A[b(q)s(q)]_k = \tilde{\Delta}_k^{(r-1)} + A[b(q)s(q)]_k$$

Since $\zeta^{(r-1)}(q)$ realizes the first $r - 1$ syndrome digits, $\tilde{\Delta}_k^{(r-1)} = 0$. To force the second term to zero, we pick $b(q)$ as any polynomial of the form $q^{r-m}\zeta^{(m-1)}(q)$, where $\zeta^{(m-1)}(q)$ is one of the previously constructed minimum length polynomials. The presence of $q^{r-m}$ enlarges the size of the corrective polynomial and provides it with sufficient corrective power. The first $r - m$ coefficients of the product $q^{r-m}\zeta^{(m-1)}(q)s(q)$ are zero due to the presence of $q^{r-m}$. The remaining $m-1$ coefficients are also zero because $\zeta^{(m-1)}(q)$ realizes the first $m-1$ terms of the syndrome sequence. We have established that the new polynomial realizes the first $r-1$ syndrome values. To accommodate the latest sample $s_r$, we observe that $\tilde{\Delta}_r^{(r-1)} = \Delta_r$ and that

$$\tilde{\Delta}_r^{(r)} = \Delta_r + A[q^{r-m}\zeta^{(m-1)}(q)s(q)]_r = \Delta_r + A\Delta_m$$

The above equation becomes zero with the choice $A = -\Delta_r/\Delta_m$ provided that $\Delta_m \neq 0$. Therefore $m$ must be selected as any of the previous steps for which $\Delta_m \neq 0$. With the above choices the new filter taps are updated by

$$\zeta^{(r)}(q) = \zeta^{(r-1)}(q) - \frac{\Delta_r}{\Delta_m} q^{r-m}\zeta^{(m-1)}(q) \tag{14.69}$$

and the corresponding shift register will generate the first $r$ terms of the given sequence. Nevertheless, the minimal length property may fail. It turns out that Eq. (14.69) will produce a minimum length shift register if $m$ is chosen as the most recent iteration for which $d_r > d_{r-1}$. This property is established below.

The sequence of shift register lengths is increasing:

$$d_r \geq d_{r-1} \tag{14.70}$$

Indeed, the shift register generating the first $r$ terms of the sequence produces, in particular, the first $r-1$ terms and $d_{r-1}$ corresponds to the smallest such register. Suppose that at step $r$ the discrepancy $\Delta_r$ is nonzero. Then

$$d_r \geq r - d_{r-1} \tag{14.71}$$

If $d_{r-1} \geq r$, the statement holds because $d_r > 0$. Suppose that $d_{r-1} < r$ and that $d_r \leq r - 1 - d_{r-1}$ or $d_r + d_{r-1} + 1 \leq r$. The value $s_r$ is generated by the register with taps $\zeta^{(r)}(q)$. On the other hand, $\zeta^{(r-1)}(q)$ fails to produce $s_r$ because the discrepancy is nonzero, by assumption. Of course $\zeta^{(r-1)}(q)$ produces the first $r-1$ syndrome digits. Putting these remarks together, we have

$$s_r = -\sum_{k=1}^{d_r} \zeta_k^{(r)} s_{r-k} = \sum_{k=1}^{d_r} \zeta_k^{(r)} \sum_{i=1}^{d_{r-1}} \zeta_i^{(r-1)} s_{r-k-i}$$

The third term is obtained from the second term upon replacing $s_{r-k}$ with the output of the register with taps $\zeta^{(r-1)}(q)$. This is permissible because we are in the range $r - d_r \leq r - k \leq r - 1$ and $r - d_r \geq d_{r-1} + 1$ by assumption. Hence Eq. (14.67) holds. Interchanging the sums, we obtain

$$s_r = \sum_{i=1}^{d_{r-1}} \zeta_i^{(r-1)} \sum_{k=1}^{d_r} \zeta_k^{(r)} s_{r-k-i} = -\sum_{i=1}^{d_{r-1}} \zeta_i^{(r-1)} s_{r-i} \neq s_r$$

a contradiction.

**TABLE 14.1   The Berlekamp-Massey algorithm**

Initialization: $\zeta^{(0)}(q) = 1$, $B^{(0)}(q) = 1$, $d_0 = 0$

For $r = 1 : 2t$,

$$\Delta_r = \sum_{i=0}^{d_{r-1}} \zeta_i^{(r-1)} s_{r-i}$$

$$\delta_r = \begin{cases} 1 & \text{if } \Delta_r \neq 0, \quad 2d_{r-1} \leq r \\ 0 & \text{otherwise} \end{cases}$$

$$d_r = \delta_r(r - d_{r-1}) + (1 - \delta_r)d_{r-1}$$

$$\begin{pmatrix} \zeta^{(r)}(q) \\ B^{(r)}(q) \end{pmatrix} = \begin{pmatrix} 1 & -\Delta_r q \\ \Delta_r^{-1}\delta_r & (1 - \delta_r)q \end{pmatrix} \begin{pmatrix} \zeta^{(r-1)}(q) \\ B^{(r-1)}(q) \end{pmatrix}$$

---

Combining the inequalities (14.70) and (14.71), we have

$$d_r \geq \max(d_{r-1}, r - d_{r-1}) \tag{14.72}$$

If $d_{r-1}$ is available at step $r$, the right-hand side of (14.72) will provide a lower bound for the minimum length $d_r$. We will show that prudent selection of $m$ achieves the lower bound and leads to the minimum length register. Suppose that $\Delta_r \neq 0$. If $2d_{r-1} \geq r$, the maximum in (14.72) is $d_{r-1}$. If we set $d_r = d_{r-1}$, namely we leave the register length unmodified, $d_r$ achieves the lower bound (14.72). If $2d_{r-1} \leq r$, we set $d_r = r - d_{r-1}$ and the lower bound is attained again. The above statements are combined by means of the flag variable $\delta_r$. $\delta_r$ is 1 if the discrepancy is nonzero and $2d_{r-1} \leq r$; otherwise, it is zero (see Table 14.1). Now we return to Eq. (14.69), and we rewrite it as

$$\zeta^{(r)}(q) = \zeta^{(r-1)}(q) - \Delta_r q B^{(r-1)}(q) \qquad B^{(r-1)}(q) = \Delta_m^{-1} q^{r-m-1} \zeta^{(m-1)}(q) \tag{14.73}$$

Finally, we pick $m$ as the most recent iteration for which $d_r > d_{r-1}$, that is, $2d_{r-1} \leq r$. Then $B^{(r)}(q)$ is updated as

$$B^{(r)}(q) = \Delta_r^{-1}\delta_r \zeta^{(r-1)}(q) + (1 - \delta_r)q B^{(r-1)}(q) \tag{14.74}$$

The development of the algorithm is complete. The list of recursions is given in Table 14.1. If $\delta_r = 0$, the first term in Eq. (14.74) is zero because the register length has not changed. $B^{(r)}(q)$ retains the previous values albeit shifted by $q$ so that the term $q^{r-m}$ is sequentially built.

We have already seen that the feedback shift register with characteristic polynomial $\zeta^{(r)}(q)$ realizes the first $r$ terms of the syndrome sequence. It will be a minimum length register if $d_r$ achieves the lower bound (14.72). This follows by induction. Indeed, let $\dim(x)$ denote the dimension of a vector $x$.

Equation (14.73) implies that

$$d_r = \dim(\zeta^{(r)}) \leq \max\left(\dim(\zeta^{(r-1)}), \dim(q^{r-m}\zeta^{(m-1)})\right)$$
$$= \max(d_{r-1}, r - m + d_{m-1})$$

By definition of $m$, we have $m - d_{m-1} = d_m$, and moreover

$$d_m = d_{m+1} = \cdots = d_{r-1}$$

Therefore

$$d_r \leq \max(d_{r-1}, r - (m - d_{m-1})) = \max(d_{r-1}, r - d_{r-1})$$

**Example 14.7**  **Decoding by the Berlekamp-Massey algorithm**
Consider the triple-error correcting BCH code of Example 14.6 and the same
received sequence $r(q) = q + q^{10}$. Using the syndromes computed in Example
14.6, we determine the error locator polynomial coefficients by the Berlekamp-
Massey algorithm. Computations are summarized in Table 14.2. The same error
locator polynomial results.                                              ∎

### 14.5.4   Spectral techniques

The BCH decoding approach outlined above involves the following steps:

1. Computation of the $2t$ syndromes.
2. Computation of the number of errors that have occurred as well as of the
   coefficients of the error locator polynomial.
3. Computation of error locations.
4. Computation of error magnitudes.

Step 1 uses Horner's rule $2t$ times. Step 2 is performed either by the Peterson-
Zierler-Görenstein scheme or by the Berlekamp-Massey algorithm. Step 3 is

**TABLE 14.2**

| $r$ | $\Delta_r$ | $\delta_r$ | $d_r$ | $B$ | $\zeta$ |
|-----|-----------|-----------|-------|-----|---------|
| 0 |        |   | 0 | $1$ | $1$ |
| 1 | $a^8$   | 1 | 1 | $a^7$ | $1 + a^8 q$ |
| 2 | $0$     | 0 | 1 | $a^7 q$ | $1 + a^8 q$ |
| 3 | $a^4$   | 1 | 2 | $a^{11} + a^4 q$ | $1 + a^8 q + a^{11} q^2$ |
| 4 | $0$     | 0 | 2 | $a^{11} q + a^4 q^2$ | $1 + a^8 q + a^{11} q^2$ |
| 5 | $0$     | 0 | 2 | $a^{11} q^2 + a^4 q^3$ | $1 + a^8 q + a^{11} q^2$ |
| 6 | $0$     | 0 | 2 | $a^{11} q^3 + a^4 q^4$ | $1 + a^8 q + a^{11} q^2$ |

carried out by exhaustive search. Finally, step 4 uses either a linear system solver or the partial fraction expansion formula (14.63).

Step 1 can alternatively be effected by the discrete Fourier transform and its fast implementations (see Sections 5.8 and 7.6.3). Let $r$, $v$, and $e$ be the received vector, the codevector and the error vector, respectively. Then $r = v + e$. Let us focus on primitive narrow sense BCH codes. Pick a primitive element $a$. Passing to the frequency domain, we have $R = V + E$, where

$$R_j = \sum_{i=0}^{n-1} r_i a^{ij} = r(a^j), \qquad 0 \leq j \leq n - 1$$

and similarly for $V$ and $E$. The syndromes are

$$s_j = r(a^j) = R_j = e(a^j) = E_j, \qquad 1 \leq j \leq 2t \qquad (14.75)$$

The syndrome vector is nothing but the first $2t$ frequency components of the error vector. Thus step 1 can be carried out by computing the DFT of $r$ and keeping its first $2t$ components. Step 2 is carried out as before. If we combine (14.75) with (14.57), we conclude that the remaining spectral components $E_j$, $2t + 1 \leq j \leq n - 1$, can be obtained at the output of the linear feedback shift register with characteristic polynomial $\zeta(q)$ initialized at the first $\nu$ syndrome values. In effect the spectral components of the error $E_j$ are the first $n$ values of the sequence $\tilde{s}(q)$. Once $E_j$ are determined, the inverse Fourier transform is applied to recover $e$ in the time domain.

## 14.6  REED-SOLOMON CODES

Reed-Solomon codes form a powerful subclass of BCH nonbinary codes. They are obtained if we set $m = 1$ in the general definition of BCH codes given in Section 14.4.2. Then $n = N - 1$ and the coefficients of each minimal polynomial reside in the same field with $a^i$, $GF(N)$. If $a$ is a primitive element in $GF(N)$, it follows that $\phi_i(q) = q - a^i$ and the generator polynomial of the Reed-Solomon code is factored as

$$g(q) = (q - a)(q - a^2) \cdots (q - a^{2t}) \qquad (14.76)$$

The coefficients of $g(q)$ are elements of $GF(N)$ and the degree of $g(q)$ equals $2t$. Therefore $n - k = 2t$.

Reed-Solomon codes satisfy the BCH bound. In fact much more is true, as we show next. Given any linear $(n, k)$ code $\mathcal{C}$, the minimum distance of the code satisfies

$$d_{\min} \leq n - k + 1 \qquad (14.77)$$

The above inequality is known as the *singleton bound*. The proof is simple. Pick a codeword in systematic form $v = [\hat{v}, u]$ such that the message $u$ has weight 1, $w(u) = 1$. Then the triangle inequality of the Hamming distance implies that

$$w(v) = w([\hat{v}, 0] + [0, u]) \leq w([\hat{v}, 0]) + w([0, u]) \leq w(\hat{v}) + w(u) \leq n - k + 1$$

Reed-Solomon codes attain the singleton bound:

$$d_{\min} = n - k + 1$$

Indeed, the BCH bound gives $d_{\min} \geq 2t + 1 = n - k + 1$. The assertion follows from the singleton bound. Thus Reed-Solomon codes offer the larger minimum distance among all linear codes of given parameters $n$ and $k$.

### Example 14.8   *A double-error Reed-Solomon code*
Consider the double-error Reed-Solomon code over GF(16). It is a (15,11) cyclic code generated by

$$g(q) = (q - a)(q - a^2)(q - a^3)(q - a^4) = q^4 + a^{13}q^3 + a^6q^2 + a^3q + a^{10}$$

Each code polynomial is a sequence of 15 digits. Each digit is an element of GF($2^4$). Thus it is represented by 4 bits (equivalent to 60 bits). Likewise the information sequence is a string of 11 symbols requiring 44 bits. Suppose that the zero codeword is transmitted and that the polynomial

$$r(q) = a^4q^2 + a^3q^7 = e(q)$$

is received. The syndromes are

$$s_1 = r(a) = a^4a^2 + a^3a^7 = a^7$$
$$s_2 = r(a^2) = a^4a^4 + a^3a^{14} = 1$$
$$s_3 = r(a^3) = a^4a^6 + a^3a^{21} = a^{13}$$
$$s_4 = r(a^4) = a^4a^8 + a^3a^{28} = a^{13}$$

We employ the Peterson-Zierler-Görenstein scheme. Let $i = 2$. We have

$$\det S(2) = \det \begin{pmatrix} s_2 & s_1 \\ s_3 & s_2 \end{pmatrix} = \det \begin{pmatrix} 1 & a^7 \\ a^{13} & 1 \end{pmatrix} = a^{10} \neq 0$$

We conclude that two errors have occurred. The coefficients of the error locator polynomial are determined by

$$\begin{pmatrix} \zeta_1 \\ \zeta_2 \end{pmatrix} = \begin{pmatrix} s_2 & s_1 \\ s_3 & s_2 \end{pmatrix}^{-1} \begin{pmatrix} s_3 \\ s_4 \end{pmatrix} = \begin{pmatrix} a^5 & a^{12} \\ a^3 & a^5 \end{pmatrix} \begin{pmatrix} a^{13} \\ a^{13} \end{pmatrix}$$

or $\zeta_1 = a^{12}$, $\zeta_2 = a^9$. Thus

$$\zeta(q) = 1 + a^{12}q + a^9 q^2$$

The roots of $\zeta(q)$ are readily found by exhaustive search to be $a^8$ and $a^{13}$. The reciprocals are $a^7$ and $a^2$, confirming the assumption. To determine the error coefficients, we use Forney's algorithm. Thus

$$s(q)\zeta(q) = (a^7 + q + a^{13}q^2 + a^{13}q^3)(1 + a^{12}q + a^9 q^2)$$

$$= a^7 + aq + a^6 q^4 + a^7 q^5$$

Hence

$$\chi(q) = s(q)\zeta(q)(\bmod\ q^{2t}) = a^7 + aq$$

Therefore

$$\gamma_1 = \frac{\beta_1^{-1}\chi(\beta_1^{-1})}{1 - \beta_2\beta_1^{-1}} = a^{13}\frac{a^7 + a^{14}}{1 + a^{20}} = a^4$$

$$\gamma_2 = \frac{\beta_2^{-1}\chi(\beta_2^{-1})}{1 - \beta_1\beta_2^{-1}} = a^8\frac{a^7 + a^9}{1 + a^{10}} = a^3$$

as expected. ∎

**Binary codes derived from Reed-Solomon codes.** Consider a $t$-error correcting Reed-Solomon code over $GF(2^m)$. Each information symbol is represented by an $m$-tuple over $GF(2)$. Using the $m$-bit representation for each information and code digit, the Reed-Solomon code gives rise to a binary linear $(n, k)$ code with $n = m(2^m - 1)$ and $2tm$ parity bits. Binary codes derived from Reed-Solomon codes are very effective against clustered errors. For example, if the original Reed-Solomon code corrects four errors, it can correct any single burst of length $3m + 1$ or less, since such a burst does not affect more than four symbols in $GF(2^m)$. Furthermore it can correct any combination of two bursts of length $m + 1$ or less.

## 14.7 CONVOLUTIONAL CODES

Convolutional encoders were first encountered in Example 3.17. An $(n, k, m)$ convolutional encoder over $GF(N)$ is a multichannel FIR filter of order $m$ with $k$

inputs and $n$ outputs taking values in $GF(N)$. The filter is preceded by a demultiplexer and followed by a multiplexer. The filter input-output expression has the form

$$\mathbf{v}_i = \sum_{j=0}^{m} \mathbf{u}_j g_{i-j} \tag{14.78}$$

Each input digit $\mathbf{u}_j$ is a $1 \times k$ row vector with entries in $GF(N)$. Likewise each output digit $\mathbf{v}_j$ is an $1 \times n$ vector. The filter coefficients $g_i$ are $k \times n$ matrices over $GF(N)$. In practice, $n$ and $k$ are small integers, while $m$ is large enough to achieve low error probabilities. Typically $k = 1$.

The encoder input sequence is formed by demultiplexing the information sequence into $k$ streams:

$$\mathbf{u}_i = \begin{pmatrix} u_i^{(1)} & u_i^{(2)} & \cdots & u_i^{(k)} \end{pmatrix}$$

$$u_i^{(1)} = \begin{pmatrix} u_0 & u_k & u_{2k} & u_{3k} & \cdots \end{pmatrix}$$

$$u_i^{(2)} = \begin{pmatrix} u_1 & u_{k+1} & u_{2k+1} & u_{3k+1} & \cdots \end{pmatrix}$$

$$u_i^{(k)} = \begin{pmatrix} u_{k-1} & u_{2k-1} & u_{3k-1} & u_{4k-1} & \cdots \end{pmatrix}$$

The multichannel output sequence of the FIR filter is serialized by the multiplexer:

$$\begin{pmatrix} v_0^{(1)} & v_0^{(2)} & \cdots & v_0^{(n)} & v_1^{(1)} & v_1^{(2)} & \cdots & v_1^{(n)} & \cdots \end{pmatrix}$$

The rate of the code is $R = k/n$. We recall from Section 3.6.5 that the multichannel convolution (14.78) reduces to $n$ ordinary scalar convolutions according to

$$v_i^{(l)} = \sum_{s=1}^{k} (u^{(s)} * g^{(sl)})_i \tag{14.79}$$

$g_i^{(sl)}$ is the $sl$ entry of the $k \times n$ matrix $g_i$. Each input sequence $u^{(s)}$ affects all output sequences through the registers with generator polynomials $g^{(s1)}$, $g^{(s2)}, \ldots, g^{(sn)}$. Let

$$K_s = \max_{1 \leq l \leq n} \{ \deg g^{(sl)} \}$$

$K_s$ stands for the maximum order shift register associated with the $s$th input. The *total encoder memory* is provided by

$$K = \sum_{s=1}^{k} K_s \tag{14.80}$$

The computation of each output sequence $v^{(l)}$ requires a parallel bank of FIR filters with coefficients $g^{(sl)}$. Invoking the series representation, Eqs. (14.78) and (14.79) take the form

$$v(q) = u(q)G(q), \quad v^{(l)}(q) = \sum_{s=1}^{k} u^{(s)}(q)g^{(sl)}(q)$$

In the most common case $k = 1$, the above expressions read as

$$G(q) = (g^{(1)}(q) \quad \cdots \quad g^{(n)}(q)), \quad v^{(l)}(q) = u(q)g^{(l)}(q)$$

In accordance with Section 3.6.6, Eq. (14.78) can alternatively be expressed as a matrix by vector product

$$v = uG$$

$G$ is the *generator matrix* of the code $G$.

Systematic encoding can also be introduced in the context of convolutional codes. In a systematic encoder the first $k$ output sequences coincide with the $k$ input sequences, ie.,

$$v^{(i)} = u^{(i)}, \qquad i = 1, 2, \ldots, k$$

The generator matrix takes the form

$$G(q) = (I \quad \hat{G}(q))$$

The size of $\hat{G}(q)$ is $k \times (n - k)$.

***Example 14.9*** *A $(2, 1, 3)$ convolutional encoder*
Consider the $(2, 1, 3)$ convolutional encoder of Fig. 14.20. The output sequences are generated by

$$v_i^{(1)} = u_i + u_{i-1} + u_{i-3}$$

$$v_i^{(2)} = u_i + u_{i-1} + u_{i-2} + u_{i-3}$$

The series representation gives

$$v^{(1)}(q) = u(q)g^{(1)}(q), \quad v^{(2)}(q) = u(q)g^{(2)}(q)$$

where

$$g^{(1)}(q) = 1 + q + q^3, \quad g^{(2)}(q) = 1 + q + q^2 + q^3$$

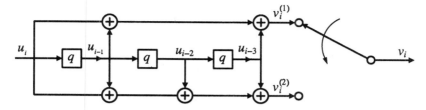

**Figure 14.20.** A (2,1,3) convolutional encoder.

The information sequence $u(q) = 1011$ or $u(q) = 1 + q^2 + q^3$ is encoded into the output sequences

$$v^{(1)}(q) = (1 + q + q^3)(1 + q^2 + q^3) = 1 + q + q^2 + q^3 + q^4 + q^5 + q^6 = 1111111$$

$$v^{(2)}(q) = (1 + q + q^2 + q^3)(1 + q^2 + q^3) = 1 + q + q^3 + q^6 = 1101001$$

The code sequence is obtained by inteleaving the above sequences

$$v = (11 \quad 11 \quad 10 \quad 11 \quad 10 \quad 10 \quad 11) \qquad\blacksquare$$

If the encoder is systematic no inversion circuit is required. In the nonsystematic case caution is required to avoid the so-called catastrophic error propagation. Inversion is warranted if the $k \times n$ generator polynomial matrix $G(q)$ has a right inverse; that is, there is a $n \times k$ matrix $\tilde{G}(q)$ such that

$$G(q)\tilde{G}(q) = q^l I \qquad (14.81)$$

$l$ is a positive integer and $I$ the identity matrix of order $k$. The right inverse is in general a rational matrix. If $\tilde{G}(q)$ is a polynomial matrix the encoder is called *noncatastrophic*. Let us take for simplicity $k = 1$. Then

$$\tilde{G}(q) = (\tilde{g}_1(q) \quad \cdots \quad \tilde{g}_n(q))^T$$

Condition (14.81) reads

$$g^{(1)}(q)\tilde{g}_1(q) + g^{(2)}(q)\tilde{g}_2(q) + \cdots + g^{(n)}(q)\tilde{g}_n(q) = q^l$$

This is a straightforward extension of the Bezout identity. It is solvable provided that the polynomials $g^{(i)}(q)$ are relatively prime:

$$\gcd(g^{(1)}(q), \ldots, g^{(n)}(q)) = 1$$

The catastrophic effects caused by the violation of the above condition are best seen via an example. Consider the (2, 1, 2) convolutional encoder generated by

$$g^{(1)}(q) = 1 + q, \quad g^{(2)}(q) = 1 + q^2$$

The above polynomials are not coprime because $1 + q$ is a common factor. Equation (14.81) has an infinite number of rational solutions $\tilde{G}(q)$ but no polynomial solution. Suppose that the input sequence is the unit step $u_i = 1$, $i \geq 0$. Then $u(q) = 1/(1 + q)$. The corresponding output sequences are

$$v^{(1)}(q) = 1 \quad \text{or} \quad v^{(1)} = 1000\cdots, \quad v^{(2)}(q) = 1 + q \quad \text{or} \quad v^{(2)} = 11000\cdots$$

Suppose that the first three bits of the code sequence are modified by the channel noise and the received sequence is the zero sequence. The minimum distance decoder will decode it to the zero codevector. The decoder fails to reconstruct the transmitted codeword making three errors. This decoding error translates to an infinite number of errors in the information sequence. Indeed, the zero codeword leads to the zero information sequence. Such deleterious effects are avoided with noncatastrophic codes.

The *minimum free distance* of a convolutional code is defined as

$$d_{\text{free}} = \min\{d(v_1, v_2) : v_1 \neq v_2\} = \min\{w(uG) : u \neq 0\}$$

The minimum is taken over all nonzero finite length information sequences $u$. Notice that lengths are finite but arbitrary. In contrast to the block case the input and output sequences of convolutional codes belong to infinite-dimensional spaces. When information sequences are restricted to polynomials of degree at most $i$ the *ith minimum distance*

$$d_i = \min\{w(uG) : u \neq 0, \deg u \leq i\}$$

is obtained. Clearly $d_i \leq d_{i+1}$. In the case of noncatastrophic codes, it can be shown that $\lim_{i\to\infty} d_i = d_{\text{free}}$. Good codes require as large distance as possible. Nonsystematic encoders offer a greater capacity in this respect than systematic coders. Good convolutional codes are constructed mainly by computer search.

## 14.8 DECODING OF CONVOLUTIONAL CODES AND THE VITERBI ALGORITHM

The Viterbi algorithm relies on the principle of optimality, the bottom line of a powerful optimization method known as dynamic programming.

The trellis diagram of a digital system or finite state machine is an expanded version of the state diagram (see Section 10.4.2) with an explicit indication of state and output transitions in time. An example is illustrated in Fig. 14.21. The horizontal axis represents time. For each time $i \geq 0$, all states that can be reached

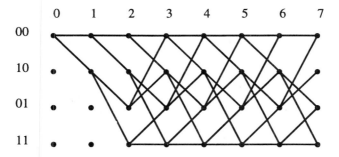

**Figure 14.21.** Trellis diagram.

from the origin at exactly $i$ time units (i.e., the reachable set at time $i$ from the origin, $\mathcal{R}(i, 0)$) are displaced on the vertical line at time $i$. Branches indicate one unit time state transitions. A branch emanates from state $s_1$ at time $n$ and terminates at state $s_2$ at time $n + 1$ if $s_2$ is reached from $s_1$ in one time unit. The input that caused the transition, and the resulting output are often indicated as labels on the branch like the state diagram (see Section 10.4.2). Each output trajectory is assigned to a path of the trellis diagram.

In the sequel we specialize to trellis diagrams of convolutional codes. Consider the convolutional code (the boldface notation is abandoned for simplicity)

$$v_i = \sum_{j=0}^{m} u_j g_{i-j}$$

The encoder state vector is

$$x_i = \begin{pmatrix} u_{i-1} & u_{i-2} & \cdots & u_{i-m} \end{pmatrix} \tag{14.82}$$

In the case of binary inputs there are at most $2^m$ possible states at each time. Two branches emanate from each state depending on whether $u_{i+1}$ is 0 or 1. Likewise two branches terminate at each state each time. Branching activities remain the same while traveling along the trellis. In other words, if a state $s_1$ is linked to the pair of states $s_2$ and $s_3$ at time $n$, the same linkage occurs throughout the trellis. Thus at all subsequent times, $s_1$ will always be connected to $s_2$ and $s_3$, and it will carry the same labels. In the general case of $k$-channel binary inputs, the number of states rises to $2^{mk}$, and the number of branches leaving a state is $2^k$.

An interesting observation about the trellis diagrams of convolutional codes (polynomial filters in general) is that the graph topology and the interconnections remain the same over all systems having the same parameters $k$ and $m$. The reason is that the state (14.82) depends only on the input. What distinguishes the trellis representation of two distinct convolutional codes are the output labels on the branches.

To facilitate the mechanics of the algorithm, we consider trellis diagrams that originate and terminate at the zero state. Zero initialization is accomplished if the input is zero at negative time instants. Termination at zero is achieved if the information sequence is appended by $m$ consecutive zeros.

***Example 14.10   A (2,1,2) convolutional encoder***
Consider the $(2,1,2)$ convolutional code with the generator polynomial

$$g(q) = \left(1 + q^2 \quad 1 + q + q^2\right)$$

The information sequence has length $L = 6$ and is given by

$$u = \left(u_0 \quad u_1 \quad \cdots \quad u_{L-1}\right) = \left(1 \quad 0 \quad 1 \quad 1 \quad 0 \quad 1\right)$$

The zero final state is imposed by appending $u$ with $m = 2$ zeros:

$$u = \left(1 \quad 0 \quad 1 \quad 1 \quad 0 \quad 1 \quad 0 \quad 0\right)$$

The resulting codeword has length $L + m$ and satisfies

$$v_i^{(1)} = u_i + u_{i-2}, \quad v_i^{(2)} = u_i + u_{i-1} + u_{i-2}$$

It is given in serialized form by

$$v = \left(11 \quad 01 \quad 00 \quad 10 \quad 10 \quad 00 \quad 01 \quad 11\right) \tag{14.83}$$

and is highlighted in Fig. 14.22.                                                   ∎

Let $u = \left(u_0 \quad u_1 \quad \cdots \quad u_{L+m-1}\right)$ be the information sequence, and let

$$v = \left(v_0 \quad v_1 \quad \cdots \quad v_{L+m-1}\right), \quad r = \left(r_0 \quad r_1 \quad \cdots \quad r_{L+m-1}\right)$$

be the code and received sequences, respectively. The entries $v_i$ and $r_i$ are binary vectors of length $n$. Each codeword is uniquely assigned to a path of the trellis

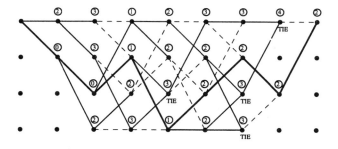

**Figure 14.22.** Illustration of the Viterbi algorithm.

diagram. The decoder is designed so that an index of the discrepancy $M(r, v)$ between the received sequence and a possible codeword is minimized. In terms of the trellis diagram the index $M(r, v)$ provides a performance indicator of the trellis paths. For a particular path $v$, $M(r, v)$ indicates the cost incurred in picking up the path $v$. We will assume that $M(r, v)$ has an additive format, namely the cost of a path is the sum of costs of the branches making up the path

$$M(r, v) = \sum_{i=0}^{L+m-1} m(r_i, v_i) \qquad (14.84)$$

Two important examples of this type is maximum likelihood decoding over discrete memoryless channels and minimum Hamming distance decoding. In the former case we have $M(r, v) = \log P(r|v)$. The memoryless structure of the channel implies that the likelihood function has the additive form (14.84) with $m(r_i, v_i) = \log P(r_i|v_i)$. In the case of minimum distance decoding, it holds that $m(r_i, v_i) = d(r_i, v_i) = \sum_{j=0}^{n-1} d(r_{ij}, v_{ij})$.

Optimum decoder design amounts to finding the path in the trellis that optimizes the cost (14.84). A trivial but prohibitively expensive solution is exhaustive search. The costs for each of the $2^{kL}$ possible paths are evaluated, and the best is selected. Fortunately there are better ways to handle the problem. The Viterbi algorithm cuts down the above search to $2^k$ alternatives utilizing the following simple observation. If $v$ is the optimal path joining the zero state to the zero state, then any subarc of $v$ joining the zero state to a state $s$ is the optimum path to $s$. In other words, suppose that $r$ is partitioned as $r = (\tilde{r} \quad \hat{r})$. The size of $\hat{r}$ is arbitrary. Let $v$ be the codeword closest to $r$. Partition $v$ accordingly as $v = (\tilde{v} \quad \hat{v})$. Then $\tilde{v}$ is closest to $\tilde{r}$. This is the principle of optimality in dynamic programming. The proof is clear. If the statement is false, there is a path joining the zero state to $s$ with better cost. Concatenation of this path with the subarc joining $s$ to the zero state leads to an admissible path with better performance than $v$, which is a contradiction.

Based on the above observation, we determine at each time $j$ the optimal path from zero to all states reached at time $j$. This is done recursively as follows: As long as $j$ stays smaller than $m$, the reachable set at time $j$ contains $2^j$ states. The full panorama of possible states appears once $j$ becomes equal to $m$ and sustains until $j$ equals $L$. In the initial period $j \leq m$, the cost paths are computed additively by (14.84). Let $M^*(s, j)$ be the cost of the optimum path joining zero with the state $s$ in $j$ time units. Let us see what happens when $j = m + 1$. For simplicity we take $k = 1$. Each state $s$ at time $j = m + 1$ is reached from two preceding states $s^0$ and $s^1$. The branches connecting $s^0$ to $s$ and $s^1$ to $s$ have branch costs $m(r_j, v_j^0)$ and $m(r_j, v_j^1)$. Therefore the two alternative paths emanating from zero and leading to $s$ at $j$ units of time have costs

$$M_0 = M^*(s^0, j - 1) + m(r_j, v_j^0), \qquad M_1 = M^*(s^1, j - 1) + m(r_j, v_j^1)$$

Clearly the optimum path from the origin to $s$ is computed as the best among the two. The path that looses is removed and does not contribute to the subsequent trellis search. This is the first manifestation of the savings offered by the Viterbi algorithm against exhaustive search. Now we set

$$M^*(s,j) = \min(M_0, M_1)$$

and we retain the optimum path leading to $s$ in $j$ units of time.

The above procedure is repeated. Suppose that at time $j$ all optimal paths to all states have been determined. Let $M^*(s,j)$ denote the optimum cost of the best path terminating at state $s$ in time $j$. To determine the optimum route to any state $s$ at time $j + 1$, we consider the two states ($2^k$ in the general case) at time $j$ that connect to $s$. The optimal paths to these two states are concatenated with the branches leading to $s$, and the resulting costs are compared. The winner is the optimal path to $s$, and the cost gives the value $M^*(s,j+1)$. The algorithm terminates after $j = L + m$ steps.

### Example 14.11    Decoding by the Viterbi algorithm

Consider the convolutional code of Example 14.10 and the codeword given by (14.83). Suppose that the received sequence is

$$r = (11 \quad 01 \quad 10 \quad 10 \quad 11 \quad 00 \quad 01 \quad 11)$$

Two errors have occurred. The decoding procedure is illustrated in Fig. 14.22. Each circled number over a state indicates the optimum cost of the path terminating at that state. Continuous lines are optimum paths linking the zero state to their terminal node. The highlighted path is the desired optimum path, and it yields the transmitted codeword. Notice that ties occur at certain states. If a state with a tie is included in the optimum path, there will be more than one optimum solutions. In such a case the convolutional code used is not capable to correct such error patterns. This problem does not arise in the specific example, since the optimum path does not pass through states with ties.    ∎

The Viterbi algorithm must reserve $2^K$ words of storage for the optimum path leading to each state and the respective optimal cost. $K$ is the total encoder memory and is given by (14.80). Each of these words must be capable to store a $kL$-bit path plus its cost. Computational complexity involves $2^K$ additions and $2^K$ comparisons. These exponential requirements make Viterbi decoding impractical for values of $K$ larger than 10. Popular alternatives include schemes searching only the high probability paths in the trellis. These methods are collectively known as *sequential decoding* and include Fano algorithm and the stack algorithm. More information can be obtained from the suggested literature.

In deep space communications and in several of planetary exploration missions, convolutional codes with maximum likelihood decoding and Reed-Solomon codes are combined in the so-called *nested* or *concatenated systems* to achieve superior performance. A convolutional code is used as an inner code to correct random errors while a Reed-Solomon code, possibly with an interleaver is used as an outer code to combat random as well as burst errors. The Voyager exposition to Uranus and Neptune used the above concatenated concept for the reliable transmission of images.

## BIBLIOGRAPHICAL NOTES

A detailed account of compression methods is given in Jayant and Noll (1984) and Gersho and Gray (1992). An extensive treatment of subband coding is provided in Vetterli and Kovačević (1995) and Vaidyanathan (1993). Error control coding is discussed in Berlekamp (1984), Blahut (1984), Lin and Costello (1983), and Peterson and Weldon (1972). Reed-Solomon codes and various applications are covered in Wicker and Bhargava (1994).

## PROBLEMS

**14.1.** Show that the computation of the DCT requires the computation of an $N$-point DFT only. *Hint:* Let $y_1(n)$ and $y_2(n)$ denote the even and odd parts of $y(n) = x(n) + x(2N - 1 - n)$. Prove that $y_2(n) = y_1(N - 1 - n)$, and then use Eq. (5.56) of Section 5.8.1.

**14.2.** Let $x(n_1, n_2)$ denote a 2-D finite support signal of $N_1 \times N_2$ points, zero outside $0 \le n_1 \le N_1 - 1, 0 \le n_2 \le N_2 - 1$. Define the $2N_1 \times 2N_2$-point sequence

$$y(n_1, n_2) = x(n_1, n_2) + x(2N_1 - 1 - n_1, n_2) + x(n_1, 2N_2 - 1 - n_2)$$
$$+ x(2N_1 - 1 - n_1, 2N_2 - 1 - n_2)$$

Let $Y(k_1, k_2)$ denote the 2-D DFT of $y(n_1, n_2)$ and

$$C(k_1, k_2) = \begin{cases} w_{2N_1}^{k_1/2} w_{2N_2}^{k_2/2} Y(k_1, k_2), & 0 \le k_1 \le N_1 - 1 \quad 0 \le k_2 \le N_2 - 1 \\ 0 & \text{otherwise} \end{cases}$$

(a) Prove that

$$C(k_1, k_2) = \sum_{n_1=0}^{N_1-1} \sum_{n_2=0}^{N_2-1} 4x(n_1, n_2) \cos \frac{\pi k_1}{2N_1} (2n_1 + 1) \cos \frac{\pi k_2}{2N_2} (2n_2 + 1)$$

for $0 \le k_1 \le N_1 - 1$, $0 \le k_2 \le N_2 - 1$ and zero otherwise. The above equation defines the DCT of $x(n_1, n_2)$.

(b) Develop an expression for the inverse DCT, and describe the algorithmic steps for its computation.

(c) Show that an $N_1 \times N_2$-point DFT rather than $2N_1 \times 2N_2$-point DFT suffices.

**14.3.** Establish the following properties of the DCT:
   (a) If $x(n_1, n_2) = x_1(n_1)x_2(n_2)$, then $C_x(k_1, k_2) = C_{x_1}(k_1)C_{x_2}(k_2)$
   (b) If $z(n_1, n_2) = ax(n_1, n_2) + by(n_1, n_2)$, then
   $C_z(k_1, k_2) = aC_x(k_1, k_2) + bC_y(k_1, k_2)$. Furthermore derive the energy relationship

$$\sum_{n_1=0}^{N_1-1} \sum_{n_2=0}^{N_2-1} |x(n_1, n_2)|^2 = \frac{1}{4N_1 N_2} \sum_{k_1=0}^{N_1-1} \sum_{k_2=0}^{N_2-1} w_1(k_1)w_2(k_2)|C_x(k_1, k_2)|^2$$

where

$$w_i(k_i) = \begin{cases} 1/2, & k_i = 0 \\ 1, & 1 \le k_i \le N_i - 1 \end{cases} \quad i = 1, 2$$

**14.4.** Given a signal $x(n)$ with $z$ transform $x(z)$, let

$$x_p(z) = \begin{pmatrix} x_0(z) & x_1(z) & \cdots & x_{M-1}(z) \end{pmatrix}^T$$

$$x_m(z) = \begin{pmatrix} x(z) & x(wz) & \cdots & x(w^{M-1}z) \end{pmatrix}^T$$

$x_p(z)$ consists of the polyphase components of $x(z)$. $x_m(z)$ is formed by the modulated versions of $x(n)$. Prove the following:
   (a) $x_m(z) = MW_M^{-1}L(z)x_p(z^M)$, $W_M$ stands for the DFT matrix of size $M$, and $L(z) = \mathrm{diag}\big(1, z^{-1}, \ldots, z^{-(M-1)}\big)$.
   (b) $H_m^T(z) = MW_M^{-1}L(z)H_p^T(z^M)$
   (c) Aliasing is canceled if and only if

$$G_p(z^M)H_p(z^M) = P(z^M)$$
$$= L(z)W_M^{-1}\mathrm{diag}\{(zw^i)^{M-1}f(zw^i)\}W_M L^{-1}(z)$$

We saw in Exercise 5.19 that circular matrices are diagonalized by the DFT. Prove that $P(z)$ is *pseudocirculant*, namely that it is circular apart from the elements located below the diagonal that are multiplied by $z^{-1}$.

**14.5.** Prove Eq. (14.7).

**14.6.** Show that the only FIR solutions of the quadrature mirror filter bank equation (14.16) are given by

$$h_0(z) = c_1 z^{-2k_1} + c_2 z^{-(2k_2+1)}.$$

**14.7.** Show that the first-order filter

$$h(z) = \frac{a^* + z^{-1}}{1 + az^{-1}}$$

is allpass. Show that a general rational transfer function is allpass if it is written as product of such first-order filters. Show that $h(z) = b(z)/a(z)$ with $a(z) = \sum_{i=0}^{N} a_i z^{-i}$, $b(z) = \sum_{i=0}^{N} b_i z^{-i}$ is allpass if $a_i = A b_{N-i}^*$. Finally, prove that $h(z)$ is allpass if the autocorrelation of $h(n)$ is a scaled unit sample, $r_h(n) = c^2 \delta(n)$.

**14.8.** Derive the analysis filter bank of the discrete Fourier transform. Show that the analysis filters are related by $h_k(z) = h_0(zw^k)$, $1 \le k \le M - 1$. Identify $h_0(z)$.

**14.9.** Suppose that the analysis filters in a two-channel QMF bank are given by

$$h_0(z) = 2 + 6z^{-1} + z^{-2} + 5z^{-3} + z^{-5}, \quad h_1(z) = h_0(-z)$$

Find a set of stable synthesis filters which achieve perfect reconstruction.

**14.10.** (a) Examine if the filters

$$h_0(z) = 1, \quad h_1(z) = 2 + z^{-1}, \quad h_2(z) = 3 + 2z^{-1} + z^{-2}$$

achieve perfect reconstruction (M = 3) or FIR perfect reconstruction. Find a set of synthesis filters, and check stability.
(b) Repeat for the filters

$$h_0(z) = 1, \quad h_1(z) = 2 + z^{-1} + z^{-5}, \quad h_2(z) = 3 + 2z^{-1} + z^{-2}$$

and the filters

$$h_0(z) = 1, \quad h_1(z) = 2 + z^{-1} + z^{-5}, \quad h_2(z) = 3 + z^{-1} + 2z^{-2}$$

**14.11.** Prove that the decoding procedure implied by the Peterson-Zierler-Görenstein scheme for double-error correcting BCH codes coincides with the technique presented in Section 14.4.1. Note that the first method uses $s_1, s_2, s_3, s_4$ syndrome digits, while the second uses $s_1, s_2$.

**14.12.** Nonbinary BCH codes may lead to better rates than binary codes of the same blocklength and designed distance. Construct the field $GF(16)$ as $F[z]/r(z)$ where $F = GF(4)$, and $r(z)$ is irreducible over $F$ of degree 2. Let $F = \{0, 1, b, b^2\}$. Show that a suitable choice is $r(z) = z^2 + z + b$. The elements of $GF(16)$ have the form $f_0 + f_1 q$, $f_i \in F$. Let $a = q$ in

$GF(16)$. Construct the polynomial representation of $GF(16)$. (*Hint:* $0 \rightarrow 0$, $a^0 \rightarrow 1$, $a^1 \rightarrow q$, $a^2 \rightarrow q+b$, $a^3 \rightarrow b^2q+b$, $a^4 \rightarrow q+1, \ldots,$ $a^{14} \rightarrow b^2q+b^2$).

Show that the minimal polynomials of $a$ and $a^2$ are $q^2+q+b$ and $q^2+q+b^2$, respectively. Show that the double error BCH code over $F$ of blocklength 15 is generated by

$$g(q) = q^6 + b^2q^5 + q^4 + q^3 + bq^2 + bq + 1$$

Demonstrate that it has higher rate than the binary double error BCH code of blocklength 15.

**14.13.** Construct the triple-error correcting $(15, 9)$ Reed-Solomon code. Decode the received sequence

$$r(q) = a^2q^8 + a^6q^4 + a^3q^2$$

**14.14.** Construct the triple-error correcting $(15, 5)$ BCH code. Decode using the Berlekamp-Massey algorithm the received sequence

$$r(q) = q^8 + q^4 + q^2$$

**14.15.** Consider the field $GF(2^4)$ generated by the primitive element $a$ satisfying $a^4 = a + 1$. Show that $b = a^7$ is also a primitive element. Construct the generator polynomial of the double and triple error correcting BCH code corresponding to $b$. Find the parity check matrix. What is the relationship of the above codes with those of Example 14.5?

**14.16.** Determine the generator polynomials of all the primitive BCH codes of length 31. Use $GF(2^5)$ spanned by the irreducible polynomial $1+q+q^5$. Decode the received polynomials $r(q) = q^9 + q^{24}$ and $r(q) = 1 + q^{15} + q^{27}$.

**14.17.** Consider the $(3,1,2)$ convolutional code with impulse response

$$g(q) = \left(1+q \quad 1+q^2 \quad 1+q+q^2\right)$$

Construct a trellis diagram. Use the Viterbi algorithm to decode the received sequence

$$r = (1 \quad 1 \quad 0,1 \quad 1 \quad 0,1 \quad 1 \quad 0,1 \quad 1 \quad 1,0 \quad 1 \quad 0,1 \quad 0 \quad 1,1 \quad 0 \quad 1)$$

# SOME FACTS FROM ANALYSIS

A metric space is a nonempty set $X$, together with a function $d : X \times X \to R$ satisfying the following properties:

1. $d(x,y) \geq 0$, $d(x,y) = 0$ if and only if $x = y$.
2. $d(x,y) \leq d(x,z) + d(z,y)$, (triangle inequality).

$d$ is called *distance* or *metric*. Take, for example, $X = C^k$, where $C$ is the set of complex numbers. Define

$$d_2(x,y) = \sqrt{\sum_{i=1}^{k} |x_i - y_i|^2}$$

It can be shown that $d_2$ satisfies the properties of a metric. Alternatively,

$$d_\infty(x,y) = \max_{1 \leq i \leq k} |x_i - y_i|$$

also defines a distance function on $C^k$. The distance $d$ provides a measure of closedness between two points in $X$. As such it enables us to define convergence in $X$. We say that the sequence $x_N$ of points in $X$ converges to $x \in X$ if the real sequence $d(x_N, x)$ converges to zero.

Very often distance functions are induced by norms. Let $X$ be a vector space over $C$. A norm is a function assigning a nonnegative real number $|x|$ to each vector $x$, such that for any vectors $x$, $y$ in $X$ and scalar $c$ the following hold:

1. $|x| \geq 0$, and $|x| = 0$ if and only if $x = 0$.
2. $|cx| = |c||x|$.
3. $|x + y| \leq |x| + |y|$.

Examples of norms on $C^k$ are

$$|x|_1 = \sum_{i=1}^{k} |x_i|, \quad |x|_2 = \sqrt{\sum_{i=1}^{k} |x_i|^2}, \quad |x|_\infty = \max_{1 \le i \le k} |x_i|$$

Let $1 \le p \le \infty$. The above norms are special cases of

$$|x|_p = \left( \sum_{i=1}^{k} |x_i|^p \right)^{1/p}$$

Given a normed space, the distance function induced by the norm is

$$d(x, y) = |x - y|$$

It is easy to check that $d$ is a a valid metric. Thus we say that a sequence $x_N$ converges to $x$ with respect to the norm if it converges with respect to the induced distance, $|x_N - x| \to 0$, $N \to \infty$. An important result states that in a finite-dimensional linear space convergence remains the same under any norm. If $x_N$ converges to $x$ with respect to one norm, it converges to $x$ with any other norm. Therefore as far as convergence is concerned, it makes no difference which norm is used. This result fails in infinite-dimensional spaces.

The set of $k \times m$ complex matrices is a vector space of dimension $km$. If we stack each column one after the other, the matrix is identified with a vector of length $km$. The various norms on $C^{km}$ can then be used as matrix norms. A very useful alternative is provided by the induced norms. These have the form

$$|A| = \sup_{x \ne 0} \frac{|Ax|}{|x|} \tag{I.1}$$

where $|Ax|$ is any norm on $C^k$ and $|x|$ any norm in $C^m$. It can be shown that (I.1) defines a legitimate norm that in addition satisfies

$$|AB| \le |A||B|$$

# ALGEBRAIC CONCEPTS

In this appendix some basic algebraic concepts are reviewed. The primary emphasis is on the theory of finite fields. This material is needed in the analysis and design of digital signal processing systems.

## II.1 BASIC ALGEBRAIC STRUCTURES

### II.1.1 Groups

The structure of a group generalizes the familiar concept of ordinary addition. Let $G$ be a nonempty set, equipped with a binary operation, that is, a function with domain the product $G \times G$ and range $G$. Let $*$ denote this operation. We say that $(G, *)$ defines a group if the following properties are satisfied:

1. *Associative property.* For any $a$, $b$, $c$ in $G$, $(a * b) * c = a * (b * c)$.
2. *Identity element.* There exists $e$ in $G$ such that for all $a$ in $G$, $a * e = e * a = a$.
3. *Inverse.* For each $a$ in $G$ there exists $a'$ in $G$ so that $a * a' = a' * a = e$.

The group $G$ is called *abelian* or *commutative* if the operation is commutative: $a * b = b * a$ for all $a$, $b$ in $G$. If $G$ has finitely many elements, it is called a *finite group*. The number of elements of $G$ is called the *order* of $G$.

The set of integers with ordinary addition is an abelian group. The identity element is 0, and the inverse of $a$ is $-a$. The set of nonzero real numbers equipped with real multiplication is an abelian group. 1 is the identity element and $a^{-1}$ is the inverse of $a$. Given an arbitrary set $S$, the collection of one-to-one and onto functions on $S$ forms a group. The group operation is the composition of maps. If $S$ is a finite set with $n$ elements, the group is a finite group of order $n!$.

The identity and the inverse of any element in a group are uniquely defined. Furthermore the cancellation property holds: For any $a$, $b$, $c$ in $G$ with

$a * b = a * c$, it holds that $b = c$. Finally, the equations $a * x = b$ and $x * a = b$ have unique solutions in $G$ given by $x = a' * b$ and $x = b * a'$, respectively.

A subset $H$ of $G$ is a *subgroup* of $G$ if for any $a, b$ in $H$, $a * b$ as well as $a'$ belong to $H$. $H$ will then contain the identity element of $G$ and will form a group with respect to $*$. Important examples of subgroups are the cyclic subgroups defined as follows: Let $a$ be an element of the group $G$. The powers of $a$ are defined by induction as $a^0 = e$, $a^n = a * a^{n-1}$, $a^{-n} = (a')^n$. It is easy to show that the set $H = \{a^n, n \in Z\}$ forms an abelian subgroup of $G$. A group $H$ arising this way is called a *cyclic subgroup with generator a*.

**Theorem II.1** **Lagrange's theorem** If $H$ is a subgroup of the finite group $G$, the order of $H$ divides the order of $G$.

PROOF   Let $H = \{h_1, h_2, \ldots, h_n\}$. Let $h$ denote the column vector with entries the elements of $H$. We form the array

$$( h \quad g_1 * h \quad \cdots \quad g_r * h )$$

where $g_1$ is any element of $G$ not contained in $H$, $g_2$ an element of $G$ not included in the first two columns, and so forth. We stop adding new columns once all entries of $G$ are scanned by the columns already constructed. Since $G$ is finite the above procedure terminates after a finite number of steps. We show now that each element of $G$ appears in exactly one entry of the array. Indeed, all entries residing on the same column are distinct because if $g_i * h_j = g_i * h_k$; then by the cancellation property, $h_j = h_k$, which cannot happen. If two entries belonging to different columns were equal, then $g_i * h_m = g_j * h_k$ with $i > j$. Therefore $g_i = g_j * (h_k * h'_m)$. Since $H$ is a group, the entry in the parenthesis belongs to $H$. Thus $g_i$ belongs to the $j + 1$ column $g_j * h$ which contradicts the way $g_i$ was selected. Thus the order of $G$ is given by the size of the array $(r + 1)n$ and hence is a multiple of the order of $H$. ∎

An element $a$ of a group $G$ has finite order if there is a nonzero integer $n$ such that $a^n = e$, $e$ being the identity of the group. The smallest such positive integer defines the *order* of $a$. In the additive group of integers the equation $a^n = e$ takes the form $na = 0$, which implies that $a = 0$. Thus each nonzero integer has infinite order. In the multiplicative group of nonzero complex numbers, the above equation takes the form $a^n = 1$, which is satisfied by the consecutive powers of $a = e^{j2\pi/n}$. Every element $a$ of a finite group $G$ has finite order. The cyclic subgroup generated by $a$ is finite; therefore there are integers $k$, $m$ with $k > m$ such that $a^k = a^m$. Then $a^{k-m} = e$. Let $n$ be the order of $a$. Apparently the cyclic subgroup of $a$ is spanned by the first $n$ successive powers of $a$,

$$\{a^l : l \in Z\} = \{a^0, a^1, a^2, \ldots, a^{n-1}\}$$

and its order coincides with $n$. Indeed, if $k > n$, we divide $k$ by $n$: $k = qn + r$,

$0 \leq r < n$, and we obtain $a^k = a^{qn} a^r = e^q a^r = a^r$. In a similar fashion we handle negative powers. The order of an element in a finite group divides the order of the group. This is a direct consequence of the Lagrange theorem.

### II.1.2  Rings

A ring is a set $R$ equipped with two binary operations, the addition, denoted by $+$, and the multiplication, denoted by $\cdot$, such that the following properties are satisfied:

1. $R$ together with addition is a commutative group.
2. Multiplication is associative and distributes over addition:

$$a \cdot (b + c) = a \cdot b + a \cdot c, \quad (a + b) \cdot c = a \cdot c + b \cdot c$$

If multiplication is commutative, the ring is a *commutative ring*. The identity element of addition is denoted by 0 and is called *zero*. The inverse of $a$ with respect to addition is denoted by $-a$ and is called *opposite*. If multiplication has an identity element, the ring is called *ring with identity*, and the identity is denoted by 1. If $a$, $b$ are in $R$ and $a \cdot b = 1$, $b$ is referred to as *right inverse* of $a$ and $a$ as *left inverse* of $b$. In the sequel we omit $\cdot$ and we write $ab$. For any $a$, $b$ in $R$ it holds that $0a = a0 = 0$ and $(-a)b = a(-b) = -ab$. If $a$ has a left as well as a right inverse, these coincide. This unique inverse of $a$, provided that it exists, is denoted by $a^{-1}$. The cancellation property for multiplication is not in general valid in a ring with identity. This gives rise to the notion of integral domain. A commutative ring with identity is an *integral domain* if for any $a$, $b$ in $R$ with $ab = 0$, either $a = 0$ or $b = 0$.

The set of integers with ordinary addition and multiplication is an integral domain. A similar statement holds for rational numbers, real and complex numbers. The set of discrete single-channel signals equipped with the pointwise operations forms a commutative ring with identity but not an integral domain.

### II.1.3  Ring of integers and Euclid's algorithm

In contrast to addition, subtraction and multiplication, division is not executable in $Z$. Instead, it is governed by the fundamental theorem of division. For any integer $s$ and any positive integer $t$, there exist unique integers $q$ and $r$, called *quotient* and *remainder*, respectively, such that $s = qt + r$, $0 \leq r \leq t - 1$. We occasionally denote the quotient by $q(s, t) = [s/t]$. We say that $t$ *divides* $s$ or that $s$ is a *multiple* of $t$, and we write $t|s$, if there is integer $a$ such that $s = at$. The *greatest common divisor* of integers $s$ and $t$, $\gcd(s, t)$, is the greatest of all positive common divisors of $s$ and $t$. Thus $\gcd(s, t)|s$, $\gcd(s, t)|t$. Moreover, if $d|s$ and $d|t$, then $|d|$ divides $\gcd(s, t)$. The *least common multiple* of $s$, $t$ is the smallest positive integer multiple of $s$ and $t$. $s$ and $t$ are called *relatively prime* if $\gcd(s, t) = 1$. Finally, an integer $a > 1$ is called *prime* if the only divisors of $a$ are $a$ and 1.

Euclid's algorithm is an extremely useful procedure for the computation of the greatest common divisor of two integers. Despite its long history it still inspires research and finds new applications. We assume, without loss of generality, that $0 < t < s$. Let

$$s = q_1 t + t_1, \qquad 0 \le t_1 < t$$
$$t = q_2 t_1 + t_2, \qquad 0 \le t_2 < t_1$$
$$t_1 = q_3 t_2 + t_3, \qquad 0 \le t_3 < t_2$$

Setting $t_{-1} = s$, $t_0 = t$, we generally have

$$t_{n-2} = q_n t_{n-1} + t_n, \qquad 0 \le t_n < t_{n-1}$$

Let $N$ be the least positive integer such that $t_{N-1} = q_{N+1} t_N$. In the first equation, division of $s$ with $t$ gives quotient $q_1$ and remainder $t_1$. Next $t$ is divided by $t_1$ to produce the remainder $t_2$. The process is repeated until perfect division is obtained. The last nonzero remainder $t_N$ gives the greatest common divisor of $s$, $t$, that is,

$$\gcd(s, t) = t_N \tag{II.1}$$

The proof follows from the relation

$$\gcd(s, t) = \gcd(s - qt, t) \qquad \text{for all } q \in Z \tag{II.2}$$

This is a direct consequence of the definition of the greatest common divisor. If we apply the above identity to Euclid's algorithm, we obtain

$$\gcd(t_1, t) = \gcd(s - q_1 t, t) = \gcd(s, t)$$

and more generally, $\gcd(t_{k+1}, t_k) = \gcd(s, t)$, $k \ge 1$. Finally, $\gcd(s, t) = \gcd(t_{N-1}, t_N)$. Since $t_N$ divides $t_{N-1}$, Eq. (II.1) follows. If we apply Euclid's recursion backward, we arrive at the important identity

$$\gcd(s, t) = as + bt \tag{II.3}$$

where $a$, $b$ are integers. Relation (II.3) is important and nontrivial. It expresses the greatest common divisor as an integer combination of $s$ and $t$.

### II.1.4    Chinese remainder theorem

Let $c$ be an integer and $m_1$, $m_2$, ..., $m_k$, pairwise relatively prime integers. Suppose that we are given the remainders (moduli) of $c$, $c_i$, with respect to each of the integers $m_i$. The Chinese remainder theorem enables us to recover $c$ from its moduli, $c_i$. Precisely we have

**Theorem II.2**    Given a set of pairwise relatively prime integers $m_1, m_2, \ldots, m_k$, and a set of integers $c_1, c_2, \ldots, c_k$ satisfying $c_j < m_j, j = 1, 2, \ldots, k$, the system of equations

$$c = c_j \ (\text{mod } m_j) \tag{II.4}$$

has a unique solution $c$ in the interval

$$0 < c < M, \quad M = \prod_{i=1}^{k} m_i \tag{II.5}$$

The solution is given by

$$c = \sum_{j=1}^{k} c_j N_j M_j (\text{mod } M) \quad M_j = \prod_{i \neq j}^{k} m_i = \frac{M}{m_j} \tag{II.6}$$

and $N_j$ is determined from the equation

$$N_j M_j + n_j m_j = 1 \tag{II.7}$$

**PROOF**    Since $m_j$ are relatively prime, $\gcd(M_j, m_j) = 1$. Hence we can find integers $N_j$ and $n_j$ such that (II.7) holds. We show next that the division of $c$ by $m_j$ gives remainder $c_j$. We write Eq. (II.6) equivalently as

$$c = c_j M_j N_j + \sum_{i \neq j} c_i M_i N_i + \lambda M$$

Now $m_j$ divides $M_i$ for every $i \neq j$, by definition of $M_i$. Moreover $m_j$ divides $M$. Therefore

$$c = c_j M_j N_j (\text{mod } m_j) \tag{II.8}$$

On the other hand, because of (II.7) we have

$$N_j M_j = 1 (\text{mod } m_j) \tag{II.9}$$

If we multiply both sides of (II.9) with $c_j$ and take into account (II.8), we get (II.4). Clearly $c$ is in the interval $(0, M)$, since it is the remainder of the division by $M$.

Next we show uniqueness of solutions. Let $d$ be another solution of (II.4). Then $c = d (\text{mod } m_j)$. Each $m_j$ divides $c - d$ and $m_j$ are pairwise relatively prime. Therefore their product $M$ also divides $c - d$. Since $0 < c, d < M$, it holds that $-M < c - d < M$. Thus $c = d$.    ∎

*Example II.1    Illustration of the Chinese remainder theorem*

Take $m_1 = 4$, $m_2 = 5$, and $m_3 = 7$. Then $M = 140$ and $M_1 = 35$, $M_2 = 28$, and $M_3 = 20$. Suppose that $c_1 = 1$, $c_2 = 1$, and $c_3 = 5$. Equation (II.7) can be solved either by the Euclidean algorithm or by inspection. Thus we readily find $N_1 = 3$, $n_1 = -26$, $N_2 = 2$, $n_2 = -11$, and $N_3 = -1$, $n_3 = 3$. Application of (II.6) gives

$$c = (1 * 3 * 35 + 1 * 2 * 28 + 5 * (-1) * 20)(\text{mod } 140) = 61$$

Given relatively prime integers $m_1$, ..., $m_k$, the Chinese remainder theorem asserts that the mapping $CR$ of each integer $c$ in the interval $0 < c < M$, $M = \prod m_i$, into the $k$-tuple $(c_1, c_2, \cdots, c_k)$ :

$$CR : c \to (c_1, c_2, \ldots, c_k), \qquad c_i = c(\text{mod } m_i)$$

is well-defined and invertible. Its inverse is specified by formula (II.6). Furthermore it is a morphism between the rings $R_M$ and $R_{m_1} \times R_{m_2} \times \cdots \times R_{m_k}$; that is, $CR(c + d) = CR(c) + CR(d)$, $(CR(cd))_i = (CR(c))_i(CR(d))_i$. Indeed, let $c$ and $d$ be transformed into $(c_1, c_2, \ldots, c_k)$ and $(d_1, d_2, \ldots, d_k)$, respectively, that is, $c_i = c(\text{mod } m_i)$, $d_i = d(\text{mod } m_i)$. Then the sum $c + d(\text{mod } m_i)$ and the product $cd \ (\text{mod } m_i)$ are transformed into $c_i + d_i = c + d \ (\text{mod } m_i)$ and $c_i d_i = cd(\text{mod } m_i)$.

### II.1.5    Numerical systems $Z_m$

For each positive integer $m$ consider the set of integers

$$Z_m = \{0, 1, 2, \ldots, m - 1\}$$

Addition modulo $m$ and multiplication modulo $m$ are defined as follows: For any $i, j$ in $Z_m$ we form the sum $i + j$ and we divide it by $m$. The remainder $R_m(i + j)$ is a uniquely defined element of $Z_m$. Thus we set $i + j = R_m(i + j)$. Likewise $ij = R_m(ij)$. It is easy to show that the above operations render $Z_m$ a commutative ring with identity. The identity elements of addition and multiplication modulo $m$ are 0 and 1, respectively. $Z_m$ is not an integral domain unless $m$ is prime. Indeed, if $m = pq$, then $pq = 0$.

The next theorem describes the invertible elements in $Z_m$. An element $a$ is invertible if the inverse $a^{-1}$ exists so that $aa^{-1} = a^{-1}a = 1$.

**Theorem II.3**    Let $m$ be a positive integer. The set of invertible elements in $Z_m$ is given by

$$\{r \in Z_m : \gcd(r, m) = 1\}$$

PROOF    Let $r$ be an invertible element. Then there is $s$ in $Z_m$ such that $rs = 1$. Since multiplication is modulo $m$, we have $rs = qm + 1$ or $rs - qm = 1$. Hence $r$, $m$ are relatively prime. Conversely, if $\gcd(r, m) = 1$, there are integers $k, j$ with

$rk + jm = 1$. If $k \in Z_m$, we conclude that $rk = 1$. If $k$ is not in $Z_m$, we divide it by $m$ to obtain $k = nm + s, s \in Z_m$. Thus $r(nm + s) + jm = 1$, or $rs + (rn + j)m = 1$, or $rs = 1, s \in Z_m$. Hence $s$ is the inverse of $r$.    ∎

As an example let $m = 15$. The set of invertible elements of $Z_{15}$ is

$$\{1, 2, 4, 7, 8, 11, 13, 14\}$$

## II.1.6 Fields

The structure of a ring enables us to add, subtract, and multiply. When we can also divide, we obtain the structure of a field. A commutative ring with identity $F$ is called a *field* if every nonzero element has an inverse. The set of nonzero elements of a field is a group with respect to multiplication. A field with a finite number of elements is called a *finite field* or *Galois field*. We easily infer that a field is an integral domain. Rational numbers, real numbers, and complex numbers with the ordinary operations are examples of fields. Let $p$ be a prime number. We saw that the ring $Z_p$ is an integral domain. Theorem II.3 implies that the nonzero elements of $Z_p$ are invertible and thus form a group with respect to multiplication. Thus $Z_p$ forms a field. In the sequel we will denote this field by $\mathrm{GF}(p)$. Addition and multiplication tables of $\mathrm{GF}(5)$ are shown below.

| + | 0 | 1 | 2 | 3 | 4 |
|---|---|---|---|---|---|
| 0 | 0 | 1 | 2 | 3 | 4 |
| 1 | 1 | 2 | 3 | 4 | 0 |
| 2 | 2 | 3 | 4 | 0 | 1 |
| 3 | 3 | 4 | 0 | 1 | 2 |
| 4 | 4 | 0 | 1 | 2 | 3 |

| · | 0 | 1 | 2 | 3 | 4 |
|---|---|---|---|---|---|
| 0 | 0 | 0 | 0 | 0 | 0 |
| 1 | 0 | 1 | 2 | 3 | 4 |
| 2 | 0 | 2 | 4 | 1 | 3 |
| 3 | 0 | 3 | 1 | 4 | 2 |
| 4 | 0 | 4 | 3 | 2 | 1 |

## II.1.7  Polynomial rings and Euclid's algorithm

Let $F$ be a field. A polynomial with coefficients in the field $F$ is an expression of the form

$$f(z) = f_0 + f_1 z + f_2 z^2 + \cdots + f_n z^n = \sum_{i=0}^{n} f_i z^i, \qquad f_i \in F$$

The greatest nonzero power is called *degree* of $f(z)$ and is denoted by $\deg f$. The sum of two polynomials of degree $n$ is defined in the usual way

$$f(z) + g(z) = \sum_{i=0}^{n} (f_i + g_i) z^i$$

Likewise the product of $f$ and $g$ of degrees $n$ and $m$ is

$$f(z)g(z) = \sum_{i=0}^{n+m} \left( \sum_{j=0}^{i} f_j g_{i-j} \right) z^i$$

The coefficients are added and multiplied by the operations of the field $F$. It holds that

$$\deg(f+g) \le \max(\deg f, \deg g), \quad \deg(fg) = \deg f + \deg g$$

The set of polynomials equipped with the above operations forms an integral domain. It is denoted by $F[z]$. As far as division is concerned, the polynomial ring behaves like the integer ring. Thus the polynomial $r(z)$ divides the polynomial $s(z)$, and we write $r(z)|s(z)$ if there exists polynomial $q(z)$ such that $s(z) = q(z)r(z)$. A nonzero polynomial that is divided only by the constant polynomials and itself is called *irreducible* or *prime* polynomial. We usually assume that the leading coefficient is one. The *greatest common divisor* of $r(z)$, $s(z)$ is denoted by $\gcd(r(z), s(z))$ and is given by the maximum degree polynomial that divides both of them and has leading coefficient equal to 1. If $\gcd(r(z), s(z)) = 1$, $r(z)$ and $s(z)$ are called *relatively prime* or *coprime*. The least common multiple $\operatorname{lcm}(r(z), s(z))$ is the minimum degree polynomial that is divided by both $r(z)$ and $s(z)$ and has leading coefficient equal to 1. Division of polynomials is described as follows: For any polynomial $s(z)$ and any nonzero polynomial $t(z)$, there exist uniquely defined polynomials $q(z)$ and $r(z)$ such that

$$s(z) = q(z)t(z) + r(z), \qquad \deg r(z) < \deg t(z) \tag{II.10}$$

where $q(z)$ is the quotient and $r(z)$ is the remainder of the division. The greatest common divisor can be obtained by successive divisions exactly as in the case of integers. Let $\deg s(z) \ge \deg t(z)$. We set

$$t_{n-2}(z) = q_n(z)t_{n-1}(z) + t_n(z), \quad t_{-1}(z) = s(z), \quad t_0(z) = t(z), \quad t_1(z) = r(z)$$

Since $\deg t_n < \deg t_{n-1}$, after a finite number of steps we obtain $t_{N+1}(z) = 0$. Then $\gcd(s(z), t(z)) = t_N(z)$. Proceeding backward, we obtain the important identity

$$\gcd(s(z), t(z)) = a(z)s(z) + b(z)t(z) \tag{II.11}$$

for suitable polynomials $a(z)$ and $b(z)$. Equation (II.11) expresses the greatest common divisor as a polynomial combination of $s(z)$ and $t(z)$. The special case of relatively prime polynomials is particularly interesting. We have the following theorem:

**Theorem II.4**    Two polynomials $a(z)$ and $b(z)$ of degrees $n_a$, $n_b$, $n_b \le n_a$, are relatively prime if and only if there exist polynomials $x(z)$, $y(z)$ such that

$$x(z)a(z) + y(z)b(z) = 1 \tag{II.12}$$

with

$$\deg x(z) < n_b, \quad \deg y(z) < n_a$$

If $a(z)$, $b(z)$ are relatively prime, the above polynomials $x(z)$, $y(z)$ are uniquely defined. Equation (II.12) is referred to as *Bezout identity*.

PROOF    We drop $z$ to simplify notation. Suppose that (II.12) holds. Since $\gcd(a, b) | ax$ and $\gcd(a, b) | by$, $\gcd(a, b) | 1$ and $\gcd(a, b) = 1$. Conversely, suppose that $a$, $b$ are relatively prime. There exist $\tilde{x}(z)$, $\tilde{y}(z)$ such that

$$\tilde{x}(z)a(z) + \tilde{y}(z)b(z) = 1 \tag{II.13}$$

Division of $\tilde{y}$ by $a$ gives $\tilde{y} = qa + r$, $\deg r < \deg a$. Equation (II.13) becomes $(\tilde{x} + bq)a + br = 1$. The polynomials $x = \tilde{x} + bq$ and $y = r$ are solutions of (II.12), and they satisfy $\deg y = \deg r < n_a$. Moreover $\deg(xa) = \deg(1 - yb) < n_b + n_a$. Therefore $\deg x < n_b + n_a - \deg a = n_b$. To prove uniqueness, we consider another pair of solutions $x_1$, $y_1$. Then

$$\frac{a}{b} = \frac{y - y_1}{x_1 - x}$$

But $\deg(y - y_1) < n_a$, $\deg(x_1 - x) < n_b$. Thus $a$, $b$ have a common divisor, which is a contradiction.    ∎

Suppose that $f(z)$ is a polynomial of degree $k$ with coefficients in the field $F$. The irreducible factorization of $f$ (see Appendix of Chapter 7) asserts that there exist unique irreducible polynomials $f_1, f_2, \ldots, f_m$ of degrees $d_i \geq 1$, respectively, integers $k_1, k_2, \ldots, k_m$, and a constant $A$ such that

$$f(z) = A f_1^{k_1}(z) f_2^{k_2}(z) \cdots f_m^{k_m}(z), \quad \sum_{i=1}^{m} d_i k_i = k \tag{II.14}$$

An element $a$ of the field $F$ is a *root* of $f(z)$ if $f(a) = 0$. A nonzero polynomial $f(z)$ with coefficients in $F$ has the element $a \in F$ as a root if and only if the polynomial $z - a$ divides $f(z)$.

**Chinese remainder theorem for polynomials.** The Chinese remainder theorem is readily extended to polynomial rings over fields. The statement of the theorem is identical to the corresponding theorem for integers, except for inequality $0 < c < M$ which is replaced by $\deg c(z) < \sum \deg m_i(z)$.

***Polynomial finite fields.*** In close analogy with the finite fields $GF(p)$, polynomial fields are constructed via addition and multiplication modulo an irreducible polynomial. The following important theorem is stated without proof:

**Theorem II.5** Let $F[z]$ be the polynomial ring over the finite field $F$. For any positive integer $m$ there exists an irreducible polynomial over $F$ of degree $m$. ∎

Let $F$ be a finite field with $N$ elements, $m$ a positive integer and $r(z)$ an irreducible polynomial of degree $m$. Let us consider the set of polynomials of degree at most $m - 1$ with coefficients in the field $F$ and with operations modulo $r(z)$

$$f(z) + g(z) = R_{r(z)}(f(z) + g(z)), \quad f(z)g(z) = R_{r(z)}(f(z)g(z))$$

Thus the sum of $f(z)$ and $g(z)$ is obtained as the remainder of the division of the ordinary sum $f(z) + g(z)$ by $r(z)$. The product is similarly defined. It follows, as in the case of integers, that the above operations define a field. We denote this field by $F[z]/r(z)$. It is a finite field with $N^m$ elements. If $F$ coincides with $GF(p)$ the polynomial field has $p^m$ elements. Thus for any positive integer $m$ and any prime $p$ we can construct a Galois field of order $p^m$. We show in the next section that every Galois field arises this way.

## II.2  STRUCTURE OF FINITE FIELDS

### II.2.1  Exponential representation of finite fields

In the sequel $F$ denotes a finite field with $N$ elements and $a$ is a nonzero element of $F$ of order $n$.

**Lemma 1**  If $n > 1$ and $m \geq n$, then $a^m = 1$ if and only if $n|m$.

**PROOF**  If $m = kn$, then $a^m = a^{kn} = (a^n)^k = 1$. Conversely, we divide $m$ by $n$ to find $m = qn + r$, $0 \leq r < n$. Then $1 = a^m = (a^n)^q a^r = a^r$, which is a contradiction unless $r = 0$. ∎

**Lemma 2**  If $n = n_1 n_2$, then $a^{n_1}$ has order $n_2$.

**PROOF**  Indeed, $(a^{n_1})^{n_2} = a^{n_1 n_2} = a^n = 1$. Moreover, if $(a^{n_1})^k = 1$, then $n_1 k \geq n$ and $k \geq n_2$. ∎

**Lemma 3**  If the nonzero element $b$ has order $m$ and $\gcd(n, m) = 1$, then $ab$ has order $mn$.

**PROOF**  $(ab)^{mn} = a^{mn} b^{mn} = (a^n)^m (b^m)^n = 1$. Conversely, if $(ab)^k = 1$, then

$(ab)^{km} = 1$ and $a^{km}b^{km} = a^{km} = 1$. Therefore $n|km$ by Lemma 1, and since $\gcd(n,m) = 1$, $n|k$. Likewise $m|k$. Hence $mn|k$. ∎

**Lemma 4**   For any $k > 1$, $a^k$ has order $n/\gcd(n,k)$.

**PROOF**   Let $d = \gcd(n,k)$ and $l = n/d$. Then $(a^k)^l = (a^n)^{k/d} = 1$. Moreover, if $a^{km} = 1$, then $n|km$ and $l|(km/d)$. Since $l$ is relatively prime to $k/d$, it divides $m$. ∎

The element $a$ is called *primitive* if the order of $a$ is $N - 1$; that is, $n$ coincides with the number of nonzero elements of $F$. In this case the exponential signal $a^k$, $k \geq 0$ has maximal period $N - 1$. Consider, for example, the Galois field $GF(3) = \{0,1,2\}$. The order of 1 is 1, since $1^1 = 1$. The order of 2 is 2, since $2^1 = 2$, $2^2 = 1$. Hence 2 is a primitive element.

*Remark 1.*   The order of any element of $F$ is a divisor of $N - 1$. This is a direct consequence of the Lagrange theorem. Indeed, the order of an element $a$ equals the order of the cyclic subgroup generated by $a$, and the latter is a subgroup of the multiplicative group formed by the nonzero field elements.

The following fundamental result states that a finite field always has a primitive element. Hence it is generated by 0 and the $N - 1$ consecutive powers of a primitive element.

**Theorem II.6**   Every finite field with $N$ elements has a primitive element. Thus it admits the representation

$$F = \{0, a^0 = 1, a, a^2, \ldots, a^{N-2}\}$$

**PROOF**   Let $n$ be the maximum of the orders of the nonzero elements of $F$, and $a$ an element in $F$ of order $n$. By the previous remark $n$ is a divisor of $N - 1$. Let $b$ be a nonzero element of $F$ different from $a$ of order $r$. Each prime factor of $r$ is also a prime factor of $n$. Indeed, if $r_2$ is a prime factor of $r$, $r = r_1 r_2$, then the greatest common divisor of $n$ and $r_2$ is either 1 or $r_2$. If $\gcd(n,r_2) = 1$, $b^{r_1}$ has order $r_2$ (Lemma 2) and $ab^{r_1}$ has order $nr_2$ (Lemma 3). Thus $ab^{r_1}$ has order greater than $n$, which contradicts the maximality of $n$. Thus $\gcd(n,r_2) = r_2$. Let $p$ be an arbitrary prime factor of $r$ that appears $t$ times in $r$ and $s$ times in $n$. Then $n = p^s n_1$, $r = p^t r_1$, and $\gcd(p,n_1) = \gcd(p,r_1) = 1$. By Lemma 2, $a^{p^s}$ has order $n_1$. Likewise $b^{r_1}$ has order $p^t$. From Lemma 3 we deduce that $a^{p^s} b^{r_1}$ has order $p^t n_1$. Since $n = p^s n_1 \geq p^t n_1$, we have $p^s \geq p^t$. We have proved that $r|n$. Therefore the order of every nonzero element of $F$ divides $n$; that is, $b^n = 1$. This means that the polynomial of degree $n$, $z^n - 1$ has at least $N - 1$ roots in $F$. Thus $N - 1 \leq n$ and $n = N - 1$. The proof is complete. ∎

*Example II.2   Primitive elements in GF($2^2$) and GF($2^3$)*

Let us determine an irreducible polynomial of degree 2 over $GF(2) = \{0, 1\}$. This will have the form $z^2 + bz + c$. Since neither 0 nor 1 is a root, the only possibility is $r(z) = z^2 + z + 1$. The corresponding polynomial field $F[z]/r(z)$ consists of the polynomials $\{0, 1, z, z + 1\}$ and the operations modulo $r(z)$. The polynomial $z$ is a primitive element because $z^2 = z + 1 \pmod{z^2 + z + 1}$ and $z^3 = 1 \pmod{z^2 + z + 1}$. Likewise $z + 1$ is also a primitive element.

Let us next consider irreducible polynomials of degree 3 over $GF(2)$. These will have the form $z^3 + bz^2 + cz + 1$ because 0 is not a root. $b$ and $c$ cannot both be 0 or 1, since then 1 would be a root. The remaining choices lead to the polynomials $z^3 + z^2 + 1$ and $z^3 + z + 1$. Both polynomials cannot have a linear factor by the preceding analysis. They cannot have a second-degree factor because, if they did, they would have a linear factor as well. Thus both polynomials are irreducible, and they are the only irreducible polynomials of degree 3 over $GF(2)$. Let $r(z) = z^3 + z + 1$. The finite field $F[z]/r(z)$ consists of all polynomials of degree 2 with coefficients 0 and 1,

$$0, 1, z, z + 1, z^2, z^2 + 1, z^2 + z, z^2 + z + 1$$

and the polynomial operations modulo $z^3 + z + 1$. The possible orders of the field elements are the divisors of 7 (see Theorem II.1). Since 7 is prime, all elements of the field different from 1 are primitive.   ∎

Motivated by the last example we give the following definition. A positive integer $m$ is called *Mersenne prime* if the number $2^m - 1$ is prime. The numbers 2, 3, 5, 7 are Mersenne primes. Let $F = \{0, 1\}$. The polynomial field $F[z]/r(z)$ over any irreducible polynomial of degree $m$ consists of $2^m$ elements. If $m$ is Mersenne prime, $2^m - 1$ is prime, and all nonconstant polynomials have order $2^m - 1$. Therefore nonconstant polynomials are primitive.

*Remark 2.* In a field $F$ of $N$ elements, the polynomial $z^N - z$ factors as $z(z^{N-1} - 1)$. The polynomial $z^{N-1} - 1$ has at most $N - 1$ roots in $F$. Moreover the order of every nonzero element of $F$ is a divisor of $N - 1$. Lemma 1 implies that every nonzero element of $F$ is a root of $z^{N-1} - 1$. Thus the following factorization holds:

$$z^N - z = \prod_{b \in F}(z - b) = z \prod_{k=0}^{N-2}(z - a^k)$$

where $a$ is a primitive element of $F$.

The Euler function of an integer $m$, $\phi(m)$ is defined as the number of invertible elements of $Z_m$. According to Theorem II.3, $\phi(m)$ is alternatively given by the number of positive integers that are relatively prime to $m$. We urge the reader to prove the formula

$$\phi(n) = n \prod_{\substack{p|n \\ p \text{ prime}}} \left(1 - \frac{1}{p}\right)$$

For any integer $m \geq 1$, the set of invertible elements in $Z_m$ forms a multiplicative group with $\phi(m)$ elements. The order of any invertible element equals the order of the cyclic subgroup generated by the element and hence divides $\phi(m)$ by the Lagrange theorem. This proves the following important theorem:

**Theorem II.7 Fermat's theorem** For any integer $m \geq 1$ and any integer $1 \leq a \leq m - 1$ such that $\gcd(a, m) = 1$, it holds that

$$a^{\phi(m)} = 1 (\text{mod } m) \qquad \blacksquare$$

Let $a$ be a primitive element of $F$. Then $a$ has order $N - 1$. By Lemma 4, every nonzero element of $F$ is written as $a^k$ and has order $(N - 1)/\gcd(N - 1, k)$. Clearly $a^k$ will be primitive if $\gcd(N - 1, k) = 1$. Thus the number of primitive elements in $F$ is given by $\phi(N - 1)$, where $\phi$ is the Euler function.

*Example II.3    Primitive elements in GF($2^4$)*
Consider the field of polynomials of degree at most 3 with coefficients in $F = \{0, 1\}$ and with operations the standard polynomial operations modulo an irreducible polynomial of degree 4. An irreducible polynomial of degree 4 will have the form $z^4 + az^3 + bz^2 + cz + 1$. Linear factors are excluded by requiring that $a + b + c = 1$. Then factors of degree 3 are ruled out. The only possibility are factors of degree 2. These factors must be irreducible. By the preceding discussion, the only such polynomial is $z^2 + z + 1$. Thus the irreducible polynomial cannot be of the form $(z^2 + z + 1)^2 = z^4 + z^2 + 1$. Hence $(a, b, c) \neq (0, 1, 0)$. The only polynomials that satisfy the above constraints are

$$z^4 + z + 1, \quad z^4 + z^3 + 1, \quad z^4 + z^3 + z^2 + z + 1$$

We take $r(z) = z^4 + z + 1$. Then $F[z]/r(z)$ consists of all polynomials of degree at most 3 with coefficients 0 or 1. In this case $N = 16$. The divisors of 15 are 1, 3, 5, 15. These are exactly the numbers that can be obtained as orders of elements of the field. The number of primitive elements is 8. Indeed, $\phi(15) = 15(1 - 1/3)(1 - 1/5) = 8$. We claim that the polynomial of degree 1, $a = z$ is a primitive element in the above field. To establish the claim, we show that the order of $a$ cannot be 1, 3, or 5. Indeed, $a^3 = z^3 \neq 1$. Furthermore $a^5 = z^5 = z^2 + z$ (mod $r(z)$), and $z^2 + z \neq 1$. Thus the order of $a$ is 15 and $a$ is primitive. There are seven more primitive elements, and the reader is urged to find them. $\blacksquare$

### II.2.2  Field integers

Let $F$ be a finite field with $N$ elements. We consider the field elements $0, 1, 1 + 1,$ $1 + 1 + 1, \ldots$, that is, the cyclic additive subgroup generated by 1. The elements of this set are denoted by $n \cdot 1, n \in Z$, and are called *field integers*. Since $F$ is finite, its integers form a finite set. The number of field integers is designated by $p$ and is referred to as the *characteristic* of the field. According to the discussion of cyclic subgroups, $p$ is the smallest positive integer $m$ with the property $m \cdot 1 = 0$. The characteristic of a finite field is prime. Indeed, let $p = mn$, with $m, n \neq 1$. Then

$$0 = \underbrace{1 + 1 + 1 + \cdots + 1}_{p \text{ times}} = \underbrace{(1 + 1 + 1 + \cdots + 1)}_{m \text{ times}} \underbrace{(1 + 1 + 1 + \cdots + 1}_{n \text{ times}}$$

Since a field is an integral domain, either $m \cdot 1 = 0$ or $n \cdot 1 = 0$, contradicting minimality of $p$. Thus the field integers are given by the set

$$\{0, 1, 2 \cdot 1, \ldots, (p - 1) \cdot 1\}$$

It is easy to show that they form a subfield of $F$; that is, they contain 0 and 1 and are closed under all field operations. Furthermore they behave isomorphically to $GF(p)$. In the sequel we will write $n$ instead of $n \cdot 1$.

**Lemma 5**  Let $F$ be a finite field of characteristic $p$, $a_1, a_2, \ldots, a_k$, field elements and $m$ positive integer. Then

$$\left( \sum_{i=1}^{k} a_i \right)^{p^m} = \sum_{i=1}^{k} a_i^{p^m} \tag{II.15}$$

PROOF   Let $k = 2$ and $m = 1$. Then

$$(a_1 + a_2)^p = \sum_{j=0}^{p} \binom{p}{j} a_1^j a_2^{p-j} = \sum_{j=0}^{p} \frac{p!}{j!(p-j)!} a_1^j a_2^{p-j}$$

and $\binom{p}{j}$ is a field integer. Since $j!(p-j)!$ divides $p! = p(p-1)!$ and $p$ is prime, $j!(p-j)!$ divides $(p-1)!$. Hence $\binom{p}{j}$ is a multiple of $p$; namely $\binom{p}{j} = 0(\text{mod } p)$ for $j = 1, 2, \ldots, p - 1$. Consequently $(a_1 + a_2)^p = a_1^p + a_2^p$. For $m = 2$ we have $[(a_1 + a_2)^p]^p = [a_1^p + a_2^p]^p = a_1^{p^2} + a_2^{p^2}$. The general case follows by induction.  ∎

**Theorem II.8**  In a field $F$ of characteristic $p$, a field element is an integer if and only if it satisfies the equation $z^p - z = 0$; that is, it is a root of the polynomial $z^p - z$.

PROOF   Let $k$ be a field integer. Lemma 5 implies that

$$k^p = (\underbrace{1 + 1 + \cdots + 1}_{k \text{ times}})^p = \sum_{i=1}^{k} 1^p = \sum_{i=1}^{k} 1 = k$$

To prove the converse statement, we simply note that the polynomial $z^p - z$ has at most $p$ roots in $F$.   ∎

*Remark 3.* If a field has $N$ elements and characteristic $p$, then $p|N$. Indeed, the field integers viewed as an additive subgroup of the field has an order that is a divisor of $N$ by the Lagrange theorem. It follows that if $N$ is prime, then $p = N$. In this case the field integers cover the entire field, which thus becomes isomorphic to $GF(p)$.

It is easy to show that in the polynomial field $F[z]/r(z)$ with $F = GF(p)$, the field integers are exactly the polynomials of zero degree, namely $GF(p)$. Consequently the field characteristic is $p$.

### II.2.3   Minimal polynomials

Let $F$ be a finite field with $N$ elements and characteristic $p$. Let $F_0$ be a subfield of $F$. Clearly $F$ and $F_0$ have the same field integers. Given an element $b$ in $F$, the *minimal polynomial* of $b$ over $F_0$ is denoted by $\phi_b(z)$ and is defined as follows:

1. The coefficients of $\phi_b(z)$ lie in $F_0$, and the leading coefficient is 1.
2. $\phi_b(z)$ has $b$ as a root; that is, $\phi_b(b) = 0$.
3. It is the lowest-degree polynomial with the above properties.

Thus any other polynomial vanishing at $b$ with coefficients in $F_0$ and leading coefficient 1 has degree greater than the degree of $\phi_b(z)$.

Each element $b$ of $F$ has a unique minimal polynomial. Indeed, $b$ is a root of $z^N - z$. This polynomial has leading term 1, and coefficients 0, 1 and $-1$. It follows that the set of polynomials satisfying properties 1 and 2 is nonempty and finite. Thus the minimal polynomial exists. If $\phi_1(z)$ also has the above properties, then $\deg \phi_b(z) = \deg \phi_1(z)$. The polynomial $\phi_b(z) - \phi_1(z)$ has coefficients in $F_0$, is zero at $b$, and has degree less than the degree of $\phi_b(z)$. This contradicts minimality of $\phi_b(z)$ unless $\phi_1 = \phi_b$.

Next we state some useful properties of minimal polynomials.

**Property 1**   The minimal polynomial $\phi_b(z)$ is irreducible over $F_0$.
Indeed, let $\phi_b(z) = \phi_1(z)\phi_2(z)$, where none of the factors is constant. Then either $\phi_1(b) = 0$, or $\phi_2(b) = 0$, contradicting minimality.

**Property 2**   If a polynomial $f(z)$ with coefficients in $F_0$ has $b$ as a root, then $\phi_b(z)$ divides $f(z)$.

Indeed, division gives $f(z) = q(z)\phi_b(z) + r(z)$, $\deg r(z) < \deg \phi_b(z)$. Now $f(b) = 0 = r(b)$. Minimality forces $r(z)$ to be the zero polynomial.

The following two properties follow directly from property 2.

**Property 3** $\phi_b(z)$ divides $z^N - z$.

**Property 4** If $f(z)$ is an irreducible polynomial with integer coefficients that vanishes at $b$, then $f(z) = \phi_b(z)$.

To prove the last and most important property, we will need the following two lemmas:

**Lemma 6** Let $f(z)$ be a polynomial with integer coefficients, and let $f(b) = 0$. Then for each positive integer $n, f(b^{p^n}) = 0$. Thus, besides $b$, all field elements of the form $b^p$, $b^{p^2}$, ... are roots of $f(z)$.

**PROOF** Let $f(z) = \sum_{i=0}^{k} f_i z^i$. Then

$$[f(b)]^p = \left[\sum_{i=0}^{k} f_i b^i\right]^p = \sum_{i=0}^{k} f_i^p b^{ip}.$$

Since $f_i$ are field integers, Theorem II.8 implies that

$$[f(b)]^p = \sum_{i=0}^{k} f_i b^{ip} = f(b^p).$$

Hence $b^p$ is a root of $f(z)$. The general case follows in a similar fashion. ∎

Motivated by the above lemma, it is reasonable to examine if the polynomial with roots $b^{p^n}$ yields the minimal polynomial of $b$ over the field integers. This is in fact true and is established after the following lemma:

**Lemma 7** There exists $m \geq 1$ such that $b^{p^m} = b$.

**PROOF** If $p = N$, then $m = 1$ because $b$ is an integer, and by theorem 11.8, it satisfies $b^p = b$. Suppose that $p \neq N$. It suffices to show that there exists $m$ such that $p^m = 1 (\mathrm{mod}(N - 1))$ or $p^m = 1 + k(N - 1)$. Indeed, in this case $b^{p^m} = b^{1+k(N-1)} = b(b^{(N-1)k}) = b$. By the Fermat theorem, $\phi(N - 1)$ does the job. ∎

The smallest integer $m \geq 1$ such that $b^{p^m} = b$ is called *degree of b*.

**Property 5** Let $F_0$ be the field integers of $F$. The minimal polynomial of $b$ over $F_0$ has degree equal to the degree of $b$ and is determined by its roots $b, b^p$,

$b^{p^2}, \ldots, b^{p^{m-1}}$ :

$$\phi_b(z) = \prod_{i=0}^{m-1} (z - b^{p^i}) \tag{II.16}$$

**PROOF**   Let $f(z) = \prod_{i=0}^{m-1}(z - b^{p^i})$. It holds that $f(b) = 0$. We show that the coefficients of $f(z)$ are field integers. Using Lemma 5 we obtain

$$[f(z)]^p = \prod_{i=0}^{m-1}(z - b^{p^i})^p = \prod_{i=0}^{m-1}(z^p - b^{p^{i+1}}) = \prod_{i=1}^{m}(z^p - b^{p^i}) = \prod_{i=1}^{m-1}(z^p - b^{p^i})(z^p - b)$$

Thus $[f(z)]^p = \prod_{i=0}^{m-1}(z^p - b^{p^i}) = f(z^p)$. Let $f(z) = \sum f_i z^i$. Then $[f(z)]^p = \sum f_i^p z^{ip} = \sum f_i^p (z^p)^i$. Furthermore $f(z^p) = \sum f_i z^{pi}$. Thus $f_i = f_i^p$ and the coefficients of $f(z)$ are field integers according to Theorem II.8.

Next we observe that $f(z)$ is irreducible over the field of integers. Indeed, if $f(z) = f_1(z)f_2(z)$, one of the factors, say, $f_1(z)$ vanishes at $b$. It follows from Lemma 6 that $f_1(z)$ vanishes at $b^{p^n}$ and consequently includes all roots of $f(z)$. Thus $f(z)$ is irreducible and, by Property 4, coincides with the minimal polynomial of $b$.    ∎

According to Property 5 the elements $b^p, b^{p^2}, \ldots, b^{p^{m-1}}$ have the same minimal polynomial. They are called *conjugate elements* in close analogy with the field of complex numbers where complex conjugate numbers are obtained as roots of a second-degree polynomial with real coefficients. If $a$ is a primitive element of $F$, then $b = a^j$ for some $j$. The exponents of the conjugate elements of $b$, $j, jp, jp^2, jp^3, \ldots, jp^{m-1}$ determine the *cyclotomic class of $j$*.

### Example II.4   Minimal polynomials in GF(2⁴)

Let us consider the field produced by the polynomials of degree at most 3 with coefficients 0 or 1, and the polynomial operations modulo the irreducible polynomial $r(z) = z^4 + z + 1$. Let us determine the minimal polynomial of $a = z$ using property 5. The characteristic of the field is 2. The cyclotomic class of 1 consists of the integers $1, 2, 4, 8$. (Note $2^4 = 16 = 1 \pmod{15}$.) The conjugate roots of $a$ are $a^2 = z^2$, $a^4 = z^4 = z + 1 \pmod{r(z)}$, $a^8 = z^8 = z^2 + 1 \pmod{r(z)}$. In the above computation we assigned $a^k = z^k$ to the remainder of the division of $z^k$ by $r(z)$. An alternative way is based on the remark that $a$ is a root of $r(z)$. Indeed, $r(a) = a^4 + a + 1 = z^4 + z + 1 = 0 \pmod{r(z)}$. Therefore $a^4 = a + 1$, $a^5 = aa^4 = a(a + 1) = a^2 + a$, $a^6 = aa^5 = a(a^2 + a) = a^3 + a^2$, $a^7 = aa^6 = a(a^3 + a^2) = a^4 + a^3 = a + 1 + a^3$, and so forth. We found that the conjugate roots of $a = z$ are the polynomials $z^2$, $z + 1$, and $z^2 + 1$. Since $r(z)$ is irreducible over $GF(2)$ and has $a$ as a root, it follows from Property 4 that it is the minimal polynomial of $a$. We can verify the factorization anticipated by Property 5:

$$z^4 + z + 1 = (z - a)(z - a^2)(z - a^4)(z - a^8)$$

Let us next compute the minimal polynomial of $a^3$. The conjugate roots are $a^3$, $a^6$, $a^{12}$, $a^{24} = a^9$. According to Property 5, the minimal polynomial is given by

$$\phi_{a^3}(z) = (z + a^3)(z + a^6)(z + a^{12})(z + a^9) = z^4 + z^3 + z^2 + z + 1$$

An alternative procedure is described later.                                      ∎

### II.2.4   Polynomial representation of finite fields

In this section we summarize some important results concerning finite fields. We establish the existence of an element $b$ in $F$ so that every field element is expressed as a polynomial in $b$ of degree smaller than the degree of $b$. We will see that the field order is necessarily a power of its characteristic and that two fields with the same order are necessarily isomorphic. As before, $F$ stands for a finite field of order $N$ and characteristic $p$.

**Theorem II.9**   Let $b$ in $F$ of degree $m$. The set of polynomial in $b$ expressions with integer coefficients

$$f_0 + f_1 b + f_2 b^2 + \cdots + f_{m-1} b^{m-1} = f(b), \qquad \deg f(z) \leq m - 1$$

forms a subfield of $F$ of characteristic $p$ and order $p^m$.

**PROOF**   The above representation defines a unique element. Indeed, if $f(b) = \tilde{f}(b)$, $b$ would be a root of the difference $f(z) - \tilde{f}(z)$, which has degree smaller that $m$. Since $m$ is the degree of $b$, it equals the degree of the minimal polynomial of $b$. We have thus found a polynomial vanishing at $b$ of order smaller than $m$, a contradiction. Therefore the above set has exactly $p^m$ elements. It is easy to see that it is a group with respect to addition. To prove that it is closed with respect to multiplication, take two elements $f(b)$, $g(b)$. Let $h(z) = f(z)g(z)(\bmod \phi_b(z))$, the product modulo the minimal polynomial of $b$. Then $\deg h(z) \leq m - 1$ and $h(b) = f(b)g(b)$ because $\phi_b(b) = 0$. To determine the inverse of $f(b)$, we consider the equation $f(z)g(z) = 1(\bmod \phi_b(z))$, namely we lift the construction to the inverse of $f(z)$ modulo $\phi_b(z)$. This is equivalent to finding $g(z)$ and $r(z)$ such that $f(z)g(z) + r(z)\phi_b(z) = 1$. This equation is always solvable because $\phi_b(z)$ is prime and $\deg f(z) < \deg \phi_b(z)$. Therefore $f(z)$ and $\phi_b(z)$ are relatively prime. If the degree of $g(z)$ exceeds $m - 1$, we divide $g(z)$ by $\phi_b(z)$. The remainder gives the desired answer.                                      ∎

**Theorem II.10**   The order of a finite field is a power of its characteristic, that is, $N = p^m$ for some positive integer $m$.

**PROOF**   Let $a$ be a primitive element of the field. Let $m$ be the degree of $a$, that is, the degree of the minimal polynomial of $a$. We have $a^{p^m} = a$. Hence $a^{p^m - 1} = 1$. Thus the order of $a$, $N - 1$ is not greater than $p^m - 1$ and $N \leq p^m$. Conversely,

Theorem II.9 implies that the set of polynomials in $a$ of degree less than $m$ is a subfield of order $p^m$. Therefore $p^m \leq N$ and the claim is proved.    ∎

The above argument proves the following:

**Theorem II.11**    In a field of order $p^m$ every primitive element $a$ has degree $m$; that is, the minimal polynomial of $a$ has degree $m$.    ∎

**Theorem II.12**    Let $a$ be an element of degree $m$ in a field $F$ of order $p^m$. Then

$$F = \{f_0 + f_1 a + f_2 a^2 + \cdots + f_{m-1} a^{m-1} : f_i \quad \text{field integers}\}$$    ∎

We complete the theoretical discussion of finite fields with the following theorem which states that a finite field of order $p^m$ is essentially unique:

**Theorem II.13**    Two finite fields with the same number of elements are isomorphic. In particular, a field of order $p^m$ is isomorphic to the polynomial field over $GF(p)$ with operations modulo an irreducible polynomial of degree $m$.

PROOF    By Theorem II.10, the order of the two fields $F$, $\tilde{F}$, is of the form $p^m$. Let $a$ and $\tilde{a}$ be two primitive elements in $F$ and $\tilde{F}$, respectively. To construct the desired isomorphism $A : F \rightarrow \tilde{F}$, we use the polynomial representation of the fields. Each element of $F$ has the form $f(a)$, where $f(z)$ is a polynomial of degree at most $m - 1$. The integers of the two fields are both isomorphic to $GF(p)$. Hence they can be identified by the isomorphism $A_0 : n \rightarrow n$. The required field isomorphism $A$ is obtained by lifting $A_0$ in the obvious way

$$A : f_0 + f_1 a + f_2 a^2 + \cdots + f_{m-1} a^{m-1} \rightarrow \tilde{f}_0 + \tilde{f}_1 \tilde{a} + \cdots + \tilde{f}_{m-1} \tilde{a}^{m-1}$$

where $\tilde{f}_i = A_0(f_i)$. The rest of the proof is left to the reader.    ∎

*Remark 4*    Equation (II.16) refers to minimal polynomials over the field integers. Analogous representations are valid for minimal polynomials over an arbitrary subfield $F_0$ of $F$. Suppose that $F_0$ has $N_0 = p^{m_0}$ elements. It follows that if a polynomial in $F_0$ has a root $b$ in $F$ all elements of the form $b^{N_0^n}$, $n = 1, 2, \ldots$, are also roots. The proof is exactly the same with that of Lemma 6. The minimal polynomial of $b$ with coefficients in $F_0$ is given by

$$\phi_b(z) = (z - b)(z - b^{N_0})(z - b^{N_0^2}) \cdots (z - b^{N_0^{m-1}})$$

*Remark 5.*    Every irreducible polynomial over the field $F$, $r(z)$ can be factored into a product of linear factors over an extension field. Indeed, consider the polynomial field $F[z]/r(z)$. This is identical to $GF(N^k)$, $N$ is the order of the field $F$ and $k$ the degree of $r(z)$. In this field $r(z)$ has exactly $k$ distinct roots, of the form $a, a^N, \ldots, a^{N^{k-1}}$. The field $F[z]/r(z)$ is called a *splitting field* of $r(z)$ because it is an extension of $F$ in which $r$ factors into linear terms. An analogous conclusion holds for a general polynomial $f(z)$ over $F$. Each irreducible factor of $f$ of degree $d_i$ is decomposed into linear factors in $GF(N^{d_i})$. Embedding these fields in an extension field, we conclude that $f(z)$ is also decomposed into linear factors.

### II.2.5  Primitive polynomials

According to Theorem II.13 every finite field with $p^m$ elements is realized by the polynomial field $F[z]/r(z)$. The irreducible polynomial $r(z)$ has degree $m$. The element $a = z$ of $F[z]/r(z)$ has degree $m$. Indeed, $a$ is a root of $r(z)$, since $r(a) = r(z) = 0 \pmod{r(z)}$. Second, $r(z)$ has integer coefficients, and by assumption it is irreducible over the integers. Thus, by Property 4, $r(z)$ is the minimal polynomial of $a = z$. This implies that $a$ has degree $m$, and Theorem II.11 applies. Of course any other root of $r(z)$ (conjugate root of $a$), such as $z^2$, can also be used for the polynomial representation.

Can $a = z$ or any other conjugate root of $a$ be used for the exponential representation of the field? In other words, can the same element be simultaneously used for the exponential and the polynomial representation? The answer is: not always. We have already seen that a primitive element has degree $m$ and thus reproduces both representations. Conversely, an element of degree $m$ is not necessarily primitive. Take, for instance, $GF(2^4)$. The polynomial $r(z) = z^4 + z^3 + z^2 + z + 1$ is irreducible over $GF(2)$. Thus $GF(2^4)$ can be constructed using the above polynomial for the modulo $r(z)$ operations. Let $a$ be a root of $r(z)$, such as $a = z$. The relation $r(a) = 0$ gives $a^4 = a^3 + a^2 + a + 1$, $a^5 = a^4 + a^3 + a^2 + a = 1$. Thus the order of $a$ is 5, and hence $a$ is not primitive. We conclude that a root of an irreducible polynomial can generate the polynomial representation of a field but not the exponential representation, since it can fail to be primitive. The extra feature that an irreducible polynomial must possess in order to guarantee that its roots are primitive elements is offered by the so-called primitive polynomials.

*Definition.* An irreducible polynomial $r(z)$ of degree $m$ over $GF(p)$ is called *primitive*, if $p^m - 1$ is the smallest positive integer $n$ with the property $r(z) | z^n - 1$.

Suppose that $r(z)$ is primitive and that $a$ is a root of $r(z)$. Then $a$ is a primitive element. Indeed, by Property 4, $r(z)$ is the minimal polynomial of $a$. Now, if the order of $a$ is $n$, then $a^n = 1$ and $a$ is a root of $z^n - 1$. By Property 3, $r(z) | z^n - 1$. The above definition implies that $n = p^m - 1$, and hence $a$ is primitive. Conversely, if $a$ is primitive, the minimal polynomial of $a$ is primitive.

If $F[z]/r(z)$ is built upon a primitive polynomial $r(z)$, any root $a$ of $r(z)$ (e.g., $a = z$) will also yield the exponential representation. The correspondence between powers of $a$ and polynomials in $a$ is readily inferred from the equation $r(a) = 0$. Consider, for instance, $GF(2^4)$. The two representations are illustrated below. The primitive polynomial chosen is $r(z) = z^4 + z + 1$. $a$ satisfies the equation $a^4 = a + 1$. Hence $a^5 = a^2 + a$, $a^6 = a^3 + a^2$, and so forth.

$$\begin{pmatrix} 1 & a & a^2 & a^3 & a^4 & a^5 & a^6 & a^7 & a^8 & a^9 \\ 1 & a & a^2 & a^3 & a+1 & a^2+a & a^3+a^2 & a^3+a+1 & a^2+1 & a^3+a \end{pmatrix}$$

$$\begin{pmatrix} a^{10} & a^{11} & a^{12} & a^{13} & a^{14} \\ a^2+a+1 & a^3+a^2+a & a^3+a^2+a+1 & a^3+a^2+1 & a^3+1 \end{pmatrix}$$

### *Example II.5    Computation of minimal polynomials*

The minimal polynomial of an element $b$ over the field integers can be constructed by the factorization (II.16). An alternative method is described next through an example. Take $GF(2^4)$ spanned by $r(z) = z^4 + z + 1$ and present the field via the exponential and polynomial representation of a root $a$ of $r(z)$. Let us determine the minimal polynomial of $a^3$. The degree of $a^3$ cannot exceed $m = 4$ (why?). Hence the desired polynomial has the form $\phi(z) = 1 + f_1 z + f_2 z^2 + f_3 z^3 + f_4 z^4$. To determine the integer coefficients, we evaluate the polynomial at $a^3$: $\phi(a^3) = 1 + f_1 a^3 + f_2 a^6 + f_3 a^9 + f_4 a^{12}$. We convert the powers of $a$ into their corresponding polynomial in $a$ expressions. Then

$$0 = 1 + f_1 a^3 + f_2(a^2 + a^3) + f_3(a + a^3) + f_4(1 + a + a^2 + a^3)$$

or

$$0 = (1 + f_4) + a(f_3 + f_4) + a^2(f_2 + f_4) + a^3(f_1 + f_2 + f_3 + f_4)$$

Uniqueness of the polynomial representation forces all coefficients to be zero. The resulting linear system of equations is readily solved. The solution is $f_4 = 1$, $f_2 = 1, f_3 = 1, f_1 = 1$. The minimal polynomial is $z^4 + z^3 + z^2 + z + 1$.

In general the resulting linear system may have more than one solution. The minimal polynomial is the unique solution of minimum degree. Take, for example, the element $a^5$. Proceeding as above, we find that

$$1 + f_1 a^5 + f_2 a^{10} + f_3 a^{15} + f_4 a^{20} = 1 + f_3 + (f_1 + f_4)(a^2 + a) + f_2(a^2 + a + 1) = 0$$

Therefore

$$1 + f_2 + f_3 = 0$$
$$f_1 + f_2 + f_4 = 0$$

The above system has 4 solutions. These are given in polynomial form by

$$1 + z^3, \quad 1 + z + z^3 + z^4, \quad 1 + z + z^2, \quad 1 + z^2 + z^4$$

The lowest-degree polynomial is $1 + z + z^2$ and yields the minimal polynomial of $a^5$. In conformity with Property 2, it divides the remaining 3 polynomials.  ∎

## II.2.6   Summary

Every finite field has necessarily $p^m$ elements, $p$ is prime, and $m$ is a positive integer. No field with $18 = 2 \cdot 3^2$ elements exists. If $m = 1$, the field can be identified with the field $GF(p) = \{0, 1, \ldots, p - 1\}$ and the operations modulo $p$. If $m > 1$, the field is identified with the polynomial field of all polynomials of degree at most $m - 1$ having coefficients in $GF(p)$ and with the polynomial operations modulo an irreducible polynomial $r(z)$ of degree $m$. Such irreducible polynomials always exist. Note that a field such as $GF(2^4)$ of order 16 has nothing to do with the ring $Z_{16}$. The elements of these sets can be represented the same way (e.g., as quadruples), but the algebraic structures are completely different. As we have pointed out, the second is not a field, since 16 is not a prime.

The algebraic structure of a field is nicely exposed through the exponential representation and the polynomial representation. The first is convenient when multiplications are performed because $a^i \cdot a^j = a^{i+j}$. The second fits better with additions, since $f(a) + g(a) = (f + g)(a)$.

Construction of a finite field can be built upon the polynomial field $F[z]/r(z)$. By far the most interesting case is the binary case, $p = 2$. The irreducible polynomial $r(z)$ is usually determined from tables. Lists of irreducible polynomials of high degrees (up to 10,000) are available. The exponential and the polynomial representation are achieved simultaneously if $r(z)$ is a primitive polynomial. In the latter case, a root $a$ of $r(z)$ provides the exponential representation by its consecutive powers and the polynomial representation through $r(a) = 0$ or $a^m = \sum_{i=0}^{m-1} r_i a^i$. Extensive lists of primitive polynomials are available.

## BIBLIOGRAPHICAL NOTES

The material of this appendix draws upon Berlekamp (1984). The reader interested in theory is referred to Van der Waerden (1949).

# APPENDIX III

# ALGORITHMS

An algorithm can be viewed as a transformation of the form $\mathcal{A} : \ell \times F^m \to \ell$, where $F$ is a field. As the domain of the transform indicates, input data are divided in two types: the input sequence and the initial state. When a given sequence $u(n)$ and a given initial state $\mathbf{x}_0$ are presented, the algorithm $\mathcal{A}$ executes the computations with the aid of a number of processors, and it outputs the results in the sequence $y = \mathcal{A}(u, \mathbf{x}_0)$. We often deal with algorithms of the form $\mathcal{A} : \ell \to \ell$, where no initial state is explicit. In other cases the algorithm has the form $\mathcal{A} : F^m \to \ell$; that is, it starts from an initial state and produces an output without need for an input sequence. The algorithms we will be concerned with in this appendix handle real or complex data via ordinary operations. An algorithm is often composed of more complex subsystems (subroutines) that are linked via the basic interconnection schemes of Section 3.5.

Design of algorithms relies on the general principles of system analysis and design. In addition it has its own autonomous and distinct identity, due to a number of intrinsic features and attributes. An algorithm is designed in order to achieve a specific task in an efficient way. FFT algorithms, for instance, aim to compute the discrete Fourier transform in a computationally efficient way. The tasks that need algorithmic design are many. Some of these tasks play a distinguished role, since they form basic building blocks upon which many other tasks depend. Such important jobs, particularly relevant for signal processing, are the solution of linear and nonlinear equations and optimization.

An important feature that algorithms dealing with real or complex data must possess is robustness to numerical errors. Algorithms operate with finite arithmetic; hence control of numerical errors caused by finite precision is necessary. Another critical issue in the specification of an algorithm is the class of admissible algorithmic architectures. This depends on the technology used to implement the algorithm. A basic distinction is between von Neumann sequential architectures and parallel architectures. Sequential machines communicate input data to central memory one at a time. A single instruction is executed each time. The speed of accessing memory and the speed of input-output interfaces decisively affect computational performance. In contrast, parallel computers are composed of several processors, each executing part of the overall task. Two important types of computing systems are *array processors* and *multiprocessors*. VLSI array processors are major representatives of the first type. They employ a significant number of similar processors with local communication (only nearby processors interact) for the execution of a specific computation. They are particularly useful in a computationally intensive environment, as in image processing. A LANDSAT satellite transmits 25–30 frames per second (video rate) with information regarding earth resources. Each frame corresponds to $1000 \times 1000$ pixels. Filtering is required to remove transmission noise. The computations involved are of the order of billion of instructions per second. High-level image processing such as image analysis increases computational requirements by several orders of magnitude. Multiprocessors, on the other hand, consist of processors that can operate autonomously, have great computational power, and a potential for general purpose operations.

Algorithm performance is assessed by the implementation cost and speed. Critical issues are then the number of processors, the time complexity, and the communication cost. The latter is assessed in conjuction with the technology used. Communication cost in VLSI array processor implementation is measured by the communication links between processors. The shorter the buses carrying data, the lower is the cost. Communication cost in general purpose parallel computers is assessed by the number of messages transmitted during algorithm execution.

### III.1    NONLINEAR EQUATIONS

#### III.1.1    The Picard algorithm

Consider the system of nonlinear equations

$$\{x \in R^m : g(x) = 0\} \tag{III.1}$$

where $g : R^m \to R^m$. It is convenient for subsequent analysis to examine the problem of fixed points of a map $f : R^m \to R^m$

$$\{x \in R^m : f(x) = x\} \tag{III.2}$$

We can pass from one task to another via the assignment $f(x) = x - g(x)$. The fixed points of $f$ lie on the intersection of the graph of $f$ with the line $y = x$. The problem may have many, one, or no solution. For a general function $f$, it is not possible to construct an algorithm that will produce a solution in a finite number of steps. For instance, if the coordinates of $f$ are polynomial functions, the solutions do not admit algebraic expressions. Nonlinear equations over a finite set is an exception. The solution in this case can be determined in a finite number of steps by exhaustive search. In the sequel we focus on iterative techniques, that is, algorithms that estimate the solution in an infinite number of steps. The Picard algorithm

$$x(n+1) = f(x(n)), \qquad n = 0, 1, \ldots \tag{III.3}$$

suggests itself as a natural try. The algorithm starts from an initial state $x_0$ and produces successive estimates by repeated application of $f$. The estimate at step $n$ is formed by the $n$th iterate of $f$:

$$x(n) = f(f(f(\cdots (f(x_o)) \cdots))) = f^n(x_o)$$

Suppose that $f$ is continuous. If we can ensure that the sequence $x(n)$ is convergent, we can deduce that the limit of $x(n)$, $x$, is a fixed point of $f$. Indeed,

$$x = \lim_{n \to \infty} x(n) = \lim_{n \to \infty} f(x(n-1)) = f(\lim_{n \to \infty} x(n-1)) = f(x)$$

The above thoughts can be substantiated if $f$ is a contraction. Let $X$ be a Banach space (e.g., $R^m$) and $A$ a closed subset of $X$. A map $f : A \to A$ is called *contraction* if there exists $0 < L < 1$ such that

$$|f(x_1) - f(x_2)| < L |x_1 - x_2|, \qquad x_1, x_2 \in A \tag{III.4}$$

The following theorem holds:

**Theorem III.1** If $f$ is a contraction on the closed subset $A$ of the Banach space $X$, the system (III.2) has a unique solution $x^*$ in $A$ and the Picard algorithm converges to $x^*$, for any initial state $x_0 \in A$.

PROOF    Let $x_0 \in A$ and $x(n)$ the sequence produced by Eq. (III.3) and initial state $x_0$. We want to show that $x(n)$ is a Cauchy sequence. Note that

$$|x(n+1) - x(n)| = |f(x(n)) - f(x(n-1))| < L |f(x(n-1)) - f(x(n-2))|$$

and that

$$|x(n+1) - x(n)| < L^n |x(1) - x_0|$$

Let $n$, $m$ be positive integers. Using the latter equation, we have

$$|x(n+m) - x(n)| \le \sum_{i=0}^{m-1} |x(n+i+1) - x(n+i)|$$

$$< \sum_{i=0}^{m-1} L^{n+i} |x(1) - x_o| = L^n \frac{L^m - 1}{L - 1} |x(1) - x_0|$$

The right-hand side tends to 0 as $m, n \to \infty$. We have proved that $x(n)$ is a Cauchy sequence. Since $X$ is a Banach space and $A$ is closed, $x(n)$ converges in $A$. Let $x^*$ denote the limit. The contraction property implies that $f$ is continuous. Thus $x^*$ is a solution of (III.2). To prove uniqueness, we consider another solution of (III.2), $\tilde{x}$ with $\tilde{x} \neq x^*$. Then $f(x^*) = x^*$, $f(\tilde{x}) = \tilde{x}$. Thus $| x^* - \tilde{x} | = | f(x^*) - f(\tilde{x}) | < L | x^* - \tilde{x} |$. This is a contradiction.  ∎

To compute the solution of Eq. (III.1), we set $f(x) = x - g(x)$. If $f$ is a contraction, the Picard algorithm takes the form $x(n + 1) = x(n) - g(x(n))$ and converges to the unique solution. We achieve further flexibility if we note that the set of solutions of (III.1) coincides with the set of solutions of (III.2), where

$$f(x) = x - K(x)g(x) \tag{III.5}$$

and $K(x)$ is an arbitrary invertible matrix. Then we search for $K(x)$ so that the corresponding $f$ is a contraction.

**Linearization.** Theorem III.1 is the simplest among a series of theorems collectively known as *fixed point theorems*. These theorems are extremely useful in a variety of applications. For instance, they are advocated in the existence and uniqueness theory of ordinary differential equations. On the other hand, the contraction property is often hard to verify. A useful criterion along this direction is the following: Suppose that $f : R^m \to R^m$ is differentiable; that is, the Jacobian matrix of partial derivatives $\partial f / \partial x$ exists and is continuous. If $|\partial f(x)/\partial x| < L < 1$ for all $x, f$ is a contraction. This is a direct consequence of the mean value theorem. Indeed, $|f(x) - f(y)| \leq |\partial f/\partial x||x - y| < L|x - y|$. The contraction property is rarely valid over the entire space. Thus we are content to a local result, whereby the contraction holds in a closed neighborhood of $x^*$. A sufficient condition is $|\partial f/\partial x|_{x^*} < L < 1$. Indeed, the latter inequality will be valid in a closed and bounded neighborhood $A$ of $x^*$ because of continuity of $\partial f/\partial x$. $f$ carries $A$ into $A$ and is contractive on $A$. Local contraction means that the algorithm converges to $x^*$ provided that the initial state is chosen near $x^*$. The above result is referred to as *linearization* because it infers convergence of (III.3) from the linear part of $f$ at $x^*$.

***Example III.1    Evaluation of functions with simple inverse***
Let $r : R^m \to R^m$ be a function and a given point $a \in R^m$. We seek to evaluate $x = r(a)$. We will assume that the inverse function of $r$, $h$ is known and is readily evaluated. A typical example is the square root $x = \sqrt{a}$, or more generally, the $k$th root $x = \sqrt[k]{a}$, for which the inverse function is $h(x) = x^k$. Determination of $x$ amounts to solving the equation $h(x) - a = 0$. If the function $f(x) = x - h(x) + a$ is contractive, the evaluation of $x$ can be carried out by the Picard algorithm

$$x(n + 1) = x(n) - h(x(n)) + a$$

If $x = r(a) = \sqrt{a}$, then

$$f(x) = x - x^2 + a, \qquad \frac{\partial f}{\partial x} = 1 - 2x$$

The derivative is bounded by 1 on every subinterval of $(0, 1)$. Hence the above scheme can be used for the computation of the square root of a number less than 1.    ∎

### III.1.2   The Newton algorithm

Let us return to (III.1) and the function $f(x) = x - K(x)g(x)$. We seek to determine $K(x)$ so that $f$ becomes a contraction. We write the above as follows: $f(x) = x - K(x)g(x) = x - \sum_{i=1}^{m} k_i(x)g_i(x)$, where

$$K(x) = [k_1(x), k_2(x), \dots, k_m(x)] \qquad g(x) = [g_1(x), \dots, g_m(x)]^T$$

Then

$$\nabla f(x) = \frac{\partial f(x)}{\partial x} = I - \sum_{i=1}^{m} k_i(x)\frac{\partial g_i(x)}{\partial x} - \sum_{i=1}^{m} \frac{\partial k_i(x)}{\partial x}g_i(x)$$

or

$$\nabla f(x) = I - K(x)\frac{\partial g}{\partial x} - \sum_{i=1}^{m} \frac{\partial k_i(x)}{\partial x}g_i(x)$$

We choose $K(x) = [\nabla g(x)]^{-1}$. Then $\nabla f(x^*) = I - I + 0 = 0$. Linearization shows that the above choice renders $f$ a contraction. The algorithm

$$x(n + 1) = x(n) - (\nabla g(x(n)))^{-1}g(x(n)) \tag{III.6}$$

is known as the *Newton algorithm*. It converges to $x^*$ provided that the initial state is near $x^*$.

In the case of square root evaluation, the Newton algorithm takes the form

$$x(n + 1) = x(n) - \frac{1}{2x(n)}(x^2(n) - a) = \frac{1}{2}\left(x(n) + \frac{a}{x(n)}\right)$$

Convergence of the Newton scheme is very fast. It can be shown that for any $L \in (0, 1)$, there is $\epsilon > 0$ such that

$$|x(n + 1) - x^*| \leq L^{n+1}|x(0) - x^*|, \quad |x(0) - x^*| \leq \epsilon$$

and any induced norm. This superior performance is compensated by an increase of computational complexity. At every step the evaluation of partial derivatives and the inverse of a matrix is required. If derivatives are not expressed in analytic form, approximate differentiation methods must be employed.

## III.2 OPTIMIZATION AND THE STEEPEST DESCENT ALGORITHM

Optimization algorithms compute the minima or maxima of a function and thus constitute a significant component of system design. Let $J : R^m \to R$ be a function. A point $x^* \in R^m$ is called a *(local) minimum* if there exists a neighborhood of $x^*$, $A$, such that for every $x \in A$, $x \neq x^*$, $J(x^*) < J(x)$. If inequality holds for any $x \in R^m$, $x^*$ is called a *global minimum*. Local and global maxima are similarly defined. $x^*$ is a maximum of $J$ if and only if it is a minimum for $-J$. Therefore we limit attention to minimization. It is known that if $x^*$ is a local minimum of the differentiable function $J$, the derivative of $J$ at $x^*$ is zero:

$$\nabla J(x^*) = 0 \tag{III.7}$$

This relation defines a set of nonlinear equations, and the algorithms of the previous section apply. In particular, the Newton algorithm takes the form

$$x(n + 1) = x(n) - \left[\nabla^2 J(x(n))\right]^{-1} \nabla J(x(n)) \tag{III.8}$$

The basic goal in steepest descent methods is to reduce the cost step by step rather than solving Eq. (III.7). It relies on recursions of the form

$$x(n + 1) = x(n) + \gamma(n)s(n) \tag{III.9}$$

The estimate $x(n + 1)$ tries to improve the previous estimate $x(n)$ by achieving a smaller cost value, that is

$$J(x(n + 1)) < J(x(n)) \tag{III.10}$$

As Eq. (III.9) indicates, $x(n + 1)$ lies on the line that passes from $x(n)$, has the direction of $s(n)$ and lies at a distance from $x(n)$ equal to the stepsize $\gamma(n)$. The choice of the parameters $\gamma(n)$ and $s(n)$ must satisfy (III.10). Thus we pick $s(n)$ so that it defines a *descent direction*; namely it has the property

$$s^T(n)\nabla J(x(n)) < 0 \tag{III.11}$$

The idea is illustrated in Fig. III.1. Suppose that $J$ is positive. Let $J(x(n)) = c$, and let $\{x \in R^m : J(x) = c\}$ be the curve (hypersurface, in general) of points having the same cost value. The vector $\nabla J(x(n))$ is orthogonal to the tangent space of the curve at $x(n)$ and is directed outward. Descent condition (III.11)

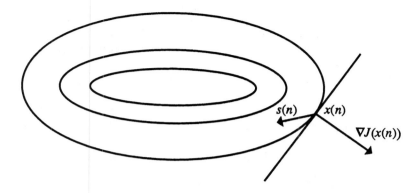

**Figure III.1.** Description of the descent direction.

states that the vector $s(n)$ forms an angle larger than 90 degrees with $\nabla J(x(n))$. Therefore, for small positive values of $\gamma(n)$, all points of the form $x(n) + \gamma(n)s(n)$ lie in the interior of the curve. Consequently they satisfy (III.10). It is worthwhile to point out that a similar argument is used in the study of stability of dynamical systems via Lyapunov functions; see Chapter 11. A formal argument goes as follows: We assume that $x(n)$ and $s(n)$ have been specified and that $s(n)$ defines a descent direction at $x(n)$; that is, $\nabla J(x(n)) \neq 0, s^T(n)\nabla J(x(n)) = -a, a > 0$. We define the real-valued function $r(\gamma) = J(x(n) + \gamma s(n))$, and we consider the Taylor expansion at $\gamma = 0$. Then $r(\gamma) = r(0) + \dot{r}(0)\gamma + o(\gamma^2)$, or

$$J(x(n) + \gamma s(n)) = J(x(n)) + \gamma s^T(n)\nabla J(x(n)) + o(\gamma^2)$$

Let $0 < \rho < a$ and $\delta = a/\rho$ such that for any $0 < \gamma < \delta, |o(\gamma^2)| < \gamma^2\rho$. Then

$$J(x(n) + \gamma s(n)) - J(x(n)) < -a\gamma + \gamma^2\rho = \gamma(\gamma\rho - a) < 0$$

Notice that $\gamma$ depends on $x(n)$ and hence on $n$.

The choice $s(n) = -\nabla J(x(n))$, clearly warrants the steepest possible descent direction (see Fig. III.1) and the algorithm

$$x(n + 1) = x(n) - \gamma(n)\nabla J(x(n)), \qquad \gamma(n) > 0 \tag{III.12}$$

forms an important representative of descent methods.

The stepsize $\gamma(n)$ indicates the distance traveled along the direction $s(n)$. Small steps represent a conservative policy and low speed of convergence. Large steps may lead to an oscillatory behavior, with the estimates bouncing round the limit. A reasonable compromise is to proceed along the direction $s(n)$, up to the point where the cost has minimum value, that is, to pick $\gamma(n)$ as the minimum of the function $r(\gamma)$

$$\gamma(n) : \min_\gamma r(\gamma) = \min_\gamma J(x(n) + \gamma s(n)) \tag{III.13}$$

The above choice is natural, albeit computationally intensive, since it requires the solution of another optimization problem. This is a one-dimensional problem and thus is simpler than the original. Constant stepsizes are often employed.

The Newton algorithm and the steepest descent algorithm can be expressed in the following unified form:

$$x(n+1) = x(n) - \gamma(n)R^{-1}(n)\nabla J(x(n)), \qquad R(n) > 0 \qquad \text{(III.14)}$$

The vector $s(n) = -R^{-1}(n)\nabla J(x(n))$ defines a descent direction, since

$$s^T(n)\nabla J(x(n)) = -\nabla J(x(n))^T R^{-1}(n)\nabla J(x(n)) < 0$$

**Example III.2    Minimization of quadratic forms**
Let

$$J(x) = x^T Rx - 2d^T x, \qquad R > 0 \qquad \text{(III.15)}$$

We easily calculate $\nabla J(x) = 2(Rx - d)$ and $\nabla^2 J(x) = 2R$. Hence the solution of the linear system

$$Rx^* = d \qquad \text{(III.16)}$$

yields the unique minimum of $J$. The steepest descent algorithm with constant stepsize has the form

$$x(n+1) = x(n) - 2\gamma(Rx(n) - d) = (I - 2\gamma R)x(n) + 2\gamma d$$

Let $\tilde{x}(n) = x(n) - x^*$. Then the latter equation becomes

$$\tilde{x}(n+1) = \tilde{x}(n) - 2\gamma(R(\tilde{x}(n) + x^*) - d)$$

Thus

$$\tilde{x}(n+1) = (I - 2\gamma R)\tilde{x}(n) = (I - 2\gamma R)^n \tilde{x}(0) \qquad \text{(III.17)}$$

The error goes to zero if and only if $A^n \to 0$, where $A = I - 2\gamma R$. The asymptotic behavior of the matrix exponential signal is studied in Section 11.5. We show there that $A^n \to 0$ if and only if all eigenvalues of $A$ lie inside the unit circle. The eigenvalues of $A$ are $1 - 2\gamma\lambda_i$, where $\lambda_i$ are the eigenvalues of $R$. Thus $|1 - 2\gamma\lambda_i| < 1$ if and only if

$$0 < \gamma < \frac{1}{\lambda_{\max}(R)} \qquad \text{(III.18)}$$

The last condition provides the range for $\gamma$. The previous algorithms can be used as a linear system solver for systems of the form (III.16), with $R > 0$.

Minimization of quadratic forms is one of the few cases where (III.13) admits an analytic expresssion. Indeed,

$$r(\gamma) = J(x(n) + \gamma s(n)) = (x(n) + \gamma s(n))^T R(x(n) + \gamma s(n)) - 2d^T(x(n) + \gamma s(n))$$

Thus

$$0 = \frac{dr(\gamma)}{d\gamma} = 2\gamma s^T(n)Rs(n) + 2s^T(n)Rx(n) - 2d^T s(n)$$

and

$$\gamma(n) = -\frac{s^T(n)Rx(n) - s^T(n)d}{s^T(n)Rs(n)} \qquad \blacksquare$$

## III.3 LINEAR EQUATIONS AND QR METHODS

Let $A$ be a $m \times k$ matrix, $m \geq k$, of rank $k$, and let $a_1, a_2, \ldots, a_k$ be the columns of $A$. As we saw in Chapter 2, the Gram-Schmidt orthogonalization constructs orthogonal vectors $q_1, q_2, \ldots, q_k$ that span the column space of $A$. The basic recursion is

$$q_i = a_i - \sum_{j=1}^{i-1} \frac{< a_i, q_j >}{< q_j, q_j >} q_j \qquad (\text{III.19})$$

The matrix

$$Q = (\,q_1 \quad q_2 \quad \cdots \quad q_k\,)$$

has $m$ rows and $k$ columns and is orthogonal because its columns are orthogonal vectors. Thus $Q^T Q = I$, and Eq. (III.19) becomes

$$a_i = q_i + \sum_{j=1}^{i-1} \frac{< a_i, q_j >}{< q_j, q_j >} q_j = q_i + \sum_{j=1}^{i-1} r_{ij} q_j$$

Putting the above equations columnwise, we have

$$(\,a_1 \quad a_2 \quad \cdots \quad a_k\,) = (\,q_1 \quad q_2 \quad \cdots \quad q_k\,) \begin{pmatrix} 1 & r_{21} & r_{31} & \cdots & r_{k1} \\ 0 & 1 & r_{32} & \cdots & r_{k2} \\ \vdots & \vdots & \vdots & \ddots & \vdots \\ 0 & 0 & 0 & \cdots & 1 \end{pmatrix}$$

Hence the Gram-Schmidt algorithm represents $A$ as

$$A = QR$$

where $Q$ is an orthogonal $m \times k$ matrix and $R$ is an upper triangular $k \times k$ matrix. If the factors $Q$, $R$ of a square matrix $A$ are known the solution of the linear system,

$$Ac = d \tag{III.20}$$

is computed exactly in two steps as follows: Eq. (III.20) becomes $QRc = d$. We set $Rc = f$. Then $Qf = d$. Since $Q$ is orthogonal, $f = Q^T d$. Therefore $f$ is computed with $O(m^2)$ multiplications and additions. The desired vector $c$ is determined by back substitution, since $R$ is triangular. Computational complexity is proportional to $m^2/2$.

The Gram-Schmidt algorithm is rarely used for the computation of the factors $Q$ and $R$. Two of the most popular methods rely on the Householder and Givens transformations. Let $w$ be a vector of unit length, $w^T w = 1$. The square matrix

$$Q = I - 2ww^T \tag{III.21}$$

is called a *Householder matrix*. $Q$ is symmetric and orthogonal. Indeed, it is clear that $Q^T = Q$, while

$$QQ^T = (I - 2ww^T)(I - 2ww^T) = I - 4ww^T + 4ww^T = I$$

The action of $Q$ is illustrated in Fig. III.2. The following lemma holds:

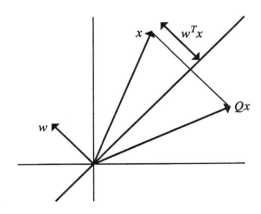

**Figure III.2.** The Householder transformation

**Lemma** For any vector $x \in R^m$, there is a vector $w = (w_1 \quad \cdots \quad w_m)^T$ such that the corresponding Householder matrix $Q$ satisfies

$$Qx = \lambda(1 \quad 0 \quad \cdots \quad 0)^T, \qquad \lambda = |x|$$

**PROOF** Let $Q$ be a matrix of the form (III.21) that satisfies the desired property. Then

$$x^T x = x^T Q Q x = \lambda^2 \quad \text{or} \quad \lambda = |x|$$

Moreover

$$x^T = (1 \quad 0 \quad \dots \quad 0)\lambda Q$$

or

$$x^T = \lambda\{(1 - 2w_1^2 \quad -2w_1 w_2 \quad \dots \quad -2w_1 w_m)\}$$

Thus

$$w_1 = \pm\left(\frac{1}{2}\left(1 - \frac{x_1}{\lambda}\right)\right)^{1/2}, \quad w_i = \frac{-x_i}{2w_1\lambda}, \quad i = 2, 3, \dots, m$$

Note that $x_1 \leq |x| = \lambda$. The case $x_1 = \lambda$ occurs when $x_2 = x_3 = \dots = x_m = 0$. Then the proposition holds trivially, with $w = 0$. In every other case $x_1 < \lambda$, the argument of the square root is positive, and $w_i$ are well-defined. ∎

We compute the $Q$, $R$ factors of $A$ by repeated application of the previous proposition. We represent $A$ by its columns as above, and we apply the proposition to the first column. Let $Q_1$ be an orthogonal matrix such that

$$Q_1 a_1 = \left(a_{11}^{(1)} \quad 0 \quad \dots \quad 0\right)^T = a_1^{(1)}$$

Let $Q_1 a_i = a_i^{(1)}$, $i = 2, 3, \cdots, k$. Then

$$Q_1 A = \left(a_1^{(1)} \quad a_2^{(1)} \quad \dots \quad a_k^{(1)}\right)$$

We consider the column $a_2^{(1)}$ and ignore its first entry. For the remaining vector $\bar{a}_2$ of length $m - 1$, we apply the previous proposition to determine an orthogonal matrix $\bar{Q}_2$ such that

$$\bar{Q}_2 \bar{a}_2 = \left(a_{22}^{(2)} \quad 0 \quad \dots \quad 0\right)^T$$

Thus the orthogonal matrix

$$Q_2 = \begin{pmatrix} 1 & 0 \\ 0 & \tilde{Q}_2 \end{pmatrix}$$

acting on $Q_1 A$ gives

$$Q_2 Q_1 A = Q_2 \begin{pmatrix} a_1^{(1)} & a_2^{(1)} & \dots & a_k^{(1)} \end{pmatrix} = \begin{pmatrix} a_{11}^{(1)} & * & \dots \\ 0 & a_{22}^{(2)} & \dots \\ 0 & 0 & \dots \\ \vdots & \vdots & \ddots \\ 0 & 0 & \dots \end{pmatrix}$$

The above construction maintains the zeros in the first column and annihilates the entries of the second column below the diagonal. After $k$ steps the above procedure leads to the upper triangular matrix:

$$Q_k Q_{k-1} \cdots Q_1 A = R$$

Givens's orthogonal matrices produce entirely analogous results.

## BIBLIOGRAPHICAL NOTES

For further reading the reader is referred to Luenberger (1969) and Golub and Van Loan (1983).

# BIBLIOGRAPHY

Ahmed, N., and Rao, K. *Orthogonal Transforms for Digital Signal Processing*. Springer-Verlag, 1975.

Anderson, B. D. O., Bitmead, R. R., Johnson, C. R., Jr., Kokotovic, P. V., Kosut, R. L., Mareels, I., Praly, L., and Riedle, B. D. *Stability of Adaptive Systems Passivity and Averaging Analysis*. MIT Press, 1986.

Anderson, B., and Moore, J. *Optimal Filtering*. Prentice Hall, 1979.

Antoniou, A. *Digital Filters Analysis and Design*. McGraw-Hill, 1979.

Arnold, V. I. *Ordinary Differential Equations*, trans. by R. Silverman. MIT Press, 1973.

Astrom, K., and Wittenmark, B. *Computer Controlled Systems Theory and Design*. Prentice Hall, 1984.

Athans, M., and Falb, P. L. *Optimal Control: An Introduction to the Theory and Its Applications*. McGraw-Hill, 1965.

Athans, M., Dertouzos, M., Spann, R., and Mason, S. *Systems, Networks and Computation, Multivariable Methods*. McGraw-Hill, 1974.

Berlekamp, E. *Algebraic Coding Theory*. McGraw-Hill, 1968. Rev. Ed., Laguna Hills, Aegean Park Press, 1984.

Bertsekas, D., and Tsitsiklis, J. *Parallel and Distributed Computation*. Prentice Hall, 1989.

Blahut, R. *Fast Algorithms for Digital Signal Processing*. Addison-Wesley, 1985.

Blahut, R. *Theory and Practice of Error Control Codes*. Addison Wesley, 1984.

Boyd, S. Volterra series engineering fundamentals. Ph.D. dissertation. University of California, Berkeley, 1985.

Bracewell, R. N. *The Fourier Transform and Its Applications*. 2d ed. McGraw-Hill, 1986.

Brauer, F., and Nohel, J. A. *Qualitative Theory of Ordinary Differential Equations*. Benjamin, 1969.

Brockett, R. *Finite Dimensional Linear Systems*. Wiley, 1970.

Burrus, C. S., and Parks, T. W. *DFT/FFT and Convolution Algorithms Theory and Implementation*. Wiley, 1985.

Caines, P. *Linear Stochastic Systems*. Wiley, 1988.

Chen, C. *Linear System Theory and Design*. Holt, Rinehart and Winston, 1984.

Cichocki, A., and Ubenauen, R. *Neural Networks for Optimization and Signal Processing.* Wiley, 1993.

Crochiere, R. E., and Rabiner, L. R. *Multirate Digital Signal Processing.* Prentice Hall, 1983.

Cybenko, G. Approximations by superposition of a sigmoidal function. *Mathematics of Control, Signals and Systems,* vol. 2, 1989, pp. 303–314.

de Moor, B. and Van Overschee, P. Numerical algorithms for subspace state space system identification. *Trends in Control, A European Perspective,* ed. by A. Isidori. Springer Verlag, 1995.

Denning, D. E. *Cryptography and Data Security.* Addison-Wesley, 1983.

Desoer, C. A., and Vidyasagar, M. *Feedback Systems, Input-Output Properties.* Academic Press, 1975.

Desoer, C., and Kuh, E. *Basic Circuit Theory.* McGraw-Hill, 1969.

Devaney, R. L. *An Introduction to Chaotic Dynamical Systems.* Benjamin Cummings, 1986.

Dickinson, B. *Systems Analysis, Design and Computation.* Prentice Hall, 1991.

Dieudonne, J. *Foundations of Modern Analysis.* Academic Press, 1969.

Doetsch, G. *Introduction to the Theory and Application of the Laplace Transformation,* Springer-Verlag, 1974.

Doob, J. L. *Stochastic Processes.* Wiley, 1960.

Doyle, J. C., Francis, B. A., and Tannenbaum, A. *Feedback Control Theory.* Macmillan, 1992.

Duda, R., and Hart, P. *Pattern Classification and Scene Analysis.* Wiley, 1973.

Dudgeon, D., and Mersereau, R. *Multidimensional Digital Signal Processing.* Prentice Hall, 1984.

Dym, H., and McKean, H. P. *Fourier Series and Integrals.* Academic Press, 1972.

Francis, B. A. *A Course in $H_\infty$ Control Theory.* Lecture Notes in Control Information Sciences, vol. 88. Springer-Verlag, 1987.

Franklin, G., and Powell, J. D. *Digital Control of Dynamic Systems.* Addison-Wesley, 1980.

Frisch, R. *Statistical Confluence Analysis by Means of Complete Regression Systems.* Pub. 5. Economic Institute, Oslo University, 1934.

Fukunaga, K. *Statistical Pattern Recognition.* 2d ed. Academic Press, 1990.

Gallagher, R. *Information Theory and Reliable Communications.* Wiley, 1968.

Gardner, W. *Introduction to Random Processes, with Applications to Signals and Systems.* Macmillan, 1986.

Gersho, A., and Gray, R. *Vector Quantization and Signal Compression.* Kluwer, 1992.

Golomb, S. W. *Shift Register Sequences.* Holden Day, 1967.

Golub, G. H., and Van Loan, C. F., *Matrix computations.* The Johns Hopkins University Press, 1983.

Goodwin, G., and Sin, K. *Adaptive Filtering and Prediction and Control.* Prentice Hall, 1984.

Gray, R. M., and Davisson, L. D. *Random Processes: a Mathematical Approach for Engineers.* Prentice Hall, 1986.

Guidorzi, R. Errors-in-variables identification and model uniqueness. *Statistical Modelling and Latent Variables*, (ed. by K. Haagen, D. J. Bartholomew, and M. Deistler.) North Holland, 1993, pp. 127–150.

Hahn, W. *Stability of Motion.* Springer-Verlag, 1967.

Hannan, E. J., and Deistler, M. *The Statistical Theory of Linear Systems.* Wiley, 1988.

Haykin, S. *Adaptive Filter Theory*, 2d ed. Prentice Hall, 1991.

Haykin, S. *Neural Networks, A Comprehensive Foundation.* Macmillan, 1994.

Heideman, M., Johnson, D., and Burrus, C. Gauss and the history of the fast Fourier transform. *IEEE ASSP* magazine, vol. 1, no. 4, Oct. 1984.

Hirsch, M., and Smale, S. *Differential Equations, Dynamical Systems, and Linear Algebra.* Academic Press, 1974.

Honig, M. L., and Messerchmitt, D. G. *Adaptive Filters, Structures, Algorithms, and Applications.* Kluwer, 1984.

Hornik, K., Stinchcombe, M., and White, H. Multilayer feedforward networks are universal approximators. *Neural Networks*, vol. 2, 1989, pp. 359–366.

Hush, D. R., and Horne, B. G. Progress in supervised neural networks. *IEEE Signal Processing*, Jan. 1993, pp. 8–39.

Isidori, A. *Nonlinear Control Systems.* 2d ed. Springer-Verlag, 1989.

Jain, A. *Fundamentals of Digital Image Processing.* Prentice Hall, 1989.

Jain, A. K., and Dubes, R. C. *Algorithms for Clustering Data.* Prentice Hall, 1988.

Jayant N. S., and Noll, P. *Digital Coding of Waveforms.* Prentice Hall, 1984.

Jian-Hua, L., Michel, A. N., and Porod, W. Analysis and synthesis of a class of neural networks: Linear systems operating on a closed hypercube, *IEEE Trans. on Circuits and Systems*, vol. 36, no. 11, Nov. 1989, pp. 1405–1422.

Kailath, T. *Linear Systems.* Prentice Hall, 1980.

Kalman, R. E. Identification from real data. *Current Developments in the Interface: Economics, Econometrics, Mathematics*, ed. by M. Hazewinkel and H. G. Rinnoy Kan. Reidel, 1982, pp. 161–96.

Kalman, R. E., and Bertram, J. E. Control system analysis and design via the second method of Lyapunov. *Trans. ASME J. of Basic Eng.*, vol. 82, 1960, pp. 371–392.

Kalman, R., Falb, P., and Arbib, M. *Topics in Mathematical System Theory.* McGraw-Hill, 1969.

Kalouptsidis, N., and Theodoridis, S. *Adaptive System Identification and Signal Processing Algorithms.* Prentice Hall, 1993.

Katznelson, Y. *An Introduction to Harmonic Analysis.* Dover, 1976.

Kohavi, Z. *Switching and Finite Automata Theory.* 2d ed. McGraw-Hill, 1978.

Kolmogorov, A. N., and Fomin, S. V. *Introductory Real Analysis.* trans. and ed. by A. Silverman. Dover, 1970.

Kreyzig, E. *Advanced Engineering Mathematics.* 4th ed. Wiley, 1979.

Kumar, P. R., and Varaya, P. *Stochastic Systems, Estimation Identification and Adaptive Control.* Prentice Hall, 1986.

Kung, S. Y. *VLSI Array Processors.* Prentice Hall, 1988.

Kwakernaak, H., and Sivan R. *Linear Optimal Control Systems.* Wiley-Interscience, 1972.

Lathi, B. *Signals, Systems and Controls.* Intext 1974.

Lee, E. B., and Marcus, L. *Foundations of Optimal Control Theory.* Wiley, 1967.

Lim, J. S. *Two Dimensional Signal and Image Processing.* Prentice Hall, 1990.

Lin, S., and Costello, D., Jr. *Error Control Coding: Fundamentals and Applications.* Prentice Hall, 1983.

Lippmann, P. R. An introduction to computing with neural nets. *IEEE Signal Processing* magazine, vol. 4, pp. 4–22.

Liu, C., and Liu, J. *Linear System Analysis.* McGraw-Hill, 1975.

Ljung, L. *System Identification Theory for the User.* Prentice-Hall, 1987.

Ljung, L., and Soderstrom, T. *Theory and Practice of Recursive Identification,* MIT Press, 1983.

Luenberger, D. *Optimization by Vector Space Methods.* Wiley, 1969.

MacWilliams, F. J., and Sloane, J. J. *The Theory of Error Correcting Codes.* North-Holland, 1977.

Manolakis, D., and Proakis, J. *Introduction to Digital Signal Processing.* Macmillan, 1989.

Marple, S. *Digital Spectral Analysis with Applications.* Prentice Hall, 1987.

McClellan, J. H., and Rader, C. M. *Number Theory in Digital Signal Processing.* Prentice Hall, 1979.

Milanese, M., and Vicino, A. Optimal estimation theory for dynamic systems with set membership uncertainty: an overview. *Automatica* 27, 1991, pp. 997–1009.

Nijmeijer, H., and Van der Schaft, A. T. *Nonlinear Dynamical Control Systems.* Springer-Verlag, 1990.

Nikias, Ch. L., and Petropoulou, A. P. *Higher-Order Spectral Analysis, a Nonlinear Signal Processing Framework.* Prentice Hall, 1993.

Norton, J. *Introduction to System Identification.* Academic Press, 1986.

Oppenheim, A., and Schafer, R. *Discrete Time Signal Processing.* Prentice Hall, 1989.

Oppenheim, A., Willsky, A., and Young, I. *Signals and Systems.* Prentice Hall, 1983.

Papoulis, A. *Probability, Random Variables and Stochastic Processes.* 3d. ed. McGraw-Hill, 1991.

Papoulis, A. *Signal Analysis.* McGraw-Hill, 1985.

Papoulis, A. *The Fourier Integral and Its Applications.* McGraw-Hill, 1962.

Peterson, W., and Weldon, E., Jr. *Error Correcting Codes.* 2d ed., MIT Press, 1972.

Phillips, C., and Nagle, H. *Digital Control System Analysis and Design.* Prentice Hall, 1984.

Pitas, L., and Venetsanopoulos, A. N. *Nonlinear Digital Filters, Principles and Applications.* Kluwer Academic Publishers, 1990.

Porat, B. *Digital Processing of Random Signals, Theory and Methods.* Prentice Hall, 1994.

Priestley, M. B. *Spectral Analysis and Time Series.* Academic Press, 1981.

Rhodes, I. B. A tutorial introduction to estimation and filtering. *IEEE Transactions on Automatic Control,* vol.16, no. 6, Dec. 1971, pp. 688–706.

Rosenblatt, M. *Stationary Sequences and Random Fields.* Birkhauser, 1985.

Rozanov, Yu. A. *Stationary Random Processes*. trans. by A. Feinstein. Holden-Day, 1967.

Rudin, W. *Real and Complex Analysis*. McGraw-Hill, 1970.

Rugh, W. J. *Nonlinear System Theory, The Volterra/Wiener Approach*. Johns Hopkins University Press, 1981.

Sandberg, I. W. Structure theorems for nonlinear systems. *Multidimensional Systems and Signal Processing*, vol. 2, 1991, pp. 267–286.

Schetzen, M. *The Volterra and Wiener Theories of Nonlinear Systems*. Wiley, 1980.

Shannon, C. E. A mathematical theory of communication. *Bell Sys. Tech. J.*, vol. 27, Oct. 1948, pp. 623–656.

Soderstrom, T., and Stoica, P. *System Identification*. Prentice Hall, 1989.

Sontag, E. *Mathematical Control Theory, Deterministic Finite Dimensional Systems*. Texts in Applied Mathematics. Springer-Verlag, 1990.

Stone, M. H. *Discrete Mathematical Structures and Their Applications*. Science Research Associates, 1973.

Strang, G. *Linear Algebra and Its Applications*. Harcourt Brace, Jovanovich, 1980.

Tsonis, A. *Chaos, from Theory to Applications*. Plenum, 1992.

Vaidyanathan, P. P. *Multirate Systems and Filter Banks*. Prentice Hall, 1993.

Van der Waerden, B. L. *Modern Algebra*. Ungar, 1949.

Vetterli, M., and Kovačević, J. *Wavelets and Subband Coding*. Prentice Hall, 1995.

Vidyasagar, M. *Control System Synthesis: A Factorization Approach*. MIT Press, 1985.

Vidyasagar, M. *Nonlinear Systems Analysis*. 2d ed. Prentice Hall, 1992.

Walter, E., and Piet-Lahanier, H. Estimation of parameter bounds from bounded error data: A survey. *Math. Comput. Simulation*, vol. 32, 1990, pp. 399–468.

Wicker, S. B., and Bhargava, V. K., eds. *Reed-Solomon Codes and Their Applications*. IEEE Press, 1994.

Widrow, B., and Stearns, S. D. *Adaptive Signal Processing*. Prentice Hall, 1985.

Willems, J. L. *Stability Theory of Dynamical Systems*. Nelson, 1970.

Wong, E., and Hajek, B. *Stochastic Processes in Engineering Systems*. Springer-Verlag, 1985.

Zadeh, L., and Desoer, C. *Linear System Theory*. McGraw-Hill, 1963.

# INDEX